Global Biogeochemical Cycles

This is Volume 50 in the
INTERNATIONAL GEOPHYSICS SERIES
A series of monographs and textbooks
Edited by RENATA DMOWSKA and JAMES R. HOLTON

A complete list of the books in this series appears at the end of this volume.

Global Biogeochemical Cycles

edited by

Samuel S. Butcher
Robert J. Charlson
Gordon H. Orians
Gordon V. Wolfe

ACADEMIC PRESS

Harcourt Brace Jovanovich, Publishers

LONDON SAN DIEGO NEW YORK BOSTON
SYDNEY TOKYO TORONTO

ACADEMIC PRESS LIMITED
24–28 Oval Road
LONDON NW1 7DX

United States Edition published by
ACADEMIC PRESS INC.
San Diego, CA 92101

This book is printed on acid-free paper

A catalogue record for this is available from the British Library
ISBN 0-12-147685-5
0-12-147686-3 *pbk*

Typeset by EJS Chemical Composition, Bath, Avon
and printed and bound in Great Britain at The Bath Press, Avon

Contents

Authors xiii
Preface xv

1 Introduction 1
1.1 An Approach to the Subject 1
1.2 This Text 5
Questions 7
References 7

2 The Origin and Early Evolution of the Earth 9
2.1 The Origin of the Elements 9
2.2 The Origin of the Solar System 14
2.3 Condensation 15
2.4 Accretion of the Planets 18
2.5 Early Evolution of the Earth 19
Questions 20
References 20

3 Evolution and the Biosphere 21
3.1 Origin and Evolution of Life 21
3.1.1 Fossil Evidence of Early Life Forms 22
3.1.2 Organic Synthesis on the Primeval Earth 25
3.1.3 Origin of Life 26
3.1.4 Early Photosynthesis 28
3.1.5 Oxygenic Photosynthesis 28
3.1.6 Impact of Oxygen on Evolution 28
3.2 The Machinery of Life 29
3.2.1 The Major Divisions of Life 37
3.2.2 The Five Kingdom System of Classification of Organisms 37
3.2.3 Endosymbiotic Theory of Evolution of Eukaryotic Organisms 40
3.2.4 Why So Many Kinds of Organisms? 41
3.3 The Functional Types of Molecules Produced by Living Organisms and Their Biogeochemical Significance 42
3.3.1 Energy Storage and Transfer Compounds 42
3.3.2 Information Storage and Transfer Molecules 43
3.3.3 Structural Molecules 44
3.3.4 Defensive Molecules 44
3.4 The Ecological Organization of the Living World 46

3.5 Overview of Major Biogeochemical Transformations 49
3.5.1 The Carbon Cycle 50
3.5.2 The Nitrogen Cycle 50
3.5.3 The Sulfur Cycle 51
3.5.4 The Phosphorus Cycle 51
3.5.5 The Metal Cycles 51
3.5.6 Concluding Remarks 51
Questions 52
References 53

4 Modeling Biogeochemical Cycles 55
4.1 Introductory Remarks 55
4.2 Reservoir Models and Cycles: Some Definitions 56
4.3 Time-scales and Single Reservoir Systems 57
4.3.1 Turnover Time 57
4.3.2 Residence Time (transit time) 58
4.3.3 Age 59
4.3.4 Relations between τ_0, τ_r and τ_a 59
4.3.5 Response time 60
4.3.6 Reservoirs in Non-steady-state 60
4.4 Coupled Reservoirs 61
4.4.1 Linear Systems 61
4.4.2 Non-linear Systems 64
4.5 Coupled Cycles 66
4.6 Note on Transport Processes 66
4.6.1 Advection, Turbulent Flux, and Molecular Diffusion 66
4.6.2 Other Transport Processes 68
4.6.3 Air–Sea Exchange 68
4.6.4 Sediment–Water Exchange 70
4.7 Note on Times of Mixing in the Atmosphere and Oceans 70
Questions 71
References 72

5 Equilibrium, Rate, and Natural Systems 73
5.1 Thermodynamics 73
5.1.1 Systems at Equilibrium: Thermodynamics 73
5.1.2 Measure of Spontaneity, K_{eq} 74
5.1.3 Condensed Phase Systems 75
5.1.4 Non-ideal Behavior 76
5.1.5 Thermodynamic Description of Isotope Effects 77
5.2 Oxidation and Reduction 77
5.2.1 Half Cell Conventions and the Nernst Equation 78
5.2.2 Stability Diagrams 79
5.2.3 Natural Systems and the Nernst Equation 81
5.3 Chemical Kinetics 82
5.3.1 Reaction Rates 82
5.3.2 Molecular Processes 82
5.3.3 Reaction Mechanisms 83
5.3.4 Kinetic Isotope Effects 85
5.4 Non-equilibrium Natural Systems 86
5.4.1 The Formation of Nitrogen Oxides 86
5.4.2 Oxygen in the Atmosphere 87

5.5 Summary 89
Questions 90
References 90

6 Tectonic Processes, Continental Freeboard, and the Rate-controlling Step for Continental Denudation 93
6.1 Introduction 93
6.2 Erosion: A Capsule Summary 94
6.2.1 Weathering and Crustal Breakdown 94
6.2.2 Weathering and the Atmospheric Gases 95
6.3 Soils and the Local Weathering Environment 96
6.3.1 Weathering Reaction Kinetics 96
6.3.2 Variable Chemistry of Soil Fluids: A Complicating Factor 97
6.3.3 Local Responses to Atmospheric Variables 98
6.4 Slope Processes and the Susceptibility of Lithologies to Erosion 98
6.4.1 Soils, Slopes, Vegetation, and Weathering Rate 99
6.4.2 Elemental Partitioning: The Role of Slope Processes 102
6.5 Landforms, Tectonism, Sea Level, and Erosion 103
6.5.1 The Effects of Erosional Regime 104
6.5.2 The Effects of Deposition, Storage, and Burial 104
6.5.3 The Effects of Tectonic Processes 104
6.5.4 Effects of Sea-level Change on Erosion 105
6.5.5 A General Concordance Between Erosion Rates and Uplift Rates 107
6.6 Erosion in Tectonically Active Areas 108
6.6.1 Erosion and Orogeny 108
6.6.2 River Chemistry and Bedrock Susceptibility in Mountainous Regions 110
6.7 Erosion of the Cratons 111
6.7.1 Erosion and Slow Uplift of the Shields 112
6.7.2 Weathering-limited Erosion on the Elevated Shield 114
6.7.3 Transport-limited Erosion on the Lowlands 114
6.8 The Effects of Transients: Continental Ice Sheets 115
6.9 Conclusion 117
Questions 117
References 118

7 Pedosphere 123
7.1 Introduction 123
7.1.1 The Pedosphere 123
7.1.2 Factors of Soil Formation 124
7.1.3 The Soil Profile 124
7.2 The Soil Constituents 126
7.2.1 General Composition 126
7.2.2 Clay Minerals and Clay Mineral Properties 128
7.2.3 Soil Organic Matter 132
7.3 Weathering and Soil-forming Processes 134
7.3.1 Physical Weathering 134
7.3.2 Chemical Weathering 135
7.3.3 Soil-forming Processes 141
7.4 Soil and Biogeochemical Cycles 145
7.4.1 Soils and River Water Chemistry 146
7.4.2 Case Studies: The Dynamics of Watersheds 148
7.4.3 Conclusions 148
Questions 148
References 148

8 Sediments: Their Interaction with Biogeochemical Cycles through Formation and Diagenesis 155
8.1 Introduction 155
8.2 Weathering and Formation of Sediments 155
8.3 The Structure of Sediments 157
8.3.1 The Solid Phase 157
8.3.2 The Liquid Phase 159
8.3.3 The Gaseous Phase 159
8.4 Physicochemical Processes in Sediments 160
8.4.1 The Liquid–Solid Interaction 160
8.4.2 Diagenesis 161
8.4.3 Diffusion 162
8.4.4 Redox Conditions 163
8.5 Transformation of Organic and Inorganic Material 165
8.5.1 Organic Matter 165
8.5.2 Sulfur 167
8.5.3 Nitrogen 167
8.5.4 Phosphorus 169
Questions 171
References 171

9 The Oceans 175
9.1 What is the Ocean? 175
9.2 Ocean Circulation 178
9.2.1 Density Stratification in the Ocean 179
9.2.2 Surface Currents 182
9.2.3 Thermocline Circulation 183
9.2.4 Abyssal Circulation 184
9.3 Biological Processes 187
9.3.1 The RKR Model 187
9.3.2 Factors Affecting the Rate of Plankton Productivity 188
9.3.3 The Geographic Distribution of Primary Productivity 190
9.3.4 Forms of Organic Matter in Seawater 190
9.3.5 Oceanic Reservoirs of Organic Carbon 192
9.3.6 An Oceanic Budget for Organic Carbon 192
9.3.7 Organic Carbon Pathways in the Ocean 193
9.4 Chemistry of the Oceans 193
9.4.1 Residence Time 196
9.4.2 Composition of Seawater 197
9.4.3 Equilibrium Models of Seawater 202
9.4.4 Kinetic Models of Seawater 204
Questions 209
References 209

10 The Atmosphere 213
10.1 Definition 213
10.2 The Vertical Structure of the Atmosphere 213
10.2.1 Hydrostatic Equation 213
10.2.2 Scale Height 214
10.2.3 Lapse Rate 215
10.2.4 Static Stability 215
10.3 Vertical Motions, Relative Humidity, and Clouds 217
10.4 The Ozone Layer and the Stratosphere 217

10.5 Horizontal Motions, Atmospheric Transport, and Dispersion 218
10.5.1 Microscale Turbulent Diffusion 219
10.5.2 Synoptic Scale Motion: The General Circulation 219
10.5.3 Geostrophic Wind 220
10.5.4 Meridional Transport of Water: The ITCZ 221
10.6 Composition 223
10.7 Atmospheric Water and Cloud Microphysics 224
10.8 Trace Atmospheric Constituents 226
10.8.1 Sulfur Compounds 226
10.8.2 Nitrogen Compounds 227
10.8.3 Carbon Compounds 227
10.8.4 Other Trace Elements 228
10.9 Chemical Interactions of Trace Atmospheric Constituents 230
10.9.1 Gas Phase Interactions 230
10.9.2 Condensed Phase Interactions 230
10.10 Physical Transformations of Trace Substances in the Atmosphere 233
10.11 Influence of Atmospheric Composition on Climate 234
10.11.1 Carbon Dioxide 234
10.11.2 Other "Greenhouse Gases" 235
10.11.3 Particles and Clouds 235
10.12 Chemical Processes and Exchanges at the Lower and Upper Boundaries of the Atmosphere 235
10.12.1 CO_2, Photosynthesis and, Nutrient Exchange 235
10.12.2 Reactions at the Surface 235
10.12.3 Cosmic Ray-induced Nuclear Reactions 235
10.12.4 Escape of H and He 238
References 238

11 The Global Carbon Cycle 239
11.1 Introduction 239
11.2 The Isotopes of Carbon 240
11.3 The Major Reservoirs of Carbon 241
11.3.1 The Atmosphere 241
11.3.2 The Hydrosphere 242
11.3.3 The Terrestrial Biosphere 246
11.3.4 The Lithosphere 248
11.4 Fluxes of Carbon between Reservoirs 249
11.5 Models of the Carbon Cycle 253
11.6 Trends in the Carbon Cycle 254
References 259

12 The Nitrogen Cycle 263
12.1 Introduction 263
12.2 Chemistry 263
12.3 Biological Transformation of Nitrogen Compounds 266
12.4 Abiotic Processes 270
12.4.1 Homogeneous Gas Phase Reactions 270
12.4.2 Heterogeneous Reactions 272
12.5 The Global Nitrogen Cycle 273
12.5.1 Nitrogen Inventories 273
12.5.2 Fluxes of Nitrogen 275
12.5.3 Anthropogenic Perturbations 279
Questions 282
References 282

13 **The Sulfur Cycle** 285
13.1 Introduction 285
13.2 Oxidation States of Sulfur 285
13.3 Sulfur Reservoirs 288
13.4 The Atmospheric Cycle of Sulfur 288
13.4.1 Transformations of Sulfur in the Atmosphere 288
13.4.2 Sources and Distribution of Atmospheric Sulfur 288
13.4.3 The Remote Marine Atmosphere 293
13.4.4 The Global Atmospheric Sulfur Budget 295
13.5 The Hydrospheric Cycle of Sulfur 295
13.5.1 Oceanic Outputs 297
13.5.2 Oceanic Inputs 298
Questions 299
References 299

14 **The Phosphorus Cycle** 301
14.1 Occurrence of Phosphorus 301
14.1.1 Dissolved Inorganic Forms of Phosphorus 302
14.1.2 Particulate Forms of Phosphorus 302
14.1.3 Organic Forms of Phosphorus 304
14.2 Sub-global Phosphorus Cycles 304
14.2.1 Freshwater Terrestrial Ecosystems 305
14.2.2 The Oceanic System 306
14.3 The Global Phosphorus Cycle 308
14.3.1 Fluxes between Reservoirs 309
14.3.2 The Steady-state Cycle 311
14.3.3 Perturbations 312
Questions 313
References 313

15 **Trace Metals** 317
15.1 Introduction 317
15.2 Metals and Geochemistry 317
15.2.1 Metal Abundance and Availability 317
15.2.2 Metal Mobilization 318
15.2.3 Human Activities as Geochemical Processes 318
15.3 An Overview of Metal Ion Chemistry 322
15.3.1 Introduction 322
15.3.2 Oxidation–Reduction Reactions 322
15.3.3 Volatilization 323
15.3.4 Complexation Reactions 324
15.3.5 Precipitation Reactions 328
15.3.6 Adsorption 330
15.3.7 Reactions Involving Organisms 334
15.3.8 Geochemical Kinetics 336
15.4 Observations of Metals in Natural Systems 338
15.4.1 Combining Physical and Chemical Information 338
15.4.2 Particle and Metal Interactions in Estuaries 339
15.4.3 Application of Chemical Principles to Evaluate Field Data 341
15.5 Examples of Global Metal Cycling 341
15.5.1 Mercury 342
15.5.2 Copper 346

15.6 Summary 349
Questions 350
References 351

16 Human Modification of Global Biogeochemical Cycles 353
16.1 Global Climate Change 353
16.2 Acid Precipitation 355
16.3 Food Production 355
16.4 Stratospheric Ozone Depletion 356
16.5 Oxidative Capacity of the Global Troposphere 357
16.6 Life and Biogeochemical Cycles 357
16.7 Conclusion 360
References 361

Glossary 363
Answers to Questions 365
Index 367

Authors

Theodore L. Anderson, *Doctoral Candidate, Department of Atmospheric Sciences, AK-40, University of Washington, Seattle, WA 98195, USA*

Mark M. Benjamin, *Jungers Professor of Civil Engineering, Department of Civil Engineering, FX-10, University of Washington, Seattle, WA 98195, USA*

Donald E. Brownlee, *Professor of Astronomy, Department of Astronomy, FM-20, University of Washington, Seattle, WA 98195, USA*

Samuel S. Butcher, *Professor of Chemistry, Department of Chemistry, Bowdoin College, Brunswick, ME 04011, USA*

Robert J. Charlson, *Professor of Atmospheric Sciences and Environmental Studies, Department of Atmospheric Sciences, AK-40, University of Washington, Seattle, WA 98195, USA*

Rolf O. Hallberg, *Professor of Microbial Geochemistry, Department of Geology and Geochemistry, Stockholm University, S-106 91 Stockholm, Sweden*

Kim Holmén, *Department of Meteorology, Stockholm University, S-106 91 Stockholm, Sweden*

Bruce D. Honeyman, *Visiting Scholar, School of Oceanography, University of Washington, Seattle, WA 98195, USA.* Present address: *Department of Environmental Sciences and Engineering Ecology, Colorado School of Mines, Golden, CO 80401, USA*

Daniel A. Jaffe, *Assistant Professor of Chemistry, Geophysical Institute and Department of Chemistry, University of Alaska Fairbanks, Fairbanks, AK 99775, USA*

Richard A. Jahnke, *Associate Professor, Skidaway Institute of Oceanography, Savannah, GA 31416, USA*

Russell E. McDuff, *Associate Professor of Oceanography, School of Oceanography, WB-10, University of Washington, Seattle, WA 98195, USA*

James W. Murray, *Associate Professor of Oceanography, School of Oceanography, WB-10, University of Washington, Seattle, WA 98195, USA*

Gordon H. Orians, *Professor of Zoology and Environmental Studies, Institute for Environmental Studies, FM-12, University of Washington, Seattle, WA 98195, USA*

Henning Rodhe, *Professor of Chemical Meteorology, Department of Meteorology, Stockholm University, S-106 91 Stockholm, Sweden*

Henri Spaltenstein, *Visiting Scholar, College of Forest Resources, AR-10, University of Washington, Seattle, WA 98195, USA.* Present address: *University of Lausanne, Valentine 18, 1400 Yverdon, Switzerland*

James T. Staley, *Professor of Microbiology, Department of Microbiology, SC-42, University of Washington, Seattle, WA 98195, USA*

Robert F. Stallard, *US4, Department of Geological and Geophysical Sciences, Princeton University, Princeton, NJ, USA.* Present address: *US Geological Survey, 325 Broadway, Boulder, CO 80303-3328, USA*

Fiorenzo C. Ugolini, *Professor of Forest Soils, College of Forest Resources, University of Washington, Seattle, WA 98195, USA.* Present address: *Dipartimento di Scienza del Suolo e Nutrizione della Pianta, Università degli Studi, Piazzale delle Cascine, 15, 50144 Firenze, Italy*

Gordon V. Wolfe, *Doctoral Candidate in the Graduate School of Arts and Sciences, FM-12, University of Washington, Seattle, WA 98195, USA*

Preface

Human activity is affecting the global environment in a profound way. Some of the changes result from losses of habitat and the associated extinctions of species. Other changes are due to high rates of addition of materials to the environment. Many of these human interventions now occur on a scale capable of changing the global biogeochemical cycles upon which life and the Earth's climate depend.

Biogeochemical cycles describe the transformation and movement of chemical substances in the global context. This text is designed for courses intended to present an integrated perspective on biogeochemical cycles. Courses related to this subject are found at the advanced undergraduate and graduate level in many colleges and universities. These courses are usually presented by a person with a specialty in one of the conventional scientific disciplines, supplemented by readings in other areas. Our goal has been to provide a comprehensive treatment under one cover so that the components are integrated and the need for additional reading is reduced.

The text springs from courses on biogeochemical cycles offered at the University of Washington and at the University of Stockholm for several years. The course at the University of Washington was started by two of the authors (Charlson and Murray), but an essential part of the course has been visits by faculty from other disciplines. Many of the chapters spring from the lecture notes of those visits. Some of the authors are former students in this course.

Much of the work important to the study of biogeochemical cycles is done in traditional disciplines, ranging from astronomy to zoology, with many disciplines that have developed fairly recently (such as chemical oceanography) playing very important roles. We assume that this will continue to be the case. None the less, given the nature of biogeochemistry, specialists need to understand what their disciplines can bring to the subject and what are the needs of other disciplines. To fully comprehend these cycles, a person must also integrate material from several disciplines.

Although we have tried to take a comprehensive approach, the text is admittedly idiosyncratic. In managing the compromise between depth of coverage and presentation of the topic in a reasonable size for a textbook, many viewpoints are given only limited space. We hope that readers can gain an appreciation of the scope of biogeochemical cycles, and will be adequately prepared to understand the growing literature in the field.

This book is about fundamental aspects of the science of biogeochemistry. As such, and while it is relevant to the major issues of global change, it is not issue oriented. Nor does this book attempt to review all of the research on these topics. It does, however, emphasize fundamental aspects of the physical, chemical, biological, and earth sciences that are of lasting importance for integrative studies of the Earth. We assume two characteristics in the reader's background. First, the reader is assumed to have a background in science comparable to someone who has completed a university level course in introductory chemistry. Second, the reader is expected to be involved in one of the disciplines that relate to the study of biogeochemical cycles. This includes the subdisciplines of chemistry, biology, and geology, and the science specialties that deal with soils, atmospheres, and oceans.

The text is organized into three major parts. The first part (Chapters 1–5) quickly surveys the basic sciences that underlie biogeochemistry. Some readers can read parts of this very quickly, depending on their background. For example, the undergraduate biology major will be familiar with much of the material in Chapter 3 and some of the material in Chapter 5.

The second part of the text, Chapters 6–10, surveys the important aspects of several "spheres" of the Earth.

Individual chapters are devoted to tectonic processes (lithosphere), soils (pedosphere), sediments, oceans (hydrosphere), and atmosphere. Most biologists, chemists, and physicists will see most of this as new material.

The final part, Chapters 11–15, integrates the material of the preceding parts in the examination of specific elemental cycles. Although the science of the first two parts of the book is fairly mature, our knowledge of many elemental cycles is still incomplete. This is one area where the instructor and student will want to consult current literature during the course. Further exploration of these elemental cycles and the examination of cycles not included here can provide many topics for student papers.

The final chapter brings our attention back to a number of important contemporary global environmental issues. These issues were the driving force for the development of the book in the first place. We hope that this synthesis will start some of its readers on pathways that improve our understanding of biogeochemical cycles and lead us toward the resolution of these global environmental issues.

Bert Bolin's visit to the University of Washington in 1976 provided a major stimulus for thinking about biogeochemical cycles at the university. Active work on this text began with a grant from the Rockefeller Foundation to the Institute for Environmental Studies at the University of Washington in 1978. Rockefeller assistance made it possible to bring several scientists to the University of Washington to discuss the role of their specialty in biogeochemical cycles. Visits from M. Alexander, P. L. Brezonik, P. J. Crutzen, R. A. Duce, R. O. Hallberg, H. D. Holland, M. L. Jackson, G. E. Likens, F. T. Mackenzie, S. Oden, H. Rodhe, and H. J. Simpson played important roles in shaping our approach to the text.

Several individuals played important roles during the final preparation of the manuscript. Most of the typing was done by Sheila Parker. The drafting of the figures was done by Kay Dewar and April Ryan. A gift from the Ford Motor Company assisted in the final preparation of the manuscript. Last, but not least, we owe thanks to the many students at the University of Washington and the University of Stockholm who explored this subject initially without a textbook and then with draft chapters. Their enthusiasm for the subject and their comments and criticism have helped maintain our interest in this manuscript over almost a decade.

1

Introduction

Samuel S. Butcher

1.1 An Approach to the Subject

The term "biogeochemical cycles" is used here to mean the study of the transport and transformation of substances in the natural environment, as seen in the global context. An example of such a global cycle is the hydrological cycle, presented in schematic form in Fig. 1-1 (Garrels and Mackenzie, 1971; Drever, 1982). A description of the cycle includes the amount of water in selected reservoirs and the rate at which water is moved from one reservoir to another. The oceans comprise the largest reservoir of water. Water from the oceans is pumped into the atmosphere reservoir by solar radiation. The evaporation of water is the largest non-radiative transfer of energy from the surface of the Earth to the atmosphere, amounting to about 20% of the solar flux incident on the Earth. The components of biogeochemical cycles are developed more fully in Chapter 4.

Water in the atmosphere may be returned to the ocean or it may be transported and deposited as precipitation on land. Once on the land, the water is returned to the ocean as run-off, although some of it may be returned to the atmosphere as evapotranspiration.

The hydrological cycle is typical of the global view that we will take in this book. In order to provide a complete description of the water cycle itself, as well as understand cycles with which it interacts, one must dig a bit deeper. The removal of water from the atmosphere as precipitation depends on many details of the behavior of water droplets in the atmosphere. This behavior in turn depends on substances in the atmosphere other than water. This topic is probed in Chapter 10.

Water is also very important as a medium for chemical reactions in which the water itself may not be changed very much at all. The properties of water as a medium are important in the ocean, fresh waters, cloud droplets, and in soils, sediments, and on rock surfaces. The relevant properties of water include its large dipole moment and ability to form hydrogen bonds. These determine its capacity to stabilize charged species (ions) and account for the large number of reactions in which a proton is transferred from one compound to another. Water and the hydrological cycle intersect with the cycles of most elements. Physical and chemical weathering of the Earth's surface and transport to the oceans by rivers and glaciers are significant processes for many elements.

Water illustrates a principle that must be kept in mind in thinking about biogeochemical cycles. Although we may often focus on just one element or class of compounds in order to simplify things, in many cases the processes involving one element may be strongly coupled with that of other elements.

Water is an obvious example of a substance that is global in scope. In this text, specific sources of pollution (stack emissions of sulfur dioxide) and small-scale natural processes (ammonia emissions from a feedlot) are considered only to the extent that they are significant in aggregate at the global scale. Further, the term "cycles" in the title should not imply that only closed, steady-state systems are considered, but should emphasize the importance of understanding where substances come from and what they are turned into.

Although biogeochemical cycles are fascinating for their own sake, they also connect with important regional and global scale issues. Consider the

Global Biogeochemical Cycles
ISBN 0-12-147685-5

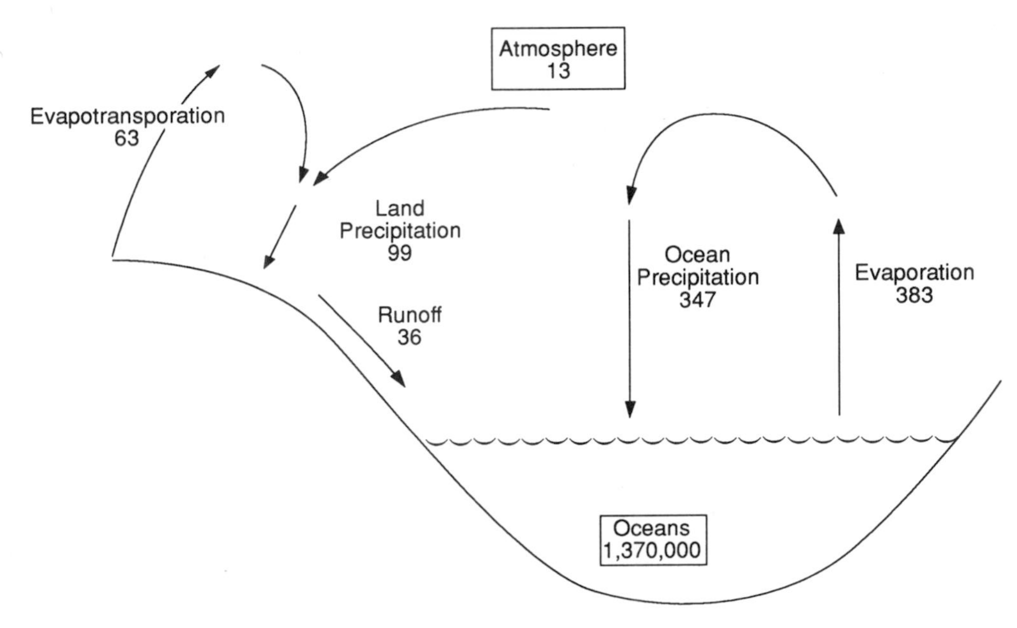

Fig. 1-1 The hydrologic cycle as an example of a global cycle. Units are 10^{18} g H_2O (burdens) and 10^{18} g H_2O/year (fluxes).

following issues:

- Food production (on land and in aquatic environments) and its dependence on temperature and other climatic factors, the availability of nutrients, and the presence of toxic agents.
- World climate and the extent to which it may be influenced by natural and cultural factors.
- Acid deposition, its effect on natural ecosystems, and the extent to which it is caused by cultural factors.
- Stratospheric ozone – its importance for human health and welfare and its sensitivity to the presence of a variety of trace gases.
- The ubiquitous nature of trace levels of DDT and other synthetic chemicals and the pathways by which these substances are dispersed and react in natural systems.

The subject offers a number of challenges that, for many different reasons, are important ones for the scientific community to meet. First, many of the issues used to define the subject of interest are important to society. That is not to say that a scientific-technical understanding of these issues will be enough to ensure a solution to those problems over which some control may be exercised. On the other hand, it seems unlikely that solutions to problems with scientific and sociopolitical components will be derived before the systems are understood.

Second, the description of these systems requires the integration of material from many disciplines. Exposure to related areas of science can be an important part of a student's scientific training. If the scientific community is not able to integrate the science necessary to describe biogeochemical systems, it seems unlikely that it will be easy for society as a whole to derive solutions for these problems.

Third, there are many aspects of biogeochemical studies that will require fresh new work in a range of traditional disciplines. To take a few examples: In some parts of the Earth, our understanding of the circulation patterns in the atmosphere and the hydrosphere does not yet permit a description of the long-range transport of substances. At the microscopic level, there are many chemical pathways in living systems for which present knowledge is far from complete. In another realm, methods are needed to handle the vast amount of data used to define these systems and to present a description of the system that is meaningful to the informed scientific layperson as well as the specialist.

Fourth, given the finite resources of most educational institutions, the development of this subject offers the challenge of thinking about how to do interdisciplinary work in a setting dominated by traditional disciplines. The establishment of a new discipline, "biogeochemistry", would sidestep the challenge of doing interdisciplinary work. Further, an increased understanding of biogeochemical systems will very likely draw heavily on new work in such areas as chemical oceanography, chemical kinetics, and analytical chemistry, to name areas which are generally chemical in nature, but which diverge substantially at the level where new methods and theories are developed.

Recognition that biogeochemical issues, such as those listed above, may be significant for mankind is not a recent event. Arrhenius (1896) began an article titled "On the influence of carbonic acid in the air on the temperature of the ground" with the statement, "A great deal has been written on the influence of the absorption of the atmosphere upon the climate. Tyndall in particular has pointed out the enormous importance of this question." The basic observations by Tyndall were made more than 30 years earlier. Arrhenius is well-known as a chemist and his influence is reflected in such concepts as Arrhenius activation energy and the Arrhenius definition of acids and bases. Arrhenius clearly expressed himself in other areas of science as well. The increased specialization of science since his time has brought many benefits but, as a result, most contemporary scientists limit their activities to a relatively narrow area. In fact, much of the need for a special treatment for biogeochemistry may stem from the development of science since the time of Arrhenius and Tyndall.

In order to locate one's own area of specialization and see how that discipline relates to the study of biogeochemical cycles, see Fig. 1-2. In this figure, the divergence of science into separate disciplines is represented as well as the limitations of any one of these disciplines in describing biogeochemical systems. Along the vertical axis, disciplines are distinguished according to whether they study complex aggregates of atoms and molecules and are limited to making relatively qualitative or descriptive statements or whether the objects of study are simpler aggregates whose behavior may be handled in more quantitative terms. In this picture, much of geology and biology are found toward the descriptive end of the spectrum, whereas physics is found toward the quantitative end. The borders between the qualita-

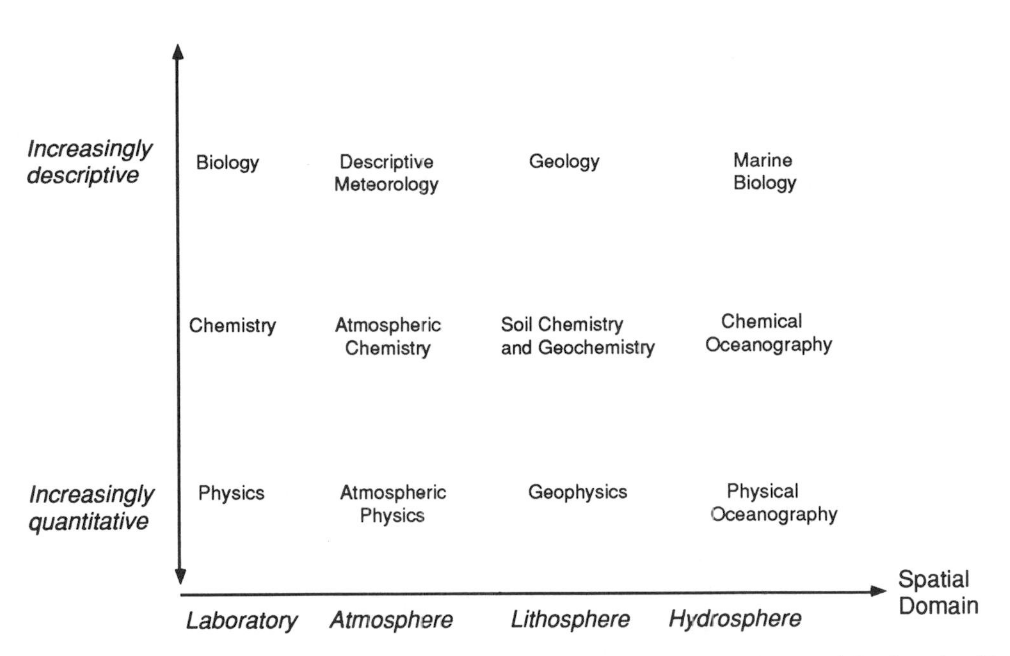

Fig. 1-2 A description of Earth science disciplines according to the approach to the subject and the domain of interest.

tive and quantitative areas are not, of course, sharp and easily described by disciplinary labels (consider biophysics).

In another dimension, as suggested by the horizontal axis, scientific disciplines diverge into fields that are domain-specific. This divergence is seen in the development of fields given to the study of oceans, the atmosphere, the lithosphere, and a fourth domain – the laboratory. Thus there are the major fields of oceanography, hydrology, atmospheric sciences, and the geological and soil sciences. In some instances, these fields have integrated the biology, chemistry, and physics necessary for a complete understanding of the domain, and in other cases they have developed into such fields as atmospheric physics, marine biology, and soil chemistry.

The laboratory has been identified as a distinct domain here to emphasize a point. In some cases, behavior of the system in the test tube or the Petri dish (*in vitro*) closely simulates the behavior of a pond or a forest ecosystem (*in vivo*). For many systems, however, care must be taken in extrapolating from the laboratory environment to the natural environment. The difference in scale of the mixing processes for the two systems and the presence of interspecies effects in the natural environment are but two factors that may be significant. This is not to say that the whole is not the sum of its parts, but simply to provide a cautionary note for the reader that it is important to recognize all of the parts and then to form the proper sum. The laboratory must be included in a description of how we go about studying natural systems for the simple reason that much experimental work of the natural sciences (biology, chemistry, and physics) occurs in the laboratory.

Contributions from many disciplines are required if one is to gain a clear understanding of the scientific-technical aspects of biogeochemical issues. The case of methane (CH_4) illustrates the roles of a range of disciplines. Methane is a constituent of natural gas and is widely used as an industrial energy source. Atmospheric levels of methane are about 1.7 ppmv and are increasing by about 1% per year. Although 1% may seem like a small rate for those concerned with double digit inflation, it is a very rapid rate of change on a geological time-scale.

Atmospheric methane is significant for a number of reasons. It is one of the gases that controls OH radical concentrations in the troposphere. Oxidation of methane is one of the chief sources of water in the stratosphere. Methane is also an active greenhouse gas. Tracing a portion of the methane picture will provide some sense of the range of disciplines involved in the study of this rather simple molecule.

Our story of methane begins with the fixation of atmospheric carbon dioxide by photosynthesis. Much of the carbohydrate (and other fixed carbon) formed is returned to the atmosphere by aerobic respiration, but a small fraction can be converted to methane by anaerobic microorganisms (Mah *et al.*, 1977). These organisms form methane in the process of obtaining energy by chopping down large carbon chains into smaller molecules. In the first step, the carbohydrate is oxidized to a carboxylic acid by reducing water to elemental hydrogen, rather than reducing elemental oxygen or sulfate. Elemental hydrogen is only an intermediate, however, as energy is also derived by the reduction of carbon dioxide by the hydrogen produced by decarboxylation of a carboxylic acid. The overall reaction may be written schematically:

$$2RCH_2OH \rightarrow RCOOH + RH + CH_4$$

As a gaseous product not soluble in water, methane will migrate toward the atmosphere, but along the way there may be organisms that derive their energy needs from the reducing power of methane. If the methane escapes oxidation, it reaches the atmosphere as a gas.

The organisms that form methane live in anaerobic (oxygen-lacking) conditions. Those significant for global methane are found in the digestive systems of ruminants (cattle, sheep, etc.) and termites and in organic carbon-rich aquatic systems (shallow freshwater sediments and rice paddies).

Once in the atmosphere, methane does not absorb visible light at wavelengths longer than 146 nm and is thus photochemically inert in the lower atmosphere. Methane does absorb in the infrared region at 3.3 and 7.7 μm and this gives rise to its role as a greenhouse gas.

The main process for the removal of methane from the atmosphere is the reaction with OH radicals. OH radicals are formed in a photochemical steady-state by processes that need not concern us here. What is important is that where there is sunlight there are OH radicals. OH radicals have a strong affinity for hydrogen atoms on other molecules. In the process, water and another radical are formed (in the case of methane):

$$CH_4 + \cdot OH \rightarrow H_2O + \cdot CH_3$$

The CH_3 radical is very reactive, and the reaction of CH_4 with OH is the first step in a series of reactions that ultimately oxidize the carbon to carbon dioxide. This reaction is not only important as a loss mechanism for methane, it also plays an important role in regulating the concentration of OH.

This small section of the methane picture thus is sketched by several disciplines that do not normally intersect. Methane formation is described by microbiologists and biochemists who have something to say about the conditions required for methane formation and use of methane by organisms. The atmospheric effects of methane are described by atmospheric scientists and the physics of radiant energy exchange. The loss of methane in the atmosphere results from a homogeneous gas phase chemical reaction that can be described in chemical kinetic terms. Putting together a global picture of methane thus involves many specialties.

It may be that development of a separate discipline of science (Fortescue, 1979) will play an important role in understanding the problems sketched above. It is probably more likely that a variety of approaches to these problems will be useful. The emphasis in this text is not on the creation of a new discipline; rather, we assume that most scientists who contribute to the field of biogeochemistry will continue to be those who devote a major effort to an area of specialization – soil microbiology, chemical kinetics, etc. At the same time, it is safe to say that our understanding of issues such as atmospheric methane will be meager until we are able to synthesize a more complete description from contributions as widely separated as chemical kinetics and microbiology. It goes without saying (almost) that the contributions to biogeochemical issues can only be enhanced when the specialists are aware of the relationship of their field to the study of biogeochemical cycles.

We do not feel that it is possible (for us), or perhaps even desirable, to present the study of biogeochemistry in the form of a series of recipes or a dogma that can be plugged into any new biogeochemical problem. We do hope to identify some of the key elements common to the study of many of these systems. We offer a variety of approaches to the problems and offer a framework within which one may critically examine any study of biogeochemistry. Finally, we hope to encourage a spirit of experimentation in the development of new ways of looking at these complex systems.

1.2 This Text

Let us briefly consider a system that will have little practical value other than to serve as a basis from which we can describe the organization of the book. The description of the Earth as a system may be undertaken by defining the biological, chemical, and physical states of all of its parts as well as the positions and kinetic energies of these parts.

Points within this many-dimensional system are defined by a location variable (denoted by X); a chemical state variable (C), and biological and physical state variables (B and P). To complete the description of the system, the densities or concentrations near a point, $N(X, C, B, P)$, will be required. These functions will in general be time-dependent. Although we might think of this system as a cube, it should actually be represented as an n- space and not bound in the sense implied by the cube. Further notes on the variables follow.

The location variables (x, y, and z in a Cartesian coordinate system) vary continuously. We may, however, choose to study a selected domain; for example, allow x, y, and z range so as to include only the oceans or only a particular estuary. The location variable may also be chosen to range only over a space that is homogeneous with respect to density or concentration. This would be the beginning of a box model (described in Chapter 4).

The chemical states of substances help to define the state of the system. In some cases, it may be important to distinguish between chemical isomers (propylene and cyclopropane) or isotopic forms (^{12}C and ^{14}C). In other cases, carbon compounds of broadly defined classes, such as organic carbon, may be lumped together. The biological state not only specifies (implicitly) a mass of material, but more importantly, it specifies a function and a potential for bringing about chemical and physical changes. The chemical and physical state of the Earth is quite different from that of Venus and Mars as a result of the biological component on Earth (see, e.g. Lovelock, 1979). Further specification of the biological state may involve descriptions of communities of organisms or even individual species. Along with specification by name and number of organisms must be included a description of how the organism relates to its surroundings. What is the organism sensitive to? Is it heterotrophic? What chemical changes does it bring about? The physical state variable includes temperature and the state of aggregation: gas, liquid, or solid including the phase (calcite or aragonite).

The physical relationships of various regions of the system and the transfer processes must also be defined. This will include a description of the important chemical reactions and their rate constants, dispersion and transport processes, and the fact that sediments and the oceans share a surface. The internal structure can be complex. For example, the population of kelp in a portion of the ocean can be coupled to the population of sea otters through the harvesting of kelp by the sea urchins and predation of sea urchins by sea otters. Exogenous inputs and outputs such as the influx of solar radiation and meteoritic matter and the efflux of infrared radiation, helium, and hydrogen are obvious examples when the system represents the entire Earth.

Mass, density, concentration, and temperature are examples of aggregate variables. Variables such as these are used because they simplify the description of the system and have applications in predicting its behavior (in this case, through the laws of conservation of mass and energy). It is likely that, for some systems, other aggregate variables will be useful. Entropy and free energy are among the thermodynamic variables that are useful for some purposes. Trophic level (as carnivore) and soil classification (as B horizon) are aggregates that serve obvious useful purposes. There are a few observations to be made concerning the use of aggregate variables. First, the student of biogeochemistry should become familiar with the use of more common aggregate variables and imaginative in the derivation of new ones. Second, one should be aware that although aggregate variables serve a useful purpose, there is a fiddler's fee to be paid. Some information about the system may be lost. The use of temperature as a measure of average energy is analogous to a description of a population by average family size. The average may be useful, but for some purposes it is also important to know something about the number of large or small families (or the fraction of highly energetic molecules).

The study of biogeochemistry involves the descriptions of the *distribution* of material throughout our system and the *rates* at which material moves from region to region. The organization of the book is divided into three major sections. The first section, consisting of Chapters 1–5, describes the processes and characteristics that are important in all parts of the system: the mix of elements present on the Earth and processes important on a cosmological scale; characteristics of biological systems; the role of equilibrium and reaction rates; and applications of box models and the role of time.

The second major section describes the characteristics of some of the major domains or "spheres" of the system. We define these subspaces as lithosphere, pedosphere, sediments, oceans, and atmosphere, and they are examined in Chapters 6 through 10.

The final section examines the biogeochemistry of selected chemical elements. Each of these chapters builds on earlier sections. Thus, the description of carbon will make use of the characteristics of biological systems, definitions of time-scales, and properties of the oceans described in earlier chapters. Five groups of substances are considered in Chapters 11 through 15. Chapter 16 brings us back to some of the questions important to society.

It is not the intent of this text to present a picture of the entire biogeochemical system. Our goals for the moment are more modest. Our aim is to break the large system down to manageable pieces and show how they can be described. At the same time, we want to keep in mind the relationship of these small pieces to the larger system. Our reduction of the larger system is presented in the final group of chapters. There are many ways to achieve this reduction, however, and the manner selected depends in large part on the questions being asked. Do we define our interest by asking: "What happens to sulfur compounds in the natural environment?" or "What effect will a reduction of anthropogenic sulfur emissions have on the fine particle concentration in the atmosphere?" The manner chosen for the reduction will also be determined by the internal structure of the system and by the nature of factors that are well-known and by those that are incompletely characterized. Through all of this we must keep in mind the caveat expressed by Wheeler (1982): "The observer's choice of what he shall look for has an inescapable consequence for what he will find."

The definition of cycles and reservoirs will also depend on the perspective of the person undertaking the study. The meteorologist will describe one portion of our large system in great detail while treating the ocean as a less interesting exogenous factor. The geologist will concentrate on a different portion of the system and the laboratory scientist will select yet another area.

To summarize, there are a few points to be carried from the preceding discussion into the rest of the book. First, there are many disciplines that have special insights to offer to the study of biogeochemical cycles. Second, there are many ways in which biogeochemical systems may be approached in defining cycles, reservoirs, and aggregate variables. The reader is encouraged to proceed with a spirit of adventure tempered by the need to examine the results critically.

Questions

1-1 The following papers might be examined as a class project. The objective is to gain an understanding of (1) the range of disciplines represented; (2) what each of the papers contributes to biogeochemistry; and (3) what linkages may exist between the subjects of these papers.

- Andreae, M. O. and H. Raemdonck (1983). Dimethylsulfide in the surface ocean and the marine atmosphere: A global view. *Science* **221**, 744–747.
- Cox, R. A. and D. Sheppard (1980). Reactions of OH radicals with gaseous sulfur compounds. *Nature* **284**, 330–331; **or** Atkinson, R., R. A. Perry, and J. N. Pitts (1978). Rate constants for the reaction of OH radicals with COS, CS_2, and CH_3SCH_3 over the temperature range 299–430 K. *Chem. Phys. Lett.* **54**, 14–18.
- Maroulis, P. J., A. L. Torres, A. B. Goldberg, and A. R. Bandy (1980). Atmospheric SO_2 measurements on Project Gametag. *J. Geophys. Res.* **85**, 7345–7349.
- Huebert, B. J. and A. L. Lazrus (1980). Bulk composition of aerosols in the remote troposphere. *J. Geophys. Res.* **85**, 7337–7344.
- Galloway, J. N., G. E. Likens, and M. F. Hawley (1984). Acid precipitation: Natural versus anthropogenic components. *Science* **226**, 829–831.
- Lindberg, R. D. and D. D. Runnells (1984). Ground water redox reactions: An analysis of equilibrium state applied to Eh measurements and geochemical modelling. *Science* **225**, 925–927.

1-2 Describe the studies in the following paper. What precautions must be taken in extrapolating from laboratory studies to a field situation? Zimmerman, P. R., J. P. Grenberg, S. O. Wandiga, and P. J. Crutzen (1982). Termites: A potentially large source of atmospheric methane, carbon dioxide, and molecular hydrogen. *Science* **218**, 563–565.

References

Arrhenius, S. (1896). On the influence of carbonic acid in the air on the temperature of the ground. *Phil. Mag., Ser. 5* **41**, 237–276.

Drever, J. I. (1982). "The Geochemistry of Natural Waters" Prentice-Hall, Englewood Cliffs, N. J.

Fortescue, J. A. C. (1979) "Environmental Geochemistry." Springer-Verlag, New York.

Garrels, R. M. and F. T. Mackenzie (1971). "Evolution of Sedimentary Rocks." W. W. Norton, New York.

Lovelock, J. E. (1979). "Gaia: A New Look at Life on Earth." Oxford University Press, Oxford.

Mah, R. A., D. M. Ward, L. Baresi, and T. L. Glass (1977). Biogenesis of methane. *Ann. Rev. Microbiol.* **31**, 309–341.

Wheeler, J. A. (1982). *In* "Mind and Nature" (R. Q. Elvee, ed.). Harper, San Francisco, Calif.

2

The Origin and Early Evolution of the Earth

D. E. Brownlee

2.1 The Origin of the Elements

Biogeochemical cycles are defined in part by the chemical elements present and their distribution in various parts of the Earth. This chapter describes how the elements that made up the solar nebula were formed. The condensation and differentiation of the elements within the nebula to form the planets are then outlined. These processes gave Earth the inventory of elements that begin to define its biogeochemistry.

The Earth is the end-product of a series of evolutionary processes that began at the time the first elements were produced 10–20 billion years (Ga) ago. The Earth's composition, its evolution, and many of its chemical and physical cycles described in this book are influenced to various degrees by processes that occurred during or before the formation of the solar system. These processes include nuclear reactions to produce the elements, gravitational collapse to produce stars and proto-planetary systems, condensation to produce solid grains, and accretion to accumulate grains into planets. The basic framework of this scheme is believed to be generally well understood, although many of the details of even the most fundamental processes such as condensation and accretion are highly uncertain. Probably the most remarkable and even haunting aspect of the Earth's genesis is that in our planetary system, the only one known, the Earth has unique properties. These unique features include a hydrosphere, vigorous plate tectonics, a large moon, and an oxygen-rich atmosphere that is in chemical disequilibrium. In tracing the development of the Earth, we will start at the very first step, the origin of the elements from which the Earth formed.

The composition of the Earth was determined both by the chemical composition of the solar nebula, from which the Sun and planets formed, and by the nature of the physical processes that concentrated materials to form planets. The bulk elemental and isotopic composition of the nebula is believed or usually assumed to be identical to that of the Sun. The few exceptions to this include elements and isotopes such as lithium and deuterium that are destroyed in the bulk of the Sun's interior by nuclear reactions. The composition of the Sun as determined by optical spectroscopy is similar to the majority of stars in our galaxy and, accordingly, the relative abundances of elements in the Sun are referred to as "cosmic abundances". Although the cosmic abundance pattern is commonly seen in other stars, there are dramatic exceptions, such as stars composed of iron or solid nuclear matter, as is the case with neutron stars. The best estimation of solar abundances is based on data from optical spectroscopy and meteorite studies and in some cases extrapolation and nuclear theory. The measured solar abundances are listed in Fig. 2-1 and Table 2-1. It is believed to be accurate to about 10% for the majority of elements. The major features of the solar abundance distribution are a strong decrease in the abundance of heavier elements, a large deficiency of Li, Be, and B, and a broad abundance peak centered near Fe. The factor of 10 higher

Global Biogeochemical Cycles
ISBN 0-12-147685-5

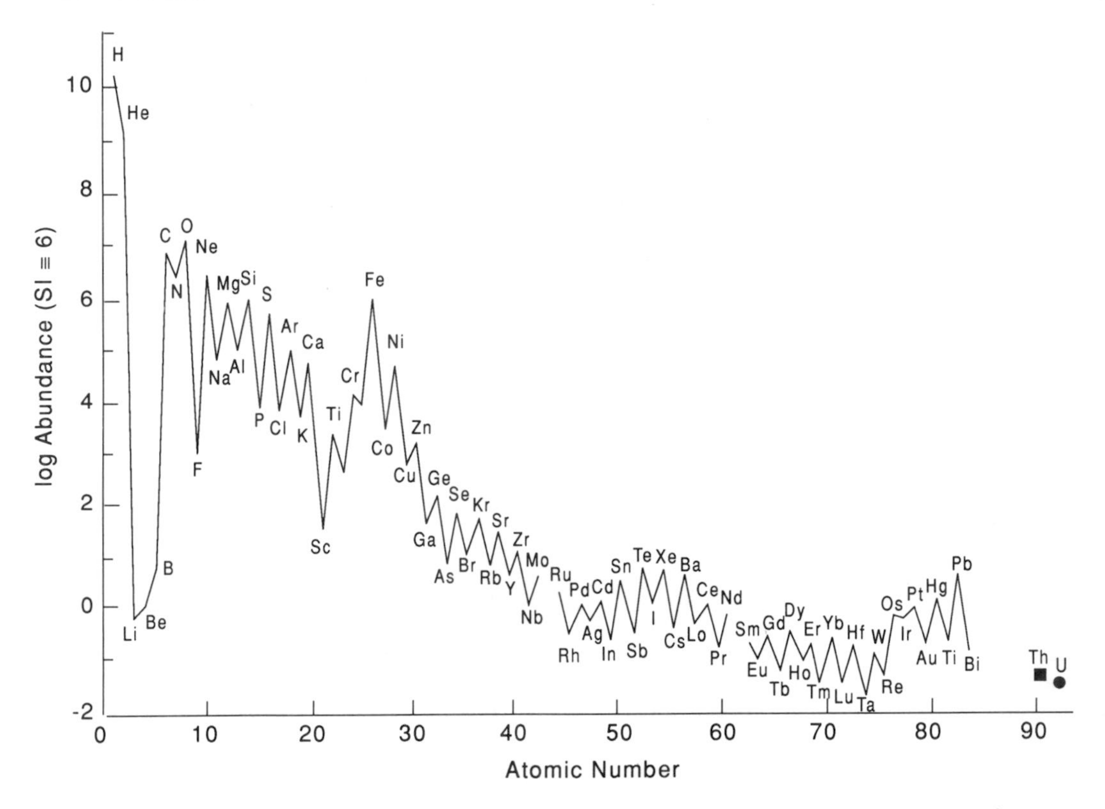

Fig. 2-1 Cosmic (solar) abundances of the elements, relative to Si, which is given the arbitrary value of 10^6.

abundance of even atomic number nuclei relative to their immediate odd atomic number neighbors is due to the higher binding energy of nuclei with even numbers of protons. The abundance curve of odd nuclei plotted against mass is a very smooth function. The cosmic abundance pattern is the net result of nuclear reactions that occurred during the origin of the universe and in the interiors of later generations of stars.

Over 98% of the atoms in the Sun are H and He that are believed to have formed in the "Big Bang", the origin of the universe. Nucleosynthesis that occurred during the Big Bang produced basically the cosmic abundance of ^{1}H and ^{4}He, and small amounts of ^{2}H, ^{3}He, and ^{7}Li, but essentially no heavier elements. The Li/H ratio produced by the Big Bang was about 10^{-9}. Element formation occurred over a 15-min time period during which temperatures dropped from very high values to less than about 10^9 K. After this brief period of expansion and cooling, the universe no longer contained matter that was hot and dense enough for nuclear reactions to occur. The fundamental nuclear reaction that occurred in the Big Bang was the fusion of hydrogen to form helium. Synthesis stopped at Li^7 because formation of the next abundant element, carbon, required higher densities than existed in the universe at the appropriate temperature range. Without future generations of hot dense matter, in the form of stars, there would never have been any elements that were heavier or more chemically interesting than H, He, and Li.

The H and He produced in the Big Bang served as "feed stock" from which all heavier elements were created. Less than 1% of the H produced in the Big Bang has been consumed by subsequent element production and thus heavy elements are rare. Essentially, all of the heavier elements were produced in stars. Following the Big Bang, the universe expanded to the point where instabilities formed galaxies, mass concentrations from which up to 14^{14} stars could form. Inside stars, the tempera-

Table 2-1 Solar abundance of the elements (atoms/10^6 atoms of Si)[a]

	Element	Abundance		Element	Abundance
1	H	2.79×10^{10}	47	Ag	0.486
2	He	2.72×10^{9}	48	Cd	1.61
3	Li	57.1	49	In	0.184
4	Be	0.73	50	Sn	3.82
5	B	21.2	51	Sb	0.309
6	C	1.01×10^{7}	52	Te	4.81
7	N	3.13×10^{6}	53	I	0.90
8	O	2.38×10^{7}	54	Xe	4.7
9	F	843	55	Cs	0.372
10	Ne	3.44×10^{6}	56	Ba	4.49
11	Na	5.7×10^{4}	57	La	0.446
12	Mg	1.075×10^{6}	58	Ce	1.16
13	Al	8.49×10^{4}	59	Pr	0.167
14	Si	1.00×10^{6}	60	Nd	0.828
15	P	1.04×10^{4}	62	Sm	0.258
16	S	5.15×10^{5}	63	Eu	0.0973
17	Cl	5.24×10^{3}	64	Gd	0.333
18	Ar	1.01×10^{5}	65	Tb	0.0603
19	K	3.77×10^{3}	66	Dy	0.394
20	Ca	6.11×10^{4}	67	Ho	0.0889
21	Sc	34.2	68	Er	0.251
22	Ti	2.40×10^{3}	69	Tm	0.0378
23	V	293	70	Yb	0.248
24	Cr	1.35×10^{4}	71	Lu	0.0367
25	Mn	9.55×10^{3}	72	Hf	0.154
26	Fe	9.0×10^{5}	73	Ta	0.027
27	Co	2.25×10^{3}	74	W	0.133
28	Ni	4.93×10^{4}	75	Re	0.0517
29	Cu	522	76	Os	0.675
30	Zn	1.26×10^{3}	77	Ir	0.661
31	Ga	37.8	78	Pt	1.34
32	Ge	119	79	Au	0.187
33	As	6.56	80	Hg	0.34
34	Se	62.1	81	Tl	0.184
35	Br	11.8	82	Pb	3.15
36	Kr	45	83	Bi	0.144
37	Rb	7.09	90	Th	0.0335
38	Sr	23.5	92	U	0.0090
39	Y	4.64			
40	Zr	11.4			
41	Nb	0.698			
42	Mo	2.55			
44	Ru	1.86			
45	Rh	0.344			
46	Pd	1.39			

[a] From Anders and Grevesse (1989).

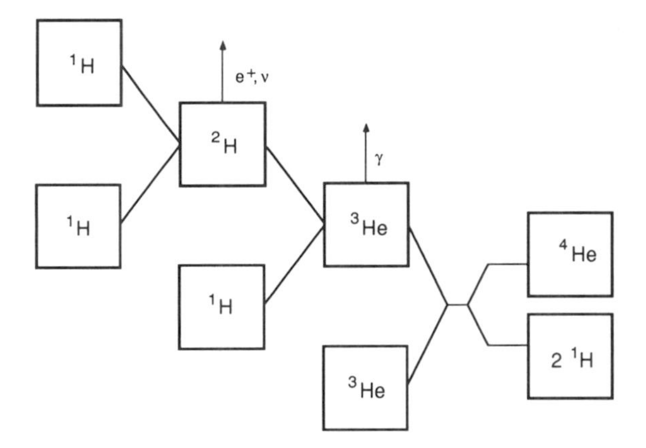

Fig. 2-2 Nuclear reactions of the H-to-He primary fusion sequence in the Sun.

tures and densities are sufficiently high for synthesis of heavy elements. The reaction that is the major energy source for stars similar to the Sun is the fusion of H to He. The basic reactions (the proton–proton chain) in the Sun are shown in Fig. 2-2. Although these reactions are the major source of energy in the solar system, they proceed at a slow and uniform rate. In the Sun's core, where the temperature is 15×10^6 K, the lifetime of a proton before it is fused to deuterium is 10 Ga. The average energy generation rate for the entire Sun is only 2 ergs/gs (200 μW/kg). Hydrogen burning occurs in stars whose interior temperatures are in the 10^7–10^8 K range, and while there are several reaction chains depending on the stellar mass, the general fusion reactions forming He are similar to those that occurred during the Big Bang.

For stars like the Sun, the burning of H to He occurs deep in stellar interiors at a fairly stable rate for over 90% of the star's lifetime. For the Sun, this "main sequence" stage will last for about 10 Ga before H depletion and rising core temperature initiate a set of more energetic nuclear reactions that occur in the final stages of its evolutionary lifetime. Throughout the geological record of the Earth, the Sun has been a main sequence star burning H to He. On the basis of theory and observations of similar stars, it is expected that the total luminosity (total radiated power) emitted from the Sun should not undergo large changes. Small changes should have occurred, however, and these may have had significant effects on the Earth and its physical and chemical cycles. The major predictable effect is a gradual increase in the solar luminosity. The increase in the mean molecular weight of nuclei in the Sun's core as H is burned results in a slow increase in interior temperature. To maintain the star in hydrostatic equilibrium, the internal pressure must remain constant to support the weight of overlying matter. To maintain constant pressure as the mean molecular weight of the gas increases, the temperature must rise. The temperature rise results in increased energy generation and it has been estimated that over the past 4 Ga, the "solar constant" has increased by at least 30%. Astronomically, such a change is minor but the stress this change places on terrestrial processes is very large. In fact, it is quite remarkable that the surface temperature of the Earth has remained constant to a level of about 3% while the solar luminosity has increased. A 30% increase in the Sun's output would increase the temperature of the sunlight side of a 300 K blackbody by 20 K. The remarkable stability of the Earth's temperature in light of solar changes and large changes in the composition of the atmosphere has led to the novel "Gaia" hypothesis, which suggests that biological organisms on the Earth collectively act to moderate the long-term atmospheric environment to the mutual benefit of terrestrial life (Lovelock, 1979). An alternative hypothesis, not involving organisms, is described by Kasting *et al.* (1988).

Following the long duration of hydrogen burning, stars enter the red giant phase where increasingly heavier elements are produced. Increasing temperatures in the stellar cores allow more massive, highly charged nuclei to collide with sufficient energy to penetrate coulomb barriers and initiate fusion reactions. The first major step is the fusion of He to form C, a reaction that takes place above 10^8 K. This occurs by the triple alpha process, an interaction that requires essentially a three-body collision between He nuclei. The nearly simultaneous collision of three particles requires high densities and is the reason why this reaction did not occur in the Big Bang. In the triple alpha sequence, two He nuclei collide to form a highly unstable ^{8}Be nucleus. The ^{8}Be must then interact with a third He to form ^{12}C on a very short time-scale because of the 10^{-16} s decay time of ^{8}Be. The reaction is very temperature-dependent and the He burning phase is violent, unstable, and relatively short-lived. At temperatures above 10^8 K, fusion reactions can produce elements up to Fe. From He to Fe the binding energy per nucleon increases with atomic number and fusion reactions are usually exothermic and provide an energy source. Beyond Fe the binding energy per nucleon decreases and exothermic reactions do not occur. Up to Fe many of the nuclei are products of alpha reactions, which involve fusion with a He nucleus. Because of this and the fact that there is high binding energy for nuclei that are multiples of ^{4}He, all of the most abundant isotopes for elements up to Fe are multiples of ^{4}He (i.e. ^{12}C, ^{16}O, ^{32}S, ^{24}Mg, ^{28}Si, etc.; see Fig. 2-1).

During the red giant phase of stellar evolution, free neutrons are generated by reactions such as ^{13}C(α,n) ^{16}O and ^{22}Ne(α,n) ^{25}Mg. The (α,n) notation signifies a nuclear reaction where an alpha particle combines with the first nucleus and a neutron is ejected to form the second nucleus. The neutrons, having no charge, can interact with nuclei of any mass at the existing temperatures and can in principle build up the elements to Bi, the heaviest stable element. The steady source of neutrons in the interiors of stable, evolved stars produces what is known as the "s process", the build-up of heavy elements by slow interaction with a low flux of neutrons. The more rapid "r process" occurs in explosive environments where the neutron flux is high. The mechanism of the s process is illustrated in Fig. 2-3. Starting with a seed isotope, successive neutron captures build up increasingly neutron-rich isotopes of the same element until an unstable isotope is reached. The typical decay of this neutron-rich isotope is beta decay, which produces the next element in the periodic table. The new element will have one more proton and one less neutron than its radioactive parent. In beta decay, a neutron disintegrates into a proton, a neutrino, and an ejected electron (beta particle). The new element created in the s process will then add new neutrons until it reaches a neutron-rich isotope that undergoes beta decay to form yet the next element. The s process can produce the isotopes along the "valley of beta stability" in the chart of the nuclides, the chart of isotopes plotted on a graph of total neutrons in nuclei *vs* atomic number. The relative abundances of isotopes produced by the s process is proportional to their binding energies and inversely proportional to their neutron capture cross-sections. Tightly bound nuclei have small cross-sections and are slower to absorb neutrons to form heavier isotopes. The s process cannot produce isotopes that are particularly neutron-rich or neutron-poor, but it can produce most of the cosmically abundant isotopes between ^{56}Fe and ^{209}Bi.

Because the path of the s process is blocked by isotopes that undergo rapid beta decay, it cannot produce neutron-rich isotopes or elements beyond Bi, the heaviest stable element. These elements can be created by the r process that is believed to occur in cataclysmic stellar explosions such as supernovae.

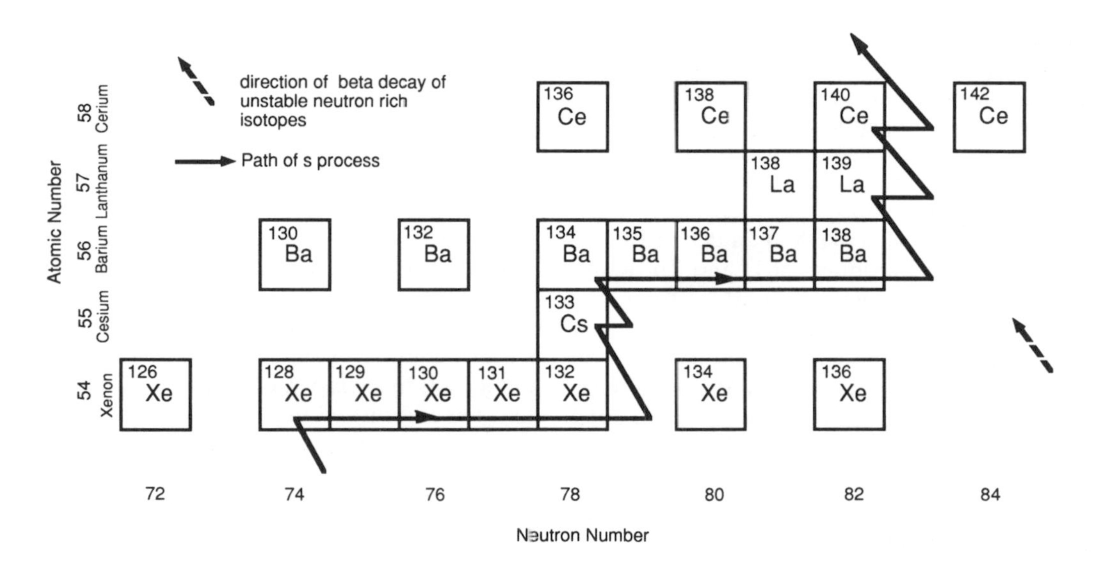

Fig. 2-3 Schematic showing the path of the s process. The isotopes ^{134}Xe, ^{136}Xe, and ^{142}Ce are beyond the reach of s process nucleosynthesis and are only produced by the r process.

In the r process, the neutron flux is so high that the interaction time between nuclei and neutrons is shorter than the beta decay lifetime of the isotopes of interest. The s process chain stops at the first unstable isotope of an element because there is time for the isotope to decay, thus forming a new element. In the r process, the reaction rate with neutrons is shorter than beta decay times, and very neutron-rich and highly unstable isotopes are created that ultimately beta decay to form stable elements. For the elements in Fig. 2-3, the decay paths of these unstable isotopes would be from the lower right following the 45° slope of β decay. The r process can produce neutron-rich isotopes such as ^{134}Xe and ^{136}Xe that cannot be reached in the s process chain (Fig. 2-3).

Thus the origin of the elements began in the Big Bang, but the formation of most of the elements important for physical and chemical processes on the Earth occurred in stars. The element production process required cycles of star formation, element formation in stellar cores, and ejection of matter to produce a gas enriched in heavy elements from which new generations of stars could form. The atoms in the Earth are products of reactions in a large number of stars; typical atoms have been cycled through several generations of stars. The synthesis of material and subsequent mixing of dust and gas between stars produced the solar mix of elements that are termed cosmic abundances. Although some isotopes are exceedingly rare, nuclear processes in stars have produced every known stable isotope.

In addition to stable elements, radioactive elements are also produced in stars. The unstable but relatively long-lived isotopes ^{40}K, ^{232}Th, ^{235}U, and ^{238}U are the internal heat source that drives volcanic activity and processes related to internal convection in the terrestrial planets. The short-lived trans-uranium elements such as Rn and Ra that are found on the Earth are all products of U and Th decay. These isotopes are sometimes used as tracers of natural terrestrial processes and cycles. Long-lived isotopes, such as ^{87}Rb and ^{147}Sm, are used for the precise dating of geological samples. When the solar system formed, it also contained several short-lived isotopes that have since decayed and are now extinct in natural systems. These include ^{26}Al, ^{244}Pu, ^{107}Pd, and ^{129}I. ^{26}Al, with a half-life of less than 1 Ma, is particularly important because it is a potentially powerful heat source for planetary bodies and because its existence in the early solar system places tight constraints on the early solar system chronology.

2.2 The Origin of the Solar System

The Sun and planets formed 4.55 Ga ago from interstellar gas and dust. This interstellar material had a bulk cosmic elemental composition, but the elements in the initial material were highly fractionated between the solid and gaseous phases. Most of the condensable materials were in the form of submicron dust grains, whereas materials that do not condense under astrophysical conditions, such as H, He, and the noble gases, were gaseous. The grains are believed to be mixtures of silicates and carbonaceous matter. In popular models for grains they have silicate cores about 100 nm in diameter coated with a 100-nm thick mantle of compounds composed primarily of H, C, N, and O. Irradiation by ultraviolet and charged particles likely leads to the formation of complex cross-linked polymers in the mantles.

The formation of the solar system is believed to have begun when a cloud of gas and dust became unstable to gravitational collapse and started an essentially unconstrained freefall. During freefall, the dimensions of the cloud decrease by a factor of nearly 1000 by the time a stable rotating lenticular nebula is established. The nebula is stable to further collapse because gravitational forces are countered by gas pressure and centrifugal forces. The condition for gravitational collapse of a cloud is that the internal gravitational potential energy exceed twice the kinetic energy. This is expressed by:

$$\frac{3GM^2}{5R} > \frac{3kTM}{\mu m_H}$$

where k is the Boltzman constant, R the initial cloud radius, G the universal gravitational constant, M the cloud mass, T the temperature, μ the mean molecular mass, and m_H the mass of the proton. The collapse of interstellar material to form the solar nebula takes on the order of 10^5–10^6 years depending on the initial conditions. During most of the collapse, the cloud is transparent to its own emitted infrared radiation and it cools to a temperature near 10 K until nebular densities are reached and the cloud becomes opaque. During this isothermal collapse phase, any condensables not originally on grains certainly condensed. The solar nebula that formed is believed to have been a stable rotating disk somewhat larger than the present planetary system. The Sun formed in the center and the planets formed from materials that accumulated in the disk. The planets Jupiter and Saturn must have formed by some variant of gravitational collapse, because to a good approximation their elemental composition matches that of the bulk nebula. Most of their mass is H and He, elements that could only have been in gaseous form. The other planets apparently formed out of solids. The outer planets Uranus and Neptune formed from icy and rocky materials, while the terrestrial planets, Mars, Earth, Venus, and Mercury, formed exclusively from rocky and metallic particles.

The solar nebula was hot and dense near its center and became cooler and more diffuse with increasing radial distance. Modern theories of evolution in the nebula indicate that the major heat source was frictional viscosity within the rotating disk of gas. The viscosity and associated redistribution of energy and angular momentum was the result of convective movement of gas in a gas disk that had differential rotation with radial distance from its center. In the outer regions of the nebula, it is likely that heating was never sufficient to vaporize pre-existing interstellar dust grains. In the inner regions of the nebula, the original grains were apparently vaporized or extensively altered and the inner (terrestrial) planets must have formed from second-generation solids that condensed from the nebular gas. This condensation process is temperature-dependent and it influenced the composition of the planets that formed. The bulk of the mass of each planet appears to have formed largely from local material in a "feeding zone", an annular ring of nebular material. Grains condensed and were then eventually swept up to form planetary bodies. The composition of material in feeding zones depended on temperature and radial distance from the center of the nebula, and this determined planetary compositions.

2.3 Condensation

The sequence of condensation of solids from a solar composition at a nebular pressure of 10 Pa (about 10^{-4} atm) is shown in Fig. 2-4. This sequence is calculated for what solids could exist in equilibrium with the solar nebula at various temperatures. Attainment of true equilibrium requires the lack of nucleation barriers, and efficient diffusion within solids so that grain interiors can maintain equilibrium with the gas phase. While strict equilibrium condensation may have occurred at higher temperatures, perfect equilibrium probably did not occur with grains larger than a few micrometers in size or at lower temperatures where diffusion is slow. At temperatures sufficiently above 1500 K, all elements were in the gas phase. The first solid grains to condense would be the highly refractory but cosmically rare elements like Pt, Os, Ir, and Re. The first abundant solids to form are oxides and silicates of Ca, Al, and Ti. Inclusions in certain meteorites are rich in these elements, although it is still not clear whether they are actual preserved condensates or refractory residues of volatilized material. Around 1400 K, compounds of the most abundant elements in the Earth – Mg, Si, and Fe – condense. At this high temperature, Mg and Si condense as Fe-free silicates and Fe condenses as an FeNi metal alloy. At lower temperatures, silicates that maintain equilibrium with the gas can incorporate FeO. At 750 K, independent of pressure, Fe metal in contact with the nebula should react with H_2S gas and form FeS, the first sulfur-bearing solid. Iron remaining in contact with gas below 450 K should react with H_2O to form magnetite, Fe_3O_4. Also at this temperature, the first water can be incorporated into solids in the form of bound and trapped water in hydrated silicates. At temperatures below 250 K, water and clathrates of methane and ammonia can form. Ultimately, if the temperature ever drops as low as 40 K, pure methane ice can exist. The approximate temperatures at which materials condensed and accreted to form the planets are indicated on the left-hand side of Fig. 2-4. While there certainly must have been complications, such as radial mixing of material and non-equilibrium condensation and grain destruction, the equilibrium condensation sequence is generally consistent with the observed properties of the planets and meteorites.

Evidence for condensation is seen in the meteorites, fragments of the asteroids that formed in the region between Mars and Jupiter. The stony meteorites that have elemental compositions that closely match those of the Sun (except for volatile elements such as H, He, N, etc.) are called *chondrites* after the presence of small spherical particles called *chondrules*. Although the chondrites generally contain close to undifferentiated solar composition, there is elemental fractionation in these objects that is related to condensation processes. Different chondrite groups are distinguished by Fe/Si ratios that vary by 50%, ratios of Ca, Al, and Ti to Si that vary by about 40%, and abundances of volatile elements, such as Cd, Bi, In, and Pb, that vary by orders of magnitude. The depletion of volatile elements is believed to be due to incomplete condensation. The correlated depletion of Ca, Al, and

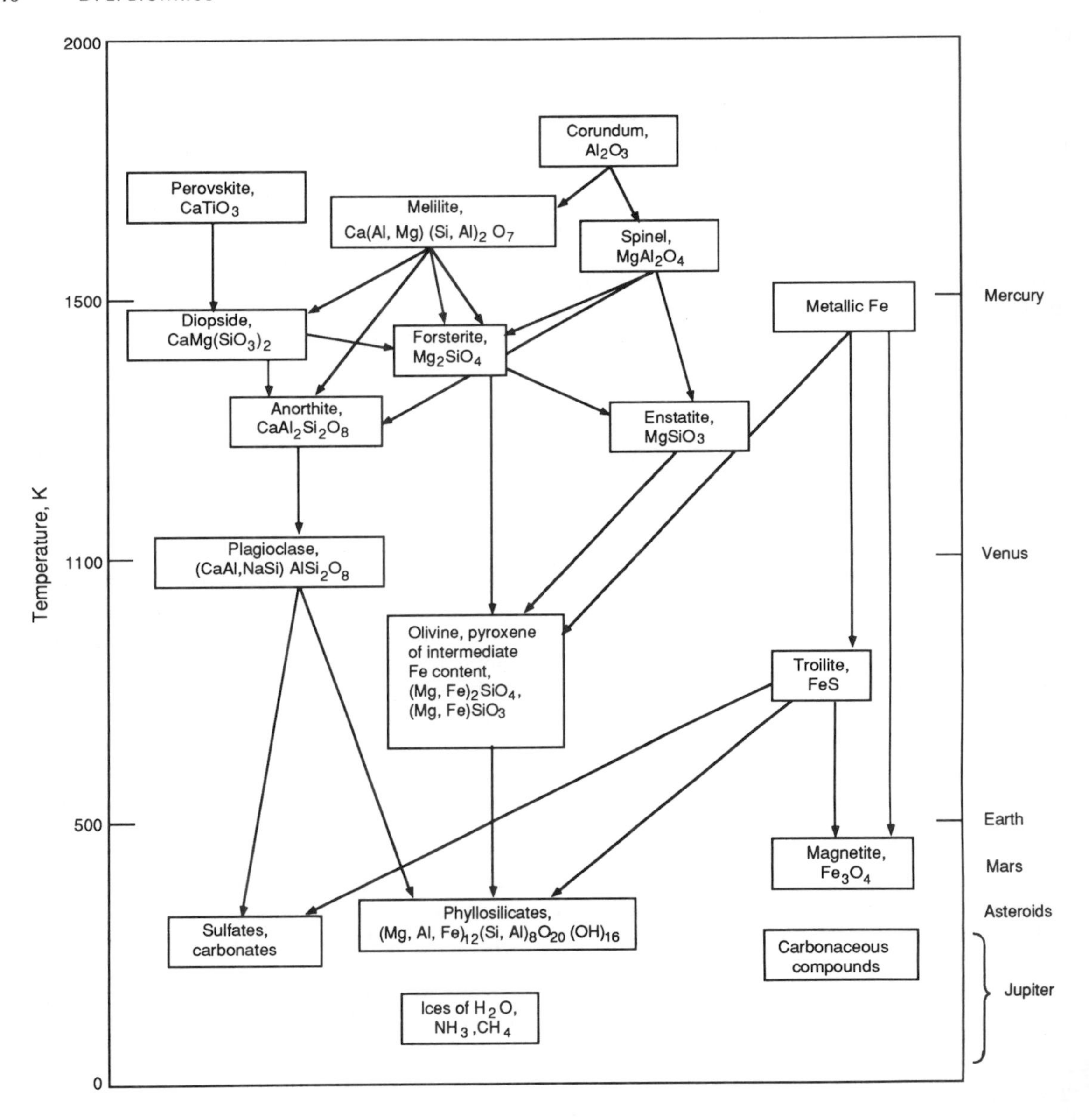

Fig. 2-4 The sequence of condensation of solids from a solar composition gas at a nebular pressure of 10 Pa (*c*. 10^{-4} atm). Adapted from Wood (1979) with the permission of Prentice-Hall, Inc.

Ti in all but the most primitive meteorites is most likely due to the loss of an early condensate that was composed of these elements. An early condensate of Ca, Al, Ti, and other highly refractory elements could have been separated by accretion from the region of the nebula where grains condensing later formed the Al-depleted chondrites. The Fe/Si fractionation may be the result of the different accretion efficiencies of metal and silicate grains, but even this is affected by condensation temperature, because it determines how Fe is divided between metal, sulfide, oxide, and silicate phases.

The effects of condensation are also seen in the bulk compositions of the planets and their satellites. The outer planets, Uranus and Neptune, have overall densities consistent with their formation from icy and stony solids. The satellites of Uranus have typical densities of $1.3\,g/cm^3$, indicative of a large ice component. The inner planets have uncompressed densities ranging from 3.4 for Mars to 5.5 for Mercury. The inner planets are composed of silicates and metal. The range in densities among the terrestrial planets is largely due to differences in the oxidation state of Fe. Most of the mass of these planets is composed of Fe, Mg, and Si, with sufficient oxygen to oxidize Si and Mg totally. The elements Fe, Mg, and Si appear to occur at solar atomic abundances of approximately 1 : 1 : 1. The oxidation state of iron ranges from almost totally oxidized in the case of Mars to totally reduced as in Mercury. This range in oxidation state is consistent with equilibration with nebular gas at low and high temperatures respectively. The oxidation state of Fe in chondrites also ranges from totally reduced in the case of the enstatite chondrites to totally oxidized in the carbonaceous chondrites. Even though the meteorites probably formed beyond the region of the terrestrial planets, the range of oxidation state is undoubtedly the result of nebular pressures and temperatures.

Water and carbon play critical roles in many of the Earth's chemical and physical cycles and yet their origin on the Earth is somewhat mysterious. Carbon and water could easily form solid compounds in the outer regions of the solar nebula, and accordingly the outer planets and many of their satellites contain abundant water and carbon. The type I carbonaceous chondrites, meteorites that presumably formed in the asteroid belt between the terrestrial and outer planets, contain 5% (m/m) carbon and 20% (m/m) water of hydration. By comparison, the terrestrial planets are depleted in carbon and water by orders of magnitude relative to the carbonaceous chondrites. Actually, it is remarkable that the Earth contains any of these compounds at all. An example of what could have happened is the case of Earth's moon, where indigenous carbon and water are undetectable. Looking at Fig. 2-4, it can be seen that no water or carbon-bearing solids should have condensed by equilibrium processes at the temperatures and pressures that probably were typical in the zone of the solar nebula that produced the Earth. Water of hydration does not occur in silicates until the temperature is below about 350 K, and ice could not exist in nebular conditions until the temperature was below 200 K. Temperatures low enough for even the formation of hydrated phases probably did not exist within the region of the terrestrial planets. The origin of carbonaceous materials is even more mysterious. In the nebula, the distribution of carbon between CO and CH_4 should be controlled by the following relation:

$$CO + 3H_2 \leftrightarrow CH_4 + H_2O$$

Above 650 K, carbon in the nebula should be primarily in CO but, below 650 K, if equilibrium persisted, nearly all carbon would be in CH_4. The condensation temperature of methane is 50 K and if all carbon was in this form, then it could not have been efficiently incorporated into solids except perhaps at the extreme outer edges of the solar nebula. It is likely, however, that equilibrium between CO and CH_4 did not occur and that CO was probably an important reservoir of carbon throughout the nebula. With abundant CO, catalytic reactions on grain surfaces could form carbonaceous coatings. It has been suggested that Fischer-Tropsch reactions, similar to the following, produced much of the carbonaceous matter in meteorites:

$$10CO + 21H_2 \rightarrow C_{10}H_{22} + 10H_2O$$

Carbon compounds, water, and other volatiles such as nitrogen compounds were probably not constituents of the bulk of the solids that formed near the Earth. They probably condensed in cooler, more distant regions and were then scattered into the region where the Earth was forming. Fragments of comets and asteroids formed in the outer solar system still fall to Earth at a rate of 1×10^7 kg/year and early in the Earth's history the rate must have been higher. Certainly, some of the Earth's volatile elements were accreted as cometary and other bodies from the outer solar system, but it is yet unknown if this was the major source of the more volatile elements that comprise the atmosphere and oceans. One possible tracer that could be used to determine the cometary input is the noble gas composition of comets and the atmosphere. Once outgassed, noble gases tend to remain in the atmosphere and are not influenced by subsequent planetary activity. Unfortunately, the noble gas composition of comets is not presently known. An argument against a common source such as comets for volatiles on the terrestrial planets is based on the differences in the ^{36}Ar contents of planetary atmospheres directly measured by spacecraft. The primordial ^{36}Ar content for Venus is nearly 100 times greater than that for Earth, which is in turn nearly 100 times the Mars value. If this volatile was carried by the same material that brought water and carbon, then all three could not have been derived from a common comet source.

2.4 Accretion of the Planets

Condensed solids, along with possible presolar solids, accumulated by the process of accretion to form planetary bodies. In the beginning, gravitational collapse may have been involved but most of the accretionary phase involved various forms of mechanical collision and sticking. The process started with low-velocity collisions of micrometer size dust and terminated with bodies as large as the Earth and Mars colliding with velocities of over 5 km/s. The first stage involved low-velocity collision of dust grains to form bodies of centimeter size. The movements of dust-sized objects were strongly coupled to gas motions but particles larger than a centimeter could decouple from local gas motions. The larger particles have surface area-to-mass ratios that are low enough that the particles are dominated by gravity and not nebular winds. Pulled by the vertical component of gravity, these rock-sized particles would settle to the central plane of the nebula forming a thin dust disk with a relatively high concentration of dust, rocks, and boulders. Energy dissipation in the disk by collisions and gas friction should produce an extremely thin disk system somewhat analogous to the rings of Saturn. The particles should have almost perfectly circular orbits about the Sun at near zero inclinations so that particles have low relative velocities. If the relative velocities are low enough and the matter density is high, then gravitational "patch" instabilities may develop in the rock disk that can very quickly form objects of kilometer size. Formation of larger bodies requires direct collisions and net accumulation of mass. As the bodies grow, gravitational perturbations from each other and from nearby forming planets increase impact velocities up into the kilometer per second range. In the higher-velocity regimes, collisions are often destructive. An impact at 4 km/s has the same energy per mass and release rate as the chemical energy in high explosives. The accretion process is complex and involves both destruction and net growth. The process is also highly competitive, with the largest body in the growing swarm having a selective advantage because of its gravitational field. Gravity expands the capture cross-section to areas much larger than the geometrical cross-section and it helps in retaining rebounded material from high-velocity impacts. At one time, the material that formed the Earth was in the form of a million or so bodies 100 km in diameter, but in the end all of these were incorporated into one object. The final stages of accretion involved collision with very large bodies perhaps as large as Mars. The tilt of the Earth's spin axis is probably the result of a single collision with a large body. This tilt gives the Earth its seasons.

A collision with a Mars-sized object may have resulted in the formation of the Moon. Our moon is by no means the largest satellite in the solar system, but it is unusual in that it and the moon of Pluto are the largest moons relative to the mass of the planets they orbit. Geochemical studies of returned lunar samples have shown that close similarities exist between the bulk composition of the Moon and the Earth's mantle. In particular, the abundances of siderophiles ("iron-loving" elements such as Ir and Au that concentrate in planetary iron cores) in the moon are similar to those in the Earth's mantle. These observations have led to the currently popular hypotheses that the Moon formed from mantle material ejected into orbit by a large impact. The collision must have occurred early in the Earth's history but after separation of its core. The Moon affects the Earth's tides in an important way. Ocean tides play a variety of roles in biological and chemical cycles. Early in the Earth's history, the Moon was closer and the tidal effects should have been larger. The tidal interactions between the Earth and the Moon result in a slowing of the Earth's spin rate and an increase in the Earth–Moon distance. Over the past 0.5 Ma, the Earth's spin rate has decreased about 10% by this effect. The existence of a massive moon also has an important stabilizing effect on the Earth's obliquity, the tilt of the spin axis relative to the Earth's orbit plane. The obliquity of the Earth has changed only a few degrees over geologic time but Mars, without a massive and stabilizing moon, has had large changes. Obliquity has major effects on climate.

After planetary accretion was complete, there remained two groups of surviving planetismals, the comets and asteroids. These populations still exist and have played a major role in the Earth's history. Asteroids from the belt between Mars and Jupiter and comets from reservoirs beyond the outer planets are stochastically perturbed into Earth-crossing orbits and have collided with Earth throughout its entire history. The impact rate for bodies of 1 km diameter is approximately 3 per million years and impacts of bodies of 10 km diameter occur on a 100 Ma time-scale. The collision of a 10-km asteroid or comet produces a crater nearly 200 km in diameter and the atmospheric and oceanic effects from shock processes and ejecta can produce severe stress on terrestrial organisms. Major global effects can include heavy particulate loading of the atmosphere, production of nitrogen oxides from shock effects, and strong infrared radiation from ejected matter. The cretaceous–tertiary extinctions that occurred 65 Ma ago are marked with a global layer that contains shocked quartz grains and the trace element signature of meteoritic material. The total mass of the layer is consistent with ejecta from the hypervelocity impact of a 10-km comet or asteroid. Although infrequent, similar and much larger impacts must have happened many times in the Earth's history. It has been hypothesized that giant impacts may have sterilized the planet due to globally high temperatures during the first few hundred million years.

2.5 Early Evolution of the Earth

When the Earth was in its early stages of accretion, it was presumably a cool object. In the final stages, accretional energy was appreciable and must have heated the upper regions of the Earth's interior. When the Earth approached its present mass, projectiles would impact at a minimum velocity of 11.2 km/s, the escape velocity. The kinetic energy at this velocity is high and at least for large objects much of the energy released can be buried and trapped. Additional heat sources were radioactive decay and formation of the core. If the Earth contained appreciable amounts of ^{26}Al, its decay would provide an early and intense source of thermal energy. The other radioactive heat sources (U, Th, and K) are longer-lived and release heat on time-scales of over 10^8 years. Partial melting of the upper layers leads to the formation of molten iron masses that sink toward the Earth's interior. The gravitational settling of dense metal releases heat that creates a one-time pulse of gravitational energy. All of these heat sources lead to the chemical differentiation of the Earth into a core, mantle, and crust, and to outgassing of volatiles to form an atmosphere. Apparently all of the terrestrial planets underwent differentiation, although there may have been significant variations in the details of crustal development and subsequent evolution. Later evolution depends on the amount of internal heat available and the thickness of crust through which geologic activity must penetrate. The Earth is still a very active body with volcanic activity and widespread plate movement. Its closest neighbors, Mars and Venus, are less active and have had different evolutionary histories.

The evolution of the Earth's atmosphere and oceans is important for many of the cycles described in the book. Both of these evolved by outgassing but there is little information about the history of these processes. Again, comets could have a main source of some of the volatiles (H_2O, CH_4, etc.); however, proof of this source is as yet lacking (see section 2.3). Unfortunately, bombardment of the inner planets by large projectiles pulverized planetary surfaces up until about 3.9 Ga ago, erasing essentially all direct records of the Earth's early history. The Earth's oceans and its atmosphere are unique in comparison with Mars and Venus. Neither of these planets have oceans and their atmospheres are composed largely of CO_2 (Table 2-2). Mars has ice on and in its surface materials, but Venus has essentially no water anywhere. With a surface temperature of 650 K, water near the surface of Venus could only exist in the gas state. It has been suggested that Venus may have had as much water as the Earth but lost it in a catastrophic blow-off process driven by extreme greenhouse heating early in its history (see Chapter 10 for a discussion of loss by escape from the upper atmosphere). Another possibility is that Venus could have lost an early atmosphere and hydrosphere as a result of large impacts, a process that may also have "eroded" the atmospheres of Earth and Mars. The CO_2 inventory of Venus is similar to that on the Earth except that it is in the atmosphere, while on the Earth all but trace amounts are trapped in carbonates. Eventually, with increased solar heating or changes in the atmosphere, the CO_2 content of the Earth's atmosphere may increase. Increased greenhouse heating and positive feedback could lead to higher surface temperatures. Mars, in contrast, is in a nearly permanent ice age with only transient periods when conditions are sufficient for the existence of liquid water. The most remarkable aspect of the Earth is its atmosphere. It is composed of oxygen and nitrogen in contact with liquid water in highly non-equilibrium proportions. Except for ^{40}Ar produced

Table 2-2 Atmospheric compositions of the terrestrial planets

Planet	Relative pressure	Principal gases (%)	Other gases (ppm)
Mercury	10^{-15}	He (*c.* 98), H (*c.* 2)	
Venus	90	CO_2 (96), N_2 (3.5)	H_2O (*c.* 100), SO_2 (150), Ar (70), CO (40), Ne (5), HCl (0.4), HF (0.01)
Earth	1	N_2 (78), O_2 (21), H_2O (1), Ar (0.93)	CO_2 (330), Ne (18), He (70), Kr (1.1), Xe (0.087), CH_4 (1.5), H_2 (5), N_2O (0.3), CO (0.12), NH_3 (0.01), NO_2 (0.001), SO_2 (0.002), H_2S (0.0002), O_3 (0.4)
Mars	0.007	CO_2 (95), N_2 (2.7)	O_2 (1300), CO (700), H_2O (300), Ne (2.5), Kr (0.3), Xe (0.08), O_3 (0.1)

by decay of ^{40}K, the atmospheric composition is controlled by biological processes that definitely do not occur on similar scales elsewhere in the solar system. Early in the Earth's history, it probably had an atmosphere dominated by CO or CO_2, like Mars and Venus. In time, the CO_2 was incorporated into carbonate rocks and nitrogen and oxygen came under the control of biological processes. The rise of oxygen due to photosynthesis started early in the Earth's history and reached modern levels before the start of the Cambrian, 600 Ma ago.

Questions

2-1 Briefly describe the general features of the distribution of different elements in the solar system and the processes that produce this distribution.

2-2 How do the compositions of the terrestrial planets differ from the average composition of the solar system?

2-3 What is unusual about the Earth's atmosphere?

2-4 What sources of energy can contribute to the high temperatures of the Earth's core?

2-5 Briefly describe compounds likely to be present that will condense at high, intermediate, and low temperatures.

References

Anders, E. and N. Grevesse (1989). Abundances of the elements: Meteorite and solar. *Geochim. Cosmochim. Acta* **53**, 197–214.

Kasting, J. F., O. B. Toon, and J. B. Pollack (1988). How climate evolved on the terrestrial planets. *Sci. Amer.* **258** (2), 90–97.

Lovelock, J. E. (1979). "Gaia: A New Look at Life on Earth." Oxford University Press, Oxford.

Penzias, A. A. (1979). The origin of the elements. *Science* **205**, 549–554.

Bibliography

Silk, J. (1989). "The Big Bang." W. H. Freeman, New York.

Wood, J. A. (1979). "The Solar System." Prentice-Hall, Englewood Cliffs, N. J.

Woolsley, S. E. and M. M. Phillips (1988). Supernova 1987A! *Science* **240**, 750–759.

3

Evolution and the Biosphere

James T. Staley and Gordon H. Orians

Although physical and chemical factors were exclusively responsible for affecting the distributions of elements 4.5 Ga ago when the Earth originated (see Chapter 2), this is no longer true. Recent evidence suggests that life originated on Earth over 3.5 Ga ago. Since that time, biological processes have become increasingly important in determining the distribution of elements and the compounds into which they are incorporated. In this chapter, we discuss the origin and evolution of life, the nature of contemporary living organisms and the molecules they synthesize, and some of the major impacts biological processes have had on the development and continued functioning of the biosphere, the shell about the Earth in which living forms reside. In addition, we present an overview of some of the important roles organisms play in the functioning of biogeochemical cycles.

3.1 Origin and Evolution of Life

People have sought to explain their own existence and the existence of other biological forms since at least the beginning of historic times. In most societies, these explanations were formulated within the framework of the prevailing religious belief system. At the time of the Renaissance, most Europeans believed in the doctrine of spontaneous generation. Proponents of this doctrine held that living organisms developed from non-living materials. For example, maggots could be "generated" by allowing meat to decay. The Italian naturalist Francesco Redi (1626–1698) was the first to conduct careful experiments on the production of maggots in decaying meat. He showed that when the meat was properly contained in a vessel covered with cheesecloth, maggots were not produced. He noted that when vessels were left exposed to air, flies would alight on the spoiling meat and lay eggs. The eggs, in turn, hatched to produce maggots, which subsequently developed into flies. Redi's experiments set to rest, at least temporarily, the doctrine of spontaneous generation.

However, this controversy was renewed with the discovery of microbes by Anton van Leeuwenhoek (1632–1723) in the seventeenth century. Experiments conducted then showed that mixtures, such as boiled aqueous extracts of meat, developed microorganisms even after prolonged heating in sealed vessels. Scientists reasonably, but incorrectly, inferred that the microorganisms were developing spontaneously from these non-living materials. Subsequent investigations have determined that temperatures greater than 100° C are necessary to sterilize (kill or remove all life from an area or a material) natural materials, because some bacteria produce heat-resistant spores that survive ordinary boiling temperatures. It was the growth of these heat-resistant spores that led many proponents of spontaneous generation to conclude that bacteria were able to develop from non-living materials.

Adequately sterilized materials do not give rise to organisms under the conditions and length of time permitted for typical scientific experimentation. Moreover, there is no convincing evidence that living organisms are developing from non-living materials anywhere on Earth today. However, if life originated on Earth, as we believe it did, it must have developed from non-living materials. The current

Global Biogeochemical Cycles
ISBN 0-12-147685-5

scientific views of when and how life might have originated and evolved are based upon studies of the fossil record in combination with appropriate dating procedures and imaginative chemical experiments in the laboratory.

It is not surprising that early students of life tended to think of its origins in terms of processes operating on familiar time-scales. At that time, the great age of the Earth was not appreciated, nor were methods available for dating ancient events. Today, suitable methods exist for determining the age of materials that are billions of years old. In addition, recently acquired knowledge of the composition of the planets aid us in identifying processes of change that may have taken place on Earth long ago.

Current evidence indicates that all major bodies in our solar system originated about the same time, approximately 4.6 Ga ago. The oldest rocks taken from the surface of the Moon and meteorites found on Earth are about 4.5 Ga old. The Moon, Mercury, Venus, and Mars have cratered surfaces that appear to be the result of the same type of meteoric activity that produced craters on Earth. The early atmosphere on Earth was probably similar to those found today on nearby planets on which life did not evolve. Hence, we can use information about those planets to help us infer the nature of the conditions on Earth under which life presumably evolved.

3.1.1 Fossil Evidence of Early Life Forms

Although fossils have been known since prehistoric times, their significance as the remains or traces of early life forms did not become fully realized until the Renaissance. The science of paleontology did not become formally established until the eighteenth and nineteenth centuries. The careful studies of William Smith (1769–1839), an English geologist, led to the recognition that certain identifiable strata in sedimentary deposits always contained the same types of fossils. Thus a relationship was found between the depth of a stratum and its age. Paleontologists also noted that the lowest strata with fossils contained fewer types of organisms and ones that were simpler in structure than strata closer to the surface. In strata formed prior to about 600 Ma ago, there was no evidence at all for living organisms. This led paleontologists to refer to this age as the Azoic (i.e. without life). More recently, however, abundant fossils have been found in rocks older than this. Those from North Pole, Australia, dated at 3.5 Ga ago contain spherical carbon-containing structures approximately the size of modern prokaryotic cells, fossils of filamentous bacteria (Fig. 3-1), and finely stratified undulating sediments (Fig. 3-2), thought to be fossilized stromatolites (Hofmann and Schopf, 1983). Stromatolites are being formed now by cyanobacteria in places where high salt concentrations or high temperatures prevent grazing on them by typical predators. Microbial fossils have even been reported in some deposits that date in excess of 3.5 Ga ago, indicating that the first microorganisms had evolved within 1 Ga after the formation of the Earth (Awramik *et al.*, 1983).

These and other findings have resulted in substantial changes in terms used to describe periods in the history of the Earth. A geological timetable typical of those now in common usage is given in Table 3-1. The Hadean Eon, which extended from the time of the origin of the Earth to about 3.8 Ga ago, was a period during which the crust formed. So much debris hit the surface of the Earth at that time, and so much crustal movement has occurred since then, that no rocks survive on the surface from this aeon. During the Archaean Eon, which extended from about 3.8 to 2.6 Ga ago, more stable crustal features developed and life evolved. The major metabolic patterns of living organisms also evolved during this eon, including fermentation, photosynthesis, and the ability of cells to convert atmospheric nitrogen into a useful form (i.e. nitrogen fixation). The Proterozoic Eon extended from 2.6 to ~0.6 Ga ago. During this time, new cell types evolved and processes of sexual reproduction were perfected, leading to the evolution of plants and animals.

The evolution of living organisms was in part possible because of important changes in the physical environment, but living organisms in turn caused profound changes in the physical nature of the Earth and its atmosphere. Ever since the evolution of life, living organisms have been major participants in biogeochemical cycles, the culmination of which are the major anthropogenic perturbations of those cycles today. The major problems that the evolution of life poses for us include: Why were living organisms confined to aquatic environments for such a large portion of the history of life on Earth? What were the environmental conditions under which life evolved? What were the changes living organisms caused in those conditions that may have prevented the further evolution of life from non-living matter? These are, clearly, difficult

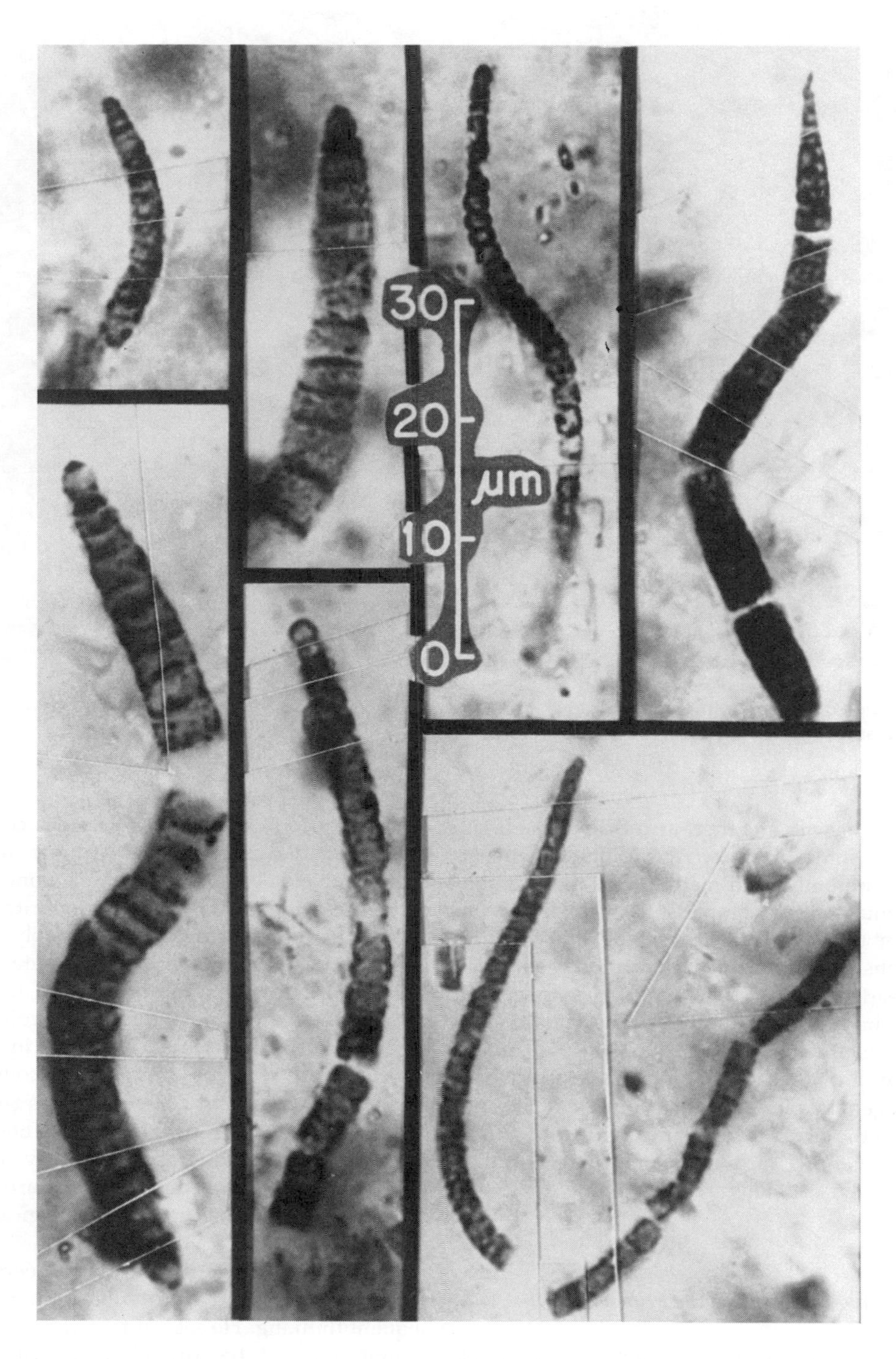

Fig. 3-1 Microfossils of several filamentous microorganisms. Bitter Springs Formation, Central Australia. Dated at 0.85 Ga (courtesy William Schopf).

3-2 A weathered outcrop with conical stromatolites from the Tooganinnie Formation (McArthur Group), Northern Territory, Australia. Dated at 1.6 Ga. A cm scale is shown at the left (courtesy William Schopf).

questions to answer, and there are no direct methods of attacking some of them. Accordingly, many of the most commonly used methods are indirect and scientists must rely on comparisons of conditions on Earth with those on planets lacking life, and on comparisons among organisms living today to reveal how the earliest forms of life might have appeared. These methods are not without their pitfalls and many details are as yet unknown. There is, in addition, substantial disagreement over details. None the less, a general picture of the evolution of life and its effect on the structure and functioning of Earth is gradually emerging, around which there is broad consensus even as details are disputed.

3.1.2 Organic Synthesis on the Primeval Earth

We are uncertain just how life originated, but a reasonable first step in the development of organisms was the non-biological synthesis of those compounds of carbon upon which living systems are all based (see Box on p. 31). There is considerable evidence indicating that these compounds can be formed abiologically. For example, one group of meteorites, the carbonaceous chondrites, contain organic matter, which was presumably formed in intergalactic space. Some of these meteorites have been analyzed chemically and found to contain a variety of organic constituents, including sugars, organic acids, and amino acids. Most interesting is the discovery that the amino acids contain equal portions of the D- and L-stereoisomers (see Section 3.2). This is further evidence that these organic materials were produced abiologically, or at least were not contaminants derived from Earth, where L-amino acids, originating from biological sources, predominate.

Stanley Miller (1953) conducted pioneering experiments that have had a profound influence on subsequent thinking. He set out to determine if organic chemicals could be formed from water and the various gases that were presumed to have existed in the atmosphere 4 Ga ago. An atmosphere containing methane, ammonia, and hydrogen as its principal constituents of biological interest was used in these experiments. Thus it differed from the current atmosphere by lacking oxygen, carbon dioxide, and molecular nitrogen. The experiments were conducted in a glass-enclosed system (Fig. 3-3). Water, representing a pre-Cambrian sea, was contained in one flask, which was connected to a condenser; the other part, representing the atmosphere, contained the gases. Electrodes inserted into the atmosphere were used to produce sparks to simulate lightning. In a relatively short time of operation, amino acids and a variety of other organic substances were produced under the anaerobic (lacking oxygen) conditions. The energy source for his experiments, electric discharges, is destructive of life as we know it, and is not used today by any bio-

Table 3-1 Geological history of the Earth[a]

Eon/era Period	Began[b]	Major events in the history of life
Hadean	4.5 Ga	
Archean	3.8 Ga	Origin of life; prokaryotes fluorish; photosynthetic cells liberate oxygen
Proterozoic	2.5 Ga	Eukaryotes evolve; several animal phyla appear
Phanerozoic	600 Ma	
Paleozoic		
Cambrian	600 Ma	Climate warms, most animal phyla present, including some that failed to survive; diverse algae and cyanobacteria
Ordovician	500 Ma	Diversification of echinoderms, other invertebrate phyla, jawless fishes; mass extinction at end of period
Silurian	440 Ma	Diversification of jawless fishes, first bony fishes; invasion of land by vascular plants, arthropods
Devonian	400 Ma	Origin and diversification of bony and cartilaginous fishes; trilobites diverse; origin of ammonoids, amphibians, insects, first forests; mass extinction late in period
Carboniferous	345 Ma	Extensive forests of early vascular plants, especially club mosses, horsetails, ferns; coal beds form; amphibians diverse; first reptiles; radiation of early insect orders
Permian	290 Ma	Reptiles, including mammal-like forms, radiate; amphibians decline; diverse orders of insects; conifers appear; continents aggregated into Pangaea; glaciations; major mass extinction, especially of marine forms at end of period
Mesozoic		
Triassic	245 Ma	Early dinosaurs; first mammals; gymnosperms become dominant; diversification of marine invertebrates; continents begin to drift; mass extinction near end of period
Jurassic	195 Ma	Diverse dinosaurs; first birds; archaic mammals; gymnosperms dominant; ammonite radiation; continents drifting
Cretaceous	138 Ma	Most continents widely separated; continued radiation of dinosaurs; angiosperms and mammals begin diversification; mass extinction at end of period
Cenozoic		
Tertiary	66 Ma	Radiation of birds, mammals, angiosperms, pollinating insects; continents nearing modern positions; drying trend in mid-Tertiary
Quaternary	3 Ma	Repeated glaciations, North and South America join; extinctions of large mammals, evolution of *Homo*; rise of civilizations, human modifications of biogeochemical cycles

[a] From Purves and Orians (1987).
[b] Time before present: 1 Ga = 1 billion years; 1 Ma = 1 million years.

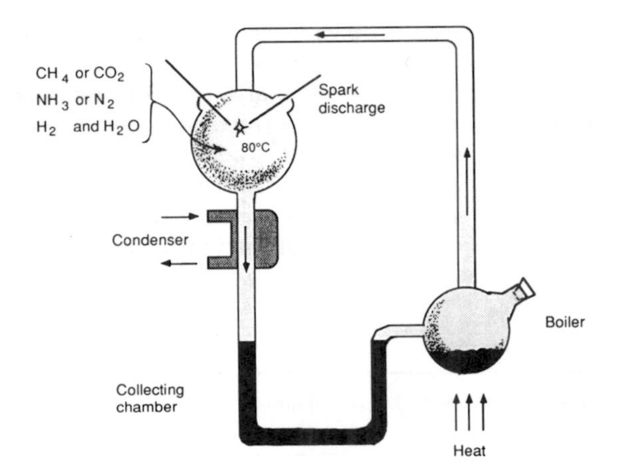

Fig. 3-3 Schematic of the apparatus Stanley Miller used to demonstrate the formation of amino acids from simple inorganic compounds under conditions similar to those of the early Earth.

logical systems. Other energy sources probably powered the evolution of life.

Rubey (1951, 1955) proposed that the early atmosphere was strongly affected by volcanic outgassing. Gases contributed to the atmosphere from this source include carbon dioxide, nitrogen, and sulfur compounds such as hydrogen sulfide, in addition to methane, hydrogen, and water. In his view, the early atmosphere was very similar to that of present-day Earth except that it was anaerobic. Rubey's view, which is supported by chemical analyses of the oceans and atmosphere, has gained wide acceptance among evolutionists and has stimulated research to determine whether organic compounds can be synthesized abiotically under such atmospheric conditions. The results from these experiments, which have been discussed by Chang *et al.* (1983), confirm that organic compounds can be synthesized by many combinations of these gases, again supporting the plausibility of prebiotic organic synthesis. Indeed, organic compounds can be produced under so many environmental conditions that the current question is which of the many possible conditions were the ones actually found on the early Earth.

3.1.3 Origin of Life

One of the major gaps in our knowledge concerns how the organic compounds that were synthesized abiogenically could have developed into the first living organism. All organisms, even the simplest of them, are extremely complex structures that carry out a variety of complex processes and are able to control precisely exchanges of materials between themselves and the environment. Thus, it is not surprising that no-one has produced in the laboratory cellular organisms or functioning prototypes of them using various mixtures of organic chemicals.

Not only is the living state complex, it is also difficult to define exactly what is meant by life. One way to define life is to state what appear to be the simplest requirements for independent existence. They include: (a) the presence of a semipermeable membrane to control the passage of materials into and out of the compartments (cells) of which all living organisms are composed; (b) chemical machinery for the synthesis and degradation of essential molecules; (c) genetic material that encodes the synthesis of the molecules required to catalyze those syntheses and degradations; (d) sufficient structure to prevent unwanted reactions from occurring; and (e) machinery to duplicate all the above capabilities in the formation of a new organism (i.e. reproduction). We do not know the minimum number of molecules required to perform these activities. Viruses are usually considered non-living. Furthermore, they are non-cellular (Section 3.2) and are able to multiply only by utilizing the complex genetic and structural machinery of more complex cellular organisms, which they parasitize. The largest viruses contain enough genetic material to encode only about 100 different genes. The smallest known organisms that are able to live independently are single-celled bacterial parasites called the Rickettsiae, which contain enough genes to encode about 200–400 proteins. Whether smaller independently living organisms were possible prior to the evolution of more complex forms of life is not known.

Most scientists speculate, however, that the first organisms were bacteria, and most agree with Oparin's (1938) hypothesis that they were heterotrophic, i.e. they obtained their energy by oxidizing organic compounds. Indeed, the first bacteria are viewed by some as being very similar to the anaerobic fermentative bacteria.

One example of fermentative bacteria is a group termed the lactic acid bacteria. These bacteria, which are commercially important in the cheese and dairy industry as well as in pickle and sauerkraut production, produce lactic acid by fermenting sugars.

This simple metabolic process occurs anaerobically by a series of enzymes involved in a pathway called glycolysis (Fig. 3-4), in which glucose, a 6-carbon sugar, is converted ultimately to pyruvic acid. Initially, the glucose is phosphorylated by the enzyme hexokinase which requires ATP (adenosine triphos-

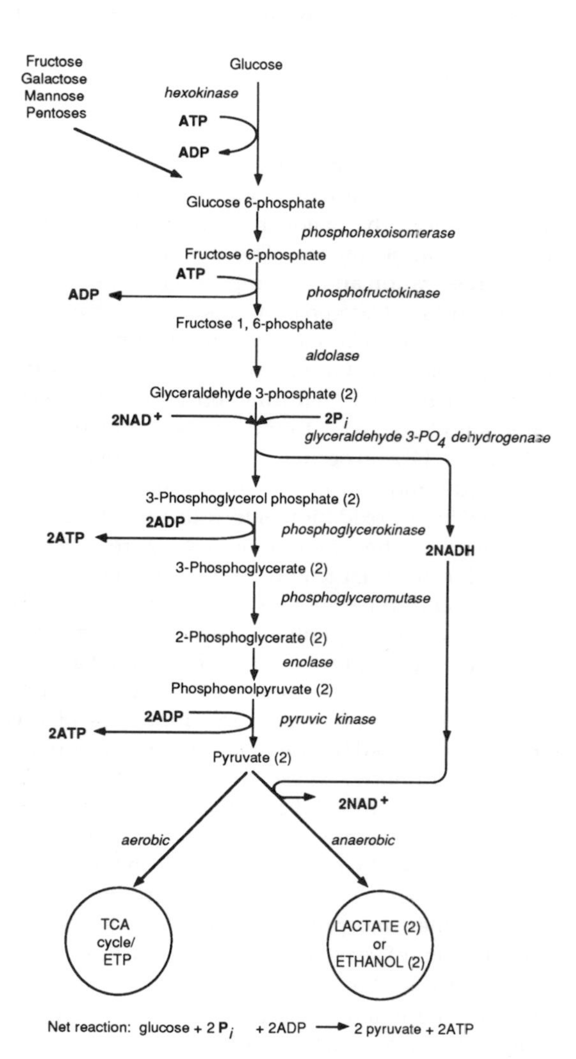

Fig. 3-4 The biochemical pathway of glycolysis, which obtains energy from the breakdown of 6-carbon sugars to a pair of 3-carbon pyruvate molecules. The enzymes at each reaction in the sequence are in italics. The energy generated is stored in the form of ATP (see Fig. 3-5). Pyruvate can be further transformed to lactate or ethanol under anaerobic conditions (fermentation), or, under aerobic conditions, it can enter the TCA cycle (see Fig. 3-10).

Fig. 3-5 Chemical diagram of ATP (adenosine triphosphate). The three functional groups are the base adenosine (upper right), a 5-carbon ribose sugar (middle), and three molecules of phosphate (left). The lines at the bottom of sugar ring indicate hydroxyl groups.

phate), a high-energy compound (Fig. 3-5). Another molecule of ATP is required to produce fructose 1,6-diphosphate. This molecule is broken down by an enzyme called aldolase to produce two 3-carbon sugars, both of which are phosphorylated. These three-carbon compounds are phosphorylated again with inorganic phosphate and each compound gives rise to two ATP before it is finally converted to pyruvic acid. Thus, a net synthesis of two molecules of ATP occurs for each molecule of glucose fermented to pyruvic acid. This ATP can be used for the synthesis of cell constituents. A subsequent step involves the formation of lactic acid, the end-product of the fermentation, from pyruvate. This is the only way these bacteria can generate energy, so they are not nearly as complex metabolically as aerobic bacteria that have additional enzymes needed to oxidize fully the pyruvate to carbon dioxide.

The Archaebacteria group, which includes the methanogenic (methane-producing) bacteria, may also have been an early group. They are so different from typical bacteria that it appears they evolved independently from them at very early times (Woese and Fox, 1977). One of the arguments favoring the early separation of methanogens from other bacteria is their metabolism. Most methanogens can utilize carbon dioxide and hydrogen gas for their carbon and energy sources, respectively, and produce methane in the process. Carbon dioxide and hydrogen are both components of the volcano-dominated atmosphere. Also, methanogens are anaerobic and therefore could have grown in the anaerobic atmosphere of the early biosphere. Interestingly, some methanogens can utilize carbon dioxide as a sole source of carbon for growth, but they use a different pathway for carbon dioxide fixation than is used by other forms of autotrophic life, including photosynthetic bacteria, cyanobacteria, higher plants, algae, and chemoautotrophic bacteria.

3.1.4 Early Photosynthesis

One of the most significant biological events affecting the Earth was the evolution of photosynthesis. Although this process is thought to have occurred before 3.0–3.5 Ga ago (Cloud, 1983), evidence for this is somewhat speculative. The first photosynthetic organisms were probably anaerobic photosynthetic bacteria, perhaps forms similar to present-day purple sulfur bacteria and green sulfur bacteria. These bacteria, which cannot photosynthesize in the presence of oxygen, carry out the anaerobic oxidation of the reduced sulfur compounds H_2S and S, with the formation of S and SO_4^{2-}, respectively:

$$2H_2S + CO_2 \rightarrow (CH_2O) + 2S^0 + H_2O \quad (1)$$

$$2S^0 + 3CO_2 + 5H_2O \rightarrow 3(CH_2O) + 2SO_4^{2-} + 4H^+ \quad (2)$$

Organic matter produced during photosynthesis is represented by (CH_2O).

Unlike photosynthesis by other organisms, including cyanobacteria, higher algae, and plants, oxygen is not evolved. Hence, this process is referred to as anoxygenic photosynthesis. Perhaps the best evidence that this process was occurring before 3.0–3.5 Ga ago is the discovery of sulfate minerals in deposits of that time (Walter *et al.*, 1980). Although small amounts of oxygen from abiotic photolysis of water could have resulted in the oxidation of reduced sulfur compounds to form sulfates, it is also possible that part or most of the sulfate was derived from anaerobic photosynthesis according to reaction (2) above.

3.1.5 Oxygenic Photosynthesis

There are some cyanobacteria, especially species of *Oscillatoria*, that can grow anaerobically and photosynthesize much like purple sulfur and green sulfur photosynthetic bacteria. They utilize hydrogen sulfide and produce sulfur granules in the process. However, these cyanobacteria are also capable of utilizing water in place of hydrogen sulfide and producing oxygen as a product. Such a process is termed oxygenic (i.e. oxygen-evolving) photosynthesis as opposed to anoxygenic photosynthesis. All cyanobacteria, algae, and higher plants carry out this type of photosynthesis:

$$CO_2 + H_2O \rightarrow (CH_2O) + O_2 \quad (3)$$

Current evidence indicates that the excess of oxygenic photosynthesis over respiration and decomposition produced the major fraction of the oxygen of the Earth's atmosphere. [The overall reaction for aerobic respiration and decomposition is the reverse of reaction (3).] Oxygenic photosynthesis has been dated to before 2 Ga ago by some distinctive geological formations. One of these formations is the extensive banded iron formations, our chief economic source of iron, found in rocks dated to 2–2.5 Ga ago. The iron occurs in the form of almost pure ferric oxide and silica, and the formations can be explained by the oxidation of ferrous iron in solution with oxygen. The amount of oxygen required to account for this is large and can be explained only if it were produced by photosynthetic organisms. The only pre-Cambrian photosynthetic organisms that could have accomplished this were cyanobacteria, and there is extensive fossil evidence of cyanobacteria-like microorganisms in formations of this age. In contrast, red bed deposits of iron oxides occur only more recently. These red beds have a more oxidized form of iron and could have been produced only if there was more oxygen available than would have been present when the banded iron formations occurred. Therefore, it appears that the oxygen content in the atmosphere has increased slowly over many millions of years. Evidently, the Earth changed over the period 2.0 to 0.5 Ga ago from an anerobic planet to one where the atmosphere now contains about 20% oxygen, as a result of oxygenic photosynthesis.

3.1.6 Impact of Oxygen on Evolution

Most complex organisms are aerobic, i.e. require oxygen for growth. Therefore, it is thought that their evolution could not have occurred until after the appearance of oxygen on Earth, i.e. not before some 2 Ga ago. Indeed, even though oxygen might have been produced prior to 2 Ga ago, considerable time would have been required before the concentration of oxygen became substantial. Initially, it would have reacted with reduced compounds such as iron, and only after these more reduced forms were fully oxidized would oxygen have accumulated to any significant levels. Moreover, when oxygen first began to accumulate on Earth, it would have been toxic to nearly all organisms then living, just as O_2 kills most organisms living in environments lacking oxygen today. This may have been the first major event of biologically caused pollution, but it provided conditions that favored both the ability to tolerate oxygen and the ability to use it in aerobic respiration.

It is interesting to note that bacteria and cyanobacteria predominated from ~3.5 to ~0.6 Ga ago, or three-quarters of the time life has existed! The tremendous variety of higher forms of life originated and evolved largely during the last 0.6 Ga. The production of oxygen is thought to have been the major event that led to the evolution of eukaryotic organisms and the resultant enormous variety of complex organisms that evolved during the Cambrian Era.

3.2 The Machinery of Life

All living organisms consist of subunits called cells. Theodore Schleiden and Theodor Schwann are attributed with the eloquent formulation of this theory in the nineteenth century. No exceptions have been discovered. Some organisms consist of a single cell and are called unicellular, whereas others, comprising many cells, are termed multicellular. Individual cells range in size, from less than 1 μm (10^{-6} m) to greater than 50 μm in diameter.

Cells are bound by a membrane referred to as the cell or cytoplasmic membrane, a semipermeable structure permitting the free passage of water and gases, but selectively permeable to organic substances such as sugars, amino acids, organic acids, and charged particles. The membrane is composed of lipids and proteins (see Box, p. 30) for descriptions of some of the general types of large organic molecules). Proteins consist of a large number of L-amino acids, covalently bound together by peptide bonds. D-Amino acids, the stereoisomers of L-amino acids, are not found in proteins. The discovery of equal concentrations of D- and L-isomers in carbonaceous chondrites suggests a non-biological or, at least, non-Earth origin for these amino acids.

Inside the cytoplasmic membrane is the cytoplasm, a viscous aqueous solution containing cations, anions, small molecules such as sugars and amino acids, and large molecules such as enzymes, a group of proteins that are responsible for catalyzing chemical reactions in cells. Also in the cytoplasm are small structures called ribosomes, the sites of protein synthesis. These, too, consist of macromolecular substances – proteins and ribonucleic acids (RNA). The subunits of RNA are the sugar ribose, the four nucleotide bases (adenine, guanine, cytosine, and uracil), and an inorganic anion (phosphate).

The hereditary material of the cell consists of deoxyribonucleic acid (DNA). DNA is similar to RNA, but the five carbon sugar, deoxyribose replaces ribose, and in place of uracil, the base thymine is found. DNA occurs as a double-stranded molecule. The two strands are held together by hydrogen bonds. DNA polymerase is one of the enzymes responsible for the process of replication whereby the double strands are separated and used as a template for the synthesis of new strands. Replication precedes cell division so that each new cell receives its own complement of double-stranded DNA.

Most plant and bacterial cells, but not animal cells, are bound by a cell wall layer external to the cell membrane. This structure confers rigidity and shape to cells. Most plants contain cellulose (a polymer of glucose) or pectin (a polymer of galacturonic acid) in their cell wall structures. Bacterial cell walls have very distinctive chemical compositions. Most contain a glycopeptide consisting of two amino sugars covalently bound to a peptide chain. The amino sugars, *N*-acetyl glucosamine and *N*-acetyl muramic acid, alternate in the formation of linear chains. Peptide bonds formed on the lactyl moiety of muramic acid permit the interconnection of the amino sugar "backbones" by short peptides. Curiously, some of the amino acids in this cross-linking peptide chain are D-amino acids. It is only in the cell walls and capsules (an outer layer that some bacteria have) of bacteria that D-amino acids occur. No other living organisms have D-amino acids. It is the uniqueness of the cell wall of bacteria that makes them selectively susceptible to the antibiotic penicillin. Penicillin resembles certain components of the cell wall glycopeptide, and therefore it interferes with an enzyme involved in assembly of the cell wall.

Large Organic Molecules

Many types of *organic molecules* are basic constituents of living organisms or are produced by them. Besides carbon and hydrogen, most naturally occurring organic molecules contain one or more of four key elements, all of which are from the second and third period of the periodic table: N, O, P, and S. Carbon atoms form most of the skeleton of these molecules. Bonding of carbon can lead to very large molecules with a variety of structures. Hydrogen may share one or more of the valence electrons of carbon and may also participate in *hydrogen bonds*. Hydrogen bonds also involve nitrogen and oxygen and are important in determining the structures of DNA and many other molecules. Much of the nitrogen is found in amino ($—NH_2$) groups, which provide the possibility of basic behavior as well as polarity, hydrophilic behavior, and, in some cases, water solubility. Oxygen, in carbonyl (C═O), carboxyl (—COOH), and hydroxyl groups (—OH), provides acidic and hydrophilic character.

Most organo-phosphorus compounds are phosphates, $R—PO_4$. The bond energy of the phosphate to the rest of the molecule is central to the flow of energy in *all* metabolic processes. Sulfur, found largely as sulfide, is central to providing three-dimensional rigidity by uniting parts of a molecule by the disulfide (—S—S—) bridge.

Amino acids, the building blocks of giant protein molecules, have a carboxyl group and an amino group attached to the same carbon atom (Fig. 3-6). A *protein* is a linear polymer of amino acids combined by peptide linkages. Twenty different amino acids are common in proteins. Their side chains, which have a variety of chemical properties, control the shapes and functions of proteins. Some of these side chains are hydrophobic (Fig. 3-6A), others are hydrophilic (Fig. 3-6B), and still others (Fig. 3-6C) occur either on the surface or the interiors of proteins.

Carbohydrates form a diverse group of compounds (Fig. 3-7) that share an approximate formula $(CH_2O)_n$. The major categories of carbohydrates are *monosaccharides* (simple sugars, Fig. 3.7A), *oligosaccharides* (small numbers of simple sugars linked together, Fig. 3-7B), and *polysaccharides* (very large molecules, Fig. 3-7C), among which are starches, glycogen, cellulose, and other important compounds. *Derivative carbohydrates*, such as sugar phosphates and amino sugars, contain additional elements. Important amino sugars are cartilage and chitin, the principal structural carbohydrate in insect skeletons and the cell walls of fungi.

Lipids (Fig. 3-8) are insoluble in water and release large amounts of energy when they are metabolized. Lipids are composed of two principal building blocks, *fatty acids* (Fig. 3-8A) and *glycerol* (Fig. 3-8B). Three fatty acids (carboxylic acids with long hydrocarbon tails) combine with one molecule of glycerol to form a triglyceride. More complex lipids are formed by the addition of other groups, the most important of which contain phosphorus (Fig. 3-8C).

Deoxyribonucleic acid (DNA, Fig. 3-9) is the genetic material of all organisms, including plants, animals, and microorganisms. [Some viruses lack DNA, but use RNA (ribonucleic acid) in its place.] DNA carries all the hereditary information of the organism and is therefore replicated and passed from parent to offspring. RNA is formed on DNA in the nucleus of the cell. The RNA carries the genetic information to the cytoplasm where it is used to produce proteins on the ribosomes. The specific proteins formed include enzymes which carry out the characteristic activities of the organism. Both RNA and DNA are formed from monomers, called nucleotides, each of which consists of a simple sugar, a phosphate group, and a nitrogen-containing base. The complex structure of DNA is founded on only four bases. A tremendous wealth of information is contained in the precise ordering of these bases to form the genetic code.

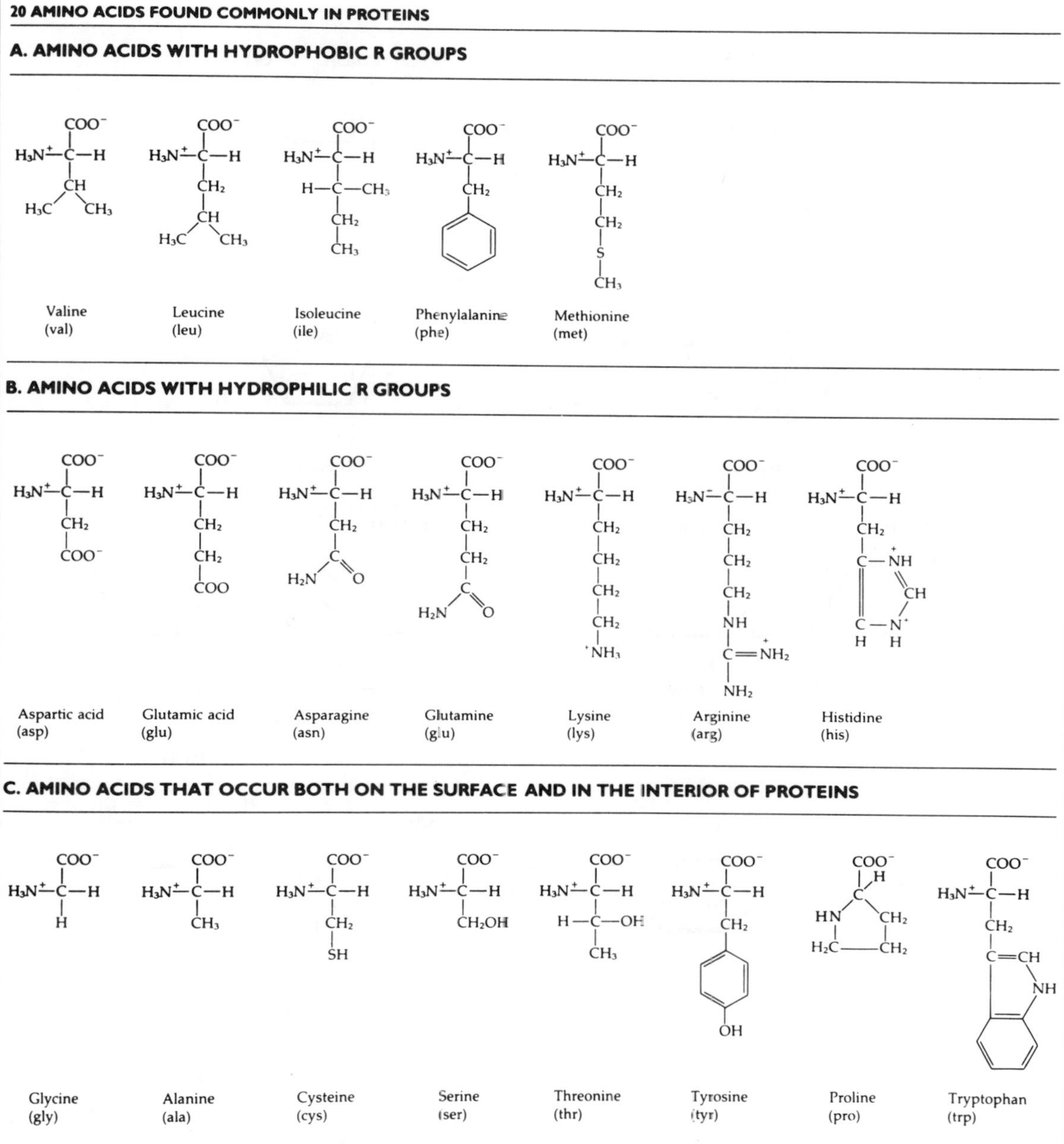

20 AMINO ACIDS FOUND COMMONLY IN PROTEINS

A. AMINO ACIDS WITH HYDROPHOBIC R GROUPS

Valine (val) | Leucine (leu) | Isoleucine (ile) | Phenylalanine (phe) | Methionine (met)

B. AMINO ACIDS WITH HYDROPHILIC R GROUPS

Aspartic acid (asp) | Glutamic acid (glu) | Asparagine (asn) | Glutamine (glu) | Lysine (lys) | Arginine (arg) | Histidine (his)

C. AMINO ACIDS THAT OCCUR BOTH ON THE SURFACE AND IN THE INTERIOR OF PROTEINS

Glycine (gly) | Alanine (ala) | Cysteine (cys) | Serine (ser) | Threonine (thr) | Tyrosine (tyr) | Proline (pro) | Tryptophan (trp)

Fig. 3-6 The 20 protein amino acids divided by R group character as (A) hydrophobic, (B) hydrophilic, and (C) mixed. Reprinted from Purves and Orians (1987) with the permission of Sinauer Associates, Inc.

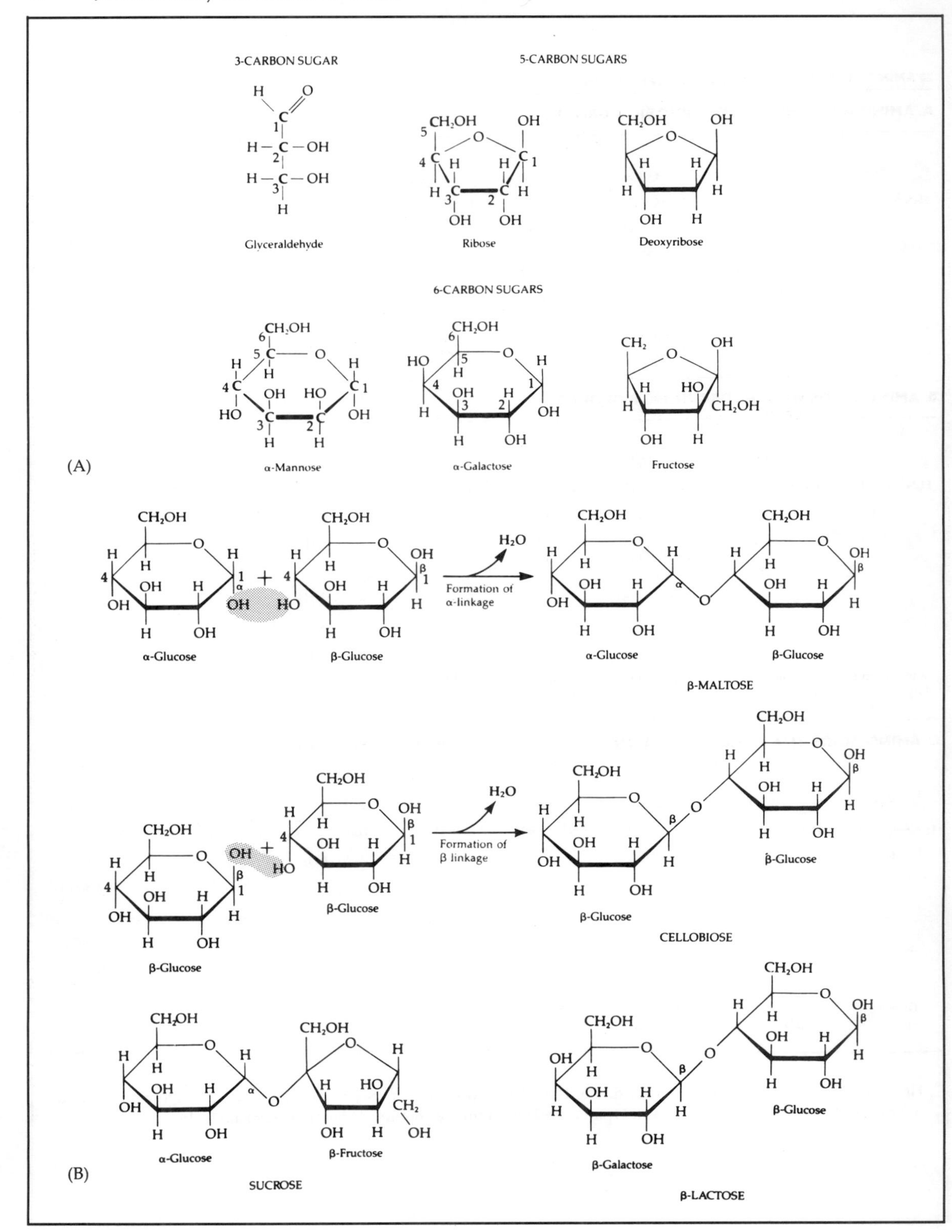
3-CARBON SUGAR
Glyceraldehyde
5-CARBON SUGARS
Ribose
Deoxyribose
6-CARBON SUGARS
α-Mannose
α-Galactose
Fructose
(A)
α-Glucose
β-Glucose
Formation of α-linkage
β-MALTOSE
Formation of β linkage
CELLOBIOSE
β-Fructose
SUCROSE
β-Galactose
β-LACTOSE
(B)

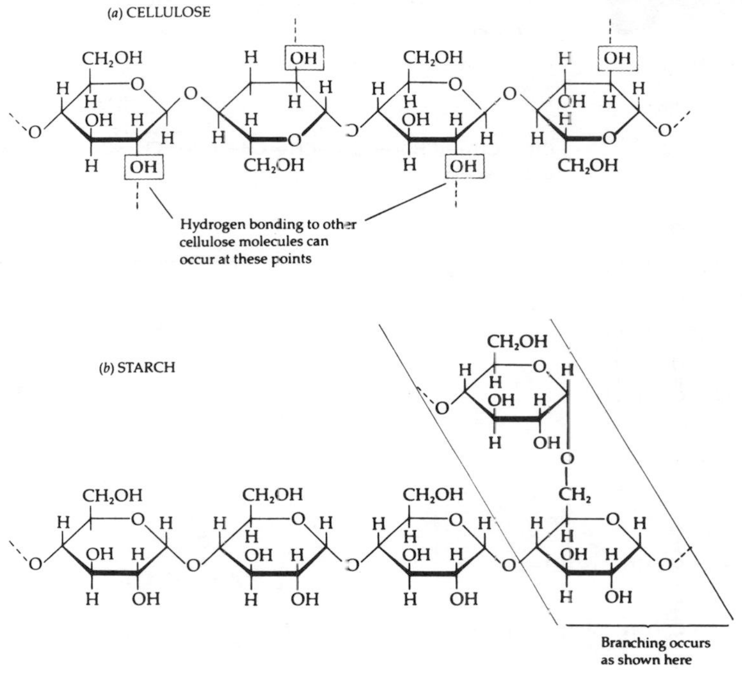

(a) CELLULOSE

Hydrogen bonding to other cellulose molecules can occur at these points

(b) STARCH

Branching occurs as shown here

(C)

Fig. 3-7 Carbohydrates in (A) 3-, 5-, and 6-carbon sugars (monosaccharides), (B) oligosaccharides, and (C) polysaccharides. Reprinted from Purves and Orians (1987) with the permission of Sinauer Associates, Inc.

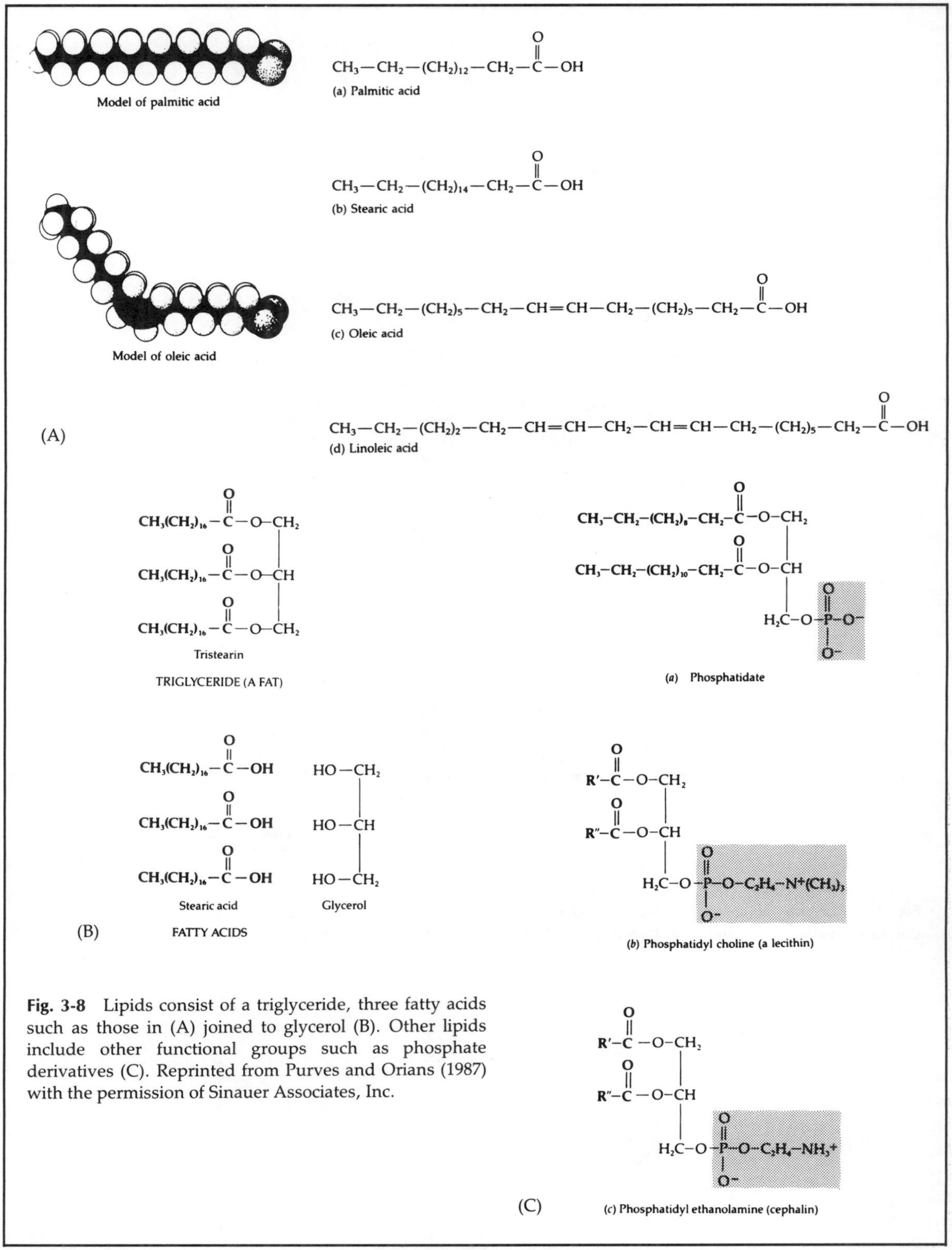

Model of palmitic acid

$CH_3-CH_2-(CH_2)_{12}-CH_2-C(=O)-OH$

(a) Palmitic acid

$CH_3-CH_2-(CH_2)_{14}-CH_2-C(=O)-OH$

(b) Stearic acid

Model of oleic acid

$CH_3-CH_2-(CH_2)_5-CH_2-CH=CH-CH_2-(CH_2)_5-CH_2-C(=O)-OH$

(c) Oleic acid

(A)

$CH_3-CH_2-(CH_2)_2-CH_2-CH=CH-CH_2-CH=CH-CH_2-(CH_2)_5-CH_2-C(=O)-OH$

(d) Linoleic acid

$CH_3(CH_2)_{16}-C(=O)-O-CH_2$

$CH_3(CH_2)_{16}-C(=O)-O-CH$

$CH_3(CH_2)_{16}-C(=O)-O-CH_2$

Tristearin

TRIGLYCERIDE (A FAT)

$CH_3-CH_2-(CH_2)_8-CH_2-C(=O)-O-CH_2$

$CH_3-CH_2-(CH_2)_{10}-CH_2-C(=O)-O-CH$

$H_2C-O-P(=O)(O^-)-O^-$

(a) Phosphatidate

$CH_3(CH_2)_{16}-C(=O)-OH$ $HO-CH_2$

$CH_3(CH_2)_{16}-C(=O)-OH$ $HO-CH$

$CH_3(CH_2)_{16}-C(=O)-OH$ $HO-CH_2$

Stearic acid Glycerol

(B) FATTY ACIDS

$R'-C(=O)-O-CH_2$

$R''-C(=O)-O-CH$

$H_2C-O-P(=O)(O^-)-O-C_2H_4-N^+(CH_3)_3$

(b) Phosphatidyl choline (a lecithin)

$R'-C(=O)-O-CH_2$

$R''-C(=O)-O-CH$

$H_2C-O-P(=O)(O^-)-O-C_2H_4-NH_3^+$

(C) (c) Phosphatidyl ethanolamine (cephalin)

Fig. 3-8 Lipids consist of a triglyceride, three fatty acids such as those in (A) joined to glycerol (B). Other lipids include other functional groups such as phosphate derivatives (C). Reprinted from Purves and Orians (1987) with the permission of Sinauer Associates, Inc.

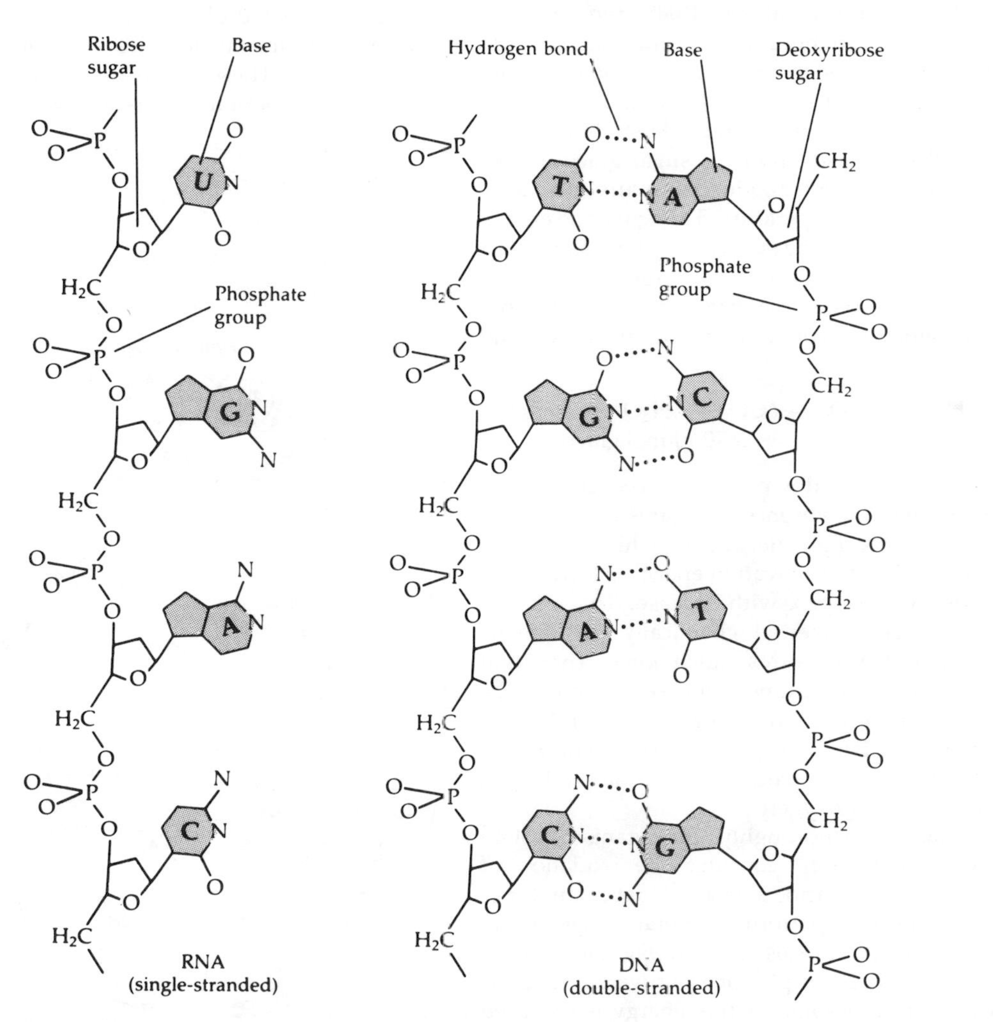

Fig. 3-9 RNA and DNA, the carriers of genetic information. Bases are denoted as: A, adenine; T, thymine; G, guanine; C, cytosine; U, uracil. Note that RNA contains U where DNA contains T. Reprinted from Purves and Orians (1987) with the permission of Sinauer Associates, Inc.

Cells are able to carry out all of the functions of life. During cell duplication (reproduction), the DNA is replicated (one double-stranded molecule is converted to two double-stranded molecules) by specific enzymatic reactions, and the newly formed DNA is separated into two separate cells produced by growth and division of the former single cell. In bacteria, this asexual process of reproduction is the primary means of multiplication to produce new individuals. In contrast, protists, plants, fungi, and animals rely primarily on sexual reproduction for the formation of new individuals. In sexual reproduction, two mating types (male and female) produce special reproductive cells termed gametes (sperm and eggs) that fuse together to form a cell that develops into a new individual organism.

All cells require energy produced and stored in the form of ATP (adenosine triphosphate) (Fig. 3-5). This energy can be released by organisms using reactions catalyzed by enzymes. Enzymes typically have molecular weights of about 30 000, though there is considerable variation. Their overall function is to lower the activation energy for reactions by complexing with the substrate for the reaction.

The action of enzymes can be illustrated by considering the metabolism of the sugar glucose. Like table sugar (i.e. sucrose), glucose occurs as a stable chemical at room temperature. The activation energy to break it down can be supplied by heat so that the glucose burns. The reaction is exergonic, and once initiated, the reaction may proceed unassisted until all of the glucose is converted to carbon dioxide and water:

$$C_6H_{12}O_6 + 6O_2 \rightarrow 6CO_2 + 6H_2O + 2870 \text{ kJ/mol glucose}$$

The chemical breakdown does not occur at room temperature in the absence of organisms, however, because the activation energy is very high.

In living cells, the activation energy is lowered by enzymes that complex with glucose. This permits glucose to be converted chemically in aqueous solutions at temperatures much lower than that required for actual burning. Glucose can actually be broken down entirely to carbon dioxide and water by many organisms. Organisms obtain energy for growth through this oxidation. Not all oxidations involve molecular oxygen.

To explain more thoroughly how organisms obtain energy by oxidation, consider the oxidation of glucose to pyruvic acid, a process that many types of organisms can perform. Several enzymes are required to carry out this reaction (Fig. 3-4). As the glucose is oxidized to produce pyruvate, energy is released. At least some of this energy is captured during the enzymatic conversions and stored in the cell as ATP by a process called substrate level phosphorylation. In yeast cells growing anaerobically, the pyruvic acid is subsequently reduced to ethanol by the appropriate enzymes in the presence of reduced pyridine nucleotides. This process is called alcoholic fermentation.

When yeast cells grow aerobically, they can oxidize the pyruvate entirely to carbon dioxide and water. These reactions take place in structures called mitochondria. The reactions form a cyclic scheme termed the tricarboxylic acid cycle (TCA cycle) or Krebs cycle in which in one cycle, the substrate, pyruvate, is converted entirely to carbon dioxide and water (Fig. 3-10). Both substrate level phosphorylation and electron transport mediated phosphorylation occur during the process. In electron transport mediated phosphorylation (Fig. 3-11), the reduced pyridine nucleotides are oxidized by dehydrogenases. These dehydrogenases along with other enzymes permit the passage of first, hydrogen

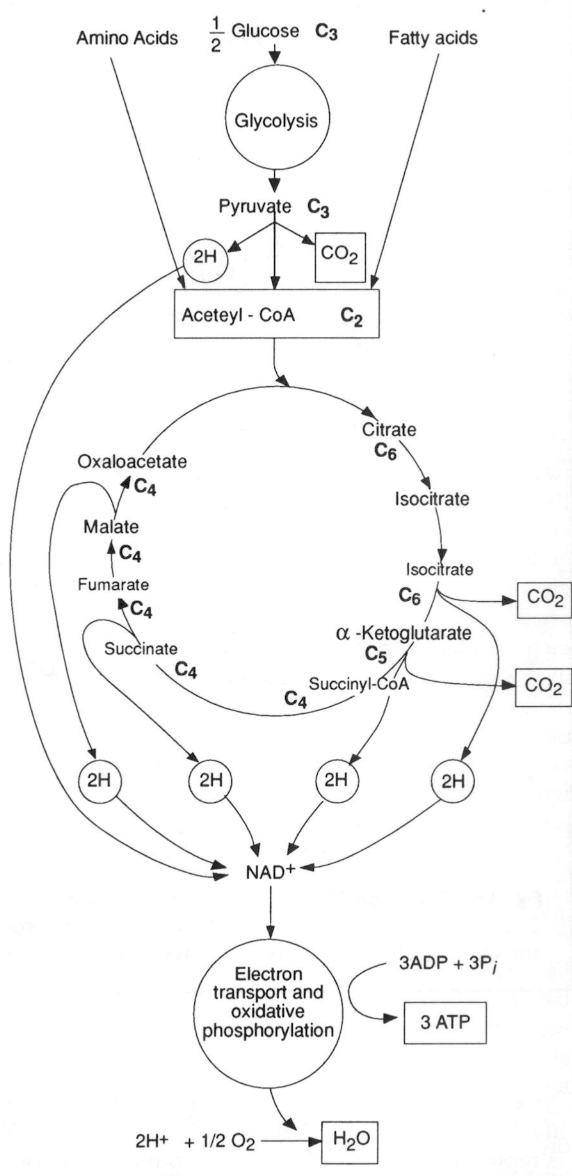

Fig. 3-10 The biochemical pathway of the tricarboxylic acid cycle (TCA cycle). Pyruvate, generated from glycolysis (Fig. 3-8), enters the cycle as acetyl-CoA (acetyl-coenzyme A), and is then degraded through a series of reactions to a 4-carbon compound, oxaloacetate. The energy resulting from the reactions is stored as ATP, which is produced through electron transport and oxidative phosphorylation (see Fig. 3-11).

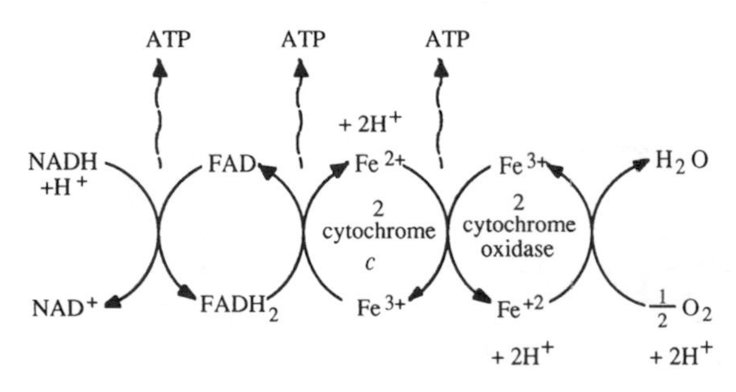

Fig. 3-11 Electron transport process schematic, showing a coupled series of oxidation-reduction reactions that terminates with the reduction of molecular oxygen to water. The three molecules of ATP shown are generated by an enzyme called ATPase, which is located in the cell membrane and forms ATP from a proton gradient created across the membrane.

atoms, and then electrons through an electron transport system of enzymes. ATP is generated by the proton gradient that is established between the cytoplasm and the environment. Much ATP can be generated in this process, because for each pair of electrons removed in the oxidation, 3 ATP molecules are produced. Thus some 30 ATP/mole glucose can be generated by electron transport phosphorylation in the TCA cycle. This corresponds to the retention of about 879 kJ of the energy from a mole of glucose, or an efficiency of about 30% of the available chemical bond energy. Substrate level phosphorylation, which also occurs, results in the formation of additional ATP.

3.2.1 The Major Divisions of Life

Organisms are placed into one of two major groups based upon the nature of their cells. Many microorganisms (i.e. the algae, protozoa, and fungi) and all complex organisms (i.e. plants and animals) are termed *eukaryotic*. In contrast, the bacteria and cyanobacteria are called *prokaryotic*. Eukaryotic organisms are distinguished from prokaryotic organisms by their cell structures and functions. In general, eukaryotic cells are more complex and usually larger than prokaryotic cells. Eukaryotic cells have their hereditary material contained in a nucleus enclosed by a membrane analogous to the cell membrane in structure (Fig. 3-12). In contrast, the genetic material of prokaryotic cells is not bound by a membrane but is in direct contact with the cytoplasm (Fig. 3-13). Although DNA is the genetic material of both cell types, in prokaryotes it is packaged into a single large molecule or chromosome, whereas eukaryotes have several chromosomes of DNA, the number being characteristic of the species. For example, human cells have 46 chromosomes in their nuclei. Eukaryotic organisms also have a nucleolus, a nuclear structure rich in RNA that is not found in prokaryotic organisms. Because prokaryotic organisms have only one chromosome, their processes of replication (DNA duplication) and cell division are simpler than in eukaryotes. In prokaryotes, the single DNA strand is replicated while attached to the cell membrane. The process of cell elongation results in the separation of the two duplicate copies of the chromosome, which ultimately reside in two separate cells after cell division. In eukaryotic organisms, the chromosomes are replicated and separated by a complex structural process called mitosis. The chromosomes are initially aligned along the center of the cell where they are subsequently replicated and separated into the two new cells.

The cytoplasm of eukaryotic cells contains mitochondria, membrane-bound structures involved in respiration, the process by which organic substances are oxidized in the presence of oxygen to yield energy by electron transport mediated phosphorylation. Photosynthetic eukaryotic organisms contain chloroplasts which, like mitochondria, are membrane-bound structures located in the cytoplasm. The ribosomes of eukaryotic cells are somewhat larger than the ribosomes of prokaryotic cells. Eukaryotic flagella responsible for motility have a cross-section consisting of 9 outlying pairs of fibrils and an inner 2 fibrils to give what is referred to as a $9+2$ arrangement. The prokaryotic flagellum appears as a single protein fibril in cross-section. These, and a number of other differences shown in Table 3-2, illustrate the fundamental differences between these two types of life forms.

3.2.2 The Five-Kingdom System of Classification of Organisms

Currently, the most widely accepted system of classification of organisms is the five-kingdom system (Table 3-3) of Whittaker (1969). Prokaryotic organisms are placed in the Kingdom Monera, which includes the bacteria and cyanobacteria whose

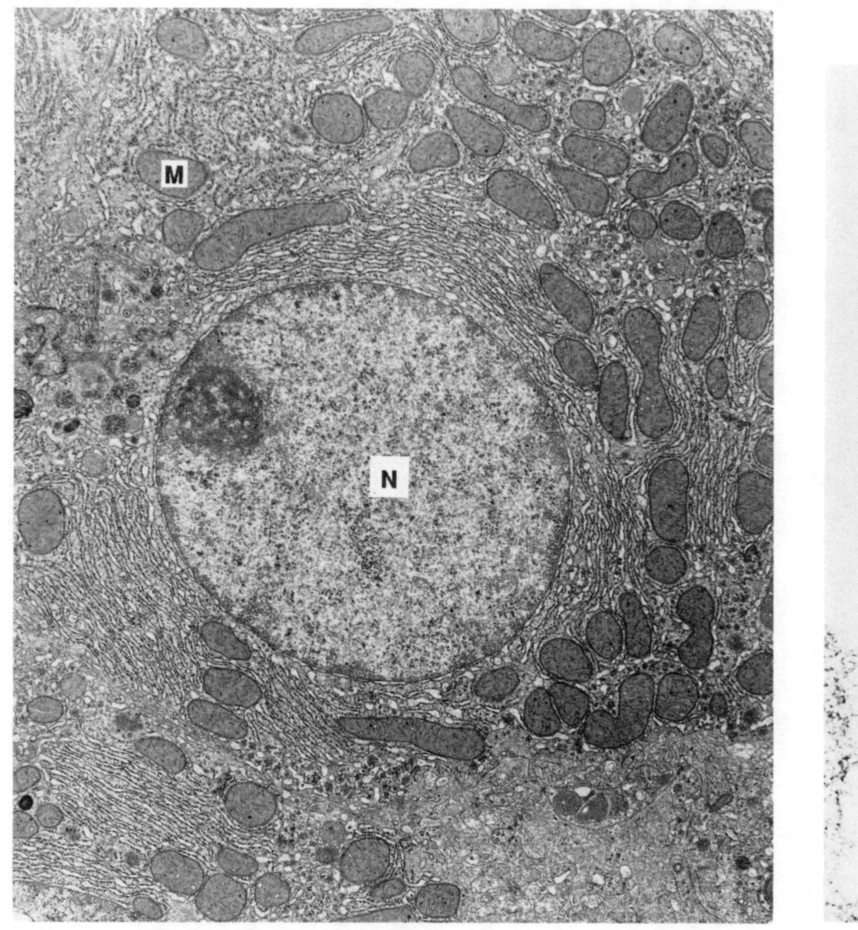

Fig. 3-12 A thin section through a eukaryotic cell. Note the nucleus (N) is bound by a nuclear membrane. In the cytoplasm of the cell, there are many mitochondria (M) and intracytoplasmic membranes. Reprinted with permission from R. Rodewald, University of Virginia/Biological Photo Service.

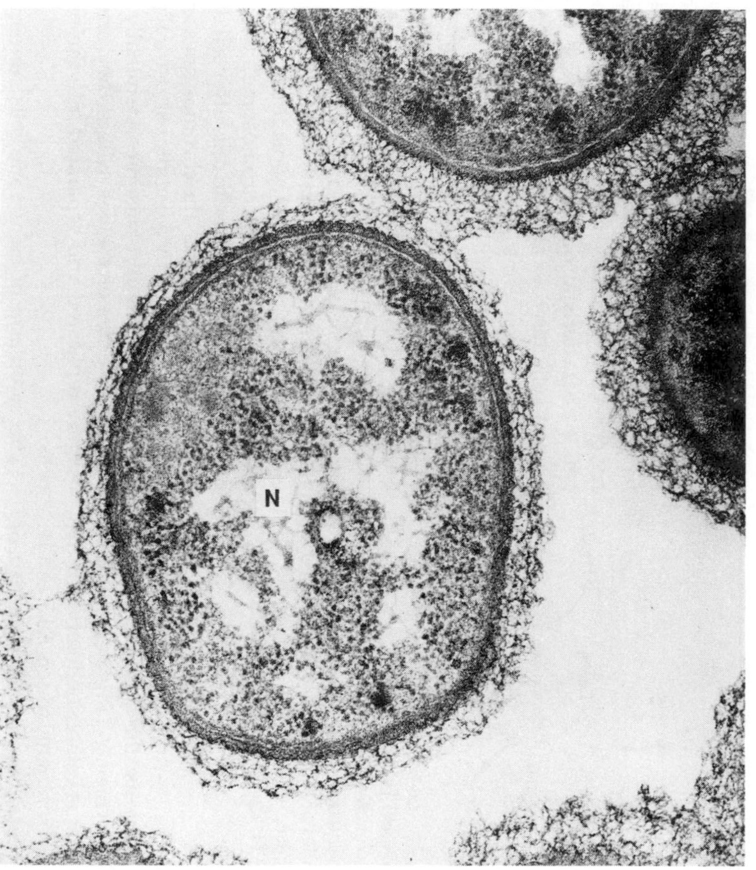

Fig. 3-13 A thin section through a prokaryotic cell. Note that the nuclear material (N) is not bound by a membrane, but is free in the cytoplasm. Mitochondria and other intracytoplasmic structures are absent. Reprinted with permission from J. J. Cardamone, Jr., University of Pittsburgh/Biological Photo Service.

Table 3-2 Comparisons of cells of prokaryotic and eukaryotic organisms

	Prokaryotic	Eukaryotic
Structure		
Nucleus		
nuclear membrane	−	+
nucleolus	−	+
number of chromosomes	1	>1
Cytoplasm		
endoplasmic reticulum	−	+
mitochondria	−	± (a few protists lack these)
chloroplasts	−	± (+ if photosynthetic)
ribosomes[a]	70 S	80 S
flagellum cross-section (not all species in either group are motile)	1 fibril	9 + 2 fibril arrangement
Cell function		
Mitosis	−	+
Cell fusion during reproduction	−	±

[a] The sedimentation coefficient, S, is directly related to the molecular mass.

cellular organization has already been outlined. All other living organisms are eukaryotes and are currently placed in four different kingdoms. The Kingdom Protista is a heterogeneous group of organisms that are either single-celled or groups of very similar cells joined together. Some are autotrophs, others are ingestive heterotrophs, and still others are absorptive heterotrophs. One phylum (Sporozoa) consists of non-motile forms, but members of the others move by means of ameboid motion, ciliary action, or flagella. Many marine species (foraminiferans) secrete skeletons of calcium carbonate and their remains are the major contributors to the formation of limestone. Others (radiolarians) secrete glassy siliceous skeletons that are responsible for sediments under many tropical seas.

The Kingdom Fungi contains mostly multicellular, heterotrophic organisms that have absorptive nutrition. They are parasites or saprophytes (living on dead matter) and they produce, at some time in their life-cycles, characteristic and often complex reproductive structures that differentiate them from protists, plants, or animals. The body of multicellular fungi is called a mycelium. Fungal cell walls contain a variety of polysaccharides, often including cellulose and, in some species, chitin. Lichens consist of a meshwork of a fungus and some photosynthetic organism, either an alga or a cyanobacterium.

Plants (Kingdom Plantae) range in size from single-celled forms to large trees and vines. Photosynthetic plants probably outweigh all other organisms by at least a factor of 10. The Rhodophyta, Chlorophyta, Pyrrophyta, Chrysophyta, and Phaeophyta, collectively known as algae, abound in fresh and marine waters and on moist terrestrial substrates. They account for over 25% of the photosynthesis occurring on Earth, and contain a variety of pigments, apparently adaptations to the very different light regimes found at different depths in the water column. They synthesize a number of different molecules for the storage of their food reserves (starch, fats, oils). The materials used in the construction of their cell walls are also highly varied (cellulose, pectin substances, silica, lignin, mucilages, and calcium carbonate). The largest species may attain lengths in excess of 35 m, but there is little differentiation of cells into distinct types. The large blades of marine kelps, for example, consist primarily of thin sheets of identical cells.

Mosses and liverworts (Bryophyta) are more complex than algae. Some of the larger species have structures that superficially appear similar to roots, stems, and leaves, but they lack the internal conducting systems present in the vascular plants (Tracheophyta). Internal transport systems (vascular systems) make possible the large sizes of terrestrial

Table 3-3 Classification of the major groups of living organisms

Taxon	Representatives
Kingdom Monera	
Phylum Bacteria	Bacteria
Phylum Cyanobacteria	Cyanobacteria
Kingdom Protista	
Phylum Mastigophora	Flagellates
Phylum Sarcodina	Amebas and their relatives
Phylum Sporozoa	Ameboid parasites
Phylum Ciliophora	Ciliates
Kingdom Fungi	
Phylum Mycomycota	Slime molds
Phylum Mastigomycota	Water molds
Phylum Eumycota	True molds, "typical" mushrooms
Kingdom Plantae	
Phylum Pyrrophyta	Dinoflagellates
Phylum Chrysophyta	Diatoms
Phylum Phaeophyta	Brown algae
Phylum Rhodophyta	Red algae
Phylum Chlorophyta	Green algae
Phylum Bryophyta	Mosses and liverworts
Phylum Tracheophyta	Vascular plants
Kingdom Animalia (only the major phyla are listed)	
Phylum Porifera	Sponges
Phylum Cnidaria	Hydras, jellyfish, corals
Phylum Ctenophora	Comb jellies
Phylum Platyhelminthes	Flatworms
Phylum Rotifera	Rotifers
Phylum Nematoda	Round worms
Phylum Mollusca	Chitons, snails, clams, squids, octopi
Phylum Annelida	Segmented worms
Phylum Arthropoda	Scorpions, spiders, crabs, insects, millipedes, centipedes
Phylum Bryozoa	Moss animals
Phylum Echinodermata	Sea lilies, seastars, sea urchins, sand dollars, sea cucumbers
Phylum Chordata	Tunicates, sharks, bony fishes, amphibians, reptiles, birds, mammals

plants where the soil is the source of some requisites for life (water, mineral nutrients) and the air is the source of others (CO_2, sunlight). The different groups of vascular plants are characterized primarily by their methods of reproduction. Vascular plants are the source of all wood.

Over 1 million species of animals (Kingdom Animalia) are known to science, and estimates of the true number range to more than 30 million because most species of insects, for example, are as yet undescribed. Because they are heterotrophs, animals represent less biomass than the autotrophs upon which they depend for their food, but because many of them construct sturdy skeletons that are durable and resistant to degradation, they are important contributors to biogeochemical cycles. And, of course, the species producing the largest contributions to and perturbations of biogeochemical cycles is an animal – *Homo sapiens*.

3.2.3 Endosymbiotic Theory of Evolution and Eukaryotic Organisms

Increasingly, our knowledge of cell structures and their molecular compositions is providing information on the evolution of organisms. For example, a great deal is known about the structure of ribosomes (Fig. 3-14). Interestingly, the structure of the archaebacterial ribosome differs considerably from that of the eubacterial ribosome, even though both are prokaryotic organisms. Furthermore, the eukaryotic ribosome shows a stronger morphological resemblance to the archaebacterial ribosome than to the eubacterial ribosome. This information suggests that the ribosome and possibly other cytoplasmic components of eukaryotic organisms may have evolved from archaebacteria.

In addition, information on ribosomal structure is providing support for a longstanding theory on evolution termed the "endosymbiotic theory". This theory holds that the membrane-bound organelles of eukaryotic cells are derived from prokaryotic organisms (Margulis, 1970). For example, the mitochondrion is viewed as being a bacterium that established an intracellular symbiosis with another prokaryotic organism. Substantial evidence exists to support the endosymbiotic theory as regards at least some of the eukaryotic organelles. For example, the mitochondrion contains DNA, which like bacterial DNA, is not bound by a nuclear membrane. Furthermore, the mitochondrion contains ribosomes that are similar to the ribosomes of prokaryotic organisms. Moreover, the process of respiration carried out by mitochondria is similar to respiration that occurs in the cell membrane of prokaryotes. Finally, the size and shape of mitochondria are typical of that of a gram-negative bacterium lacking a cell wall. Indirect supporting evidence for the contention that endosymbiosis occurs between bacteria and other

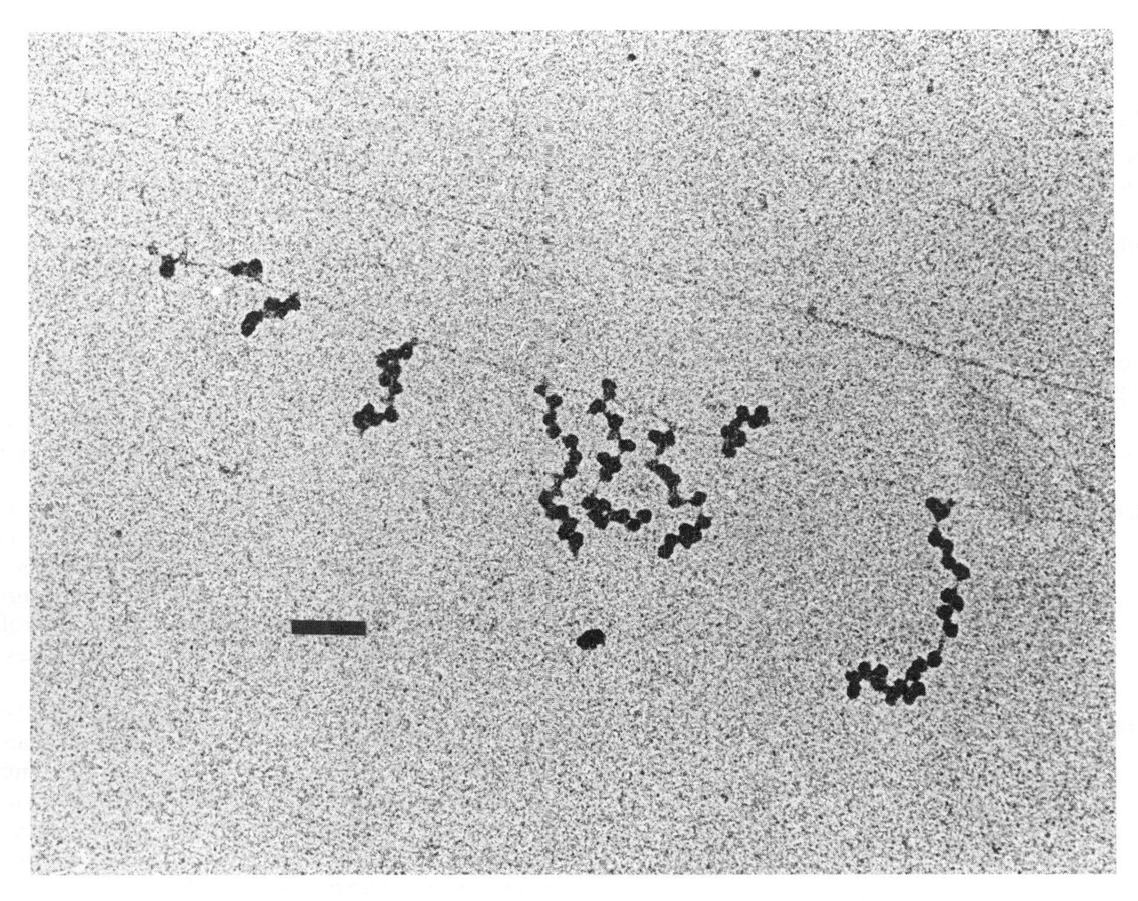

Fig. 3-14 Electron micrograph showing polyribosomes from the bacterium *Escherichia coli*. Filaments are DNA molecules. Ribosomes are attached to mRNA which they are translating. Bar = 0.1 μm. Photograph courtesy of Dr. O. L. Miller, Jr. (from Miller *et al.*, 1970).

microbes is the endosymbiotic association of bacteria similar to the genus *Cytophaga* with the protozoan *Paramecium*. Similarly, the chloroplast of eukaryotic organisms is thought to be derived from an association between a cyanobacterium and other organisms.

3.2.4 Why So Many Kinds of Organisms?

We are accustomed to the fact that there are many species of living organisms, but it is not obvious why so many different species should have evolved or why natural communities of living organisms should be so complex. The general answer is that many different species have evolved because no organism is efficient at making a living in all environments. In colloquial language, "The jack of all trades is the master of none." Moreover, as new forms of life evolve, they make possible still others. The evolution of autotrophs provide food for heterotrophs. The availability of hosts leads to the evolution of parasites.

The details of the structure and functioning of any organism are best thought of as representing compromise solutions, molded by natural selection over long periods of evolutionary time, to the many different problems posed for it by its physical and biological environments. Rarely is a solution to one problem a suitable solution to another. A plant can either have leaves that allow for rapid entry and exit of CO_2 and water or leaves that can resist water loss in dry environments, but not both. A bird can have wings designed for either speed or maneuverability, but not both. The kinds and abundances of molecules synthesized by different organisms are also the result of conflicting selective pressures. These molecules influence biogeochemical cycles primarily as byproducts of selection for dealing with problems at the level of the individual organism. It is highly unlikely that these molecules, however important they are for bio-geochemistry, evolved because of their geochemical influences. The details of the process of natural selection are not our focus here. More information is available in the books of Dawkins (1982, 1987) and Futuyma (1986). Our purpose is to point out the many uses to which organisms put the molecules they synthesize and why these uses determine the quality and quantity of chemicals produced, and hence their potential biogeochemical significance.

3.3 The Functional Types of Molecules Produced by Living Organisms and Their Biogeochemical Significance

The enormous variety of molecules synthesized by living organisms serves a rather limited number of functions. Basically, molecules are used for (a) structural support, (b) metabolism, energy storage, and energy transfer, (c) information storage and information transfer, (d) modifiers of other chemicals (enzymes), and (e) defense against predators, parasites, and competitors. A given molecule may, of course, serve more than one function. Because of the functional requirements of molecules for these different purposes, some types are important in biogeochemical cycles whereas others are not. The key factors influencing the biogeochemical significance of a molecule are its *per capita* rate of production, the abundance of its producers, its rate of chemical decomposition, the nature of its degradation products, and the mobility of the molecule in its original and transformed states.

3.3.1 Energy Storage and Transfer Compounds

The means of storing energy for varying periods of time are essential for survival because, in general, organisms are not able to absorb food continuously at exactly the same rate at which it is being used. Photosynthetic organisms produce carbohydrates only during daylight but require energy throughout the night. Heterotrophic organisms consume many types of food, some of which must be actively pursued. At times, they experience high food intake rates, whereas at other times they experience low intake rates and periods of starvation.

The energy derived from the breakdown of cellular fuels is transferred eventually to ATP. ATP occurs in all cells of all living organisms at concentrations of 0.001–0.005 mol/L. The hydrolysis of ATP to ADP releases 29 288 J/mol of free energy at pH 7.0 and 25° C. However, the concentrations of the reactants in cells are not equimolar and the presence of Mg in living cells may enable living cells to obtain as much as 72 300 J/mol from ATP hydrolysis. Normally, however, the turnover of ATP is very rapid, and the supply of ATP in most cells is sufficient for only a very short time of activity. For example, a cell of the bacterium *Escherichia coli*, dividing once every 20 min, must convert at a minimum 2.5 million molecules of ATP to ADP and phosphate every second. However, it is estimated that an *E. coli* cell contains only about 1 million molecules of ATP, less than enough for 0.5 s of work.

Cells store energy in the form of more complex molecules, primarily carbohydrates and fats. In plant cells, most energy is stored in the form of starch granules. In most animal cells, energy is stored as fat droplets, but liver and muscle cells store energy as glycogen granules. Even with these storage mechanisms, however, most cells have energy stores sufficient to support metabolic processes for no more than a few hours.

Long-term energy storage usually involves special organs in which surplus energy is stored and precise systems for regulating the deposition of materials in storage and their recovery when needed. The differences in the storage molecules and organs of plants and animals are related to the mobility of most animals and the sedentary existence of plants. A motile organism must carry its stored energy at considerable metabolic cost. A sedentary organism, on the other hand, can deposit energy in any out-of-the-way place and never need carry it around. For motile organisms, the quantity of energy that can be stored per unit weight of storage tissue is an important factor influencing the value of different storage molecules. The oxidation of 1 g of carbohydrate or protein yields about 17 730 J, whereas the oxidation of 1 g of fat yields about 39 750 J. Although proteins can be used as fuel, the extraction of energy from them is more complicated. Proteins are normally used for energy only when there are shortages of fats and carbohydrates. Because fats yield over twice the energy per unit weight as carbohydrates and proteins, they are the best storage molecules for motile organisms. Most animals store excess energy as fats. For plants, however, the ease with which energy can be extracted from carbohydrates makes them the best overall storage molecules; long-term storage of energy in most plants occurs as starch and related compounds. The main exception is seeds, where weight and size economy are important for dispersal and fitting into irregularities in the soil where germination takes place. Plant seeds are the major sources of oils for human consumption.

The amounts of materials involved in energy transformations by living organisms are enormous, but many of these processes are less significant biogeochemically than might be supposed. Basic metabolic processes are the same in all living eukaryotes. They extract energy from food by glycolysis, followed by either fermentation or aerobic respiration. Prokaryotes, on the other hand, are much more diverse. In addition to using molecular oxygen as the agent for cellular oxidation, some prokaryotes can use sulfuric acid, carbon dioxide, or nitrate ions. The requirements that energy be readily extractable from any storage molecule and the value, for motile organisms, of high caloric densities of storage molecules, results in rapid turnover of these molecules while the organism is living. On the death of an organism, its energy stores are rapidly eaten by its predators or utilized by detritivores. None the less, because of the great volume of elements involved in basic energy metabolism, these processes are major contributors to the cycles of some elements, such as carbon, where the entire cycle is dominated by the photosynthesis and respiration of living organisms. The same is true for sulfur, because of its importance in proteins and its use in osmoregulation, and nitrogen, because of its importance in proteins. The bacteria that reduce carbon dioxide to obtain their energy release about 2 billion metric tons of methane into the atmosphere each year, about one-third of which comes from methanogens living in the guts of mammalian herbivores.

3.3.2 *Information Storage and Transfer Molecules*

Information is stored in living organisms primarily in the form of nucleic acids (DNA and RNA) and the enzymes they form. One of the basic principles of information storage and control is that the information itself is capable of triggering processes that consume much more material and energy than is found in the informational molecules themselves. A given DNA molecule may be copied many times and the enzyme it produces is used repeatedly to catalyze molecular interactions without actually being consumed in the process. The total amount of materials committed to the storage of information itself is rather modest, and there is no functional requirement for informational molecules to be highly resistant to attack.

Informational molecules may also convey messages among organisms. The first animal communication signals to evolve were probably chemical ones. Chemical signals are important in the lives of most organisms and have been extensively studied in such organisms as the cellular slime molds, where they are responsible for the remarkable aggregations of cells that precede the formation of their interesting fruiting bodies (Bonner, 1967, 1970). It is even possible, as first suggested by Haldane (1955), that hormones were involved later in evolution as the intercellular equivalents of the chemicals that governed communication among single-celled organisms.

Because of the differences between air and water as media for the transmission of chemical signals, the structure of airborne signals differs from that of water-borne ones. Most airborne signals have from 5 to 20 carbon molecules and a molecular weight of between 80 and 300. In water, however, molecular weight has a major impact on diffusivity. The time to reach the maximum radius of the active space of a signal (the space over which it is present above threshold level) and the fade-out time (the time between the release of the signal and the total disappearance of the active space) are approximately 10 000 times greater in water than in air. As a consequence, many water-borne chemical signals are much larger in size than the largest airborne ones (Wilson, 1970).

Chemical signals have two major properties that reduce their biogeochemical significance. First, they are produced in very small quantities. The thresholds for their detection are often remarkably low, sometimes lower than the minimum concentrations detectable by our most sophisticated equipment. Second, chemical signals are designed to last only a short time. Indeed, a short life is essential to their functioning. Most of the events being signaled by chemicals are short-term ones (the presence of a food item, selection of a mate), and production of a signal that remains above threshold for long periods of time is rarely advantageous. Indeed, it is normally highly disadvantageous. Therefore, these molecules do not accumulate in places where they may be locked up for long periods of time, and the quantities circulating through the air and water are trivial in comparison with the quantities of defensive and structural molecules.

3.3.3 Structural Molecules

Structural molecules enable living organisms to maintain their shapes and move. A skeleton is any device for transmitting forces from one part of an organism to another. If the organism is motile, this transmission of forces is the basis for locomotion. If the organism is stationary, the skeleton may provide support to maintain shape against the pull of gravity and against other physical forces such as wind and water movement. It may also serve as a defense against predators.

The cells of most plants maintain their shapes by means of turgor pressure against a cell wall of cellulose, strengthened by the deposition of lignins. Lignins are phenolic heteropolymers with molecular weights greater than 5000. They are found extensively in woody tissues. The concentration of lignins in cells varies from nearly zero in submerged parts of aquatic plants to over 40% in the woody cells of gymnosperms and angiosperms. Next to cellulose, lignins are the most abundant of all natural polymers (Swain, 1979). The wood of some plants consists of mixed lignins, whereas others are nearly pure polymers, but the ecological significance of these differences is not clear. All lignins are highly resistant to biological degradation.

Microorganism and animal skeletons are chemically more diverse, ranging from the rigid shells of molluscs to the flexible chitin of the exoskeletons of arthropods. Many unicellular organisms have rigid external skeletons and, although these organisms are very small, their abundance may result in large depositions of materials. For example, the foraminiferans in the protist phylum Sarcodina, secrete shells of calcium carbonate. Shells of different species have distinctive shapes and, because they are readily preserved as fossils in marine sediments, they serve as indicators of the geological period in which they lived. Another important component of marine depositions is the shells of members of the Phylum Mollusca. Except for the cephalopods (squids and octopuses), nearly all members of this diverse phylum (chitons, bivalves, gastropods) produce calcium carbonate shells, which are often very thick in species that inhabit rocky shorelines where they are subjected to very strong pressures from waves.

A very complicated exoskeleton is excreted by arthropods, consisting of many layers of proteins and lipids and a nitrogen-containing polysaccharide known as *chitin*. The more massive, rigid parts of arthropod exoskeletons are strengthened by the deposition of calcium carbonate in their middle layers or by the tanning of the proteins by cross-linking of the protein chains by orthoquinones.

The existence of complex, long-lived organisms depends on their ability to synthesize sturdy structural materials whose longevity is commensurate with the life of the organism they are supporting. This requires that the materials be resistant to biodegradation. Because of their abundance and resistance to degradation, these materials play important roles in biogeochemical cycles. However, since they are usually also involved with defenses against pathogens and predators, we delay the discussion of their biogeochemical significance until we have discussed defensive molecules.

3.3.4 Defensive Molecules

The energy-rich tissues of living organisms are excellent sources of energy and materials for other organisms. Among the major problems faced by all living organisms is avoiding becoming a meal for some predator or parasite, while at the same time extracting enough energy from the bodies of other organisms to meet metabolic needs. Organisms, particularly plants, also use chemicals in warfare against competitors. Defensive molecules are generally referred to as allelochemicals. Chemical suppression of competition by plants is termed allelopathy.

The defensive molecules of plants, often called secondary substances, are probably derived evolutionarily from materials with a basic metabolic function. Some continue to serve more than one function, e.g. defense, providing structure and, perhaps, controlling water loss (some plant resins). Unlike molecules that are necessary to basic cellular metabolism, secondary substances are strikingly different from species to species. Many are found in only a single species of plant or a few closely related species. Some are characteristic of particular families of plants and provide good keys to evolutionary relationships at that level. Others appear sporadically in quite unrelated plants, indicating that they evolved independently, perhaps from similar precursors, although in most cases the origins of secondary substances are unknown. Every species of plant has an assemblage of secondary substances that is unique to it. Detailed treatments of the types of secondary plant substances and the status of current theory concerning their evolution, distribution within plants, and the extent of plant investment in defensive substances, are provided by Harborne (1988) and Rosenthal and Janzen (1979).

The type of defense employed and the extent of commitment to it are believed to have evolved in relation to the risks of discovery of the particular plants and tissues by herbivores. The rationale for this belief is that because production of defensive molecules is costly and generally serves no other metabolic function, allocation of energy to their production and maintenance will not be beneficial unless it results in a reduction in attack rates by herbivores sufficient to compensate for the costs. Plants differ in the probability that they will be found and eaten, independently of their defenses. For example, ephemeral plants and tissues have a higher probability of escaping being found by herbivores than do longer-lived plants and tissues. Moreover, herbivores differ in their abilities to utilize the tissues of different plants. Specialist herbivores, closely adapted to their food plants, are expected to be better able to handle defenses than are generalist herbivores that eat a wide variety of plants but are not specifically adapted to dealing with the defenses of any one of them.

Based on this type of reasoning, investigators have divided defensive substances into (1) the *acute toxins* (qualitative defenses) that are present in very low concentrations in plant tissues and which exert their effects on herbivores by interfering with some basic metabolic process such as transmission of nervous impulses, and (2) *digestibility-reducing substances* (quantitative defenses) that are present in higher concentrations in plant tissues, that act in the gut of the animal to reduce its ability to utilize its food, particularly proteins, and whose effectiveness increases directly with their concentration (Cates and Rhoades, 1977; Feeny, 1970, 1976; McKey, 1974; Rhoades and Cates, 1976). Qualitative defenses appear to be most characteristic of plants and tissues that are relatively difficult for herbivores to locate, whereas quantitative defenses are best developed in plants and tissues that are easy to locate. About 80% of woody perennial dicotyledonous plants contain tannins (the most important group of digestibility-reducing molecules) compared to 15% of annuals and herbaceous perennials (Bate-Smith and Metcalf, 1957; Rhoades and Cates, 1976). Plants also increase their commitment to defensive substances as a result of being attacked. Substances synthesized in response to attacks by pathogens, especially fungi, are known as phytoalexins. They are of varied chemical nature, their only general property being a degree of lipid solubility (Harborne, 1988).

Animals also defend themselves chemically against predators but this is less widespread than might be expected, probably because large predators on most animals are closely related to their prey, and hence are biochemically much more similar than animals are to plants. Therefore, effective defenses are likely to be autotoxic (Orians and Janzen, 1974). Plants, however, are able to evolve chemicals that attack nerves, muscles, and hormonal systems.

Many plants release chemicals that inhibit the germination and growth of other plants. Included among them are some of the terpenes responsible for the fragrance of many plants. They may travel from the leaves of plants to the soil as the leaves decompose. Many allelopathic substances are released from the roots of plants and have important effects on soil bacteria and fungi. The role that they play in organizing plant communities in nature is still controversial.

The qualitative defenses of plants (acute toxins) are present in minute quantities in plant tissues, and they contribute little to the flux of materials through ecosystems. Quantitative defenses (digestibility-reducing substances), however, are often present in large quantities. For example, the protein-complexing resin coating the leaves of creosote bush (*Larrea cuneifolia*) may be as much as 44% of the dry weight of the leaves (Rhoades, 1977). Moreover, the complexes these substances form with proteins are highly resistant to decay by macro- and micro-organisms, and hence they may accumulate in large quantities in ecosystems. The dark color of black-water rivers is due to high concentrations of humic and tannic acids in the water, evidently derived from the heavily defended leaves of trees growing on the impoverished soils characteristic of the watersheds of these rivers (Janzen, 1974). Resistance to decay of wood is strongly related to its lignin content. Many of these compounds decompose slowly in aerobic environments and hardly at all in anaerobic ones. Tannins and lignins are important components of materials in terrestrial ecosystems, with major effects on the carbon cycle.

The structural molecules of the skeletons and shells of invertebrates, which function as physical defenses against predation, are important in marine environments, where they produce carbonate and silicate rocks. Deposition in anaerobic environments has also been the basis for the formation of the extensive deposits of gas and oil that now fuel modern industrial societies. Removal of carbon from the biosphere by organisms to produce carbonate rocks, coal, oil, and hydrocarbon gases has been responsible for the presence of oxygen in the atmosphere of the Earth. Reversal of this process by human consumption of fossil fuels has already produced a detectable increase in atmospheric carbon dioxide.

The potential biogeochemical role of stable structural molecules can also be appreciated by considering the consequences of a failure to degrade the common structural molecules. For example, if it were not for the presence of chitin-splitting bacteria, the loss of nitrogen from the atmosphere through the formation of chitin, the main structural molecules of arthropods and one of the most stable of the organic nitrogen compounds, would be very serious. About 34 mg of carbon are fixed annually per square centimeter of ocean surface, corresponding to a production of about 50 mg of phytoplanktonic dry matter. About 5 mg of dry zooplanktonic material, containing about 5% chitin or 0.35% chitin nitrogen, are generally produced from that amount of phytoplankton, yielding an annual utilization of about 0.018 mg/cm^2 nitrogen chitin in the ocean. Production of chitin on land is probably at least that great. Thus, if chitin-splitting organisms had not evolved and no inorganic decomposition had taken place, the estimated current rate of chitin production would have exhausted the atmosphere of its nitrogen in 41 Ma, provided that adequate nitrogen fixation had occurred (Hutchinson, 1944).

3.4 The Ecological Organization of the Living World

The bodies of living organisms are excellent sources of energy-rich molecules that can be used to fuel the machinery of other organisms. Since early in the evolution of life, some organisms have obtained their energy and materials by consuming others. The major dynamics of the living world today are dominated by eating and being eaten. Interrelations among the millions of species of living organisms are very complex. Much of the work of ecologists is devoted to untangling the webs of connections that form the bases of ecological communities. For the purposes of the study of biogeochemical cycles, much of this richness of detail can be ignored and living organisms can be grouped into major categories based on the sources of their energy (Fig. 3-15).

The entry of materials and energy into the living world is overwhelmingly by means of photosynthesis. The fraction of solar energy falling on the surface used in photosynthesis is small, even though many refinements have been made in the course of millions of years of evolution. The Earth intercepts 5×10^{24} J of energy per year from the sun. Of this quantity, only about 3×10^{20} J are captured by photosynthesis. An important factor contributing to the small percentage captured is the absorption spectrum of chlorophyll and its associated pigments, which capture only a small fraction of the total energy present in sunlight. Much of the incident solar radiation is reflected from the surface of the Earth. Much is converted to heat and some of this is used to evaporate water, and hence to drive the global hydrological cycle.

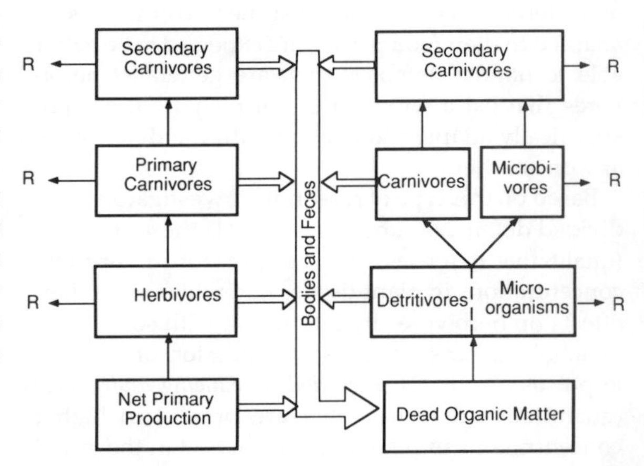

Fig. 3-15 Trophic levels in ecosystems. Thin arrows show flow of energy up the food chain (through living biomass) and the broad arrows show the complementary flow of dead organic matter (detritus) back down. R indicates respiration.

The chemical composition of all organisms is similar. For example, the elemental composition of algae in terms of carbon, nitrogen, and phosphorus is given by the following ratio of atoms (Redfield *et al.*, 1963): $C_{106}N_{16}P$.

Organisms at other levels in the food chain, from primary consumer to mineralizing bacteria, have similar elemental compositions. The stoichiometry of this ratio poses certain constraints on primary producers. They must have in their diet adequate levels and appropriate forms of these and other essential elements in order to grow. Frequently, phosphorus and nitrogen can be limiting to the growth of algal phytoplankton because of their low abundance in a habitat, especially following blooms in which growth has depleted them.

The groupings of organisms according to source of energy is called a classification by *trophic level*. The major trophic levels in ecological communities are photosynthesizers (primary producers), herbivores (eaters of plants or parts of them), primary carnivores (eaters of herbivores), secondary carnivores (eaters of primary carnivores), tertiary carnivores, and detritivores (eaters of the dead remains of once living organisms). These categories are artificial in the sense that many organisms, such as human beings, do not fit into any one of the categories, but rather obtain their energy from several different trophic levels. The groupings are also artificial in that they group together organisms that eat very different types of tissues. For example, the trophic level "herbivores" includes eaters of wood, roots, leaves, flowers, fruits, seeds, nectar, and pollen, all very different resources, so different in fact that few organisms eat all of them. Indeed, the exploiters of pollen and nectar are highly specialized morphologically for exploiting those unusual resources, and they probably interact very little with all other herbivores. These problems notwithstanding, trophic levels are, as we shall see, of considerable utility for the student of biogeochemical cycles.

Net primary production is defined as gross primary production (total photosynthesis) minus respiration. The global distribution of net primary production in the Earth's major ecological zones is shown in Table 3-4. Oceans, despite their much larger surface area, contribute much less than half of the net primary production of the Earth. This is because surface waters, where photosynthesis can take place, are highly deficient in nutrients over most of the oceans. Oceanic production is concentrated in coastal areas, especially where upwelling of deep water brings nutrients into the photic zone. On land, photosynthesis is often limited by dryness and cold temperatures, as well as by nutrient shortages, which is why subtropical and tropical areas con-

Table 3-4 Net primary production of the Earth's major ecological zones[a]

Zone	Area (10^6 km^2)	Mass of plants (10^6 tonnes)	Primary production (10^6 tonnes C)
Polar	8.1	13.8	1.3
Coniferous forests	23.2	439.1	15.2
Temperate	22.5	278.7	18.0
Subtropical	24.3	323.9	34.6
Tropical	55.9	1347.1	102.5
Total land	133.9	2402.5	171.5
Glaciers	13.9	0	0
Lakes and rivers	2.0	0.04	1.0
Total continents	149.3	2402.5	172.5
Oceans	361.0	0.2	60.0
Earth total	510.3	2402.7	232.5

[a] After Rodin *et al.* (1975).

tribute much more than their proportional share to global primary production (Table 3-4).

The energy captured by an individual organism is used to support basic metabolism and maintenance for growth, and for reproduction. Moreover, much of the energy potentially available to the next higher trophic level is in fact not captured by it. Many organisms or parts of them die and are consumed by detritivores which dominate the flow of energy from green plants and are the primary pathway for return of nutrients in a form useful for assimilation by green plants. On average, about 10% of the energy captured by one trophic level is taken in by the next level, but there are wide variations around this overall average. For example, the small, unicellular algae that dominate marine photosynthesis have high growth and cell division rates, possess easy-to-digest tissues, and can be eaten by very small animals. As a result, a higher proportion of the primary products of photosynthesis in the ocean are eaten by herbivores than is the case on land where materials may accumulate for centuries in the form of wood. In many marine areas, there may actually be, for parts of the year, a larger standing crop of herbivores than of the plants they feed upon due to the very high algal growth and reproduction rates.

Animals differ markedly in the way in which the energy they take in is allocated to different activities. The relationships between energy assimilation, utilization, and expenditure may be summarized by the following energy budget equation:

$$A = P + R = C - FU$$

where A = assimilated energy, P = energy of production due to growth (P_g) and reproduction (P_r), R = respiratory energy, C = consumed energy, and FU = egested (fecal) and urinary energy. The relationships between annual production (P) and respiration (R) have been analyzed by Humphreys (1979) using published data from over 200 population energy budgets. Some of his results, grouped by categories of animals, are shown in Table 3-5. These results demonstrate the high cost, in terms of energy consumption, of maintaining the constant high body temperatures of birds and mammals, compared to the much lower maintenance costs of other groups. For example, it takes about eight times as much energy per day to sustain a small bird as to support a lizard of the same weight. As a result, ecosystems dominated by large vertebrates, such as African savannas, tend to have very short food chains, and much energy in the system is dissipated in the high metabolic rates of the dominant species.

The consequences of the massive "loss" of energy accompanying passage from one trophic level to another include the relative shortness of food chains and the fact that organisms low in the trophic ladder tend to dominate the cycling of elements through the biosphere. This is especially true on land where vascular plants dominate both the physical structure and the flow of energy in ecosystems. In the oceans, because the major photosynthesizers are small and structurally simple, the cycling of many elements is much more strongly influenced by herbivores and carnivores.

Table 3-5 Mean values for Respiration (R), Production (P), and Assimilation (A) in kJ/m^2 year for different groups of animals[a]

Group	R	P	A	R/P	R/A	P/A
Multicellular ectotherms						
non-social insects	11.9	8.6	20.5	1.4	0.6	0.4
non-insect invertebrates	135.6	41.6	177.2	3.3	0.8	0.2
social insects and fish	23.1	1.9	24.9	12.5	0.9	0.1
Multicellular endotherms						
birds	41.8	0.5	42.3	78.5	1.0	0.01
mammals	11.7	0.4	12.1	30.2	1.0	0.03

[a] After Humphreys (1979).

3.5 Overview of Major Biogeochemical Transformations

The continuing processes of evolution and natural selection have resulted in the development of complex ecosystems containing a myriad of species interacting with one another and their environment. These organisms could not exist and would not have evolved were it not for their ability to obtain energy (i.e. produce a net ATP yield) from their life processes. Organisms can be viewed as complex chemical catalysts that have evolved to exploit energy-yielding chemical transformations that would proceed much less rapidly in their absence. However, whereas there must be a net formation of ATP by organisms, some of the reactions they perform are not, of themselves, energetically spontaneous. None the less, they are essential for the organism's overall metabolism and they are important to the global ecosystem. For example, oxygenic photosynthetic organisms derive their energy from sunlight. Yet they also have enzymes that "split" the water molecule to produce oxygen and hydrogen atoms and other enzymes that fix carbon dioxide to produce organic compounds. Both of these processes require energy and are extremely significant geochemically. Similar examples can be cited for other organisms. In sulfate-reducing bacteria, the primary energy source is the anaerobic oxidation of organic compounds. Sulfate receives the electrons from this process and is reduced to hydrogen sulfide.

In this section, we summarize some of the major geochemical transformations in which organisms partake (Table 3-6). In subsequent chapters, more

Table 3-6 Elemental cycles and selected important transformations in which organisms play a role

Cycle	Process (chemical transformation)	Biological group responsible
Carbon	Carbon dioxide fixation ($CO_2 \rightarrow$ organic material)	Photosynthetic organisms: plants, algae, bacteria Chemoautotrophic organisms: nitrifying bacteria, some sulfur oxidizers, iron oxidizers, hydrogen oxidizers
	Aerobic respiration (organic material + $O_2 \rightarrow CO_2 + H_2O$)	All plants, animals, and strictly aerobic microbes
	Organic decomposition (or mineralization) (organic material $\pm O_2 \rightarrow\rightarrow$ inorganic material)	Microorganisms, especially fungi and bacteria
	Methane production [$CO_2 + H_2$ (or simple organic compound such as acetate) $\rightarrow CH_4 + H_2O$]	Methane producing bacteria
Nitrogen	Nitrogen fixation [$N_2 \rightarrow\rightarrow RNH_2$ (organic amino group)]	Free living prokaryotes: *Azotobacter* spp., some *Clostridium* spp., some Cyanobacteria, photosynthetic bacteria Symbiotic prokaryotes: *Rhizobium* spp. and others
	Nitrification ($NH_3 \rightarrow NO_2^-$) ($NO_2^- \rightarrow NO_3^-$)	Chemoautotrophic nitrifying bacteria
	Dissimilatory denitrification[a] ($NO_3^- \rightarrow N_2, N_2O$)	Anaerobic respiring bacteria
	Ammonification (organic nitrogen $\rightarrow NH_3$)	Many microbes, especially bacteria
Sulfur	Sulfur oxidation ($H_2S \rightarrow\rightarrow S \rightarrow S_2O_3^{2-} \rightarrow\rightarrow SO_4^{2-}$)	Purple and green sulfur photosynthetic bacteria, some cyanobacteria Chemoautotrophic sulfur oxidizers
	Dissimilatory sulfate ($SO_4^{2-} \rightarrow\rightarrow H_2S$)	Sulfate reducing bacteria[b]
	Dimethyl sulfide production $SO_4^{2-} \rightarrow (CH_3)_2S$	Certain marine algae
Metal cycles	Iron and manganese oxidation and reduction	Iron bacteria and manganese bacteria

[a] Assimilatory denitrifiers reduce nitrate to the amino acid level where it is incorporated into protein. Many plants and bacteria can do this and, therefore, use nitrate as a nitrogen source.

[b] Assimilatory sulfate reducers reduce sulfate to the sulfhydryl level where it is incorporated into the sulfur amino acids of protein. Many plants and bacteria can do this.

detailed information will be given on each of these cycles.

3.5.1 The Carbon Cycle

A variety of autotrophic organisms fix carbon dioxide with the formation of organic material. The anoxygenic photosynthetic bacteria use sunlight as an energy source and reduced sulfur compounds as ultimate sources of hydrogen for the reduction of CO_2 to organic material. In contrast, the oxygenic photosynthetic organisms (cyanobacteria, algae, and plants) derive hydrogen for the reduction of CO_2 by splitting water, which also results in the formation of oxygen. Thus, photosynthetic organisms are involved not only in the carbon cycle but also the sulfur cycle and the oxygen cycle. Indeed, like other organisms, they must obtain other nutrients for growth, such as nitrogen and phosphorus, and therefore they are involved in these and other cycles as well.

The nitrifying bacteria, the hydrogen bacteria, some of the sulfur-oxidizing bacteria, iron bacteria, and methanogenic bacteria derive their energy not from sunlight, but from the oxidation of inorganic compounds. All of these organisms carry out carbon dioxide fixation with the formation of organic carbon.

All plants, almost all animals, and aerobic microorganisms engage in aerobic respiration, in which energy is derived from the oxidation of stored organic compounds. The result is the formation of carbon dioxide and water and the removal of oxygen from the atmosphere. This process is similar to organic decomposition. Organic decomposition, however, is carried out by saprophytic fungi and bacteria which obtain the organic material from non-living sources. Thus, saprophytic fungi and bacteria cause the decay of dead trees and other non-living plant and animal material. Animals, too, are important in the decomposition of plant and animal organic material.

The methane-producing bacteria derive their energy from the oxidation of simple organic com-pounds such as methanol and acetate, or from molecular hydrogen. Methane is the reduced product of their metabolism. These are among the most strictly anaerobic organisms. They are responsible for swamp gas (methane) production in the sediments of aquatic habitats. Some reside as symbionts in the rumen of cattle and hindguts of termites.

3.5.2 The Nitrogen Cycle

Some prokaryotic organisms can take molecular nitrogen from the atmosphere and reduce it to form the amino group of an amino acid. The process, known as nitrogen fixation, requires much energy, and the complex enzyme responsible for the process, nitrogenase, has been found only in certain prokaryotic organisms. Many of the bacteria that carry out this process are free-living. These include some cyanobacteria which are important nitrogen fixers in aquatic habitats, as well as soil nitrogen fixers in the genera *Azotobacter* and *Clostridium*. Interestingly, all of the anoxygenic photosynthetic bacteria that have been tested have been found to be nitrogen fixers. This is one of the reasons the process is thought to be extremely ancient.

From a geochemical standpoint, the most important nitrogen fixers are those that grow symbiotically in association with higher plants. For example, the bacterial genus *Rhizobium* occurs in special root nodules of leguminous (and some other) plants, including clover, alfalfa, beans, and peas. The plant supplies organic nutrients for growth of the bacterium and the bacterium produces nitrogen in a usable form for the plant.

The nitrifying bacteria, universally found in aerobic soil and aquatic environments, derive energy from the oxidation of reduced inorganic nitrogen compounds (ammonia and nitrite). As do autotrophic bacteria, they obtain carbon from carbon dioxide in the atmosphere.

The process of dissimilatory dentrification occurs anaerobically and is mediated by bacteria that use nitrate in place of oxygen as an acceptor of electrons during respiration. The result is the formation of molecular nitrogen and nitrous oxide. The nitrous oxide plays a role in the chemistry of stratospheric ozone and is, therefore, extremely important biogeochemically. These bacteria are heterotrophic and derive energy from the anaerobic oxidation of organic compounds.

Many organisms, especially bacteria, decompose organic material with the release of ammonia, a process referred to as ammonification.

3.5.3 The Sulfur Cycle

Reduced sulfur compounds serve as hydrogen donors for the anoxygenic photosynthetic bacteria such as the green and purple sulfur bacteria and some cyanobacteria. In contrast, chemoautotrophic sulfur bacteria obtain energy from the oxidation of reduced sulfur compounds including hydrogen sulfide, sulfur, and thiosulfate. As with the nitrifying bacteria, these bacteria are primarily aerobic and use carbon dioxide as their source of carbon. The ultimate product of their metabolism is sulfuric acid. These bacteria are responsible for the production of acid mine waters in areas where strip mining has exposed pyrite minerals to rainfall and oxygen. Some of these bacteria can grow at pH values as low as 1.0 and pH values of 3.0 and 4.0 are common in run-off streams from mining areas. Fish cannot live in these waters and most plants cannot grow in exposed soils of such high acidity.

Dissimilatory sulfate reducers such as *Desulfovibrio* are heterotrophic, deriving energy for their growth from the anaerobic oxidation of organic compounds such as lactic acid and acetic acid. Sulfate is reduced and large amounts of hydrogen sulfide are generated in this process. The black sediments of aquatic habitats that smell of sulfide are due to the activities of these bacteria. The black coloration is caused by the formation of metal sulfides, primarily iron sulfide. These bacteria are especially important in marine habitats because of the high concentrations of sulfate that exists there.

Dimethylsulfide (DMS) is the major volatile sulfur compound of biogenic origin emitted from the oceans into the atmosphere. It is estimated that the annual global sea-to-air flux is 39 million metric tons of sulfur per year. DMS is produced by the enzymatic cleavage of dimethylpropiothetin (DMPT). The function of DMPT in these algae is not totally resolved, but there is strong evidence (Andreae and Bernard, 1984; Vairavamurthy *et al.*, 1985) that it may function as an osmotic solute. The dipolar ionic nature of DMPT gives the molecule a very low membrane permeability. The osmotic role of DMPT is also suggested by the fact that, except for the cyanophyta, freshwater algae produce little or no DMS. The dimethylsulfide produced by marine algae reacts in the atmosphere to form sulfuric acid as a natural component of acid rain.

3.5.4 The Phosphorus Cycle

Phosphorus is not an abundant constituent of the biosphere. It occurs primarily either as inorganic phosphate or as various organic forms of phosphate. Phosphorus occurs principally in the form of phosphate, which is non-volatile, and is not transported in the atmosphere as a gas to any significant degree. Its mobilization in soil and water is affected by precipitation with divalent metals such as calcium and magnesium. None the less, it is an essential component of living organisms where it is found in their nucleic acids and high-energy compounds such as ATP. Thus, organisms have developed mechanisms for concentrating phosphorus, which is often a limiting nutrient, from soil and water. Phosphorus is a major component of plant fertilizers. In freshwaters, algal blooms are frequently controlled by the availability of phosphate. Microorganisms are able to store phosphate as a polymer inside their cells. This is especially important to aquatic forms, such as algae and cyanobacteria, during periods of phosphorus limitation.

3.5.5 The Metal Cycles

Many bacteria can oxidize and reduce metals and metallic ions. Some can even derive energy from those oxidative processes. For example, iron oxidizers such as *Thiobacillus* spp. can grow on reduced iron compounds and obtain energy from their oxidation if the pH is sufficiently low, as in pyrite oxidation in acid mine waters. At this time, however, there are no known chemoautotrophs that can derive energy from the oxidation of manganese, although many heterotrophic bacteria can perform this activity. Many bacteria are also involved in the deposition of oxidized iron and manganese compounds, thereby immobilizing these elements. Trace amounts of metals can be very important to organisms either through toxic effects or the roles that metals play in enzymes and energy transfer compounds.

3.5.6 Concluding Remarks

This brief survey is illustrative rather than comprehensive. Organisms are involved in many other transformations that have important environmental consequences. Microbial processes are especially diverse and important. Pesticide degradation, cellulose decomposition, toxin production, deposition of limestone, rock and mineral weathering, formation of sulfur deposits, antibiotic production, and methyl mercury formation are just a few of the many processes of ecological and geochemical significance that are mediated, at least in part, by microorganisms (see Table 3-7). Many further details of important transformations are brought out in the chapters on specific elements.

The long-term influences of living organisms on the atmosphere of the Earth have been enormous. Had the Earth remained lifeless, concentrations of carbon dioxide in the atmosphere probably would have remained very high, and the temperature would have been very different from that of today. Oxygen would have slowly increased due to the splitting of water by sunlight, but it would not have reached concentrations more than 1% of the present concentration. The atmosphere of the Earth also differs in the chemical interactions taking place in it. If life were to disappear today, nitrogen in the atmosphere would eventually be transformed into nitrate, which would be transferred to the oceans, lowering their pH considerably from present values.

The mean temperature of the Earth is a result of input of energy from the Sun and loss of energy by emission of radiation. The input of energy is a function of the reflectivity (albedo) of the Earth. Ice reflects 80–95% of incident light, dry grassland 30–40%, and a conifer forest 10–15%. Therefore, seasonal changes in vegetation substantially alter the amount of radiation absorbed by the surface. Similarly, major changes in vegetation due to human activity have the same effect. The surface temperature of the Earth also depends on atmospheric concentrations of carbon dioxide, nitrous oxide, and methane, greenhouse gases that also have important connections to the biosphere.

Questions

3-1 Since the Earth was originally anaerobic, anaerobic microorganisms evolved early in the Earth's history. Discuss the evidence for and implications of an anaerobic origin of life.

3-2 Louis Pasteur first described organisms that could live without air (namely oxygen). This was a sensation to biologists of the day. How many types of anaerobic organisms are known today? How does each obtain energy?

3-3 Evidence from analyses of ribosomal RNA has led many biologists to believe in a three- rather than a five-kingdom system of classification of organisms. The three kingdoms consist of eubacteria, archaebacteria, and eukaryotes. Compare this classification to the five-kingdom system and indicate where you would place the protozoa, cyanobacteria, methane-producing bacteria, and fungi in a three-kingdom system.

3-4 The unique microbial processes of the nitrogen cycle are nitrogen fixation and nitrification. Describe each of these by a chemical equation and briefly discuss the energy-yielding metabolism of the microorganisms that are responsible for these transformations.

Table 3-7 Biological sources of the Earth's major atmospheric gases[a]

Gas	Principal biological source	Residence time in the atmosphere
Nitrogen	Bacteria	10^7 to 10^9 years
Oxygen	Photosynthesis	Thousands of years
Carbon dioxide	Organism respiration, fuel combustion	About 100 years
Carbon monoxide	Bacterial processes, incomplete combustion	A few months
Methane	Bacteria in anaerobic environments	A few years
Nitrous oxide	Bacteria and fungi	About 100 years
Ammonia	Bacteria and fungi	A few days
NO_x	Reaction of pollutants in sunlight	A few days
Hydrogen sulfide	Anaerobic bacteria	A few days
Hydrogen	Photosynthetic bacteria, methane oxidation	A few years

[a] Modified from Margulis and Lovelock (1974).

3-5 Over most of evolutionary history all living organisms were single-celled and mostly lacked hard parts. How would a living world of this structure influence biogeochemical cycles? In what ways would their effects differ from that provided by today's living world?

3-6 The rise of predation as a major form of interaction among organisms led to the evolution of hard protective coverings on many organisms. Outline the major ways that this ecological development might have affected the size and nature of biogeochemical cycles.

3-7 The invasion of land occurred relatively recently in the history of life. How did this event influence the types and quantities of molecules produced by living organisms?

3-8 Carbon constitutes about 11% of the Earth's crust and approximately the same proportion of animal bodies. If these compositions are so similar, how could living organisms have altered the carbon composition of the atmosphere so strikingly?

3-9 In 1948, the Swiss chemist Paul Mueller was awarded a Nobel Prize for his discovery of the insecticidal properties of DDT. If Mueller had made his discovery more recently, would he have been as likely to receive the Nobel Prize? How might thinking about biogeochemical cycles have influenced recent judgements?

3-10 To what extent are the attributes of ecosystems collective properties, that is, the summation of the characteristics of components of the system, as opposed to being emergent attributes, that is, properties that are totally unpredictable from observations of the components of those systems? How does your answer influence the way you think about biogeochemical cycles?

References

Andreae, M. O. and W. R. Bernard (1984). The marine chemistry of dimethyl sulfide. *Mar. Chem.* **14**, 267–269.

Awramik, S. M., J. W. Schopf, and M. R. Walter (1983). Filamentous fossil bacteria 3.5×10^9 years old from the Archaen of Western Australia. *Precambrian Res.* **20**, 357–374.

Bate-Smith, E. C. and C. R. Metcalf (1957). Laucoanthocyanins 3. The nature and distribution of tannins in dicotyledonous plants. *J. Linn. Soc. (Lond.)* **55**, 669–705.

Bonner, J. T. (1967). "The Cellular Slime Molds," 2nd edn. Princeton University Press, Princeton, N. J.

Bonner, J. T. (1970) The chemical ecology of cells in the soil. *In* "Chemical Ecology" (E. Sondheimer and J. B. Simeone, eds.), pp. 133–155. Academic Press, New York.

Cates, R. G. and D. F. Rhoades (1977). Prosopis leaves as resource for insects. *In* "Mesquite: Its Biology in Two Desert Scrub Ecosystems" (B. Simpson, ed.), US/IBP Synthesis Series, Vol. 4, pp. 61–83. Dowden, Hutchinson and Ross, Stroudsberg, Penn.

Chang, S., D. Desmarias, R. Mack, S. L. Miller, and G. E. Strathern (1983). Prebiotic organic synthesis and the origin of life. *In* "Earth's Earliest Biosphere, Its Origin and Evolution" (J. W. Schopf, ed.), pp. 53–88. Princeton University Press, Princeton, N. J.

Cloud, P. (1983) Early biogeologic history: The emergence of a paradigm. *In* "Earth's Earliest Biosphere, Its Origin and Evolution" (J. W. Schopf, ed.), pp. 14–29. Princeton University Press, Princeton, N. J.

Dawkins, R. (1982). "The Extended Phenotype." Oxford University Press, Oxford.

Dawkins, R. (1987). "The Blind Watchmaker." W. W. Norton, New York.

Feeny, P. O. (1970) Seasonal changes in oakleaf tannins and nutrients as a cause of spring feeding by winter moth caterpillars. *Ecology* **51**, 656–681.

Feeny, P. O. (1976). Plant apparency and chemical defense. *Rec. Adv. Phytochem.* **10**, 1–40.

Futuyma, D. J. (1986). "Evolutionary Biology," 2nd edn. Sinauer Associates, Sunderland, Mass.

Haldane, J. B. S. (1955). Animal communication and the origin of human language. *Sci. Progr.* (Lond.) **43**, 385–410.

Harborne, J. B. (1988). "Introduction to Ecological Biochemistry," 3rd edn. Academic Press, New York.

Hofmann, H. J. and J. W. Schopf (1983). Early Proterozoic microfossils. *In* "Earth's Earliest Biosphere, Its Origin and Evolution" (J. W. Schopf, ed.), pp. 321–359. Princeton University Press, Princeton, N. J.

Humphreys, W. F. (1979). Production and respiration in animal populations. *J. Anim. Ecol.* **48**, 427–453.

Hutchinson, G. E. (1944). Nitrogen in the biogeochemistry of the atmosphere. *Amer. Sci.* **32**, 178–195.

Janzen, D. H. (1974). Tropical blackwater rivers, animals, and mass fruiting by the Dipterocarpaceae. *Biotropica* **6**, 69–103.

Margulis, L. (1970). "Origin of Eucaryotic Cells." Yale University Press, New Haven, Conn.

Margulis, L. and J. E. Lovelock (1974). Biological modulation of the Earth's atmosphere. *Icarus* **21**, 471–489.

McKey, D. (1974). Adaptive patterns in alkaloid physiology. *Amer. Nat.* **108**, 305–320.

Miller, O. L., B. A. Hamkalo, and C. A. Thomas (1970). Visualization of bacterial genes in action. *Science* **169**, 392–395.

Miller, S. L. (1953). A production of amino acids under possible primitive Earth conditions. *Science* **117**, 528–529.

Oparin, A. I. (1938). "The Origin of Life." Macmillan, New York (reprinted in 1953, Dover, New York).

Orians, G. H. and D. H. Janzen (1974). Why are embryos so tasty? *Amer. Nat.* **108**, 581–592.

Purves, W. K. and G. H. Orians (1987). "Life: The Science of Biology," 2nd edn. Sinauer Associates, Sunderland, Mass.

Redfield, A. C., B. H. Ketchum, and F. A. Richards (1963). The influence of organisms on the composition of seawater. *In* "The Sea" (M. N. Hill, ed.), Vol. 2, pp. 26–77. Wiley-Interscience, New York.

Rhoades, D. F. (1977). The antiherbivore chemistry of *Larrea*. *In* "Creosote Bush" (T. J. Mabry, J. H. Hunziker, and D. R. DiFeo, Jr., eds.), US/IBP Synthesis Series, Vol. 4, pp. 135–175. Dowden, Hutchinson, & Ross, Stroudsberg, Penn.

Rhoades, D. F. and R. G. Cates (1976). Toward a general theory of plant antiherbivore chemistry. *In* "Biochemical Interaction Between Plants and Insects" (J. W. Wallace, ed.), pp. 168–212. Plenum Press, New York.

Rodin, L. E. *et al.* (1975). *Proceedings of the First International Congress on Ecology*, Wageningen.

Rosenthal, G. A. and D. H. Janzen (1979). "Herbivores: Their Interaction with Secondary Plant Metabolites." Academic Press, New York.

Rubey, W. W. (1951). Geological history of sea water: An attempt to state the problem. *Geol. Soc. Amer. Bull.* **62**, 1111–1148.

Rubey, W. W. (1955). Development of the hygrosphere and atmosphere, with special reference to probable composition of the early atmosphere. *In* "Crust of the Earth" (A. Poldenvaart, ed.), pp. 631–650. Geological Society of America, New York.

Swain, T. (1979). Tannins and lignins. *In* "Herbivores: Their Interaction with Secondary Plant Metabolites" (G. A. Rosenthal and D. H. Janzen, eds.), pp. 658–682. Academic Press, New York.

Vairavamurthy, A., M. O. Andreae, and R. L. Iverson (1985). Biosynthesis of dimethylpropionthetin by *Hymenomnonas carterae* in relation to sulfur source and salinity variations. *Limnol. Oceanog.* **30**, 59–70.

Walter, M. R., R. Buick, and J. S. R. Dunlop (1980). Stromatolites 3 400–3 500 Myr old from the North Pole area. Western Australia. *Nature* **284**, 443–445.

Whittaker, R. H. (1969). New concepts of kingdoms of organisms. *Science* **163**, 150–160.

Wilson, E. O. (1970). Chemical communication within animal species. *In* "Chemical Ecology" (E. Sondheimer and J. B. Simeone, eds.), pp. 133–155. Academic Press, New York.

Woese, C. and G. E. Fox (1977). Phylogenetic structure of the prokaryotic domain: The primary kingdoms. *Proc. Natl. Acad. Sci. USA* **74**, 5088–5090.

4

Modeling Biogeochemical Cycles

Henning Rodhe

4.1 Introductory Remarks

The chemical and physical status of the Earth is characterized by transport and transformation processes, many of which are of a cyclical nature. The circulation of water between oceans, atmosphere, and continents is an example of such a cyclic process. The basic characteristics of a cycle of a particular element or compound are often described in terms of the content in the various reservoirs and the fluxes between them. In our example, the reservoirs could be "the oceans", "the water in the atmosphere", "the ground water", etc. A fundamental question in the cycle approach is the determination of how the rates of transfer between the reservoirs depend on the content of the reservoirs and on other, external, factors. In many cases, the details of the distribution of the element within each of the reservoirs are disregarded.

The cycle approach to describe the physiochemical environment on Earth has advantages as well as disadvantages. The advantages include the following.

1. It provides an overview of fluxes, reservoir contents, and turnover times.
2. It gives a basis for quantitative modeling.
3. It helps to estimate the relative magnitudes of anthropogenic and natural fluxes.
4. It stimulates questions such as: Where is the material coming from? Where is it going next?
5. It helps to identify gaps in knowledge.

The following are some of the disadvantages.

1. The analysis is by necessity superficial. It provides little or no insight into what goes on inside the reservoirs or into the nature of the fluxes between them.
2. It gives a false impression of certainty. Very often, at least, one of the fluxes in a budget is derived from balance considerations. Such estimates may erroneously be taken to represent solid knowledge.
3. The analysis is based on averaged quantities that cannot always be easily measured because of spatial variation and other complicating factors.

Many important geophysical problems cannot be studied using the simplified cycle approach. Weather forecasting, for example, requires a detailed knowledge about the distribution of winds, temperature, etc., within the atmosphere. In this case, it is obviously not possible to use a reservoir model with the atmosphere as one of the reservoirs. It would not be much better even if one divided the atmosphere into a few reservoirs. A forecast model requires a resolution fine enough to resolve explicitly the structure of the most important weather phenomena such as cyclones, anticyclones, and wave patterns on spatial scales from a few hundred kilometers upwards. Models of this kind are based either on a division of the physical space into a large but finite number of grid cubes (grid point models) or on a separation of the variables into different wave numbers (spectral models). It is possible to look at such models as reservoir models consisting of very many reservoirs. However, this is normally not done.

The purpose of this chapter is to introduce and define the basic concepts used in the description and modeling of biogeochemical cycles. The last two

Global Biogeochemical Cycles
ISBN 0-12-147685-5

sections contain a brief summary of transport processes and the time-scales characterizing exchange processes in the atmosphere and oceans.

4.2 Reservoir Models and Cycles: Some Definitions

The following definitions are applicable to studies of biogeochemical cycles.

1. *Reservoir (box, compartment)*. An amount of material defined by certain physical, chemical, or biological characteristics that, under the particular consideration, can be considered as reasonably homogeneous. For example:

- oxygen in the atmosphere;
- carbon monoxide in the southern hemisphere;
- carbon in living organic matter in the ocean surface layer;
- ocean water having a density between ϱ_1 and ϱ_2;
- sulfur in sedimentary rocks.

In situations where the reservoir is defined by its physical boundaries, it is not uncommon to refer to its content of the specific element as its *burden*. We will denote the content of the reservoir by M. The dimension of M would normally be mass, although it could also be, for example, moles.

2. *Flux*. The amount of material transferred from one reservoir to another per unit time, in general denoted by F (mass per time). For example:

- the rate of evaporation of water from the ocean surface to the atmosphere;
- the rate of oxidation of N_2O in the stratosphere (i.e. flux from the atmospheric N_2O-nitrogen reservoir to the stratospheric NO_x-nitrogen reservoir);
- the rate of deposition of phosphorus on marine sediments.

In more specific studies of transport processes, the flux is normally defined as the amount of material transferred per unit area per unit time. To distinguish between these two conflicting usages, we shall refer to the latter as "flux density".

3. *Source*. A flux (Q) of material *into* a reservoir.
4. *Sink*. A flux (S) of material *out of* a reservoir. Very often this flux is assumed to be proportional to the content of the reservoir ($S = kM$). In such cases, the sink flux is referred to as a first-order process. If the sink flux is constant, independent of the reservoir content, the process is of zero order. Higher-order fluxes, i.e. $S = kM^\alpha$ with $\alpha > 1$, also occur.
5. *Budget*. A balance sheet of all sources and sinks of a reservoir. If sources and sinks balance and do not change with time, the reservoir is in *steady-state*, i.e. M does not change with time. It is common in many budget estimates that some fluxes are better known than others. If steady-state prevails, a flux that is unknown *a priori* can be estimated by its difference from the other fluxes. If this is done, it should be made very clear in the presentation of the budget which of the fluxes is estimated as a difference.
6. *Turnover time*. The turnover time of a reservoir is the ratio of the content M of the reservoir to the sum of its sinks S or the ratio of M to the sources Q (see Section 4.3). The turnover time is the time it will take to empty the reservoir in the absence of sources if the sinks remain constant. It is also a measure of the average of the times spent by individual molecules or atoms in the reservoir (more about this in the next section).
7. *Cycle*. A system consisting of two or more connected reservoirs, where a large part of the material is transferred through the system in a cyclic fashion. If all material cycles *within* the system, the system is *closed*. In many situations, one may consider systems of connected reservoirs that are not cyclic but where material flows unidirectionally. In this connection, some reservoirs (at the end of the chain) may be *accumulative*, whereas others remain unbalanced (*non-accumulative*) (cf. Holland, 1978).
8. *Geochemical cycle*. In geology and geochemistry, the reservoirs and fluxes depicted in Fig. 4-1 are often referred to as the geochemical cycle. It might as well have been called the geophysiochemical cycle.
9. *Biogeochemical cycle*. This term is often used to describe the global or regional cycles of the "life elements" C, O, N, S, and P with reservoirs including the whole or part of the atmosphere, the ocean, the sediments, and the living organisms. Figure 4-2 shows the principal reservoirs and fluxes in the global biogeochemical carbon cycle as an example. The term can be applied to the corresponding cycles of other elements or compounds.

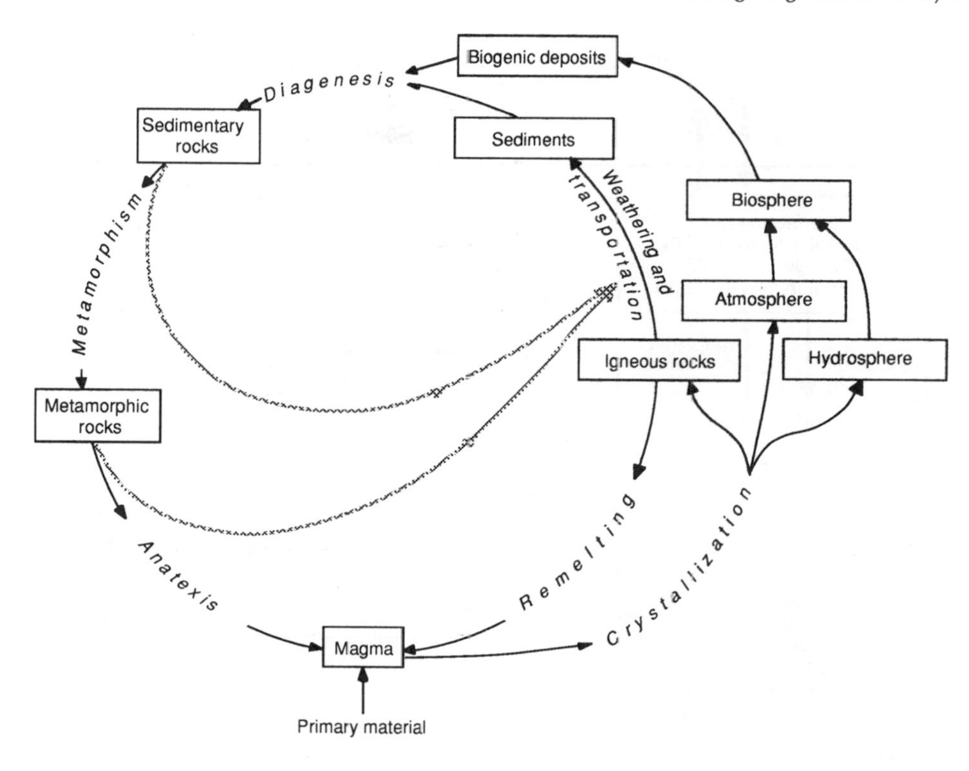

Fig. 4-1 The geochemical cycle showing the flux of material between various reservoirs in the geosphere. Adapted from Mason (1984) with the permission of John Wiley and Sons, Inc.

General comments

Budgets and cycles can be considered on very different *spatial scales*. In this book, we concentrate on global, hemispheric, and regional scales. The choice of a suitable scale (i.e. the size of the reservoirs) should be determined by the degree of ambition of the analysis as well as by the homogeneity of the spatial distribution. For example, in carbon cycle models it is reasonable to consider the atmosphere as one reservoir (the concentration of CO_2 in the atmosphere is fairly uniform). On the other hand, oceanic carbon content and carbon exchange processes exhibit large spatial variations and it is reasonable to separate the surface layer from the deeper layers, the Atlantic from the Pacific, etc. Many sulfur and nitrogen compounds in the atmosphere occur in very different concentrations in different regions of the world. For these compounds, regional budgets tell us more about the real situation at any one place than global budgets.

We shall in the main be concerned with processes that occur on *time-scales* longer than, or equal to, a season (3 months). This implies that we may consider the variables and the parameters of our models as time-averaged quantities with an averaging time of at least a season.

4.3 Time-scales and Single Reservoir Systems

4.3.1 Turnover Time

The turnover time is the ratio between the content (M) of a reservoir and the total flux out of it (S):

$$\tau_0 = M/S \tag{1}$$

The turnover time may be thought of as the time it would take to empty the reservoir if the sink (S) remained constant while the sources were zero ($\tau_0 S = M$). In fluid reservoirs like the atmosphere or

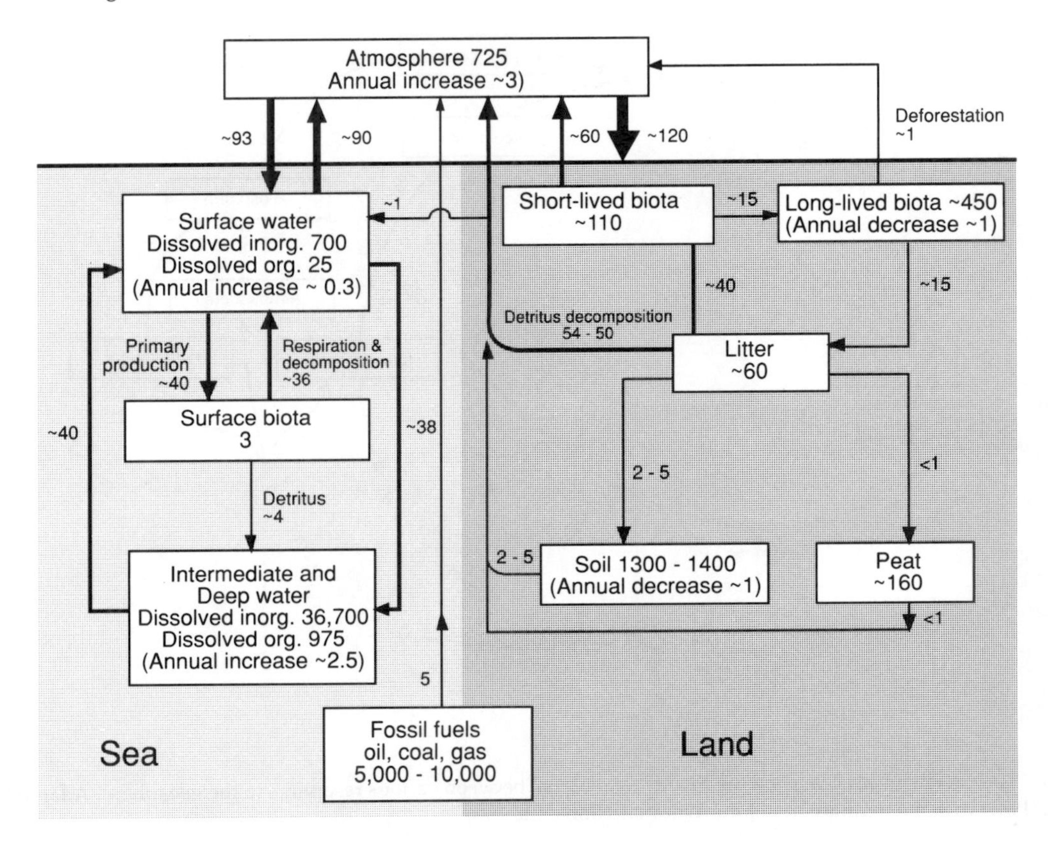

Fig. 4-2 Principal reservoirs and fluxes in the carbon cycle. Units are 10^{15} g C (burdens) and 10^{15} g C/year (fluxes). From Bolin (1986) with permission from John Wiley and Sons.

the ocean, the turnover time is also related to the spatial variability of the tracer concentration within the reservoir: a long turnover time corresponds to a small variability and vice versa (Junge, 1974; Hamrud, 1983).

If material is removed from the reservoir by two or more separate processes, each with a flux S_i, one can define turnover times with respect to each such process as:

$$\tau_{0i} = M/S_i \qquad (2)$$

Since $\Sigma S_i = S$, these time-scales are related to the turnover time of the reservoir, τ_0, by

$$\tau_0^{-1} = \Sigma \tau_{0\,i}^{-1} \qquad (3)$$

The equation describing the rate of change of the content of a reservoir can be written as:

$$\frac{dM}{dt} = Q - S = Q - M/\tau_0 \qquad (4)$$

If the reservoir is in steady-state ($dM/dt = 0$), sources (Q) and sinks (S) must balance. In this case, S can be replaced by Q in equation (1).

4.3.2 *Residence Time (transit time)*

The *residence time* is the time spent in a reservoir by an *individual* atom or molecule. It is also the age of a molecule when it leaves the reservoir. If the pathway of the atom from the source to the sink is characterized by a physical transport, the term *transit time* can be used as an alternative. Even for the same element (or compound), different atoms (or molecules) will have different residence times in a given reservoir. The probability density function of residence times is denoted by $\phi(\tau)$, where $\phi(\tau)\,d\tau$ describes the fraction of the atoms (molecules) having a residence time in the interval to τ to $\tau + d\tau$. The probability density function may have very

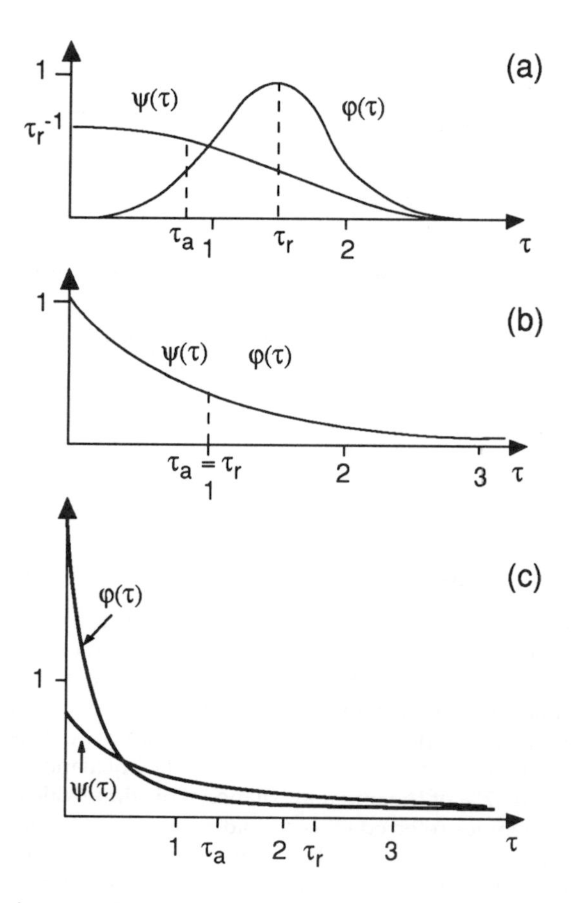

Fig. 4-3 The *age* frequency function $\psi(\tau)$ and the *transit time* frequency function $\phi(\tau)$ and the corresponding average values τ_a and τ_r for the three cases described in the text: (**a**) $\tau_a < \tau_r$; (b) $\tau_a = \tau_r$; (c) $\tau_a > \tau_r$. Adapted from Bolin and Rodhe (1973) with the permission of the Swedish Geophysical Society, Stockholm.

different shapes. Some examples are shown in Fig. 4-3. This figure also contains the frequency function of age. These terms are further illustrated below.

Figure 4-3a might correspond to a lake with inlet and outlet on opposite sides of the lake. Most water molecules will then have a residence time in the lake roughly equal to the time it takes for the mean current to carry the water from the inlet to the outlet. Another example is a human population where most people live to a mature age. Figure 4-3b illustrates the common situation with exponential decay or with a well-mixed reservoir. A simple example could be the reservoir of all ^{238}U on Earth. The half-life of this nuclide is 4.5×10^9 years, implying that the content of this reservoir today is about half of what it was when the planet Earth was formed 4.5 Ga ago. The probability density function of residence time of the uranium atoms originally present is an exponential decay function. The average residence time is 6.5×10^9 years. (The average value of time for an exponential decay function is the half-life divided by ln 2.) In the reservoir corresponding to Fig. 4-3c, the removal is biased toward "young" particles. This might occur when the sink is located close to the source (the "short circuit" case).

The *average residence time* (average transit time) τ_r is defined by

$$\tau_r = \int_0^\infty \tau\phi(\tau)\,d\tau \tag{5}$$

In many cases, the word "average" is left out and this quantity is simply referred to as "residence time".

4.3.3 Age

The age of an atom or molecule in a reservoir is the time since it entered the reservoir. Age is defined for all molecules, whether they are leaving the reservoir or not. As with residence times, the probability density function of ages $[\psi(\tau)]$ can have different shapes. In a steady-state reservoir, however, $\psi(\tau)$ is always a non-increasing function. The shapes of $\psi(\tau)$ corresponding to the three residence time distributions discussed above are included in Fig. 4-3.

The *average age* of atoms in a reservoir is given by:

$$\tau_a = \int_0^\infty \tau\psi(\tau)\cdot d\tau \tag{6}$$

4.3.4 Relations between τ_0, τ_r and τ_a

For a reservoir in steady-state, τ_0 is equal to τ_r, i.e. the turnover time is equal to the average residence time spent in the reservoir by individual particles (Eriksson, 1971; Bolin and Rodhe, 1973).

This may seem to be a trivial result, but it is actually of great significance. For example, if τ_0 can be estimated from budget considerations by comparing fluxes and burdens and if the average transport velocity (V) within the reservoir is known, the average distance ($L = V\tau_r$) over which the transport takes place in the reservoir can be estimated.

The relation between τ_0 and τ_a is less simple. τ_a can be larger or less than τ_0 depending upon the shape of the age distribution (cf. Fig. 4-3). For a well-mixed reservoir, or one with a first-order removal process, $\tau_a = \tau_0$.

In the case of a human population (Fig. 4-3a), τ_a is only about half of τ_0: the average age of all Swedes is between 35 and 40 years, whereas the average residence time, i.e. the average length of life (average age at death), is just over 70 years.

In the situation where most atoms leave the reservoir quickly but a few of them survive very long, τ_a is larger than τ_0 (the "short circuit" case). Some further examples of age distributions and relations between τ_a and τ_0 are given in Lerman (1979).

When equating τ_0 and τ_r, it must be made clear that the flux (S) defining τ_0 is the *gross flux* and not a net flux. For example, the removal of water from the atmosphere is brought about by both precipitation and dry deposition (direct uptake by diffusion to the surface). The dry deposition is normally not explicitly evaluated but subtracted from the gross evaporation flux to yield the net evaporation from the surface. The turnover time of water in the atmosphere calculated as the ratio between the atmospheric content and the precipitation rate (10 days) is thus *not* equal to the average residence time of water molecules in the atmosphere. The actual value of the average residence time of individual water molecules is substantially shorter.

4.3.5 *Response Time*

The response time of a reservoir is a time-scale that characterizes the adjustment to equilibrium after a sudden change in the system. A precise definition is not easy to give except in special circumstances as in the following example.

Consider a single reservoir for which the sink is proportional to the content ($S = kM$) and which is initially at equilibrium with fluxes $Q_0 = S_0$ and content M_0. The turnover time of this reservoir is:

$$\tau_0 = M_0/S_0 = 1/k \qquad (7)$$

Suppose now that the source strength is suddenly changed to a new value Q_1. How long would it take for the reservoir to reach a new equilibrium? The adjustment process is described by the differential equation

$$\frac{dM}{dt} = Q_1 - S = Q_1 - kM \qquad (8)$$

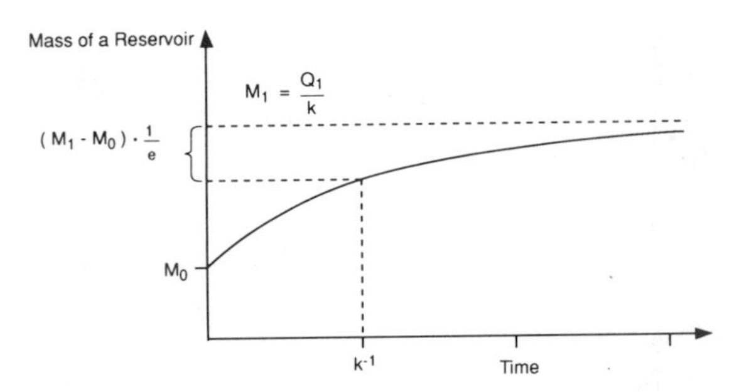

Fig. 4-4 Illustration of an exponential adjustment process. In this case, the response time is equal to k^{-1}.

with the initial condition $M(t = 0) = M_0$.

The solution

$$M(t) = M_1 - (M_1 - M_0)\exp(-kt) \qquad (9)$$

approaches the new equilibrium value ($M_1 = Q_1/k$) with a response time equal to k^{-1} or τ_0. The change of the reservoir mass from the initial value M_0 to the final value M_1 is illustrated in Fig. 4-4. In this case, with an exponential adjustment, the response time is defined as the time it takes to reduce the imbalance to $e^{-1} = 37\%$ of the initial imbalance. This time-scale is sometimes referred to as "e-folding time". Thus, for a single reservoir with a sink proportional to its content, the response time equals the turnover time.

As a specific example, consider oceanic sulfate as the reservoir. Its main source is river run-off (pre-industrial value: 100 Tg S/year) and the sink is probably incorporation into the lithosphere by hydrogeothermal circulation in mid-ocean ridges (100 Tg S/year, McDuff and Morel, 1980; cf. Chapter 13). The content of sulfate in the oceans is about 1.3×10^9 Tg S. If we make the (unrealistic) assumption that the present run-off (200 Tg S/year) would continue indefinitely, how long would it take for the sulfate concentration in the ocean to adjust to this new flux? The answer is $\tau_0 \sim 1.3 \times 10^9$ Tg/10^2 Tg/year $\sim 10^7$ years. A more detailed treatment of a similar problem can be found in Southam and Hay (1976).

4.3.6 *Reservoirs in Non-steady-state*

Let us analyze the situation when one observes a change in reservoir content and wants to draw conclusions regarding the sources and sinks. We rewrite equation (8) as:

$$\frac{1}{M}\frac{dM}{dt} = \frac{Q}{M} - \frac{1}{\tau_0}$$

where $\tau_0 = 1/k$ is the turnover time in the steady-state situation. Let us denote the left-hand side of the equation (the observed rate of change of the reservoir content) by τ_{obs}^{-1}. If the mass is observed to increase by, say, 1% per year, τ_{obs} would be 100 years. Two limiting cases can be singled out:

1. $\tau_{obs} \gg \tau_0$. In this case, there has to be an approximate balance between the two terms on the right-hand side of the equation:

$$\frac{Q}{M} \approx \frac{1}{\tau_0} \text{ or } Q \approx \frac{M}{\tau_0}$$

This means that the observed change in M mainly reflects a change in the source flux Q or the sink function. As an example, we may take the methane concentration in the atmosphere, which is now increasing by about 1% per year. The turnover time is estimated to be about 10 years, i.e. much less than τ_{obs} of 100 years (Cicerone and Oremland, 1988). Consequently, the observed rate of increase in atmospheric methane is a direct consequence of a similar rate of increase of emissions into the atmosphere. (In fact, this is not quite true. A fraction of the observed increase is probably due to a decrease in sink strength caused by a decrease in the concentration of hydroxyl radicals responsible for the decomposition of methane in the atmosphere.)

2. $\tau_{obs} \ll \tau_0$. In this case, the balance is $dM/dt \approx Q$, which means that there is an increase in reservoir content about equal to the source flux with little influence of the sink. The reservoir is then in an accumulative stage and its mass is increasing with time largely as a function of Q. The content of CFC-12 in the atmosphere is an interesting example of this situation. The observed increase is about 4% per year ($\tau_{obs} \approx 25$ years) and the turnover time is about 120 years. The rate of increase is thus a reflection of an imbalance between sources and sinks rather than an increase in the source flux Q. The emissions of CFC-12 actually remained essentially constant between 1974 and 1988.

In situations where τ_{obs} is comparable in magnitude to τ_0, a more complex relation prevails between Q, S, and M. Atmospheric CO_2 falls in this last category, although its turnover time (3–4 years, cf. Fig. 4.2) is much shorter than τ_{obs} (about 300 years). This is because the atmospheric CO_2 reservoir is closely coupled to the carbon reservoir in the biota and in the surface layer of the oceans (Section 4.4). The effective turnover time of the combined system is actually several hundred years (Rodhe and Björkström, 1979).

4.4 Coupled Reservoirs

The treatment of time-scales and dynamic behavior of single reservoirs given in the previous section can easily be generalized to systems of two or more reservoirs. While the simple system analyzed in the previous section illustrates many important characteristics of cycles, most natural cycles are more complex. The matrix method described in Section 4.4.1 provides an approach to systems with very large numbers of reservoirs that is, at least, notationally simple. The treatments in the preceding section and in Section 4.4.1 are still limited to linear systems. In many cases, we assume linearity because our knowledge is not adequate to assume any other dependence and because the solution of linear systems is straightforward. There are, however, some important cases where non-linearities are reasonably well understood. A few of these cases are described in Section 4.4.2.

As important as coupled reservoirs and non-linear systems are, the less mathematically inclined may want to read this section only for its qualitative material. The treatment described here is not essential for understanding the reading later in the book.

4.4.1 Linear Systems

A linear system of reservoirs is one where the fluxes between the reservoirs is linearly related to the reservoir contents. A special case, that is commonly assumed to apply, is one where the fluxes between reservoirs are proportional to the content of the reservoirs where they originate. Under this proportionality assumption, the flux F_{ij} from reservoir i to reservoir j is given by:

$$F_{ij} = k_{ij} M_i \tag{10}$$

The rate of change of the amount M_i in reservoir i is

thus:

$$\frac{dM_i}{dt} = \sum_{j=1}^{n} k_{ji} M_j - M_i \sum_{j=1}^{n} k_{ij} \text{ for } j \neq i \quad (11)$$

where n is the total number of reservoirs in the system.

This system of differential equations can be written in matrix form as

$$\frac{d\mathbf{M}}{dt} = \mathbf{kM} \quad (12)$$

where the vector $\mathbf{M}$ is equal to $(M_1, M_2 \ldots M_n)$ and the elements of matrix $\mathbf{k}$ are linear combinations of the coefficients k_{ij}. The solution to equation (12) describes the adjustment of all reservoirs to a steady-state by a finite sum of exponential decay functions (Lasaga, 1980). The time-scales of the exponential decay factors correspond to the non-zero eigenvalues of the matrix $\mathbf{k}$. The response time of the system, τ_{cycle}, may be defined by:

$$\tau_{cycle} = \frac{1}{|E_1|} \quad (13)$$

where E_1 is the non-zero eigenvalue with smallest absolute value (Lasaga, 1980). The treatment can be generalized by adding an external forcing function on the right-hand side of equations (11) and (12).

As an illustration of the concept introduced above, let us consider a coupled two-reservoir system with no external forcing (Fig. 4-5). The dynamic behavior of this system is governed by the two differential equations:

$$\frac{dM_1}{dt} = -k_{12}M_1 + k_{21}M_2$$

$$\frac{dM_2}{dt} = k_{12}M_1 - k_{21}M_2 \quad (14)$$

the expression of conservation of mass

$$M_1 + M_2 = M_T \quad (15)$$

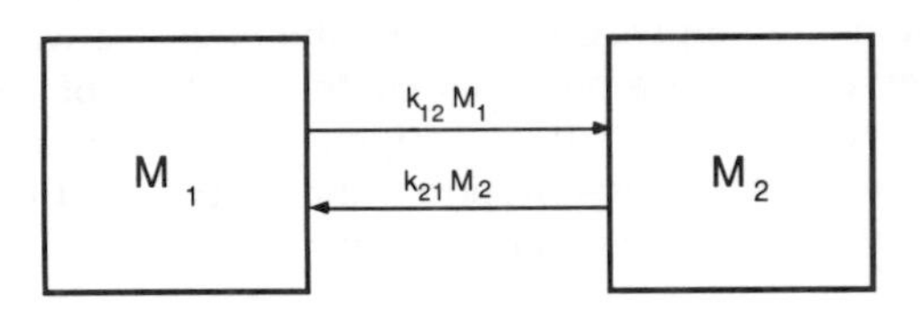

Fig. 4-5 A coupled two-reservoir system with fluxes proportional to the content of the emitting reservoirs.

and the initial condition

$$M_1(t=0) = M_1^0$$
$$M_2(t=0) = M_2^0 = M_T - M_1^0 \quad (16)$$

Equations (14) can be written in matrix form as:

$$\frac{d\mathbf{M}}{dt} = \mathbf{kM} \quad (17)$$

where $\mathbf{M}$ is the vector (M_1, M_2) describing the contents of the two reservoirs and $\mathbf{k}$ the matrix:

$$\begin{pmatrix} -k_{12} & k_{21} \\ k_{12} & -k_{21} \end{pmatrix}$$

The eigenvalues of $\mathbf{k}$ are the solutions to the equation

$$\begin{vmatrix} -k_{12}-\lambda & k_{21} \\ k_{12} & -k_{21}-\lambda \end{vmatrix} = (-k_{12}-\lambda)(-k_{21}-\lambda) - k_{12}k_{21} = 0 \quad (18)$$

$\lambda_1 = 0$ and $\lambda_2 = -(k_{12}+k_{21})$. The general solution to equation (17) can be written as:

$$\mathbf{M}(t) = \psi_1 \exp(\lambda_1 t) + \psi_2 \exp(\lambda_2 t) \quad (19)$$

where ψ_1 and ψ_2 are the eigenvectors of the matrix $\mathbf{k}$. In our case, we have:

$$\mathbf{M}(t) = \psi_1 + \psi_2 \exp(-(k_{12}+k_{21})t) \quad (20)$$

or, in component form and in terms of the initial conditions:

$$M_1(t) = \frac{k_{21}}{k_{12}+k_{21}} M_T + \left(M_{10} - \frac{k_{21}M_T}{k_{12}+k_{21}}\right) \times \exp[-(k_{12}+k_{21})t]$$

$$M_2(t) = \frac{k_{12}}{k_{12}+k_{21}} M_T + \left(M_T - M_{10} - \frac{k_{12}M_T}{k_{12}+k_{21}}\right) \times \exp[-(k_{12}+k_{21})t] \quad (21)$$

It is seen that in the steady-state the total mass is distributed between the two reservoirs in proportion to the sink coefficients (in reverse proportion to the turnover times), independent of the initial distribution.

In this simple case, there is only one time-scale characterizing the adjustment process, that is $(k_{12}+k_{21})^{-1}$. This is also the response time, τ_{cycle}, as defined by equation (13):

$$\tau_{cycle} = \frac{1}{k_{12}+k_{21}} \quad (22)$$

or if expressed in terms of the turnover times of the two reservoirs:

$$\tau_{cycle}^{-1} = \tau_{01}^{-1} + \tau_{02}^{-1} \quad (23)$$

The response time in this simple model will depend on the turnover times of both reservoirs and will always be shorter than the shortest of the two turnover times. If τ_{01} is equal to τ_{02}, τ_{cycle} will be equal to half of this value.

An investigation of the dynamic behavior of a coupled three-reservoir system using the techniques described above is included in the list of problems at the end of the chapter.

It may be noted that the steady-state solution of equation (12) is not necessarily unique. This can easily be seen in the case of the four-reservoir system shown in Fig. 4-6. In the steady-state, all material will end up in the two accumulating reservoirs at the bottom. However, the distribution between these two reservoirs will depend on the amount initially located in the two upper reservoirs.

Before turning to non-linear situations, let us consider two specific examples of coupled linear systems. The first describes the dynamic behavior of a multi-reservoir system, the second represents a steady-state situation of an open two-reservoir system.

Example 1

As a specific example of a time-dependent linear system, we may take the model of the phosphorus cycle as formulated by Lerman *et al.* (1975; cf. Fig. 4-7). These authors used a computer to solve the system of equation (11) with a time-dependent source term added to represent the transient situation with an exponentially increasing industrial mining input (7% increase per year). The same situation was studied in a more elegant way by Lasaga (1980) with the aid of matrix algebra. The evolution of the phosphorus content of the various reservoirs (except the sediments) during the first 70 years is shown in Table 4-1. Within this time-frame, the only noticeable change is seen to occur in the land reservoir. Lasaga showed that the adjustment time-scale of the system, τ_{cycle}, is 53 000 years. This is much shorter than the turnover time of the sediment reservoir (2×10^8 years) but much longer than the turnover times of all other reservoirs. This cycle is described in greater detail in Chapter 14.

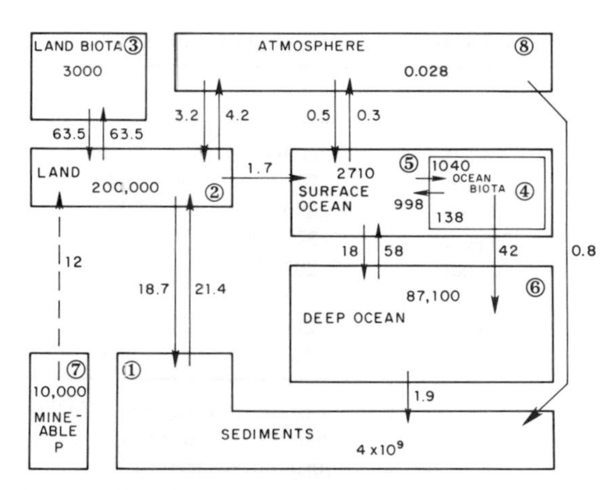

Fig. 4-7 The phosphorus cycle: Phosphorus contents are in units of Tg P, and transfer rates are in units of Tg P/year. Adapted from Lerman *et al.* (1975) with the permission of the Geological Society of America.

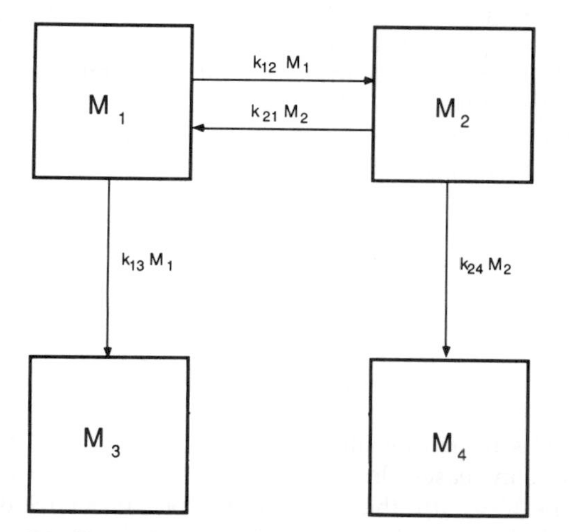

Fig. 4-6 Example of a coupled reservoir system where the steady-state distribution of mass is not uniquely determined by the parameters describing the fluxes within the system but also by the initial conditions (see text).

Example 2

As a much simpler example, let us consider a system consisting of two connected reservoirs (Fig. 4-8). Steady-state is assumed to prevail. Material is introduced at a constant rate Q in reservoir 1. Some of this material is removed (S_1) and the rest (T) is transferred to reservoir 2, from which it is removed at a rate S_2. The turnover times (average residence times) of the two reservoirs and of the combined reservoir (defined as the sum of the two reservoirs)

Table 4-1 Response of phosphorus cycle to mining output

Time	Land	Land biota	Oceanic biota	Surface ocean	Deep ocean
0 years	200 000	3000	138	2710	87 100
10 years	200 173	3000	138	2710	87 100
20 years	200 522	3001	138	2710	87 100
30 years	201 224	3003	138	2710	87 100
40 years	202 636	3008	138	2710	87 100
50 years	205 481	3018	138	2710	87 100
60 years	211 208	3018	138	2711	87 100
70 years	222 741	3078	138	2712	87 101

Notes: Phosphorus amounts are given in Tg P (1 Tg $= 10^{12}$ g). Initial contents and fluxes as in Fig. 4-7 (system at steady-state). In addition, a perturbation is introduced by the flux from reservoir 7 (mineable phosphorus) to reservoir 2 (land phosphorus), which is given by 12 exp $(0.07t)$ in units of Tg P/year.

are easily calculated to be:

$$\tau_{01} = \frac{M_1}{T+S_1} = \frac{M_1}{Q} \tag{24}$$

$$\tau_{02} = \frac{M_2}{S_2} = \frac{M_2}{T} \tag{25}$$

$$\tau_0 = \frac{M_1+M_2}{S_1+S_2} = \frac{M_1+M_2}{Q} = \frac{M_1}{Q} + \frac{M_2}{S_2}\cdot\frac{S_2}{Q}$$

$$= \frac{M_1}{Q} + \frac{M_2}{S_2}\cdot\frac{T}{Q} = \tau_{01} + \alpha\tau_{02} \tag{26}$$

where $\alpha = T/Q$ is the fraction of the material passing through reservoir 1 that is transferred to reservoir 2.

In the special case where $S_1 = 0$, $T = Q$, and $\alpha = 1$ (all material introduced in reservoir 1 is transferred to reservoir 2), the turnover time of the combined reservoir equals the sum of the turnover times of the individual reservoirs.

Among other situations, this two-reservoir model has been applied to sulfur in the atmosphere with SO_2-sulfur as one reservoir and sulfate-sulfur as the other (Rodhe, 1978).

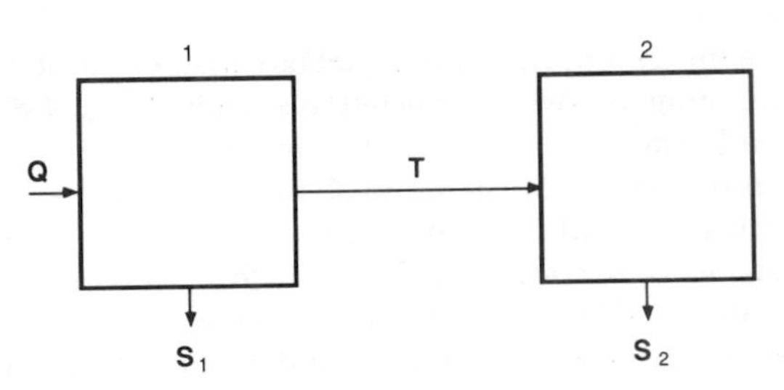

Fig. 4-8 An open two-reservoir system.

4.4.2 *Non-linear Systems*

In many situations, the assumption about linear relations between removal rates and reservoir contents is invalid and more complex relations have to be assumed. No simple theory exists for treating the various possible non-linear situations. The following discussion will be limited to a few examples of non-linear reservoir/flux relations and cycles. For a more comprehensive discussion, see the review by Lasaga (1980).

Consider a single reservoir with a constant rate of supply and a removal rate proportional to the square of the reservoir content. The equation governing the rate of change of the reservoir content is:

$$\frac{dM}{dt} = Q - BM^2 \tag{27}$$

If $M(0) = 0$, the solution to this equation is:

$$M = \sqrt{\frac{Q}{B}\cdot\frac{1-\exp(-2\sqrt{QB}t)}{1+\exp(-2\sqrt{QB}t)}} \tag{28}$$

This is graphically illustrated in Fig. 4-9. Initially, the mass increases almost linearly with time. After the time $1/2\sqrt{(QB)}$, the removal term becomes effective and the mass levels off. M eventually reaches a steady-state equal to $\sqrt{(Q/B)}$, but the response time-scale is not as easily defined as in the linear case. Relative to a simple exponential relaxation process,

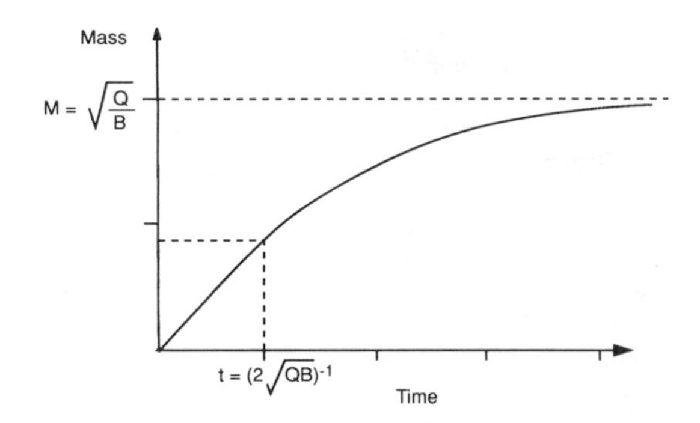

Fig. 4-9 The shape of the function given in equation (28).

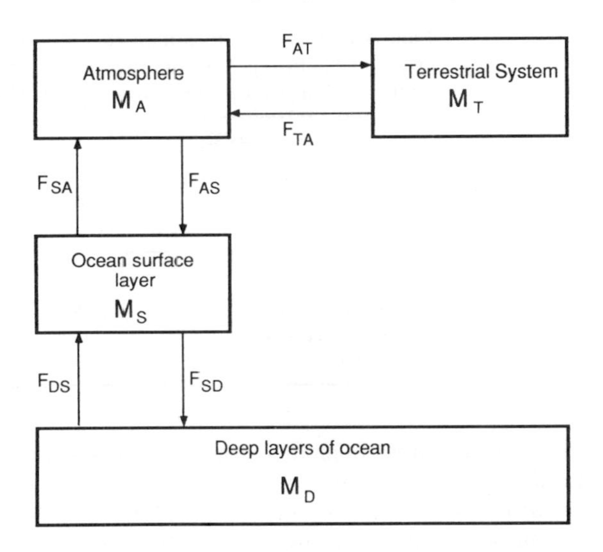

Fig. 4-11 Simplified model of the biogeochemical carbon cycle. Adapted from Rodhe and Björkström (1979) with the permission of the Swedish Geophysical Society.

the adjustment given by equation (28) is more rapid initially, and slower as time progresses.

In general, if the removal flux is dependent upon the reservoir content raised to the power α ($\alpha \neq 1$), i.e. $S = BM^{\alpha}$, the adjustment process will be faster or slower than the steady-state turnover time depending on whether α is larger or smaller than 1 (Rodhe and Björkström, 1979).

A similar simple non-linear adjustment process is described by the equation:

$$\frac{dM}{dt} = AM - BM^2 \tag{29}$$

which is a common model for the growth of biological systems (it is called logistical growth). The term AM represents exponential growth (unlimited supply of space and nutrients) and the term BM^2 is a removal term, a negative feedback effect of "crowdedness". Initially (where $AM_0 \gg BM_0^2$), the growth will first be close to exponential and will then gradually level off to the equilibrium value A/B (Fig. 4-10).

An important example of non-linearity in a biogeochemical cycle is the exchange of carbon dioxide between the ocean surface water and the atmosphere and between the atmosphere and the terrestrial system. To illustrate some effects of these non-linearities, let us consider the simplified model of the carbon cycle shown in Fig. 4-11. M_s represents the sum of all forms of dissolved carbon (CO_2, H_2CO_3, HCO_3^-, and CO_3^{2-}). The ocean to atmosphere flux is related to the dissolved species $CO_2(aq)$, and this flux is related to the total carbon content in the surface layer (M_S) by:

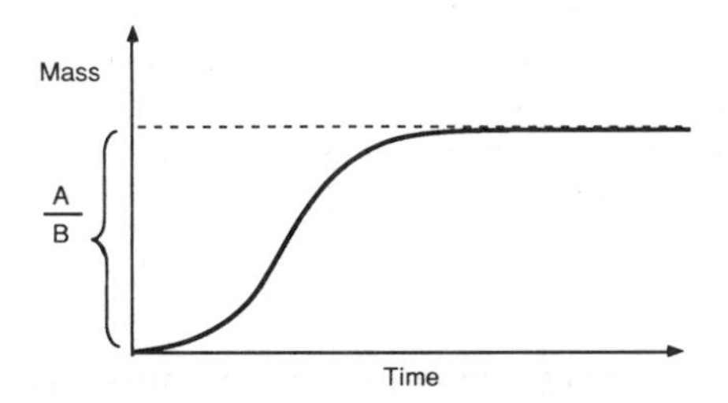

Fig. 4-10 Shape of "logistical growth". The rate of change increases slowly initially. The rate of growth reaches a maximum and eventually drops to zero as the mass levels off, approaching the value A/B.

$$F_{SA} = k_{SA} M_S^{\alpha_{SA}} \tag{30}$$

where the exponent α_{SA} (the buffer factor) is about 9. The buffer factor results from the equilibrium between CO_2 (aq) and the more prevalent forms of dissolved carbon. This effect is discussed further in Chapter 11. As a consequence of this strong dependence of F_{SA} on M_S, a substantial increase in CO_2 in the atmosphere is balanced by a small increase of M_S.

Similarly, the flux from the atmosphere to the terrestrial system may be represented by the expression:

$$F_{AT} = k_{AT} M_A^{\alpha_{AT}} \tag{31}$$

The exponent α_{AT} is considerably less than unity owing to the fact that CO_2 generally is not the limiting factor for vegetation growth. This means

Table 4-2 Steady-state carbon contents (unit: Pg = 10^{15} g) for the four-reservoir model of Fig. 4-11 during the unperturbed (pre-industrial) situation, after the introduction of 1000 Pg carbon, and after the introduction of 6000 Pg carbon

	Pre-industrial content (Pg)	After 1000 Pg		After 6000 Pg	
		Content (Pg)	% increase	Content (Pg)	% increase
Atmosphere	700	840	20	1 880	170
Terrestrial system	3 000	3 110	4	3 655	22
Ocean surface layer	1 000	1 020	2	1 115	12
Deep ocean	35 000	35 730	2	39 050	12

that even a substantial increase in M_A does not produce a corresponding increase in F_{AT}.

Assuming that the carbon cycle of Fig. 4-11 will remain a closed system over several thousands of years, we may ask the question of how the equilibrium distribution within the system would change after the introduction of a certain amount of fossil carbon. Table 4-2 contains the answer for two different assumptions about the total input. The first (1000 Pg) corresponds to the total input from fossil fuel up to about the year 2000; the second (6000 Pg) is roughly equal to the now known accessible reserves of fossil carbon (Keeling and Bacastow, 1977).

If all fluxes are proportional to the reservoir contents, the percentage change in reservoir content will be equal for all the reservoirs. The non-linear relations discussed above give rise to substantial variations between the reservoirs. Note that the atmospheric reservoir is much more significantly perturbed than any of the other three reservoirs. Even in the case with a 6000-Pg input, the carbon content of the oceans does not increase by more than 12% in the steady-state.

On the other hand, with "only" 1000 Pg emitted into the system, i.e. less than 3% of the total amount of carbon in the four reservoirs, the atmospheric reservoir would still remain significantly affected (20%) in the steady-state. In this case, the change in oceanic carbon would be only 2% and hardly noticeable. The steady-state distributions are independent of where the addition occurs. If the CO_2 from fossil fuel combustion were collected and dumped into the ocean, the final distribution would still be the same.

If all fluxes were proportional to the reservoir content, i.e. if the buffer factor had been unity, all reservoirs would be equally affected: 15% in the 6000-Pg case and 2.5% in the 1000-Pg case.

4.5 Coupled Cycles

An important class of cycles with non-linear behavior is represented by situations when coupling occurs between cycles of different elements. The behavior of coupled systems of this type has been studied in detail by, for example, Prigogine (1967). In these systems, multiple equilibria are sometimes possible and oscillatory behavior can occur. There have been recent suggestions that atmospheric systems of chemical species, coupled by chemical reactions, could exhibit multiple equilibria under realistic ranges of concentration (Fox *et al.*, 1982; White, 1984). However, no such situations have been confirmed by measurements.

4.6 Note on Transport Processes

In this chapter to date, little consideration has been given to the nature of the transport processes responsible for fluxes of material between and within reservoirs. This appendix includes a very brief discussion of some of the processes that are important in the context of global biogeochemical cycles. More comprehensive treatments can be found in textbooks on geology, oceanography, and meteorology, and in reviews such as Lerman (1979) and Liss and Slinn (1983).

4.6.1 Advection, Turbulent Flux, and Molecular Diffusion

Let us consider a fluid in which a tracer *i* is mixed. A flux of the tracer within the fluid can be brought about either by organized fluid motion or by molecular diffusion. These two flux processes can be written as:

$$\mathbf{F}_{i1} = \mathbf{V}q_i\varrho = \mathbf{V}c_i \quad (32)$$

and

$$\mathbf{F}_{i2} = -D_i\varrho\nabla q_i \quad (33)$$

where $\mathbf{F}_{i1}$ and $\mathbf{F}_{i2}$ denote the flux vectors of the tracer (dimensions: mass/length^2 time), $\mathbf{V}$ the fluid velocity vector (length/time), ϱ the density of the fluid (mass/length^3), q_i the tracer mixing ratio (mass/mass), c_i the mass concentration of the tracer (mass/length^3), D the molecular diffusivity (length^2/time), and ∇ the gradient operator (as length^{-1}). The expression ∇q_i denotes the vector ($\partial q_i/\partial x$, $\partial q_i/\partial y$, $\partial q_i/\partial z$).

The continuity of tracer mass is expressed by the equation

$$\frac{\partial c_i}{\partial t} = -\nabla\cdot\mathbf{F}_i + Q - S = -\nabla\cdot(\mathbf{F}_{i1} + \mathbf{F}_{i2}) + Q - S$$
$$= -\nabla\cdot(\mathbf{V}c_i) + \nabla\cdot(D_i\varrho\nabla q_i) + Q - S \quad (34)$$

where Q and S represent production and removal of the tracer (mass/length^3 time). Here $\nabla\cdot\mathbf{F}_i$ denotes the scalar quantity:

$$\frac{\partial F_{ix}}{\partial x} + \frac{\partial F_{iy}}{\partial y} + \frac{\partial F_{iz}}{\partial z}$$

If variations in fluid density and diffusivity can be neglected, we have:

$$\frac{\partial c_i}{\partial t} = -\nabla\cdot(\mathbf{V}c_i) + D\nabla^2 c_i + Q - S \quad (35)$$

In most situations, a fluid would be turbulent, implying that the velocity vector, as well as the concentration c_i, exhibits considerable variability on time-scales smaller than those of prime interest. This situation can be described by writing these quantities as the sum of an average quantity (normally a time average) and a perturbation:

$$\mathbf{V} = \bar{\mathbf{V}} + \mathbf{V}'$$
$$c_i = \bar{c}_i + c_i'$$

From equation (32), the transport flux $\mathbf{F}_{i1}$ then becomes:

$$\mathbf{F}_{i1} = (\bar{\mathbf{V}} + \mathbf{V}')(\bar{c}_i + c_i')$$
$$= \bar{\mathbf{V}}\bar{c}_i + \bar{\mathbf{V}}c_i' + \mathbf{V}'\bar{c}_i + \mathbf{V}'c_i' \quad (36)$$

and its average value

$$\bar{\mathbf{F}}_{i1} = \bar{\mathbf{V}}\bar{c}_i + \overline{\mathbf{V}'c'} \quad (37)$$

Note that the averages of $\mathbf{V}'$ and c' are equal to zero. The continuity equation can now be written as:

$$\frac{\partial \bar{c}_i}{\partial t} = -\nabla\cdot(\bar{\mathbf{V}}\bar{c}_i) - \nabla\cdot(\overline{\mathbf{V}'c_i'}) + D\nabla^2\bar{c}_i + Q - S \quad (38)$$

The two terms on the right-hand side of equation (37) describe transport by *advection* and by *turbulent flux*, respectively. The separation of the motion flux into advection and turbulent flux is somewhat arbitrary; depending upon the circumstances, the averaging time can be anything from a few minutes to a year or even more.

Because in most situations the perturbation quantities ($\mathbf{V}'$ and c_i') are not explicitly resolved, it is not possible to evaluate the turbulent flux term directly. Instead, it has to be related to the distribution of averaged quantities – a process referred to as parameterization. A common assumption is to relate the turbulent flux vector to the gradient of the averaged tracer distribution [in analogy with the molecular diffusion expression in equation (33)]:

$$(\mathbf{F}_{i2})_{\text{turb}} = \bar{\mathbf{V}}'\bar{c}' = -k_{\text{turb}}\nabla\bar{c}_i \quad (39)$$

The coefficient k_t thus introduced (dimension: length^2/time) is called the turbulent, or eddy, diffusivity. In the general case, the eddy diffusivity is given separate values for the three spatial dimensions.

It must be remembered that the eddy diffusivities are not constants in any real sense (like the molecular diffusivities) and that their numerical values are very uncertain. The assumption underlying equation (39) is therefore open to question.

In most cases, the term expressing the divergence of the molecular flux in equation (38) ($D\nabla^2\bar{c}_i$) can be neglected compared to the other two transport terms. Important exceptions occur, e.g. in a thin layer of the atmosphere close to the surface and in similar layers of the oceans close to the bottom and to the surface (viscous sublayers). Molecular diffusion is also an important transport process in the upper atmosphere, at heights above 100 km.

Order of magnitude values for vertical eddy diffusivity in the atmosphere and the ocean are shown in Fig. 4-12. The values for the viscous layers represent molecular diffusivities of a typical air molecule like N_2.

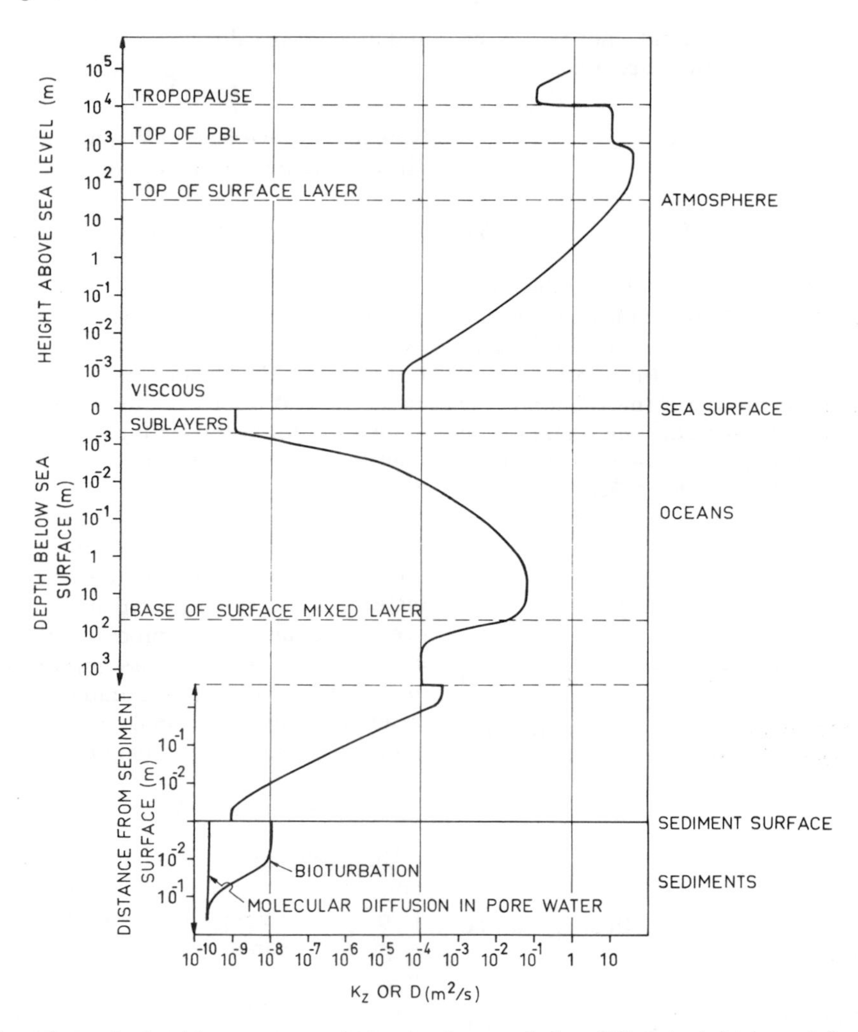

Fig. 4-12 Orders of magnitude of the average vertical molecular or turbulent diffusivity (whichever is largest) through the atmosphere, oceans, and uppermost layer of ocean sediments.

4.6.2 Other Transport Processes

Under some circumstances, transport processes other than fluid motion and molecular diffusion can be important. One important example is *sedimentation* due to gravity acting on particulate matter submerged in a fluid, e.g. removal of dissolved sulfur from the atmosphere by precipitation scavenging, transport of organic carbon from the surface waters to the deep layers and to the sediment by settling detritus. The rate of transport by sedimentation is determined essentially by the size and density of the particles and by the drag exerted by the fluid.

Geochemically significant mixing and transport can sometimes be accomplished by biological processes. An interesting example is redistribution of sediment material caused by the movements of worms and other organisms (*bioturbation*). Exchange processes between the atmosphere and oceans and between the oceans and the sediments are treated below in separate sections.

4.6.3 Air–Sea Exchange

4.6.3.1 Gas transfer

The magnitude and direction of the net flux, *F*, of any gaseous species across an air–water interface, counted positive if the flux is directed from the atmosphere to the ocean, is related to the difference in concentration, Δc, in the two phases by the relation:

$$F = K\Delta c \tag{40}$$

Here $\Delta c = c_a - K_H c_w$ with c_a and c_w representing the concentrations in the air and water respectively and K_H the Henry's Law constant. The parameter K, linking the flux and the concentration difference, has the dimension of a velocity. It is often referred to as the transfer (or piston) velocity. The reciprocal of the transfer velocity corresponds to a resistance to transfer across the surface. The total resistance ($R = K^{-1}$) can be viewed as the sum of an air resistance (R_a) and a water resistance (R_w):

$$R = R_a + R_w = \frac{1}{k_a} + \frac{K_H}{\alpha k_1} \tag{41}$$

The parameters k_a and k_1 are the transfer velocities for chemically unreactive gases through the viscous sublayers in the air and water, respectively. They relate the flux F to the concentration gradients across the viscous sublayers through expressions similar to equation (40):

$$F = k_a(c_a - c_{a,i})$$
$$F = k_1(c_{w,i} - c_w) \tag{42}$$

Here $c_{a,i}$ and $c_{w,i}$ are the concentrations right at the interface (cf. Fig. 4-13). They are related by $c_{a,i} = K_H c_{w,i}$.

The parameter α in equation (41) quantifies any enhancement in the value of k_1 due to chemical reactivity of the gas in the water. Its value is unity for an unreactive gas; for gases with rapid aqueous phase reactions (e.g. SO_2), much higher values can occur.

A comparison of the resistance in air and water for different gases shows that the resistance in the water dominates for gases with low solubility that are unreactive in the aqueous phase (e.g. O_2, N_2, CO_2, CH_4). For gases of high solubility or rapid aqueous chemistry (e.g. H_2O, SO_2, NH_3), processes in the air control the interfacial transfer.

The numerical values of the transfer velocity K for the different gases are not well established. Its magnitude depends on such factors as wind speed, surface waves, bubbles, and heat transfer. A globally averaged value of K often used for CO_2 is about 10 cm/h. Transport at the sea–air interface is also discussed in Chapter 9 (for a review, see Liss, 1983).

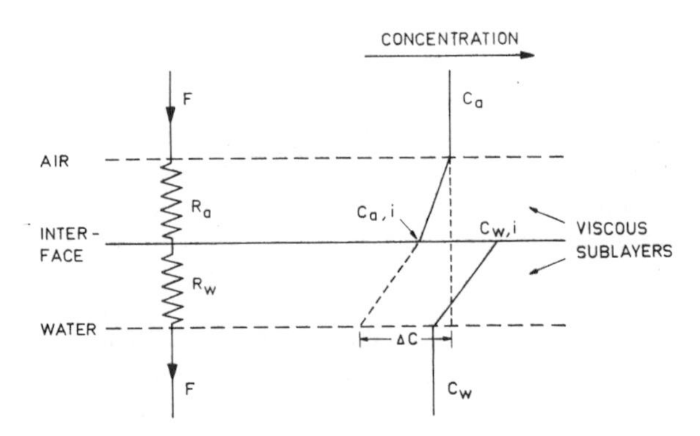

Fig. 4-13 A simplified model of flux resistances and concentration gradients in the viscous sublayers at the air–sea interface.

4.6.3.2 *Transfer of particles*

Liquid water, including its soluble and insoluble constituents, is transferred from the oceans to the atmosphere when air bubbles in the water rise to the surface. These bubbles form from air trapped by breaking waves called "whitecaps". As the bubbles burst at the surface, water droplets are injected into the atmosphere. These water droplets are small enough to remain airborne for several hours. Whitecaps begin to form in winds common over the oceans, and significant quantities of sea salt made airborne in this way are transported to the continents and deposited in coastal areas.

The flux of particles in the other direction, deposition on the ocean surface, occurs intermittently in precipitation (*wet* deposition) and more continuously as a direct uptake by the surface (*dry* deposition). These fluxes may be represented by a product of the concentration of particulate matter in air close to the surface and parameters often referred to as deposition velocities:

$$F = F_w + F_d = (v_w + v_d)\, c_a$$

The deposition velocities depend on the size distribution of the particulate matter, on the frequency of occurrence and intensity of precipitation, the chemical composition of the particles, the wind speed, nature of the surface, etc. Typical values of v_w and v_d for particles below about 1 μm in diameter are in the range 0.1–1.0 cm/s (Slinn, 1983). The average residence time in the atmosphere for such particles is of the order of a few days.

4.6.4 Sediment–Water Exchange

The sediment surface separates a mixture of solid sediment and interstitial water from the overlying water. Growth of the sediment results from accumulation of solid particles and inclusion of water in the pore space between the particles. The rates of sediment deposition vary from a few millimeters per 1000 years in the pelagic ocean up to centimeters per year in lakes and coastal areas. The resulting flux of solid particles to sediment surface is normally in the range 0.006–6.0 kg/m^2 year (Lerman, 1979). The corresponding flux of materials dissolved in the trapped water is 10^{-6}–10^{-3} kg/m^2 year. Chemical species can also be transported across the sediment surface by other transport processes. The main processes are (Lerman, 1979):

1. Sedimentation of solids (mineral, skeletal, and organic materials).
2. Flux of dissolved material and water into sediment, owing to the growth of the sediment column.
3. Upward flow of pore water and dissolved material caused by pressure gradients.
4. Molecular diffusional fluxes in pore water.
5. Mixing of sediment and water at the interface (bioturbation and water turbulence).

An estimate of the advective fluxes (processes 1, 2, and 3) requires knowledge of the concentration of the species in solution and in the solid particles as well as of the rates of sedimentation and pore water flow. The diffusive type processes 4 and 5 depend on vertical gradients of the concentrations of the species as well as on the diffusivities. In regions where bioturbation occurs, the effective diffusivity in the uppermost centimeters of the sediments can be more than that due to molecular diffusion in the pore water alone (cf. Fig. 4-12).

4.7 Note on Times of Mixing in the Atmosphere and Oceans

It is often important to know how long an element spends in one environment before it is transported somewhere else. For example, if a time-scale characterizing a chemical or physical transformation process in a region has been estimated, a comparison with the time-scale characterizing the transport away from this region will tell which process is likely to dominate.

The question of residence time and its definition in a steady-state reservoir was discussed earlier in this chapter. In such cases, the average residence time in the reservoir was shown to be equal to the turnover time $\tau_0 = M/S$, where M is the mass of the reservoir and S the total flux out of it. It is important to note that if one considers the exchange between two reservoirs of different mass, the time-scale of exchange will be different depending upon whether the perspective is from the small or the big reservoir. An interesting example is that of mixing between the troposphere and stratosphere in the atmosphere. Studies of radioactive nuclides injected into the lower stratosphere by bomb testing have shown that the time-scale characterizing the exchange between the lower stratosphere and troposphere is one to a few years. This means that a "particle" injected in the lower stratosphere will stay for this time, on average, before entering the troposphere. On the other hand, a gas molecule like N_2O, which is chemically stable in the troposphere, will spend several decades in the troposphere before it is mixed up into the lower stratosphere, where it is decomposed by photochemical processes. Thus, even though the gross flux of air from the troposphere to the stratosphere, F, is equal to the gross flux of air from the stratosphere to the troposphere, the time-scale of mixing between these two reservoirs is several decades from the tropospheric point of view, τ_T, but only a few years as seen from the stratosphere, τ_S. The reason for the difference is, of course, the small mass of the stratosphere, M_S, as compared to the troposphere, M_T. Formally, we can write:

$$\frac{\tau_S}{\tau_T} = \frac{M_S/F}{M_T/F} = \frac{M_S}{M_T} \approx 0.1$$

Time-scales of transport can also be applied to situations when no well-defined reservoirs can be defined. If the dominant transport process is advection by mean flow or sedimentation by gravity, the time-scale characterizing the transport between two places is simply $\tau_{adv} = L/V$, where L is the distance and V the transport velocity. Given a typical wind speed of 20 m/s in the mid-latitude tropospheric westerlies, the time of transport round the globe would be about 2 weeks.

In situations where the transport is governed by diffusive processes, a time-scale of transport can be defined as:

$$\tau_{turb} = \frac{L^2}{D}$$

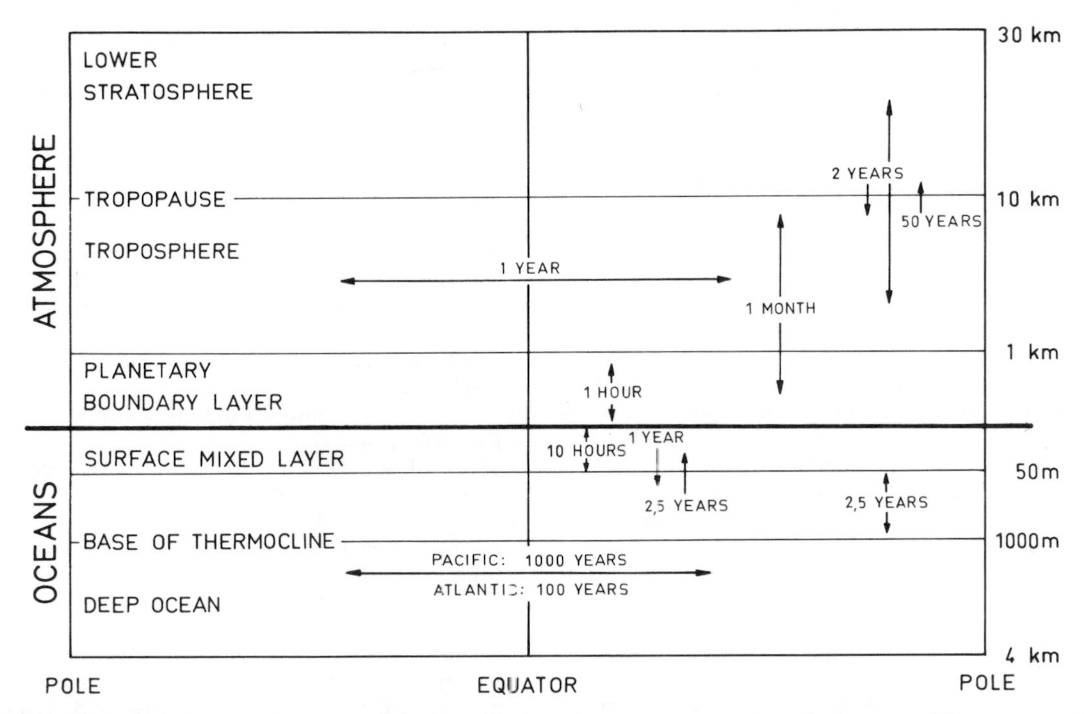

Fig. 4-14 Rough estimates of characteristic time for exchange of air and water respectively, between different parts of the atmosphere and oceans. All heights and depths are variable.

where L is the distance and D the diffusivity (molecular or turbulent). Applying this definition to the vertical mixing through the surface mixed layer of the ocean, assuming the depth of the layer to be 50 m and the turbulent diffusivity 0.1 m²/s we get:

$$\tau_{\text{turb}} = \frac{(50)^2}{0.1} \text{ s} \approx 7 \text{ h}$$

Some important time-scales characterizing the transport within the oceanic and atmospheric environments are summarized in Fig. 4-14. In view of the somewhat ambiguous nature of the definitions of these time scales, the numbers should not be considered as more than indications of the magnitudes.

Acknowledgement. I wish to thank Anders Björkström for valuable comments on the manuscript.

Questions

4-1 Consider a reservoir with two separate sources Q_1 and Q_2 and a single sink S. The magnitudes of Q_1 and S and their uncertainties have been estimated to be 75 ± 20 and 100 ± 30 (arbitrary units). Assuming that there is no direct way of estimating Q_2, how would you derive its magnitude and uncertainty range from budget considerations? What assumption must be made regarding the reservoir?

4-2 Calculate the turnover time of carbon in the various reservoirs given in Fig. 4-2.

4-3 What is the relation between the turnover time τ_0, the average transit time τ_r, and the average age τ_a in a reservoir where all "particles" spend an equal time in the reservoir?

4-4 Consider a reservoir with a source flux Q and two sink fluxes S_1 and S_2. S_1 and S_2 are proportional to the reservoir content M with proportionality constants k_1 and k_2. The values of k_1 and k_2 are (1/year) and (0.2/year), respectively. The system is initially in steady-state with $M = M_0$ and $Q = S_{01} + S_{02}$. Describe the change in time of M if the source is suddenly reduced to half its initial value. What is the response time of the reservoir?

4-5 Consider the water balance of a lake with a constant source flux Q. This outlet is of the "threshold" type where the sink is proportional to the mass of water above a threshold value M_1; $S = k(M - M_1)$. Calculate the turnover time of water at steady-state and the response time relative to changes in Q.

4-6 For the more mathematically inclined, investigate the dynamic behavior of a coupled linear three-reservoir model using the technique outlined in Section 4.4.

Answers can be found on p. 365.

References

Bolin, B. (1986). How much CO_2 will remain in the atmosphere? *In* "The Greenhouse Effect, Climate Change, and Ecosystems." SCOPE Report No. 29 (B. Bolin, B. R. Döös, J. Jäger, and R. A. Warrick, eds.), pp. 93–155. John Wiley, Chichester.

Bolin, B. and H. Rodhe (1973). A note on the concepts of age distribution and transit time in natural reservoirs. *Tellus* **25**, 58–62.

Cicerone, R. J. and R. S. Oremland (1988). Biogeochemical aspects of atmospheric methane. *Global Biogeochem. Cycles* **2**, 299–327.

Eriksson, E. (1971). Compartment models and reservoir theory. *Ann. Rev. Ecol. Syst.* **2**, 67–84.

Fox, J. L., S. C. Wofsy, M. B. McElroy, and M. J. Prather (1982). A stratospheric chemical instability. *J. Geophys. Res.* **87**, 11126–11132.

Hamrud, M. (1983). Residence time and spatial variability for gases in the atmosphere. *Tellus* **35B**, 295–303.

Holland, H. D. (1978). "The Chemistry of the Atmosphere and Oceans." John Wiley, New York.

Junge, C. E. (1974). Residence time and variability of tropospheric gases. *Tellus* **26**, 477–488.

Keeling, C. D. and R. B. Bacastow (1977). Impact of individual gases on climate. *In* "Energy and Climate", pp. 72–95. US National Research Council, Washington, D.C.

Lasaga, A. C. (1980). The kinetic treatment of geochemical cycles. *Geochim. Cosmochim. Acta* **44**, 815–828.

Lerman, A. (1979). "Geochemical Processes: Water and Sediment Environments." John Wiley, New York.

Lerman, A., F. T. Mackenzie, and R. M. Garrels (1975). Modelling of geochemical cycles: Phosphorus as an example. *Geol. Soc. Amer. Mem.* **142**, 205–218.

Liss, P. S. (1983). Gas transfer: Experiments and geochemical implications. *In* "Air–sea Exchange of Gases and Particles" (P. S. Linn and W. G. N. Slinn, eds.), pp. 241–298. Reidel, Dordrecht, The Netherlands.

Liss, P. S. and W. G. N. Slinn (eds.) (1983). "Air–sea Exchange of Gases and Particles." Reidel, Dordrecht, The Netherlands.

Mason, B. (1966). "Principles of Geochemistry." 3rd edn. John Wiley, New York.

McDuff, R. E. and F. M. M. Morel (1980). The geochemical control of seawater (Sillén revisited). *Environ. Sci. Technol.* **14**, 1182–1186.

Prigogine, I. (1967). "Introduction to Thermodynamics of Irreversible Processes." Wiley-Interscience, New York.

Rodhe, H. (1978). Budgets and turn-over times of atmospheric sulfur compounds. *Atmos. Environ.* **12**, 671–680.

Rodhe, H. and A. Björkström (1979). Some consequences of non-proportionality between fluxes and reservoir contents in natural systems. *Tellus* **31**, 269–278.

Slinn, W. G. N. (1983). Air–sea transfer of particles. *In* "Air–Sea Exchange of Gases and Particles" (P. S. Liss and W. G. N. Slinn, eds.), pp. 299–405. Reidel, Dordrecht, The Netherlands.

Southam, J. R. and W. W. Hay (1976). Dynamical formulations of Broecker's model for marine cycles of biologically incorporated elements. *Math. Geol.* **8**, 511–527.

White, W. H. (1984). Does the photochemistry of the troposphere admit more than one steady state? *Nature* **309**, 242–244.

5

Equilibrium, Rate, and Natural Systems

Samuel S. Butcher

5.1 Thermodynamics

Simple chemical and physical processes play important roles in many biogeochemical systems. For example, removal of oxygen by the oxidation of surface materials and other reactions in the atmosphere and in surface waters may be described by the following chemical reactions:

Oxidation of sediments:

$$\tfrac{1}{2}H_2O + FeS_2 + \tfrac{15}{4}O_2 \rightarrow Fe^{3+} + 2SO_4^{2-} + H^+$$

Oxidation in the atmosphere:

$$CH_4 + 2O_2 \rightarrow CO_2 + 2H_2O$$

Although the processes may be described by rather simple equations, there are many factors that can affect the reaction rate and whether or not equilibrium is attained. Each process may occur under a wide variety of conditions of temperature, oxygen partial pressure, and presence of organisms and catalysts. Many of the mechanisms may not be understood in detail and the concentrations of key reactants may be poorly characterized. It is often necessary to parameterize the rates of these processes rather than base them on fundamental thermodynamic and kinetic data.

Despite these shortcomings, it is useful to understand the fundamental processes on which the more complex biogeochemical systems are based. Understanding the basic processes enables one to quantify the sensitivity of the overall process to such variables as temperature, pressure, radiation intensity, organisms and physical substrates, and the concentrations of a wide range of other substances, which may include primary reactants, catalysts, and toxins.

This chapter examines the physical principles used to describe chemically reactive systems. The conditions under which systems are considered fall into two categories: (1) the system may be at thermodynamic equilibrium with no input or output of energy and (2) energy inputs do exist and the system may or may not be changing.

We first review the thermodynamic principles necessary to describe equilibrium systems. A discussion of electrochemistry is also included. Next, the rates of chemical changes, or chemical kinetics, are examined. Finally, we examine selected natural systems in which thermodynamic and kinetic factors are important.

5.1.1 *Systems at Equilibrium: Thermodynamics*

The laws of thermodynamics are the cornerstones of any description of a system at equilibrium. The First Law (the energy conservation law) introduces H, the enthalpy, as a *state function* of a system. State functions depend only on the state of the system and not on how it managed to arrive at that state.

The Second Law (the entropy law) introduces entropy, S, as a state function. More importantly, this law also describes the equilibrium state of a system as one of maximum entropy. For a system at constant temperature and pressure, the *equilibrium condition* is expressed as:

$$\Delta G = \Delta G_{\text{prod}} - \Delta G_{\text{react}} = 0 \qquad (1)$$

G, the free energy, or the Gibbs free energy, is related to enthalpy and entropy by:

$$G = H - TS$$

Global Biogeochemical Cycles
ISBN 0-12-147685-5

When the temperature of reactants and products are equal, ΔG is given by:

$$\Delta G = \Delta H - T\Delta S$$

[We represent a change in any chemical reaction as reactant − product, where *reactant* is the collection of substances appearing on the left-hand side of the reaction equation and *products* appear on the right-hand side. Reactant and product bear no reference to whether or not the substance is actually being formed or is disappearing.]

Equation (1) defines the equilibrium condition under the constraint that temperature and pressure are constant. A related consequence of the Second Law states that if $\Delta G < 0$, the reaction reactant to product is *thermodynamically* spontaneous. (If $\Delta G > 0$, the reaction product to reactant is spontaneous.)

Although ΔG is the overall determinant of spontaneity, it is convenient to examine the two thermodynamic components of ΔG. The ΔH term is largely a function of the strengths of the chemical bonds in reactant and product. To the extent that there are more strong bonds (and strong associations between molecules) in the product than in the reactant, ΔH will be negative. A negative ΔH, therefore, contributes to the overall spontaneity of the reaction.

The ΔS term is a measure of the relative degree of disorder in reactant and product. To the extent that the product has greater disorder than the reactant, ΔS will be positive. A positive ΔS will also contribute to the overall spontaneity of the reaction. For gas phase systems in which the number of independent molecules and atoms is greater in the product, ΔS will be positive. If there is no change in the number of atoms and molecules it will be difficult to say (without evaluating ΔS from tables of data) whether ΔS is positive or negative. In the liquid phase, such simple rules for ΔS are less easily applied. The formation of ions in the aqueous phase may be associated with an increase of order in the liquid through complexes between the ions and water molecules.

In practice, G and H for a substance are defined *relative* to the G and H for the constituent elements. These relative values are known as *free energy of formation* and *enthalpy of formation* for standard conditions and referred to as ΔG^0 and ΔH^0. Values for these functions may be obtained from standard tables of thermodynamic data, usually for the reference temperature of 298.2 K. The Chemical Rubber Company Handbook (Weast, 1987) is one of the more commonly available sources. More complete sources, including some with data for a range of temperatures, are listed in the references at the end of the chapter. Note that many tabulations still represent these energy functions in calories and that it may be necessary to make the conversion to Joules (1 cal = 4.1840 J). Because of the definition of the energy of formation, elements in their standard state (carbon as graphite, chlorine as Cl_2 gas at one bar, bromine as Br_2 liquid, etc.) have free energies and enthalpies of formation equal to zero. If needed, the absolute entropies of substances (from which ΔS may be evaluated) are also available in standard sources.

For aqueous phase ions, there is an additional degree of freedom to be dealt with in defining G and H. In this case, the change in G and H for the formation of an ion from its constituent element [say $\frac{1}{2}Cl_2(gas) + e^- \rightarrow Cl^-(aq)$] is measured *relative* to the change in G and H for the formation of H^+ from H_2(gas). This relative change in G and H is termed the *standard free energy* or *enthalpy of formation* for ions. As a result of this definition, ΔG^0 and ΔH^0 for formation for H^+ are zero. In practice, the definitions given above lead to the following algorithm for determining ΔG^0 and ΔH^0 for a reaction. ΔG^0 for a reaction may be obtained by simply adding the ΔG^0 (formation) values for the product species and subtracting the sum of ΔG^0 (formation) for all reactant species. Where an element in its standard state or H^+ is involved, substitute zero.

5.1.2 Measures of Spontaneity, K_{eq}

As an example reaction, consider the gas phase equilibrium:

$$N_2(g) + 2O_2(g) \leftrightarrow 2NO_2(g)$$

The extent to which products or reactants are favored in this reaction is described by the equilibrium constant, K_{eq}:

$$K_{eq} = \frac{(P_{NO_2}/P^0)^2}{(P_{O_2}/P^0)^2(P_{N_2}/P^0)}$$

P^0 is reference pressure, commonly chosen as 1 bar, or (1/1.01325) atmosphere. The exponents for O_2 and NO_2 result from the stoichiometric coefficients in the balanced equation. If we agree to measure pressure in bars, then K_{eq} is given numerically by:

$$K_{eq} = \frac{P^2_{NO_2}}{P^2_{O_2} P_{N_2}}$$

In order to establish a relationship between the equilibrium constant and the thermodynamic variables, a connection between pressure or concentration and free energy is needed. The dependence of G on pressure for a gas, A, is

$$G = G^0 + RT \ln (P_A/P^0)$$

For a nearly pure liquid (often the case for water in aqueous solutions), G is given by the following expression where X is the mole fraction of the liquid:

$$G = G^0 + RT \ln (X)$$

For substances present in the liquid phase in small amounts (compared to the number of moles of liquid), the following expression is used, where c is the concentration of the substance (expressed in molarity units):

$$G = G^0 + RT \ln (c)$$

By combining the above relationships with the general condition for equilibrium, the following relationship is obtained:

$$\Delta G^0 = -RT \ln (K_{eq}) \qquad (2)$$

The *temperature dependence* of the equilibrium constant is given by the van't Hoff equation:

$$\frac{d \ln (K_{eq})}{dT} = \frac{\Delta H^0}{RT^2} \qquad (3)$$

Thus the degree of spontaneity at a given temperature depends on ΔG^0 (or ΔH^0 and ΔS^0) through equation (2). The *change* in spontaneity with temperature is given by equation (3) and depends only on ΔH^0.

K_{eq} for a gas phase reaction is defined in terms of the partial pressures of reactants and products. At times it is more convenient to express the reactivity of a substance in terms of moles or molecules per unit volume. The equilibrium constant in concentration terms is obtained from K_{eq} by using the Ideal Gas equation:

$$K_{eq} = K(c)(RT)^{\Delta n} = K'(c')(RT/N_0)^{\Delta n}$$

where $K(c)$ is the equilibrium constant in mole/volume units, $K'(c')$ is the constant in molecule/volume units, N_0 is Avogadro's number, and Δn is the change in the number of moles of gas for the reaction. In the case of the NO_2 equilibrium, $\Delta n = -1$.

Explanatory Example

Take as an example the reaction $N_2(g) + 2O_2 \rightarrow 2NO_2(g)$:

$$\Delta G^0(298 \text{ K}) = +103.68 \text{ kJ}$$
$$\Delta H^0(298 \text{ K}) = +67.70 \text{ kJ}$$

Thus, from equation (2), K_{eq} for this reaction is 6.7×10^{-19} bar at 298 K. If we consider the equilibrium in air and take 0.79 and 0.21 bar for the pressures of N_2 and O_2, respectively, the equilibrium pressure for NO_2 will be 1.5×10^{-10} bar. We characterize the reaction $N_2 + 2O_2 \rightarrow 2NO_2$ as non-spontaneous when the pressures of reactants and products are all 1 bar.

The positive value for ΔH^0 indicates that the *degree* of spontaneity will *increase* as T increases. (In this case, it means that the reaction will become less non-spontaneous with increasing temperature.)

5.1.3 Condensed Phase Systems

The results obtained above may also be extended to cases where reactants are in phases other than the gas phase. The equilibrium between ammonia in a large cloud droplet and in the gas phase, $NH_3(aq) \leftrightarrow NH_3(g)$, is described by the equilibrium constant expression:

$$K_{eq} = K_H = \frac{(P_{NH_3}/P^0)}{c_{NH_3}}$$

Here c_{NH_3} is the concentration of undissociated ammonia in water. The equilibrium constant for this class of equilibria is also known as a Henry's Law constant. The ΔG^0 value for $NH_3(aq)$ should be for dilute aqueous solutions. It is common to express Henry's Law in the form $P_{NH_3} = K_H c_{NH_3}$. It is important to note the units of K_H.

The solubility equilibrium for $CaCO_3$ (calcite) $\leftrightarrow Ca^{2+} + CO_3^{2-}$ is defined by:

$$K_{eq} = K_{sp} = c(Ca^{2+})\, c(CO_3^{2-})$$

This expression is the familiar solubility product. This equilibrium constant does not contain a term for $CaCO_3$ (calcite). To the extent that the calcite is pure (does not contain dissolved impurities) and consists of particles large enough that surface effects are unimportant, this equilibrium does not depend on the "concentration" of calcite or particle size. Exceptions to this case are important in the formation of cloud droplets (where the particle size dependence is known as the Kelvin Effect) and in the solubility of finely divided solids.

Many different complexes of elements in a given oxidation state may exist in water. The amphoteric nature of Al(III) and Fe(III) results from the formation of a series of dissolved species, MOH^{2+}, $M(OH)_2^+$, $M(OH)_3$, and other forms, in addition to the more common M^{3+}. The speciation of soluble Al and Fe is thus a sensitive function of pH.

Acid–base and precipitation equilibrium in aqueous media often leads to the use of aggregate variables. The variables are often useful in characterizing alkalinity and conservation of mass conditions. The first type of variable springs from the conservation of mass conditions. Dissolved aluminum may be present in any of the forms described above. The total dissolved aluminum is given by:

$$Al_T = [Al^{3+}] + [AlOH^{2+}] + [Al(OH)_2^+] + \ldots$$

If no aluminum is added or removed, Al_T is conserved.

A similar case occurs in carbonate equilibria, which leads to the formation of H_2CO_3, HCO_3^-, and CO_3^{2-}. Total dissolved carbon is represented by

$$C_T = [CO_2(aq)] + [H_2CO_3] + [HCO_3^-] + [CO_3^{2-}]$$

In the case of CO_2, one must consider the hydrated form (H_2CO_3) and the unhydrated form [$CO_2(aq)$]. This latter form is in equilibrium with the atmosphere:

$$CO_2(gas) \leftrightarrow CO_2(aq)$$

The $CO_2(aq)$ is in turn in equilibrium with H_2CO_3:

$$H_2CO_3(aq) \leftrightarrow CO_2(aq) + H_2O \quad K_2$$

For many purposes, $CO_2(aq)$ and H_2CO_3 are lumped together as "carbonic acid", or $H_2CO_3^*$. The Henry's Law and acid base equilibria are often written in terms of $H_2CO_3^*$ (Stumm and Morgan, 1981). K_2 is about 650 and so most "carbonic acid" is in fact $CO_2(aq)$. The actual carbonic acid concentration is given by:

$$[H_2CO_3] = [H_2CO_3^*]/(1 + K_2) \cong [H_2CO_3^*]/650$$

This is the reason that the ionization constant for "H_2CO_3" is smaller than would be expected for an oxyacid with this formula. ($pK_1 = 6.4$ when written in terms of $H_2CO_3^*$, and $pK_1 = 3.6$ in terms of H_2CO_3.)

Alkalinity is often used to describe the acid neutralizing ability of an aqueous solution. It is defined in the following manner. In any ionic equilibrium, there is a conservation of charge condition:

$$\Sigma(+\text{ charges}) = \Sigma(-\text{ charges})$$

This may also be written:

$$\begin{aligned}&\Sigma(+\text{ charges for conservative ions})\\&\quad + \Sigma(+\text{ charges for non-conservative ions})\\&\quad\quad = \Sigma(-\text{ charges for conservative ions})\\&\quad\quad\quad + \Sigma(-\text{ charges for non-conservative ions})\end{aligned}$$

Conservative ions are ones that do *not* undergo acid–base reactions at the pH values of interest. These include Na^+, K^+, Ca^{2+}, SO_4^{2+}, Cl^-, etc. Non-conservative ions *do* undergo acid–base reactions. These include H_3O^+, OH^-, HCO_3^- and CO_3^{2-}. Alkalinity is defined as follows:

$$\begin{aligned}&\text{Alkalinity}\\&\quad = \Sigma(+\text{ charges for conservative ions})\\&\quad\quad - \Sigma(-\text{ charges for conservative ions})\\&\quad = \Sigma(-\text{ charges for non-conservative ions})\\&\quad\quad - \Sigma(+\text{ charges for non-conservative ions})\end{aligned}$$

If, for example, we make a 0.10 M solution of KOH, the only conservative ions would be K^+ at 0.10 M and the alkalinity would be 0.10 M. The alkalinity would be −0.10 M for a 0.10 M solution of HCl. Negative alkalinity is also known as acidity.

Solutions to complex ionic equilibrium problems may be obtained by a graphical log concentration method first used by Sillen (1959) and more recently described by Butler (1964) and Morel (1983). Computer-based numerical methods are also used to solve these problems (Morel, 1983).

5.1.4 Non-ideal Behavior

Departures from ideality have been studied extensively for gases and gas mixtures. For most conditions of interest in the Earth's atmosphere, the assumption of ideal behavior is a reasonable approximation. The two most prevalent gases (N_2 and O_2) are non-polar and have critical temperatures (126 and 154 K, respectively) far below most temperatures of environmental interest. These gases behave fairly ideally even though their pressures are high. For other gases, the partial pressures common in the atmosphere are so low that ideal behavior is a good approximation.

In the condensed phase, departures from ideality are much more common. A significant departure from ideality results from the effect of the ionic strength of aqueous solutions on the energies of ions. This effect should be considered in any quantitative consideration of ionic equilibria in seawater (Stumm and Morgan, 1981).

5.1.5 Thermodynamic Description of Isotope Effects

Thermodynamic energy terms (and equilibrium constants) may differ for isotopic species of an element. This effect is described in theoretical detail by Urey (1947), and applications to geochemistry are discussed by Broecker and Oversby (1971) and Faure (1977). In the case of the vapor/liquid equilibrium for water, the vapor is enriched relative to the liquid in the lighter isotopic species, $^1H_2{}^{16}O$, relative to heavier species, $^1H^2H^{16}O$ (or $HD^{16}O$), $^1H_2{}^{18}O$, and other isotopic species.

Small variations in isotopic composition are usually described by comparing the ratio of isotopes in the sample material to the ratio of isotopes in a reference material. The standard measure of isotopic composition is $\delta X'$, defined in parts per thousand (per mil or ‰) by:

$$\delta X' = \left(\frac{(X'/X)}{(X'/X)_{ref}} - 1 \right) \times 1000 \qquad (4)$$

The amount of the standard isotopic species and the tracer isotopic species are represented by X and X' for the sample and the reference material. The reference substance is chosen arbitrarily, but is a substance that is homogeneous, available in reasonable amounts, and treatable by standard analytical techniques for measuring isotopes (generally mass spectrometry). For instance, a sample of ocean water known as Standard Mean Ocean Water (SMOW) is used as a reference for ^2H and ^{18}O. Calcium carbonate from the Pee Dee sedimentary formation in North Carolina, USA (PDB) is used for the carbon isotopes.

If the sample has less of the tracer isotopic species than the reference material, δ will be negative. Since many of the tracer species are heavier than the reference species (^{14}C, ^{13}C *vs* ^{12}C or ^{34}S *vs* ^{32}S), this has led to use of the term "light" for substances having less of a tracer species than the standard species.

As an example, the ratio of the equilibrium vapor pressures for ^{16}O water (p_1) and ^{18}O water (p_2) depends on temperature and is expressed by the following equation (see Faure, 1977). The relationship of this expression to the van't Hoff equation can be seen:

$$\ln\left(\frac{p_1}{p_2}\right) = \frac{7.88}{T} - 0.0177$$

At 25° C, the ratio of equilibrium pressures is 1.0088 (from the above equation). This means that pure $H_2{}^{16}O$ has a slightly greater vapor pressure than pure $H_2{}^{18}O$. The ratio of ^{18}O to ^{16}O in the atmosphere is equal to the ratio in the ocean × 1/1.0088. Using the expression for $\delta^{18}O$ and the fact that ocean water is used as the reference, a value of −8.7‰ is obtained for vapor in equilibrium with a 25° C ocean.

Water vapor formed at low latitudes by evaporation from the ocean at 25° C is "lighter" than the ocean water from which it evaporated. As this water moves to higher latitudes and the first bit of vapor condenses and rains out, it has a composition very close to the low-latitude ocean, but the vapor remaining in the atmosphere becomes even lighter. Further precipitation at still higher latitudes makes $\delta^{18}O$ for the remaining vapor even more negative. Precipitation at high latitudes is thus isotopically lighter than low-latitude precipitation.

As may be seen from the equation given above, the degree of equilibrium fractionation of water will increase as temperature decreases. $\delta^{18}O$ in ice cores from Greenland and Antarctica is more negative for those periods in which evaporation leading to precipitation occurred at a lower temperature. The $\delta^{18}O$ and δ^2H signals can thus be used as a means of tracking changes in mean temperatures (Saigne and Legrande, 1987; Faure, 1977). The kinetic basis for the isotope effect will be discussed later.

5.2 Oxidation and Reduction

Reactions in which electrons are transferred from one reactant to another form a subset of systems treated by thermodynamics. One advantage of focusing on the electrons when examining conditions for spontaneity and equilibrium for oxidation reduction systems is that the electrons represent a common currency for comparing different reactions. The potential for electrons to be transferred in one direction or another reflects the thermodynamic spontaneity of the reaction.

5.2.1 Half Cell Conventions and the Nernst Equation

The elemental reaction used to describe a redox reaction is the *half reaction*, usually written as a reduction:

$$\tfrac{1}{2}O_2(\text{gas}) + 2e^- + 2H^+ \rightarrow H_2O(\text{liq}) \qquad E^0 = 1.2290\,\text{V}$$

The relative tendency for this reaction to occur, or the driving force for the reaction, is described by the corresponding *half cell potential*, E^0, or the half cell free energy change, ΔG^0. ΔG^0 is the free energy change for the given reaction at standard conditions *relative* to the reduction for the *standard hydrogen electrode*, for which the half reaction is:

$$H^+ + e^- \rightarrow \tfrac{1}{2}H_2(\text{gas})$$

The half cell potential is related to the free energy change by the following relationship for reversible conditions:

$$\Delta G^0 = -zFE^0 \tag{5}$$

where F is the Faraday constant (= 96 485 coulomb/mol or 96 485 J/V mol) and z is the number of electrons appearing in the balanced reduction half cell. A negative value for ΔG^0 (or a positive value for E^0) means that the given reduction has a greater tendency to occur than does the reduction of H^+ to H_2. ΔG^0 values are easily obtained from tables of thermodynamic data. As mentioned above, ΔG^0 is zero by convention for elements in their standard states and for H^+. To consider ΔG^0 for the reduction of oxygen above we need only consider ΔG^0 for the formation of H_2O. $\Delta G(H_2O) = -237.18$ kJ, and thus E^0 for the reduction of oxygen is 1.2290 V.

The dependence of the relative potential for the half reaction on concentration is given by the Nernst equation:

$$E = E^0 - (2.303RT/zF)\log\frac{[\text{products}]}{[\text{reactants}]} \tag{6}$$

[Products] represents the product of activities of all products raised to a power that is the stoichiometric factor in the balanced equation; [reactants] is the analogous expression for the reactants. Eh is used to describe the reduction potential relative to the standard hydrogen electrode. The factor $2.303RT/F$ has the value 0.05916 at the common reference temperature of 298.2 K. The factor 2.303 results from the conversion from a natural logarithm (used to express the concentration dependence of thermodynamic functions) to the common log (more often used in electrochemistry and in measurements of hydrogen ion activity). The Nernst equation for the reduction of oxygen will then be:

$$Eh_1 = 1.2290 - 0.02958\log\frac{1}{P^{1/2}(O_2)\,[H^+]^2}$$

As an alternative, the tendency for a reduction to occur may also be expressed in terms of a hypothetical electron activity. Even though free electrons do not occur in aqueous solutions, it is convenient to relate the reaction tendency to this hypothetical activity. The electron activity for a reduction electrode X may be defined relative to the electron activity for a standard hydrogen electrode by the following relationship:

$$\Delta G = zRT\ln\frac{a_x}{a_H} \tag{7}$$

The activity of electrons in equilibrium with the standard hydrogen electrode is arbitrarily defined to be unity and $-\log(a_x)$ is defined as pE in a manner analogous to the definition of pH as $-\log[H^+]$. Hostettler (1984) discusses the difficulty of associating pE with an electron activity. By substitution, an expression similar to the Nernst equation is obtained for pE. The term pE is $-\Delta G/(2.303zRT)$:

$$\text{pE} = \text{pE}^0 - (1/z)\log\frac{[\text{products}]}{[\text{reactants}]} \tag{8}$$

In this chapter, equilibrium relationships will be expressed in terms of Eh. Some authors (see Stumm and Morgan, 1981; Drever, 1988) use pE instead. The two descriptions are equivalent and we briefly compare them.

The half cell for the reduction of sulfur (solid) to H_2S is written below:

$$S(\text{solid}) + 2e^- + 2H^+ \rightarrow H_2S(\text{aq}) \qquad \text{pE}^0 = +0.1444$$

$$Eh_2 = 0.1444 - 0.02958\log([H_2S/[H^+]^2)$$

For the oxygen and sulfur reductions, the pE

equations will be:

$$pE_1 = 20.77 - ½\log \frac{1}{P^{1/2}(O_2)[H^+]^2}$$

$$pE_2 = 2.44 - ½\log \frac{[H_2S]}{[H^+]^2}$$

Since Eh and pE compare potential and activity with respect to the same reference (the standard hydrogen electrode), we may compare the oxygen electrode with the sulfur electrode by subtracting the expression for sulfur from the expression for oxygen

$$Eh_1 - Eh_2 = 1.0846 - 0.02958\log \frac{1}{P^{1/2}(O_2)[H_2S]}$$

and

$$pE_1 - pE_2 = 18.33 - ½\log \frac{1}{P^{1/2}(O_2)[H_2S]}$$

If we take the case in which the quantity under the log operator is equal to 1, then Eh and pE are both positive. In the case of the potential picture given by Eh, the electrons will flow *to* the electrode with the higher potential (more positive), i.e. the oxygen electrode. In the electron activity picture represented by pE, the electrons will flow from the electrode of higher activity to the electrode with the lower activity. Since pE for oxygen is positive relative to that for sulfur, the electrons in equilibrium with oxygen have a *lower* activity. In either case, the thermodynamically spontaneous reaction is one in which oxygen is reduced and hydrogen sulfide is oxidized.

5.2.2 Stability Diagrams

The half cell reactions for hydrogen and oxygen form a starting point from which to consider redox systems in water. The Nernst equation for the reduction of oxygen may be written in terms of pH:

$$Eh_1 = 1.2290 + \frac{0.05916}{4}\log[P(O_2)] - 0.05916\text{pH}$$

The equation for the reduction of H^+ is

$$Eh_2 = 0 - \frac{0.05916}{2}\log[P(H_2)] - 0.05916\text{pH}$$

A plot of the two half cell potentials as a function of pH is shown in Fig. 5-1 for the case in which O_2 and H_2 are at unit activity.

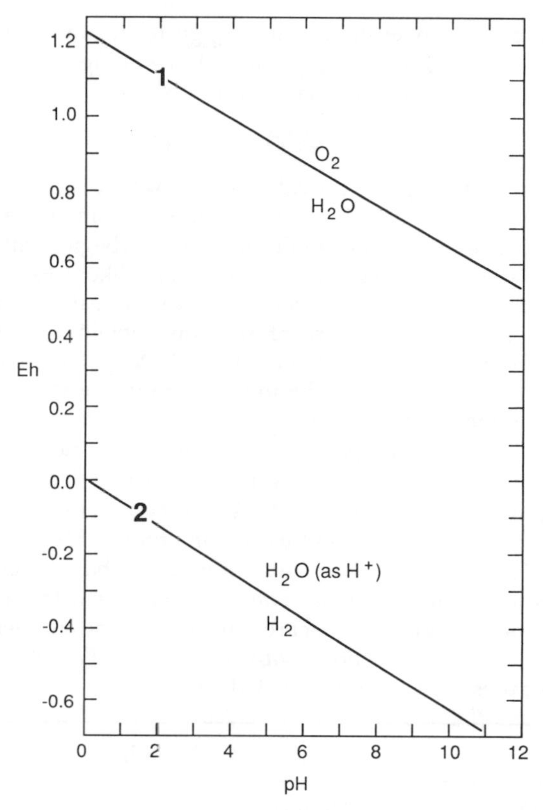

Fig. 5-1 Eh as a function of pH diagram for the reduction of oxygen and water.

Considering conditions for the O_2/H_2O system, if the oxygen gas activity is greater than unity, Eh_1 will be greater than that described by line 1. Conversely, if Eh_1 lies above line 1, the oxygen activity for the equilibrium system is greater than unity. Thus, the region above line 1 is said to represent the stability field for O_2 and the region below the line is the stability field for H_2O for the O_2/H_2O half cell. Similarly, line 2 represents the boundary between the stability field for H_2O (in the form of H^+), which lies above the line and the stability field for H_2, which lies below. It is convenient to label these curves by the stability fields that they define. In each case, we will have the oxidized state of an element above the line and the reduced state below the line.

The direction of a thermodynamically spontaneous reaction may also be examined using diagrams such as Fig. 5-1. Suppose that we have H_2, O_2, and H_2O all at unit activity (this is a thought experiment only!) at some arbitrary pH. Since Eh for the oxygen reduction is always greater than that for

H^+ reduction, electrons can be expected to flow from the lower Eh to the higher Eh resulting in the following spontaneous reaction:

$$2H_2(g) + O_2(g) \rightarrow 2H_2O(liq)$$

Eh/pH diagrams provide a simple way to describe the stabilities of various compounds as a function of pH. The construction of Eh/pH diagrams begins with a list of those species and reactions likely to be important. It is not essential to have exactly the right list at the start. As more information is obtained, that information can be added to the list. A limited view of the sulfur likely to be found in natural waters is examined as an example.

The list of compounds of sulfur will include the following: In group 1, S(VI) in the form of SO_4^{2-}; in group 2, S(IV) as SO_3^{2-}, HSO_3^-, and H_2SO_3; in group 3, S(0), S(elemental sulfur); and in group 4, S(−II) as S^{2-}, HS^-, and H_2S. The compounds have been grouped by oxidation state. Compounds with the same oxidation number are converted to one another by acid–base reactions with water. The ΔG^0 values for these species are tabulated below:

	ΔG^0 (kJ/mol)
SO_4^{2-}	−744.6
SO_3^{2-}	−486.6
HSO_3^-	−527.8
H_2SO_3	−537.9 $pK_1 = 1.8, pK_2 = 7.2$
S(s)	0 (by definition)
S^{2-}	85.8
HS^-	12.05
H_2S	−27.87 $pK_1 = 7.1, pK_2 = 14$

The pK values describing the equilibria of the acids H_2SO_3 and H_2S are also included. We assume for the present that hydrolysis of SO_4^{2-} is not important and that no metal sulfides are precipitated.

We make a start by considering a solution with a total sulfur concentration of 1.0 M. Where ions or other dissolved species are present, we also assume that the solutions are ideal; that is, activity is equal to the concentration in molarity units. As a first step, consider the reduction of S(s) to the −2 oxidation state represented by S^{2-}, HS^-, and H_2S. The prevalent dissolved species will depend on the pH and the acid–base equilibrium constants.

For pH < pK_1, the dominant S(−II) species is H_2S. The reduction half cell for the formation of H_2S is:

$$S(s) + 2H^+ + 2e^- \rightarrow H_2S \qquad E^0 = +0.1444$$

and

$$Eh = 0.1444 - 0.05916\,pH - 0.02958 \log [H_2S]$$

Since $[H_2S]$ will be about 1 M in this region, the Eh curve is drawn for the region 0 < pH < 7.1 in Fig. 5-2. This curve should not be extended beyond 7.1 because H_2S is not the dominant S(−II) species for higher pH values.

For 7.1 < pH < 14, HS^- will be the dominant sulfur species and the half reaction data are given below:

$$S(s) + H^+ + 2e^- \rightarrow HS^- \qquad E^0 = -0.0624$$

$$Eh = -0.0624 - 0.02958 \log [HS^-] - 0.02958\,pH$$

The curve describing this reduction is then drawn in Fig. 5-2 as curve 1 for this pH region. We assume that $HS^- = 1$ M. Note the change in slope relative to the S/H_2S reduction. A vertical line separates areas where H_2S and HS^- are dominant species. In fact, the function of these two curves at pH = 7.1 should

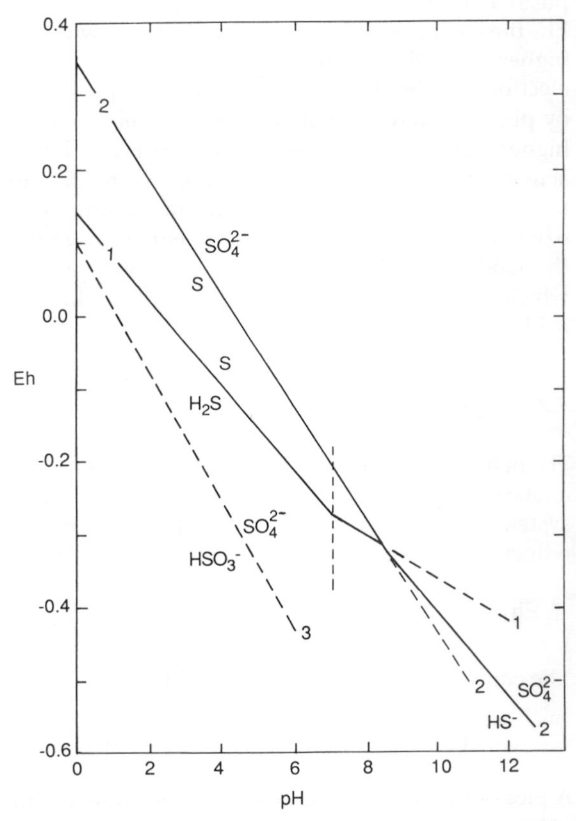

Fig. 5-2 Working diagram of Eh as a function of pH for selected aqueous sulfur species.

not be represented by the intersection of straight lines, but by a smooth curve. This detail will not alter the basic picture offered by these diagrams. If we wanted to consider pH values greater than 14, the reduction of S to S^{2-} would be examined.

We next consider the reduction of SO_4^{2-} to S(s), represented by the following equations and curve 2:

$$SO_4^{2-} + 8H^+ + 6e^- \rightarrow S(s) + 4H_2O \qquad E^0 = +0.3526$$

$$Eh = +0.3526 - 0.0789\,pH + 0.0099 \log [SO_4^{2-}]$$

The first special case to be dealt with is the crossing of curves 1 and 2 at about pH 8. Consider the relative positions of curves 1 and 2 for pH > 8. This indicates that S(s) is not stable in this region and will undergo a spontaneous autoredox reaction, $S(s) \rightarrow SO_4^{2-} + HS^-$ (not balanced). The consequence is that S(s) is not stable for pH > 8 and the relevant redox couple will be that for SO_4^{2-}/HS^-:

$$SO_4^{2-} + 8e^- + 9H^+ \rightarrow HS^- + 4H_2O \qquad E^0 = +0.2486$$

The solid line 2 for the above reduction thus replaces the dashed portions of lines 1 and 2 for pH > 8.

The S(IV) species are considered next. For $1.8 < pH < 7.2$, the dominant species is HSO_3^-. An examination of the curve for the SO_4^{2-}/HSO_3^- half reaction (shown as dashed curve 3) indicates that HSO_3^- is unstable with respect to SO_4^{2-} and S(s). HSO_3^- is not thermodynamically stable at all. For HSO_3^- to be stable would require the $[HSO_3^-]/[SO_4^{2-}]$ ratio to be less than 1×10^{-10}. Such small ratios are not normally of interest in geochemical considerations. (Small amounts of a substance may, however, play an important role in the rates of processes.) It may be shown that the same results are obtained for the other S(IV) species and thus they do not appear on the *equilibrium* diagram at all. Saying that HSO_3^- is not stable does not mean that S(IV) species do not exist in natural waters. S(IV) species are *thermodynamically* unstable, but the reactions by which they decompose may be very slow. The final Eh/pH diagram for this limited sulfur system is shown in Fig. 5-3. The S(0) and S(−II) species are stable only when O_2 is not present.

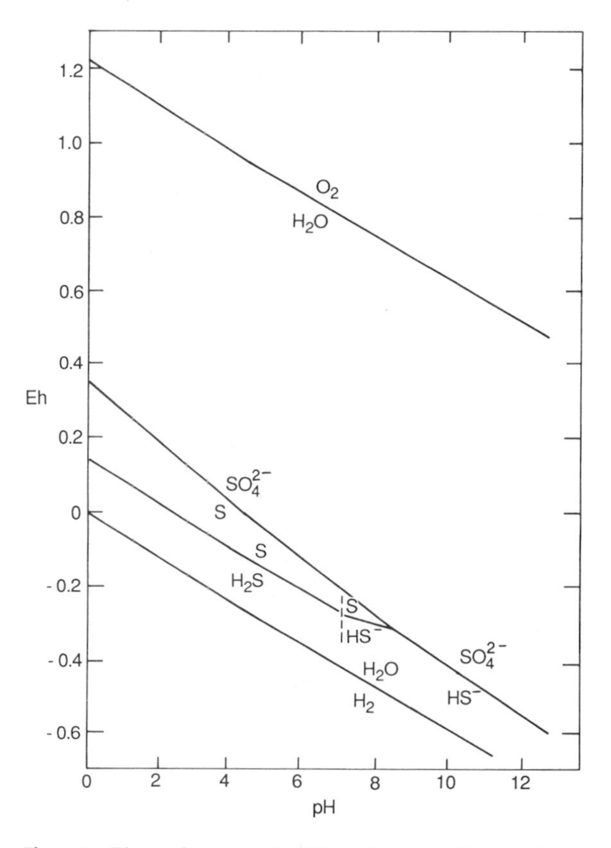

Fig. 5-3 Eh as a function of pH for selected sulfur species.

5.2.3 *Natural Systems and the Nernst Equation*

The extent to which natural systems are described by the Nernst equation depends on the relative rates at which electrons are transferred to and from various substances. These rates vary over several orders of magnitude. For example, we would not expect the electron transfer for the reduction of the hydrated ferric ion, $Fe^{3+} + e^- \rightarrow Fe^{2+}$, to be described by the same kinetic factors used to describe the reduction of nitrate, $2NO_3^- + 8H^+ + 8e^- \rightarrow N_2O + 5H_2O$. The latter reaction must involve a large number of molecular steps and will probably be a much slower process. The mechanisms of a few inorganic electron transfer processes have been summarized by Taube (1968). The presence of these kinetic factors when several redox couples are possible means that the Eh value measured with an instrument may not be related in a simple way to the concentrations of species present, and different redox couples may not be in equilibrium with one another. Lindberg and Runnells (1984) have presented data on the extent of disequilibrium in ground waters. Bockris and Reddy (1970), Stumm and Morgan (1981), and Hostettler (1984) describe factors that determine the potential of non-equilibrium redox systems.

5.3 Chemical Kinetics

5.3.1 Reaction Rates

The rates of chemical reactions are described at two levels. In the first, an empirical relationship is used to relate the overall rate of a process to the concentrations of various reactants. A common form for this expression is shown below:

$$\frac{dA}{dt} = -kA^m B^n C^p$$

This equation is known as the *rate law* for the reaction. The concentration of a reactant is described by A; dA/dt is the rate of change of A. The *rate constant* is represented by k. The parameters m, n, and p represent the order of the reaction with respect to A, B, and C, respectively. The exponents do not have to be integers in an empirical rate law. The units of k will depend on the units of the concentrations and on the values of m, n, and p.

Other types of expressions are assumed for the rate law. For instance, enzyme-catalyzed reactions are often described by the Michaelis-Menten model, which leads to a rate law in the following form:

$$\frac{d[S]}{dt} = \frac{k[E]_0[S]}{K_s + [S]}$$

$[S]$ is the substrate concentration, $[E]_0$ is the total enzyme concentration, and k and K_s are empirical constants.

The *temperature dependence* of a rate is often represented by the temperature dependence of the rate constant, k. This dependence is often represented by the Arrhenius equation, $k = A\exp(-E_a/RT)$. For some reactions, this relationship may be written as $k = AT^n \exp(-E_a/RT)$. In either case, E_a is known as the *activation energy* for the reaction (see also Benson, 1960; Moore and Pearson, 1981).

Time-scales for chemical reactions studied in both the laboratory and the environment range from about 10^{-12} to 10^6 s. These reactions take place in the gas phase or in the condensed phase. A wide variety of laboratory experimental methods are used to determine rate parameters. Methods for determining rate parameters are described in physical chemistry texts and other references listed at the end of this chapter.

Reactions occurring in beakers, or in the natural world, result from a series of simple processes between atoms and molecules. For instance, in the formation of NO from N_2 and O_2, the direct process in which a molecule of N_2 collides with a molecule of O_2 to produce two molecules of NO is an extremely slow process relative to other pathways for producing NO. Instead, the formation of NO at high temperature, as in combustion or lightning, may result from a series of simple steps comprising the Zeldovich mechanism (Bagg, 1971). M may be any of several molecules:

$$M + O_2 \rightarrow O + O + M$$
$$O + O + M \rightarrow O_2 + M$$
$$O + N_2 \rightarrow NO + N$$
$$N + O_2 \rightarrow NO + O$$

The study of reaction mechanisms comprises the second level of the examination of the rate of change of chemical systems. A reaction mechanism is a series of simple molecular processes, such as the Zeldovich mechanism, that lead to the formation of the product. The combination of these simple steps and their rate constants defines the overall reaction rate expressed by the rate law. In one of the most recently defined areas of chemical dynamics, the mechanical details of the molecular steps are examined.

5.3.2 Molecular Processes

The process of assembling individual molecular steps to describe complex reactions has probably enjoyed its greatest success for gas phase reactions in the atmosphere. In the condensed phase, molecules spend a substantial fraction of the time in association with other molecules and it has proved difficult to characterize these associations. Three basic types of fundamental processes are recognized.

Unimolecular processes are reactions involving only one reactant molecule. Photolytic reactions such as the decomposition of ozone by light and radioactive decay processes are examples of unimolecular processes:

$$O_3 + (h\nu) \rightarrow O_2 + O$$
$$^{14}C \rightarrow {}^{14}N + e^-$$

The rates of these processes depend *only* on the concentration of reactant:

$$d[O_3]/dt = -d[O]/dt = -J[O_3] \text{ or } d[^{14}C]/dt = -k[^{14}C]$$

Rate constants for photolytic reactions are commonly represented by the symbol J.

A complete description of the rate of a photolytic reaction should include the electronic states of the product molecules. J contains contributions from photons of all wavelengths that lead to a reaction. The flux of photons and the likelihood that photon absorption will lead to a reaction depend on the wavelength of the radiation. In the ozone example, J is sensitive to the flux of ultraviolet [λ (wavelength) < 320 nm] radiation and thus J increases with height in the atmosphere, varies with time of day, and becomes zero at night.

Bimolecular processes are reactions in which two reactant molecules collide to form two or more product molecules. In most cases, the reaction involves a rather simple rearrangement of bonds in the two molecules:

$$NO + O_3 \rightarrow NO_2 + O_2$$

Often, a single atom is transferred from one molecule to another and one bond is formed as another is broken. The rate of a bimolecular process depends on the product of concentrations of the two reactants. In this case:

$$\frac{d[NO]}{dt} = -\frac{d[NO_2]}{dt} = -k[NO][O_3]$$

Termolecular processes are common when two reactant molecules combine to form a single small molecule. Such reactions are often exothermic and the role of the third body is to carry away some of the energy released and thus stabilize the product molecule:

$$O_2 + O + M \rightarrow O_3 + M$$

The rate of the reaction depends on the product of reactant concentrations, including the third body:

$$\frac{d[O_3]}{dt} = k[O_2][O][M]$$

Almost any molecule can act as a third body, although the rate constant may depend on the nature of the third body. For the example reaction in air, the most important third body molecules are N_2 and O_2.

5.3.3 Reaction Mechanisms

The chemistry of the stratospheric ozone will be sketched with a very broad brush in order to illustrate some of the characteristics of complex reactions. A model for the formation of ozone in the atmosphere was proposed by Chapman and may be represented by the following "oxygen only" mechanism (other aspects of stratospheric ozone are discussed in Chapters 10 and 12):

$$O_2 \rightarrow O + O \qquad J_1$$
$$O + O_2 + M \rightarrow O_3 + M \qquad k_2$$
$$O_3 \rightarrow O_2 + O \qquad J_3$$
$$O + O_3 \rightarrow 2O_2 \qquad k_4$$

Reactions 2 and 3 regulate the balance of O and O_3, but do not materially affect the O_3 concentration. Any ozone destroyed in the photolysis step (3) is quickly reformed in reaction (2). The amount of ozone present results from a balance between reaction (1), which generates the O atoms that rapidly form ozone, and reaction (4), which eliminates an oxygen atom and an ozone molecule. The concentrations of O and O_3 that result under conditions of constant sunlight (constant J_1 and J_3) are termed steady-state concentrations. These are the concentrations of O and O_3 defined by the equations $d[O]/dt = 0$ and $d[O_3]/dt = 0$. Two equations result that may be solved for [O] and $[O_3]$:

$$0 = 2J_1[O_2] - k_2[O][O_2][M] + J_3[O_3] - k_4[O][O_3]$$
$$0 = k_2[O][O_2][M] - J_3[O_3] - k_4[O][O_3]$$

Although this simple mechanism accounts for the presence of O_3 in the stratosphere, close examination of the stratospheric ozone concentration and the rate constants indicates that the loss term for ozone (reaction 4) is not large enough to balance the rate at which ozone is produced.

The reason for the discrepancy is found in a series of *catalytic* reactions such as the following:

$$NO + O_3 \rightarrow NO_2 + O_2 \qquad k_5$$
$$NO_2 + O \rightarrow NO + O_2 \qquad k_6$$
$$O_3 + O \rightarrow 2O_2 \qquad \text{net reaction}$$

The net effect of this pair of reactions produces the same end result as reaction (4), O and O_3 are destroyed. The NO/NO_2 pair of compounds is referred to as a catalyst because it enhances the rate of the reaction ($O + O_3 \rightarrow 2O_2$) without being changed in the process (Crutzen, 1971; Johnston, 1971). (These and other catalysts do have finite lifetimes. In this case, it appears that NO_2 finally forms nitric acid, $HONO_2$, which may be removed by condensation on ice crystals followed by precipitation.) Catalysts are an "invisible hand" and can greatly increase the rate of a reaction. The pair of chlorine radical species Cl and ClO undergoes an analogous set of reactions (Stolarski and Cicerone, 1974; Molina and Rowland, 1974):

$$Cl + O_3 \rightarrow ClO + O_2$$
$$ClO + O \rightarrow Cl + O_2$$

Catalysts are also very important in biological systems where enzymes effectively regulate the rate of reactions and determine which products are formed. The Michaelis-Menten mechanism is represented by the following steps:

$$S + E \rightarrow ES^* \quad k_1$$
$$ES^* \rightarrow E + S \quad k_2$$
$$ES^* \rightarrow E + P \quad k_3$$

In this case, S represents the substrate, or the substance undergoing change, and P represents the product molecule(s). E is the enzyme, which forms a complex, ES^*. This complex is capable of undergoing further reaction, with a lowered activation energy, to form a specific product molecule. The overall reaction is $S \rightarrow P$ with the enzyme being regenerated.

Rate limiting steps are the molecular processes that effectively limit the rate of the overall reaction. They are the slowest step in the series of steps leading to product formation. The rate limiting step may be examined for an enzyme-catalyzed reaction. For the Michaelis-Menten model, the rate of product formation (or the negative of the rate of substrate loss) is given by:

$$\frac{dP}{dt} = \frac{k_3[E_0][S]}{\left(\frac{k_2 + k_3}{k_1} + [S]\right)}$$

Compare this expression with the empirical form given in equations in the first part of this section. This rate expression may be examined for two limiting cases of substrate concentration:

Case 1: $[S] \ll \frac{k_2 + k_3}{k_1}$

Considering only the largest term in the denominator of the rate expression, the rate is now given by:

$$\frac{dP}{dt} = \left(\frac{k_1 k_3}{k_2 + k_3}\right)[E]_0[S] = k'[E]_0[S]$$

The rate is second-order overall, first-order with respect to the total enzyme concentration and the substrate concentration. The overall reaction is limited by the slowest step in the series, which is reaction 1. Since this reaction involves bringing E and S together, it is a second-order reaction.

Case 2: $[S] \gg \frac{k_2 + k_3}{k_1}$

The rate law now becomes $dP/dt = k_3[E]_0$. The rate is independent of the substrate concentration and first order with respect to enzyme concentration. In this case, reaction 3, in which the complex decomposes to form the product, is the slowest step. Although the discussion above assumes that we have an isolated enzyme reacting with a substrate, the same principles may be applied to the more complex case when an entire organism, or a series of organisms, is consuming a substrate.

The case of bacterial reduction of sulfate to sulfide described by Berner (1984) provides a useful example. The dependence of sulfate reduction on sulfate concentration is shown in Fig. 5-4. Here we see that for $[SO_4] < 5$ mM, the rate is a linear function of sulfate concentration. For $[SO_4^{2-}] > 10$ mM, the rate is reasonably independent of sulfate concentration. The sulfate concentration in the ocean is about 28 mM and thus in shallow marine sediments the reduction rate does not depend on sulfate concentration. (The rate *will* depend on the concentration of organisms and the concentration of other necessary reactants – organic carbon in this case.) In freshwaters, the sulfate concentration is much less than 5 mM and the sulfate reduction rate does depend on the sulfate

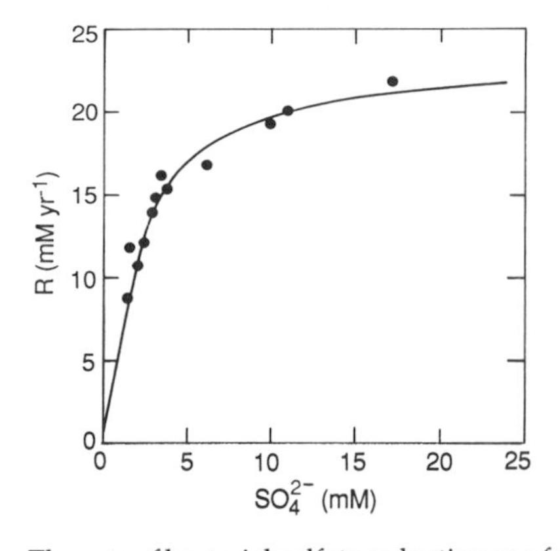

Fig. 5-4 The rate of bacterial sulfate reduction as a function of sulfate concentration. Adapted from Berner (1984) with the permission of Pergamon Press.

concentration (and may be independent of the concentration of organic carbon).

The catalytic destruction of ozone by NO/NO_2 also illustrates rate-limiting steps. At 30 km altitude, the rate constants for three bimolecular ozone destruction reactions are given below. The units are all cm^3/molec s:

$$O_3 + O \rightarrow 2\,O_2 \qquad k_4 = 1.0 \times 10^{-15}$$

$$NO + O_3 \rightarrow NO_2 + O_2 \qquad k_5 = 4.7 \times 10^{-15}$$

$$NO_2 + O \rightarrow NO + O_2 \qquad k_6 = 9.3 \times 10^{-12}$$

Reactions 1, 2 and 3 given earlier are included to complete the mechanism and it is also assumed that the sum $[NO] + [NO_2]$ is a constant ($6.4 \times 10^9/cm^3$ is used here). Values for J_1, J_3, and k_2 are 8×10^{-11}/s, 6.5×10^{-4}/s, and $1.0 \times 10^{-33}\ cm^6/molec^2s$, respectively. The steady-state concentrations of species for this limited mechanism are:

$$[O_3] = 4.0 \times 10^{12}/cm^3$$

$$[O] = 9.8 \times 10^{7}/cm^3$$

$$[NO] = 2.9 \times 10^{8}/cm^3$$

$$[NO_2] = 6.1 \times 10^{9}/cm^3$$

We can empirically test for the rate-limiting step by examining how the ozone concentration is changed as we change each rate constant in turn. In this case, doubling k_5 reduces the ozone concentration about 2%. On the other hand, doubling k_6 reduces the ozone concentration by nearly 50%. In this example, the reaction between NO_2 and O atoms is the rate-limiting step in the catalytic process. The relative concentrations of NO and NO_2 also provide a clue to the rate-limiting step. Over 90% of the nitrogen atoms in NO/NO_2 are present as NO_2. Thus, increasing k_5 cannot increase the concentration of NO_2 significantly.

The power of the catalyst may also be seen in the relative rate constants. The rate constant for the loss of ozone due to reaction with NO is five times the loss of ozone in the reaction with O atoms. The constant for loss of O atoms in the reaction with NO_2 is nearly four orders of magnitude faster than that due to reaction with ozone. These numbers are brought out to indicate the very large range in the values for rate constants. This range results mainly from differences in activation energies and the steric requirements for the reactions. (Steric requirements define the orientations of atoms necessary to form the new bonds.) Although the concentrations of the catalytic species in this example are lower than the ozone concentration by three or four orders of magnitude, their effect is magnified when each NO and NO_2 molecule goes through the catalytic cycle hundreds or thousands of times before being removed.

Atmospheric reactions have been successfully represented as a sum of molecular reactions and mixing processes. Rate constants for a large number of atmospheric reactions have been tabulated by Baulch *et al.* (1982, 1984) and Atkinson and Lloyd (1984). Rates of reactions on solid surfaces have been examined in selected cases for solids suspended in air or water or in sediments. Reactions for the atmosphere as a whole and for cases involving aquatic systems, soils, and surface systems are often parameterized by the methods of Chapter 4. That is, the rate is taken to be a linear function or a power of some limiting reactant – often the compound of interest. As an example, the global uptake of CO_2 by photosynthesis is often represented in the empirical form $d[CO_2]/dt = -k[CO_2]^m$.

5.3.4 Kinetic Isotope Effects

Molecules containing different isotopic species of an element may react at different rates. The effect is noticeable if the rate-limiting step in a reaction is one in which a bond to the element in question is formed or broken. The primary kinetic isotope effect results from the lower vibrational frequencies of heavier isotopic species. The quantum mechanical zero point vibrational energies are also lower for heavier isotopes and thus slightly more energy is required to break a bond to ^{13}C, for instance, than to ^{12}C. Isotope ratios may be measured with high accuracy on small amounts of geochemical samples and thus one may infer something about the processes that form a compound from the observed isotope distribution.

For a kinetic isotope effect to exist, the reverse reaction must not occur to a significant extent (in which case we would have a thermodynamic isotope effect), and the molecules undergoing reaction must be drawn from a larger pool of molecules that do not react. If all the molecules are going to undergo reaction, there will be no discrimination between isotopes and no kinetic isotopic effect. For these reasons, the kinetic isotope effect for ^{13}C in living matter occurs in the early stages of the photosynthetic process when a small fraction of the CO_2 (or HCO_3^-) available is added to organic substrates to form carboxylic acids.

Photosynthesis begins with the transfer of CO_2 into the cell from the atmosphere or water. Photosynthetic enzymes then transfer the inorganic carbon to a 5-carbon organic compound to form two 3-carbon carboxylic acid molecules. (This is the case for the C3 photosynthetic mechanism.) In these steps, reaction of $^{12}CO_2$ occurs slightly faster than reaction of $^{13}CO_2$ and the organism has a more negative $\delta^{13}C$ than the atmosphere or ocean from which it grows. Thus, while marine inorganic carbon has $\delta^{13}C$ of about 0‰ compared to the reference and atmospheric carbon has $\delta^{13}C$ of about −7‰, the $\delta^{13}C$ values for plants range from −10 to −30‰ (Schidlowski, 1988). The isotope distribution depends somewhat on the species of plant and it is a strong function of whether the plant fixes carbon by the C3 mechanism (most plants) or the C4 mechanism (a smaller group including corn, sugar cane, and some tropical plants).

Isotope effects also play an important role in the distribution of sulfur isotopes. The common state of sulfur in the oceans is sulfate and the most prevalent sulfur isotopes are ^{32}S (95.0%) and ^{34}S (4.2%). Sulfur is involved in a wide range of biologically driven and abiotic processes that include at least three oxidation states [S(VI), S(0), and S(−II)]. Although sulfur isotope distributions are complex, it is possible to learn something of the processes that form sulfur compounds and the environment in which the compounds are formed.

5.4 Non-equilibrium Natural Systems

5.4.1 The Formation of Nitrogen Oxides

The formation of the nitrogen oxides, nitric oxide (NO) and nitrogen dioxide (NO_2) from N_2 and O_2, provides an example of the interplay between thermodynamics and kinetics. The calculation of the equilibrium concentrations of NO and NO_2 is straightforward if we assume that these are the only compounds formed and that the initial pressures of N_2 and O_2 are known.

The 298 K values for ΔG^0 and ΔH^0 are given below in kJ/mol:

	ΔG^0	ΔH^0
NO(g)	86.596	90.291
NO_2(g)	51.241	33.095

The enthalpies and free energies of formation of NO and NO_2 are just the values given above, since N_2 and O_2 are the constituent elements. There are two things to note about these values. First, ΔG^0 for the formation of NO is more positive than that for NO_2 at 298 K. Second, ΔH^0 for NO formation is much more positive than ΔH^0 for NO_2 formation. Thus, according to the van't Hoff equation, the equilibrium constant for NO formation will increase much more rapidly as the temperature is increased.

We may describe the temperature dependence of ΔG and ΔH in terms of the functions $(G_T - H_{298})/T$ and $H_T - H_{298}$. These functions are easily interpolated to obtain values at temperatures intermediate to the points given in the tables. The equilibrium pressures of NO and NO_2 are given in Fig. 5-5 for initial N_2 and O_2 pressures equal to 0.79 and 0.21 bar.

The equilibrium pressure for NO_2 is much greater than the NO pressure for temperatures less than about 500 K. Above 1000 K, the NO pressure is much greater than the NO_2 pressure. The following discussion will focus on NO formation.

The reason for considering this equilibrium at high temperatures is that there are many processes that

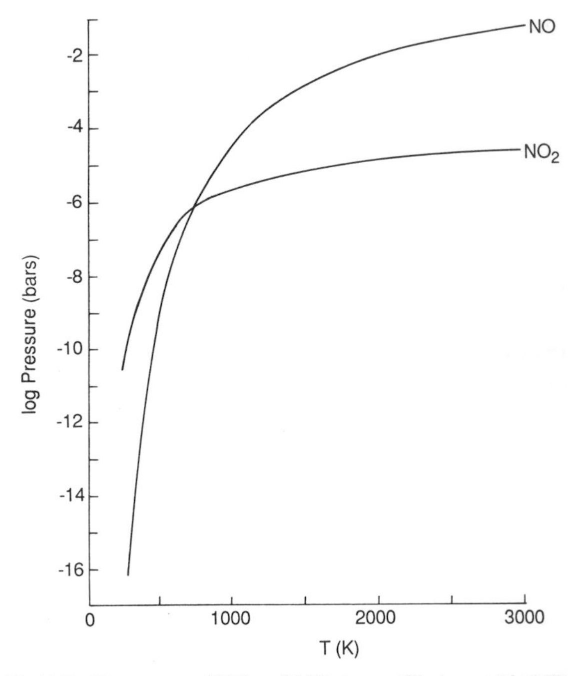

Fig. 5-5 Pressures of NO and NO_2 in equilibrium with 0.79 bar of N_2 and 0.21 bar of O_2 as a function of temperature.

raise air to a high temperature. Hydrocarbon or biomass combustion can produce temperatures of 1500–3000 K and lightning discharges can produce temperatures of the order of 30 000 K (Yung and McElroy, 1979). Other processes capable of producing high temperatures include shock waves from comet or meteorite impacts (Prinn and Fegley, 1987) and nuclear bomb explosions (Goldsmith *et al.* 1973).

The actual mechanism by which NO is formed does not include the direct reaction of N_2 and O_2 to any significant extent. This direct reaction would involve the breaking of two strong bonds and the formation of two new bonds, an unlikely event. Rather, the oxidation of nitrogen begins with a simple reaction:

(i) $$M + O_2 \rightarrow O + O + M$$

This reaction occurs to only a small extent, but the oxygen atoms thus formed may form NO through the following catalytic cycle:

(ii) $$O + N_2 \rightarrow NO + N$$

(iii) $$N + O_2 \rightarrow NO + O$$

net $$N_2 + O_2 \rightarrow 2NO$$

The reverse reactions of (i), (ii), and (iii) are also important in establishing the equilibrium between N_2, O_2, and NO.

At low temperatures, the rates of these reactions are very slow either because the rate constants are very small or because the concentrations of O and N are very small. For these reasons, equilibrium is not maintained at low temperatures. The activation energies for the critical steps are quite large, 494 kJ/mol for reaction (i) and 316 kJ/mol for reaction (ii). The rate constants increase sharply with temperature and equilibrium is maintained at higher temperatures. The time-scale for equilibrium for the overall reaction $N_2 + O_2 \leftrightarrow 2NO$ is less than 1 s for $T > 2000$ K.

As the temperature of an N_2/O_2 mixture is increased above 2000 K, the observed concentration of NO (as well as those for NO_2, N, O, and other species) will approach the equilibrium values appropriate for that temperature. As the temperature of the mixture of these gases decreases, the concentrations will follow the equilibrium values. Equilibrium will be maintained as long as the time-scale for the chemical reaction is shorter than the time-scale for the temperature change, i.e. the chemical reaction is more rapid than the temperature change. The time-scale for the chemical reaction increases rapidly as the temperature decreases because of the large activation energies.

The time-scale for the chemical reaction will be examined using one of the reactions for the loss of NO, the direct reaction between NO molecules.

(iv) $$NO + NO \rightarrow N_2O + O$$
$$k_4 = 1.56 \times 10^{-12} \exp(-33\,125/T)\ \text{cm}^3/\text{molec s}$$

The time-scale for adjustment of NO to equilibrium is rather difficult to evaluate for the complete mechanism. We can obtain an approximation and an upper limit to the time-scale by just considering this reaction. The exponential time-scale for this reaction is given by:

$$1/\tau = 4k[NO]_{eq}$$

For the equilibrium conditions described above, this time-scale increases from 1 s at 2500 K to 100 000 s at 1500 K.

As the reaction mixture cools, a point is reached when the time-scale for the chemical reaction is longer than the time-scale for the temperature change. When this occurs, the change in the actual concentration begins to lag behind the change in the equilibrium concentration. When temperatures fall to low values (less than about 1000 K in this case), the reactions are so slow that little further change occurs.

The amount of NO in the mixture when temperatures are returned to ambient values is not so much a function of the peak temperature reached, but depends on the temperature at which the chemical time-scale equals the cooling time-scale. If the cooling rate is very fast, the lowest temperature at which equilibrium is maintained will be higher – and the amount of NO present will be greater. The total amount of NO formed will also depend on the amount of air raised to high temperature and the extent to which other air is heated by entrainment, compression, radiation, and other factors. For example, in an automobile, the time-scale for cooling of the combustion mixture is about 0.05 s. The pressure of NO emitted from the tailpipe is about 0.0006 bar, far in excess of the ambient equilibrium value of about 3×10^{-16} bar.

5.4.2 Oxygen in the Atmosphere

The presence of oxygen in the atmosphere flies in the face of conventional thermodynamic logic. Oxygen is not stable with respect to reduced organic matter (with which it should react to form carbon dioxide) and it is not stable with respect to reaction with dinitrogen to form a dilute solution of nitric acid in the oceans.

The total amount of oxygen in the atmosphere is about 1.2×10^{21} g(O). The processes affecting atmospheric oxygen in the short-term are photosynthesis and the combination of respiration by living organisms and decomposition of fixed carbon by abiotic processes. The discussion of the control of atmospheric oxygen is taken from the treatment by Holland (1978) and the reader should refer to this work for a more detailed discussion. We will represent photosynthesis by the following reaction:

$$CO_2 + H_2O \xrightarrow{\text{light}} CH_2O + O_2$$

(The process may also involve the oxidation of H_2S instead of H_2O.) ΔG for this process is about + 480 kJ/mol (carbon). The energy is provided by absorption of visible light, which supplies about 230 kJ/mol photons. The rate at which oxygen is produced by photosynthesis for the entire Earth is about 1.9×10^{17} g(O)/year. Most of this oxygen is removed rather quickly by respiration/decomposition and thus photosynthesis by itself does not account for the net production of oxygen. The production of oxygen results from the physical removal of some of the reduced carbon from contact with oxygen before it has a chance to decompose. This process is known as *carbon burial* and it represents the difference between photosynthesis and respiration/decomposition. Carbon burial is the removal of fixed carbon to anaerobic sediments where reaction with atmospheric oxygen does not occur until the sediments are returned to the surface by tectonic processes. The rate of carbon burial corresponds to 3.2×10^{14} g(O)/year, less than 1% of the amount of oxygen formed by photosynthesis.

That there must be some mechanism for the control of atmospheric oxygen is demonstrated by two observations. First, there is a great potential for a very different amount of oxygen in the atmosphere. Atmospheric oxygen could be much less than it is today and it might be substantially higher than present levels. Atmospheric oxygen is a small fraction of the 1.1×10^{25} g of reduced oxygen in the 2.4×10^{25} g of crustal matter (Mason, 1966) and 1.3×10^{24} g of reduced oxygen in the hydrosphere. Of the $\sim 2 \times 10^{22}$ g of carbon in sedimentary rocks, only about 20% is reduced carbon (Schidlowski, 1988). Of course, increased amounts of buried carbon may be limited by the nutrient supply as well. The second point is that the average residence time for oxygen in the atmosphere, with respect to carbon burial, is about 4 Ma [1.2×10^{21} g(O)/3.2×10^{14} g(O)/year] and the oxygen concentration appears to have been quite constant over a time-scale much longer than this. No change in the amount of atmospheric oxygen has occurred since direct measurements have been made. Machta and Hughes (1970) have not been able to account for a measureable change during the period 1920–70. This analysis indicates that the time-scale for change is greater than 200 000 year. The presence of mammals during the past 180 Ma suggests that large changes of oxygen have not occurred during this time (e.g. Walker, 1974).

In our analysis, we will examine only the factors that affect atmospheric oxygen on a time-scale greater than ~ 100 years and neglect the annual "panting" associated with the rapid photosynthetic and respiration/decomposition cycle.

The rate of carbon burial depends on the rate of carbon fixation and the probability that the fixed carbon will escape oxidation. The rate of carbon fixation is independent of the amount of oxygen in the atmosphere (to first order) but the probability that fixed carbon will escape oxidation must decrease as the amount of oxygen increases. Oxidation of fixed carbon by oxygen can be expected to become more likely as the oxygen pressure increases, whether the carbon is exposed to the atmosphere on the surface or dissolved in water which is in contact with the atmosphere. The rate of carbon burial must approach very low values as the oxygen pressure in the atmosphere becomes large.

If the oxygen pressure decreases, the carbon burial rate will increase, but not without bound. Although the probability of burial will approach unity, the rate of carbon fixation will be limited by the availability of nutrients buried with the carbon or other factors affecting the rate of photosynthesis. The overall curve describing the rate of oxygen *production* is described by curve P in Fig. 5-6. Even though the oxidation of carbon is favored thermodynamically [$\Delta G = -480$ kJ/mol(C)], the reaction cannot occur if the carbon is removed from chemical contact with the oxygen.

The removal of atmospheric oxygen results from the oxidation of reduced materials brought to the surface. The main sources of reduced material in decreasing order of importance are: reduced carbon, sulfur as sulfides and pyrites, and ferrous iron.

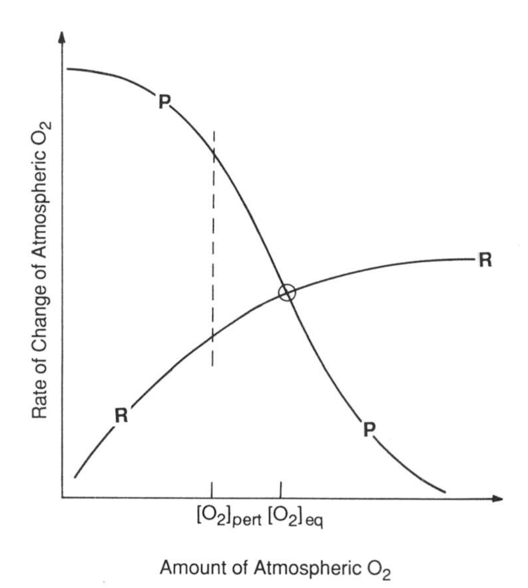

Fig. 5-6 Rate of production (P) and removal (R) of atmospheric oxygen as a function of total atmospheric oxygen. Dashed line indicates perturbed level.

These reduced compounds are brought to the surface by the uplift of rocks and by volcanic activity. (Anthropogenic processes also play a small role.) The reduced materials combine with oxygen at the surface in the process of chemical weathering. Chemically reduced volcanic emissions (e.g. H_2S, SO_2, etc.) combine with oxygen in the atmosphere. At low oxygen pressures, the oxidation rate can be expected to depend on the oxygen pressure. (At low pressures, the rate-limiting step should involve the supply of oxygen.) At high oxygen pressures, the rate of removal may depend on the rate of uplift, or other physical geologic factors. Since most reduced material in freshly exposed rock becomes completely oxidized on Earth (Holland, 1978), the removal process seems to be operating near the level part of the curve. The curve describing the *removal* of atmospheric oxygen is shown as curve R in Fig. 5-6. Where the P and R curves cross, production and removal rate are equal and the amount of oxygen in the atmosphere is constant.

We can examine the *stability* of this model by imagining a perturbation of the system from its equilibrium position. Suppose that a brief interval of increased tectonic activity brought a large amount of reduced material to the surface in a short time. This would decrease atmospheric oxygen from $[O_2]_{eq}$ to a lower value $[O_2]_{pert}$. As may be seen in Fig. 5-6, at the $[O_2]_{pert}$ concentration the production rate will exceed the removal rate. Thus $[O_2]$ can be expected to increase until $[O_2]_{eq}$ is reached (assuming that curves P and R still apply).

The formation of reduced carbon can account for the qualitative features of the oxygen cycle. The fate of the reduced carbon in sediments is, however, coupled with the sulfur cycle, as described by Berner (1984) and Garrels and Lerman (1981). Some of the reduced carbon may be oxidized by bacteria under anoxic conditions with sulfate acting as the oxidizing agent. The carbon is fully oxidized to carbonate and the sulfur is reduced in several stages to a sulfide or to iron pyrite. The "stripped-down", balanced, chemist's redox equation is:

$$15CH_2O + 8SO_4^{2-} + 2Fe_2O_3 \rightarrow 15CO_3^{2-} + 4FeS_2 + 8H_2O + 14H^+$$

The carbonate is present in sediments as magnesium and calcium carbonates. The calcium and magnesium also appear as the sulfate and silicate, respectively, so that a more descriptive redox equation is:

$$15CH_2O + 8CaSO_4 + 7MgSiO_3 + 2Fe_2O_3 \rightarrow 8CaCO_3 + 7MgCO_3 + 7SiO_2 + 4FeS_2 + 15H_2O$$

The spontaneity of this reaction is reflected in the ΔG^0 value of -125 kJ/mol(C). The result is that the total equivalents of reduced matter produced by carbon burial are redistributed and now include some pyrite. Sulfate is removed from the water permeating the sediments and magnesium and calcium carbonates are formed. Thus, the carbon, oxygen, and sulfur cycles are coupled by a redox reaction which also includes iron. The extent of the coupling depends on the supplies of reactive organic carbon, sulfate, and reactive iron. When the reduced material is returned to contact with the atmosphere, it is oxidized and an amount of oxygen equivalent to that formed when the carbon was first buried is removed.

5.5 Summary

Thermodynamics establishes the boundaries of what is possible in the natural world. In many cases, enough information is available to enable one to say something about the spontaneity of processes, i.e. thermodynamic spontaneity: what reaction will occur if a reaction does take place? We often know much less about *how fast* the reaction is. Will the spontaneous reaction occur within our lifetime? The question of *rate* involves many issues that are much less well understood. We probably know more about gas phase rates than we do about rates in other media. Even so, it has taken many years to sort out the puzzle of stratospheric ozone depletion. Rates in the aqueous phase and other condensed media are much less well understood. Biological processes involving metabolic pathways in countless organisms, catalytic effects involving trace species, and processes occurring at phase boundaries are among the factors that complicate an understanding of rates in the natural world.

Although the natural systems described above are far from equilibrium, many localized regions of natural systems are well-described in thermodynamic and equilibrium terms. In brief, if the reactions that redistribute compounds between the reactant and product states are fast, then equilibrium conditions may be applied. In some cases, part of a system may be described in equilibrium terms and part cannot be. As an example, ammonia is generally not in equilibrium with oxidized nitrogen in the atmosphere and surface waters and ammonia is not distributed between the oceans and atmosphere according to equilibrium expressions (Quinn *et al.*, 1988). None the less, the distribution between NH_4^+ and NH_3(aq) is described by the usual equilibrium expression because the proton exchange reaction is very fast. Generally speaking, one must be cautious in applying equilibrium relationships unless it is known that the underlying reactions are fast.

Questions

5-1 The free energy of formation of ozone is +163.2 kJ/mol and the enthalpy of formation is 142.7 kJ/mol at 298 K. Estimate the equilibrium pressure of ozone at 25 km. The total pressure at 25 km is about 0.025 bar and the atmosphere is 21% O_2. The temperature is about 217 K and you may assume that ΔH and ΔS are constant over this temperature interval.

5-2 The Eh measured in a groundwater sample was −0.067 V at a pH of 8.83 (Hostettler, 1984). The concentrations of SO_4^{2-} and HS^- were 2.29 and 0.003 mM, respectively. Is this measured Eh consistent with the observed SO_4^{2-} and HS^- concentrations? Discuss briefly. Relevant data for the reduction of SO_4^{2-} is given in this chapter (assume $T = 298$ K).

5-3 For problem 2 above, what would the ratio of Fe^{2+} to Fe^{3+} be if this reduction couple is in equilibrium with Eh = −0.067? E^0 for $Fe^{3+} + e^- \rightarrow Fe^{2+}$ is +0.77.

5-4 At 30 km, the O_3 and O concentrations are about 2×10^{12} and $5 \times 10^7/cm^3$, respectively. The principle reactions with Cl and ClO are:

$$Cl + O_3 \rightarrow ClO + O_2 \qquad k_1 = 9.0 \times 10^{-12}\ cm^3/molec\ s$$

$$ClO + O \rightarrow Cl + O_2 \qquad k_2 = 4.5 \times 10^{-11}\ cm^3/molec\ s$$

The total concentration of Cl + ClO is $5 \times 10^8/cm^3$ and the rate of formation and removal of Cl and ClO is 50 molec/cm^3 s. Estimate the Cl/ClO ratio and the actual concentrations of these species. (You may assume that reactions 1 and 2 are much faster than the rates of formation and removal.) Approximately how many catalytic cycles does Cl go through before being removed from the stratosphere?

5-5 The atmospheric lifetime for methylchloroform, CH_3CCl_3, is about 9.7 years. Assume that the only reaction destroying methylchloroform is reaction with OH in the troposphere. (In fact, a sizeable fraction is transported to the stratosphere.) The rate constant for the OH + CH_3CCl_3 reaction is given by $k = 5.1 \times 10^{-12} \exp(-1800/T)\ cm^3/molec\ s$. Make any reasonable assumptions and estimate an average tropospheric OH concentration.

5-6 The total living biomass amounts to about 5.6×10^{17} g C and has an approximate formula CH_2O. Suppose that half of this biomass was suddenly burned to CO_2 and H_2O. How much would the partial pressure of O_2 change? How might the control system described in Section 5.4 respond in order to restore atmospheric oxygen to its 21% level? (The total amount of oxygen in the atmosphere is 1.2×10^{21} g.)

Answers can be found on p. 365.

References

Atkinson, R. and A. C. Lloyd (1984). Evaluation of kinetic and mechanistic data for modeling of photochemical smog. *J. Phys. Chem. Ref. Data* **13**, 315–444.

Bagg, J. (1971). The formation and control of oxides of nitrogen in air pollution. *In* "Air Pollution Control," Part One (W. Strauss, ed.), pp. 35–94. John Wiley, New York.

Baulch, D. L., R. A. Cox, P. J. Crutzen, R. F. Hampson, J. A. Kerr, J. Troe, and R. T. Watson (1982). Evaluated kinetic and photochemical data for atmospheric chemistry. *J. Phys. Chem. Ref. Data* **11**, 327–496 (suppl. I).

Baulch, D. L., R. A. Cox, R. F. Hampson, J. A. Kerr, J. Troe, and R. T. Watson (1984). Evaluated kinetic and photochemical data for atmospheric chemistry. *J. Phys. Chem. Ref. Data* **13**, 1259–1380 (suppl. II).

Benson, S. W. (1960). "The Foundations of Chemical Thermodynamics." McGraw-Hill, New York.

Berner, R. A. (1984). Sedimentary pyrite formation: An update. *Geochim. Cosmochim. Acta* **48**, 605–615.

Bockris, J. O. and A. K. N. Reddy (1970). "Modern Electrochemistry." Plenum Press, New York.

Broecker, W. S. and V. M. Oversby (1971). "Chemical Equilibrium in the Earth." McGraw-Hill, New York.

Butler, J. N. (1964). "Ionic Equilibrium." Addison-Wesley, Reading, Ma.

Crutzen, P. J. (1971). Ozone production rates in an oxygen–hydrogen–nitrogen oxide atmosphere. *J. Geophys. Res.* **76**, 7311–7327.

Drever, J. I. (1988). "The Geochemistry of Natural Waters." Prentice-Hall, Englewood Cliffs, N. J.

Faure, G. (1977). "Principles of Isotope Geology." John Wiley, New York.

Garrels, R. M. and A. Lerman (1981). Phanerozoic cycles of sedimentary carbon and sulfur. *Proc. Natl. Acad. Sci. USA* **78**, 4652–4656.

Goldsmith, P., A. F. Tuck, J. S. Foot, E. L. Simmons, and R. L. Newson (1973). Nitrogen oxides, nuclear weapon testing, Concorde, and stratospheric ozone. *Nature* **244**, 545–551.

Holland, H. D. (1978). "The Chemistry of Atmosphere and Oceans." John Wiley, New York.

Hostettler, J. D. (1984). Electrode reaction, aqueous electrons, and redox potentials in natural waters. *Amer. J. Sci.* **284**, 734–759.

Johnston, H. S. (1971). Reduction of stratospheric ozone by nitrogen oxide catalysts from supersonic transports. *Science* **173**, 517–522.

Lindberg, R. D. and D. D. Runnells (1984). Groundwater redox reactions: An analysis of equilibrium state applied to Eh measurements and geochemical modeling. *Science* **225**, 925–927.

Machta, L. and E. Hughes (1970). Atmospheric oxygen in 1967 to 1970. *Science* **168**, 1582–1584.

Mason, B. (1966). "Principles of Geochemistry," p. 41. John Wiley, New York.

Molina, M. J. and F. S. Rowland (1974). Stratospheric sink for chlorofluoromethanes: Chlorine atom-catalyzed destruction of ozone. *Nature* **249**, 810–812.

Moore, W. J. (1972). "Physical Chemistry," 4th edn. Prentice-Hall, Englewood Cliffs, N. J.

Morel, F. M. M. (1983). "Principles of Aquatic Chemistry." John Wiley, New York.

Prinn, R. G. and B. Fegley (1987). Bolide impacts, acid rain, and biospheric traumas at the Cretacious–Tertiary boundary. *Earth Planet. Sci. Lett.* **83**, 1–15.

Quinn, P. K., R. J. Charlson, and T. S. Bates (1988). Simultaneous measurements of ammonia in the atmosphere and ocean. *Nature* **335**, 336–338.

Saigne, C. and M. Legrande (1987). Measurements of methanesulfonic acid in Antarctic ice. *Nature* **330**, 240–242.

Schidlowski, M. (1988). A 3,800 million year isotopic record of life from carbon in sedimentary rocks. *Nature* **333**, 313–318.

Sillén, L. G. (1959). Graphical presentation of equilibrium data. *In* "Treatise in Analytical Chemistry," Part I, Vol. 1, (I. M. Kolthoff, P. J. Elving and E. B. Sandell, eds.), pp. 277–317. Intersceince, New York.

Stolarski, R. S. and R. J. Cicerone (1974). Stratospheric chlorine: A possible sink for ozone. *Can. J. Chem.* **52**, 1610–1615.

Stumm, W. and J. J. Morgan (1981). "Aquatic Chemistry," 2nd edn. John Wiley, New York.

Taube, H. (1968). Mechanisms of oxidation–reduction reactions. *J. Chem. Educ.* **45**, 453–461.

Urey, H. C. (1947). The thermodynamic properties of isotopic substances. *J. Chem. Soc.* **1947**, 562–581.

Walker, J. C. G. (1974). Stability of atmospheric oxygen. *Amer. J. Sci.* **168**, 193–214.

Wayne, R. P. (1985). "Chemistry of Atmospheres." Clarendon Press, Oxford.

Weast, R. C. (ed.) (1987). "The Handbook of Chemistry and Physics." Chemical Rubber Co. Press, Cleveland, Ohio.

Yung, Y. L. and M. B. McElroy (1979). Fixation of nitrogen in the prebiotic atmosphere. *Science* **203**, 1002–1004.

Bibliography

Adamson, A. W. (1986). "Physical Chemistry," 3rd edn. Academic Press, Orlando, Florida.

Atkins, P. W. (1986). "Physical Chemistry," 3rd edn. W. H. Freeman, New York.

Bard, A. J. and L. R. Faulkner (1980). "Electrochemical Methods." John Wiley, New York.

Calvert, J. G. and J. N. Pitts (1966). "Photochemistry." John Wiley, New York.

Campbell, I. M. (1986). "Energy and the Atmosphere," 2nd edn. John Wiley, New York.

Chase, M. W., C. A. Davies, J. R. Downey, D. J. Frurip, R. A. McDonald, and A. N. Syverud (1985). JANAF thermodynamic tables. *J. Phys. Chem. Ref. Data* **14**, 1–1856 (suppl. 1).

Finlayson-Pitts, B. J. and J. N. Pitts (1986). "Atmospheric Chemistry." John Wiley, New York.

Garrells, R. M. and C. L. Christ (1965). "Solutions, Minerals, and Equilibria." Harper and Row, New York.

Krauskopf, K. B. (1967). "Introduction to Geochemistry." McGraw-Hill, New York.

Laidler, K. J. (1965). "Chemical Kinetics," 2nd edn. McGraw-Hill, New York.

Lerman, A. (1979). "Geochemical Processes." John Wiley, New York.

Levine, I. N. (1983). "Physical Chemistry." McGraw-Hill, New York.

Moore, J. W. and R. G. Pearson (1981). "Kinetics and Mechanism." John Wiley, New York.

National Bureau of Standards (1952). "Selected Values of Chemical Thermodynamic Properties," Circular 500, US Department of Commerce, Washington, D.C.

National Bureau of Standards (1956). "Tables of Chemical Kinetics," Circular 510, US Department of Commerce, Washington, D.C.

National Research Council (1926). "International Critical Tables." McGraw-Hill, New York.

Seinfeld, J. H. (1986). "Atmospheric Chemistry and Physics of Air Pollution." John Wiley, New York.

Stull, D. R., E. F. Westrum, and G. C. Sinke (1969). "The Chemical Thermodynamics of Organic Compounds." John Wiley, New York.

Trotman-Dickenson, A. F. and G. S. Milne (1967). "Tables of Bimolecular Gas Reactions," National Standard Reference Data Series, NBS 9, US Department of Commerce, Washington, D.C.

Wagman, D. D., W. H. Evans, V. B. Parker, R. H. Schumm, I. Halow, S. M. Bailey, K. L. Churney, and R. L. Nuttall (1982). The NBS tables of chemical thermodynamic properties. *J. Phys. Chem. Ref. Data* **11** (suppl. 2).

Walker, J. C. G. (1977). "Evolution of the Atmosphere." Macmillan, New York.

SCOTT W. STARRATT

6

Tectonic Processes, Continental Freeboard, and the Rate-controlling Step for Continental Denudation

Robert F. Stallard

6.1 Introduction

It is now widely accepted that the compositions of the atmosphere and world ocean are dynamically controlled. The atmosphere and the ocean are homogeneous with respect to most major chemical constituents. Each can be viewed as a reservoir for which processes add material, remove material, and alter the compositions of substances internally. The history of the relative rates of these processes determines the concentrations of substances within a reservoir and the rate at which concentrations change. Commonly, only a few processes predominate in determining the flux of a substance between reservoirs. In turn, particular features of a predominant process are often critical in controlling the flux of a phase through that process. These are rate-controlling steps.

Weathering and erosion of bedrock are fundamental to the geochemical cycles that control the composition of the atmosphere, the oceans, and sedimentary rock. Consequently, identification of rate-controlling aspects of the erosion process is crucial to the analysis of global biogeochemical cycles. This chapter sketches how rates of erosion are controlled at different spatial and temporal scales, starting with the surfaces of mineral grains and expanding to whole continents. Most of the discussion focuses on how tectonic processes, continental freeboard, climate, and the susceptibility of various lithologies to erosion influence the rate of continental denudation. Furthermore, to understand the development of the Earth through time, atmospheric properties that might affect erosion rates, such as temperature, moisture availability, oxygen and carbon dioxide partial pressure, etc., are examined.

Most of the examples in the following discussion are from the humid tropics. There are several fundamental reasons for having chosen the humid tropics as a study environment. Weathering and erosion are important to global geochemical cycles because of the chemical reactions that occur during weathering. Most of these reactions involve water, and although some are simple inorganic reactions, many are biologically mediated. The high temperatures, moist conditions, and luxuriant vegetation of the humid tropics are ideal for rapid chemical weathering. Moreover, erosional processes in this zone are particularly important on a global scale. According to Meybeck (1979), the humid tropics presently occupy about 25% of the Earth's land surface and supply about 65% of the dissolved silica and 38% of the ionic load delivered by rivers to the oceans. Data compiled by Milliman and Meade (1983) indicate that the same region contributes about 50% of the total river solid load to the ocean. Both studies demonstrate that the bulk of this material is derived from active orogenic belts and island arcs. Finally, the equatorial region was least affected by the climatic fluctuations of the ice ages.

The study of chemical weathering in drier and

Global Biogeochemical Cycles
ISBN 0-12-147685-5

cooler regions of the Earth is beset by numerous complications and ambiguities. In dry regions, chemical weathering is very slow, and commonly many of the characteristics of landforms and soils seem to have been inherited from earlier, moister times. Moreover, materials derived from chemical weathering and atmospheric deposition often precipitate out as mineral deposits in soils. Cooler climates have been strongly affected by the repeated glaciations of the Pleistocene. In glaciated areas, soils are exceedingly young, and soils of adjacent non-glaciated regions were affected by a variety of periglacial processes. Still wider regions near ice sheets were veneered with loess.

This chapter examines climatic and tectonic controls on erosion in the tropics and the implications of these observations regarding the composition of erosion products in general. The role of glaciations in continental denudation will briefly be examined.

6.2 Erosion: A Capsule Summary

The energy that powers terrestrial processes is derived primarily from the Sun and from the Earth's internal heat production (mostly radioactive decay). Solar energy drives atmospheric motions, ocean circulation (tidal energy is minor), the hydrologic cycle, and photosynthesis. The Earth's internal heat drives convection which is largely manifested at the Earth's surface by the characteristic deformation and volcanism associated with plate tectonics, and by the hot-spot volcanism associated with rising plumes of especially hot mantle material.

Erosion is the process that tears down the subaerial landforms constructed by crustal deformation and volcanism. Matter derived from the Earth's crust generally moves from high elevation on land to low elevations in the ocean along pathways (rivers, glaciers, wind, etc.) that are often long and complex. There are many pauses on the way (e.g. alluvial deposits and lakes, glacial moraines, and sand dunes), during which compositions can change. Erosion is typically most rapid in mountainous regions and coastal areas, and is slowest in flatlands. Matter is transported largely by rivers and to a lesser extent by winds and glaciers. During erosion, crustal material is initially mobilized by weathering.

6.2.1 Weathering and Crustal Breakdown

Weathering involves the chemical and physical breakdown of bedrock through its interaction with the hydrologic and atmospheric cycles. Chemical weathering is the breakdown of bedrock by chemical reactions. The products may include dissolved solids and new minerals (usually clays). Physical weathering is the physical breakdown of unweathered or partially chemically weathered bedrock; some old (primary) minerals remain intact. During erosion, these two types of weathering frequently operate in tandem. Chemical weathering weakens the rock; physical weathering finishes it off, making it available for transport processes. There are a few "rules of thumb":

1. Chemical weathering is more important in warm, moist regions, whereas physical weathering is more important in cold, dry areas.
2. Contributions made by chemical weathering are greater in regions where there is much vegetation.
3. Contributions made by physical weathering are much greater in steep terrains (i.e. more primary minerals remain) and overall weathering rates are higher.

The compositions of dissolved and solid erosion products are initially determined by the stability of the bedrock minerals at the site of weathering. The composition will change during transport to the ocean as the result of further weathering during storage in alluvial deposits or during authigenic mineral formation in lakes, alluvial soils, and groundwaters.

All minerals can be weathered chemically. The susceptibility of minerals varies considerably, however. This is illustrated nicely by the relative mineral stability for weathering under tropical conditions (Table 6-1). Note how stability appears to be closely related to composition. Similar mineral groups cluster together. For igneous and metamorphic minerals, stability is almost the reverse of the Bowen's reaction series (Goldich, 1938); the first minerals to crystallize out of a magma are the most susceptible to weathering. Note also that the vulnerability to chemical weathering of ionically bonded chemical sediments is in reverse order to the typical marine evaporite sequence (see Holland, 1974).

Minerals may weather congruently to produce

Table 6-1 Mineral stability in tropical soils[a]

MOST STABLE	
Quartz >	Pure silica
K-Feldspar, Micas > Na-Feldspar > Ca-Feldspar, Amphiboles > Pyroxenes, Chlorite >	Igneous and metamorphic aluminosilicates
Dolomite > Calcite >	Carbonate minerals
Gypsum, Anhydrite ≫ Halite	Evaporite minerals
	LEAST STABLE

[a] Adapted from Stallard (1985).
Note: Na-feldspar + Ca-feldspar = plagioclase.

only dissolved weathering products or incongruently to produce both dissolved cations (Na^+, K^+, Mg^{2+}, Ca^{2+}), silica [$Si(OH)_4$], and solid products that are cation-depleted and usually silica-depleted. Common minerals that always weather congruently are halite, anhydrite, gypsum, aragonite, calcite, dolomite, and quartz. Halite, anhydrite, and gypsum are so unstable that they almost never occur in the solid load of rivers. Calcite and dolomite sometimes occur in alkaline rivers that are supersaturated with respect to these minerals, e.g. the Yellow River of China. Quartz is the most persistent of all common primary minerals and occurs in most river sediments. Minerals that contain iron and aluminium usually weather incongruently to produce clays and iron/aluminium sesquioxides (oxides and hydroxides of Fe^{3+} and Al^{3+}). Magnesium, potassium, and to a lesser extent calcium and sodium are retained in cation-rich clays such as the smectites, vermiculites, illites, and chlorites.

The most stable minerals are often physically eroded before they have a chance to decompose chemically. Minerals that decompose contribute to the dissolved load in rivers, and their solid chemical-weathering products contribute to the secondary minerals in the solid load. The secondary minerals and the more stable primary minerals are the most important constituents of clastic sedimentary rocks. Consequently, the secondary minerals of one cycle of erosion are often the primary minerals of a subsequent cycle. When a weakly cemented sedimentary rock consisting of chemically stable minerals is exposed to weathering, it often breaks down physically, and although erosion might be very rapid, the solid products undergo little chemical alteration. If physical erosion processes are not very intense, most of the erosion products are dissolved, and stable primary minerals and secondary minerals will accumulate over the bedrock to form soil. Additional details on mineral transformations by weathering are provided in Chapter 7.

6.2.2 Weathering and the Atmospheric Gases

It is chemical weathering that makes continental denudation so important to geochemical cycles. Solution transport by rivers into the ocean is the largest single flux of many elements into the seawater reservoir. Carbon dioxide and molecular oxygen, compounds of fundamental biogeochemical importance, are consumed by weathering reactions involving primary minerals. The rate of consumption by weathering reactions, although much slower than the rate of the cycling through organisms, may be important in controlling the long-term concentrations of CO_2 and O_2 in the atmosphere and ocean (Garrels and Lerman, 1981; Berner *et al.*, 1984).

The consumption of CO_2 during weathering is indirect. Most members of two important classes of minerals, the silicates and the carbonates, consume hydrogen ions and release alkali (Na^+, K^+) and alkaline earth (Ca^{2+}, Mg^{2+}) cations during weathering. The primary proton donor is carbonic acid H_2CO_3, formed by the hydrolysis of carbon dioxide. An accumulation of carbonate and bicarbonate in solution results. Organic acids, which are photosynthetic derivatives of CO_2, are often important in shallow soil horizons, and mineral acids (sulfuric, nitric, hydrochloric) from atmospheric or rock weathering sources also can be important. The net effect of the reactions is to consume protons and, when weak acids are the proton donors, to produce alkalinity. Carbon dioxide is returned to the atmosphere primarily by carbonate deposition, metamorphic reactions, volcanism, and mid-ocean ridge hydrothermal circulation.

Oxygen is consumed largely by the oxidation of reduced iron in silicates and sulfides, of sulfur in sulfides, and of organic carbon in sedimentary rocks. Oxygen is returned to the atmosphere as a result of biological fixation of reduced carbon, sulfur, and iron and the subsequent burial of these compounds. Not all chemical weathering reactions involve CO_2 or O_2. Certain minerals, such as quartz (SiO_2) and the major evaporite minerals (NaCl, $CaSO_4$, and $CaSO_4 \cdot H_2O$) dissolve without reacting with anything except water.

6.3 Soils and the Local Weathering Environment

The initial partitioning of elements into dissolved and solid phases during continental denudation is controlled by the chemical and physical processes associated with soil genesis. The description of weathering processes as they occur in soil profiles is discussed in Chapter 7. A vast body of observational data, not to be reviewed here, has shown that soils are remarkably complex on the local scale. Soil systems are neither physically nor chemically homogeneous. Both solid and dissolved matter is washed down through soil profiles; capillary action operating in tandem with evaporation can lead to upward transport of dissolved material; frost action, root growth, tree falls, and burrowing can mix material within a profile; transport of fluids in soils on non-porous substrates is largely lateral; soil porosity is controlled by both chemical and physical processes; freezing can have profound effects on the structure of soil and the composition of residual fluids; biochemical processes generate many thermodynamically unstable reactants; the chemical activities of many chemical species (e.g. H_2O, CO_2, O_2) change tremendously on time-scales that are infinitesimally short geologically (minutes to days); and reactions frequently occur in thin films and on grain services.

6.3.1 Weathering Reaction Kinetics

It has been said that one of the reasons why weathering reaction kinetics has not been studied in more detail is that the reactions are too slow to be studied by graduate students. The as yet insurmountable problem of experimentally or computationally modeling the soil environment is another factor.

Several simple experimental systems that simulate some aspect of the groundwater environment have been used to study the breakdown of individual minerals. These kinetics studies have encompassed quartz (Brantley *et al.*, 1986), feldspars (Holdren and Berner, 1979; Holdren and Speyer, 1985), pyroxenes and amphiboles (Berner and Schott, 1982; Schott and Berner, 1985), carbonates (Berner, 1978), and glasses (White, 1983). The relative stability observed in laboratory weathering is consistent with field-based observations; however, experimental rates appear to be faster than those in natural systems.

Electron microscopy of mineral grains from laboratory experiments and from natural soils provides some interesting observations about reaction mechanisms and rate controls during weathering. A continuum of rate-limiting mechanisms can be defined between reactions that are transport (diffusion)-controlled and those that are surface reaction-controlled (Berner, 1978). In the former case, reactions are limited by diffusion of solution products away from the crystal surfaces. This situation occurs where reactions are sufficiently rapid that fluid concentrations near the primary mineral are close to equilibrium values for the driving reaction, which need not be the final equilibrium for the system. Under a microscope, grain surfaces appear smooth with rounded edges (Fig. 6-1). In the latter case, reaction rates are

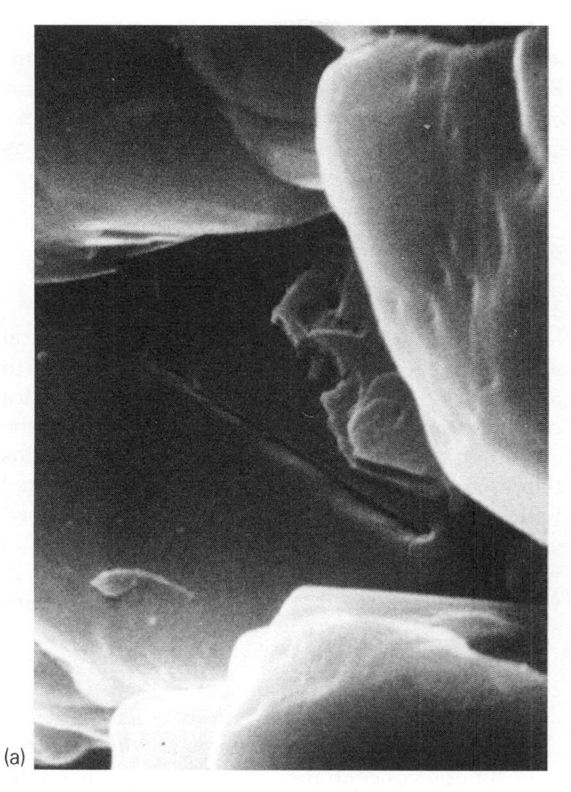

Fig. 6-1 Electron micrographs of quartz grains from a soil profile described in (a) Stallard (1985) and (b) Brantley *et al.* (1986). The quartz grains in (a) are from a sample close to fresh bedrock near the bottom of the profile; the sample in (b) is from near the top of the profile. Note that the surfaces of the grains in (a) are smooth, whereas those in (b) are deeply etched. Presumably, the etched grains are dissolving in soil waters that are strongly undersaturated with respect to quartz. Such etching indicates that reaction rates are affected by variations in surface energy. The smooth grains are probably dissolving at or close to equilibrium conditions. Reprinted with permission from the author.

controlled by local surface energy on the crystals. Surface reaction control is important under circumstances where surrounding fluids are decidedly undersaturated relative to the equilibria that are driving the weathering reaction. Surfaces appear rough and pitted, commonly on zones of obvious crystallographic defects (Fig. 6-1; see also Gilkes *et al.*, 1980; and Eggleton and Buseck, 1980). Most studies of primary minerals in natural soils show extensive pitting, which is suggestive of surface reaction control.

There are several important caveats to the use of this observational model. There is, for example, a considerable range of undersaturation, down to 50% in the case of the quartz, where surface defects are not especially reactive, and crystals appear smooth (Brantly *et al.*, 1986). The formation of weathering rinds and the build-up of clays and other secondary weathering products in the immediate vicinity of the weathering grains can strongly affect the transport processes. This aspect of weathering has yet to be realistically examined in the laboratory. Finally, in soils, fluid compositions change rapidly and a wide range of reactions may be important during weathering (Drever and Smith, 1978; Eberl *et al.*, 1986; Herbillon and Nahon, 1988; Trolard and Tardy, 1989; van Breeman, 1988a, b).

6.3.2 *Variable Chemistry of Soil Fluids: A Complicating Factor*

Weathering rates are most sensitive to the throughput of water. In soils, this is a decidedly discontinuous process. Typically, water flows through soil following rainfall or snowmelt. Once saturated, the flux of water is largely dependent on the physical properties of the soil and not on the rate of supply. Water that cannot be accommodated by flow through the soil, because of soil saturation or a high rate of supply, is re-routed into overland run-off and interacts minimally with primary minerals. Following wetting, typical soils drain rapidly. At most times, soils are drying. This scenario is evident in discharge curves for small river catchments, where precipitation and snowmelt events show up as spikes over a very long background discharge which reflects slow groundwater inputs.

Groundwater environments can be represented as a simple flow-through system. For the situation where chemical weathering of mineral grains is transport-controlled, the weathering rate of a mineral should be directly dependent on the rate of throughput of water. For the situation where rates are controlled by surface reactions, weathering rate should be independent of the rate of throughput of water.

Soil water flow is decidedly episodic. During dry times, the water solutions in the soil are probably fairly concentrated and not very reactive. Time-averaged reaction rates should be roughly proportional to the fraction of time reacting minerals are in contact with thermodynamically undersaturated (and reactive) water. In a study of the relationship between denudation rate and run-off for rivers draining igneous and metamorphic rock in Kenya, Dunne (1978) obtains the relationship (denudation rate in tons/km^2 year) = 0.28 (run-off in mm/year)$^{0.66}$.

The fact that soils dry between episodes of water flow complicates weathering reaction scenarios. During drying, solute concentrations in water films increase, and the areal extent of the films decrease. The chemical activity of water drops. As a result, chemical equilibria that might be important for controlling weathering reactions within wet soils are replaced by new equilibria reflecting the elevated concentration of solutes. A different suite of clays and sesquioxides might become stable and silica (opal), calcite, and various evaporite minerals can precipitate. During subsequent wetting, both primary minerals and secondary minerals formed under drying conditions may react. Features such as etch pits may form during episodes of wetting and thermodynamic instability, even if they would not normally form under average or typical conditions. Moreover, secondary minerals that formed under dry conditions may persist through wetting cycles. The formation of calcium carbonate nodules in soils (caliche) is an example of this. The mineral constituents of such soils may not be at equilibrium. Freezing, which also produces residual fluids with elevated concentrations of dissolved solutes, presumably does not have as significant an effect as drying because lower temperatures are involved.

6.3.3 *Local Responses to Atmospheric Variables*

The influence that variations of temperature and levels of atmospheric CO_2 and O_2 have on chemical weathering are more subtle. Temperature appears to have a direct effect on weathering rate. The silica concentration of rivers (Meybeck, 1979, 1987) and the alkalinity of groundwaters in carbonate terrains (Harmon *et al.*, 1975) are both positively correlated with temperature variations. It is not clear, however, whether temperature-related variations in weathering rates are largely due to variations in vegetational activity that parallel temperature variations.

Partial pressures of O_2 and CO_2 in soils are controlled largely by soil biology. Oxygen is consumed and CO_2 is produced in soils by decay and root respiration. Plausible variations of the atmospheric concentration of these gases has little effect on their partial pressures in soils. The rate of the hydrologic cycle, however, is thought to respond directly to global mean temperature and is therefore indirectly sensitive to the partial pressure of atmospheric CO_2, which as a "greenhouse" gas can affect global mean temperatures (see Berner *et al.*, 1983).

In summary, of all the local variables that can affect weathering rate, the supply of water is clearly the most important. Biology is very important as the supplier of proton donors and complexing agents for weathering reactions, as a mediator in the moisture budget for the soil, and as a controller of soil structure. Beyond suggesting that surface reactions are important in controlling the weathering rate for most minerals and confirming mineral stability sequences, laboratory models of weathering chemistry have offered little to the study of weathering processes.

6.4 Slope Processes and the Susceptibility of Lithologies to Erosion

Weathering, atmospheric deposition, and the fixation of atmospheric gases are the ultimate sources of the material transported by rivers. These processes operate over the surface of the river catchment, and the resultant water and weathering products must be transported downslope before arriving in a channel. Examination of erosion processes on hill slopes provides insight into how chemical weathering rates are controlled at an intermediate spatial scale and how chemical elements are partitioned between the dissolved and solid loads of rivers.

The erosion process on slopes can be envisioned as a continuum between the weathering-limited and transport-limited extremes (Carson and Kirkby, 1972; Stallard, 1985). Erosion is classified as transport-limited when the rate of supply of material by weathering exceeds the capacity of transport processes to remove the material. Erosion is weathering-limited when the capacity of the transport process exceeds the rate at which material is generated by weathering. These two styles of erosion represent an interesting parallel to controls of weathering reaction rates on mineral surfaces, discussed earlier, wherein a similar continuum was defined between surface reaction control and transport (diffusion) control (Stallard, 1988).

The weathering and transport processes associated with either end of the continuum are quite different. Where erosion is weathering-limited, erosion rate is controlled by the rate at which chemical and physical weathering can supply dissolved or loose particulate material. In essence, erosion rates are controlled by susceptibility to weathering. Soils are thin, because loose material moves rapidly downslope. Much of this material is only partially weathered, because most rocks lose their structural integrity before they are completely chemically decomposed. Processes characteristic of weathering-limited regions include rockfalls, landslides, or anything that tends to maintain a fresh or slightly weathered rock surface (Table 6-2). These processes often operate at a threshold slope angle. In humid climates, weathering-limited conditions are associated with thin soils and steep straight slopes which often undergo parallel retreat at a threshold angle (Fig. 6-2).

In contrast, under transport-limited conditions, weathering rates are ultimately limited by the formation of soils that are sufficiently thick or impermeable to restrict free access by water to unweathered material. Erosion rates are low, and soils and solid weathering products are cation-deficient. In regions where transport-limited erosion predominates, soils are thick and slopes are slight and convexo-concave (Fig. 6-2b). With time, these slopes tend toward increasing flatness. Soil creep is a process typical of transport-limited situations. Most soil mass movement and wash processes, however, are intermediate between weathering-limited and transport-limited in character.

Table 6-2 Erosion regimes: Features and processes associated with transport-limited and weathering-limited erosion

Transport-limited	Weather-limited
Thick soils	Thin or no soils
Slight slopes that are convexo-concave	Steep slopes that are straight and at a threshold angle
Erosion rates independent of lithology	Erosion rates depend on bedrock susceptibility
Processes:	Processes:
soil creep	rock slides
removal of dissolved phases	strong sheet wash
	soil avalanches
	removal of dissolved phases

Notes: Weathering-limited – potential transport processes greater than weathering supply. Transport-limited – supply by weathering greater than the capacity of transport processes to remove material (until weathering is slowed by feedback).

Erosion associated with chemical weathering caused by circulating soil/ground water is intermediate between being transport-limited and weathering-limited in character. Chemical erosion would be transport-limited whenever reactions at the mineral-grain level are also transport-limited. As discussed earlier, this occurs when the flushing rate for water is sufficiently slow that equilibrium is reached with respect to some controlling reaction. Chemical weathering of carbonate rocks is probably transport-limited as soil and groundwaters are nearly saturated with respect to carbonate minerals (Holland *et al.*, 1964; Langmuir, 1971). In the case of silicate weathering, transport-limited erosion would occur where the silicates are particularly unstable or where water movement is restricted by low porosities or lack of hydraulic head in soils or bedrock.

6.4.1 Soils, Slopes, Vegetation, and Weathering Rate

For a given set of conditions (lithology, climate, slope, etc.), there is presumably an optimum soil

(a)

(b)

Fig. 6-2 Photographs illustrating (a) long straight slopes in the Andes and (b) convexo-concave slopes transitional into very flat terrain. The view of the Andes is taken from Huayna Picchu, the small peak next to Machu Picchu. The view is up the valley of a small tributary to the Urubamba River. Note that the V-shaped valley formed by fluvial erosion becomes a U-shaped glacial valley at its highest end. The glacier was active during the last ice age. (b) Taken from an airplane, a photograph of the Guayana Shield in southern Venezuela. In the foreground is the Orinoco River downriver from the Casiquiare Canal, a natural channel that connects the Orinoco and the Amazon River systems. The hills in the foreground are granitic and about 200 m high. Reprinted with permission from the author.

thickness that maximizes the rate of bedrock weathering (Fig. 6-3); (Carson and Kirkby, 1972; Stallard, 1985). For less than optimum soil thicknesses, there is insufficient pore volume in the soil to accept all the water supplied by precipitation and downhill flow. Excess water runs off and does not interact with the subsurface soil and bedrock. In contrast, water infiltrates and circulates slowly through thick soils (especially where forested). If profile thicknesses greatly exceed the optimum, long residence times for water at the base of the profile lead to a reduction of weathering rate as a result of equilibration with respect to secondary phases. With increasing thickness, soil profiles also become more structured with definite horizons. Some of these can be rather impermeable. Water is routed through weathered material, and weathering rates are thereby reduced.

Figure 6-3 portrays a hypothetical model of how chemical weathering and transport processes interact to control soil thicknesses. The relationship between soil thickness and rate at which chemical weathering can generate loose solid material is indicated by the solid curve. The rate at which transport processes can potentially remove loose solid weathering products is indicated by horizontal dotted lines. The rate of generation by chemical weathering initially increases as more water has the opportunity to interact with bedrock in the soil. As soil thickens, the optimum thickness for maximum rates of chemical weathering is exceeded, and weathering rates decrease with increasing soil thickness (E–F). The decrease presumably occurs because the rate of chemical weathering is limited by diminished access of reactive waters to unweathered materials. If the potential transport rate exceeds weathering rate, loose soil material should not remain on the slope. If weathering rate exceeds potential transport rate, soils develop. As soils develop, weathering rates vary, leading to an evolution of the soil profile.

For soil profiles that are less than the optimum thickness, there is a destabilizing feedback between soil thickness and weathering rate. Assume that a

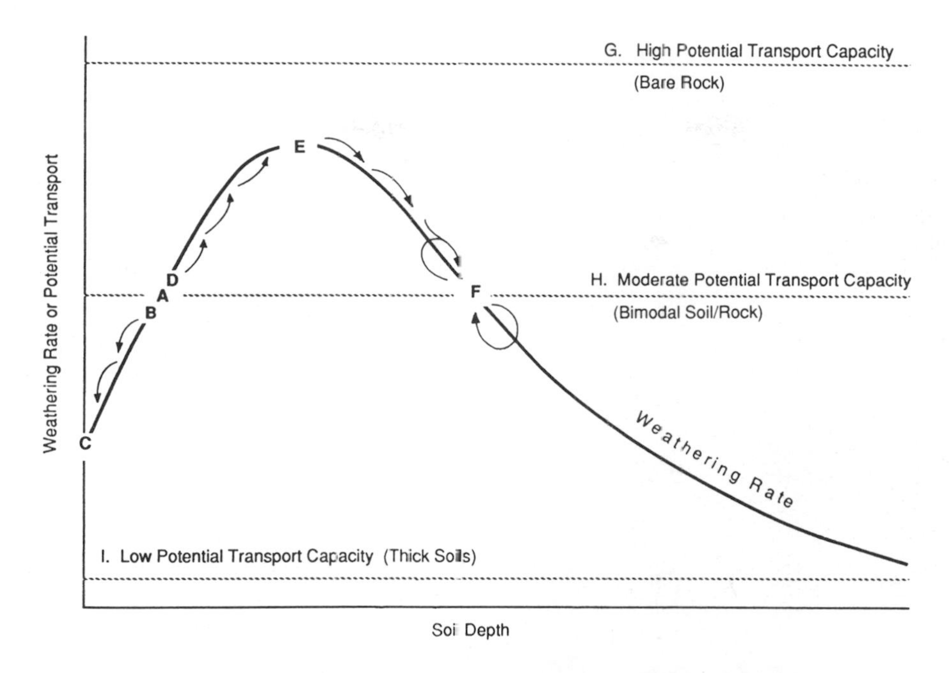

Fig. 6-3 Conceptual model to describe the interaction between chemical weathering of bedrock and downslope transport of solid erosion products. It is assumed that chemical weathering is required to generate loose solid erosion products of the bedrock. The solid curve portrays a hypothetical relationship between soil thickness and rate of chemical weathering of bedrock; the dotted lines correspond to different potential transport capacities. Low potential transport capacity is expected on a flat terrain, whereas high transport is expected on steep terrain. For moderate capacity, C and F are equilibrium points. Modified from Stallard (1985) with the permission of Kluwer Academic Publishers.

thin soil is in a dynamic equilibrium such that weathering inputs balance transport losses (Fig. 6–3, A). Weathering rate can be reduced relative to transport rate by either increasing the strength of transport processes or by thinning the soil (B). Either way, transport removal would exceed weathering supply, and the soil would continue to thin. Eventually, only a hard cohesive saprolite or bedrock would remain (C). If the soil is thickened, or if the capacity of transport processes is reduced, the soil would tend to accumulate (D–E). Finally, a situation involving stabilizing feedback occurs (F), and a thick soil forms such that transport of loose material balances weathering inputs. If soil thickens beyond point F, weathering rates decrease, and transport processes restore the soil thickness back to the stable value. This model suggests that soil distributions should be distinctly bimodal – either thin, weathering-limited (G), or thick, transport-limited (I). For some intermediate values of potential transport rate (H), either hard rock or a moderately thick soil could exist.

When soil thickness is at the stable value (F), erosion is transport-limited. Chemical weathering is also transport-limited. This is, however, not because of reaction kinetics; instead, this limitation is primarily controlled by physical factors, specifically restricted access of water to the primary minerals.

The effects of vegetation are complex. Vegetation reduces short-term physical erosion by sheltering and anchoring soils. This effect is equivalent to reducing potential transport in Fig. 6-3. A cover of vegetation does not necessarily reduce denudation rates. Vegetation can maintain a thin veneer of soil on steep slopes, particularly under wet conditions. As the soil thickens, it often becomes unstable, detaches, and slides down slope (Garwood *et al.*, 1979; Pain, 1972; Scott, 1975a; Scott and Street, 1976; Stallard, 1985; Wentworth, 1943; see Fig. 6-4). Under such circumstances, weathering rates can be

Fig. 6-4 Photograph of landslides (soil avalanches) that occurred following earthquakes in Panama on July 17, 1976, near Jaque. In the background is a bay of the Pacific Ocean. The effects of this earthquake are described by Garwood *et al.* (1979), who estimated that about 42 km^2 (about 10%) of the region near the epicenter of the earthquake was devegetated. The bedrock is mostly island-arc basalts and andesites. (Photography by N. C. Garwood.)

exceptionally high because of the extra moisture and bioacids; likewise, denudation rates are very high because of the continuous resupply of fresh rock. The effect of erosion following fires, tree falls, and land clearing on slopes is similar to that of slides (Scott, 1975b; Stallard, 1985). On slight slopes, over extended periods of time (up to millions of years), vegetation may reduce weathering rates by allowing very thick soils to accumulate. For a given soil thickness, however, weathering should be faster because of the supply of bioacids.

6.4.2 Elemental Partitioning: The Role of Slope Processes

There are two principal ways to selectively partition different elements between the dissolved and solid loads in rivers: by selective chemical weathering of particular primary minerals, and by the formation of secondary phases that are enriched or depleted in certain elements, relative to bedrock (Stallard, 1985).

Different styles of erosion are associated with different degrees of partitioning of elements between dissolved and solid load. As rocks weather chemically, they lose their structural integrity. When only kaolinite, gibbsite or other cation-depleted phases form, as commonly happens with transport-limited erosion, cation ratios in solution should match those in bedrock. During weathering-limited erosion, unstable primary minerals are selectively removed, causing elemental partitioning. For example, in moist vegetated areas on crystalline rocks or indurated sediments, solifluction, soil avalanching, and sheet run-off remove weakly cohesive material (solum and soft saprolite), leaving a cohesive hard saprolite behind (Stallard and Edmond, 1983; Stallard, 1985). Where transport processes are sufficiently intense, more stable minerals such as zircon, quartz, potassium feldspars, and micas survive chemical weathering and are eroded. Some of these resistant minerals contain substantial K and Mg, but none contain much Na or Ca. Thus, K and Mg are enriched relative to Na and Ca in solid erosion products; Na and Ca are enriched in solution relative to K and Mg (Fig. 6-5). If Mg is incorporated into the lattice of many secondary clays, as often happens, this further accentuates its retention in bulk solids.

Dissolved phases are assumed to best reflect the weathering processes occurring at the erosion site, because water is not usually stored for long periods (many years) in soils or during fluvial transport. Solid products, however, can weather chemically during transport downslope and through fluvial systems. Solids degrade when they accumulate at the base of slopes (colluvium) or during storage on floodplains (alluvium); this obviously affects dissolved components. Soil solutions also evolve as they flow downslope. Contact with fresher materials is prolonged, and evapotranspiration can concentrate solutions (Carson and Kirkby, 1972; Stallard, 1988). Both of these phenomena can cause the formation of new suites of clays and sesquioxides, and the precipitation of carbonates.

Aggradation or degradation of biomass or soil reservoirs may also produce effects that appear to be fractionation. This is because the elemental ratios in vegetation or soil reservoirs can be very different from those of bedrock. Sufficiently large and rapid changes in these reservoirs are sometimes evident in

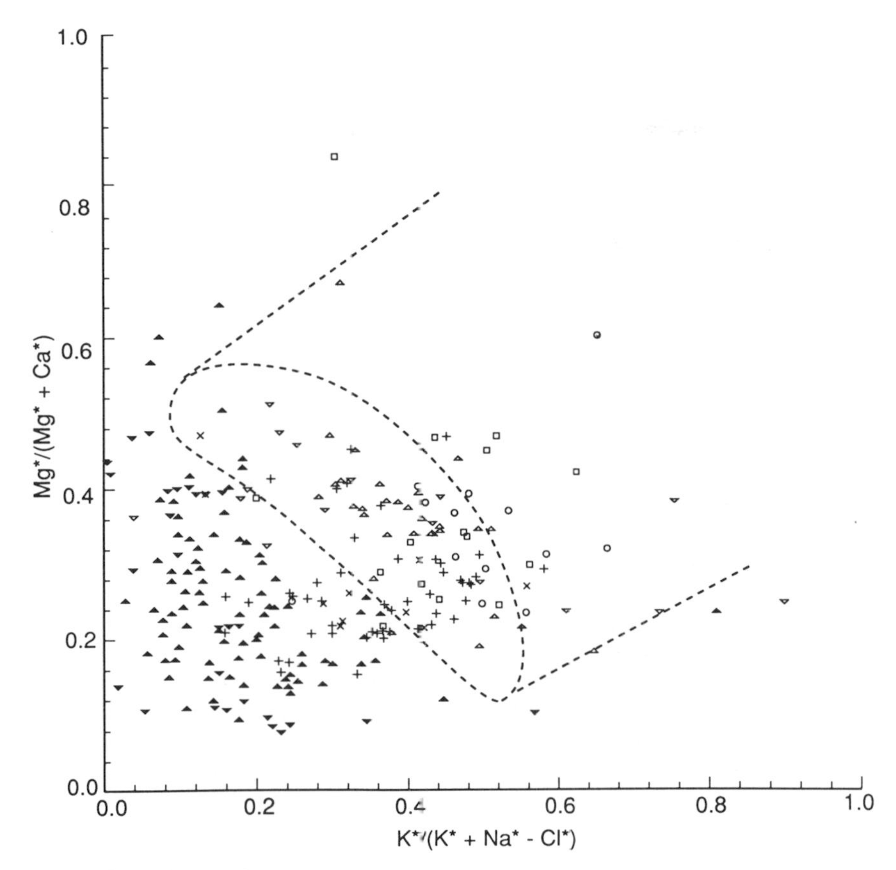

Fig. 6-5 Relation of $Mg^{2+}/Mg^{2+} + Ca^{2+}$) to $K^+/(K^+ + Na^+ - Cl^-)$ for dissolved material in surface waters of the Orinoco River and Amazon River basins, the Isthmus of Panama, and the Island of Taiwan and for rock types representative of bedrock in these basins. Symbols: ▲, samples from rivers that drain mostly mountain belts developed on felsic continental crust; ▼, samples from rivers that drain mostly island-arc mountain belts developed on mafic oceanic crust; +, samples from rivers that originate in continental mountain belts, but that drain large areas of craton; x, rivers that drain alluvial and non-alluvial sedimentary rocks in foreland and intracratonic basins; □, rivers that drain only alluvial sediments in foreland and intracratonic basins; △, rivers that drain hilly to mountainous shield on felsic crust; ▽, rivers that drain hilly to mountainous shield on mafic crust or deeply eroded island arc; ○, rivers that drain peneplaned shield. The dashed oval represents the range of analyses for common igneous rocks; the dashed lines that extend away from the igneous rock field represent the composition field for common sedimentary rocks. Adapted from Stallard (1988) with the permission of Kluwer Academic Publishers.

river chemistry. For example, the uptake and release of potassium in association with the seasonal growth and loss of leaves can affect the composition of streams that drain temperate deciduous forests (Likens *et al.*, 1977).

6.5 Landforms, Tectonism, Sea Level, and Erosion

On a larger scale, landscape development reflects those mechanisms that expose bedrock, weather it, and transport the weathering products away. Present and past tectonism, geology, climate, soils, and vegetation are all important to landscape evolution. These factors often operate in tandem to produce characteristic landforms that presumably integrate the effects of both episodic and continuous processes over considerable periods of time.

Many important erosion-related phenomena are episodic and infrequent, such as flash floods, landslides, and glaciations, whereas others such as orogenesis and soil formation involve time-scales that exceed those of major climate fluctuations. In either case, the time-scale of human existence has been too short to make adequate observations. Consequently, it is difficult to estimate directly the rates or characterize the effects of such phenomena on erosion products. The key to understanding weathering and erosion, on a continental scale, is to decipher the relationship between landforms, the processes that produce them, and the chemistry and discharge of river-borne materials.

To a first approximation, landscape formation processes can be viewed as being controlled by climatic and tectonic factors. Climate delimits where different types of weathering can occur, while tectonics controls how the effects of weathering are expressed. The role of tectonic setting is emphasized in this discussion because tectonic history is often critical to controlling denudation rates and the composition of erosion products. The object of the discussion is not to focus on the particulars of landscape development. Instead, landforms are primarily used to distinguish between different erosional regimes and to identify features that are useful for characterizing denudation and uplift rates, especially indicators of past sea level. The linkage between tectonic processes and landforms is embodied in the concept of a morphotectonic region.

6.5.1 The Effects of Erosional Regime

Many landforms have a convex upper slope, a straight main slope, and a concave lower slope. Carson and Kirkby (1972) argue that the main slope is dominated by weathering-limited processes, whereas the upper and lower slopes are primarily areas of transport-limited erosion or even deposition in the case of the lower slope. If the overall slope is largely transport-limited, it will undergo parallel retreat at the threshold angle. With parallel retreat, topographic form is maintained and characteristic landscapes are thereby generated.

The weathering regime exerts a major control on the production rate and the composition of erosion products from different lithologies (Stallard, 1985, 1988). For weathering-limited conditions, erosion rate is controlled by the susceptibility of the bedrock to weathering. The tectonic history, physical properties (porosity, shear strength, jointing, etc.), and chemical properties of the bedrock exert a major influence. For a given climate, within a specific catchment, each rock type should contribute an amount of material to river transport that is proportional to both its extent of exposure and its susceptibility to weathering. (Deep groundwater transport important in karst terrains would contribute additionally.) Solid erosion products should include abundant partially weathered (cation-rich) rock and mineral fragments. If the bedrock is especially physically unstable, it will contribute strongly to the solid load of rivers; likewise, chemically unstable bedrock will contribute strongly to the dissolved load.

In transport-limited conditions, however, susceptibility is not so important due to the isolating effect of thick soils. In the extreme situation of a flat landscape, the erosional contribution by a particular rock type should be related only to the area exposed. Solids would be cation-deficient.

6.5.2 The Effects of Deposition, Storage, and Burial

Sediment-laden rivers flowing over flat terrain commonly develop extensive floodplains. At times, floodplains coalesce into broad depositional alluvial plains such as the Llanos of South America. The sediments in those deposits weather chemically. Less stable minerals in the sediment are broken down and alluvial soils develop. Eventually, only the most stable minerals such as quartz remain, and the clays are transformed into cation-deficient varieties. Sediment in such rivers, especially the sand, may go through many cycles of deposition, weathering, and erosion before it is transported out of the system. Compositionally, this sediment resembles that derived from transport-limited erosion. Elemental fractionation between the original bedrock and erosion products still occurs because of the permanent burial of some cation-rich material and the uninterrupted transport of much of the fine-grained suspended sediment out of the system (Johnsson *et al.*, 1988; Stallard, 1985, 1988).

6.5.3 The Effects of Tectonic Processes

Tectonic processes provide the material of which landscapes are made. Tectonism produces uplift followed by erosion, or subsidence followed by deposition. Sediment storage makes the calculation of denudation rates rather time-scale-dependent.

Sediment deposited at one time can be re-eroded through increases in river discharge caused by wetter climates or lowering of base level, perhaps brought about by uplift. What might be considered eroding bedrock on one scale can be just stored sediment in transit to the ocean on another, longer scale. In the following sections, tectonically active and tectonically quiet areas are discussed separately to highlight the differences in erosional style between these two types of settings. An important aspect of this difference is the degree to which erosion is influenced by eustatic sea-level fluctuations.

To a first approximation, fast tectonic processes occur at plate boundaries (Fig. 6-6). The primary exceptions to this are incipient rifts and the so-called "hot spots or mantle plumes", both of which are associated with uplift and volcanic activity in plate interiors. Collisional plate boundaries are of greatest interest, because these are the sites of the intense faulting, folding, and volcanism associated with the world's great mountain belts. Most divergent plate boundaries are underwater mid-ocean ridges, and are of no direct concern. Incipient divergent boundaries on land are associated with rift zones and their characteristic volcanism and block-faulted mountains. Transcurrent shearing (neither convergent nor divergent) plate boundaries are characterized by intense seismic activity, but usually little impressive mountain building. Hot spots are evidenced by local, sometimes intense volcanism, apparently fixed with respect to the mantle and not with the plate upon which it is occurring.

The term "craton" is generally used to refer to tectonically quiet or stable continental areas. Major subdivisions include "shields" where long-term erosion has exposed extensive areas of old crystalline basement, "platforms" where shields have a flat-lying sedimentary veneer, intracratonic basins where slow long-term subsidence has led to thicker sedimentary deposits on the craton, and "passive continental margins" where continental crust has rifted, separated, cooled, and subsided (Pitman, 1978; Sloss and Speed, 1974). Between cratons and mountain belts there is often a "foreland basin" where basement has slowly subsided as a result of sedimentary loading and tectonic downwarping. Intracratonic basins, passive margins, and foreland basins represent loci of long-term sediment accumulation and storage (Ronov *et al.*, 1969; Sloss, 1963, 1979; Soares *et al.*, 1978; Vail *et al.*, 1977; Vail and Herdenbol, 1979). Intracratonic basins subside through especially long stretches of geologic time. The Amazon Trough, for example, has been active for all of the Phanerozoic (Soares *et al.*, 1978).

6.5.4 Effects of Sea-level Change on Erosion

The ocean surface represents the master base level for continental erosion and sedimentation. Given a sufficient period of time, in the absence of tectonic processes, continents would presumably be eroded flat to about sea level. It is not surprising, therefore, that most tectonically quiet areas on continents tend to have low elevations and are often flat, whereas tectonically active areas, mostly mountain belts, have high elevations and steep slopes (Fig. 6-2 and 6-4).

High and low stands of sea level are directly recorded as sedimentary coastal onlap sequences and as erosional terraces. These records are complicated in regions of crustal instability, and the rate and nature of crustal deformation determines whether evidence of short-term or long-term sea-level fluctuations are preserved and how easily this evidence is interpreted. Because continental basement warps and fractures through time, and because evidence of sea level is erased by erosion, the interpretation of this evidence to produce sea-level curves for the Phanerozoic is a subject of considerable debate.

Eustatic sea-level fluctuations occur on a wide range of time-scales (Vail *et al.*, 1977; Vail and Herdenbol, 1979). The shortest duration changes appear to be associated with cyclic glacio-eustatic sea-level changes. Amplitudes are on the order of 100 m and appear to involve the superposition of several cycles ranging in period from 20 000 to 400 000 years. These cycles are related to changes of glacial ice and seawater volume apparently driven by climatic processes sensitive to the Earth's orbital and rotation motions. On a thousand-fold longer time-scale are sea-level fluctuations associated with the development of coastal onlap sequences and depositional sequences in the intracratonic basins of North America, the Russian Platform, and Brazil. These fluctuations have amplitudes of 100 to 200 m and periods ranging from 10 to 80 Ma. Within regions of sluggish tectonics, the effects of sea-level change on these time-scales seem to be especially important in affecting both landscape morphology and the nature of sedimentary deposits. All of the

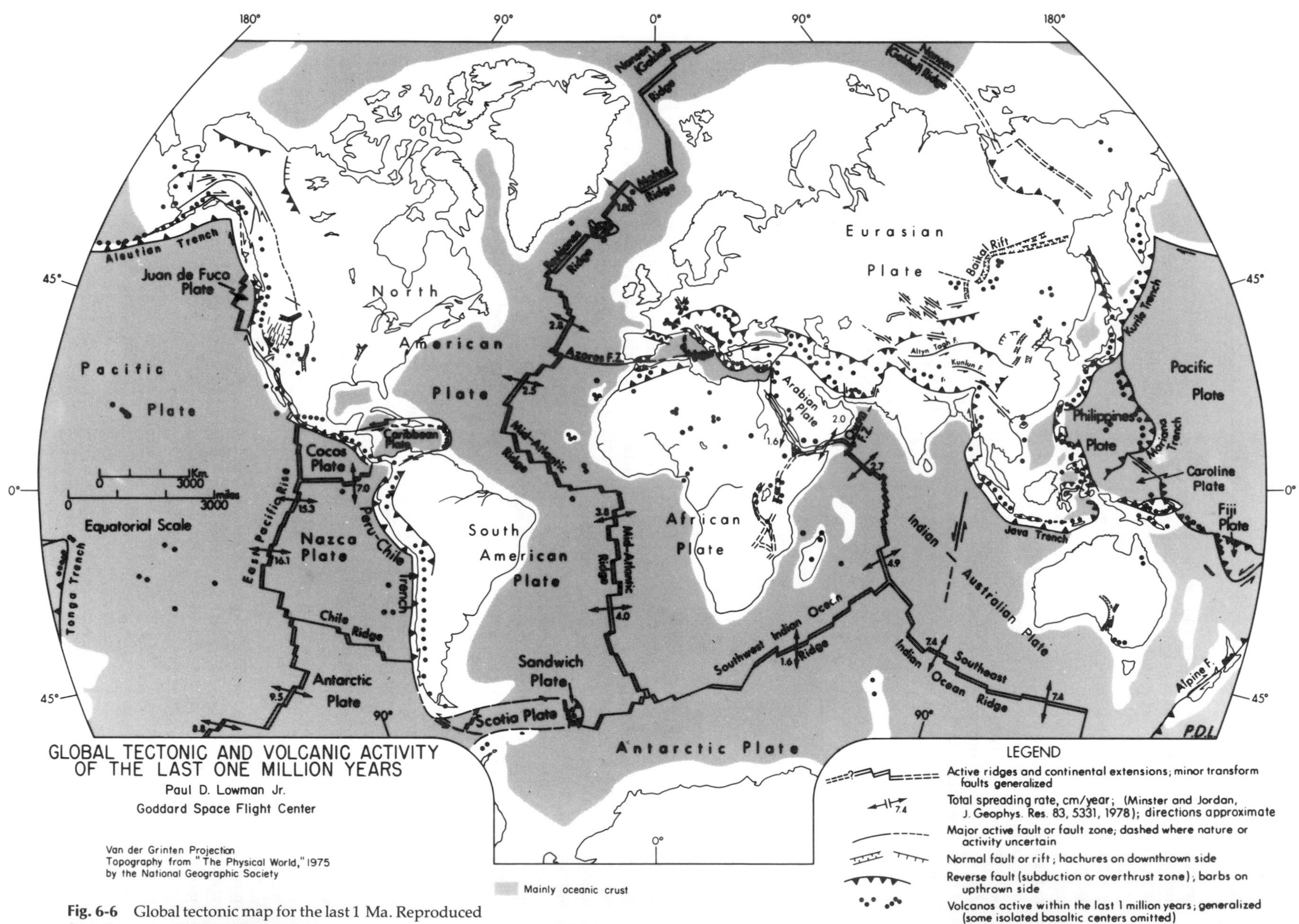

Fig. 6-6 Global tectonic map for the last 1 Ma. Reproduced from Lowman (1981) with permission.

above variations are superimposed on two very long-term sea-level fluctuations that occurred over the last 700 Ma (Fischer, 1983). These have amplitudes of 300–400 m with low stands in Eocambrian and Triassic-Jurassic times. Lows apparently coincide with major episodes of continental break-up. Sea level again seems to be approaching such a minimum.

The clearest sedimentary records of sea-level change occur where there is a good and steady supply of sediment and where the land is slowly and steadily sinking relative to mean sea level (Pitman, 1978). Areas that commonly satisfy this requirement are passive margins, intracratonic basins, and the foreland basins of major orogenic belts. When sea level rises, thicker sedimentary units are deposited more inland, generating a coastal onlap. With sufficiently high sea level, epeiric seas begin to flood the cratons, often filling foreland and intracratonic basins before spilling out over normally high ground; this is a major transgression. At low sea level, very little deposition occurs on the cratons. Instead, most of the previously deposited sediments are removed by erosion and redeposited along the continental margins. Passive margins seem to have the most easily interpreted subsidence history, but the longest records are found in intracratonic basins. In the tectonically active areas, deformation and subsidence are typically too rapid to preserve information about eustatic sea level.

Several investigators have argued that there is a major tectonic component to the subsidence histories of passive margins and intracratonic basins. Sloss and Speed (1974) and Sloss (1979) note that subsidence episodes appear to be globally synchronous and coincident with high sea levels. They argue that simple sedimentary loading caused by deposition at times of high sea levels is not an adequate explanation and suggests that some common deep driving mechanism controls both sea level and subsidence.

In regions where land is steadily rising relative to mean sea level, the effects of sea-level fluctuations are sometimes recorded as erosional features on land. Whenever the rate of sea-level rise matches the rate of uplift, there is an apparent sea level stand still. Both deposition and erosion are controlled by this almost fixed base level, and a terrace may form. If sea level falls and again rises, the terrace will have risen sufficiently so that it is preserved upslope. Episodic uplift can produce terraces, but these should not be synchronous with similar terraces developed in other regions, unless episodes of uplift were being controlled by some deep-seated process. The best-formed terraces from the late Pleistocene-Holocene sea-level fluctuations occur on coasts where uplift has been rapid (Fig. 6-7). These take the form of combined wave-cut benches, beach ridges, and carbonate banks. Deformation and erosion in these areas is often so rapid, however, that no evidence of longer-term (even early Pleistocene) base-level changes is preserved. The erosional effects of long-term sea-level fluctuations, however, are spectacularly recorded on some cratons.

6.5.5 A General Concordance Between Erosion Rates and Uplift Rates

Figure 6-7 displays uplift curves, representing the elevation of various sea-level indicators *vs* time, for tectonically active (A, B, C) and quiet (D, E, F) regions. Note that uplift rates for Taiwan, a fold and thrust belt, exceed estimates for shield regions in South America by almost three orders of magnitude. If we assume that the hypsographic curves for any of these regions have remained fixed through time, which implies that the region has had a steady-state appearance, then uplift rates equate with denudation rates. However, because the climatic and tectonic processes that govern landform development do not operate constantly and continuously, current denudation rates may differ significantly from these values.

Dissolved solid concentrations (dissolved phases derived from the weathering of bedrock) can be used to estimate ranges of denudation rates for particular regions. Histograms of dissólved solid concentrations for different morphotectonic regions of the Amazon and Orinoco basins are presented in Fig. 6-8. Solution chemistry of rivers provides the best gauge of current weathering for a particular terrain because sediment compositions are more difficult to interpret because of sediment storage as alluvium or colluvium. These denudation rates are in general agreement with more rigorous calculations (Gibbs, 1967; Lewis *et al.*, 1987; Paolini, 1986; Stallard, 1980). Comparison with Fig. 6-7 shows a reasonable match between denudation rates and uplift rates for a particular type of terrain. The most concentrated water samples and highest denudation rates are observed in river basins in tectonically active areas.

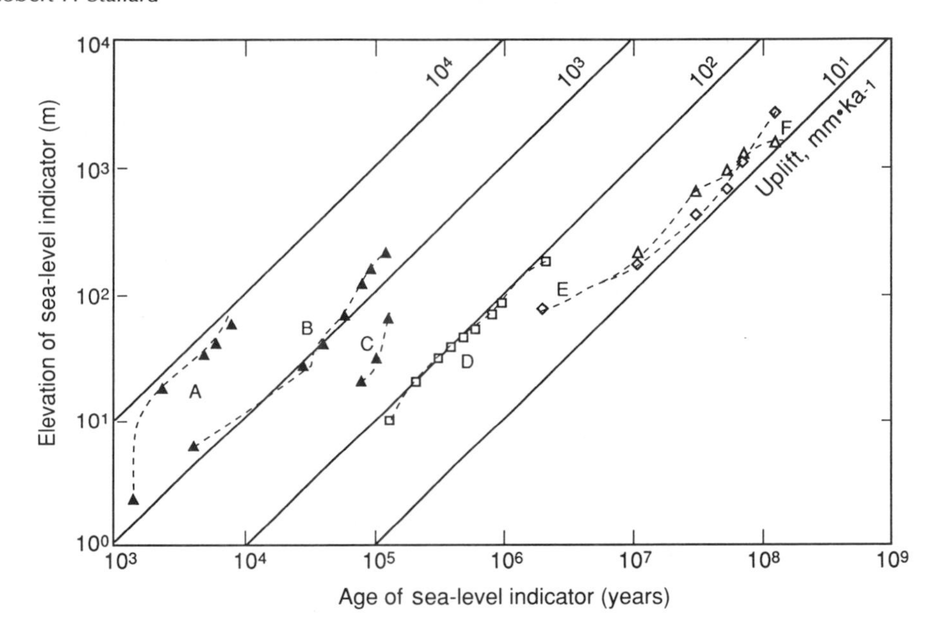

Fig. 6-7 The elevation of the terraces and other sea-level indicators compared to their age for several locations (Stallard, 1988). To derive actual uplift rates, one needs to know the age of the terrace and the eustatic sea level at that time. (A) Elevation data from the Hengchun Peninsula, Taiwan, and Eastern Coast Range of Taiwan, corrected for glacial-eustatic effects and grouped by intervals of about 1500 years. Uplift seems to have been discontinuous, but has averaged about 5000 m/Ma (Peng *et al.*, 1977). (B) Calculated composite terrace elevations for the Huon Peninsula, New Guinea, using data from Bloom *et al.* (1974). Uplift has been continuous and has ranged from 940 to 2560 m/Ma; the calculation used 1620 m/Ma. (C) Terrace elevations for the Island of Barbados (Clermont traverse) from Matthews (1973). Uplift rate has been about 400 m/Ma. (D) Depositional terraces in the Amazon Trough from Klammer (1984). Ages are estimated from those of interglacials on the standard oxygen isotope curve. Uplift is calculated to have been about 80 m/Ma. Klammer argues that the apparent change in elevation has been caused by drop in sea level since the Pliocene rather than by uplift. (E) and (F) Erosion surface from the Guayana and Brazilian Shields, respectively (King, 1957; McConnell, 1968; Aleva, 1984). Adapted from Stallard (1988) with the permission of Kluwer Academic Publishers.

6.6 Erosion in Tectonically Active Areas

In many active mountain belts, both rapid long-term uplift and erosion are sustained by tectonic recycling of sediments. For example, the island of Taiwan (Fig. 6-7, A), which is a fold-and-thrust belt, has been formed over the last several million years during an ongoing collision between the Luzon island arc and the Eurasian continental margin (Chi *et al.*, 1981; Covey, 1984; Suppe, 1981). During this orogeny, metamorphosed and diagenetically altered sediments have been uplifted, eroded, and deposited, only to be reincorporated into the cycle once more (Covey, 1984; Manias *et al.*, 1985). The central range of Taiwan has some of the highest denudation rates (13 000 tonne/km^2 year solid + 650 tonne/km^2 year dissolved = 5150 mm/year $\times$ 10^3) measured for river catchments anywhere on Earth (Li, 1976). The high denudation rate is a reflection of the poorly lithified, highly tectonized nature of the sedimentary rocks that compose the island. Sediment-yield data compiled by Milliman and Meade (1983) indicate that island arcs and mountain belts in the tropical and subtropical west Pacific may contribute as much as 22% of all solid material discharged by rivers into the ocean. Furthermore, the tropical mountainous areas in southeast Asia and India may contribute another 33%.

6.6.1 Erosion and Orogeny

Mountain building usually involves compressional deformation of the crust. Recent studies of the

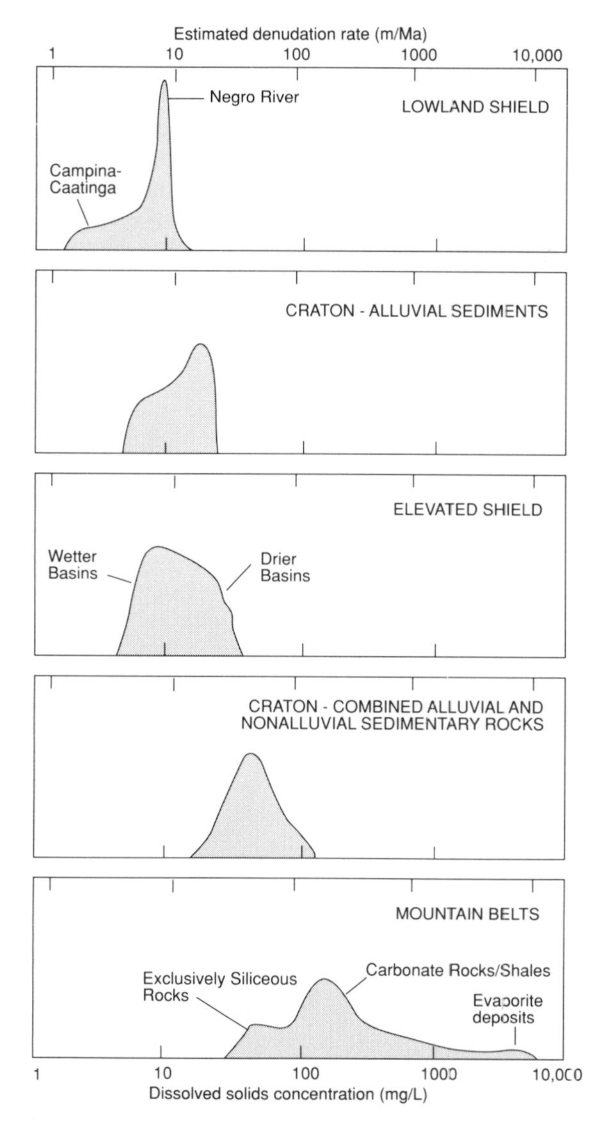

Fig. 6-8 Histogram of dissolved solids of samples from the Orinoco and Amazon river basins and corresponding denudation rates for morpho-tectonic regions in the humid tropics of South America (Stallard, 1985). The approximate denudation scale is calculated as the product of dissolved solid concentrations, mean annual run-off (1 m/year), and a correction factor to account for large ratios of suspended load in rivers that drain mountain belts and for the greater than average annual precipitation in the lowlands close to the equator. The correction factor was treated as a linear function of dissolved solids and ranged from 2 for the most dilute rivers (dissolved solids less than 10 mg/L) to 4 for the most concentrated rivers (dissolved solids more than 1000 mg/L). Bedrock density is assumed to be 2.65 g/cm^3. Adapted from Stallard (1988) with the permission of Kluwer Academic Publishers.

physics of orogeny suggest that there is a feedback between the nature of the building process and denudation rates. Suppe (1981), Davis *et al.* (1983), and Dahlen *et al.* (1984) have modeled the effects of brittle deformation in accretionary fold–thrust mountain belts such as Taiwan and the Andes. The basis of their model is the hypothesis that rock deformation is governed by pressure-dependent, time-dependent brittle fracture or frictional sliding – "Coulomb behavior". In such belts, sediments are scraped off of the subducting lithosphere, and a large wedge of deformed sediment, separated from underriding crust by a basal decollement, is built. The process resembles the behavior of snow being pushed in front of a bulldozer blade. The mountain belt develops a regional profile that tapers toward the subducting plate. If the angle of the taper (0–6°) is less than a critical value, the wedge sticks to the basal decollement. Compression, due to the addition of material at the toe of the wedge, is taken up by the internal deformation of the wedge such that the wedge steepens. Once the angle of regional taper reaches the critical value, the entire wedge can slide over the basal decollement, deforming as material is added so as to maintain the critical taper. If too much material (~15 km) piles up, the deepest material starts to deform ductily or plastically, greatly reducing basal resistance to sliding. Flat high plateaus, such as the Andean Altiplano and the Tibetan Plateau, may develop over the areas that are deforming in this manner.

Models indicate that the overall topographic profile of such belts evolves into a stable form such that erosional outputs balance accretional inputs (Suppe, 1981; Davis *et al.*, 1983). A mountain belt cannot reach unlimited height as more material is added. Two factors mitigate against this: (1) erosion and (2) changes in rock deformation from brittle to ductile with increasing depth. There should be a continuous supply of easily eroded material so long as accretion continues. Since erosion thins the wedge and reduces the taper, the intensity of deformation should increase with increasing erosion rates. It seems reasonable, therefore, that the lithologic susceptibility to erosion and climate-related erosional intensity may in turn influence the form of the entire mountain range, not just the form of the slopes. Suppe (1981) argues that the width and regional slope of mountain belts is susceptible to the climate regime. For example, mountain belts might tend to be wider where erosion rates are reduced, such as in a region of dry climates. There must be some minimum regional denudation rate for a steady-state profile to form. Under the limiting case of no erosion, an accretionary mountain belt should continue to widen so long as accretion continues.

6.6.2 River Chemistry and Bedrock Susceptibility in Mountainous Regions

The composition of dissolved and solid material transported by rivers in mountain belts of the humid tropics is consistent with weathering-limited erosion. In the Andes, sediments constitute the principal basement lithology, and the river chemistry correlates with catchment geology (Stallard, 1980, 1985, 1988; Stallard and Edmond, 1983). For example, black shales have particularly high Mg : Ca ratios, and Bolivian rivers that drain black shales are exceptionally Mg-rich. Rivers that drain evaporites have the highest total cation (TZ+) concentration, followed by those that drain carbonates, and finally by those that drain only siliceous rocks. This is

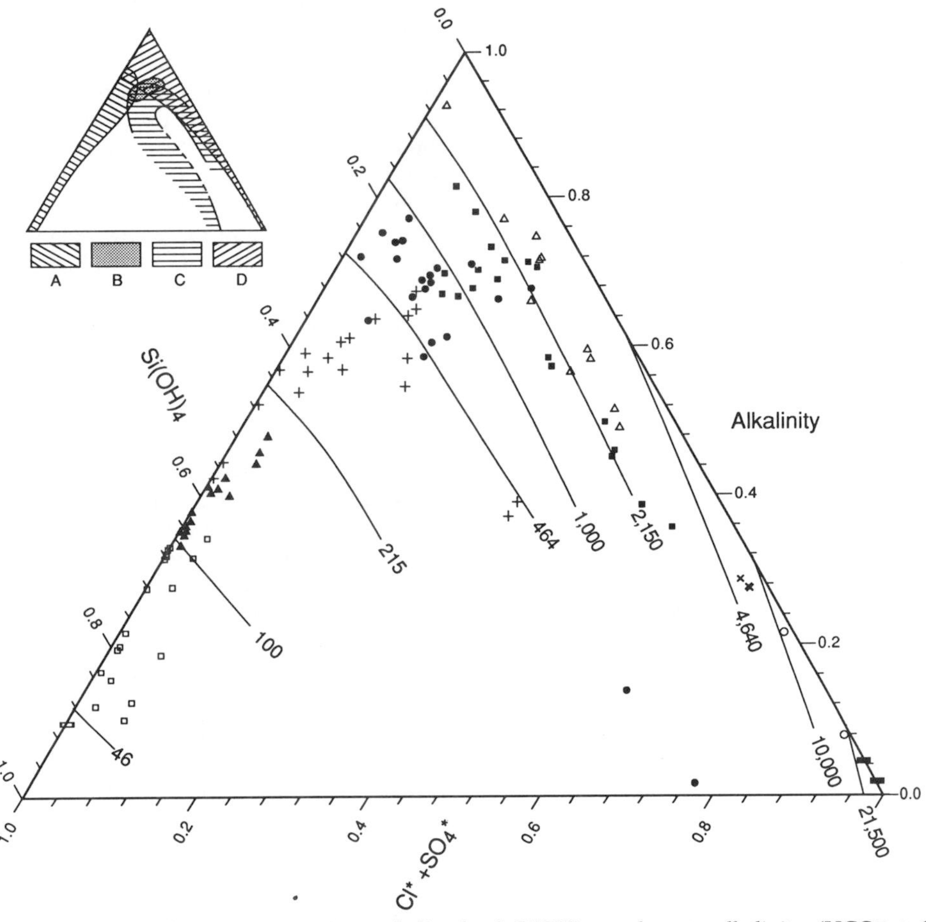

Fig. 6-9 Ternary diagram showing proportions of dissolved $Si(OH)_4$, carbonate alkalinity ($HCO_3^- + CO_3^{2-}$), and ($Cl^- + SO_4^{2-}$) in the Orinoco River and Amazon River basins. Charged species are in equivalents; $Si(OH)_4$ is in mole units. The curves in the larger figure are numbered in total cation concentration (mEq/L). Unlike previous figures, symbols represent the total cation concentration interval that includes the sample's concentration. The predominant symbol within each interval corresponds to samples whose concentrations plot within that interval. In the smaller figure, the patterned areas correspond to the predominant source of samples whose concentrations plot within the areas: (A) streams that drain cratonic areas; (B) streams that originate in mountain belts, but that drain large areas of cratons; (C) streams that drain mountain belts with extensive black shales; (D) streams that drain mountain belts with extensive carbonate rocks and evaporite deposits. Adapted from Stallard (1988) with the permission of Kluwer Academic Publishers.

illustrated by the trimodal nature of the histogram of dissolved solids from Andean rivers in Fig. 6-8 and by the ternary diagram in Fig. 6-9. Here, $Si(OH)_4$, alkalinity, and $Cl^- + SO_4^{2-}$) are used as input markers for the respective lithologies, bearing in mind that alkalinity is also produced by weathering of silicate minerals and SO_4^{2-} by weathering of pyrite and other sulfides. Total cations increase systematically from silica to alkalinity to (chloride + sulfate). Many Andean rivers are slightly supersaturated with respect to calcite (Stallard, 1988; Stallard and Edmond, 1987); this saturation ultimately limits the supply of alkalinity to the rivers. Similar limits do not apply to Na, K, Mg, and Ca because of additional inputs from silicates and evaporites. Erosional contributions by the weathering of carbonates and especially by evaporite minerals greatly exceed those expected if inputs are assumed to be proportional to the fraction of the catchment area in which these lithologies are exposed. Much of the evaporite input is sustained by actively extruding salt diapirs (Benavides, 1968; Stallard, 1980; Stallard and Edmond, 1983).

The presence of unstable and cation-rich minerals in the suspended load and bed material of rivers that drain the Andes indicates that erosion is extraordinarily rapid. In these rivers, the solution enrichment of Na^+ relative to K^+ and Ca^{2+} relative to Mg^{2+}, when compared to bedrock ratios (Fig. 6-5), indicates that K- and Mg-rich solid phases are being eroded. Moreover, the chemical and mineralogical compositions of the sediment correlate with geology. The sands in tributaries that have their headwaters in the mountain belts are commonly litharenites (DeCelles and Hertel, 1989; Franzinelli and Potter, 1983; Johnsson, 1989; Johnsson *et al.*, 1988; Potter, 1978; Stallard *et al.*, 1990). Especially unstable minerals such as calcite, amphiboles, and pyroxenes are present in samples from some of these rivers. Micas and 2 : 1 clays, including illites, vermiculites, and smectites are abundant in the fine-grained fraction (Gibbs, 1967; Irion, 1976; Koehnken, 1990; Stallard *et al.*, 1990). Chlorites (2 : 1 : 1 clays), quartz, feldspars, and amphiboles are also common in the silt-size fraction. Pyrophyllite occurs in some rivers. When compared to average igneous rocks, fine-grained sediments typically are enriched in Al relative to soluble cations, in Mg relative to Ca, and in K relative to Na (Stallard, 1985; Stallard *et al.*, 1990). Andean rivers often acquire an intense red color only after crossing regions with red beds. Furthermore, the suspended load of rivers in the Peruvian Andes that have not yet crossed the red beds is enriched in vermiculite and mica, whereas in the lower courses of the same rivers, it is enriched in smectite and kaolinite. Finally, the marine shales, which are abundant in the Bolivian, Colombian, and Venezuelan Andes, are very micaceous, and illite is particularly abundant in rivers that drain these regions.

6.7 Erosion of the Cratons

Erosion on cratons has not been well described. This is in part because denudation rates are very low and because sea-level fluctuation may be important to the erosion process. Cratons seem to undergo major episodes of erosion following drops of sea level. When the level drops to a stable stand of several million years, much of the landscape is eroded down to the new level and an erosion surface or a planation surface forms.

Two principal models have been put forward to describe the development of extremely eroded topography following a drop in base level in a region that had once been planed flat (James, 1959). One is the classic Davis cycle (Davis, 1932), the other is a model developed by King after Penck (King, 1953, 1967). These differ in their prediction of how the raised topography is dissected to form hills and valleys. According to Davis, rivers first cut steep valleys into the landscape, then their valleys broaden, and regional slopes are flattened until the last remnants of the original surface are worn away at interfluves. At this point, the landscape is said to be mature. Slopes continue to flatten until formerly elevated interfluves become low swells. This is a peneplain. In King's model, initial slopes are formed along the edges of the uplift and the sides of large penetrating river valleys. Slopes then evolve into steepened equilibrium forms that separate the old erosion surface from an incipient new surface. As time proceeds, these slopes undergo parallel retreat into the older surface which is consumed while the younger surface is extended. The end-product is a pediplain. The steep slopes are the locus of most of the erosion and the remaining terrain is almost flat, being drained by rivers having little erosive capacity. In the Davis model, no remnants of previous surfaces can exist. In King's model, however, successive erosional levels can remain stacked and can even actively expand through parallel retreat into the older surfaces long after the initiating change in base level has been superceded.

A search for stacked erosion surfaces became the centerpiece of work by King (1967), who compiled spectacular continental-scale examples from South America, Africa, India, and Australia. In a sense, these surfaces are rather like super-terraces, occurring within a narrow range of altitudes for a particular region. Surfaces nearest sea level are well defined and undissected, whereas the most elevated surfaces are usually remnants of small extent or are simply delineated by a large number of hills having peaks of similar height topped by deeply weathered soils (Fig. 6-10). In South America (Fig. 6-7E, F), Africa, and other tropical cratons, at least five levels seem to be identifiable. King argued that to a first approximation the surfaces are globally synchronous (within about 10 Ma) and that the oldest ones are of great age, perhaps predating the rifting of the South Atlantic (late Jurassic to early Cretaceous).

Subsequent geomorphic studies and work related to bauxite exploration have produced better dating and descriptions of relationships among erosion surfaces in South America (Aleva, 1979, 1984; Krook, 1979; McConnell, 1968; Menendez and Sarmentaro, 1984; Zonneveld, 1969). The most economic bauxite deposits occur on the Neogene (55 Ma) surfaces. Although bauxites from older surfaces are of high grade, they are of small extent and very dissected. The bauxites become progressively more ferruginous on younger surfaces. Where the substrate is quartzose, quartz persists through the entire weathering profile, which may be as much as 50 m thick on the 55-Ma surface (Menendez and Sarmentaro, 1984).

These surfaces are dated by classical stratigraphic techniques. Deposits of fluvial, lacustrine, or eolian sediments that contain dateable pollen or vertebrates are sometimes preserved on the surface. This is not common and dating usually is accomplished by tracing a surface into an area of subsidence such as a passive margin or an intracratonic basin. Occasionally, the surfaces have been sufficiently downwarped to be buried under marine sediments. More often, such as off the northeastern South American coast, erosion surfaces coincide with hiatuses in the sedimentary section; these are frequently overlain by deposits of quartz and bauxite gravels and sub-arkosic sands. This is thought to be indicative of the early dissection of thick and highly weathered soils following a drop in base level. Such hiatuses are not unlike those associated with sea-level high stands seen in coastal onlap sequences. Accordingly, Aleva (1984) notes that the major surfaces appear to coincide with major high stands on the late Mesozoic-Cenozoic sea-level curve. The most recent and youngest surfaces are 10 or so glacio-eustacic terraces from the Amazon valley described by Klammer (1984) (Fig. 6-7d).

The Brazilian and Guayana shields are among the best exposures of elevated crystalline basement in the humid tropics. The topography can be spectacular with steep slopes, either of bare rock or thinly vegetated, often topped by high plateaus, some of which are quite expansive. The highest landforms are 2000- to 3000-m table-like mountains called "Tepuis" (Fig. 6-11). These are topped by thick orthoquartzites. Inselbergs are locally common. Great talus piles do not accumulate below cliffs and at the base of slopes. Much of this material probably just dissolves. A karst-like topography developed on the orthoquartzite, gneisses, and granites is found on some of the higher and presumably most ancient areas (Blancaneaux and Pouyllau, 1977; Szczerban, 1976; Urbani, 1986). The karst on quartzite is established to such a degree that stalactites consisting of opal, cristobalite, tridymite, and quartz have formed locally (Chalcraft and Pye, 1984). Aerial and radar photographs show that all but the youngest surfaces are dissected to various degrees, and that this dissection is geologically controlled. Many rivers appear to follow fault systems and dike swarms (Fig. 6-10; Schubert *et al.*, 1986; Stallard *et al.*, 1989). This would indicate that the older surfaces are no longer expanding in the manner described by King, since the dissected surfaces can no longer serve as fixed base levels for overlying scarps. Especially resistant lithologies such as quartzites and some intrusives persist as high elevations. Thus, even though weathering rates are very slow (Fig. 6-8), the observation that lithology plays an important role in landscape morphology indicates that weathering-limited erosion is occurring.

6.7.1 Erosion and Slow Uplift of the Shields

The development of topography consisting of stacked erosion surfaces on many shields indicates that vast areas of cratons are undergoing a slow but sustained rise relative to sea level. The remnants of erosion surfaces are evidence of the former thickness of the cratons. It is interesting to speculate that this tendency to rise has held true for many shield areas

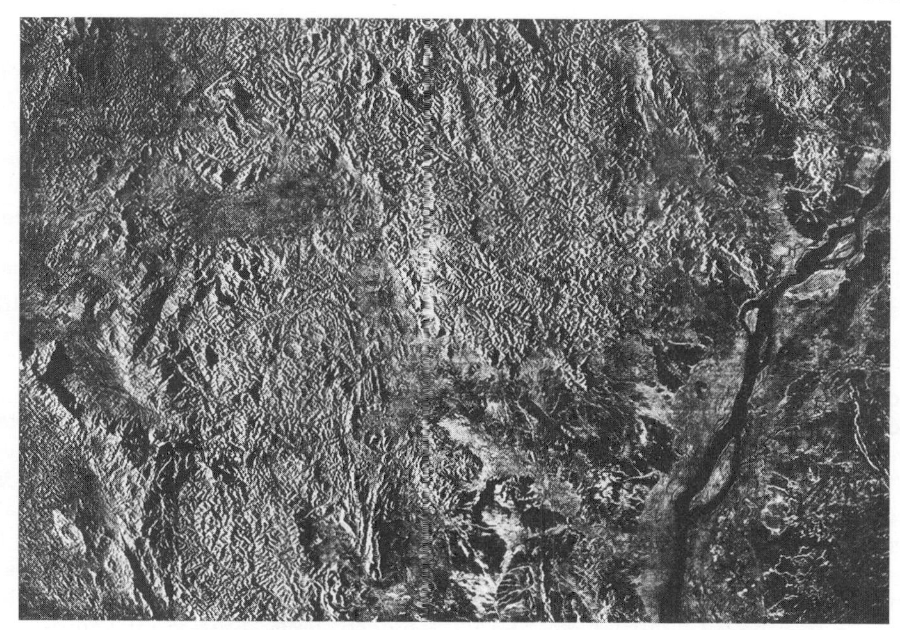

Fig. 6-10 Radar image of erosion surface in Venezuela (Petroleos de Venezuela S.A., 1977). The region bounded by this photograph is approximately 50 × 100 km. The Orinoco river crosses the left side of the image. The confluence of the Meta is just off the lower left corner of the image. To the west of the Orinoco are the Llanos of the Andean Foreland Basin; to the east of the Orinoco is the Guayana Shield. The erosion surfaces appear as areas of slightly dissected, raised topography in the Guayana Shield. An irregular NNW–SSE trending escarpment that starts to the left of center and runs to the lower edge separates the two surfaces.

Fig. 6-11 Photograph of Ayun Tepui, taken from Canaima, Venezuela. The cliffs, which are about 2000 m high, consist of orthoquartzite. The irregular top is a karst topography formed by the slow dissolution of the quartz. It is estimated that the top of this tepui, and similar ones elsewhere on the Guayana Shield, have been exposed since the Jurassic. Reprinted with permission from the author.

throughout much of geologic time, thereby explaining the high metamorphic grade of rocks exposed on the surface of most shield areas. In this sense, shields would be dynamical opposites to intracratonic basins. In this context, continental platforms are areas that neither rise nor sink dramatically. Tectonic modes change with time. For example, the thick quartzites that top many of the highest peaks of the Guayana shield are evidence that this area once behaved as either a platform or a basin.

Part of the rising and sinking of cratons may be due to isostatic adjustments in response to erosion and to sediment accumulation. Part, however, may be due to deep-seated processes. Burke and Wilson (1972), Crough (1979), and Morgan (1983) have argued that uplift and erosion of shields is caused by random repeated passages of continental crust over mantle plumes. As a segment of crust passes over a mantle plume, it is heated and becomes more buoyant. This event produces a fairly rapid uplift followed by a slow reduction of elevation as a result of combined erosion and cooling. If erosion is sufficiently intense, the formerly elevated segment of crust becomes a depositional basin once it has completely cooled. Morgan (1983) used this idea to examine the relationship between hot spots and regional geology in the continents bracketing the Atlantic. He hypothesized the recent passage of a hot plume under the Guayana shield to explain its presently elevated nature.

The globally synchronous development of erosion surfaces and slow uplift rates of shields indicates that models based on localized heating of the crust and mantle are not adequate to explain the uplift process. To produce near simultaneous surfaces on separate continents, King (1967) invoked global epeirogeny – uplift driven by some sort of deep-seated process. The change in base level does not have to act continuously once scarps are initiated. They may become self-sustaining to some degree, for as they retreat, isostatic adjustment would raise the topography (King, 1956), thereby stabilizing the effects of the initial base-level change. The isostatic uplift would be cumulative, thus explaining the stacked topography. Global sea-level drop seems to be a more reasonable initiating event, but it is still a matter of speculation as to whether the uplift is continuous or episodic and isostatically self-sustaining or driven by deep-seated, global tectonic processes. There is, however, an interesting parallel between King's view of erosion surfaces and the idea of Sloss and Speed (1974) and Sloss (1979) that there is a tectonic component to the subsidence of sedimentary basins at times of high sea level.

6.7.2 Weathering-limited Erosion on the Elevated Shield

Rivers from the shields transport very little dissolved or solid load, and the solids tend to be more resistant mineral phases (Franzinelli and Potter, 1983; Gibbs, 1967: Irion, 1976; Johnsson, 1989; Johnsson *et al.*, 1988; Koehnken, 1990; Potter, 1978; Stallard *et al.*, 1989). The major cations of the dissolved load are in bedrock proportions (Fig. 6-5), indicating that bedrock is strongly leached during weathering and most of the cation load is in solution. River sands include more potassic feldspars, as might be expected if weather-limited erosion is occurring on the deep slopes. Erosion rates in these terrains are much lower than in active mountain belts like the Andes, even though elevations are as much as 3000 m. Both lithology and basement structure contribute to the low erosion rates in elevated crystallizing basement terrains (Stallard, 1988; Stallard *et al.*, 1989). Rocks are frequently massive and composed of minerals like quartz, potassium and sodium feldspars, and micas, which are somewhat resistant to weathering. Uplift apparently involved ductile (plastic) deformation and was followed by a long history of weathering that may have removed all unstable lithologies. The basement of active organic belts is not so massive owing to faulting, brittle deformation and volcanism. Furthermore, active uplift provides a continuous supply of fresh unstable lithologies.

6.7.3 Transport-limited Erosion on the Lowlands

Vast tracts of the South American and West African lowlands have predominantly transport-limited denudation regimes (Stallard, 1988). These regions represent the flattest and youngest erosional/depositional surfaces of the late Neogene. On the shield, the substrate is crystalline basement, while in the intracratonic basins, the Andean foreland basin, and the coastal plains, much of the sedimentary substrate consists of strongly "pre-weathered" fluvio-lacustrine sediment (Stallard, 1985, 1988;

Stallard and Edmond, 1983; Stallard *et al.*, 1989). Soils and solid loads of rivers are rich in quartz, kaolinite, and iron sesquioxides, all of which are cation-depleted phases. River sands are quartzose (Franzinelli and Potter, 1983; Johnsson *et al.*, 1988) and the suspended load is largely kaolinite plus quartz. Aluminium sesquioxides (gibbsite) are much less common. In rivers with pH > 5, particles have iron sesquioxide coatings; at lower pH values, coatings are absent (Stallard *et al.*, 1989). Dissolved major cations are in bedrock proportions. Dissolved solid concentrations, and by inference denudation rates, are exceedingly low (Fig. 6-8). This agrees with the diminution of weathering rates in conjunction with the development of thick soils. Dissolved loads tend to be lower on the shields than on the sediments, even though rocks of the shield are composed of less stable minerals; however, the low permeabilities of the rocks and many of the soils of the shields compared to those of the sediments apparently counterbalances this.

Where easily weathered lithologies such as carbonates and evaporites are near the surface, such as in the lower Amazon valley, their contribution to the rivers appears minor, probably because thick residual soil covers have developed. This indicates that susceptibility to weathering is indeed less important in controlling lowland erosion rates. Carbonate platforms, such as southern Florida and the Yucatan in North America, are exceptions because of deep-water circulation throughout the carbonate karst.

6.8 The Effects of Transients: Continental Ice Sheets

As seen in the previous discussion, susceptibility of bedrock to erosion is the primary factor controlling erosion rate in steep terrains, whereas supply of reactants to reactive minerals limits erosion in flat terrains. Geochemists are frequently called upon to assess the impact of some human-induced environmental pertubation. Would pulverized granite make a good fertilizer in areas of thick, cation-depleted, tropical soils? What effect does permanent deforestation of a mountain slope have on erosion on a short time-scale? On a time-scale that is 10 times longer? What about acid rain? What about long-term climatic warming or increased atmospheric carbon dioxide? What happens when a formerly moist region dries out? The Earth has recently undergone a remarkable perturbation, continental glaciation. Glaciations are especially interesting to evaluate using concepts of erosion regime.

During the past 2.4 Ma, the Earth has gone through several episodes of continental glaciation (see Hughes, 1985). Each of the last seven of these episodes (600 000 years) show a characteristic pattern of waxing and waning (Broecker and Denton, 1989). Each episode started slowly. Extensive ice sheets gradually built up in Canada (the Laurentide ice sheet), northern Europe (the European or Fenno-Scandian ice sheet), and across northern Asia. Smaller ice sheets formed in Iceland and southern South America. Temperatures cooled globally, and the altitude of mountain glacial formation descended about 1000 m. The ice caps of Greenland and Antarctica became more active. The glacial build-up lasted about 100 000 years and ended abruptly, over a few hundred years. Within the build-ups and terminations there were short episodes of glacial advance and recession. The last ice age ended about 13 500 years ago.

Glaciers are powerful agents of physical erosion. In a detailed geomorphological study of the glaciated Canadian Shield, Sugden (1978) concluded that erosional style was related primarily to the basal thermal regime of the ice. From the center of divergence on an ice gap, Sugden identified five idealized zones. Erosion is minimal under the cold-based ice at the center of the cap – shear from ice flow is minimal and the ice is anchored to the bedrock. Surrounding this was a zone of basal melting, then a zone of basal freezing, and finally a zone of melting. Various processes drive erosion in these zones (Fig. 6-12). Sugden concludes that the excavation of debris is an important factor influencing the amount of erosion accomplished by an ice sheet. The excavation process constrasts strongly with fluvial systems or alpine glaciers, for which downslope transport is important. Sugden argues that landforms on the Canadian Shield are equilibrium forms with respect to the thermal regime of the ice sheet at maximum glaciation.

The estimation of depth of erosion by continental ice sheets has been a controversial endeavor. In mountainous regions such as the North American Rockies, alpine glaciers have scoured deep valleys into a topography that was, prior to glaciation, much smoother and more rolling. For continental ice sheets, it is much harder to find a convenient frame of reference, such as pre-existing topography, from which depth of erosion can be judged. White (1972) argued that the ice sheets had removed much of the

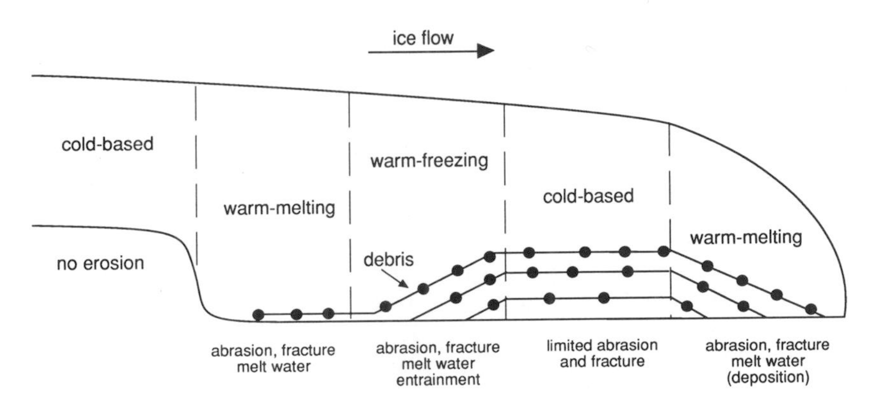

Fig. 6-12 Idealized model of the relationship between styles of glacial erosion and the basal thermal regime of ice. Many factors affect the basal thermal regime, the most important of which are surface temperature, ice thickness (pressure melting), and the rate of ice deformation (frictional heating). Cold-based ice is generally protective and has little erosive power unless it contains debris inherited from an up-glacial basal freezing zone. In warm-melted zones, basal slip between ice and rock promotes fracture, plucking, and abrasion. If ice subsequently passes into a freezing zone, bedrock particles can be entrained during the freezing process. This further enhances excavation of bedrock. During a glaciation, patterns of erosion may shift as ice sheets develop and retreat, but peak erosion seems to occur near glacial maxima. The top surface of an ice sheet is relatively flat. Erosion is more pronounced at basement topographic lows (either valleys or lowlands), relative to adjacent basement highs, because the thicker ice over the lows promotes pressure melting, and because low areas tend to be zones of ice convergence. Less porous rocks such as crystalline shield may be more susceptible to glacial erosion than porous rocks such as sediments, because diversion of meltwater through the substrate may promote binding of the ice to the bedrock. Such ice would act more like cold-based ice. Adapted from Sugden (1978) with the permission of the International Glaciological Society.

sedimentary cover of the Canadian Shield and drew a cross-section indicating that there may have been as much as 1000 m of erosion. This conflicted with geological, pedological, and geomorphological evidence that some areas of the Canadian Shield had been minimally eroded during the ice ages and that the oldest glacial deposits are rich in material from crystalline basement (Gravenor, 1975; Sugden, 1976). More recently, careful compilations of the quantities of glacial sediment deposited in the oceans around North America and in depositional basins on North America by Laine (1980) and Bell and Laine (1985) indicate that about 120 or more meters of physical erosion were caused by continental ice sheets. If present river yields from the Canadian Shield are used, about 20 m of chemical and 20 m of physical erosion would have occurred over the same time period (3 Ma assumed). Clearly, continental ice sheets are potent agents of erosion – on shields, ice is evidently more powerful than the processes associated with the humid tropical weathering.

The above-mentioned river yields are undoubtedly affected by the recent glacial scouring transient. Throughout the glaciated region, there is still an abundance of loose glacial debris, much of which is fresh bedrock and mineral grains. This great abundance of fresh bedrock minerals at the ground surface contrasts markedly with the situation in flat regions undergoing transport-limited erosion where soil minerals are cation-deficient. Gravenor (1975) presents data that indicate that prior to glaciation the shield may once have had a soil mantle that even included bauxites. Presumably, both chemical and physical erosion must be proceeding much faster now than under steady-state conditions involving a thick soil. Thus, Bell and Laine (1985) used worst-case values – generally a good practice. What might be a more realistic estimate of weathering rates without glaciations?

Investigators studying glaciated areas in the Rocky Mountains have noted that in granitic areas, waters are particularly high in Ca^{2+} ions relative to other major soluble cations (Mast *et al.*, in press). One source appears to be intergranular calcite, one of the last minerals to crystallize out of the cooling granite melt. Volumetrically, this calcite is a minor constituent. Calcite is also far more susceptible than other minerals in granite (Table 6-1). A primary reason for the Ca^{2+} abundance may be that the calcite is making a disproportionate contribution to the steam waters because of the recent exposure of fresh granite by glacial activity.

6.9 Conclusion

Weathering processes exert a major control on many aspects of the chemistry of the Earth's surface environment. Considerations of temporal and spatial scale are very important in evaluating how the weathering processes affect the chemical response of the atmosphere and ocean reservoirs. Continental denudation, the transfer of material from the land surface to the ocean, is a particularly important aspect of this response. In small stream basins, solution chemistry is controlled by processes that act on a hillslope scale. There is commonly little storage of water in most large river systems; thus for large river systems, hillslope processes remain important. Storage on a time-scale of months to millions of years must be considered for solid material. Long-term burial results in a net loss to the river, and chemical weathering during storage can be important for large river systems.

In regions where the erosion regime is weathering-limited, susceptibility of the bedrock to chemical and physical weathering controls erosion rates. This susceptibility relates directly to the chemical and physical properties of the rock. Susceptibility also depends on local climate. Moreover, weathering rates are affected by the soils that form on the rock, and the nature of the vegetation that grows on the soil. Vegetation, by supplying bio-acids that chemically degrade the rock, can increase the rate of erosion of rocks that are resistant to physical erosion.

Susceptibility of bedrock to erosion is not necessarily controlled by the reactivity of the mineral grains that make up the bedrock. In the case of limestones, the weathering reactions proceed until equilibrium is reached. The rate is controlled by the supply of reactants (protons). Rocks composed of silicate minerals are different. For example, the massive, poorly jointed granites of the Guayana Shield erode more slowly than the tectonized, highly cracked and sheared granites of the Andes. The difference is not caused by contrasting mineralogies, but by differences in the permeability of water and stability of steep slopes. Again, the supply of reactants is very important, but because the weathering does not proceed to equilibrium or to completion, the reactivity of individual mineral grains is also important. Thus, when compared to the bedrock proportions, K is enriched relative to Na and Mg relative to Ca in solid erosion products. Likewise, Na^+ and Ca^{2+} are enriched in solution relative to K^+ and Mg^{2+}. In flat terrains, these enrichments do not occur, because the mobilization of major soluble cations from bedrock is complete.

In regions where erosion is transport-limited, weathering rates are controlled by the supply of reactive fluids to unstable minerals. This is controlled by soil properties, regional base level, and, ultimately, sea level.

The concept of weathering regime may be useful for interpeting the effects of important phenomena in Earth history. These include the evolution of life, changes in the patterns of plate tectonic interactions, major meteorite impacts, glaciations, and so on. Continental chemical denudation depends on the proportions of the Earth's surface that are eroding under the two types of regimes: weathering-limitation associated with steeper slopes; transport-limitation associated with flat areas. Susceptibility has to be evaluated for each climate region.

On the time-scale of the entire history of the Earth, denudation rate is controlled by the tectonic processes that supply fresh bedrock to the subaerial Earth surface environment. There are two important aspects to this. One is the uplift of mountains and volcanism to produce steep terrains that undergo weathering-limited erosion. The other is the rise and fall of global sea level that affects erosion and sedimentation on the cratons. Low sea level equates with erosion of the cratons; high sea level equates with cessation of erosion and even sedimentation on the cratons. Glaciers are the most potent agents of physical erosion on cratons and perhaps in mountain terrains where the bedrock is resistant to chemical weathering. The role of continental ice sheets in excavating shield over the history of the Earth deserves further consideration.

Questions

6-1 Provide some simple reasons for the rules of thumb for physical and chemical weathering given in Section 6.2.1.

6-2 Describe how vegetation can increase and decrease the weathering rate.

6-3 What is the distinction between physical and chemical weathering?

6-4 Describe the factors that limit mountain height. How does this relate to the areas of the Earth that deliver the largest dissolved and suspended loads to the oceans?

References

Aleva, G. J. J. (1979). Bauxite and other duricrusts in Surinam: A review. *Geologie en Mijnbouw* **58**, 321–336.

Aleva, G. J. J. (1984). Laterization, bauxitization and cyclic landscape development in the Guiana Shield. *In* "Proceedings of the 1984 Bauxite Symposium" (L. Jacob Jr., ed.), pp. 297–318. American Institute of Mining, Metallurgical, and Petroleum Engineers, New York.

Bell, M. and E. P. Laine (1985). Erosion of the Laurentide region of North America by glacial and glaciofluvial processes. *Quarternary Res.* **23**, 154–174.

Benavides, V. (1968). Saline deposits of South America. *Geol. Soc. Am. Spec. Pap.* **88**, 249–290.

Berner, R. A. (1978). Rate control of mineral dissolution under earth surface conditions. *Amer. J. Sci.* **278**, 1235–1252.

Berner, R. A. and S. Schott (1982). Mechanism of pyroxene and amphibole weathering II. Observations of soil grains. *Amer. J. Sci.* **282**, 1214–1231.

Berner, R. A., A. C. Lasaga, and R. M. Garrels (1983). The carbonate-silicate geochemical cycle and its effect on atmospheric carbon dioxide over the past 100 million years. *Amer. J. Sci.* **283**, 641–683.

Blancaneaux, P. and M. Pouyllau (1977). Formes d'altération pseudokarstiques en relation avec la geomorphologie des granites precambriens du type Rapakivi dans le territoire Federal de l'Amazone. Venezuela Cah. *Orstom Sér. Pédol.* **15**, 131–142.

Bloom, A. L., W. S. Broecker, J. M. A. Chappell, R. K. Matthews, and K. J. Mesolella (1974). Quarternary sea level fluctuations on a tectonic coast: New ^{230}Th/^{234}U dates from the Huon Peninsula, New Guinea. *Quarternary Res.* **4**, 185–205.

Brantley, S. L., S. R. Crane, D. A. Crerar, R. Hellmann, and R. Stallard (1986). Dissolution at dislocation etch pits in quartz. *Geochim. Cosmochim. Acta* **50**, 2349–2361.

Broecker, W. S. and G. H. Denton (1989). The role of ocean-atmosphere reorganization in glacial cycles. *Geochim. Cosmochim. Acta* **53**, 2465–2501.

Burke, K. and J. T. Wilson (1972). Is the African Plate stationary? *Nature* **239**, 387–390.

Carson, M. A. and M. J. Kirkby (1972). "Hillslope, Form and Process." Cambridge University Press, Cambridge.

Chalcraft, D. and K. Pye (1984). Humid tropical weathering of quartzite in southeastern Venezuela. *Zeitschrift für Geomorphologie, N. F.* **28**, 321–332.

Chi, W. R., J. Namson, and J. Suppe (1981). Stratigraphic record of plate interactions in the coastal range of eastern Taiwan. *Geol. Soc. China Mem.* **4**, 491–530.

Covey, M. C. (1984). Sedimentary and tectonic evolution of the western Taiwan foredeep. Ph.D. thesis, Princeton University, Department of Geological and Geophysical Sciences, Princeton, N.J.

Crough, S. T. (1979). Hotspot epeirogeny. *Tectonophysics* **61**, 321–333.

Dahlen, F. A., J. Suppe, and D. Davis (1984). Mechanics of fold-and-thrust belts and accretionary wedges: Cohesive Coulomb theory. *J. Geophys. Res.* **89**, 10,087–10,101.

Davis, D., J. Suppe, and F. A. Dahlen (1983). Mechanics of fold-and-thrust belts and accretionary wedges. *J. Geophys. Res.* **88**, 1153–1172.

Davis, W. M. (1932). Piedmont Benchlands and Primärrümpfe. *Geol. Soc. Amer. Bull.* **43**, 399–440.

DeCelles, P. G. and F. Hertel (1989). Petrology of fluvial sands from the Amazonian foreland basin, Peru and Bolivia. *Geol. Soc. Amer. Bull.* **101**, 1552–1562.

Drever, J. I. and C. L. Smith (1978). Cyclic wetting and drying of the soil zone as an influence on the chemistry of ground water in arid terrains. *Amer. J. Sci.* **278**, 1448–1454.

Dunne, T. (1978). Rates of chemical denudation of silicate rocks in tropical catchments. *Nature* **274**, 244–246.

Eberl, D. D., J. Srodon, and H. R. Northrop (1986). Potassium fixation in smectite by wetting and drying. *In* "Geological Processes at Mineral Surfaces" (J. A. Davis and K. F. Hayes, eds.), Series No. 323, pp. 296–325. American Chemical Society, Washington, D.C.

Eggleton, R. A. and P. R. Buseck (1980). High resolution electron microscopy of feldspar weathering. *Clays and Clay Minerals* **28**, 173–178.

Fischer, A. G. (1983). The two Phanerozoic supercycles. *In* "Catastrophies in Earth History: The New Uniformitarianism" (W. Berggren and J. Van Couvering, eds.), pp. 129–150. Princeton University Press, Princeton, N. J.

Franzinelli, E. and P. E. Potter (1983). Petrology, chemistry, and texture of modern river sands, Amazon River system. *J. Geol.* **91**, 23–39.

Garrels, R. M. and A. Lerman (1981). Phanerozoic cycles of sedimentary carbon and sulfur. *Proc. Natl. Acad. Sci. USA* **78**, 4652–4656.

Garwood, N. C., D. P. Janos, and N. Brokaw (1979). Earthquake-caused landslides: A major disturbance to tropical forests. *Science* **205**, 997–999.

Gibbs, R. J. (1967). The geochemistry of the Amazon River system: Part I, The factors that control the salinity and composition and concentration of suspended solids. *Geol. Soc. Amer. Bull.* **78**, 1203–1232.

Gilkes, R. J., A. Suddhiprakarn, and T. M. Armitage (1980). Scanning electron microscope morphology of deeply weathered granite. *Clay Minerals* **28**, 29–34.

Goldich, S. S. (1938). A study in rock weathering. *J. Geol.* **46**, 17–58.

Gravenor, C. P. (1975). Erosion by continental ice sheets. *Amer. J. Sci.* **275**, 594–604.

Harmon, R. S., W. B. White, J. J. Drake, and J. W. Hess

(1975). Regional hydrochemistry of North American carbonate terrains. *Water Resources Res.* **11**, 963–967.

Herbillon, A. J. and D. Nahon (1988). Laterites and laterization processes. *In* "Iron in Soils and Clay Minerals" (J. W. Stucki, B. A. Goodman, and U. Schwertmann, eds.). *NATO ASI Series C: Mathematical and Physical Sciences* **217**, 251–266.

Holdren, Jr., G. R. and R. A. Berner (1979). Mechanism of feldspar weathering – I. Experimental studies. *Geochim. Cosmochim. Acta* **43**, 1161–1171.

Holdren, Jr., G. R. and P. M. Speyer (1985). Reaction rates–surface area relationships during the early stages of weathering – I. Initial observations. *Geochim. Cosmochim. Acta* **49**, 675–681.

Holland, H. D. (1974). Marine evaporites and the composition of sea water during the Phanerozoic. *In* "Studies in Paleo-oceanography". (W. W. Hay, ed.), pp. 187–192. Special Publication No. 20, Society Econ. Peleon. Min., Tulsa, Oklahoma.

Holland, H. D., T. V. Kirsipu, J. S. Huebner, and U. M. Oxburgh (1964). On some aspects of the chemical evolution of cave waters. *J. Geol.* **72**, 36–67.

Hughes, T.J. (1985). The great Cenozoic ice sheet: Palaeogeography, palaeoclimatology. *Palaeoecology* **50**, 9–43.

Irion, G. (1976). Mineralogisch-geochemische Unterschungen an der pelitschen Fraktion amazonischer Oberboden und Sedimente. *Biogeographica* **7**, 7–25.

James, P. E. (1959). The geomorphology of eastern Brazil as interpreted by Lester C. King. *Geographical Rev.* **49**, 240–246.

Johnsson, M. J. (1989). Chemical weathering controls on the composition of modern fluvial sands in tropical weathering environments. Ph.D. thesis, Princeton University. Department of Geological and Geophysical Sciences, Princeton.

Johnsson, M. J., R. F. Stallard, and R. H. Meade (1988). First-cycle quartz arenites in the Orinoco River basin, Venezuela and Colombia. *J. Geol.* **96**, 263–277.

King, L. C. (1953). Canons of landscape evolution. *Geol. Soc. Amer. Bull.* **64**, 721–751.

King, L. C. (1956). A geomorfologia do Brasil oriental. *Rev. Brasileira de Geogr.* **18**, 147–265.

King, L. C. (1957). A geomorphological comparison between eastern Brazil and Africa (central and southern). *Geol. Soc. Lond. Quart. J.* **112**, 445–474.

King, L. C. (1967). "The Morphology of the Earth", 2nd edn. Oliver and Boyd, Edinburgh.

Klammer, G. (1984). The relief of the extra-Andean Amazon basin. *In* "The Amazon, Limnology and Landscape Ecology of a Mighty Tropical River and its Basin" (H. Sioli, ed.), pp. 47–83. W. Junk, Dordrecht. The Netherlands.

Koehnken, L. (1990). The composition of fine-grained weathering products in a large tropical river system, and the transport of metals in fine-grained sediments in a temperate estuary, Ph.D. Thesis, January 1990. Princeton University, Department of Geological and Geophysical Sciences, 246p.

Krook, L. (1979). Sediment petrographical studies in northern Surinam. Ph.D. thesis, Free University, Amsterdam, The Netherlands.

Laine, E. P. (1980). New evidence from beneath the western North Atlantic for the depth of glacial erosion in Greenland and North America. *Quaternary Res.* **14**, 188–198.

Langmuir, D. (1971). The geochemistry of some carbonate ground waters in central Pennsylvania. *Geochim. Cosmochim. Acta* **35**, 1023–1045.

Lewis, Jr., W. M., S. K. Hamilton, S. L. Jones, and D. D. Runnels (1987). Major element chemistry, weathering and element yields for the Caura River drainage, Venezuela. *Biogeochemistry* **4**, 159–181.

Li, Y.-H. (1976). Denudation of Taiwan Island since the Pliocene epoch. *Geology* **4**, 105–107.

Likens, G. E., F. H. Bormann, R. S. Pierce, J. S. Eaton, and N. M. Johnson (1977). "Biogeochemistry of a Forested Ecosystem." Springer-Verlag, New York.

Lowman, P. D., Jr. (1981). A Global Activity Map. *Bull. Int. Assoc. Engineer. Geol.* **23**, 37–49.

Manias, W. G., M. Covey, and R. Stallard (1985). The effects of provenance and diagenesis on clay content and crystallinity in Miocene through Pleistocene deposits, southwestern Taiwan. *Petrol. Geol. Taiwan* **21**, 173–185.

Mast, M. A., J. I. Drever, and J. Baron (1990). Chemical weathering in the Loch Vale watershed, Rocky Mountain National Park, Colorado. *Water Resources Res.* **26**, 2971–2978.

Mathews, W. H. (1975). Cenozoic erosion and erosion surfaces of eastern North America. *Amer. J. Sci.* **275**, 818–824.

Matthews, R. K. (1973). Relative elevation of late Pleistocene high sea level stands: Barbados uplift rates and their implications. *Quarternary Res.* **3**, 147–153.

McConnell, R. B. (1968). Planation surfaces in Guyana. *Geographical J.* **134**, 506–520.

Menendez, A. and A. Sarmentero (1984). Geology of the Los Pijiguaos bauxite deposits, Venezuela. *In* "Proceedings of the 1984 Bauxite Symposium" (L. Jacob Jr., ed.), pp. 387–406. American Institute of Mining, Metallurgical, and Petroleum Engineers, New York.

Meybeck, M. (1979). Concentrations des eaux fluviales en éléments apports en solution aux océans. *Rev. Géogr. dynam. Géol. phys.* **21**, 215–246.

Meybeck, M. (1987). Global chemical weathering of superficial rocks estimated from river dissolved loads. *Amer. J. Sci.* **287**, 401–428.

Milliman, J. D. and R. H. Meade (1983). World-wide delivery of river sediment to the oceans. *J. Geol.* **91**, 1–21.

Morgan, W. J. (1983). Hotspot tracks and the early rifting of the Atlantic. *Tectonophysics* **94**, 123–139.

Pain, C. F. (1972). Characteristics and geomorphic effects of earthquake initiated landslides in the Albert range of Papua New Guinea. *Engineering Geol.* **6**, 261–274.

Paolini, J. (1986). Transporte de carbono y minerales en el río Caroní. *Interciencia* **11**, 295–297.

Peng, T.-H., Y.-H. Li, and F. T. Wu (1977). Tectonic uplift rates of the Taiwan Island since the early Holocene. *Geol. Soc. China Mem.* **2**, 57–69.

Petroleos de Venezuela, S. A. (MARAVEN) (1977). *"Mosaico de Imágenes de Radar de Visión Lateral."* Ministerio de Energía y Minas, Dirección General Sectoral de Geolgía y Minas, Caracas, Venezuela, Scale 1 : 2 500 000, Sheet No. NB 19-8.

Pitman III, W. (1978). Relationship between eustacy and stratigraphic sequences of passive margins. *Geol. Soc. Amer. Bull.* **89**, 1389–1403.

Potter, P. E. (1978). Petrology and chemistry of modern big river sands. *J. Geol.* **86**, 423–449.

Ronov, A. B., A. A. Migdisov, and N. V. Barskaya (1969). Tectonic cycles and regularities in the development of sedimentary rocks and paleogeographic environments of sedimentation of the Russian Platform (an approach to a quantitative study). *Sedimentology* **13**, 179–212.

Schott, J. and R. A. Berner (1985). Dissolution mechanisms of pyroxenes and olivenes during weathering. *The Chemistry of Weathering* (J. I. Drever, ed.), pp. 35–54. D. Reidel Publishing Company: Dordrecht, The Netherlands.

Schubert, C., H. O. Briceño, and P. Fritz (1986). Paleoenvironmental aspects of the Caroní-Paragua river basin (southeastern Venezuela). *Interciencia* **11**, 278–289.

Scott, G. A. J. (1975a). Relationships between vegetation cover and soil avalanching in Hawaii. *Proc. Assoc. Amer. Geogr.* **7**, 208–212.

Scott, G. A. J. (1975b). Soil profile changes resulting from the conversion of forest to grassland in the montaña of Peru. *Great Plains-Rocky Mountain Geogr. J.* **4**, 124–130.

Scott, G. A. J. and J. M. Street (1976). The role of chemical weathering in the formation of Hawaiian amphitheatre-headed valleys. *Zeitschrift für Geomorphologie, N. F.* **20**, 171–189.

Sloss, L. L. (1963). Sequences on the cratonic interior of North America. *Geol. Soc. Amer. Bull.* **74**, 93–114.

Sloss, L. L. (1964). Tectonic cycles of the North American craton. *Kansas Geol. Survey Bull.* **169**, 449–460.

Sloss, L. L. and R. C. Speed (1974). Relationships of cratonic and continental-margin tectonic episodes. *In* "Tectonics and sedimentation, SEPM Spec. Publ. 22" (W. R. Dickenson, ed.), pp. 98–119. Society of Economic Paleontologists and Mineralogists, Tulsa, Oklahoma, USA.

Sloss, L. L. (1979). Global sea level changes: A view from the craton. *In* "Geological and Geophysical Investigations of Continental Margins" (J. Watkins, S. L. Montardert, and P. W. Dickerson, eds.). *Amer. Assoc. Petrol. Geologists Mem.* **29**.

Soares, P. C., P. M. B. Landim, and V. J. Fulfaro (1978). Tectonic cycles and sedimentary sequences in the Brazilian intracratonic basins. *Geol. Soc. Amer. Bull.* **89**, 181–191.

Stallard, R. F. (1980). Major element geochemistry of the Amazon River system, Ph.D. Thesis, June 1980, Massachusetts Institute of Technology – Woods Hole Oceanographic Institution Joint Program in Oceanography. *In* "WHOI-80-29", 366 p. Woods Hole Oceanographic Institution, Woods Hole Massachusetts, USA.

Stallard, R. F. (1985). River chemistry, geology, geomorphology, and soils in the Amazon and Orinoco basins. *In* "The Chemistry of Weathering" (J. I. Drever, ed.), pp. 293–316. Reidel, Dordrecht, The Netherlands.

Stallard, R. F. (1988). Weathering and erosion in the humid tropics. *In* "Physical and Chemical Weathering in Geochemical Cycles" (A. Lerman and M. Meybeck, eds.), pp. 225–246. Kluwer Academic Publishers, Dordrecht, The Netherlands.

Stallard, R. F. and J. M. Edmond (1981). Geochemistry of the Amazon I. Precipitation chemistry and the marine contribution to the dissolved load at the time of peak discharge. *J. Geophys. Res.* **86**, 9844–9858.

Stallard, R. F. and J. M. Edmond (1983). Geochemistry of the Amazon 2: The influence of the geology and weathering environment on the dissolved load. *J. Geophys. Res.* **88**, 9671–9688.

Stallard, R. F. and J. M. Edmond (1987). Geochemistry of the Amazon 3: Weathering chemistry and limits to dissolved inputs. *J. Geophys. Res.* **92**, 8293–8302.

Stallard, R. F., L. Koehnken, and M. J. Johnsson (1990). Weathering processes and the composition of inorganic material transported through the Orinoco River system, Venezuela and Colombia. *In* "El Río Orinoco como Ecosistema/The Orinoco River as an Ecosystem" (F. H. Weibezahn, H. Alvarez, W. M. Lewis, Jr, eds), pp. 81–119. Impresos Rubel, C.A.: Caracas, Venezuela.

Sugden, D. E. (1976). A case against deep erosion of shields by ice sheets. *Geology* **4**, 580–582.

Sugden, D. E. (1978). Glacial erosion by the Laurentide ice sheet. *J. Glaciology* **83**, 367–391.

Suppe, J. (1981). Mechanics of mountain building in Taiwan. *Geol. Soc. China Mem.* **4**, 67–89.

Szczerban, E. (1976). Cavernas y simas en areniscas precambricas del Territorio Federal Amazonas y Estado Bolívar. *Venez. Dir. Geol. Bol. Geol. Pub. Esp.* **7** (2), 1055–1072.

Trolard, F. and Y. Tardy (1989). An ideal solid solution model for calculating solubility of clay minerals. *Clay Minerals* **24**, 1–21.

Urbani, P. F. (1986). Notas sobre el origen de las cavidades en rocas cuarcíferas precámbricas del Grup Roraima, Venezuela. *Interciencia* **11**, 298–300.

Vail, P. R. and J. Herdenbol (1979). Sea-level changes during the Tertiary. *Oceanus* **22**(3), 71–79.

Vail, P. R., R. M. Mitchum, R. G. Todd, J. M. Widmier, III, S. Thompson, J. B. Scngree, J. N. Bubb, and W. G. Hatlelid (1977). Seismic stratigraphy and global changes of sea-level. *In* "Seismic-stratigraphy Applications to Hydrocarbon." *Amer. Assoc. Petrol. Geologists Mem.* **26**.

van Breeman, N. (1988a). Effects of seasonal redox processes involving iron on the chemistry of periodically reduced soils. *In* "Iron in Soils and Clay Minerals" (J. W. Stucki, B. A. Goodman, and U. Schwertmann, eds.). *NATO ASI Series C: Mathematical and Physical Sciences* **217**, 797–824.

van Breeman, N. (1988b). Long-term chemical, mineralogical, and morphological effects of iron-redox processes in periodically flooded soils. *In* "Iron in Soils and Clay Minerals" (J. W. Stucki, B. A. Goodman, and U. Schwertmann, eds.). *NATO ASI Series C: Mathematical and Physical Sciences* **217**, 825–842.

Wentworth, C. K. (1943). Soil avalanches on Oahu, Hawaii. *Geol. Soc. Amer. Bull.* **54**, 53–64.

White, A. F. (1983). Surface chemistry and dissolution kinetics of glassy rocks at 25°. *Geochim. Cosmochim. Acta* **47**, 805–815.

White, W. A. (1972). Deep erosion by continental ice sheets. *Geol. Soc. Amer. Bull.* **83**, 1037–1096.

Zonneveld, J. I. S. (1969). Preliminary remarks on summit levels and the evolution of the relief in Surinam (S. America). *Verhandelingen Kon. Ned. Geol. Mijnbouwk. Gen.* **27**, 53–60.

Zonneveld, J. I. S. (1975). Some problems of tropical geomorphology. *Zeitschrift für Geomorphologie, N. F.* **19**, 377–392.

7

Pedosphere

Fiorenzo C. Ugolini and Henri Spaltenstein

7.1 Introduction

7.1.1 The Pedosphere

The pedosphere is the envelope of the Earth where soils occur and where the soil-forming processes are active. Soils are physical entities formed at the Earth's surface that have acquired chemical and physical properties in response to environmental factors. This concept of soils was introduced by Dokuchaev, a Russian scientist at the end of the last century (Joffe, 1949).

The biosphere, that part of the Earth in which life exists (Hutchinson, 1970), overlaps the pedosphere, but the two are not coincident. Soil can develop in the absence of life. Abiotic or virtually abiotic conditions may exist in the high Arctic and in the ice-free areas of continental Antarctica. Although these polar regions experience the most severe and harsh climates on Earth, soil formation occurs in frozen and ultraxeric soils as evidenced by ion migration along unfrozen films of water covering the soil particles (Ugolini and Anderson, 1973). Current weathering and synthesis of minerals in Antarctic soils proves that soil processes are not exclusively in the realm of biological systems (Claridge, 1965; Jackson *et al.*, 1976; Ugolini and Jackson, 1982; Claridge and Campbell, 1984; Campbell and Claridge, 1987).

Nevertheless, most soil formation does occur in the presence of life, and soil formation proceeds at a faster rate in the presence of biota (Ugolini and Edmonds, 1983). Although green plants can build organic molecules by photosynthesis, elements other than C, O, and H must be obtained from the soil. The cations and anions needed for building organic molecules are sequestered in the crystal structure of minerals. They become available to photosynthetic plants and other organisms via the processes of soil formation. Soil is a source of nutrients as the minerals weather. In addition, it functions as the repository of wastes and the burial place for dead organisms. Decomposition processes, favored by the ability of soils to hold moisture and to moderate temperatures, are responsible for recycling nutrients.

By promoting decomposition and by retaining nitrogen, phosphorus, and nutrient cations, the soil prevents the rapid loss of essential elements to groundwater and rivers. The result of this process is a net accumulation of available nutrients in the soil and, thus, an enhanced productivity at all trophic levels of the ecosystem.

Chemical weathering is essential to the formation of soil. Weathering is mediated by the porous nature of the soil, which allows the retention of moisture and the exchange of gases. Synthesis of new minerals, especially the clay minerals and oxides, occurs, and a dynamic equilibrium is established between the clay minerals and the ions in the soil solution. These exchangeable processes provide nutrients to the green plants that grow at the interface between the soil and the atmosphere. Nitrogen, abundant in the atmosphere but very rare in the lithosphere, is fixed from the air by symbiotic and asymbiotic organisms in the soil. This fixed N is made available to plants. Biomass grown on the soil serves as a sink for many of the products of weathering, especially Ca, Mg, K, and P. These ions, liberated through abiotic and biotic weathering processes, are temporarily stored in the biomass, but eventually return to the soil upon the death

Global Biogeochemical Cycles
ISBN 0-12-147685-5

and decomposition of the living organisms. The mineralization of the organic debris recycles many of the nutrients needed by the green plants. Thus the pedological cycle is part of the larger global biogeochemical cycle.

7.1.2 Factors of Soil Formation

Soil processes are a function of environmental factors to the extent that soil properties can be predicted in a general way if the factors are known. Jenny (1941) attempted to relate the factors of soil formation (climate, organisms or biota, topography, parent material (the geological substratum), and time) to any soil property such as N, C, clay, depth of leaching of carbonates, and others. Jenny saw these factors not as soil formers but as parameters defining the state of the soil. Furthermore, for the purpose of providing numerical relationships, Jenny considered each of the factors as an independent variable capable of varying while the others are held constant. Determining the role of each environmental factor in influencing the development of any particular soil is difficult. Many soils have been resting on the landscape for a long time; these soils are therefore polygenetic. They have acquired some of their properties under a constellation of soil-forming factors different from those currently in operation. In addition, the factors are not always independent in their actions. This is particularly true for the biota which both influences and is influenced by climate, parent material, time, and relief. Consequently, there is a problem in holding the biotic factor as an independent variable. Quantitative solutions have been attempted and numerical values have been obtained for topo-, chrono-, and climofunctions (Yaalon, 1975; Bockheim, 1980). Jenny's approach is a valuable tool for assessing the role of different environmental factors on the formation of soil, but its applicability is restricted to selected ecosystems where nature has cooperated with Jenny's design! Notwithstanding these limitations, Jenny's approach is used in our discussion as a conceptual framework to relate the soil-forming processes to the environments of the Earth. Consequently, in the graphic display of these processes (see Fig. 7-4), we have assumed that soil formation, along the transect from the poles to the equator, is occurring under uniform parent material, similar topography, and equivalent time, and the significant variables are climate and the biota.

7.1.3 The Soil Profile

The soil profile is the manifestation of the soil-forming or pedogenic processes on the geological substratum or parent material (Fig. 7-1). The soil profile is an assemblage of horizons displayed by a vertical cut at the surface. Horizons are layer-like, more or less parallel to the surface, and differ from each other in morphology, composition, and consistency (Table 7-1).

Soil profiles form because of the endless migration of ions, molecules, and particles into the soil material. Meteoric inputs include H_2O, CO_2, O_2, nitrogenous compounds, pollutants, salts, and dust. These molecules and compounds come from space, from the atmosphere and the ocean, and other terrestrial ecosystems.

The composition of the meteoritic input can be drastically changed as it passes through the canopy. Cronan (1984) summarized the predominant canopy processes and their effects on the composition of the throughfall. A specific example is given where the effects of two different canopies on the incident precipitation were examined. A considerable difference was found between the *Abies mariesii* (Maries' fir) and *Miscanthus sinensis* (Japanese pampas grass) canopies (Ugolini *et al.*, 1988). Whereas in the case of *A. mariesii* the acidity of the precipitation is only partially neutralized by the canopy, in the case of *M. sinensis* the neutralization is complete causing

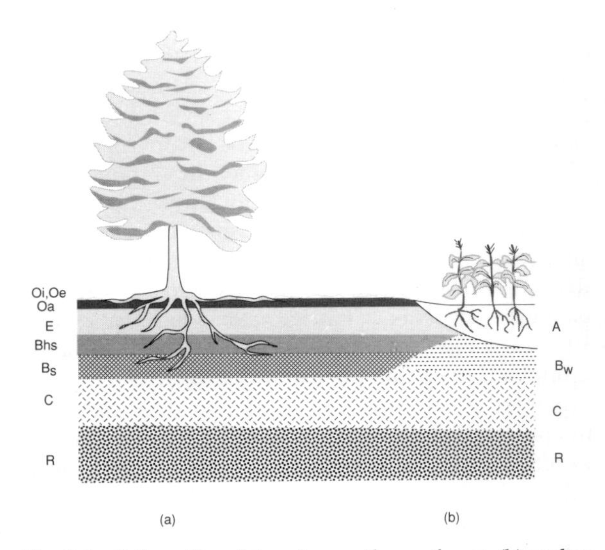

Fig. 7-1 Soil profiles: (a) under coniferous forest; (b) under herbaceous cover.

Table 7-1 Soil profile and major soil horizons[a]

1. *Soil profile*: a vertical cut at the surface of the continental earth crust displaying an assemblage of genetic horizons

2. *Horizons*: layers, more or less parallel to the surface, that differ from each other in morphology, composition, and consistence (Fig. 7-1)

3. *How the soil profile develops*: continuous or sporadic migration of ions, molecules, particles and aerosols. Water is the vehicle for transportational processes

4. *Soil morphology* (Fig. 7-1):
   Organic horizons (O)
   - Oi – fibric (partly decomposed)
   - Oe – hemic (intermediately decomposed)
   - Oa – sapric (highly decomposed)

   Mineral horizons
   - A – mineral mixed with humus, dark-colored; topmost horizon
   - E – horizon of maximum eluviation of clay, iron, aluminum, and showing a concentration of resistant minerals, such as quartz. It is generally lighter in colour than the A horizon
   - B – zones of accumulations or illuviation. Illuviation means wash into. This horizon represents the region of maximum accumulation of material such as Fe, Al, and silicate clays. In arid regions, $CaCO_3$, $CaSO_4$, or other salts may accumulate
   - C – unconsolidated material underlying the solum (A and B). It may or may not be made of the same parent material from which the solum formed. It is the least weathered and also little affected by the soil-forming process, although it may be weathered
   - R – underlying consolidated rock

5. *Subordinate distinctions* (selected)
   - a – highly decomposed organic matter
   - e – intermediately decomposed organic matter
   - h – illuvial accumulation of organic matter
   - i – slightly decomposed organic matter
   - k – accumulation of carbonates
   - o – residual accumulation of sesquioxides
   - s – illuvial accumulation of sesquioxides
   - t – accumulation of clay
   - w – development of color or structural change

[a] After the US Department of Agriculture (1986).

a rise in pH. Also, while NH_4^+ and NO_3^- concentrations are slightly enhanced in the throughfall of the *A. mariesii* canopy, these nitrogen compounds are completely adsorbed by *M. sinensis*. The leaching of the basic cations of Ca, Mg, K, and Na is even more dramatic. Leaching of basic cations from *M. sinensis* exceeds leaching from *A. mariesii* by a factor of 5. *M. sinensis* throughfall is also richer in dissolved organic carbon, HCO_3^-, SO_4^{2-}, and Cl^- than the *A. mariesii* throughfall. In conclusion, the interaction between the meteoric inputs and the vegetative canopy substantially changes the composition of the water entering the soil. Water movement into the soil is the essence of soil formation. As water percolates through the soil, it transports a myriad of cations and anions, radicals, molecules, and particulate matter. Many of these substances participate in a number of reactions that alter the soil material or the parent material. As the soil solution migrates through the soil, the products at one level become the reactants for the next level. The result of localized reactions occurring over time imparts changes to the original material; these changes are manifested in the appearance of soil horizons (Fig. 7-1). By considering the ecosystems previously discussed – *A. mariesii* and *M. sinensis* – it is possible to outline briefly the processes occurring under the two different species.

7.1.3.1 *Soil Profile 1:* A. mariesii

In the case of *A. mariesii*, as the throughfall enters the surface of the soil, it penetrates through an organic layer where organic acids produced by decomposition of the litter layer are present. These organic acids break down the minerals and form complexes of metals such as Al and Fe. The organo-metal complexes are water-soluble and able to migrate into the soil, provided the carbon-to-metal ratio is large. The solubility of these complexes is reduced as the solution migrates downward and more metals are complexed. When the carbon : metal ratio reaches a critical value, the complexes are arrested. The sequence of events – weathering, complexation, and leaching of poly-valent metals from the surface and their arrest at depth – results in the formation of a bleached and light in color E horizon at the surface and a dark Fe, Al, and C rich, Bhs horizon, below (see Fig. 7-1a). While organic acids dominate in the E and Bhs horizons, H_2CO_3 dominates the lowest part of the profile (the Bs, BC, and C horizons). The low temperatures and the low rate of diffusion through the dense Bhs help maintain a high PCO_2 in the Bs and lower horizons. Carbonic acid is the major weathering agent, but this acid does not form complexes with the polyvalent metals (Al and Fe), and consequently the latter are not mobilized, but tend to remain in the horizons. Reactions of these metals, specifically aluminum, with silicic acid (H_4SiO_4) results in the formation of secondary minerals. The Bs is the horizon where this weathering regimen occurs; here the soil is reddish because of the presence of iron oxides or hydroxides that precipitate as a separate phase. Details of these processes have been provided for similar soils developed under conifers in the central Cascades of Washington State, USA (Ugolini *et al.*, 1977; Ugolini and Dehlgren, 1987).

7.1.3.2 *Soil Profile 2:* M. sinensis

Under *M. sinensis* the surface of the soil is not covered by a layer of organic matter. The major input of organic matter into the soil occurs through the myriads of fine roots typical of herbaceous species. The products of microbial decomposition of the below-ground biomass are water-insoluble humic substances that do not migrate in the soil. These humic substances impart a dark color to the soil and are responsible for the formation of the A horizon. The organo-metal complexes formed in the A horizon are not water-soluble and cannot migrate into the horizon below, the B horizon (see Fig. 7-1b). This latter horizon is formed by *in situ* weathering by H_2CO_3, which acts as the major proton donor. Genesis of this horizon is equivalent to the Bs horizon of the *A. mariesii* profile.

In the examples given above, two dramatically different profiles have developed from the same volcanic ash material that has been exposed to contrasting environmental conditions such as different climate and vegetation. The two examples of profile development indicate that different plant species are responsible for the production of the different proton donors or acids. Soluble and insoluble organic acids and carbonic acid are the major proton donors associated with *A. mariesii* and *M. sinensis*. These proton donors are responsible for the weathering regimen and for the transport of ions in the soil. In the final analysis, they are responsible for the different soil-forming processes and therefore for the different profiles. Since the plant assemblage is related to climate, we have the concept that climate determines the biota, which in turn dictates the nature of the proton donors and thus the chemical regimen of the soil (Ugolini, 1986a, 1987). This concept will be further extended to other ecosystems of the Earth.

7.2 The Soil Constituents

7.2.1 *General Composition*

Soil is a multi-phase system consisting of solids, liquids, and gases. In a typical soil, solids, liquids, and gases occupy 50, 15–35, and 15–35% of the total volume respectively (Jackson, 1964). As an example, a soil sampled in a New Jersey field during the autumn showed 55% of its volume occupied by solids and the remaining 45% by pores. Of the 45% pore volume, 12% was occupied by airspace and 33% by water (Ugolini, unpublished).

About 40% of the solid phase is inorganic and about 10% is organic, either dead or alive. The inorganic constituents of the soil are dominated by four elements: O, Si, Al and Fe (Jackson, 1964).

Table 7-2 Elemental composition by atoms of the Earth's crust[a]

Element	%	Cumulative %	Element	%	Cumulative %
O	60.4		Ti	0.19	
Si	20.5	80.9%	P	0.07	
Al	6.3		F	0.07	
H	2.9	90.9%	Mn	0.04	
Na	2.6		C	0.04	
Ca	1.9		S	0.02	
Fe	1.9		Sr	0.009	
Mg	1.8		Ba	0.006	99.9%
K	1.4	99.5%			

[a] From Mason and Moore (1982).

Together with Mg, Ca, Na, and K, they constitute 99% of the soil mineral matter. This elemental composition and distribution does not differ much from the average composition of the major volcanic and plutonic rocks (Table 7-2).

Of approximately 80 elements found in soils, 17 are used by plants to grow. These are called the essential elements and include H, C, O, N, P, S, K, Mg, Ca, B, Cl, Mn, Fe, Cu, Zn, Mo, and Co. Hydrogen and O are derived from the hydrosphere, C from the atmosphere, N from the biosphere, and the remaining 13 from the lithosphere or the mineral soil. Soil is mainly an O-Si-Al-Fe matrix containing relatively small amounts of the other essential elements. Plant growth depends on the rapid recycling of many nutrients and, in the case of commercial crops, to the application of fertilizers.

Inorganic components of soils occur as minerals with definite crystal structure or as hydrous oxides of Al, Fe, Mn, and Ti displaying varying degrees of crystallinity. Non-crystalline alumino silicates and non-crystalline silica may be important in some soils. Minerals in soils are divided into primary and secondary minerals. Primary minerals, which occur in igneous, metamorphic, and sedimentary rocks, are formed during magmatic differentiation. A list of these minerals is given in Table 7-3. Quartz and feldspars are the most abundant primary minerals in

Table 7-3 Primary minerals commonly found in soils

	Weatherability	Cations useful for plants	Other elements useful for plants or liberated at a high rate
Quartz	–	–	–
Feldspar-K	+	K	–
Plagioclase	+	Ca–Na	–
Mica	(+)	K	B
Biotite	++	K–Mg	Fe
Chlorite–Mg	++	K–Mg	Fe
Amphibole	+(+)	Mg–Ca	Fe, minor elements
Pyroxene	++	Mg–Ca	Fe, minor elements
Volcanic glass	++	Ca–Mg–K–Na	Si–Al
Magnetite	–	–	Fe
Apatite	(+)	Ca	P
Other minerals not of igneous or metamorphic origin, but found in the soil inherited from the parent material (carbonated rock):			
Calcite	++	Ca	–
Dolomite	++	Mg–Ca	–

soils. Micas, olivine, pyroxenes, amphiboles, and others are present but in smaller quantities. Primary minerals are mostly present in the sand fraction and coarse silt and have small specific surface areas.

Secondary minerals are generally formed in near-surface conditions. Secondary minerals include layer silicates or clay minerals, carbonates, phosphates, sulfur minerals, and different hydroxides and oxy-hydroxides of Al, Fe, Mn, Ti, and Si. Non-crystalline minerals such as allophane and imogolite are also included among the secondary minerals. Secondary minerals, such as the clay minerals, may show specific surface areas in the range 20–800 m^2/g and up to 1000 m^2/g in the case of imogolite (Wada, 1985). Surface area is very important because most reactions in soil are surface reactions at the solid and liquid interface. A brief examination of layer silicates and soil colloids is useful for understanding the phenomena of adsorption, fixation, and weathering.

7.2.2 Clay Minerals and Clay Mineral Properties

Layer silicate structure, as for the rest of silicate minerals, is dominated by the strong Si—O bond which accounts for the insolubility of these minerals. Other elements involved in the building of layer silicates are Al, Mg, or Fe coordinated with O and OH groups. The spatial arrangement of Si and the above metals with O and OH results in the formation of the tetrahedral and octahedral sheets. The combination of the tetrahedral and octahedral sheets forms a layer, or the units of layer silicates. A number of layer silicate structures can be generated with different arrangements of the tetrahedral and octahedral sheets or other hydroxide layers (Table 7-4).

A characteristic common to all the layer silicates is the electric charge present at their surfaces. This charge, for most layer silicates in soils, is predominantly negative. The negative charge tends to attract cations in the soil solution; these cations are electrostatically held and are exchangeable. Cation exchange capacity (CEC) measures the ability of layer silicates, or other colloidal substances, to hold cations. In a fertile agricultural soil, these exchangeable cations represent most of the nutrients needed by plants: NH_4^+, Na^+, K^+, Ca^{2+}, and Mg^{2+}. In acid soils, part or all of the cation exchange capacity is satisfied by Al^{3+} and H^+ ions, neither one being an essential nutrient for plants.

It is fortunate for humans that the phenomenon of cation exchange capacity exists. Without this property, the nutrients released by weathering and mineralization would be easily lost to the rivers and ocean, leaving an infertile land. According to Jackson (1969), life that had originated in the sea was able to move onto the land because the clay produced by weathering had the capacity (CEC) for holding cations and making them available to plants.

7.2.2.1 The 1 : 1 minerals-kaolinite group

The 1 : 1 type layer silicates (Table 7-4) consist of a tetrahedral silica sheet and an octahedral sheet of hydroxy aluminum; the two combine to form a layer. The individual layers are about 0.7 nm thick. The 1 : 1 minerals belong to the kaolinite-serpentine group. Kaolinite [$Al_2Si_2O_5(OH)_4$] is probably the most abundant mineral of this group and it represents one of the most common clay minerals in soils. It prevails in soils of subtropical and tropical regions. The surface area of kaolinite is small (Table 7-4); furthermore, this mineral shows no plasticity and cohesion and has low swelling and shrinkage properties (Bohn *et al.*, 1985). It has a low cation exchange capacity. The small amount of negative charge is derived from deprotonation of the OH groups. As a result, the cation exchange capacity is pH-dependent.

7.2.2.2 The 2 : 1 mineral group

The 2 : 1 layer silicates consist of two tetrahedral sheets with an octahedral sheet in between. A characteristic of these minerals is the isomorphic substitution in both the tetrahedral and octahedral sheets. These substitutions produce unbalanced charges in the clay structure. Most of the 2 : 1 clay minerals in soil display net negative charges. These charges are compensated by the adsorption of cations giving rise to the phenomenon of cation exchange capacity. Charge density on the basal plane of 2 : 1 layer silicates is important in distinguishing among the several mineral species.

7.2.2.2.1 *Micas.* Micas or muscovites are 2 : 1 minerals with strong layer charges due to the isomorphous substitution of Al^{3+} for Si^{4+} in the tetrahedral sheets. The excess of net unbalanced negative charges, ideally 1 charge per formula unit, is compensated by K^+ ions strongly bonded between adjacent basal planes. Consequently, micas have a relatively low CEC and a low surface area (Table 7-4)

Table 7-4 Selected properties of phyllosilicates, allophane, and imogolite[a]

Type	Species	Cation charge per unit formula	Exchange capacity, $cmol(p^+)/kg$	Basal spacing		Surface area (m^2/g)	Expandability
				Contract. (nm)	Expand. (nm)		
1:1	Kaolinite	~0	1–10	0.72		10–20	Non-expandable
2:1	Mica	~1	20–40	1.0		70–120	Non-expandable
2:1	Vermiculite	~0.6–0.9	120–150	1.0	1.5	600–800	Expandable (variable)
2:1	Montmorillonite	~0.6–0.25	80–120	1.0	1.9	600–800	Expandable (variable)
2:1:1	Chlorite	Similar to mica	10–40	1.4		70–150	Non-expandable
	Allophane	Variable	10–150	[b]		~1000	
	Imogolite	Variable	10–150	[b]		~1000	

[a] Adapted from Soderman and Quigley (1965) and Bohn *et al.* (1985).

[b] X-ray diffraction pattern for allophane shows the characteristics of non-crystalline material (Wada, 1985); imogolite patterns consist of a number of broad peaks (Wada, 1985).

because they display only the external surface of the crystal (Bolt and Bruggenwert, 1976; Fanning and Keramidas, 1977). The thickness of the mica layer is about 1.0 nm. Micas are common in soils and are largely of primary origin, although evidence for authigenic soil mica has been presented (Jackson *et al.*, 1976). Micas are abundant in shales, slates, phyllites, schists, gneiss, granites, and sediments derived from these rocks. Micas are important as "parent material" for other 2 : 1 layer silicates, such as vermiculite and montmorillonite, with which they are often interstratified (Jackson, 1964). The weathering of micas releases K, which may become available to plants.

7.2.2.2.2 *Vermiculites.* Vermiculites are 2 : 1 expanding minerals with a structure similar to micas (Table 7-4). They are considered to be derived from the alteration of micas (Douglas, 1977). Cation exchange capacity is high, as is the surface area. Potassium or NH_4^+ in solution tends to be strongly fixed by vermiculites. Upon fixation of these ions, the CEC decreases and the properties of vermiculite become like those of mica. In acid soil, hydroxy aluminum polymers can be fixed in the interlayer position to form an "island-like" structure (Jackson, 1964). These hydroxy aluminum polymers dissolve under acidic conditions, such as those produced by acid rains.

7.2.2.2.3 *Smectite.* Montmorillonite belongs to the smectite or montmorillonite-saponite group (Borchardt, 1977). Montmorillonite is a 2:1 layer silicate that can expand along the *c* axis from 0.96 to 1.8 nm and more depending on the exchangeable cation and the degree of interlayer solvation (Jackson, 1964; Grim, 1968). The charge density for montmorillonite is low; both the cation exchange capacity and total surface area are high. This clay maximizes all the characteristics associated with colloidal material, such as small particle size, adsorption, plasticity, cohesion, shrinkage, and swelling. Montmorillonite and smectite minerals are widely distributed in soils and sediments (Velde, 1985). These minerals seem to be stable in neutral, slightly leached soils where silica and basic cation activities are high. However, montmorillonite has also been found in acid and highly leached soil horizons of Podzols or Spodosols (Brown and Jackson, 1958; Coen and Arnold, 1972; Zachara, 1979; Ross, 1980). In this case, it may contain hydroxy aluminum polymers in the interlayer position.

7.2.2.3 The 2 : 1 : 1 mineral group, chlorite

Chlorites in soil occur as primary minerals derived from mafic rocks and as secondary minerals from the weathering of biotite, hornblende, and other amphiboles and minerals (Barnhisel, 1977). Chlorites are 2 : 1 : 1 minerals consisting of 2 : 1 mica structure in addition to an interlayer hydroxide sheet. Chlorites have low CEC and surface areas.

7.2.2.4 Non-crystalline aluminosilicates

7.2.2.4.1 *Allophane.* Allophane is a hydrous aluminosilicate with a variable composition (Wada and Harward, 1974; Wada, 1977, 1985). Electron micrographs show that allophanes are made of "hollow spherules" with a diameter of 3.5–5 nm (Wada, 1985). Chemically, allophane consists of Al_2O_3, Fe_2O_3, and SiO_2 with minor amounts of Mg^{2+}, Ca^{2+}, K^+, and Na^+. Degrees of hydration and pH are important in determining the exchange properties of this mineral colloid. High cation exchange capacity is displayed in neutral and slightly alkaline conditions, whereas low values are measured in acid environments (Table 7-4). At low pH values, allophane and allophanic soils display positive charges and are able to bond to large amounts of phosphate and sulfate. Allophane is present in soils derived from tephra and other pyroclastic materials. Accordingly, areas covered by volcanic soils (about 1% of the Earth's land surface, Leamy *et al.*, 1980), may show phosphate and sulfate deficiency (Wada, 1977).

7.2.2.4.2 *Imogolite.* Imogolite is a hydrous aluminosilicate with a thread-like morphology found mostly, but not exclusively, in soils derived from volcanic ash (Wada, 1977). Imogolite shows a more definite structure than allophane (Cradwick *et al.*, 1972) and has an approximate composition of $SiO_2Al_2O_32.5H_2O$. In humid temperate climates, allophane and imogolite prevail in volcanic soils 5000–10 000 years old; after 10 000 years, these minerals are changed into others (Wada, 1977, 1980; Fig. 7-2).

7.2.2.5 Oxides, hydroxides, and oxyhydroxides

Oxides, hydroxides, and oxyhydroxides of Al, Fe, Si, and Ti in soils are mostly secondary minerals formed as weathering progresses (Jackson, 1964). These minerals vary from amorphous to crystalline and often display variable charges. This property de-

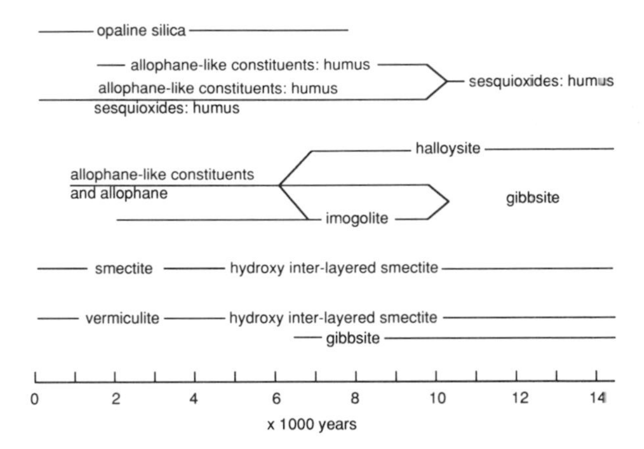

Fig. 7-2 Formation and transformation of clay minerals and their organic complexes in soils developed from volcanic ash in humid, temporate climatic zones. Horizontal lines indicate approximate duration of the respective constituents. Modified from Wada and Harward (1974) with the permission of Academic Press, Inc.

pends on the composition of soil colloids and on the ionic character of the surrounding soil solution (Bowden *et al.*, 1980).

Most soil particles have a net negative charge because it is the dominant charge on layer silicates and organic matter. However, some highly weathered soils and soils dominated by volcanic ash containing allophane and hydrous oxides may actually have a net positive charge.

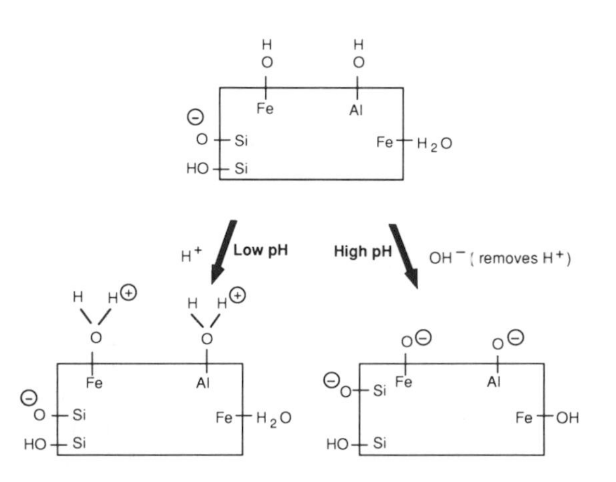

Fig. 7-3 Diagram illustrating the development of positively and negatively charged sites on surfaces of soil constituents, at low and high pH. Adapted from Parfitt (1980) with the permission of the New Zealand Society of Soil Science Offset Publications.

Variable charges arise due to the gain or loss of H^+ from functional groups on the surface of solids (Fig. 7-3). The charges developed on these groups depend on: (1) the pH of the ambient solution, which regulates the degree of protonation and deporotonation of the functional groups; (2) the nature and concentration of cations and anions in solution; and (3) temperature (Wada, 1977, 1978). The soil components that contain functional groups are oxides and hydroxides, allophane, imogolite, and, to a much lesser extent, the layer silicates. Functional groups present in organic matter, such as OH and COOH, also display variable charges.

7.2.2.5.1 *Aluminum hydroxides.* Gibbsite, $Al(OH)_3$, the most abundant aluminum hydroxide in soils (Jackson, 1964; Hsu, 1977), is especially abundant in old soils affected by intensive weathering under tropical climates. Under these circumstances, the desilication of the minerals leaves aluminum hydroxide as a residue. Aluminum hydroxides and aluminum polymers occurring in cryptocrystalline or non-crystalline states display variable charges and become positively charged under acid conditions, adsorbing anions such as SO_4^{2-}, NO_3^-, Cl^-, and PO_4^{3-} (Hsu, 1977). Table 7-5a shows major Al minerals.

Table 7-5 Most commonly occurring Fe, Al, and Ti oxides, oxide-hydroxides, and hydroxides

(a)	*Al-oxides*: Greek prefixes according to Weiser and Milligan		
	Corundum	αAl_2O_3	(Rhombohedral)
	Boehmite	$\gamma AlOOH$	(Orthorhombic)
	Diaspore	$\alpha AlOOH$	(Orthorhombic)
	Hydrargillite	$\gamma Al(OH)_3$	(Monoclinic)
	Bayerite	$\alpha Al(OH)_3$	
	Gibbsite	$\gamma Al(OH)_3$	(Monoclinic)
(b)	*Fe-oxides*		
	Goethite	$\alpha FeOOH$	(Orthorhombic)
	Hematite	αFe_2O_3	(Trigonal)
	Maghemite	γFe_2O_3	(Cubic or tetragonal)
	Lepidocrocite	$\gamma FeOOH$	(Orthorhombic)
	Magnetite	Fe_3O_4	(Cubic)
	Ferrihydrite	$5Fe_2O_3 \cdot 9H_2O$	(Hexagonal)
(c)	*Ti-oxides*		
	Rutile	TiO_2	(Tetragonal)
	Anatase	TiO_2	(Tetragonal)
	Ilmenite	$FeTiO_3$	

7.2.2.5.2 Iron oxyhydroxides, oxides, and others. Goethite ($\alpha FeOOH$) is the most common iron-oxide in soils. It occurs under many climatic conditions but is preferentially found in cool humid areas (Schwertmann, 1988) and it imparts a yellow-brown color to many soils (Schwertmann and Taylor, 1977; Schwertmann, 1985). Hematite (αFe_2O_3) is also frequently seen, but is not as widespread as goethite; it occurs in warmer environments than goethite (Schwertmann, 1988). Soils rich in hematite are red. Geothite and hematite are both associated with intense weathering (Jackson and Sherman, 1953; Jackson, 1964).

Lepidocrocite ($\gamma FeOOH$) is less frequent than goethite or hematite and it occurs in poorly drained soils where it forms from oxidation of Fe^{2+} hydroxy compounds (Schwertmann and Taylor, 1977). It displays an orange color (Schwertmann, 1988).

Ferrihydrite, $[5Fe_2O_3 \cdot 9H_2O]$, once called "amorphous ferric hydroxide", has been reported in bog iron and drainage ditches and in environments where other compounds prevent the formation of goethite and lepidocrocite (Schwertmann, 1988). It occurs as bulky precipitates containing water, adsorbed ions, and organic material. It appears as small (5.0–10.0 nm), spherical particles with a high surface area (200–350 m^2/g; Schwertmann and Taylor, 1977). Ferrihydrite is probably the form of iron present in the B horizon of true Podzols.

Magnetite (Fe_3O_4) is a detrital mineral derived from the parent rock and it is mostly segregated in the sand fraction. When oxidized, it changes into maghemite (Bohn *et al.*, 1985). Table 7-5b shows the major Fe minerals.

Under well drained conditions, iron in soil, unless chelated, is insoluble and tends to be concentrated through weathering as the more soluble ions are removed. In poorly drained soils, iron is reduced to the soluble ferrous state and is either removed from the soil or precipitated as sulfide, phosphate, or carbonate minerals. Other processes, such as chelation, are also effective in mobilizing the iron in the soil profile. The bulk of secondary iron in soils is in the oxide forms.

Iron oxide and oxyhydroxides, especially when displaying large surface areas, are important in developing pH-dependent charges.

7.2.2.5.3 Titanium oxides. Titanium in soils is present in finely divided crystals of primary minerals such as rutile and anatase (TiO_2), sphene ($CaTiSiO_5$), and ilmenite ($FeTiO_3$). Titanium minerals are resistant to weathering and tend to persist and accumulate in soils. Deeply weathered soils of Hawaii may contain 5–20% or more of TiO_2 (Sherman, 1952; Tamura *et al.*, 1955). Table 7-5c shows the major Ti minerals.

7.2.2.5.4 Silicon oxides. Among the silicon oxides, quartz (SiO_2) is the most abundant mineral in soil. Other minerals (tridymite, cristobalite, and glass) are common in soils derived from pyroclastic material. Opal is associated with diatomaceous rocks, and amorphous silica is of hydrothermal origin (Wilding *et al.*, 1977). Plant opal or phytoliths are of plant origin and form as silica accumulates in the cells of plants. Grasses take up considerable amounts of silicon and often contain 3–5% silica on a dry weight basis. However, values as high as 20% have been recorded (Norgren, 1973). Silicon is taken up in monomeric form as $Si(OH)_4$.

During the process of soil development in humid temperate climates, the soil material undergoes desilication and silica is readily lost as monomeric silica (Acquaye and Tinsley, 1965; Ugolini *et al.*, 1977a). Monomeric silica can react with metal ions and leads to the synthesis of secondary minerals such as clay minerals.

7.2.3 Soil Organic Matter

7.2.3.1 Origin, production, and decomposition

Organic matter in soils is made up of plant and animal remains partially decomposed, substances synthesized during the decomposition, and microbial bodies. Organic matter is primarily carbon, oxygen, and hydrogen derived directly or indirectly from photosynthesis. Decomposition processes incompletely break down the complex organic molecules, so that organic matter tends to accumulate in soils. Decomposition is an extremely important process in terms of releasing energy and nutrients to the soil system.

Plant litter consists mainly of sugars, cellulose, hemicellulose, lignin, waxes, and polyphenols, and to a smaller extent proteins and cations (Alexander, 1977). The rate of litter decomposition is a function of climate and the composition of the litter. In general, decomposition is most rapid in well-aerated, moist, mesic, and near neutral soils. Cold, humid environments with high water-tables and acidic conditions favor accumulation of undecomposed or partially decomposed plant debris rather than decomposition. The nitrogen level in the plant litter is often a limiting factor in decomposition, but lignin content may also control decomposition rates, especially after long periods of decomposition (Fogel and Cromack, 1977; Edmonds, 1979).

7.2.3.2 *Humic substances*

The components of soil organic matter are most usefully described in terms of broad solubility classes, collectively referred to as humic substances. Humic substances "are amorphous, dark colored, hydrophilic, acidic, partly aromatic, chemically complex organic substances that range in molecular weight from a few hundred to several thousand" (Schnitzer and Kodama, 1977). Humic substances are fractionated into three main groups on the basis of their solubility in acid and base: (1) humic acid (HA), soluble in base, but precipitated by acidification of the alkaline extract at pH 2; (2) fulvic acid (FA), soluble in both alkaline and acid; (3) humin, not soluble either in acid or base (Kononova, 1966, Schnitzer and Khan, 1972). Whereas these three humic fractions are structurally similar, their molecular weight and functional groups differ. Humic acid and humin contain more H, N, and S, but less O than FA; the total acidity and the number of carboxyl groups of FA are greater than those of HA and humin, and the ratio of carboxyl to phenolic hydroxyl groups is almost 3 for the FA but near 2 for HA and humin. Table 7-6 shows additional analytical features of the three humic substances (Schnitzer and Kodama, 1977).

Table 7-6 Analytical characteristics of HA, FA, and humin[a,b]

	HA	FA	Humin
Elementary composition % (dry, ash-free)			
Carbon	56.4	50.9	55.4
Hydrogen	5.5	3.3	5.5
Nitrogen	4.1	0.7	4.6
Sulfur	1.1	0.3	0.7
Oxygen	32.9	44.8	33.8
Oxygen-containing groups (meq/g dry, ash-free)			
Total acidity	6.6	12.4	5.9
Carboxyl	4.5	9.1	3.9
Total hydroxyl	4.9	6.9	—
Phenolic hydroxyl	2.1	3.3	2.0
Alcoholic hydroxyl	2.8	3.6	—
Total carbonyl	4.4	3.1	4.8
Quinone	2.5	0.6	—
Ketonic carbonyl	1.9	2.5	—
Methoxyl	0.3	0.1	0.1

[a] From Schnitzer and Kodama (1977).
[b] HA and humin from the Al horizon of a haploboroll and FA from the Bh horizon of a spodosol.

Humic substances are involved in many reactions, many of which are a consequence of their colloidal properties. Humic substances have surface areas as high as 800–900 m^2/g, and have adsorptive capacities greater than the layer silicates: 150–300 cmol (p^+)/kg (1.50–3.00 meq/g) of material. Humic substances in the soil are therefore important in water retention, in ameliorating the adverse properties of clayey and silty soils by favoring aggregation, in changing the albedo because of their dark color, in providing buffering capacity, and in supplying nutrients (Brady, 1974; Bohn *et al.*, 1985).

In the pedosphere, the FA and the low molecular weight HA, which are water-soluble, are of particular interest. The interaction of these humic substances is important both in the processes of weathering and soil formation. According to Schnitzer (1971) FA constitutes between 25 and 75% of the total organic matter in soil. In swamp water, up to 85% of the organic matter content may be FA.

In terms of pedological processes, the FA and low molecular weight HA fractions are significant in attacking and weathering minerals via complexation, dissolution, and transport (Schnitzer and Kodama, 1977; Stevenson, 1982). Experiments conducted by Kodama and Schnitzer (1973) on the dissolution of chlorite minerals confirm the effectiveness of the FA in decomposing these minerals. They found that FA solutions were more effective than acidified distilled water in dissolving Al, Fe, and Mg, especially from Fe-rich chlorite. Infrared analysis of the reacted FA demonstrated the participation of the COOH groups in these reactions. The solvent reactions of humic substances with silicates and with metal oxides and hydroxides are important in rendering otherwise insoluble metals, such as Al, Fe, and Mn, soluble and mobile (Schnitzer and Skinner, 1963a,b; Rosell and Babcock, 1968; Baker, 1973; Stevenson, 1982). In the soil profile, Al–FA and Fe–FA complexes are either water-soluble or insoluble depending on the ratio between FA and the metals. When the ratio is large, these complexes are soluble and migrate; when the ratio is small, they are insoluble and remain within the soil profile (Schnitzer, 1969; Schnitzer and Desjardins, 1969; Dawson *et al.*, 1978; Stevenson, 1982, 1985; Ugolini *et al.*, 1990).

7.3 Weathering and Soil-forming Processes

Weathering occurs because rocks and minerals become exposed to physical and chemical conditions different from those under which they formed. Weathering in the pedosphere takes place close to the surface where the overburden pressure and temperature are low, and H_2O is plentiful. Because most soils harbor large numbers of microorganisms, their metabolic activity results in the consumption of O_2 and production of CO_2. The partial pressure of CO_2 in soil may be 10–100 times greater than that of the atmosphere (Holland, 1978). Under waterlogged conditions, where gas diffusion between the soil and the atmosphere is impeded, the oxygen partial pressure may approach zero and the soil solution may contain CO_2, CH_4, and H_2S. Under these conditions, Fe and Mn may become soluble.

The inorganic solid phase of any soil consists of a number of minerals displaying different degrees of weathering susceptibility. The extent of weathering of these minerals depends on the stabilities of the minerals and the physical and chemical environment in which the minerals are immersed in the soil, including the supply of water and the removal or transport of the products of weathering (see Chapter 6; Garrels and Christ, 1965; Kittrick, 1977; Colman and Dethier, 1986).

Physical weathering involves changes in the degree of consolidation with little or no chemical and mineralogical changes to rocks and minerals. Chemical weathering involves changes in the chemical and mineralogical composition of rocks and minerals. In nature, the two processes occur concurrently and are difficult to separate (Reiche, 1950; Jackson and Sherman, 1953; Birkeland, 1984).

7.3.1 *Physical Weathering*

When rocks and minerals are stressed above their tensile strength they break. Commonly, rocks fracture along joints, fissures, or planes that have developed during cooling, tectonism, sedimentary processes, or along lines of weakness at the mineral grain boundaries. When buried rock masses are exposed at the surface by erosion or uplifting, the lowering of the overburden pressure, or unloading, allows the rocks to expand. This expansion includes fracturing.

Crystal growth involves the formation of ice and salt crystals. Frost wedging is the prying apart of materials by expansion of water when it freezes (Washburn, 1980). The unidirectional growth of ice crystals may be as destructive as the phase change (Washburn, 1980). The pressure produced on freezing water is well above the tensile strength of many rocks; however, this pressure may not be commonly attained in nature because rocks are not completely saturated but contain air gaps. Hydration shattering, the ordering and disordering of water molecules adsorbed at the surface of rocks, may be responsible for processes ascribed to frost wedging (Dunn and Hudec, 1966, 1972; Hudec, 1974). Washburn (1980), McGreevy (1981), and Ugolini (1986b) have reviewed frost shattering. The presence of shattered bedrock, block slope, block fields, and generally angular rock debris in cold environments provides sufficient proof that frost wedging is at work. Laboratory experiments suggest that repeated freeze–thaw cycles can produce clay-sized particles (Lautridou and Ozouf, 1982).

In arid environments, where the soluble products of weathering are not completely removed from the soil, saline solutions may circulate in the soil as well as in rock fractures. If upon evaporation the salt concentration increases above its saturation point, salt crystals form and grow (Goudie *et al.*, 1970). Wellman and Wilson (1965) presented a theoretical description relating the pressure developed by the growing salt crystals to volume and surface area. Salt weathering occurs in cold or hot deserts or areas where salts accumulate (Cooke and Smalley, 1968). Frost and salt weathering combined have a synergistic effect that could be more damaging than salt or frost alone (Williams and Robinson, 1981). Boulders, blocks, and cliffs affected by salt weathering display cavities and holes and sometimes acquire grotesque forms, as observed in the cold desert of Antarctica (Ugolini, 1986b).

Thermal expansion induced by insolation may be of importance in desert areas where rocky outcrops and soil surfaces are barren. In a desert, daily temperature excursions are wide and rocks are heated and cooled rapidly. Minerals have different coefficients of thermal expansion. Consequently, when a rock is heated or cooled, its minerals differentially expand and contract inducing stresses and strains in the rock. Ollier (1969) provided examples of rock weathering due to insolation. Fire can develop temperatures far in excess of insolation and be effective in fracturing rocks (Blackwelder, 1927).

Plants and animals disrupt and disaggregate rocks and fracture or abrade individual grains or minerals. Endolithic algae growing in deserts may be capable of disintegrating rocks through shrinking and swelling (Friedman, 1971). Lichens are effective agents in physical weathering by extending fungal hyphae into rocks and by expansion and contraction of the thalli (Fry, 1927; Yarilova, 1947; Syers and Iskandar, 1973). Higher plants grow roots in rock crevices and eventually the increased pressure breaks and disrupts the substratum. Earthworms, as mentioned by Darwin (1896), digest and abrade a considerable amount of soil. Insectivores such as moles, or rodents such as ground squirrels, hamsters, lemmings, marmots, mice and others, tunnel, excavate earth, and build dens. In this process, rodents comminute the rocks and create fine particles (Ugolini and Edmonds, 1983).

7.3.2 Chemical Weathering

Chemical weathering involves chemical changes of rocks and minerals under near-surface conditions. Mineral grains in the soils are bathed by a film of water. This condition exists even in the intense cold and extremely dry environment of continental Antarctica (Ugolini and Anderson, 1973). The dissolution of these minerals in the surrounding water depends on a number of factors: (1) the solubility of the ions of the minerals in water, which is related to the bonding strength of ions, and (2) the composition, pH, redox potential, and frequency of removal of the surrounding water. Taking these factors into consideration, it is possible to predict, in some cases, the course of weathering of minerals. The ions that tend to dissolve are those that form weak chemical bonds in the mineral. Those that precipitate are those that form strong bonds.

In chemical weathering, proton donors or acids of organic and inorganic origin are involved, where the proton acceptors are the minerals in the soil (Stumm *et al.*, 1985). In the last 100 years, anthropogenic pollution has contributed to the natural source of protons and this in turn has affected weathering rates and types.

7.3.2.1 Proton donors and other chemical agents

As briefly mentioned in Section 7.1.1, the nature, abundance, and strength of proton donors are responsible for the soil processes and the formation of the different horizons. Figure 7-4 further illustrates this principle. The major proton sources and weathering reactions are shown in Table 7-7.

7.3.2.1.1 Water. Although the dissociation constant for water at 25° C is low ($K_w = 10^{-14}$), water is so ubiquitous on this planet that, as Keller (1957) states, the reaction is inescapable. Even under continuous low temperatures and paucity of water, soil particles are surrounded by a liquid film of water and cannot escape hydrolysis reactions (Ugolini and Anderson, 1973).

7.3.2.1.2 Carbonic acid. Water in equilibrium with atmospheric CO_2 at 25° C has pH 5.65. Natural acidification of precipitation to pH less than 5.65 can occur in remote maritime air masses due to the presence of NH_4HSO_4 aerosols (Charlson and Rodhe, 1982). In the soil, the partial pressure of CO_2 may be 10–100 and up to 400 times greater than that of the atmosphere (Holland, 1978). Given a CO_2 pressure 10 or 100 times that of the atmosphere, pH values of 5.15 and 4.65, respectively, are expected. If additional sources of protons are introduced into the soils, say, from organic or inorganic acids, the pH is depressed further.

7.3.2.1.3 Organic acids. Organic acids are an important source of protons and ligands in some soil systems. They may be either metabolic products of plants and microbes or residual products of organic decomposition. Low molecular weight aliphatic acids, sugar acids, amino acids, and phenols have been detected in soil or soil solution by different investigators (Graustein *et al.*, 1977; Vedy and Bruckert, 1982; Stevenson, 1982; Dawson *et al.*, 1984; McColl *et al.*, in press). Among the humic substances, fulvic acid and low molecular weight humic acid fractions, in addition to other organic acids, are significant not only as proton donors but also in providing the ligand in the processes of complexation, dissolution, and transport (Schnitzer and Kodama, 1977; Robert *et al.*, 1987). Protons are dissociated from the organic acids during the complexation reaction, but if the organic acids are completely decomposed, then no proton transfer occurs (Ulrich, 1980; Van Breemen *et al.*, 1983).

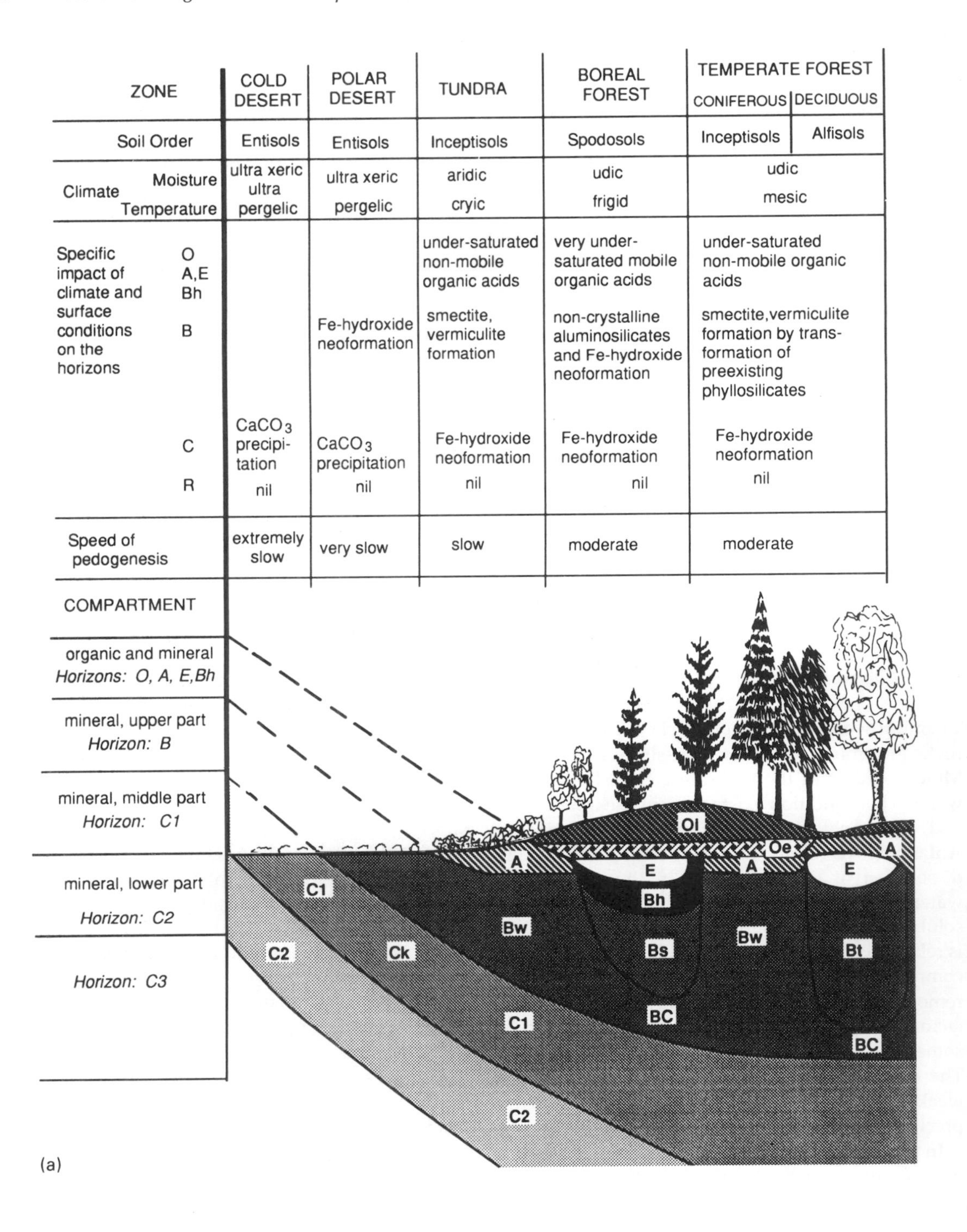

ZONE		COLD DESERT	POLAR DESERT	TUNDRA	BOREAL FOREST	TEMPERATE FOREST CONIFEROUS	TEMPERATE FOREST DECIDUOUS
Soil Order		Entisols	Entisols	Inceptisols	Spodosols	Inceptisols	Alfisols
Climate	Moisture	ultra xeric	ultra xeric	aridic	udic	udic	
	Temperature	ultra pergelic	pergelic	cryic	frigid	mesic	
Specific impact of climate and surface conditions on the horizons	O, A,E, Bh			under-saturated non-mobile organic acids	very under-saturated mobile organic acids	under-saturated non-mobile organic acids	
	B		Fe-hydroxide neoformation	smectite, vermiculite formation	non-crystalline aluminosilicates and Fe-hydroxide neoformation	smectite, vermiculite formation by transformation of preexisting phyllosilicates	
	C	$CaCO_3$ precipitation	$CaCO_3$ precipitation	Fe-hydroxide neoformation	Fe-hydroxide neoformation	Fe-hydroxide neoformation	
	R	nil	nil	nil	nil	nil	
Speed of pedogenesis		extremely slow	very slow	slow	moderate	moderate	

(a)

GRASSLAND	DESERT	SAVANNA TREELESS	SAVANNA ARBOREAL	TROPICAL RAINFOREST
Mollisols	Aridisols	Vertisols	Ultisols	Oxisols
ustic	aridic	ustic	udic	perudic
mesic	thermic	isothermic		isohyperthermic
saturated non-mobile organic acids		saturated non-mobile organic acids	under-saturated organic acids	very under-saturated mobile organic acids
Fe-hydroxide neoformation $CaCO_3$ precipitation	Fe-hydroxide neoformation $CaCO_3$ precipitation	smectite neoformation	kaolinite neoformation	Fe- and Al-hydroxide neoformation
$CaCO_3$ precipitation	$CaCO_3$ precipitation	$CaCO_3$ precipitation	Fe-hydroxide neoformation	kaolinite neoformation
nil	nil	nil		Fe-hydroxide neoformation
moderate	slow	fast		very fast

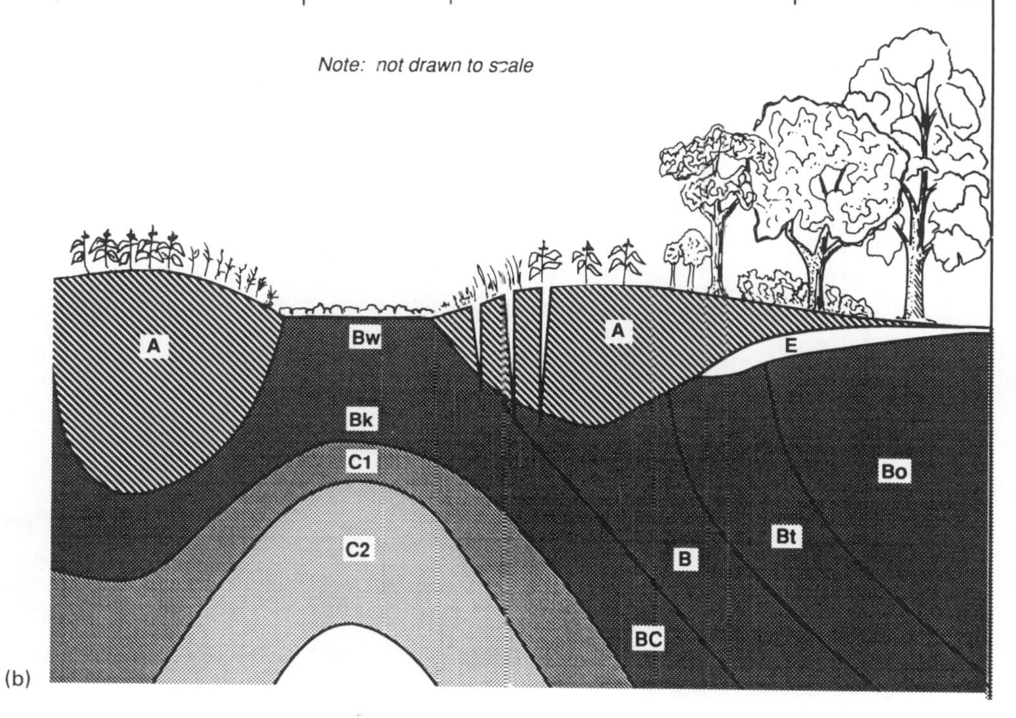

Fig. 7-4 Soil and soil-forming processes: A global view. The processes in different climates are further described in Section 7.3.3. The moisture and temperature regimes are generalized and intended only to show major pedoclimatic environments. Spodosols, for example, can also occur in a cryic regime and even in equatorial regions. Other orders could also occur in more than one moisture and temperature environment. See also Table 7-7.

Table 7-7 Major proton donors and weathering reactions occurring in different soil compartments[a]

Compartments	Horizons	Mineral species weathered	Major proton source	Major weathering reactions	Kinetics
Organic and mineral	O, A, E, Bh	Quartz/ muscovite	Organic acids	(Primary/secondary Al-silicate) + (organic acids) → (Al-organo-complex) + (Mg, Ca, Na, K)$^+$ + (organic anions)$^-$ + H_4SiO_4	Very fast congruent dissolution
Mineral, upper part	B	Feldspar	Carbonic acid high CO_2 partial pressure	(Primary/secondary Al-silicate) + H_2CO_3 + H_2O → (secondary Al-silicate/hydroxide) + (Mg, Ca, Na, K)$^+$ + HCO_3^- + H_4SiO_4	Fast incongruent dissolution
Mineral, middle part	C1	Plagioclase	Carbonic acid low CO_2 partial pressure	(Primary/secondary Al-silicate) + H_2CO_3 + H_2O → (secondary Al-silicate/hydroxide) + (Mg, Ca, Na, K)$^+$ + HCO_3^- + H_4SiO_4	Slow incongruent dissolution
Mineral, lower part			Oxidation and hydrolysis of iron Fe(II–III) + H_2O + O_2 → FeOOH + H^+	(Primary Fe–Al-silicate) + H^+ + H_2O → (secondary Al-silicate) + (Fe-oxide/ hydroxide) + H^+ + H_4SiO_4	Very slow incongruent dissolution
	C2	Biotite	Oxidation and hydrolysis of iron Fe(II–III) + H_2O + O_2 → FeOOH + H^+	(Primary Fe–Al-silicate) + H^+ + H_2O → (secondary Al-silicate) + (Fe-oxide/ hydroxide) + H^+ + H_4SiO_4	Very slow incongruent dissolution
	C3	Amphibole	Oxidation and hydrolysis of iron Fe(II–III) + H_2O + O_2 → FeOOH + H^+	(Primary Fe-silicate) + H^+ + H_2O → (Fe-oxide/hydroxide) + H^+ + H_4SiO_4	Extremely slow incongruent dissolution

[a] This table is part of Fig. 7-4. It shows the major proton donors and the major weathering reactions occurring in the different compartments (horizons) of a soil profile. It is assumed that the parent material consists of unconsolidated granite, materials are freely drained, and pedogenic impeding layers are absent. Also, no perched or fluctuating water is present.

7.3.2.1.4 Cation uptake. Uptake of cations by plants or microorganisms must satisfy the electroneutrality principle. Uptake of cations by living organisms is accompanied by release of H^+ from the organisms to the soil solution (Ulrich, 1980). The liberated H ions are neutralized by weathering reactions or ion exchange processes (Van Breeman *et al.*, 1983).

7.3.2.1.5 Nitrification. Protons are liberated during the oxidation of NH_3 or NH_4^+. This oxidation reaction is mediated by microorganisms. When active N-fixation occurs and nitrogen is fixed in excess of what can be taken up, mineralization and oxidation of the amino groups result in proton production according to the reactions:

$$R—NH_2 + H_2O \rightarrow R—OH + NH_3(g)$$
$$NH_3(g) + H_2O \rightarrow NH_4^+ + OH^- \quad (1)$$
$$2NH_4^+ + 3O_2 \rightarrow 2NO_2^- + 4H^+ + 2H_2O$$
$$2NO_2^- + O_2 \rightarrow 2NO_3^-$$

N-fixation occurring under red alder and accompanied by nitrification is responsible for the acidification of the soil (Van Miegroet and Cole, 1984). Also, human practices such as clear cutting, excessive N fertilization, or drainage of wetlands, can induce high NO_3^- production. Nitrate salts are fairly soluble. Furthermore, this anion does not react with Al or Fe, like phosphate, to form insoluble compounds. Reducing the nitrate content of the soil solution requires uptake by a growing biomass, or transformation by the microorganisms into gaseous N_2 and N_2O or NO_x, following the very schematic equation:

$$2NO_3^- + 2H^+ \rightarrow N_2 + {}^5\!/_2O_2 + H_2O \quad (2)$$

If these processes do not occur, the nitrates reach the groundwater and the rivers. Nitrate contamination can be very dramatic under cultivated lands and drained wetlands.

7.3.2.1.6 Mineralization and oxidation. Elements other than N are involved in mineralization and oxidation. In the case of sulfur, when oxidation is complete, H_2SO_4 is formed and protons are produced. The following schematic reactions are involved:

$$R—SH + H_2O \rightarrow R—OH + H_2S$$
$$H_2S + 2O_2 \rightarrow H_2SO_4 \rightarrow 2H^+ + SO_4^{2-} \quad (3)$$

In the case of P, mineralization may produce a strong acid (Ulrich, 1980):

$$R—OH—PO(OH)_2 + H_2O \rightarrow R—OH + H_3PO_4 \quad (4)$$

7.3.2.1.7 Oxidation reactions. Compounds present in the soil in a reduced state are a potential source of protons when they become oxidized. Soils under reducing conditions may contain H_2S that upon improvement in the drainage may be oxidized to H_2SO_4 (Van Breemen and Brinkman, 1976).

7.3.2.1.8 Polyvalent metals. Protons can be liberated by hydrolysis of hydrated polyvalent metals such as Al^{3+}, Fe(II), and Fe(III). Aluminum ions in the weathering environment are coordinated with six molecules of water (Jackson, 1963); the water in this hydration shell is more acidic than bulk water (Stumm and Morgan, 1981). Trivalent hexahydronium Al ion, by reacting with water, liberates protons and acts as an acid (Jackson, 1960, 1963) as shown in the following reaction:

$$Al(H_2O)_6^{3+} + H_2O \leftrightarrow Al(OH)(H_2O)_5^{2+} + H_3O^+ \quad (5)$$

The pK_1 for this reaction is 5 and therefore comparable to the pK_1 for acetic acid. Polymerized hydroxy aluminum structures also can act as weak acids.

Iron(III) acts as a proton donor when liberated from silicate minerals (Jackson, 1963) in an analogous reaction having a pK_1 of 2.2:

$$Fe(H_2O)_6^{3+} + H_2O \leftrightarrow Fe(OH)(H_2O)_5^{2+} + H_3O^+ \quad (6)$$

7.3.2.1.9 Electron donors or acceptors (oxidation–reduction): Oxidation and reduction reactions are important in weathering because they affect the solubility and the mobility of a number of elements. The solubilities of Fe, Mn, N, and S are strongly dependent on oxidation state. The transfer of electrons is often mediated by microorganisms.

Soil organic matter harbors a high density of microbes that obtain their energy via the oxidation of the organic compounds. Organic matter, especially fresh plant material, is the major electron donor (Bohn *et al.*, 1985). The following reaction depicts the oxidation of a simplified carbohydrate:

$$CH_2O + O_2 \rightarrow CO_2 + H_2O + \text{energy} \quad (7)$$

Oxygen is the common electron acceptor. When oxygen is not present, other compounds become electron acceptors. Iron(III) is thus reduced to Fe(II), Mn(III-IV) to Mn(II), SO_4^{2-} to H_2S, NO_3^- to NO_2^- or, by further reduction, to NH_3 and to gaseous N_2.

Iron(II) and Mn(II) are more soluble than in the higher oxidation states, whereas the reduction of NO_3^- can involve loss of N from the system.

In soil and weathering environments, where chemical reactions take place in an aqueous medium, the reduction potential is limited by the reduction potential of H_2O. Consequently, redox reactions in soils occur in the Eh range of 0.8 V in well-drained soils to −0.5 V in extremely reduced soils (Jackson, 1969).

7.3.2.2 *Weathering reactions*

Following these general considerations of the source of protons and the role of oxidation–reduction reactions in chemical weathering, general weathering reactions can be written for silicate and carbonate minerals. Four situations are discussed:

1. An incongruent dissolution of a silicate mineral reaction with H_2CO_3.
2. A congruent dissolution of a silicate mineral reacting with organic chelating agents.
3. A congruent dissolution of carbonate.
4. An oxidation–reduction reaction.

7.3.2.2.1 An incongruent dissolution of a silicate mineral reacting with H_2CO_3. A general weathering reaction for an alumino silicate may be represented schematically as follows (Stumm and Morgan, 1981):

$$\text{cation Al-silicate} + H_2CO_3 + H_2O \rightarrow HCO_3^- + \text{cation} + H_2SiO_4 + \text{Al-silicate} \quad (8)$$

Specifically, the reaction of a K-feldspar with H_2CO_3 and H_2O can lead to mica as a secondary mineral (modified from Pedro, 1982):

$$3KAlSi_3O_8(\text{feldspar}) + 2H_2CO_3 + 12H_2O \rightarrow KAl_3Si_3O_{10}(OH)_2(\text{mica}) + 6H_4SiO_4 + 2K^+ + 2HCO_3^- \quad (9)$$

The secondary mineral changes with increasing quantities of H_2CO_3 and H_2O involved in the reaction. Thus mica can be further weathered to kaolinite (Pedro, 1982):

$$KAl_3Si_3O_{10}(OH)_2(\text{mica}) + H_2CO_3 + {}^3\!/_2H_2O \rightarrow {}^3\!/_2Al_2Si_2O_5(OH)_4(\text{kaolinite}) + K^+ + HCO_3^- \quad (10)$$

Kaolinite, in turn, can be weathered to gibbsite (Pedro, 1982):

$$^3\!/_2\ Al_2Si_2O_5(OH)_4(\text{kaolinite}) + {}^{15}\!/_2H_2O \rightarrow 3Al(OH)_3(\text{gibbsite}) + 3H_4SiO_4 \quad (11)$$

These incongruent dissolutions result in the synthesis of clay minerals, release of monomeric silica and cations, and formation of bicarbonate.

These reactions proceed toward the right, provided some of the products are removed. Since H_2CO_3 is not a chelating agent, the soluble cations (K, Na, and Ca) and silicon are removed, but polyvalent cations such as Fe and Al are only slightly soluble and tend to accumulate. The net result is the retention and accumulation of Fe and Al due to negative enrichment.

7.3.2.2.2 A congruent dissolution of a silicate mineral reacting with a chelating agent:

$$\text{cation Al-silicate} + \text{chelating agent} + H_2O \rightarrow \text{Al-organo ligand} + H_4SiO_4 + \text{cation} - \text{ligand} \quad (12)$$

Congruent dissolution results in the formation of a polyvalent metal chelate, some divalent metal chelate, and release of monomeric silica. The chelating agent, an organic acid, could be either FA or a low molecular weight aliphatic or phenolic acid (Robert and Berthelin, 1986). Iron and aluminum, otherwise sparingly soluble in most weathering environments, acquire solubilities above those predicted by their solubility products when chelated (van Schuylenborgh and Bruggenwert, 1965). Chelated Fe and Al can be translocated into the soil profile.

7.3.2.2.3 Conguent dissolution of carbonate. The dissolution of $CaCO_3$ is regulated by the following reactions:

$$CaCO_3(s) \leftrightarrow Ca^{2+} + CO_3^{2-} \qquad pK = 8.40$$
$$H_2O + CO_2(g) \leftrightarrow H_2CO_3 \qquad pK = 1.46$$
$$H_2CO_3 \leftrightarrow HCO_3^- + H^+ \qquad pK = 6.35$$
$$H^+ + CO_3^{2-} \leftrightarrow HCO_3^- \qquad pK = -10.33$$
$$\textit{Overall:}\ CaCO_3(s) + H_2O + CO_2 \leftrightarrow Ca^{2+} + 2HCO_3^- \qquad pK = 5.8 \quad (13)$$

Dissolution of $CaCO_3$ is a congruent reaction that results in the formation of soluble products. The above reaction is driven to the right by an increase of CO_2 partial pressure and the removal of the calcium bicarbonate by water. If the calcareous rock contains impurities such as silicates, oxides, organic compounds, and others, the latter are left as residue. This residue is the substratum upon which soils develop.

7.3.2.2.4 An oxidation–reduction reaction. A common reaction that occurs in soil involves both the oxidation and reduction of Fe. Here the oxidation of Fe(II) is shown; the reduction reaction for Fe(III) would be just the reverse. Iron(II) present in the parent material oxidizes slowly in well-drained and aerated soils (Bohn *et al.*, 1985; Birkeland, 1984). Reduction of Fe(III) requires an electron donor.

$$O_2 + 4Fe^{2+} + 4H^+ \rightarrow 4Fe^{3+} + 2H_2O \quad (14)$$

An important aspect of the oxidation of Fe(II) into Fe(III) is the simultaneous hydrolysis of the oxidized Fe(III). Iron is present in primary iron-bearing minerals as Fe(II). It must first be liberated before being oxidized and hydrolyzed. Protons are needed for the weathering and liberation of the iron. The following reaction shows that subsequent to the oxidation of Fe(II) to Fe(III), the hydrolysis that follows causes the production of protons:

$$Fe^{2+} + \tfrac{1}{4}O_2 + \tfrac{3}{2}H_2O \rightarrow \alpha FeOOH(s) + 2H^+ \quad (15)$$

The H^+ can, in turn, attack the minerals causing more Fe(II) to be released that once oxidized and hydrolyzed could produce more protons, thus initiating a self-sustaining reaction, provided enough O_2 is present.

The different oxidation states and the different stable solid phases for iron can be predicted from an Eh–pH diagram (Chapter 5; Garrels and Christ, 1965). Fe(II) dominates at low pH and high Eh values, Fe(III) at low and intermediate pH and low Eh. Fe(II) is more soluble than Fe(III) and thus capable of migrating within the soil under reducing conditions. Goethite (FeOOH) is more stable under a wide pH range, but only under slightly reducing conditions. Magnetite (Fe_3O_4), siderite ($FeCO_3$), and pyrite (FeS_2) are stable in reducing environments.

7.3.3 Soil-forming Processes

As a result of weathering reactions presented in the above section, it is apparent that ions, molecules, and compounds are either mobilized and removed from the weathering environment or are insoluble, and are retained and eventually accumulate.

The retention or removal of ions, molecules, and compounds in the weathering and soil environment depends on the nature, strength, and abundance of the proton donors. The biota directly or indirectly affects the production of the proton donors. Weathering and soil-forming processes are therefore linked to climate, topography, and time, or to the soil-forming factors. Biotic and abiotic processes are responsible for the rate of weathering and for the losses of elements into the groundwater and streams. The biogeochemical cycle of an individual chemical element in the terrestrial ecosystems is controlled by processes occurring in the soil and the above- and below-ground biomass.

Water and temperature play paramount roles in weathering and soil formation. Water is not only the universal solvent in the weathering environment, but is also the vehicle for the redistribution of the products of weathering. An obvious consequence of the transport role of water can be appreciated in those cases where soluble products of weathering are retained in the system because of lack of leaching. Therefore, to examine weathering and soil-forming processes, it is necessary to consider the distribution of the percolating water and temperature. When this relationship is plotted, a scheme as shown in Fig. 7-4 can be generated. This diagram considers an ideal transect from the pole to the equator. Here, following Jenny's approach, one can examine weathering and soil formation for a level surface with uniform siliceous rock exposed to different regional climates and covered by different vegetation assemblages. Following this scheme, nine major weathering and soil-formation environments are recognized. These nine situations will be discussed briefly, starting from the extremely dry and cold environments of the cold desert to the hot and hyperhumid regions of the tropical rainforest (Fig. 7-4).

7.3.3.1 Cold desert

Mostly centered in the ice-free areas of Antarctica but also in northern Greenland, the cold desert experiences a severe climate. Physical weathering is caused mainly by frost wedging and insolation, both favored by the barren surfaces. Salt weathering is spectacularly manifested by cavernous weathering (Ugolini, 1986b). Chemical weathering is restricted by the persistent low temperatures and paucity of liquid water (McCraw, 1960; Ugolini, 1963; Tedrow and Ugolini, 1966; Campbell and Claridge, 1969, 1987).

Inorganic and organic proton donors are limited, although chemical weathering does occur (Claridge 1965; Ugolini and Jackson, 1982; Claridge and Campbell, 1984). Mica is altered to vermiculite and eventually to smectite (Claridge, 1965; Ugolini and Jackson, 1982). The liberation of Fe from the Fe-silicate minerals and its hydrolysis reactions and precipitation as oxyhydroxide liberates protons (see reaction 15). The liberated protons in turn attack the minerals and free more Fe, thus maintaining a self-sustaining weathering reaction (Ugolini, 1986b). An unfrozen film of water surrounding the soil particles provides additional protons for the weathering of the minerals (Ugolini and Anderson, 1973). These processes are very slow, and consequently weathering in Antarctica is apparent only in very old surfaces; the Bw horizon requires over 135 000 years to form (Bockheim, 1979). These soils, when well developed, show a simple morphology consisting of a desert pavement (D), and Bw and C horizons.

In the virtual absence of percolating water, salts liberated through weathering or accreted by meteoric input are removed slowly. Consequently, chlorides, sulfates, nitrates, and other soluble salts are retained in the soil. Soils formed in this environment are called Entisols (Soil Survey Staff, 1975; Fig. 7-4).

7.3.3.2 *Polar desert*

This region occupies the northernmost sectors of the Arctic. Precipitation and temperature are low, but higher than in the cold desert. Although barren conditions prevail (Bliss *et al.*, 1984), vascular plants are present, but restricted to small patches. Physical weathering is evidenced by the frost-shattered rocks that cover the barren landscape. Given sufficient moisture, algae are found in the interstices of the desert pavement, while lichens cover pebbles and cobbles (Bliss *et al.*, 1984). Proton donors include carbonic acid and a few low molecular weight acids and lichen acids. Dissociation of the unfrozen water and the hydrolysis of Fe are still relatively important (see Section 7.3.2.2). The rate of chemical weathering remains at a low level but higher than in the cold desert. Bw horizons form in about 10 000 years (Mann *et al.*, 1986). The soil morphology consists of a D, and Bw and C horizons. Carbonate rocks are dissolved and secondary carbonates accumulate; alkaline pH values are common (Tedrow, 1977; Mann *et al.*, 1986). Polar desert soils are Entisols.

7.3.3.3 *Tundra*

The tundra region stretches south of the polar desert to the margin of the tree line. Precipitation and temperature are low, but none the less most of the terrain is waterlogged during the growing season because of the permafrost – the continuous impermeable frozen layer. A small percentage of the landscape is well drained and supports soils that have received the full impact of the soil-forming factors. Organic acids are active at the surface and carbonic acid and Fe-hydrolysis reactions occur in the subsoil. There is some movement of aluminum and alteration of minerals. Mica is altered to vermiculite and smectite (Brown and Tedrow, 1964). Secondary iron and aluminum oxides are formed. Secondary carbonate is present at shallow depths, but the upper part of the soil is acid (Tedrow, 1977). The morphology of these soils consists of A, Bw, and C horizons. The well-drained soils of this region are Inceptisols (Soil Survey Staff, 1975; Fig. 7-4).

In the tundra, where the permafrost table is close to the surface, waterlogging conditions exist during the thaw season. Waterlogging is associated with reduction induced by depletion of O_2. Reducing conditions are manifested by the appearance of a unique morphological characteristic, the gley horizon. This horizon is discolored with prevailing gray color. It may display poor structure and it may show the presence of Fe(II), Mn(II), and sulfides. With the exclusion of molecular O_2, organisms involved in the decomposition or oxidation of soil organic matter cannot use oxygen as an electron acceptor. Oxidizing agents with a lower electrical potential than O_2 (Table 7-8) must be used. Nitrate is utilized as electron acceptor after the O_2 is exhausted. Nitrate can be reduced to NH_4^+ or to N_2 or N_2O by denitrifying bacteria. Reduction of Fe(III) follows. When Fe(III) accepts an electron and is reduced to Fe(II), the iron solubility increases. If the water-table drops and the soil is oxidized again, protons are liberated by the oxidation of the reduced Fe; this reaction can strongly acidify the soil (Brinkman, 1970; Van Breemen and Brinkman, 1976). Following the reduction of Fe(III), the solubility of phosphate increases and different minerals may be synthesized: pyrite (FeS_2), siderite ($FeCO_3$), magnetite (Fe_3O_4), and vivianite $(Fe_3PO_4)_2 \cdot 8H_2O$. Reduction of Mn(III-IV) to Mn(II) may lead to Mn toxicity.

Sulfate is reduced to sulfide when the redox potential in soil drops to 0 to -0.15 V. Both organic and inorganic sulfate are biochemically reduced by anaerobic bacteria (Starkey, 1966). Upon returning to oxidizing conditions, sulfide may be oxidized to H_2SO_4, producing very acidic conditions. Further

Table 7-8 Order of utilization of principal electron acceptors in soils, equilibrium potentials of these half-reactions at pH 7, and measured potentials of these reactions in soil[a]

Reactions	Eh at pH 7 (V)	Measured redox potentials in soils (V)
O_2 disappearance		
$\frac{1}{2}O_2 + 2e^- + 2H^+ \rightarrow H_2O$	0.82	0.6 to 0.4
NO_3^- disappearance		
$NO_3^- + 2e^- + 2H^- \rightarrow NO_2^- + H_2O$	0.54	0.5 to 0.2
Mn^{2+} formation		
$MnO_2 + 2e^- + 4H^+ \rightarrow Mn^{2+} + 2H_2O$	0.40	0.4 to 0.2
Fe^{2+} formation		
$FeOOH + e^- + 3H^+ \rightarrow Fe^{2+} + 2H_2O$	0.17	0.3 to 0.1
HS^- formation		
$SO_4^{2-} + 9H^+ + 6e^- \rightarrow HS^- + 4H_2O$	−0.16	0.0 to −0.15
H_2 formation		
$H^+ + e^- \rightarrow \frac{1}{2}H_2$	−0.41	−0.15 to −0.22
CH_4 formation		
$(CH_2O)_n \rightarrow \frac{n}{2}CO_2 + \frac{n}{2}CH_4$	—	−0.15 to −0.22

[a] After Bohn *et al.* (1985).

water stagnation and the lowering of oxidation potential to values of −0.15 to −0.22 V may lead to fermentation and formation of methane (CH_4) and other strongly reduced compounds.

Reduction processes are not limited to the tundra zone, but occur in other regions under different climates. In the absence of permafrost, permanent saturation of the soil may result from impermeable layers or low topographic positions on the landscape. Clay-rich B horizons may maintain a seasonal perched water-table and cause permanent or temporary saturation of the top soil. Details of these conditions and others, including the chemical processes, have been discussed by Van Breemen and Brinkman (1976), Duchaufour (1982), Bouma (1983), and Blume (1988).

7.3.3.4 Boreal forest

South of the Arctic tree-line and into the forest, temperature and precipitation increase. There is much more biomass than in the tundra and the differentiation of the soil profile in an upper and a lower compartment is greatly accentuated. The upper profile is dominated by organic acids, the lower by carbonic acids; congruent dissolution is thus expected in the upper horizon (see Section 7.3.2.2) and incongruent dissolution in the lower one. Organic acids act both as proton donors and chelating agents. They attack the minerals of the upper compartment, which results in a chromatographic-like redistribution of fulvic acids, iron, and aluminum. This upper compartment is differentiated in very distinct horizons (O, E, and Bh horizons). In this upper compartment, the humic acid : fulvic acid ratio is low (Kononova, 1966). The lower compartment (Bs and BC horizons) is dominated by carbonic acid (Ugolini *et al.*, 1987). Minerals in the E horizon are dissolved; on the other hand, smectite appears to be stable in this environment. In the B horizon, non-crystalline alumino silicate, allophane, and imogolite may form as well as oxides and hydroxides of iron and aluminum. Vermiculite with interlayer Al is also present. This intense weathering regimen is limited to the upper 1 m of soil (Duchaufour, 1982; Ugolini and Dahlgren, 1987). Spodosols are the soils formed in this region (Soil Survey Staff, 1975; Fig. 7-4).

7.3.3.5 Temperate deciduous forest

This climatological zone is found south of the boreal forest belt. Plants produce litter rich in nitrogen and cations and poor in lignin. Temperatures and rates of decomposition are higher than in the boreal forest belt. Earthworms mix the humic substances with the surficial mineral soil, and the chromatographic-like redistribution of the organic acids and trivalent metals does not occur (Duchaufour, 1982). The upper compartment of the soil profile consists of a single horizon, the A horizon. The humic acid : fulvic acid ratio in this horizon is relatively low (Kononova, 1966). The lower compartment is dominated by carbonic acid weathering and possibly low molecular weight organic acids. Migration of clay can occur under specific conditions. In this case, a leached E horizon, clay poor, and a Bt clay-rich horizon, are formed. The morphology consists of an A or E, Bw or Bt, and C horizons. Inceptisols and Alfisols are the soils developed in this region (Soil Survey Staff, 1975; Fig. 7-4).

7.3.3.6 *Grassland*

Grassland or steppe experiences a continental climate with hot, dry summers and cold winters. Rains are prevalent during the spring and fall. These conditions favor retention of organic matter in the soil, high CO_2 partial pressures, and formation of carbonic acids. Carbonic acid is the major proton donor; however, where nitrification occurs, HNO_3 also plays a role. Melanization, the darkening of the soil due to organic matter accumulation, is the prevailing soil-forming process.

Melanization is well expressed under grassland vegetation growing in a continental climate characterized by cold, dry winters, moist springs, and dry summers punctuated by thunderstorms. These conditions are typical for the steppe of Ukraine where the annual precipitation is about 450 mm and the mean temperatures for the coldest and warmest months are −7° C and 21.5° C respectively (Joffe, 1949; Duchaufour, 1978). Similar conditions exist in the Great Plains of North America and Eastern Europe. Melanization also occurs in the Argentine pampas where the winters are milder and the summers less dry than in continental North America and Eurasia. Soils developed under these conditions display a thick A horizon, i.e. dark, well structured and full of roots. The B horizon may be poorly developed, while the C horizon may contain soft concretions of $CaCO_3$ (Fig. 7-4).

In this system, herbaceous species produce fine roots that pervade the soil mass and provide a steady input of relatively easily degradable organic matter. The below-ground biomass is rich in protein and cellulose and favors a prolific growth of microorganisms. Polysaccharides are some of the metabolic products of plant and microbial activity. The presence of well-humified organic matter, polysaccharides, and clay minerals such as vermiculite and smectite, promote the formation of stable aggregates. The bonding between the minerals and organic surfaces involves a number of mechanisms such as ion-exchange, hydrogen bonding, protonation, water bridging, van der Waal's forces, and other factors (Harter, 1977). The rich invertebrate and small rodent fauna rework and homogenize the mineral soil and organic matter (Joffe, 1949; Baxter and Hole, 1967). Earthworms rework and ingest the soil and excrete casts that are very stable and rich in nutrients.

The pH values of these soils are neutral to slightly alkaline; Fe and Al are stable, and the clay migrates in the profile only at high moisture levels (Anderson, 1987). Humic acids (HA) tend to prevail over the fulvic acids (FA), because during the dry season the fulvic soils tend to polymerize and form large molecules; consequently, the HA : FA ratio is high (Kononova, 1966). Where nitrogen-fixing plants are present, nitric acid can also participate in the weathering and transport processes. Under different chemical regimes, melanization also occurs in alpine areas in poorly drained soils and lowland hardwood (Buol *et al.*, 1980).

Weathering is less intense than in the forest; none the less, 2 : 1 clay minerals such as vermiculite and smectite are formed. Secondary carbonates accumulate in the profile in the more arid sectors of the steppes (Dachaufour, 1982). Soil profiles consist of thick and homogeneous organic-rich A and Bw or Bt horizons. Soils common in the grassland are called Mollisols (Soil Survey Staff, 1975; Fig. 7-4).

7.3.3.7 *Hot desert*

Scant precipitation and high temperatures reduce leaching and favor the retention of the soluble products of weathering. High temperature increases the rate of chemical weathering, but because of the short growing season, biomass production is reduced and organic matter content in the soil is generally low. The organic compartment is thin. 2 : 1 clay minerals are formed and iron oxides are present. Carbonic acid weathering plays an important role in addition to Fe-hydrolysis. Dissolution and precipitation of carbonate occurs, and thus secondary carbonates accumulate in the soil (Gile *et al.*, 1966; Birkeland, 1984). Clay may accumulate in the B horizon in response to migration due to high subsurface flow or to weathering *in situ*. The soils of the hot desert are the Aridisols (Soil Survey Staff, 1975; Fig. 7-4).

7.3.3.8 Savanna

Savannas occur in tropical or subtropical climates with a marked dry season (Duchaufour, 1984). Carbonic acid is the prevailing proton donor. During the wet season, weathering occurs and kaolinite is formed in the well-drained soils. The soil solution, charged with H_4SiO_4 and basic cations, migrates to lowlands and depressions where the soil solution is concentrated during the dry season. This leads to an intensive synthesis of 2:1 expanding clay minerals, mostly montmorillonite. Organic matter tends to accumulate all through the soil, and the clay content is high. These soils are poorly drained during the wet season but become strongly desiccated during the dry season. Common soils for the savanna are the Vertisols (Soil Survey Staff, 1975; Fig. 7-4).

7.3.3.9 Tropical rainforest

As precipitation increases, the savanna is replaced by tropical rainforest. The increased rainfall and high temperatures favor a high biomass production, strong weathering, and intense leaching. Luxurious vegetation grows under these conditions. Mineralization of organic matter is rapid due to high temperatures, while the soil organic matter content is low. Hydrolysis reactions are intense, favored by the large volume of water that penetrates the soil, while an aggressive carbonic acid weathering regimen occurs favored by a high CO_2 partial pressure. Monomeric silica and bases are removed by leaching, whereas iron and aluminum oxides are retained (Pedro, 1982). Clay minerals of the 1:1 type, kaolinite, are formed *in situ* (see reaction 10). However, under intense leaching conditions, gibbsite may predominate as the 1:1 clay minerals become unstable (see reaction 11). Iron, not being involved to the extent that Al is in the synthesis of clay minerals, becomes segregated as a separate phase and forms intensively red oxides and oxyhydroxides such as hematite (Fe_2O_3). Intense red colors betray the presence of hematite (Schwertmann, 1988). One important result of the iron accumulation is the formation of an iron crust at the surface. This hard crust forms at the surface when the iron oxides crystallize in response to desiccation caused by changes in climate or ecological conditions. The weathering profile is deep, up to 100 m in thickness, and it is difficult to distinguish soil profiles and soil horizons. The most intensely weathered conditions are found at the surface (Duchaufour, 1982). Oxisols are unique to this environment (Soil Survey Staff, 1975; Fig. 7-4).

7.4 Soil and Biogeochemical Cycles

The soil, in the context of biogeochemical cycles, is an open system, which receives inputs and outputs of C and N, and mineral elements, e.g. Si, Mg, Ca, Na, K, and P. Soils are at the base of the primary production of terrestrial ecosystems. The soil is also the connecting link between rainfall and river flows, and thus it regulates the load in the hydrological system. Thus, the soil occupies a key position and it plays an extremely complex role in biogeochemical cycles. On the other hand, the link between the soil and the hydrological system can be reduced to a few simple "rules". Before looking at these rules, we need to assess the importance of the soil within geochemical cycles.

The amount of carbon present in the soil is closely connected to the CO_2 concentration of the atmosphere. But atmospheric CO_2 is regulated mainly by the ocean rather than by the soil (see Chapters 9 and 11). The amount of N in the soil also does not influence the N in the atmosphere because the atmosphere is a huge reservoir regulated mainly by the ocean (see Chapter 12). Nevertheless, the soil has a tremendous influence on the nitrate load of the rivers.

Additional elements found in rivers (Si, Al, Fe, Mg, Ca, K, P, and others), present as soluble salts or as particles, originate within the soil via weathering of the rocks. One could predict that the quality and quantity of this load would depend strongly upon the soils surrounding the river, but this is not always the case. In fact, the chemistry of rivers flowing on granitic terrains in Western Europe and tropical Africa appear rather similar, whereas the soils in these two regions are dramatically different (see Chapter 6; Tardy, 1969). These results are expected because the link between the soil cover and the river load obeys a few simple rules, which can largely offset the complexity of the soil cover and the associated environmental factors.

7.4.1 Soils and River Water Chemistry

7.4.1.1 The soil compartments

Figure 7-4 shows that most soil orders can be divided into two major compartments: an organo-mineral compartment on top, followed by the mineral compartment, further divided into upper, middle, and lower. The organo-mineral compartment depends strongly on the complex interactions among the climate, biota, parent material, topography, and time, whereas in the mineral compartment these interactions are less important. The middle, and especially the lower, compartments are very similar throughout the different soil orders. The chemistry of a river depends mostly on the level at which the water exits the soil. There are four main levels: (1) above the soil surface (run-off), (2) the organo-mineral compartment, (3) the upper mineral compartment, and (4) the lower mineral compartment. The soil solution, collected *in situ* within the soil profile, is an elegant way to illustrate the different compartments. Figure 7-5 displays the soil solution composition of a Spodosol; Spodosols show a dramatic differentiation between the soil compartments.

In the following discussion, the same conditions as presented in Fig. 7-4 are considered: same granitic parent material, same age, same elevation, and good drainage. The terms "low", "very low", etc., describing concentrations of an element in solution, refer to its concentration relative to the other elements in the parent material.

7.4.1.1.1 The run-off. Overland flow carries material to the river essentially as particles: sand, silt, clay, and organic matter. This material consists of plant debris, more or less decomposed, and aggregates of humified organic material and mineral particles. Beside the typical mineral elements carried to the rivers such as Si, Ca, and others, C, N, and P are also present. Run-off from Spodosols consists of very acidic organic matter; a Mollisol yields aggregates of neutral humic acids and clay. In all cases, the load carried to the rivers has a high nutrient content.

7.4.1.1.2 The organo-mineral compartment. The load of the water leaving the soil at this level consists mainly of soluble compounds because the matrix of the soil acts as a filter retaining the particles. In some cases, clay particles are also carried away. The load strongly reflects whether the soil is affected by intense or mild weathering. Intense weathering is due to very acidic conditions, as in the case of Spodosols, or to acidic conditions and high temperature as for Ultisols, or to a very weatherable parent material, such as volcanic glass in the case of Andisols. In all these cases, the soil solution contains relatively high amounts of silica as H_4SiO_4. This silica cannot react with Al or Fe, because the latter

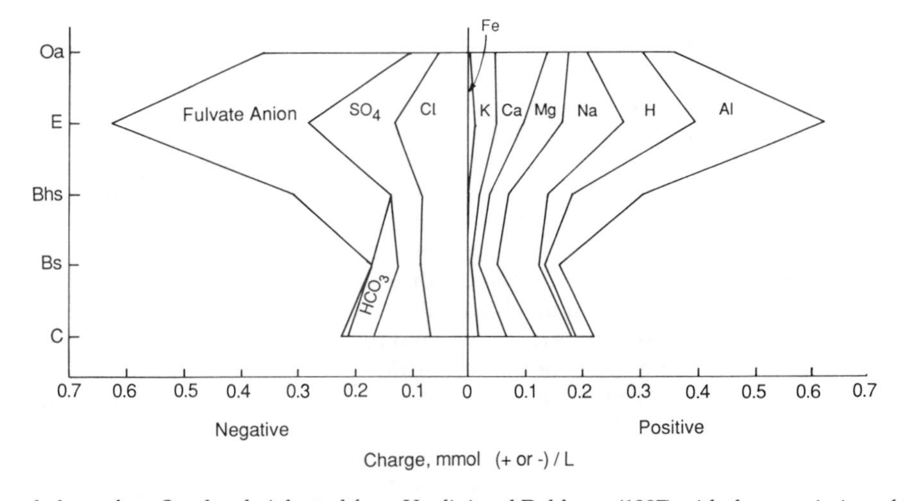

Fig. 7-5 Charge balance for a Spodosol. Adapted from Ugolini and Dahlgren (1987) with the permission of the author.

elements are effectively complexed by the organic matter.

Soils affected by mild weathering include the Mollisols, Alfisols, Vertisols, and the Aridisols. Here the soil solution contains little H_4SiO_4, due to the low rate of weathering. From these considerations one can deduce, for example, that Spodosols will yield a very acidic and brown solution containing fulvic acids partially saturated with Al and Fe, and containing some nitrogen and phosphorus. On the other hand, a Mollisol yields a neutral solution, clear of or containing only small amounts of organic matter. The cations in solution are Ca, Mg, K, and Na, accompanied by HCO_3^- and NO_3^- anions, small amounts of phosphate, and very little Fe and Al. In all cases, the water exiting the soil at this level of the profile is likely to contain nutrients such as N and P.

7.4.1.1.3 The upper mineral compartment. The solution exiting the soil at this level contains very little organic matter, if any. For this reason, organic C and N, as well as Al and Fe, are not present, or only in very small amounts. In this mineral environment, pH is a critical factor affecting the mobility of P because it is often too alkaline or too acid to allow the existence of P as a soluble anion. In even slightly alkaline pH, P precipitates with Ca in an insoluble form; in acid pH, it precipitates with Al and Fe. The load carried to the river consists mostly of alkaline cations (Mg^{2+}, Ca^{2+}, Na^+, K^+), the inorganic anion HCO_3^-, and little H_4SiO_4. Potassium can be retained within the soil if vermiculitic clay minerals are present, which is true for most alfisols. In this case, the amount of K reaching the river is low. The same behavior can be expected for Mg if the soil contains smectites (vertisols). Silicic acid losses are strongly attenuated under mild weathering conditions. In soils of intense weathering, Si reacts with Al to form allophanes and imogolite (Spodosols, Andisols) or kaolinite (Utisols). In soils of mild weathering, silica may participate in the formation of vermiculite of smectite. Therefore, the Si concentration of the water reaching a river is relatively low.

7.4.1.1.4 The lower mineral compartment. In this compartment, the pH is broadly the same in all soils, ranging from slightly alkaline to alkaline (Fig. 7-4). Where smectite is formed, Mg is sequestered within this compartment. The pH is too high to allow soluble forms of Al, Fe, and P. The load carried to the river consists mostly of Ca^{2+} and Na^+, HCO_3^-, and little H_4SiO_4. In dry environments, Ca precipitates within the soil as $CaCO_3$, and only Na, this time accompanied by the CO_3^{2-} anion, reaches the river.

To summarize, under conditions presented in Fig. 7-4, when the water entering a river comes from below the organo-mineral compartment, it is similar in all the soil orders. For example, the soil solution in the C horizon of soils, as of different orders such as Spodosols and Inceptisols, can be very similar. From these observations, the link between the complex processes occurring within a soil and the hydrological system can be reduced to a "few simple rules". The chemical composition of the river load is reduced to a few components: it contains little H_4SiO_4, variable amounts of alkaline cations (Ca, Mg, Na, K), and the HCO_3^- anion. There is little organic C or N, and a virtual absence of Al, Fe, and P. The climate, rather than the soil has more impact on the composition of the river water.

7.4.1.2 The climate

In wet climates, the concentration of soil-derived elements in river water is generally low. In humid and hyper-humid climates, the concentration of H_4SiO_4 is dictated by the equilibrium with kaolinite and tends to be low (Tardy, 1969; Kittrick, 1977; Velbel, 1985; Clayton, 1986; Pavich, 1986). In parent materials that are well drained and coarse textured, the H_4SiO_4 concentration in the solution going through the lower mineral compartment can be low enough to lead to gibbsite neoformation, even in temperate climates (Dejou *et al.*, 1972; Green and Eden, 1971; Macias-Vaquez *et al.*, 1987). In dry climates, the concentration of elements in the river water is generally high; H_4SiO_4 concentration is higher than that supported by the equilibrium with kaolinite. Montmorillonite is the mineral controlling this equilibrium (Kittrick, 1977).

If some Ca is retained within the soil by $CaCO_3$ precipitation, one would expect the soil solution to be Ca-saturated in regard to calcite; therefore, the Ca concentration in the river water would also be high.

While the schematic approach presented here is relatively simple, in reality things are more complicated. This is mainly due to two factors: (1) the complexity of the hydrological dynamics of the watersheds, and (2) some unique soil processes, either natural or human-induced.

7.4.2 Case Studies: The Dynamics of Watersheds

The watershed or drainage basin, at any scale, is the natural unit for the study of the relationships between the geology, the terrestrial ecosystems, and the soil cover on the one hand, and the hydrological system on the other. The scale ranges from watersheds of a few hectares to basins of huge rivers that cover large portions of continents (Johnson *et al.*, 1968; Likens, 1985; Stallard, 1985; Velbel, 1985; Clayton, 1986; Dethier, 1986; Paces, 1986; Pavich, 1986). Often the water feeding a river can come simultaneously from different soil compartments, as a result of factors such as topography, lithology, the thickness of the soil, and others. At any given time, it can come from the lower mineral compartment from one area, and from the organo-mineral one from another. Run-off can occur during a storm or during a dramatic snowmelt, whereas most of the water for the rest of the year may originate in the mineral compartment.

7.4.3 Conclusions

In natural ecosystems, with the exclusion of swamps, arid areas, and young tectonic reliefs, the water leaving the soil and entering rivers contains mainly the bicarbonate anion HCO_3^-, alkaline cations, mostly Ca and Na, and little H_4SiO_4. Nutrients such as N and P are added to the rivers in small amounts by other processes – soil erosion at the edge of the river, run-off – and from direct inputs such as leaves and other organic debris. Therefore, the nutrient content of rivers and lakes is generally low. In a heavily forested and still pristine watershed of southeast Alaska, Stednick (1981) measured the output of N and P to be about 4.5 and 0.8 kg/ha year, whereas it reaches 185, 275 and 40 in kg/ha year for HCO_3^-, Ca, and Na respectively. Silicon is exported at the rate of about 80 kg/ha year (measured as H_4SiO_4). One of the most important anthropogenic impacts for these ecosystems is a huge input of nutrients such as N and P by clear-cutting, drainage of wetlands, fertilizer application, and agricultural practices that induce erosion of the topsoil.

In arid areas, run-off is often the main source of the water reaching the bottom of valleys. The rivers carry a load consisting mainly of particles, and the nutrient content, N and P, is high. Throughout the world, estuaries of rivers draining arid lands, or the lakes they empty into, are incredibly rich in aquatic life, which in turn sustains a very abundant wildlife. Flood plains located downstream of arid areas are also known for their rich soils. The Yellow River in China, the Colorado in the USA, the Nile in Africa, are only some of the most famous examples. The construction of dams along these rivers allows flood control, and water for irrigation and power; but by retaining the silts behind the dams it severely impacts some of the richest and most productive ecosystems in the world.

In wetlands, river loads originate within the organo-mineral compartment. Therefore, the rivers are rich in nutrients. But in numerous cases, especially in cold climates, the acidity of the water does not allow the development of an important aquatic biomass and wildlife.

Questions

7-1 Discuss the nature of the physical weathering operating in each of the following environments: (1) seashore, (2) hot desert, (3) temperate forest.

7-2 In some ecosystems, such as those dominated by red alder (*Aldus rubra* Bong.), there is a large production of HNO_3. Under these conditions, where the HNO_3 is the major proton donor, would you expect podzolization to occur?

7-3 Rewrite the weathering reactions shown in Section 7.3.2.2 using HNO_3 in place of H_2CO_3.

References

Acquaye, D. K. and J. Tinsley (1965). Soluble silica in soils. *In* "Experimental Pedology" (E. G. Hallsworth and D. V. Crawford, eds.), pp. 126–148. Butterworth, London.

Alexander, M. (1977). "Introduction of Soil Microbiology," 2nd edn. John Wiley, New York.

Anderson, D. W. (1987). Pedogenisis in the grassland and adjacent forests of the Great Plains. *Adv. Soil Sci.* **7**, 53–93.

Baker, W. E. (1973). Role of humic acids from Tasmanian podzolic soils in mineral degradation and metal mobilization. *Geochim. Cosmochim. Acta* **37**, 269–281.

Barnhisel, R. I. (1977). Chlorites and hydroxy interlayered vermiculite and smectite. *In* "Minerals in Soil Environments" (J. B. Dixon and S. B. Weed, eds.), pp. 331–356. Soil Science Society of America, Madison, Wisconsin.

Baxter, F. P. and F. D. Hole (1967). Ant (*Formica cinerea*) pedoturbation in a prairie soil. *Soil Sci. Soc. Amer. Proc.* **31**, 425–428.

Birkeland, P. W. (1984). "Soils and Geomorphology." Oxford University Press, New York.

Blackwelder, E. B. (1927). Fire as an agent in rock weathering. *J. Geol.* **35**, 134–140.

Bliss, L. C., J. Svoboda, and D. I. Bliss (1984). Polar deserts, their plant cover and plant production in the Canadian High Arctic. *Holarctic Ecol.* **7**, 305–324.

Blume, H. P. (1988). The fate of iron during soil formation in humid-temperate environments. *In* "Iron in Soils and Clay Minerals" (J. W. Stucki, B. A. Goodman, and U. Schwertmann, eds.), pp. 749–777. Reidel, Dordrecht, The Netherlands.

Bockheim, J. G. (1979). Relative age and origin of soils in eastern Wright Valley, Antarctica. *Soil Sci.* **128**, 142–152.

Bockheim, J. G. (1980). Solution and use of chronofunctions in studying soil development. *Geoderma* **24**, 71–85.

Bohn, H. L., B. L. McNeal, and G. A. O'Connor (1985). "Soil Chemistry," 2nd edn. John Wiley, New York.

Bolt, G. H. and G. M. Bruggenwert (1976). Composition of the soil. *In* "Soil Chemistry: A. Basic Elements" (G. H. Bolt and M. G. M. Bruggenwert, eds.), pp. 1–12. Elsevier, New York.

Borchardt, G. A. (1977). Montmorillonite and other smectite minerals. *In* "Minerals in Soil Environments" (J. B. Dixon and S. B. Weed, eds.), pp. 293–356. Soil Science Society of America, Madison, Wisconsin.

Bouma, J. (1983). Hydrology and soil genesis of soils with aquic moisture regimes. *In* "Pedogenesis and Soil Taxonomy: I. Concepts and Interactions" (L. P. Walding, N. E. Smeck, and G. F. Hall, eds.), pp. 253–281. Elsevier, Amsterdam, The Netherlands.

Bowden, J. W., A. M. Posner, and J. P. Quirk (1980). Adsorption and charging phenomena in variable charge soils. *In* "Soils with Variable Charge" (B. K. G. Theng, ed.), pp. 147–166. New Zealand Society of Soil Science. Offset Publications, Palmerston North, New Zealand.

Brady, N. C. (1974). "The Nature and Properties of Soils." 8th edn. Macmillan, New York.

Brinkman, R. (1970). Ferrolysis, a hydromorphic soil forming process. *Geoderma* **3**, 199–206.

Brown, B. E. and M. L. Jackson (1958). Clay mineral distribution in the Hiawatha sandy soils of northern Wisconsin. *Clays and Clay Minerals* **71**, 213–266.

Brown, J. and J. C. F. Tedrow (1964). Soils of the northern Brooks Range, Alaska: 4. Well-drained soils of the glaciated valleys. *Soil Sci.* **95**, 187–195.

Buol, S. W., F. D. Hole, and R. J. McCracken (1980). "Soil Genesis and Classification," 2nd edn. Iowa State University Press, Ames, Iowa.

Campbell, I. B. and G. G. C. Claridge (1969). A classification of frigic soils – the zonal soils of the Antarctic continent. *Soil Sci.* **107**, 75–85.

Campbell, I. B. and G. G. C. Claridge (1987). "Antarctica: Soils, Weathering Processes and Environment." Development in Soil Science Series, Vol. 16. Elsevier, Amsterdam, The Netherlands.

Charlson, R. J. and H. Rodhe (1982). Factors controlling the acidity of natural rainwater. *Nature* **295**, 683–685.

Claridge, G. G. C. (1965). The clay mineralogy and chemistry of some soils from the Ross Dependency, Antarctica. *N.Z. J. Geol. Geophys.* **8**, 186–220.

Claridge, G. G. C. and I. B. Campbell (1984). Mineral transformation during weathering of dolerite under cold arid conditions. *N.Z. J. Geol. Geophys.* **27**, 537–545.

Clayton, J. L. (1986). An estimate of plagioclase weathering rate in the Idaho batholith based upon geochemical transport rates. *In* "Rates of Chemical Weathering of Rocks and Minerals" (S. M. Coleman and D. P. Dethier, eds.), pp. 453–466. Academic Press, New York.

Coen, G. M. and R. W. Arnold (1972). Clay mineral genesis of some New York Spodosols. *Soil Sci. Soc. Amer. Proc.* **36**, 342–350.

Colman, S. M. and D. P. Dethier (eds.) (1986). "Rates of Chemical Weathering of Rocks and Minerals." Academic Press, New York.

Cooke, R. U. and I. J. Smalley (1968). Salt weathering in deserts. *Nature* **220**, 1226–1227.

Cradwick, P. D. G., V. C. Farmer, J. D. Russel, C. R. Masson, K. Wada, and N. Yoshinaga (1972). Imogolite, a hydrated aluminum silicate of tubular structure. *Nature* **240**, 187–189.

Cronan, C. S. (1984). Biogeochemical responses of forest canopies to acid precipitation. *In* "Direct and Indirect Effects of Acidic Deposition on Vegetation" (R. A. Linthurst, ed.), pp. 65–79. Butterworth, Boston, Mass.

Darwin, C. (1896). "The Formation of Vegetable Mould Through the Action of Worms with Observations of their Habitats," pp. 305–313. D. Appleton, New York.

Dawson, H., F. C. Ugolini, B. F. Hrutfiord, and J. Zachara (1978). Role of soluble organics in the soil process of a Podzol, central Cascades, Washington. *Soil Sci.* **126**, 290–296.

Dawson, H., B. Hrutfiord, and F. C. Ugolini (1984). Mobility of lichen compounds from *Cladonia mitis* in Arctic soils. *Soil Sci.* **38**, 40–45.

Dejou, J., J. Guyot, and C. Chaumont (1972). La gibbsite, minéral banal d'altération des formations superficielles et des sols dévelopés sur roches cristallines dans les zones tempérees humides. *Proc. 24th Int. Geol. Cong.*, Section 10, pp. 417–425.

Dethier, D. P. (1986). Weathering rates and the chemical flux from catchments in the Pacific Northwest, U.S.A. *In* "Rates of Chemical Weathering of Rocks and Minerals" (S. M. Coleman and D. P. Dethier, eds.), pp. 503–530. Academic Press, New York.

Douglas, L. A. (1977). Vermiculites. *In* "Minerals in Soil Environments" (J. B. Dixon and S. B. Weed, eds.), pp. 259–292. Soil Science Society of America, Madison, Wisconsin.

Duchaufour, P. (1978). "Ecological Atlas of Soils of the World." Masson, New York.

Duchaufour, P. (1982). "Pedology" (translated by T. R. Patan). George Allen and Unwin, London.

Duchaufour, P. (1984). "Pedologie." Masson, Paris.

Dunn, J. R. and P. P. Hudec (1966). Frost deterioration: Ice or ordered water? *Geol. Soc. Amer. Spec. Pap.* **101**, 256 (abstract).

Dunn, J. R. and P. P. Hudec (1972). Frost and sorbtion effects in argillaceous rocks. *In* "Frost Action in Soils." Highway Research Board, *Natl. Acad. Sci.-Natl. Acad. Eng.* **393**, 65–78.

Edmonds, R. L. (1979). Decomposition and nutrient release in Douglas-fir needle litter in relation to stand development. *Can. J. For. Res.* **9**, 132–140.

Fanning, D. S. and V. Z. Keramidas (1977). Micas. *In* "Minerals in Soil Environments" (J. B. Dixon and S. B. Weed, eds.), pp. 195–258. Soil Science Society of America, Madison, Wisconsin.

Fogel, R. and K. Cromack, Jr. (1977). Effects of habitat and substrate quality on Douglas-fir litter decomposition in western Oregon. *Can. J. Bot.* **55**, 1632–1640.

Friedman, E. I. (1971). Light and scanning electron microscopy of the endolithic desert algal habitat. *Phycologia* **10**, 411–428.

Fry, E. J. (1927). The mechanical action of crustaceous lichens on substrata of shale, schist, gneiss, limestone, and obsidian. *Ann. Bot.* **41**, 437–460.

Garrels, R. M. and C. L. Christ (1965). "Solutions, Minerals, and Equilibria." Harper and Row, New York.

Gile, L. H., F. F. Peterson, and R. B. Grossman (1966). Morphological and genetic sequences of carbonate accumulation in desert soils. *Soil Sci.* **101**, 347–360.

Goudie, A., R. Cooke, and I. Evans (1970). Experimental investigation of rock weathering by salts. *Area* **4**, 42–48.

Graustein, W. C., K. Cromack, Jr., and P. Sollins (1977). Calcium oxalate: Occurrence in soil and effects on nutrients and geochemical cycle. *Science* **198**, 1252–1254.

Green, C. P. and M. J. Eden (1971). Gibbsite in weathered Dartmoor granite. *Geoderma* **6**, 315–317.

Grim, R. E. (1968). "Clay Mineralogy," 2nd edn. McGraw-Hill, New York.

Harter, R. D. (1977). Reactions of minerals with organic compounds in the soil. *In* "Minerals in Soil Environments" (J. B. Dixon and S. B. Weed, eds.), pp. 709–739. Soil Science Society of America, Madison, Wisconsin.

Holland, H. D. (1978). "The Chemistry of the Atmosphere and Oceans." John Wiley, New York.

Hsu, P. H. (1977). Aluminum hydroxides and oxyhydroxides. *In* "Minerals and Soil Environments" (J. B. Dixon and S. B. Weed, eds.), pp. 99–143. Soil Science Society of America, Madison, Wisconsin.

Hudec, P. P. (1974). Weathering of rocks in arctic and subarctic environments. *In* "Canadian Arctic Geology: Symposium on the Geology of the Canadian Arctic, Saskatoon 1973" (J. D. Aitken and D. J. Glass, eds.), pp. 313–335. Geological Association of Canada/ Canadian Society for Petrology and Geology, Calgary, Alberta.

Hutchinson, G. E. (1970). "The Biosphere." W. H. Freeman, San Francisco, Calif.

Jackson, M. L. (1960). Structural role of hydronium in layer silicates during soil genesis. *Proc. 7th Int. Cong. Soil Sci.*, Madison, Wisconsin, pp. 445–455.

Jackson, M. L. (1963). Aluminum bonding in soils: A unifying principle in soil science. *Soil Sci. Soc. Amer. Proc.* **27**, 1–10.

Jackson, M. L. (1964). Chemical composition of soils. *In* "Chemistry of Soil" (F. E. Bear, ed.), pp. 71–141. Reinholt, New York.

Jackson, M. L. (1969). "Soil Chemical Analysis: Advanced Course," 2nd edn. 7th printing, 1973. Department of Soil Science, University of Wisconsin, Madison, Wisconsin.

Jackson, M. L. and G. D. Sherman (1953). Chemical weathering of minerals in soils. *Adv. Agron.* **5**, 219–318.

Jackson, M. L., S. Y. Lee, F. C. Ugolini, and P. A. Helmke (1976). Age and uranium content of soil micas from Antarctica by the fission track replica method. *Soil Sci.* **123**, 241–248.

Jenny, H. (1941). "Factors of Soil Formation: A System of Quantitative Pedology." McGraw-Hill, New York.

Joffe, J. S. (1949). "Pedology," 2nd edn. Pedology Publications, New Brunswick, N.J.

Johnson, N. M., G. E. Likens, F. H. Borman, and R. S. Pierce (1968). Rate of chemical weathering of silicate minerals in New Hampshire. *Geochim. Cosmochim. Acta* **32**, 531–545.

Keller, W. D. (1957). "The Principles of Chemical Weathering." Lucas Brothers, Columbia, Missouri.

Kittrick, J. A. (1977). Mineral equilibria and the soil system. *In* "Minerals in Soil Environments" (J. B. Dixon and S. B. Weed, eds.), pp. 1–25. Soil Science Society of America, Madison, Wisconsin.

Kodama, H. and Schnitzer (1973). Dissolution of chlorite by fulvic acid. *Can. J. Soil Sci.* **53**, 240–243.

Kononova, M. M. (1966). "Soil Organic Matter: Its Nature, Its Role in Soil Formation and Soil Fertility," 2nd edn. Pergamon Press, Oxford.

Lautridou, J. P. and J. C. Ozouf (1982). Experimental frost shattering: 15 years of research at the Center de Geomorphologie du CNRS. *Prog. Phys. Geog.* **6**, 215–232.

Leamy, M. L., G. D. Smith, F. Colmet-Daage, and M. Otowa (1980). The morphological characteristics of Andisols. *In* "Soils with Variable Charges" (B. K. G. Theng, ed.), pp. 17–34. New Zealand Society of Soil Science. Offset Publications, Palmerston North, New Zealand.

Likens, G. E. (ed.) (1985). "An Ecosystem Approach to Aquatic Ecology: Mirror Lake and Its Environment." Springer-Verlag, New York.

Macias-Vasquez, F., M. L. Fernandez-Marcos, and W. Chesworth (1987). Transformations mineralogique dans les podzols et les sols podzolique de Galice (NW Espagna). *In* "Podzols et Podzolisation" (D. Righi and A. Chauvel, eds.), pp. 163–177. Institute National de la Recherche Agronomique, Plaisir et Paris.

Mann, D. H., R. S. Sletten, and F. C. Ugolini (1986). Soil development at Kongsfjorden, Spitsbergen. *Polar Res.* **4**, 1–16.

Mason, B. and C. B. Moore (1982). "Principles of Geochemistry," 4th ed. John Wiley, New York.
McColl, J. G., A. A. Pohlman, J. M. Jersak, S. C. Tam, and R. R. Northrup (in press). Organics and metals solubility in forest soils. *Proc. 7th Nth Amer. Forest Soil Conf.*, July, 24–28, 1988, University of British Columbia, Vancouver, B.C., Canada.
McCraw, J. D. (1960). Soils of the Ross Dependency, Antarctica. N.Z. *Soc. Soil Sci. Proc.* **4**, 30–35.
McGreevy, J. P. (1981). Some perspectives on frost shattering. *Prog. Phys. Geog.* **5**, 56–75.
Norgren, A. (1973). Opal phytoliths as indicators of soil age and vegetative history. Ph.D. Thesis, Oregon State University, Corvallis, Oregon. University Microfilms, Ann Arbor, Michigan, Diss. Abstr. Int. 33, 3421B.
Ollier, C. D. (1969). "Weathering." Oliver and Boyd, Edinburgh.
Paces, T. (1986). Rates of weathering and erosion derived from mass balance in small drainage basins. *In* "Rates of Chemical Weathering of Rocks and Minerals" (S. M. Coleman and D. P. Dethier, eds.), pp. 531–550. Academic Press, New York.
Parfitt, R. L. (1980). Chemical properties of variable charge soils. *In* "Soils with Variable Charge" (B. K. G. Theng, ed.), pp. 167–194. New Zealand Society of Soil Science Offset Publications, Palmerston North, New Zealand.
Pavich, M. J. (1986). Processes and rates of saprolite production and erosion on a foliated granite rock of the Virginia Piedmont. *In* "Rates of Chemical Weathering of Rocks and Minerals" (S. M. Coleman and D. P. Dethier, eds.), pp. 551–590. Academic Press, New York.
Pedro, G. (1982). The conditions of formation of secondary constituents. *In* "Constituents and Properties of Soils" (M. Bonneau and B. Souchier, eds.), pp. 63–81. Academic Press, New York.
Reiche, P. (1950). "A Survey of Weathering Processes and Products." University of New Mexico Publications in Geology, No. 3, revised edn. The University of New Mexico Press, Alburquerque, New Mexico.
Robert, M. and J. Berthelin (1986). Role of biological and biochemical factors in soil mineral weathering. *In* "Interactions of Soil Minerals with Natural Organics and Microbes" (P. M. Huang and M. Schnitzer, eds.), pp. 453–496. Soil Science Society of America, Madison, Wisconsin.
Robert, M., M. H. Razzaghe, and J. Ranger (1987). Role du facteur biochimique dans la podzolisation. *In* "Podzols et Podzolisation" (D. Righi and A. Chauvel, eds.), pp. 207–223. Institute National Recherche Agronomique, Plaisir et Paris.
Rosell, R. A. and K. L. Babcock (1968). Precipitated manganese isotopically exchanged with ^{54}Mn and chelated by soil organic matter. *In* "Isotopes and Radiation in Soil Organic Matter Studies," pp. 453–469. International Atomic Energy Agency, Vienna, Austria.
Ross, G. J. (1980). The mineralogy of Spodosols. *In* "Soils with Variable Charge" (B. K. G. Theng, ed.), pp. 127–143. New Zealand Society of Soil Science. Offset Publications, Palmerston North, New Zealand.
Schnitzer, M. (1969). Reaction between fulvic acid, a humic compound and inorganic soil constituents. *Soil Sci. Soc. Amer. Proc.* **33**, 75–81.
Schnitzer, M. (1971). Metal–organic interactions in soils and waters. *In* "Organic Compounds in Aquatic Environments" (S. J. Faust and J. V. Hunter, eds.), pp. 297–315. Marcel Dekker, New York.
Schnitzer, M. and J. G. Desjardins (1969). Chemical characteristics of a natural soil leachate from a humic podzol. *Can. J. Soil Sci.* **49**, 151–158.
Schnitzer, M. and S. U. Khan (1972). "Humic Substances in the Environment." Marcel Dekker, New York.
Schnitzer, M. and H. Kodama (1977). Reactions of minerals with soil humic substances. *In* "Minerals in Soil Environments" (J. B. Dixon and S. B. Weed, eds.), pp. 741–770. Soil Science Society of America, Madison, Wisconsin.
Schnitzer, M. and S. I. M. Skinner (1963a). Organic–metallic interactions in soils: 1. Reactions between number of metal ions and the organic matter of a Podzol Bh-horizon. *Soil Sci.* **96**, 86–94.
Schnitzer, M. and S. I. M. Skinner (1963b). Organic–metallic interactions in soils: 2. Reactions between different forms of iron and aluminum and the organic matter of a Podzol Bh-horizon. *Soil Sci.* **96**, 181–187.
Schwertmann, U. (1985). The effect of pedogenic environments on iron oxide minerals. *Adv. Soil Sci.* **1**, 171–200.
Schwertmann, U. (1988). Occurrence and formation of iron oxides in various pedoenvironments. *In* "Iron in Soils and Clay Minerals" (J. W. Stucki, B. A. Goodman, and U. Schwertmann, eds.), pp. 267–308. Reidel, Dordrecht, The Netherlands.
Schwertmann, U. and R. M. Taylor (1977). Iron oxides. *In* "Minerals in Soil Environments" (J. B. Dixon and S. B. Weed, eds.), pp. 145–180. Soil Science Society of America, Madison, Wisconsin.
Sherman, G. D. (1952). The titanium content of Hawaiian soils and its significance. *Soil Sci. Soc. Amer. Proc.* **16**, 15–18.
Soderman, L. G. and K. M. Quigley (1965). Geotechnical properties of three Ontario clays. *Can. Geotech. J.* **II**, 167–189.
Soil Survey Staff (1975). "Soil Taxonomy." Agricultural Handbook 436. US Government Printing Office, Washington, D.C.
Stallard, R. F. (1985). River chemistry, geology, geomorphology, and soils in the Amazon and Orinoco basins. *In* "The Chemistry of Weathering" (J. L. Drever, ed.), pp. 293–316. Reidel, Dordrecht, The Netherlands.
Starkey, R. L. (1966). Oxidation and reduction of sulfur compounds in soils. *Soil Sci.* **101**, 297–306.
Stednick, J. D. (1981). Precipitation and streamwater chemistry in an undisturbed watershed in southeast Alaska. *Res. Pap. PNW-291*, US Department of

Agriculture, Forest Service, Pacific Northwest and Range Experiment Station, Portland, Oregon.

Stevenson, F. J. (1982). "Humus Chemistry." John Wiley, New York.

Stevenson, F. J. (1985). Geochemistry of soil humic substances. *In* "Humic Substances in Soil, Sediment, and Water" (G. R. Aiken, D. M. McKnight, R. L. Wershaw, and P. MacCarthy, eds.), pp. 13–52. John Wiley, New York.

Stumm, W. and J. J. Morgan (1981). "Aquatic Chemistry," 2nd edn. Wiley-Interscience, New York.

Stumm, W., G. Furrer, E. Wieland, and B. Zinder (1985). The effects of complex-forming ligands on the dissolution of oxides and alumino-silicates. *In* "The Chemistry of Weathering" (J. I. Drever, ed.). Reidel, Dordrecht, The Netherlands.

Syers, J. K. and I. K. Iskandar (1973). Pedogenic significance of lichens. *In* "The Lichens" (V. Ahmadjian and M. Hare, eds.), pp. 225–248. Academic Press, New York.

Tamura, T., M. L. Jackson, and G. D. Sherman (1955). Mineral content of latosolic brown forest soil and a humic ferruginous latosol of Hawaii. *Soil Sci. Soc. Amer. Proc.* **19**, 435–439.

Tardy, Y. (1969). Geochimie des alterations. Etude des arenes et des eaux de quelques massifs cristallins d'Europe et d'Afrique. These Doc. Etat, University of Strasbourg, Strasbourg, France.

Tedrow, J. C. F. (1977). "Soils of the Polar Landscape." Rutgers University Press, New Brunswick, N.J.

Tedrow, J. C. F. and F. C. Ugolini (1966). Antarctic soils. *Amer. Geophys. Union Antarctic Res. Ser.* **8**, 161–177.

Ugolini, F. C. (1963). Pedological investigation in the lower Wright Valley, Antarctica. *Proc. Int. Conf. Permafrost.* NAS-NRC Publication No. 1278, pp. 55–61, Washington, D.C.

Ugolini, F. C. (1986a). Pedogenic zonation in the well-drained soils of the Arctic regions. *Quat. Res.* **26**, 100–120.

Ugolini, F. C. (1986b). Processes and rates of weathering in cold and polar desert environments. *In* "Rates of Chemical Weathering of Rocks and Minerals" (S. M. Colman and D. P. Dethier, eds.), pp. 193–235. Academic Press, New York.

Ugolini, F. C. (1987). The proton donor theory Part I and II. *Soil Sci. Amer. Ann. Meet., Agronomy,* November 28–December 4, abstract, p. 233, Atlanta, Georgia.

Ugolini, F. C. and D. M. Anderson (1973). Ionic migration and weathering in frozen Antarctic soils. *Soil Sci.* **115**, 461–470.

Ugolini, F. C. and R. A. Dahlgren (1987). The mechanism of podzolization as revealed by soil solution studies. *In* "Podzols et Podzolisation" (D. Righi and A. Chauvel, eds.), pp. 195–203. Institute National Recherche Agronomique, Plaisir et Paris.

Ugolini, F. C. and R. L. Edmonds (1983). Soil biology. *In* "Pedogenesis and Soil Taxonomy I. Concepts and Interactions" (L. P. Walding, N. E. Smeck, and G. F. Hall, eds.), pp. 193–231. Elsevier, Amsterdam, The Netherlands.

Ugolini, F. C. and M. L. Jackson, (1982). Weathering and mineral synthesis in Antarctic soils. *In* "Antarctic Geoscience" (C. Craddock, ed.), pp. 1101–1108. University of Wisconsin Press, Madison, Wisconsin.

Ugolini, F. C., R. Minden, H. Dawson, and J. Zachara (1977a). An example of soil processes in the *Abies amabilis* zone of central Cascades, Washington. *Soil Sci.* **124**, 291–302.

Ugolini, F. C., H. Dawson, and J. Zachara (1977b). Direct evidence of particle migration in the soil solution of a podzol. *Science* **198**, 603–605.

Ugolini, F. C., M. G. Stoner, and D. J. Marrett (1987). Arctic pedogenesis: 1. Evidence for contemporary podzolization. *Soil Sci.* **144**, 90–100.

Ugolini, F. C., R. A. Dahlgren, S. Shoji, and T. Ito (1988). An example of andolization and podzolization as revealed by soil solution studies, southern Hakkoda, northeastern Japan. *Soil Sci.* **145**, 111–125.

Ugolini, F. C., R. A. Dahlgren, and K. Vogt (1990). The genesis of spodosols and the role of vegetation in the Cascades Range of Washington, U.S.A. *In* "Classification and Management of Spodosols," *Proc. Vth Int. Soil Correlation Meet., Northeastern USA and Canada,* (J. M. Kimble and R. D. Yeck, eds.), pp. 370–380. Soil Conservation Service, USDA and Soil Management Support Service, USAID.

Ulrich, B. (1980). Production and consumption of hydrogen ions in the ecosphere. *In* "Effect of Acid Precipitation on Terrestrial Ecosystems" (T. C. Hutchinson and M. Havas, eds.), pp. 255–282. Plenum Press, New York.

US Department of Agriculture, Agency for International Development (1986). "Designations for Master Horizons and Layers in Soils." Department of Agronomy, College of Agriculture and Life Sciences, Cornell University, Ithaca, New York.

Van Breemen, N. and R. Brinkman (1976). Chemical equilibrium and soil formation. *In* "Soil Chemistry" (G. H. Bolt and M. G. M. Bruggenwert, eds.), pp. 141–170. Elsevier, Amsterdam, The Netherlands.

Van Breemen, N., J. Mulder, and C. T. Driscoll (1983). Acidification and alkalinization of soils. *Plant and Soil* **75**, 283–308.

Van Miegroet, H. and D. W. Cole (1984). The impact of nitrification on soil acidification and cation leaching in a red alder ecosystem. *J. Environ. Qual.* **13**, 586–590.

Van Schuylenborgh, J. and M. G. M. Bruggenwert (1965). On soil genesis in temperate humid climate. V. The formation of "albic" and "spodic" horizon. *Neth. J. Agric. Sci.* **13**, 267–279.

Vedy, J. C. and S. Bruckert (1982). Soil solution: Composition and pedogenic significance. *In* "Pedology, Constituents and Properties of Soil" (M. Bonneau and B. Souchier, eds.) (V. C. Farmer, trans. ed.), pp. 184–213. Academic Press, New York.

Velbel, M. A. (1985). Hydrogeochemical constraints on mass balances in forested watersheds of the southern Appalachians. *In* "The Chemistry of Weathering" (J. I. Drever, ed.), pp. 231–247. Reidel, Dordrecht, The Netherlands.

Velde, B. (1985). "Clay Minerals – A Physical-Chemical Explanation of Their Occurrence." Developments in Sedimentology No. 40. Elsevier, Amsterdam, The Netherlands.

Wada, K. (1977). Allophane and imogolite. *In* "Minerals in Soil Environments" (J. B. Dixon and S. B. Weed, eds.), pp. 603–638. Soil Science Society of America, Madison, Wisconsin.

Wada, K. (1978). Allophane and imogolite. *Dev. Sedimentol.* **26**, 147–185.

Wada, K. (1980). Mineralogical characteristics of andisols. *In* "Soils with Variable Charge" (B. K. G. Theng, ed.), pp. 89–107. New Zealand Society of Soil Science. Offset Publications, Palmerston North, New Zealand.

Wada, K. (1985). The distinctive properties of andosols. *Adv. Soil Sci.* **2**, 173–229.

Wada, K. and M. E. Harward (1974). Amorphous clay constituents of soils. *Adv. Agron.* **25**, 211–260.

Washburn, A. L. (1980). "Geocryology: A Survey of Periglacial Processes and Environments." John Wiley, New York.

Wellman, H. W. and A. T. Wilson (1965). Salt weathering, neglected geological erosive agent in coastal and arid environments. *Nature* **205**, 1091–1098.

Wilding, L. P., N. E. Smeck, and L. R. Drees (1977). Silica in soils: Quartz, cristobalite, tridymite, and opal. *In* "Minerals in Soil Environments" (J. B. Dixon and S. B. Weed, eds.), pp. 471–552. Soil Science Society of America, Madison, Wisconsin.

Williams, R. B. G. and D. A. Robinson (1981). Weathering of sandstone by the combined action of frost and salt. *Earth Surf. Proc. Landforms* **6**, 1–9.

Yaalon, D. H. (1975). Conceptual models in pedogenesis: Can soil forming functions be solved? *Geoderma* **14**, 189–205.

Yarilova, E. A. (1947). The role of lithophilous lichens in the weathering of massive crystalline rocks. *Pochvovedeniye* **3**, 533–548.

Zachara, J. M. (1979). Clay genesis and alteration in two tephritic subalpine podzols of the Central Cascades, Washington. Unpublished M.S. Thesis, University of Washington, Seattle, Wash.

8

Sediments: Their Interaction with Biogeochemical Cycles through Formation and Diagenesis

Rolf O. Hallberg

8.1 Introduction

Sediments are the key to ancient and historical environments. By understanding the processes behind the formation and diagenesis of sediments, we can use structure and composition to reveal the depositional environment. A sequence of sedimentary layers can tell us about environmental changes over time. The recent sedimentary record reveals cultural impacts on the environment during the industrial era, while for example, ancient sedimentary rocks give insight into the evolution of free oxygen in the atmosphere during the Precambrian Era. Sediments should not only be looked upon as inactive leftovers from yesterday. During formation and diagenesis, they take an active part in the biogeochemical cycles of elements. It is this ongoing interaction with biogeochemical cycles that is treated in this chapter. This interaction is, however, a very comprehensive topic, which involves many variables and complicated physical, chemical, and biological activities. In reviewing such a broad subject, which extends into many disciplines and where in many cases existing knowledge is fragmentary and incomplete, there are inevitably many aspects that have to be left out or mentioned only briefly. For example, in the sediment, the cycles of many elements are largely the result of microbial activity, and the many factors affecting this activity also alter the reaction rates and equilibria. Too few investigations, however, have led to reliable estimates of the actual influence of the biota on physicochemical processes and their role in biogeochemical cycles. The main examples in this chapter are from the marine and brackish environment. Freshwater habitats will not be specifically dealt with, though some references will be made to lake investigations (Mortimer, 1941, 1942) since processes in sediments of freshwater and marine environments have much in common.

8.2 Weathering and Formation of Sediments

Marine sediments are formed by the deposition of particles transported to the ocean by rivers and by wind. These particles come from the disintegration of rocks by weathering processes. Physical, chemical, and biological action on rocks result in three types of weathering products: detrital material from the rock and vegetation; solutes from the dissolution of minerals and organic matter; and new minerals from chemical reactions between solutes and minerals. (The breakdown of the parent rock is discussed in Chapter 6 and the formation of soil is discussed in Chapter 7.) In between the material closest to the rock and topsoil we can usually distinguish between several soil (or weathering) horizons. When the soil is eroded, transported, and deposited we have formed a sediment. Thus a sediment is laid down in *layers*, while a soil exhibits

Global Biogeochemical Cycles
ISBN 0-12-147685-5

weathering *horizons*. The mineralogical composition of the detrital material depends on the type of source rock, and also on the duration of weathering and transport. As a "rule of thumb", we can say that the longer the period of transportation, the higher the concentration of quartz particles. The reason for this is that quartz has a high resistance to weathering relative to olivine and biotite, for example.

The formation of sediments through weathering interacts with biogeochemical cycles by chemical weathering. We can distinguish between three types of chemical transformation of rocks by reaction with water, oxygen, and carbon dioxide:

- H_2O – hydratization and hydrolysis,
- O_2 – oxidation,
- CO_2 – carbonatization.

We can thus conclude that weathering processes take an active part in the cycling of oxygen and carbon, but does weathering affect these cycles to a significant extent? Consider the following examples.

Oxygen is formed by photosynthesis according to the reaction:

$$CO_2 + H_2O \rightarrow O_2 + CH_2O \text{ (cell material)} \quad (1)$$

The annual primary production of organic carbon through photosynthesis is about 7.3×10^{16} g/year. The major part of this carbon is decomposed or respired in a process that also involves the biogeochemical transformation of nitrogen, sulfur, and many other elements. Only a small part of the annual primary production of carbon escapes decomposition and is buried in the sediments. On average, sediments and sedimentary rocks contain 3% carbon, which corresponds to a net production of oxygen of about 3×10^{14} g/year. This production has the potential to double atmospheric oxygen on a time-scale of 4 Ma. In contrast, the fossil record indicates that the partial pressure of oxygen in the atmosphere has fluctuated very little, at least during the last 600 Ma (Conway, 1943). There must obviously be a sink that can accommodate the net annual production of oxygen. This sink is the annual weathering of rocks, which can be estimated from the mass of sediments formed during the last 600 Ma, which is $18\,000 \times 10^{20}$ g (Gregor, 1968). During this period, soil formation and sediment formation have fluctuated greatly. The average weathering rate is 30×10^{14} kg/year. The average continental rock contains ferrous iron and sulfide sulfur (mainly as pyrite) and organic carbon. During the weathering processes, these constituents are oxidized and consume oxygen. The amounts of oxygen necessary for each of these reactions are given in Table 8-1.

Table 8-1 Average oxygen consumption during weathering of rocks[a]

$C + O_2 \rightarrow CO_2$	12 ± 3 g O_2/kg
$S^{2-} + 2O_2 \rightarrow SO_4^{2-}$	6 ± 2 g O_2/kg
$4\text{"FeO"} + O_2 \rightarrow 2Fe_2O_3$	2 ± 1 g O_2/kg

[a] After Holland (1978, p. 285).

The total annual consumption of oxygen by weathering can be estimated (Holland, 1978) as follows:

$$(20 \pm 6\text{ g } O_2/\text{kg rock}) \times (2 \times 10^{13}\text{ kg rock}) = \sim 4 \times 10^{14}\text{ g } O_2$$

This means that weathering of rocks and burial of organic carbon in sediments during their formation are important processes for the balance of oxygen content of the atmosphere.

Weathering of rocks is also a sink for CO_2. Garrels and Mackenzie (1971) have estimated that the formation of 1500 g of sedimentary rocks requires 100 g of CO_2. If we use the same number as above for the annual formation of sedimentary rocks, we have a sink of 2×10^{14} g CO_2/year. About half of that comes from the oxidation of organic carbon in the weathering rock. The rest is from the atmosphere. The burning of fossil fuel increases the partial pressure of CO_2 in the atmosphere, but increased CO_2 levels increase the weathering rate. Thus the weathering rate and sediment formation ultimately control the carbon dioxide content of the atmosphere.

Weathering, erosion, transport, and deposition take material from the continental rocks to the sediments in the oceans. If there was no sink for these sediments, the ocean should fill up in less than 100 Ma. On the other hand, if there was no source for rocks on the continents, they should be degraded to ocean level in less than 50 Ma (Holland, 1978, p. 146). Neither of these figures make sense, and the reason is that they do not take continental plate tectonics into account. Sediments are metamorphosed and returned from the oceans back to the continents either by overthrusting (continental plate/continental plate convergence) or subduction (continental plate/ocean plate convergence). In the former case we are dealing mainly with sediments from the continental shelf, and in the latter case

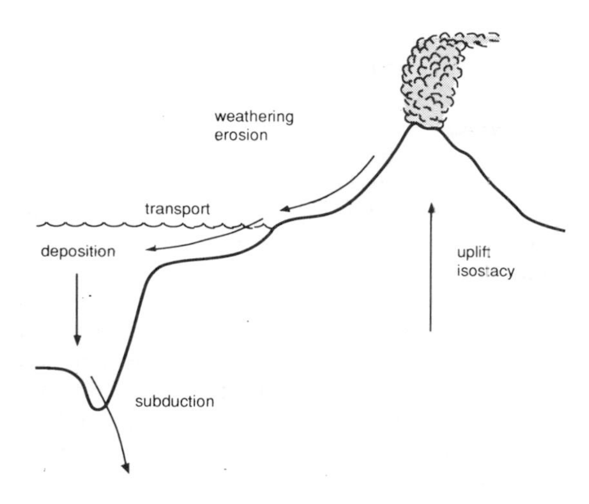

Fig. 8-1 Generalized cycle of sediments and sedimentary rocks.

mainly deep ocean sediments. Thus mountains are built up by sediments from the continental shelf, while the magmatic bedrocks originally were sediments deposited outside the shelf area. The continental plates are buoyed by lithospheric magma and are consequently uplifted to compensate for the loss of mass by weathering and erosion. Finally, the metamorphosed sediments from the ocean floor are uplifted to such an extent that they are exposed to weathering processes. Subduction and uplift are thus two important processes to close our rock/sediment cycle (Fig. 8-1).

8.3 The Structure of Sediments

8.3.1 The Solid Phase

Sediments are classified in a number of ways. Grain size conveys some basic information for exchange processes and correlates to a certain extent with the minerogenic distribution. Clay minerals and organic matter, for example, occur in the $<20 \mu m$ size fraction. For biogeochemical purposes, however, the granulometric description is far from sufficient. A more satisfactory classification is based on the material and genetic differences of the sediments as represented in Table 8-2. The arrangement of the particles is of even greater importance as it governs the permeability and porosity of the sediments. A sediment (with few exceptions) is composed of relatively porous clay aggregates and randomly distributed coarser grains of detrital material (Fig. 8-2).

The aggregates of sedimentary particles are usually arranged before deposition. Within these

Table 8-2 Material-genetic classification of sediments[a]

Sediment type	% CaCO	% clastic and clayey material	% amorphous silica	% pelagic sed.	Composition
Detrital or epiclastic	<30	>50			Denudation products of continental rocks
Biogenic					
A. Calcareous	>30			~48	Foraminifera, coccoliths, calcareous algae, molluscs, bryozoa, and corals
B. Siliceous			>30	~14	Diatoms and radiolaria
Chemogenic					Iron-manganese nodules, glauconite, phosphorite, nodules, phillipsite, palagonite, celestobarite, and evaporites
Volcanogenic					Pyroclastic material
Polygenic	<10	>50	<10	~38	Red clay

[a] Modified from Lisitzin (1972) and Sverdrup *et al*. (1942).

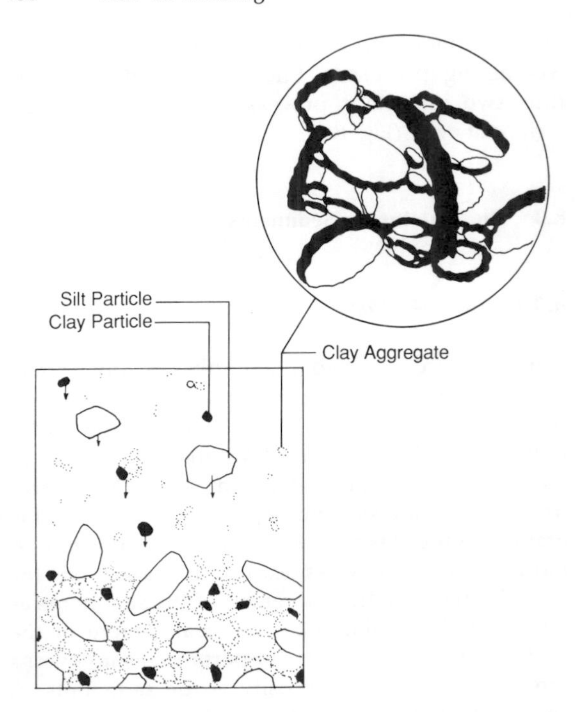

Fig. 8-2 Hypothetical particle arrangement in a sediment and a clay aggregate. Adapted from Casagrande (1940).

aggregates, forces are set up between atoms, molecules, and ions that depend on the ionic strength of the electrolyte. As a result, the arrangements of the particles are different in brackish and marine environments. Clays deposited in freshwater form relatively porous aggregates and small voids, while marine clays form large dense aggregates separated by large voids. Salinity variations in interstitial water are most significant in coastal areas and different degrees of permeability and porosity may therefore be expected in sediments of the continental margins as compared to deep-sea sediments of the same grain size.

The net electric charge of the solid surfaces influences the interaction between particles. Clay minerals are of special relevance in a description of the solid phase because of their relatively high specific surface. The specific surface of a montmorillonite clay has been found to be $800\,m^2/g$. Kaolinites, on the other hand, have accessible specific areas ranging from 10 to $20\,m^2/g$.

Clay minerals have silicate layer structures of either type 1:1 or 2:1. Kaolinite is of type 1:1 and consists of alternating layers of silicon tetraheda and aluminum magnesium octahedra as shown in Fig. 8-3a. Montmorillonites and micas are typical representatives of the 2:1 type with an aluminum–magnesium octahedral layer sandwiched between two silica tetrahedral layers. Two models have been suggested for the montmorillonite structure. According to one model (see Fig. 8-3b), the silicon atoms are situated in one plane only (*cis*-coordination) (Hofmann *et al.*, 1933). The other (see Fig. 8-3c) places the silicon atoms in each layer on two planes (*trans*-coordination) (Edelman and Favejee, 1940). Ionic substitutions, usually aluminum and iron for silicon in the tetrahedral layer and magnesium and iron for aluminum in the

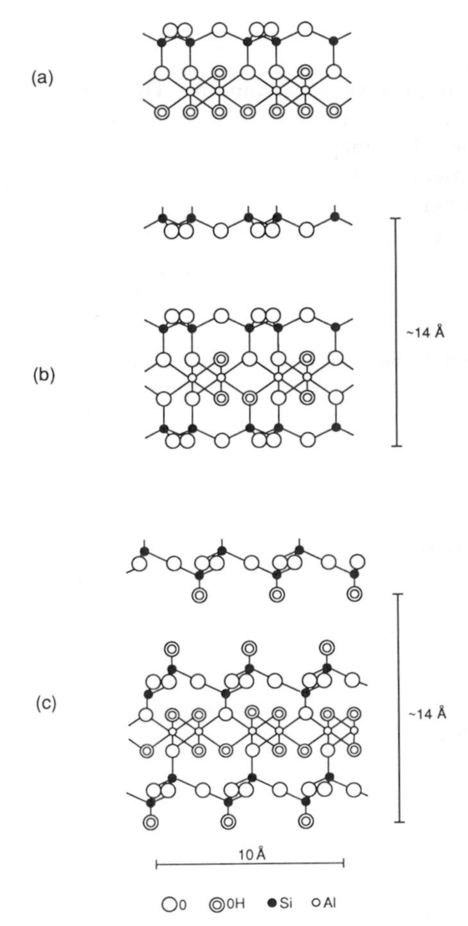

Fig. 8-3 A schematic illustration of the 1 : 1 (a) and 2 : 1 (b and c) layer lattice silicates.

octahedral layer, cause an imbalance of electrical charge that is compensated for by the presence of easily exchangeable cations at the mineral surface.

8.3.2 The Liquid Phase

The liquid phase of the solid–water interface consists of three different types of aqueous electrolytic solutions:

1. "Normal" water of random ionic ordering at some distance from a solid surface.
2. Adsorbed water, essentially free from ions, on a solid surface.
3. "Structured" water with a layering of ionic ordering (1) and (2) near a solid surface.

Because of the close relationship to the mineral particles in the sediment, interlamellar water is usually of types (2) and (3). A special type of water, so-called "polywater" or "superwater", which has been reviewed and considered by Kamb (1971) and Henniker (1949), is a modification of water solution, as a result of impurities in a water solution. It is an interesting phenomenon, provided such solutions occur in nature. "Polywater" has been observed to have a density of about 1.4 g/cm^3 and a viscosity about 15 times greater than normal water. Capillaries with "polywater" might be expected in a fine-grained sediment (Low and White, 1970) with a considerable amount of clay mineral particles like mud. The significance of the various types of water in sediments must not be underestimated, as they may influence other processes taking place in the aqueous phase.

8.3.3 The Gaseous Phase

In the gaseous phase of the sediment–water interface, oxygen, nitrogen, hydrogen sulfide, methane, carbon dioxide, and ammonia are found. The latter occurs principally as ammonium ion at pH values in the range 5–8. Oxygen and hydrogen sulfide react with each other spontaneously, and consequently these two gases are usually regarded as non-co-existent. In a sediment environment, however, heterogeneity of the sediment with regard to organic matter and permeability gives rise to niche formation. Such niches can be microenvironments of different chemical composition and exist close to one another as more or less isolated chemical and biological systems. The presence of hydrogen sulfide is not therefore a reliable indicator of entirely anoxic conditions within the sediment. Microniches were invoked by Hallberg (1974) to explain very low redox values in intertidal sediments with some oxygen still present (Fig. 8-4). A theoretical model for the size of a spherical-induced microniche in an oxic medium has been evaluated by Jörgensen (1977), who states "that if the surrounding medium is not essentially anoxic, the minimum effective size of a reduced microniche must be in the order of 100 μm to several mm". Fecal pellets may serve as such microniches at the uppermost part of the sediment (Fig. 8-5), while burrowing organisms may distribute oxygen to the reduced part of the sediment.

Methane is produced by bacteria in the strongly reduced part of the sediment. The methane-producing bacteria are obligatory anaerobes and are most active below the horizon of hydrogen sulfide production. Methane is used as a carbon source by heterotrophic microorganisms in the overlaying sediments. Methane concentrations decrease rapidly below the sediment–water interface. Thus, its importance for the biogeochemical cycles of carbon in the marine system is negligible.

Carbon dioxide is produced as a result of the metabolism of all heterotrophic organisms. The concentrations of CO_2 in pore water of reduced

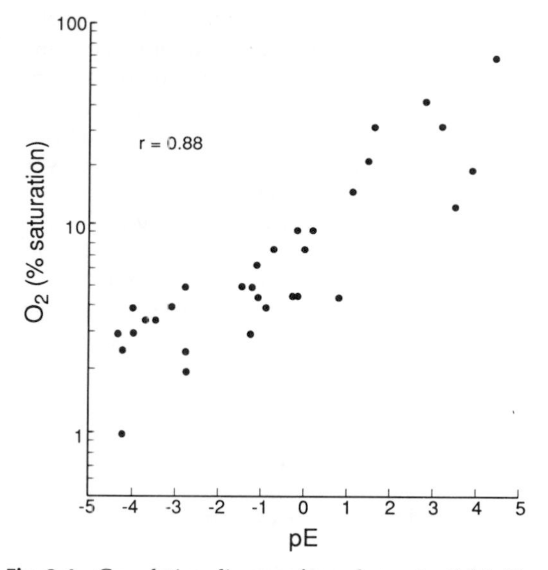

Fig. 8-4 Correlation diagram for redox potential (pE) and oxygen saturation (note the log scale). *In situ* electrode measurements in a Dutch intertidal area. Reprinted with permission from the author.

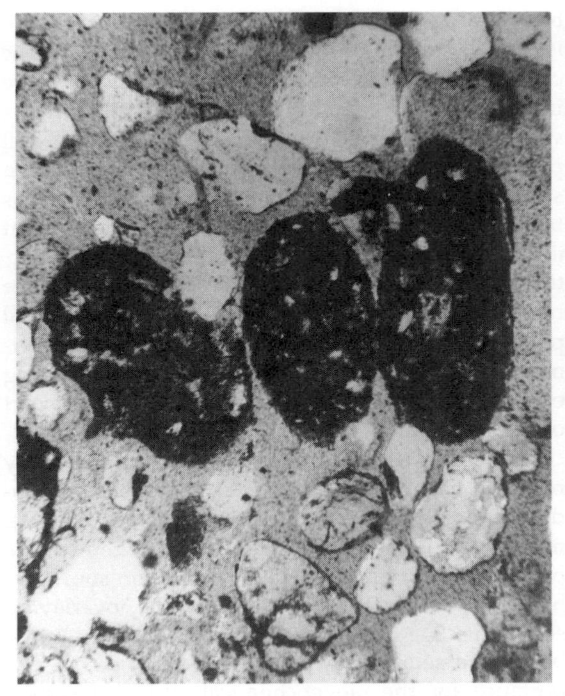

Fig. 8-5 Micrograph from a Dutch intertidal area showing fecal pellets and quartz grains. Reprinted with permission from the author.

sediments are therefore high. Autotrophic microorganisms consume CO_2 in the oxidized part of the sediment, which can vary in depth from 1 m in deep-sea sediments to a few mm in organic-rich sediments of the continental margins. The buffer capacity of ocean water is very low and the addition of 0.5 mmol/kg of CO_2 will reduce the pH of seawater by more than 1 pH unit. It is therefore surprising that the concentration of CO_2 in pore water of marine sediments may reach values of 60 mmol/kg (Presley, 1969), and yet show pH shifts less than 1 unit. It is evident that the increase in CO_2 must be counterbalanced by other processes that tend to increase the pH (such as production of ammonia and increase in total alkalinity) so that the result is a fairly constant pH.

8.4 Physicochemical Processes in Sediments

8.4.1 The Liquid–Solid Interaction

The most dynamic processes are found where the exchange of energy is most dominant. That is often at gas–liquid and liquid–solid interfaces. An understanding of the nature of the arrangement of water molecules and solvated ions close to a surface is a prerequisite for an understanding of the equilibria between water and the solid phase of a sediment. Pollutants like heavy metals and chlorinated hydrocarbons are adsorbed on the clay fraction and on organic material in the sediments. The amount and type of clay minerals are the principal determinants of liquid–solid exchange capacities and surface activities.

The reactions between clay minerals and ocean water were used by Sillén (1961) to explain the relative abundance of the major cations in ocean water and also pH of the ocean. Sillén used a thermodynamic approach where clay minerals and quartz are in equilibrium with each other. The following equilibria define the ratios $(K^+)/(H^+)$ and $(Mg^+)/(H^+)^2$:

$$KAl_3Si_3O_{10}(OH)_2 \text{ (illite)} + H^+ + \tfrac{3}{2}H_2O \leftrightarrow \tfrac{3}{2}Al_2Si_2O_5(OH)_4 \text{ (kaolinite)} + K^+ \quad (2)$$

$$\tfrac{1}{5}Mg_5Al_2Si_3O_{10}(OH)_8 \text{ (chlorite)} + 2H^+ \leftrightarrow \tfrac{1}{3}Al_2Si_2O_5(OH)_4 \text{ (kaolinite)} + \tfrac{1}{5}SiO_2 + \tfrac{7}{5}H_2O + Mg^{2+} \quad (3)$$

In a similar way, the coexistence of kaolinite, smectite, and quartz fix the ratio $(Na^+)/(H^+)$. Although Sillén's predicted data agree very well with the cation ratios of ocean water, it has not been possible to prove that these "reverse weathering" reactions between clay minerals and quartz take place to a significant extent. The interaction between clay minerals and ambient water is complicated and therefore difficult to study under natural conditions.

Our understanding of the solid–water interaction is thus further impaired by our limited knowledge of clays. This has to be taken into consideration when interpreting the results of different kinds of interlamellar water extraction methods. One procedure is the pressure-extraction method. The effect of this squeezing technique on the chemistry of the solutions is not clear. Manheim and Kryukow, Komarova, Zhuchkova and other coworkers cited in Manheim (1966) propose that the chemistry of solutions does not change with increasing extraction pressure until a certain point of water content (50–60% water content of wet weight at a pressure of 60 MPa). Other scientists (Englehardt and Gaida, 1963; Chilingarian and Ricke, 1968, 1969) found significant changes in experiments with montmorillonite clays (high cation-exchange capacity) Kaolinite clays, which have comparatively low cation-exchange capacities, gave no concentration change.

It is important to make a distinction between hydrophilic and hydrophobic clay minerals. Such properties will affect the ionic strength in the interstitial spacing and the cation exchange capacity of the clay mineral (Table 8-3). Montmorillonite, representative of the hydrophilic type, shows a characteristic swelling dependent on the amount of water in the interlamellar spacings. Interlamellar spacings as large as 60 Å have been reported in aqueous suspensions (Grim, 1953).

As a consequence of the fact that a hydrophilic clay interacts with water molecules through the formation of hydrogen bonds, a negative charge is taken on by the clay particles in aqueous solution. This is because of proton transfer in the hydrogen bond, which leads to a certain amount of proton expulsion. In a clay aggregate, adsorbed water molecules and fixed cations compensate for the charge defects in the clay crystallites. The equilibrium between an inner absorbed and outer non-absorbed solution of a clay crystallite will depend on the reintroduction of water lattice defects. These defects raise the negative surface charge of the crystallite due to a facilitated protolysis, initiating a diffusion process. If the outer solution changes its composition, the inner solution will also change in order to maintain equilibrium. The kinetics of this equilibrium are rapid, and inner–outer solution equilibration can be assumed to be maintained at all times. This is called the Donnan equilibrium interaction and applies particularly to minerals with an extensive absorbed ion population. A useful way of defining *in situ* inner and outer solution concentrations has been suggested by Mangelsdorf *et al.* (1969) and Murthy and Ferrell (1972).

The Donnan equilibrium theory implies that dilution of a clay/water system containing monovalent and divalent cations displaces the equilibrium in such a manner that the absorption of divalent ions increases, whereas the absorption of monovalent ions decreases. The ionic charge is not the only determining factor in the absorption effect. Factors such as temperature, pH, and specific ions also play important roles. Hydration energy, which appears to be one of the most important factors for the absorption and fixation of cations, displaces the ionic equilibria in a manner opposing the Donnan equilibrium theory. According to Sawhney (1972), "cations with low hydration energy such as Ca^{2+}, Mg^{2+} and Sr^{2+}, produce expanded interlayers and are not fixed".

In marine sediments, the clay minerals are not always monomineralic but frequently constitute mixed phases and are termed "mixed-layer clays". Weaver (1958) has shown that a mica that has been depleted of K^+ ("stripped mica") and only slightly altered in its alumino-silicate content, will readily adsorb K^+ from seawater and thereby reconstitute the 10 Å mica-like structure. He also states that "those layers which have a charge density between 120–150 meq/100 g will selectively adsorb and fix enough K^+ to contract a significant proportion of these layers to a 10 Å basal spacing. The result will be a 'mixed-layer' phase. This 'mixed-layer' phase is recognizable in that it will give rise to a (001) X-ray diffraction maximum over a broad range of angles because of interstratification of 10 Å and 17 Å layers." Such diffraction patterns from the clay mineral fraction of marine sediments may lead to the erroneous interpretation that montmorillonite or vermiculite have been transformed into illite.

Table 8-3 Cation exchange capacity for clay minerals at pH 7 (meq/100 g dry wt.)[a]

Mineral	CEC
Kaolinite	3–15
Halloysite · $2H_2O$	5–10
Halloysite · $4H_2O$	40–50
Chlorite	10–40
Hydromuscovite (Illite)	10–40
Attapulgite	20–30
Vermiculite	100–150
Montmorillonite	80–150

[a] From Grim (1953).

8.4.2 Diagenesis

Physicochemical reactions within the sea–sediment sphere tend to reach equilibrium. Those reactions that are so rapid that they occur prior to burial in the bottom sediments are referred to as "halmyrolysis" (e.g. formation of clay aggregates), while those that take place in the upper part of the sediment are termed "early diagenesis". The diagenetic processes include cementation, compaction, diffusion, redox reactions, transformation of organic and inorganic material, and ion exchange phenomena. A short survey of some of the diagenetic processes is presented here, although it should be emphasized that many questions are left to be answered as some of them (e.g. ion exchange) have only begun to receive adequate attention. For a more detailed presentation of diagenetic processes, the reader is referred to Berner (1980).

8.4.2.1 Cementation

Cementation is the precipitation of a binding material around grains, thereby filling the pores of a sediment. Among the processes mentioned, Berner (1971, p. 97) states that:

> ... cementation by silica must be predominantly a phenomenon of later diagenesis because almost no examples are found in recent marine sediments. By contrast, cementation by calcium carbonate may occur rapidly after deposition. A good example is beachrock. Beachrock is beach and intertidal sand (usually carbonate and skeletal fragments) cemented by $CaCO_3$ in subtropical to tropical climate zones. The cement forms so rapidly that human artifacts only a few decades old are commonly found cemented into the beachrock. Other examples of early diagenetic $CaCO_3$ cementation are provided by subtidal lithified calcarenites, reefs, and pelagic oozes (for a summary of numerous occurrences see Mackenzie *et al.*, 1969). In all cases the cements are aragonite or high magnesium calcite indicating formation from sea water or other magnesium rich solutions.

8.4.2.2 Compaction

Compaction is the decrease in volume of a sediment resulting primarily from an expulsion of water due to compression by the deposition of overlying sediment. Rosenqvist (1958, 1962) stated on the basis of a study of approximately 100 stereoscopic micrographs of undisturbed marine clays, that the particle arrangements within the clay aggregates were of the "corner/plane cardhouse" type suggested by Tan (1957) and Lambe (1958). From microstructural investigations of these "cardhouse" aggregates (Pusch, 1970), it can be concluded that during the compaction process, the aggregates approach one another assuming new positions of equilibrium.

Thus the initial state of compaction results in a collapse of the more unstable original structures and a rearrangement of the particles so that a tighter packing is obtained. Increased packing will expel the interstitial water from the sediment and decrease the porosity. As different clay minerals exhibit different packing characteristics, which are in turn dependent upon the chemical composition of the interstitial water, relationships become too complex for a simple generalization (Meade, 1966). The total compaction that occurs during accumulation of a sediment layer can be calculated from the following formula with the assumption that steady-state compaction has existed since deposition (the rate of burial of sediment and the diagenetic processes, including bioturbation, are constant):

$$\text{compaction} = \frac{h_0 - h}{h_0} = \frac{n_0 - n}{1 - n} \tag{4}$$

where h_0 = thickness of the layer at deposition, h = observed thickness of the layer, n_0 = porosity of the top-most sedimentary layer, and n = porosity of the observed layer.

For the upper part of the sediment, where the sea–sediment interaction is most predominant, steady-state compaction is an acceptable assumption as porosity and compaction undergo a linear change during burial.

8.4.3 Diffusion

Diffusion is referred to here as molecular diffusion in interstitial water. During early diagenesis, the chemical transformation in a sediment is dependent on the reactivity and concentration of the components taking part in the reaction. When transformation occurs, the original concentration of these compounds is depleted, thereby setting up a gradient in the interstitial water. This gradient will be the driving force for molecular diffusion. Diffusional transport and the kinetics of the transformation reactions determine the net effectiveness of the chemical reaction.

The process of diffusion is discussed in Chapter 4 and in detail by Crank (1956). Diffusion in a sediment is complicated by the presence of particles in the fluid medium. Diffusion is thus retarded, and a calculation of sediment diffusion must also include the terms porosity, represented by n, and tortuosity. Since the value of tortuosity for natural sediments is seldom known, it is more convenient to use the term "formation factor" or "lithological factor", denoted L, which takes into account everything but porosity. Fick's diffusion constant D, is replaced by the whole sediment diffusion constant D_s, where $D_s < D$:

$$D_s = \frac{Dn}{L} \tag{5}$$

According to Horne *et al.* (1969), pressures up to several thousand atmospheres seem to have no significant effect on D_s. A confusing variety of D_s data are found in the literature, probably because of a misinterpretation of results or lumping of "net ion flux" and "pure diffusion processes". A critical review of the subject is presented in a paper by Manheim (1970). Table 8-4 provides a list of D_s values found in the literature for the most common dissolved chemical species in sediment. Values for D_s are sometimes not available or are difficult to estimate but can be obtained indirectly by means of electrical conductivity measurements (Klinkenberg, 1951).

Not only diffusion, but interactions between diffusion and chemical transformation, determine the performance of a transformation process. Weisz (1973) described an approach to the mathematical description of the diffusion–transformation interaction for catalytic reactions. A similar approach can probably be applied to sediments. The Weisz dimensionless factor compares the time-scales of diffusion and chemical reaction:

$$\phi = \frac{\tau_{\text{diffusion}}}{\tau_{\text{reaction}}} = \frac{(R^2/D)}{[c/(dc/dt)]} \tag{6}$$

where R is a characteristic size of sediment particles. Following the discussion of Chapter 4 and Weisz, we can identify the following cases:

- When $\phi < 1$, the time-scale for diffusion is short compared with that for chemical reaction and reactivity will determine the rate of change of concentration.
- For $\phi > 1$, the time-scale for diffusion is long and diffusion will play an important role in determining the reaction rate.

Table 8-4 Diffusion constants of sediments (D_s) for some common species[a]

Na^+, K^+, NH_4^+, Cl^-, NO_3^-	5–7×10^{-6} cm^2/s
Ca^{2+}, Mg^{2+}, Mn^{2+}, Fe^{2+}, SO_4^{2-}	3–5×10^{-6} cm^2/s
Fe^{3+}, PO_4^{3-}, amorphous silica	1–3×10^{-6} cm^2/s

[a] The data given should serve only as reference values following the rule, the higher the ionic potential, the thicker the hydration layer of the water molecules around the ion, and the slower the ionic diffusion. Cations generally diffuse more rapidly than anions.

Finally, it must be stressed that diffusion of dissolved species in solutions is a key physicochemical process for the sea–sediment interaction and energy exchange at the sediment–water interface. For a more comprehensive presentation of diffusion, the reader is referred to Cussler (1984).

8.4.4 Redox Conditions

The reducing and oxidizing conditions in a sediment determine the chemical stability of the solid compounds and the direction of the spontaneous reactions. The redox state can be recognized as a voltage potential measured with a platinum electrode. This voltage potential is usually referred to as E or Eh defined by the Nernst equation, as described in Chapter 5:

$$\text{Eh} = \text{Eh}^0 - (RT/nF)\ln\frac{[\text{products}]}{[\text{reactants}]} \tag{7}$$

The Nernst equation is applicable only if the redox reaction is reversible. Not all reactions are completely reversible in natural systems; activities of reacting components may be too low or equilibrium may be reached very slowly. In a sediment, the biotic microenvironment may create a redox potential that is different from the surrounding macroenvironment. For this reason, measurements of Eh in natural systems must be cautiously evaluated and not used strictly for calculations of chemical equilibria. Calculations of redox equilibria are in some cases valuable, in the sense that they will give information about the direction of chemical reactions.

If a system is not at equilibrium, which is common for natural systems, each reaction has its own Eh value and the observed electrode potential is a mixed potential depending on the kinetics of several reactions. A redox pair with relatively high ion activity and whose electron exchange process is fast tends to dominate the registered Eh. Thus, measurements in a natural environment may not reveal information about all redox reactions but only from those reactions that are active enough to create a measurable potential difference on the electrode surface.

Eh gradients down a sediment core will cluster in a few rather distinct voltage regions. These regions are correlated with specific redox environments. The redox capacity of the natural environment is very low between these regions due to a lack of dominating redox pairs. Intermediate values will, therefore, be extremely difficult to register, as demonstrated by Bågander and Niemistö (1978). During their investigation, they were able to prove a good reproducibility between two separate measurements on 107 parallel sediment samples. Their values fell into three distinct clusters leaving areas without observed values between them (Fig. 8-6):

Eh = 0.1 to 0.4	Mn^{4+}/Mn^{2+}; O_2/H_2O_2; IO_3^-/I^-; NO_3^-/N
Eh = 0.2 to 0.0	Fe^{3+}/Fe^{2+}
Eh = −0.1 to −0.3	S/S^{2-}; S/S_n^{2-} $(n<5)$

The reader is referred to Berner (1963), Sillén (1965a,b), Breck (1972), Liss *et al.* (1973), Whitfield (1969, 1974), and Parsons (1975) for further information.

In marine sediments, usually only the uppermost layer of the sediment exhibits oxidizing conditions while the rest is reduced. The thickness of the oxidized layer and the reducing capacity of the sediment below depend on:

1. The concentration of oxygen in the ambient bottom water.
2. The rate of oxygen penetration into the sediment.
3. The accessibility of utilizable organic matter for the bacterial activity.

The depositional rate of organic matter is higher in the coastal areas than in the open sea. Even though the exchange of water is higher in shallow areas of the ocean, the deep water is not deficient of oxygen. We should thus expect to find a general trend of thicker oxidized sediments with increasing distance from the shoreline. This has been confirmed by, for example, Lynn and Bonatti (1965). They found that the thickness of the oxidized upper sediment in the Pacific Ocean between 15°S and 20°N increased with distance from the continent. Thicknesses were <1 cm near the shore and 8–15 cm at distances of 800 km offshore. This effect is also discussed in Section 9.3.

Oxygen concentration as such cannot explain the oxidizing potentials. A decrease in oxygen concentration from 100 to 1% saturation will only affect Eh by 30 mV, which lies well within the normal fluctuations of a natural system. The presence of an oxygen concentration higher than 10% saturation (about 1 mg O_2/L), however, has a retarding effect on the sulfate-reducing reactions of anaerobic bacteria, as they do not begin to proliferate until the oxygen concentration is almost undetectable.

The inflection point of the redox gradient, constituting the boundary between oxidizing and reducing environments in sediments, lies at around +250 mV. Mortimer (1942) has prescribed +200 mV as the inflection value, though due to natural variations, a range of +250 mV ±50 mV would probably be more correct. This boundary, the redoxcline (Hallberg, 1972), is recognized as comparable to other natural boundaries, such as the halocline and thermocline. The redoxcline is usually directly associated with the transformation of iron:

$$Fe(OH)_3 + 3H^+ + e^- \leftrightarrow Fe^{2+} + 3H_2O \quad (8)$$

The normal range in ferrous iron concentrations leads to an Eh variation between 200 and 300 mV.

The redoxcline is usually situated close to the sediment–water interface resulting in a redox turnover from oxidizing to reducing conditions during early diagenesis. The redox turnover will, in

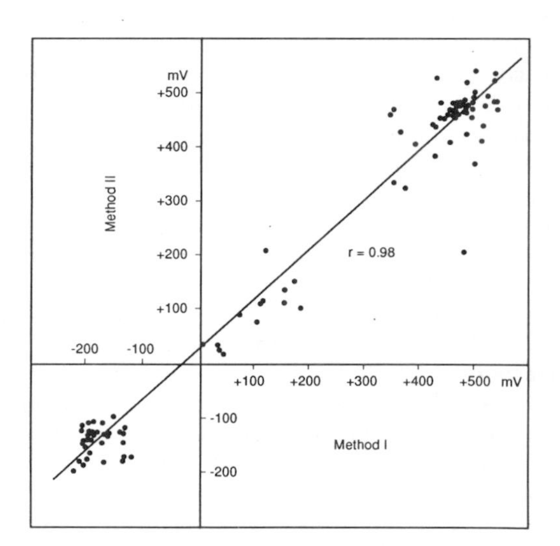

Fig. 8-6 The figure shows the reproducibility between two separate synoptic investigations of electrode potential based on 107 parallel sediment samples from 17 localities in the Baltic. Note the three distinct clusters of plotted data. For further information, see the text. Adapted from Bågander and Niemistö (1978) with the permission of Academic Press, Inc.

turn, produce disequilibrium within the sediment, causing dissolution of certain minerals and compounds and increased ion exchange across the redoxcline. The change of redox potential is a useful tool for describing the sedimentary environment.

8.5 Transformation of Organic and Inorganic Material

8.5.1 Organic Matter

In the previous sections, we examined the physical and physicochemical parameters of a sediment. They constitute the basis of the sedimentary ecosystem and are, except for the redox potential, only to a minor extent affected by the biomass. Before going into the transformation of organic and inorganic material, a brief discussion should be given on the input of organic matter to the sediments. Organic matter constitutes the energy source for almost all transformations taking place in a sediment, though usually only a few percent of the organic matter produced in the sea is available to the microorganisms in the sediments. This is illustrated in Fig. 9-16. Below the euphotic zone, the flux of carbon to the sediment can be described with the following general formula (see Chapter 9; Suess, 1980).

$$C_{\text{flux}} = \frac{C_{\text{prod}}}{0.0238z + 0.212} \qquad (9?)$$

C_{prod} is the primary production rate of carbon at the surface, and C_{flux} is the organic carbon flux at a water depth of z meters (>50 m), both in g/m^2 year.

If we disregard the chemosynthesis occurring near "black smokers" along the rift zones of the oceans, we can think of all organic matter as a direct or indirect result of photosynthetic primary production. Generally speaking, about one-third of the global primary production occurs in the sea. Although the continental margins and estuaries constitute less than 10% of the marine area, they are responsible for 25% of the marine primary production (Whittaker and Likens, 1973). Together with the input of material from the continents, these areas have a dominating influence on the biogeochemical cycling of elements. They will, therefore, be the target for our discussions in the following sections. The distribution of primary production on Earth is shown on the inside cover, in Table 9-8, and in Figs 9-14 and 11-12.

A general formula for the composition of planktonic organic matter (Fleming, 1940; Redfield *et al.*, 1963; and see Chapters 3 and 9) may be given as:

$$(CH_2O)_{106}(NH_3)_{16}(PO_4)$$

The longer the residence time for the organic debris in the water mass, the more degraded it will be before it reaches the bottom. Toth and Lerman (1977) show that the refractoriness of sediment organic matter and its rate of decomposition are functions of the sedimentation rate. The C:N:P ratios of the organic debris will thus change due to the different activation energies that must be overcome in order to break the C—N, O—P, C—C, and C—H bonds (Toth and Lerman, 1977). Duursma (1961) found an increase in the C:P ratio of particulate oceanic matter with depth. Nitrogen and phosphorus are released more readily than carbon during the decomposition of organic matter. This has been elaborated on in the paper by Toth and Lerman (1977), who have also modeled the early diagenetic processes for sulfate, ammonia, and phosphorus in sediments. This subject is also discussed in Section 9.4.4.

Ion exchange at the sediment–water interface is governed by redox conditions which, in turn, are greatly influenced by microbial activity. The driving force for their activity is the organic debris from the primary production. It should be stressed that *microbial activity* is correlated with the *input of organic matter* to the sediment rather than with the number of bacteria. Volkmann and Oppenheimer (1962), for instance, did not find any correlation between observed decrease of organic carbon and bacterial plate counts from surface sediments of the central Texas Gulf coast. If the loss in organic matter is due largely to bacterial activity, then such an observed loss during storage should provide a better indication of bacterial activity.

Bacteria can be divided into two main groups, the heterotrophic and the autotrophic bacteria. The activity of the heterotrophic organisms, which is dependent on the amount and refractoriness of *organic matter as carbon source*, is of great importance for the exchange of ions between the sea and the sediment. These bacteria perform many transformations that cannot be brought about by larger organisms or by inorganic reactions within a reasonable time-span. Denitrification and sulfate reduction are examples of such reactions (Fig. 8-7). Most of the bacteria in a sediment are anaerobes.

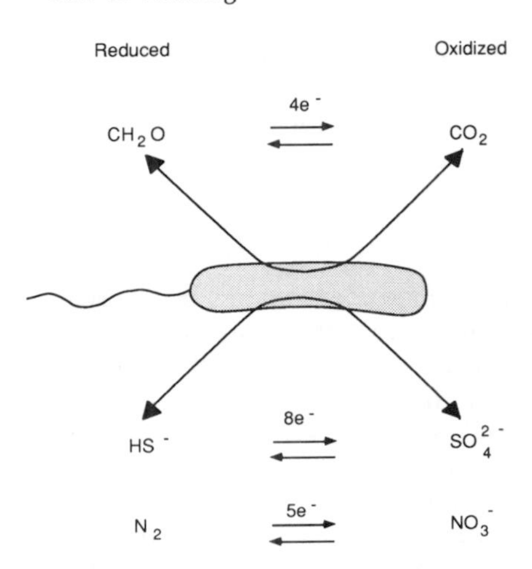

Fig. 8-7 Bacterial transformation of carbon coupled to redox reactions of sulfur and nitrogen. If the carbon transformation goes in one direction, the other reactions have to go in the opposite direction.

They are affected adversely by oxygen and therefore try to create anoxic and reducing conditions. They live in sediments that have been exhausted of oxygen by aerobic organisms. Significant numbers of viable bacteria ($>10^5$/g sediment wet weight) can be found even at great water depths. Generally, their abundance in homogeneous sediments decreases with depth.

The microbiological transformation of organic matter in an anoxic sediment can be represented by the following generalized formula:

$$(C,O,N,P) + HSO_4^- \rightarrow HCO_3^- + NH_4^+ + HPO_4^{2-} + HS^- \quad (10)$$

This reaction gives rise to HCO_3^-, which will affect the chemical sea–sediment interaction with regard to pH and alkalinity. Moreover, it demonstrates an impact on exchange and turnover processes for sulfur, nitrogen, and phosphorus. The fluxes of these components from a Baltic sediment are illustrated in Fig. 8-8.

The degradation of organic matter is very important for marine life as it facilitates the recirculation of nutrients, which are removed from surface waters by primary production. The release of nitrogen and phosphorus from the sediments and their transport to the surface water with upwelling bottom water along the western margins of the continents is an important process for the fishing industries in these areas. The remineralization of phosphate from the sediments of the Peru continental margin may generate up to 10% of the nutrient pool in the waters of the Peru undercurrent (Suess, 1981). The high productivity of these waters, on the other hand, creates a heavy rain of organic matter to the sediments off these coastal areas where it becomes the energy source for the anaerobic processes. The cycle is closed. Cultural inputs of nutrients to this cycle may be hazardous in waters with restricted circulation. In such waters, hydrogen sulfide is accumulated and life is restricted to specific types of bacteria (e.g. the deep waters of the Black Sea, Baltic Sea, and Norwegian fjords). Anthropogenic nutrients in such waters increase the eutrophication and enhance the accumulation of hydrogen sulfide.

In oceanic sediments, macro- (>1 mm) and microorganisms adapt to the existing environment. They play an important role in the mixing of surface sediment layers. Burrowing by organisms in marine sediments is so common that it is the preservation of depositional structures that requires explanation, not their destruction (Arrhenius, 1952, p. 86).

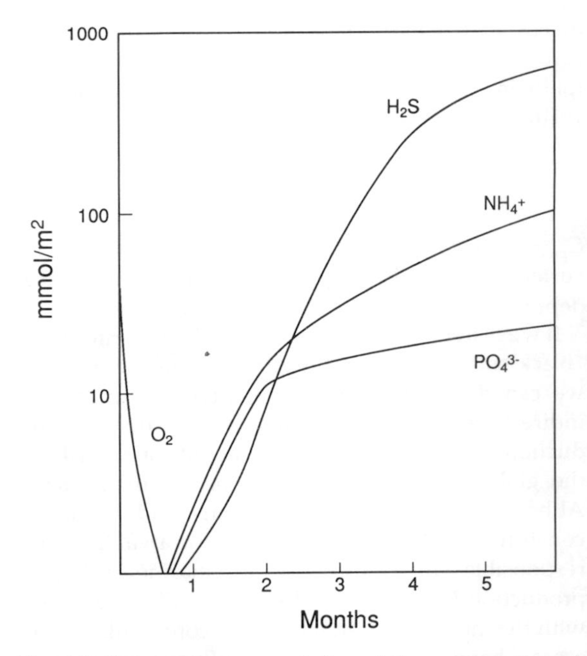

Fig. 8-8 Schematic representation of the consumption of oxygen and production of H_2S, NH_4^+, and PO_4^{3-} in a CISES. This is aimed to represent the first part of the "life-cycle" in a stagnant basin of the Baltic Sea.

Bioturbation is the mixing of sediments by the biological infauna. It is usually attributed to macro- and meiofauna (0.1–1.0 mm of sieved sediments) and very little is said about the role of microfauna. The mixing of sediments by the former can easily be observed in coastal areas where their abundance is high and the resuspension and cycling of the annual sediment influx can be as high as 99% (Young, 1971). In anaerobic sediments, however, where macro- and meiofauna are absent due to lack of oxygen, there is still a large community of microorganisms: 10^7 individuals/cm^3 of sediments is not an uncommon value. Several types are motile and have been observed to swim 3 mm/day (Oppenheimer, 1960). If all bacteria in a cubic centimeter of sediment moved 1 mm/day, they would cover a total distance of 10 km/day. These organisms cannot move particles but may have a significant effect on the mixing of the interstitial water, and thus also on the exchange between water and sediments (Östlund *et al.*, 1989). The relation between organisms and sediments is well described in a paper by Rhoads (1974). Quantitative estimates of bioturbation have been made by, for example, Guinasso and Schink (1975) and Berner (1980). Guinasso and Schink introduce a dimensionless parameter D/Lv, where D is the eddy diffusivity, L is the depth to which mixing takes place, and v is sedimentation rate. They propose in their model that if D/Lv is greater than 10, the surface layer becomes homogeneous; if D/Lv is less than 0.1, little mixing can take place before the sediments are buried.

8.5.2 Sulfur

Sulfur is treated in Chapter 13 and discussed only briefly here. The dominant reaction in the sedimentary sulfur cycle is microbial sulfate reduction. This gives rise to the formation of hydrogen sulfide which, by precipitating iron as "black unstable sulfide", will give the reduced sediment its characteristic blackish color:

$$2CH_2O + HSO_4^- \rightarrow H_2S + 2HCO_3^- + H^+ \quad (11)$$

The sulfate reducers, mainly *Desulfovibrio* spp. and *Desulfotomaculum* spp., are heterotrophs and the amount of utilizable organic matter determines their activity. The activity of sulfate-reducing organisms is generally rather low and will vary with water depth as verified by investigations in the Black Sea (Sorokin, 1964). Sorokin found that in the shallow areas along the coast with a relatively good supply of organic matter, the production of H_2S in the bottom water was about 115 mg/Lday, whereas the deeper basins produced only 1.5 mg/Lday. Marine sediments in general show the same tendency, though the data are somewhat scattered. This is demonstrated in Fig. 8-9 (Goldhaber and Kaplan, 1975). Bacterial sulfate reduction contributes reduced sulfur species to the atmosphere. As this process is most active in shallow water areas, intertidal areas in particular should be considered because of their direct contact between sediment and atmosphere. Photosynthetic sulfide oxidizing bacteria in the topmost layer of sediments act as an active filter for the transport of reduced sulfur species to the atmosphere. We should, therefore, expect the peak hours for transport of sulfur when the sediments are subject to little sunlight.

8.5.3 Nitrogen

Most of the nitrogen (>90%) contributed to oceanic sediments comes from organic compounds. Nitrate

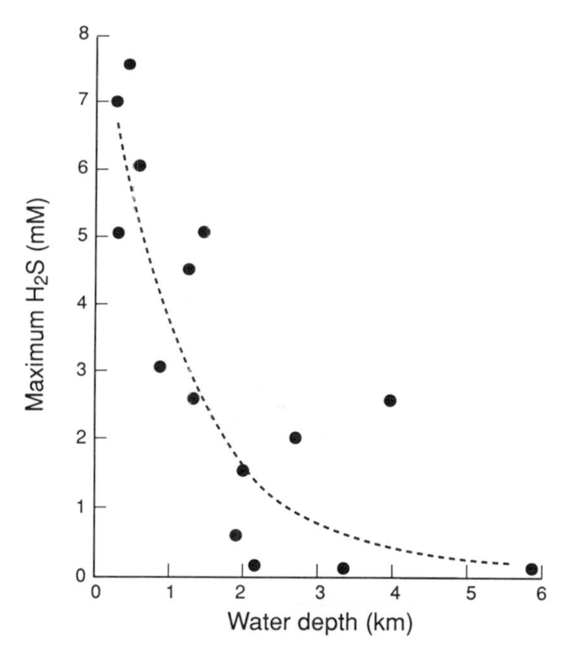

Fig. 8-9 Plot of maximum dissolved sulfide (for cores with essentially complete sulfate reduction) *vs* water depth of the sediment–water interface. The line drawn through the data is arbitrary. Adapted from Goldhaber and Kaplan (1975) with the permission of Williams and Wilkins.

and nitrite contribute only a minor portion of the nitrogen. Emery *et al.* (1964) state that a high proportion of total Kjeldahl nitrogen in surface sediment consists of nitrogen in amino acids. (The Kjeldahl method converts most of the amino nitrogen of organic matter to ammonium ion, which is then determined colorimetrically. It fails to account for nitrogen in the form of azide, azine, azo, nitrate, nitrite, nitro, nitroso, oxime, and semi-carbazone.) There is a very good correlation between amino acid-N and total N in the topmost 10 cm of Baltic sediments (Fig. 8-10; Engvall, 1978). Organic matter and total N are also well correlated. The nitrogen contents of these sediments were 3.3% of total organic matter dry wt. This is about half of what should be expected from the general formula for organic matter with C:N:P being 106:16:1. Deamination is a comparatively rapid process, however, and the C:N ratio should increase in the residue during the degradation process. Engvall (1977) found that the annual nitrogen release from anaerobic sediments was 1% of total nitrogen. The low nitrogen content can thus not be explained by release during anaerobic processes but must mainly be caused by loss of nitrogen before deposition. The C : N ratio of the organic debris also varies with season. During spring, it consists of phytoplankton (diatoms) with a C:N ratio <10, whereas in the autumn the dominating organic matter is higher algae (e.g. *Cladophora*) with a C:N ratio of 20. The ammonium flux from the Baltic sediments is accordingly higher during the spring by a factor of 4–5. Bader (1955) reports a range of 5–25 for the C:N ratio of superficial marine sediments, depending on the depositional environment, while Hartmann *et al.* (1973) have reported a value of 13.25 for a number of sediment samples along the continental margin of West Africa. The amino acid/Kjeldahl N ratio decreases rapidly with depth, and most of the nitrogen at any depth exists in an organic form other than proteins or other materials which are hydrolyzable to amino acids.

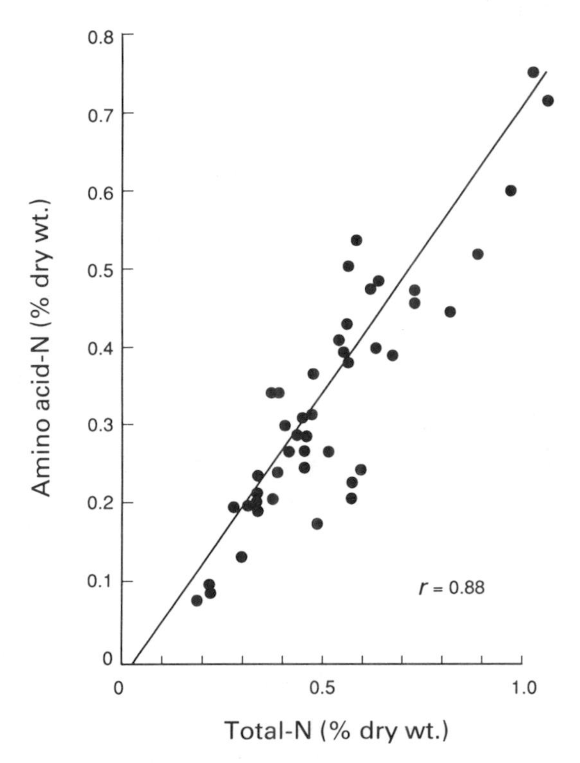

Fig. 8-10 Relation between amino acid-N and total-N in the top 10 cm of 10 sampling sites in the Baltic Proper. Reproduced from Engvall (1978) with permission.

Nitrogen transformation in the sediment is governed primarily by bacteria (Wood, 1965), and the decomposition of proteinaceous material takes place in aerobic as well as in the anaerobic environment due to various deamination reactions (or ammoniafication) performed by a variety of microorganisms. The final product of the deamination process is ammonium. Ammonium, as opposed to nitrate, is easily adsorbed by the clay minerals, and the sediment may act as a sink for nitrogen. Deamination of an amino acid is shown below:

$$RHC(NH_2)COOH + H_2O + FP \rightarrow RCOCOO^- + NH_4^+ + FPH_2 \quad (12)$$

The flavoprotein enzyme is indicated by FP; R represents an organic group. In anaerobic sediments, we would expect a good correlation between the release of ammonium and sulfide (Fig. 8-8). Richards (1965) reported sulfur/nitrogen correlations in the anoxic water of Lake Nitinat that agree well with the theoretical value 3.3 that can be calculated from reaction (10) and the general formula for organic matter. Corresponding figures for the Black Sea and Cariaco Trench are 4.45 and 1.92 respectively (Redfield *et al.*, 1963). Closed *in situ* experimental systems (CISES; see box) show a very good correlation between the release of sulfide and ammonium. However, some data show seasonal variations dependent on the composition of the organic matter. Spring and

CISES

Closed *in situ* experimental systems (CISES) in the form of plexiglass boxes are used to study sediment–water interactions (Hallberg *et al.*, 1972) and to obtain experimental data for computer simulations. The experiments can be performed with much less disturbance of the natural character of the sediment, and the data are more readily reproduced and are more representative of natural conditions than laboratory studies. Natural systems are, in contrast, open to the surrounding environment, and it can be argued that the box systems are not representative. It can also be argued that for nearly all kinds of systems, there are regions in which equilibrium is approached even though gradients exist throughout the system as a whole. The concept of local equilibrium is a fundamental assumption in theories of irreversible processes. Such local equilibrium conditions may be expected to develop for species involved in rapid processes at the sediment–water interface (Stumm and Morgan, 1970, p. 67).

The plexiglass box for *in situ* experiments is open at the bottom and forms a closed system when pushed into the sediment. Sampling of the enclosed water is done with syringes through self-sealing rubber membranes. The boxes are equipped with electrodes for the measurement of pH, Eh, and temperature. The water loss from sampling is compensated for by a simultaneous addition of seawater and is considered in the final calculations (Hallberg *et al.*, 1972).

autumn values were 6.6 and 54.6 respectively and differed by an order of magnitude.

The rate of ammonium formation in sediments is discussed in theoretical terms by Berner (1974). He has applied his model to two environmentally different sediment cores. Berner assumes the rate of ammonium formation to be a pseudo-zero-order reaction, since the ammonium concentration in cores is a linear function of depth. A pseudo-zero-order reaction for ammonium was also found to be true in the CISES studies.

There are some reports, especially from limnic environments, about gaseous nitrogen in sediments. This is probably due to denitrification processes:

$$5CH_2O + 4NO_3^- + 4H^+ \rightarrow 5CO_2 + 7H_2O + 2N_2 \quad (13)$$

Nitrogen may exist as dissolved N_2 in equilibrium with ammonium ion at Eh and pH levels encountered in natural environments. This subject should be investigated further, as the occurrence of dissolved N_2 could explain losses in N-budgets, and necessitate special precautions in the treatment of sediment samples (in order to avoid significant losses of N_2).

The main factors governing the turnover of nitrogen at the sediment–water interface are the total amount of proteinaceous matter, diffusion, ion exchange with clay minerals, and bioturbation.

8.5.4 Phosphorus

The most stable and predominant form of dissolved phosphate in marine sediments is orthophosphate (henceforth referred to as phosphate) with an oxidation state of +5. Phosphate, unlike nitrogen (oxidation state varying between −3 and +5) and sulfur (oxidation state −2 to +6), is not directly involved in redox reactions. Phosphate in solution is chiefly present as ion pairs with the major cations of seawater (Kester and Pytkowicz, 1967). In sediments it is also found in organic matter, adsorbed on hydrous ferric oxides (Mortimer, 1941, 1971; Mackereth, 1966; Stumm and Leckie, 1970), adsorbed on clay minerals (Chen, 1972), and as various forms of apatite.

Apatite – $Ca_5(PO_4)_3X$, where X stands for F, OH, and Cl – is usually not found in sediments except where calcite is also present at high concentrations. At marine pH values (7–8), the precipitation of apatite is greatly accelerated by the presence of calcite, the surface of which acts as a nucleating agent for crystallization (topotactic effect). This has been demonstrated by Leckie (1969) and Stumm and Leckie (1970), who also state that in the absence of calcite, high degrees of supersaturation of apatite can be maintained indefinitely.

Vivianite, $Fe_3(PO_4)_2 \cdot 8H_2O$, is formed during reduced conditions, as is easily demonstrated in batch cultures of sulfate-reducing bacteria where an environment similar to that of reduced sediments is simulated. According to Nriagu (1972), "Vivianite is probably the most stable Fe(II) orthophosphate solid phase encountered in sedimentary environments." Vivianite is reported from reduced freshwater (Nembrini *et al.*, 1983) and estuarine sediments (Bray *et al.*, 1973), but has not been firmly identified and reported from normal saline environments.

In a continuous culture of sulfate-reducing bacteria, Hallberg and Wadsten (1980) found a new compound, which is stable under natural conditions and therefore may be an important mineral for phosphate fixation in sediments.

During oxidizing conditions, the sediment acts as a sink for phosphate (e.g. Mortimer, 1942). This has also been verified in the CISES studies where different amounts of phosphate were added to closed oxidized systems. Not all of the phosphate added was taken up by the sediment. The 20% left in solution may be explained by the lack of free cations (e.g. calcium and iron), since phosphate is normally fixed in the sediment by these ions.

During reducing conditions, phosphate is released. At the beginning of a redox turnover, the reaction is fast and probably of pseudo–zero-order but changes then into a first-order reaction. The factor influencing the release, as for ammonium and sulfide, is the activity of the sulfate-reducing bacteria, which is directly related to the amount and reactivity of the organic matter. Thus, during the spring, the release rate and total amount of phosphate is higher than during the autumn (Fig. 8-11).

Phosphate release is supposed to be retarded, if the bottom water, enriched in phosphate, is not exchanged by hydrodynamic movements. Lee (1970), for instance, states that "It appears that the hydrodynamics of the system is often the rate-controlling step in exchange reactions." Opposite results have been demonstrated in identical CISES, which except for the initial concentration of phosphate (a difference between zero and 1200 μg/L PO_4-P) revealed the same types of reactions and amounts of totally released phosphate. The sediment was affected by phosphate exchange processes down to a depth of about 2 cm (see also Hayes, 1964).

Phosphorus in sediments can be divided into several different fractions according to chemical separation methods. Williams *et al.* (1976) outlined a scheme of three fractions: apatite-P, organic P, and

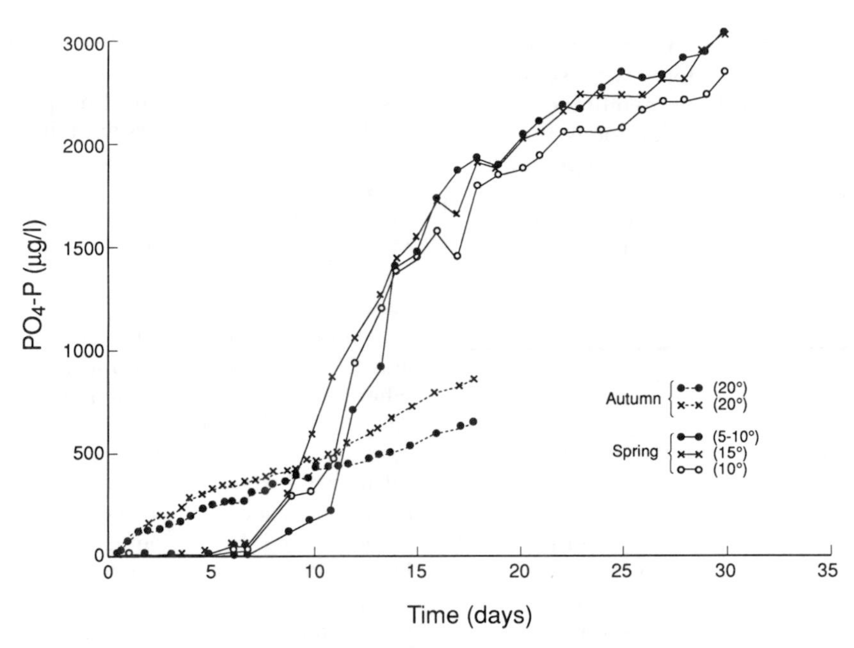

Fig. 8-11 The release of phosphate from sediments in CISES. The early part of the release is at a much higher rate during spring than during autumn. Reproduced from Holm (1978) with permission.

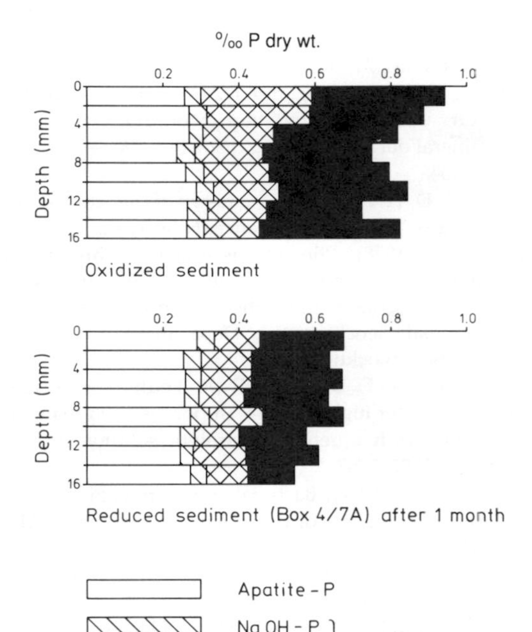

Fig. 8-12 Changes of phosphorus in a Baltic sediment during a redox turnover from oxidizing toward reducing conditions. The phosphorus has been separated into three different groups where NAI-P denotes non-apatite inorganic phosphorus (see text). The phosphorus released from the sediment comes solely from the organic-P and NAI-P fractions. Reproduced from Holm (1978) with permission.

non-apatite inorganic P (NAI-P). The latter consists, to a large extent, of phosphate associated with ferric oxide gel. In CISES studies, no significant change occurred in the apatite-P fraction. The phosphate released during the redox turnover was totally related to changes in the other two fractions (see Fig. 8-12). Because of the association with ferric oxide gel, phosphate can easily accumulate in the oxidized upper layer of the sediment and not be released to the bottom water until after a redox turnover.

Questions

8-1 Why do we speak in terms of soil horizons and sediment layers?

8-2 Discuss the significance of clay minerals in a description of the solid phase of a sediment.

8-3 Give examples of early diagenetic processes.

8-4 What kind of information can be obtained from Eh measurements in a sediment?

8-5 Discuss the processes important in defining the redoxcline.

8-6 Why do continental margins play a dominant role in the biogeochemical cycling of elements?

8-7 Give examples of bacterial transformations in a sediment that are of special importance for biogeochemical cycles.

8-8 Why should one expect a good correlation between the release of ammonium and sulfide from an anaerobic sediment?

8-9 Under what conditions might a sediment act as a sink for phosphate? What cations are normally associated with phosphate in a sediment?

References

Arrhenius G. (1952). *Rep. Swedish Deep-Sea Expedition (1947–1948)* **5**, 227.

Bågander, L. E. and L. Niemistö (1978). An evaluation of the use of redox measurements for characterizing recent sediments. *Estaurine Coastal Mar. Sci.* **6**, 127–134.

Bader, R. G. (1955). Carbon and nitrogen relations in surface marine sediments. *Geochim. Cosmochim. Acta* **7**, 205–211.

Berner, R. A. (1963). Electrode studies of hydrogen sulfide in marine sediments. *Geochim. Cosmochim. Acta* **27**, 563–575.

Berner, R. A. (1971). "Principles of Chemical Sedimentology." McGraw-Hill, New York.

Berner, R. A. (1974). Kinetic models for the early diagenesis of nitrogen, sulfur, phosphorus, and silicon in anoxic marine sediments. *In* "The Sea" (E. D. Goldberg, ed.), pp. 427–450. John Wiley, New York.

Berner, R. A. (1980). "Early Diagenesis: A Theoretical Approach." Princeton University Press, Princeton, N.J.

Bray, J. T., O. P. Bricker, and B. N. Troup (1973). Phosphate in interstitial waters of anoxic sediments, oxidation effects during sampling procedure. *Science* **180**, 1362–1364.

Breck, W. G. (1972). Redox potentials by equilibration. *J. Marine Res.* **30**, 121.

Casagrande, A. (1940). The structure of clay and its importance in foundation engineering. *J. Boston Soc. Civ. Engrs* **19** (4), 168–209.

Chen, Y. (1972). Phosphate interaction with aluminum oxide, kaolinite, and sediments. Doctoral dissertation in Engineering (unpublished), Harvard University, Cambridge, Mass.

Chilingarian, G. V. and H. H. Ricke III (1968). Data on consolidation of fine-grained sediments. *J. Sediment. Petrol.* **38**, 811–816.

Chilingarian, G. V. and H. H. Ricke III (1969). Some chemical alterations of subsurface waters during diagenesis. *Chem. Geol.* **4**, 235–252.

Conway, E. J. (1943). The chemical evolution of the ocean. *Proc. RIA* **48B**, 161–212.

Crank, J. (1956). "The Mathematics of Diffusion." Clarendon Press, Oxford.

Cussler, E. L. (1984). "Diffusion–Mass Transfer in Fluid Systems." Cambridge University Press, Cambridge.

Duursma, E. (1961). Dissolved organic carbon, nitrogen and phosphorus in the sea. *Neth. J. Sea Res.* **1**, 1–147.

Edelman, C. H. and J. C. L. Favejee (1940). On the crystal structure of montmorillonite and halloysite. *Zeitschrift für Kristallographie* **102**, 417–431.

Emergy, K. O., C. Stitt, and P. Saltman (1964). Amino acids in basin sediments. *J. Sediment. Petrol.* **34**, 433–437.

Englehardt, W., von and K. H. Gaida (1963). Concentration changes of pore solutions during compaction of clay sediments. *J. Sediment. Petrol.* **33**, 919–930.

Engvall, A.-G. (1977). Nitrogen exchange at the sediment–water interface. *Ambio Sp. Rep.* **5**, 141–146.

Engvall, A.-G. (1978). The fate of nitrogen in early diagenesis of Baltic sediments. Ph.D. thesis, Contr. in Microbial Geochemistry, Department of Geology, University of Stockholm.

Fleming, R. H. (1940). The composition of plankton and units for reporting population and production. *Proc. Sixth Pacific Sci. Cong. Calif.* **3**, 535–540.

Garrels, R. M. and F. T. Mackenzie (1971). "Evolution of Sedimentary Rocks." W. W. Norton, New York.

Goldhaber, M. B. and I. R. Kaplan (1975). Controls and consequences of sulfate reduction rates in recent marine sediments. *Soil Sci.* **119**, 42–55.

Gregor, C. B. (1968). The rate of denudation in Post-Algonkian time. *Koninkl. Ned. Akad. Wetenschap. Proc.* **71**, 22.

Grim, R. E. (1953). "Clay Mineralogy." McGraw-Hill, New York.

Guinasso, Jr, N. L. and D. R. Schink (1975). Quantitative estimates of biological mixing rates in abyssal sediments. *J. Geophys. Res.* **80**, 3032–3043.

Hallberg, R. O. (1972). Sedimentary sulfide mineral formation: An energy circuit system approach. *Mineral. Deposita* **7**, 189–201.

Hallberg, R. O. (1974). Paleoredox conditions in the eastern Gotland basin during the recent centuries. *Havsforskningsinst. Skr.* **238**, 3–16.

Hallberg, R. O. and T. Wadsten (1980). Crystal data of a new phosphate compound from microbial experiments on iron sulfide mineralization. *Am. Mineral.* **65**, 200–204.

Hallberg, R. O., L. E. Bågander, A. G. Engvall, and F. A. Schippel (1972). Method for studying geochemistry of sediment–water interface. *Ambio* **1** (2), 71–72.

Hartmann, M., P. Müller, E. Suess, and C. H. van der Weijden (1973). Oxidation of organic matter in recent marine sediments. *"Meteor" Forsch.-Ergebnisse* **C12**, 74–86.

Hayes, F. R. (1964). The mud–water interface. *Oceanogr. Mar. Biol. Ann. Rev.* **2**, 121–145.

Henniker, J. C. (1949). Depth of the surface zone of a liquid. *Rev. Mod. Phys.* **21**, 322.

Hofmann, C., K. Endell, and D. Wilm (1933). Kristallstruktur und Quellung von Montmorillonit. (Das Tonmineral der Bentonite). *Zeitschrift für Kristallographie* **86**, 340–348.

Holland, H. D. (1978). "The Chemistry of the Atmosphere and Oceans." Wiley-Interscience, New York.

Holm, N. G. (1978). Phosphorus exchange through the sediment–water interface: Mechanism studies of dynamic processes in the Baltic Sea. Ph.D. thesis, Contr. in Microbial Geochemistry, Department of Geology, University of Stockholm.

Horne, R. A., A. F. Day, and R. P. Young (1969). Ionic diffusion under high pressure in porous solid materials permeated with aqueous, electrolytic solution. *J. Phys. Chem.* **73**, 2782–2783.

Jörgensen, B. B. (1977). Bacterial sulfate reduction within reduced microniches of oxide marine sediments. *Mar. Biol.* **41**, 7–17.

Kamb, B. (1971). Hydrogen-bond stereochemistry and "anomalous water." *Science* **172**, 231.

Kester, D. R. and R. M. Pytkowicz (1967). Determination of the apparent dissociation constants of phosphoric acid in seawater. *Limnol. Oceanogr.* **12**, 243–252.

Klinkenberg, L. J. (1951). Analogy between diffusion and electrical conductivity in porous rocks. *Geol. Soc. Amer. Bull.* **62**, 559–563.

Lambe, T. W. (1958). The structure of compacted clay. *J. Soil Mech. Found. Div., Proc. ASCE*, SM2, 84, Part 1, paper 1654.

Leckie, J. O. (1969). Interaction of calcium and phosphate at calcite surfaces. Doctoral dissertation in engineering (unpublished), Harvard University, Cambridge, Mass.

Lee, F. G. (1970). Factors affecting the transfer of material between water and sediments. Literature Review No. 1, Eutrophication Information Program, Water Resources Center, University of Wisconsin, Madison, Wisconsin.

Lisitzin, A. P. (1972). Sedimentation in the world ocean. *Soc. Econ. Paleontol. Mineral. Spec. Publ.* **17**, 1–218.

Liss, P. S., J. R. Herring, and E. D. Goldberg (1973). The iodide/iodate system in sea water as a possible measure of redoxpotential. *Nature* **242**, 108.

Low, P. F. and J. L. White (1970). Hydrogen bonding and polywater in clay–water systems. *Clays and Clay Minerals* **18**, 63–66.

Lynn, D. C. and E. Bonatti (1965). Mobility of manganese in diagenesis of deep-sea sediments. *Mar. Geol.* **3**, 457–474.

Mackenzie, F. T., R. N. Ginsburg, L. S. Land, and O. P. Bricker (1969). Carbonate cements. *Bermuda Biological Station Res. Spec. Publ.* No. 3, 325.

Mackereth, F. J. H. (1966). Some chemical observations on post-glacial lake sediments. *Phil. Trans. Roy. Soc.* **B250**, 165–220.

Mangelsdorf, Jr., P. C., T. R. S. Wilson, and E. Daniell

(1969). Potassium enrichment in interstitial waters of recent marine sediments. *Science* **165**, 171–174.

Manheim, F. T. (1966). A hydraulic squeezer for obtaining interstitial water from consolidated and unconsolidated sediments. *U.S. Geol. Survey Prof. Pap.* 550-C, 256–261.

Manheim, F. T. (1970). The diffusion of ions in unconsolidated sediments. *Earth Planet. Sci. Lett.* **9**, 307–309.

Meade, R. H. (1966). Factors influencing the early stages of the compaction of clays and sands: review. *J. Sediment. Petrol.* **36**, 1085–1101.

Mortimer, C. H. (1941). The exchange of dissolved substances between mud and water, I and II. *J. Ecol.* **29**, 280–329.

Mortimer, C. H. (1942). The exchange of dissolved substances between mud and water in lakes III and IV *J. Ecol.* **30**, 147–201.

Mortimer, C. H. (1971). Chemical exchanges between sediments and water in the Great Lakes: Speculations on probable regulator mechanisms. *Limnol. Oceanogr.* **16**, 387–404.

Murthy, A. S. P. and R. E. Ferrell Jr. (1972). Comparative chemical composition of sediment interstitial waters. *Clays and Clay Minerals* **20**, 317–321.

Nembrini, G. P., J. A. Capobianco, M. Viel, and A. F. Williams (1983). A Mössbauer and chemical study of the formation of vivianite in sediments of Logo Maggiore (Italy). *Geochim. Cosmochim. Acta.* **47**, 1459–1464.

Nriagu, J. O. (1972). Stability of vivianite and ion-pair formation in the system $Fe_3(PO_4)_2 \cdot H_3PO_4 \cdot H_2O$. *Geochim. Cosmochim. Acta* **36**, 459–470.

Oppenheimer, C. H. (1960). Bacterial activity in sediments of shallow marine bays. *Geochim. Cosmochim. Acta* **19**, 244–260.

Östlund, P., L. Hallstadius, and R. O. Hallberg (1989). Porewater mixing by microorganisms monitored by radiotracer method. *Geomicrobiol. J.* **7**, 253–264.

Parsons, R. (1975). The role of oxygen in redox processes in aqueous solution. *Proc. Dahlem Workshop on the Nature of Seawater*, Berlin, March 10–15.

Presley, B. J. (1969). Chemistry of interstitial water from marine sediments. Ph.D. thesis, University of California, Los Angeles, Calif.

Pusch, R. (1970). Clay microstructure. *National Swedish Building Research Document* D8.

Redfield, A. C., B. H. Ketchum, and F. A. Richards (1963). The influence of organisms on the composition of seawater. *In* "The Sea" (M. N. Hill, ed.), pp. 26–27. Wiley-Interscience, New York.

Richards, F. A., J. D. Cline, W. W. Broenkow and L. P. Atkinsson (1965). Some consequences of the decomposition of organic matter in Lake Nitinat, an anoxic fjord. *Limnol. Oceaonog.* **10**, R185–R201.

Rhoads, D. C. (1974). Organic–sediment relations on the muddy sea floor. *Oceanogr. Mar. Biol. Ann. Rev.* **12**, 263–300.

Rosenqvist, I. T. (1958). Remarks to the mechanical properties of soil water systems. *Geol. Fren. Stockholm, Frh.* **80**, 435–457.

Rosenqvist, I. T. (1962). The influence of physico-chemical factors upon the mechanical properties of clays. *Proc. 9th Nat. Conf. Clays and Clay Minerals*, pp. 12–27.

Sawhney, B. L. (1972). Selective sorption and fixation of cations by clay minerals: A review. *Clays and Clay Minerals* **20**, 93–100.

Sillén, L. G. (1961). The physical chemistry of sea water. *In* "Oceanography" (M. Sears, ed.), pp. 549–581. American Association for the Advancement of Science, Washington D.C.

Sillén, L. G. (1965a). Oxidation state of earth's ocean and atmosphere. I. *Arkiv. Kemi* **24**, 431–456.

Sillén, L. G. (1965b). Oxidation state of earth's ocean and atmosphere, II. *Arkiv Kemi* **25**, 159–176.

Sorokin, J. L. (1964). On the primary production and bacterial activities in the Black Sea. *J. Cons. Int. Explor. Mer.* **29**, 41–60.

Stumm, W. and J. D. Lockie (1970). Phosphate exchange with sediments: Its role in the productivity of surface waters. *Adv. Water Pollution Res.* **2**, 1–26.

Stumm, W. and J. J. Morgan (1970). "Aquatic Chemistry." John Wiley, New York.

Suess, E. (1980). Particulate organic carbon flux in the oceans: Surface productivity and oxygen utilization. *Nature* **288**, 260–263.

Suess, E. (1981). Phosphate regeneration from sediments of the Peru continental margin by dissolution of fish debris. *Geochim. Cosmochim. Acta* **45**, 577–588.

Sverdrup, H. U., M. W. Johnson and R. H. Fleming (1942). "The Oceans." Prentice-Hall, Englewood Cliffs, N.J.

Tan, T. K. (1957). Discussion on soil properties and their measurement. *Proc. 4th Int. Soil Mech. Found. Engng.* **3**, 87–89.

Toth, D. J. and A. Lerman (1977). Organic matter reactivity and sedimentation rates in the ocean. *Am. J. Sci.* **277**, 465–485.

Volkmann, C. M. and C. H. Oppenheimer (1962). The microbial decomposition of organic carbon in surface sediments of marine bays of the central Texas Gulf coast. *Publ. Inst. Mar. Sci.* **8**, 80–96.

Weaver, C. E. (1958). The effects and geologic significance of potassium "fixation" by expandable clay minerals derived from muscovite, biotite, chlorite, and volcanic material. *Am. Mineral.* **43**, 839–861.

Weisz, P. B. (1973). Diffusion and chemical transformation. *Science* **179**, 433–440.

Whitfield, M. (1969). Eh as an operational parameter in estaurine studies. *Limnol. Oceanogr.* **14**, 547–558.

Whitfield, M. (1974). Thermodynamic limitations on the use of the platinum electrode in Eh measurements. *Limnol. Oceanogr.* **19**, 857–865.

Whittaker, R. H. and G. E. Likens (1973). Carbon in the biota. *In* "Carbon and the Biosphere" (G. M. Woodwell and E. V. Pecan, eds.), pp. 281–302. US Atomic Energy Commission, CONF-720510, Virginia, USA.

Williams, J. D. H., J. M. Jaquet, and R. L. Thomas (1976). Forms of phosphorus in the surficial sediments of Lake Erie. *J. Fish. Res. Board Can.* **33**, 413–429.

Wood, E. J. F. (1965). "Marine Microbial Ecology." Chapman and Hall, London.

Young, D. K. (1971). Effect of infauna on sediment and seston of a subtidal environment. *Vie et Milieu* **22**, 557–571 (suppl.).

9

The Oceans

James W. Murray

The oceans play a major role in the global cycles of most elements. There are several reasons for this. As is evident in images from space, most of the Earth's surface is ocean. When viewed from space, we see mostly water because oceans cover 71% of the Earth's surface. The oceans are in interactive contact with the lithosphere, atmosphere, and biosphere, and virtually all elements pass through the ocean at some point in their cycles. Given sufficient time, the water and sediments of the ocean are the receptacle of most natural and anthropogenic elements and compounds.

Transport processes across the ocean boundaries and within the ocean are central to studies of the global cycles. Such processes as air–sea exchange of gases and aerosols, biological production of particles within the sea, and sedimentation need to be considered. The productivity of the ocean and even climate are influenced by wind-generated surface currents and thermohaline circulation in the deep ocean. The complicated and diverse processes in estuaries influence how much material of riverine origin reaches the sea.

The ocean is also by far the largest reservoir for most of the elements in the atmosphere–biosphere–ocean system. Perturbations caused by our increase in population and industrialization are passing through the ocean, and because the time-scale for ocean circulation is long (about 2000 years) relative to the time-scale of modern society, a new steady-state of quasi-equilibrium will slowly be established. Until that time, local concentrations of toxic chemicals, especially in estuaries and bays with restricted circulation, will be a major concern for mankind.

In this chapter, we first review some of the basic descriptive aspects of the ocean and its physiographic domains and show briefly how the ocean fits into the global water balance. We then present a brief review of surface and abyssal ocean circulation. The superposition of the biological cycle on ocean circulation is what controls the distribution of a large number of elements within the ocean, so spatial variations and the stoichiometry of biological productivity and the transport of biologically produced particles are reviewed. Ocean sediments are the main site of deposition for most elements and thus they record the course of events over geological time. Sediments are considered in Chapter 8 and are not considered in detail here. Finally, we review the basic properties of ocean chemistry and attempt to classify the elements into groups according to the mechanisms that control their distribution.

9.1 What is the Ocean?

The topography and structure of the ocean floor are highly variable from place to place and reflect tectonic processes within the Earth's interior. These major features are shown in Fig. 9-1. These features have varied in the past so that the ocean bottom of today is undoubtedly not like the ocean bottom of 50 Ma ago. The major topographic systems, common to all oceans, are the continental margins, the ocean-basin floors, and the oceanic ridge systems. Tectonic features such as fracture zones, plateaus, trenches, and mid-ocean ridges act to subdivide the main oceans into a larger number of smaller basins.

The continental margin regions are the transition zones between the continents and ocean basins. Though the features may vary, the general features shown occur in all ocean basins in the form of either

Global Biogeochemical Cycles
ISBN 0-12-147685-5

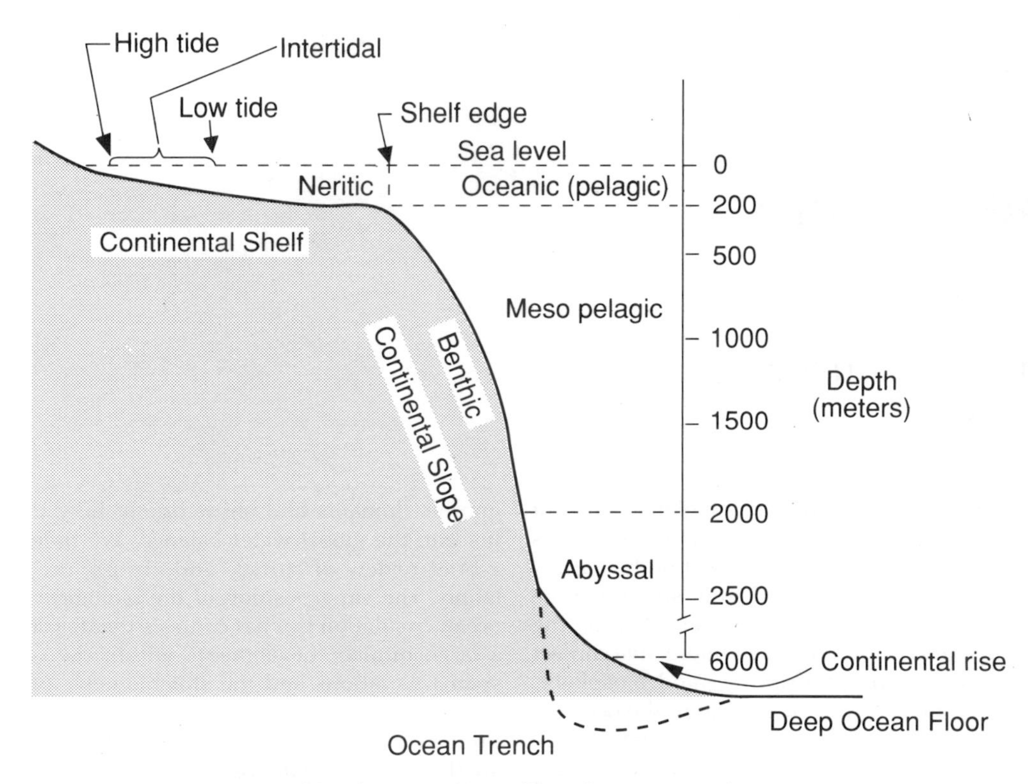

Fig. 9-1 Schematic representation of the physiographic profile at the ocean margins.

two sequences: shelf–slope–rise–basin or shelf–slope–trench–basin. The continental shelf is the submerged continuation of the adjacent land, modified in part by marine erosion or sediment deposition. The seaward edge of the continental shelf can frequently be clearly seen and it is called the shelf break. The shelf break tends to occur at a depth of about 130 m over most of the ocean. It is thought to have formed when sea level stood at its lowest during Pleistocene glacial times. At that time, the shoreline was at the edge of the present continental shelf, which was then a coastal plain. On the average, the continental shelf is about 70 km wide, although it can vary widely (compare the east coast of China with the west coast of Peru). The continental slope is characterized as the region where the gradient of the topography changes from 1:1000 on the shelf to greater than 1:40. Thus continental slopes are the relatively narrow, steeply inclined submerged edges of the continents. Though the angle changes, it would certainly not be called precipitous. The continental slope may form one side of an ocean trench as it does off the west coast of Mexico or Peru, or it may grade into the continental rise. The ocean trenches are the topographic reflection of the subduction of the oceanic plates beneath the continents. The greatest ocean depths occur in such trenches. The deepest is the Challenger Deep, which descends to 11 035 m in the Marianas Trench. The continental rises are mainly depositional features that are the result of the coalescing of thick wedges of sedimentary deposits carried by turbidity currents down the slope. Deposition is caused by the reduction in current speed when it flows out onto the gently sloping rise. Gradually, the continental rise grades into the ocean basins and the abyssal plains.

The relationships between ocean depths and land elevations are shown in Fig. 9-2. On the average, the continents are 840 m above sea level, while the average depth of the oceans is 3730 m. If the Earth were a smooth sphere with the land planed off to fill the ocean basins, it would be uniformly covered by water to a depth of 2430 m. The hypsographic curve

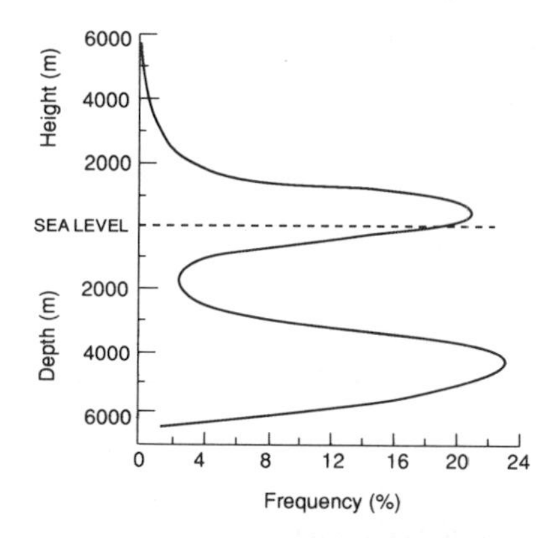

Fig. 9-2 A frequency distribution of elevation intervals of the Earth's surface.

shows the relative amount of area in each kilometer elevation interval.

The area, volume, and average depth of the ocean basins and some marginal seas are given in Table 9-1. The Pacific Ocean is obviously the largest and contains more than one-half of the Earth's water. It also receives the least river water per area of the major oceans (Table 9-2). Paradoxically, it is also the least salty. The land area of the entire Earth is strongly skewed toward the northern hemisphere.

The ocean contains the bulk of the Earth's water (1.37×10^{24} g) and moderates the global water cycle. The distribution of the mass of water is about 80% in the ocean and about 20% as pore water in sediments and sedimentary rocks. The reservoir of water in rivers, lakes, and the atmosphere is trivial (0.003%). Disregarding the pore water because it is not in free circulation, we find that 97% of the world's cycling water is in the ocean (Table 9-3). The unit of 10^{20} g is so common in geochemical cycles that it is sometimes called a geogram. The average residence time of water in the atmosphere with respect to net transfer (evaporation minus precipitation over oceans) from the oceans to the continents is about one-third of a year [0.13×10^{20} g/($3.83 - 3.47 \times 10^{20}$ g/year) = 0.33 years). The ocean's role in controlling the water content of the atmosphere has important implications for the Earth's climate.

Table 9-1 Area, mean depth, and volume of oceans and seas

Region	Area (10^6 km^2)	Mean depth (m)	Volume (10^6 km^3)
Pacific Ocean	155.25	4282	707.56
Atlantic Ocean	32.44	3926	323.61
Indian Ocean	73.44	3963	291.03
Three oceans only	321.13	4117	1322.20
Arctic Mediterranean	14.09	1205	16.98
American Mediterranean	4.32	2216	9.57
Mediterranean Sea and Black Sea	2.97	1429	4.24
Asiatic Mediterranean	8.14	1212	9.87
Baltic Sea	0.42	55	0.02
Hudson Bay	1.23	128	0.16
Red Sea	0.44	491	0.21
Persian Gulf	0.24	25	0.01
Marginal seas	8.08	874	7.06
Three oceans plus adjacent seas	361.06	3795	1370.32
Pacific Ocean including adjacent seas	179.68	4028	723.70
Atlantic Ocean including adjacent seas	106.46	3332	354.68
Indian Ocean including adjacent seas	74.92	3897	291.94

Table 9-2 A breakdown of the water balance for the four main ocean basins (cm/year for the area of the respective basins)[a]

Ocean	Precipitation	Run-off from adjoining land areas	Evaporation	Water exchange with other oceans
Atlantic	78	20	104	6
Arctic	24	23	12	35
Indian	101	7	138	30
Pacific	121	6	114	13

[a] From Budyko (1958).

Table 9-3 A detailed breakdown of the water volume in various reservoirs [a]

Environment	Water volume (10^3 km^3)	Percentage of total
Surface water		
Freshwater lakes	125	0.009
Saline lakes and inland seas	104	0.008
Rivers and streams	1.3	0.0001
Total	230	0.017
Subsurface water		
Soil moisture	67	0.005
Ground water	8000	0.62
Total	8067	0.625
Ice caps and glaciers	29 000	2.15
Atmosphere	13	0.001
Oceans	1 330 000	97.2
Totals (approx.)	1 364 000	100

[a] Data from Leopold (1974).

9.2 Ocean Circulation

The chemistry and biology of the ocean are superimposed on the ocean's circulation, thus it is important to review briefly the forces driving this circulation and give some estimates of the transport rates. There are many reasons why it is important to understand the basics of the circulation. Two examples are given as an illustration:

1. Atmospheric testing of nuclear bombs resulted in the contamination of the surface of the ocean with various isotopes including ^{14}C, ^{3}H, ^{90}Sr, ^{239}Pu, and ^{240}Pu. These isotopes are slowly being mixed through the ocean. Other processes like biological uptake and adsorption onto particles are also important but cannot be adequately resolved without first understanding the basic circulation.
2. The atmospheric CO_2 concentration has been increasing since the beginning of the industrial age, but the increase is less than anthropogenic emissions. Some of the missing CO_2 has gone into the ocean. All CO_2 taken up by the ocean is by the process of gas-exchange. Some of the excess CO_2 has been transported into the intermediate and deep water by the subduction of water masses. Circulation replenishes the surface with water undersaturated with respect to the anthropogenically perturbed CO_2 levels.

In this section, we briefly review what controls the density of seawater and the vertical density stratification of the ocean. Surface currents, abyssal circulation, and thermocline circulation will be considered individually.

9.2.1 Density Stratification in the Ocean

The density of seawater is controlled by its salt content or salinity and its temperature. The salinity is defined as the total salt content and the units are given as grams of salt per kilogram of seawater or parts per thousand (‰). Salinity is expressed on a mass of seawater basis because mass is conserved as the temperature or pressure change. For more details on the formal definition of salinity and on the preparation of very accurate standards, see the UNESCO reports (UNESCO, 1981). The salinity of surface seawater is controlled primarily by the balance between evaporation and precipitation. As a result, the highest salinities are found in the so-called central gyre regions centred at about 20° north and 20° south, where evaporation is extensive but rainfall is minimal. Suprisingly, they are not found at the equator where evaporation is great but so is rainfall.

The temperature of seawater is fixed at the sea surface by heat exchange with the atmosphere. The average incoming energy from the Sun at the Earth's surface is about four times higher at the equator than at the poles. The average infrared radiation heat loss to space is more constant with latitude. As a result, there is a net input of heat into the tropical regions and this is where we find the warmest surface seawater. Heat is then transferred from low to high latitudes by winds and by ocean currents. The geothermal heat flux from the interior of the Earth is generally insignificant except in the vicinity of hydrothermal vents at spreading ridges and in the oldest abyssal waters of the northern North Pacific (Joyce *et al.*, 1986).

Because the seawater signatures of temperature and salinity are acquired by processes occurring at the air–sea interface, we can also state that the density characteristics of a parcel of seawater are determined when it is at the sea surface. This density signature is locked into the water when it sinks. The density may be modified by mixing with other parcels of water, but if the density signatures of all the end member water masses are known, and there aren't too many of them, this mixing can be unraveled and the proportions of the different source waters to a given parcel can be determined.

To a first approximation, the vertical density distribution of the ocean can be described as a three-layered structure. The surface layer is the region from the sea surface to the depth having a temperature of about 10° C. The transition region where the temperature decreases from 10 to 4° C is called the thermocline. The deep sea is the region below the thermocline.

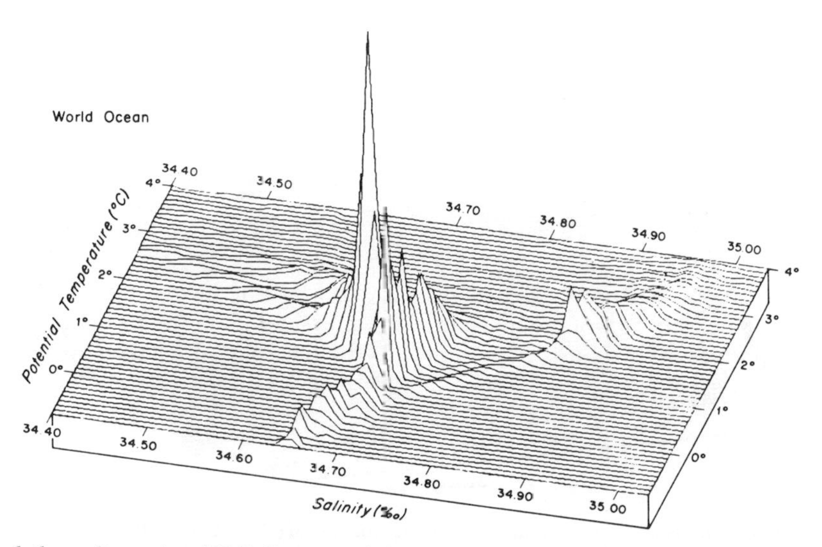

Fig. 9-3 Simulated three-dimensional T-S diagram of the water masses of the world ocean. Apparent elevation is proportional to volume. Elevation of highest peak corresponds to 26.0×10^6 km^3 per bivariate class 0.1 °C × 0.01‰. Reproduced from Worthington (1981) with the permission of MIT Press.

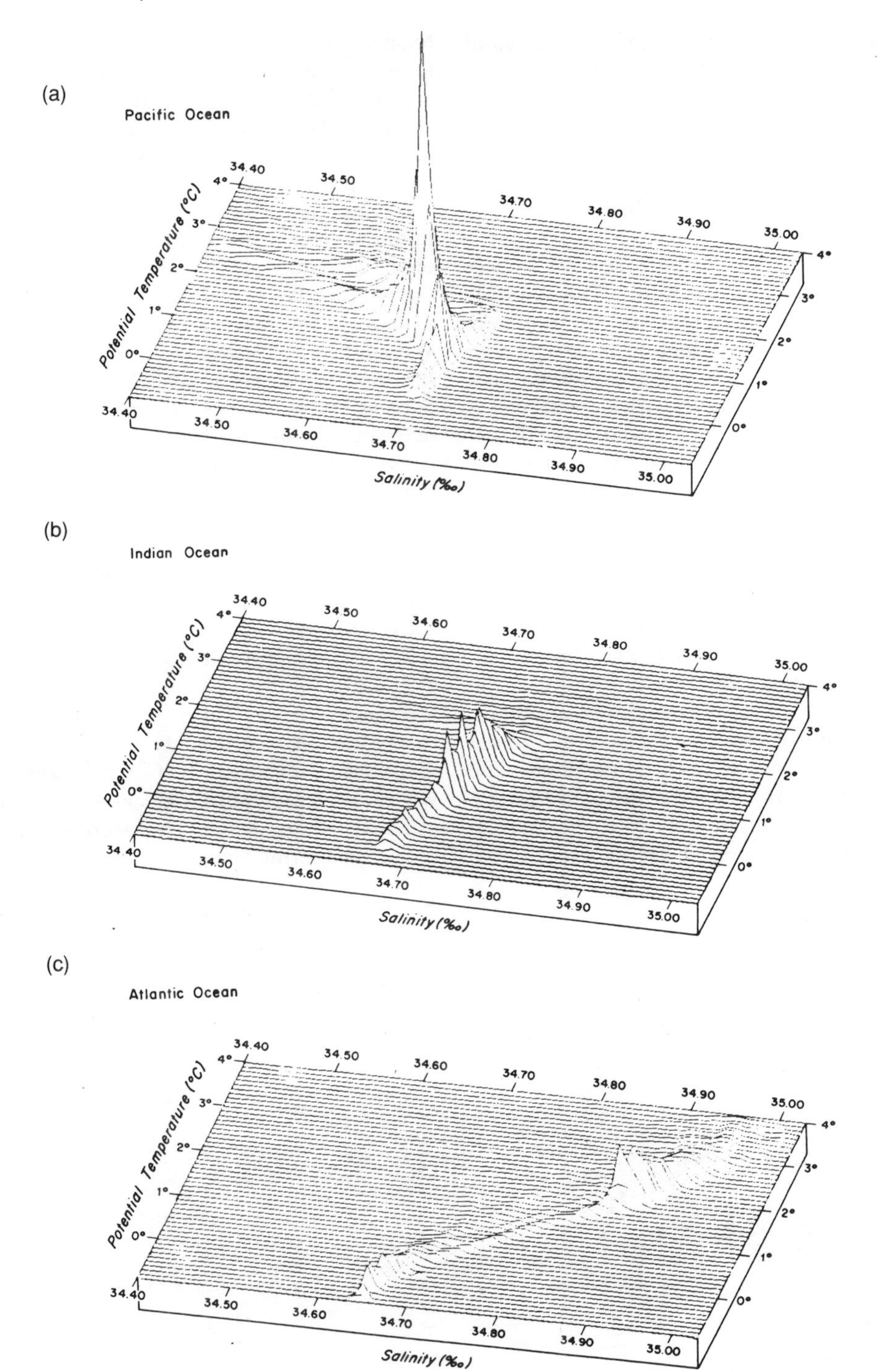

Fig. 9-4 Simulated three-dimensional T-S diagram of the water masses of (a) the Pacific, (b) the Indian, and (c) the Atlantic oceans. Apparent elevation is proportional to volume. Elevation of highest peak corresponds to (a) 26.0×10^6, (b) 6.0×10^6, and (c) 4.6×10^6 km^3 per bivariate class 0.1 °C × 0.01‰. Reproduced from Worthington (1981) with the permission of MIT Press.

Table 9-4 Average temperatures and salinity of the oceans, excluding adjacent seas[a]

	Temperature (°C)	Salinity (‰)
Pacific (total)	3.14	34.60
North Pacific	3.13	34.57
South Pacific	3.50	34.63
Indian (total)	3.88	34.78
Atlantic (total)	3.99	34.92
North Atlantic	5.08	35.09
South Atlantic	3.81	34.84
Southern Ocean[b]	0.71	34.65
World ocean (total)	3.51	34.72

[a] After Worthington (1981).
[b] Ocean area surrounding Antartica, south of 55°S.

Because temperature (T) and salinity (S) are the main factors controlling density, oceanographers use T-S diagrams to describe the features of the different water masses. The average temperature and salinity of the world ocean and various parts of the ocean are given in Figs. 9-3 and 9-4 and Table 9-4. The North Atlantic contains the warmest and saltiest water of the major oceans. The Southern Ocean (the region around Antarctica) is the coldest and the North Pacific has the lowest average salinity. These variations can be seen with more resolution in simulated three-dimensional T-S diagrams where the elevation is proportional to volume for the entire ocean (Fig. 9-3) and the Pacific (Fig. 9-4a), Indian (Fig. 9-4b), and Atlantic (Fig. 9-4c) Oceans.

Conventional T-S diagrams for specific locations in the individual oceans are shown in Fig. 9-5. The inflections in the curves reflect the inputs of water from different sources. For example, in the Atlantic Ocean, the curves reflect input from Antarctic Bottom Water (AABW), North Atlantic Deep Water (NADW), Antarctic Intermediate Water (AIW), Mediterranean Water (MW), and Warm Surface Water (WSW).

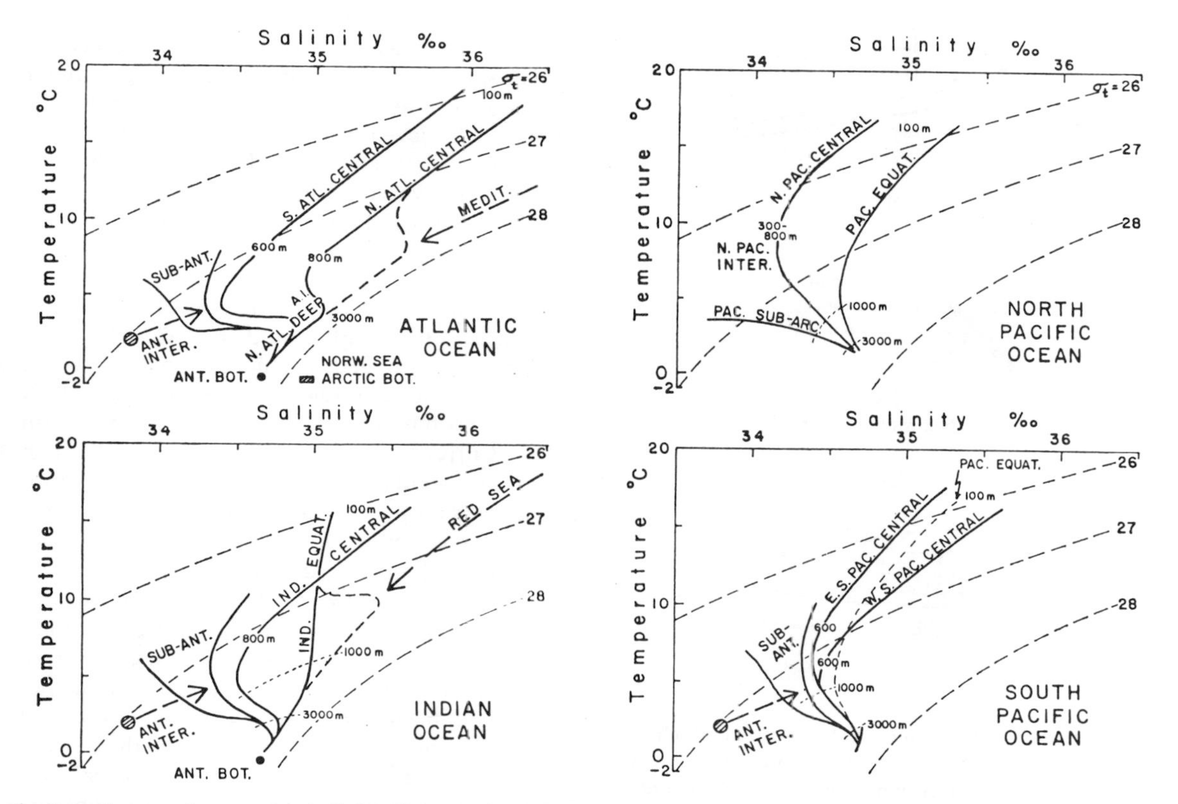

Fig. 9-5 Average temperature/salinity diagrams for the main water masses of the Atlantic, Indian, and Pacific Oceans. Reproduced from Pickard and Emery (1982) with the permission of Pergamon Press.

9.2.2 Surface Currents

Surface ocean currents respond primarily to the climatic wind field. The prevailing winds supply much of the energy that drives surface water movements. This becomes clear when charts of the surface winds and ocean surface currents are superimposed. The *wind-driven circulation* occurs principally in the upper few hundred meters and is therefore primarily a horizontal circulation, although vertical motions can be induced when the geometry of surface circulation results in convergences (downwelling) or divergences (upwelling). The depth to which the surface circulation penetrates is dependent on the water column stratification. In the equatorial region, the currents extend to 30–500 m, whereas in the circumpolar region where stratification is weak the surface circulation extends to the sea floor.

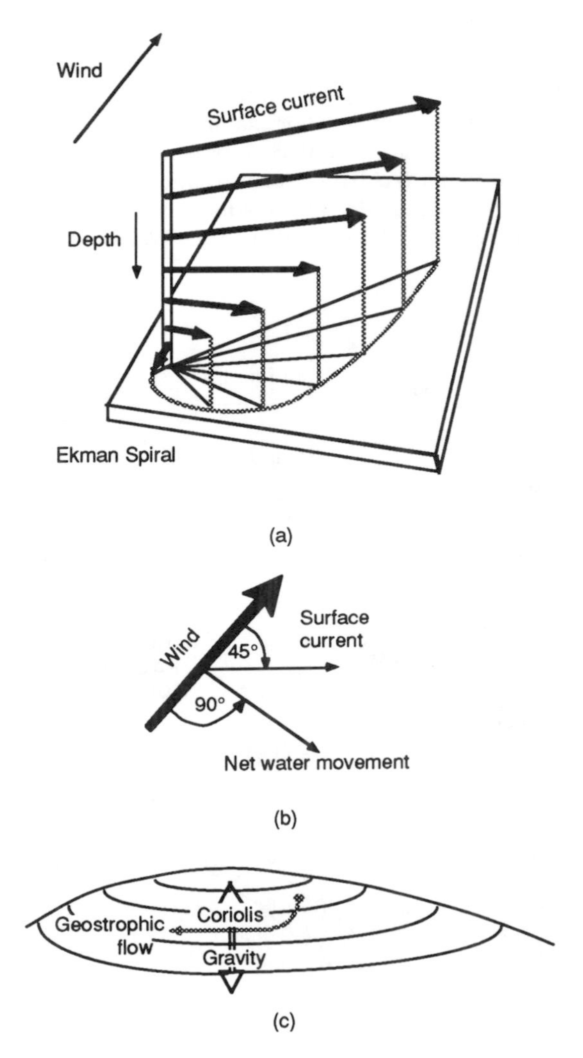

Fig. 9-6 Sketch of (a) current vectors with depth characteristic of an Eckman spiral; (b) relationship between wind, surface current, and net water movement vectors; and (c) production of circular gyres from the net interaction of the Coriolis force and Eckman transport.

The net direction of motion of the water is not always the same as the wind, because other factors come into play. These are shown schematically in Fig. 9-6. The wind blowing across the sea surface drags the surface along and sets this thin layer in motion. The surface drags the next layer and the process continues downward, involving successively deeper layers. As a result of friction between the layers, each deeper layer moves more slowly than the one above and its motion is deflected to the right (clockwise) in the northern hemisphere by the Coriolis force (see Chapter 10). If this effect is represented by arrows (vectors) whose direction indicates current direction and length indicates speed, the change in current direction and speed with depth forms a spiral. This feature is called the Ekman spiral. If the wind blew continuously in one direction for a few days, a well-developed Ekman spiral would develop. Under these conditions, the integrated net transport over the entire depth of the Ekman spiral would be at 90° to the wind direction. Normally, the wind direction is variable so that the net actual transport is some angle less than 90° to the wind direction.

As a result of the Ekman transport, changes in the sea surface topography and the Coriolis force combine to form geostrophic currents. Take the North Pacific for example. The westerlies at ~40°N and the northeast trades (~10°N) set the North Pacific Current and North Equatorial Current in motion as a circular gyre. Because of the Ekman drift, surface water is pushed toward the center of the gyre (~25°N) and piles up to form a sea surface "topographic high". As a result of the elevated sea surface, water tends to flow "downhill" in response to gravity. As it flows, however, the Coriolis force deflects the water to the right (in the northern hemisphere). When the current is constant and results from balance between the pressure gradient force due to the elevated sea surface and the Coriolis force, the flow is said to be in geostrophic balance. The actual flow is then nearly parallel to the contours of the elevated sea surface and clockwise. The sea surface topography of the Pacific Ocean was determined by Tai and Wunsch (1983) from satellite altimetry. The absolute elevation of the subtropical gyre can be clearly seen and fits the schematic description given above.

As a result of these factors (wind, Ekman transport, Coriolis force), the surface ocean circulation in the mid-latitudes is characterized by clockwise gyres in the northern hemisphere and counterclockwise gyres in the southern hemisphere. The main surface currents around these gyres for the world's oceans are shown in Fig. 9-7. The regions where Ekman transport tends to push water together are called convergences. Divergences result when surface waters are pushed apart.

Total transport by the surface currents varies greatly and reflects the mean currents and cross-sectional areas. Some representative examples will illustrate the scale. The transport around the subtropical gyre in the North Pacific is about $70 \times 10^6\, m^3/s$. The Gulf Stream, which is a major northward flow off the east coast of North America, increases from $30 \times 10^6\, m^3/s$ in the Florida Straits to $150 \times 10^6\, m^3/s$ at 64°30′W, or 2000 km downstream.

9.2.3 Thermocline Circulation

The transition region between the surface and deep ocean is referred to as the thermocline. This is also a pycnocline zone where the density increases appreciably with increasing depth. Most of the density change results from the decrease in temperature (hence thermocline).

Concepts regarding the existence of the thermocline are changing. An older view held that the thermocline represented a vertical balance between the downward diffusion of heat and the upward advection of cold abyssal water that originated at high latitudes (see Section 9.2.4). It was presumed that these two processes are in long-term balance resulting in a steady-state. A one-dimensional vertical model can be fit to most of the data; however, the vertical diffusivity parameters required to fit the data are unsatisfactory from several points of view (i.e. they are too large; see Jenkins, 1980).

Recently, a simple but physically more realistic model has evolved based on lateral transport. According to this view, the interior of the ocean is ventilated by rapid mixing and advection *along* isopycnal surfaces. The density surfaces that lie in the thermocline at 200–1000 m in the equatorial region shoal and outcrop at high latitudes. The argument is that water acquires its T and S (and chemical tracer) signature while at the sea surface and then sinks and is transported horizontally as shown in Fig. 9-8. A map showing the winter outcrops of isopycnal surfaces in the North Atlantic Ocean is shown in Fig. 9-9. Characteristic values of the horizontal eddy diffusion coefficient (K) are of

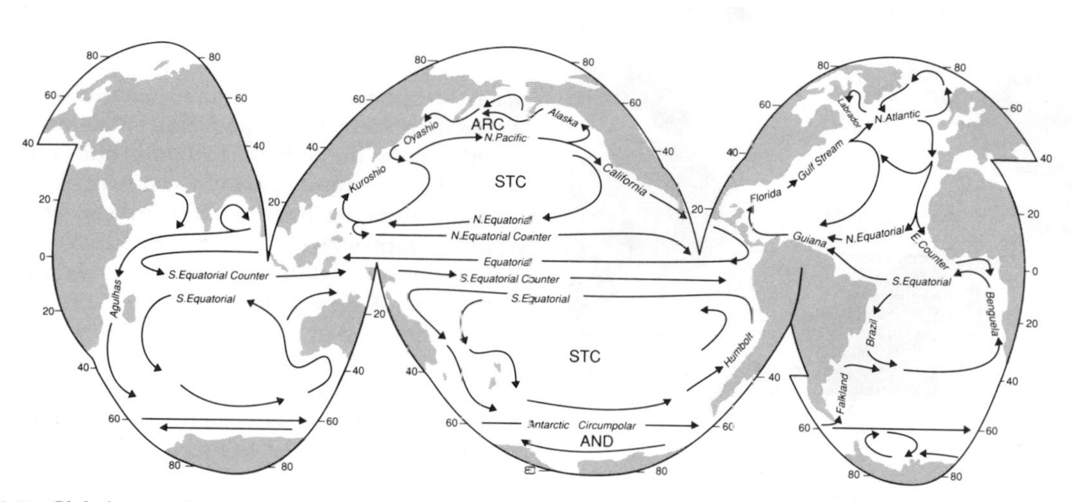

Fig. 9-7 Global map of major ocean currents. AND, Antarctic divergence; STC, subtropical convergence; ARC, Arctic convergence.

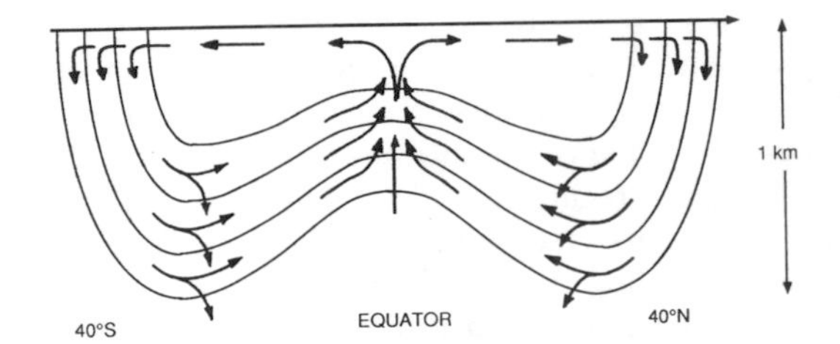

Fig. 9-8 The pathways followed by the water ventilating the main oceanic thermocline. Adapted from Broecker and Peng (1982) with the permission of Eldigio Press.

the order of 10^7 cm^2/s. Assuming a distance (L) of the order of 2000 km (30°) and assuming the characteristic time is $\tau = L^2/K$, we obtain a characteristic ventilation time for the main thermocline of about 130 years.

The horizontal isopycnal thermocline model is important for determining the fate of the excess atmospheric CO_2. The increase of CO_2 in the atmosphere is modulated by transport of excess CO_2 from the atmosphere into the interior of the ocean. The direct ventilation of the thermocline in its outcropping regions at high latitudes appears to play an important role in removing CO_2 from the atmosphere (Brewer, 1978; Siegenthaler, 1983).

Nuclear bomb-produced $^{14}CO_2$ and ^3H (as HTO) have been used to describe and model this rapid thermocline ventilation (Ostlund *et al.*, 1974; Sarmiento *et al.*, 1982; Fine *et al.*, 1983). For example, the distributions of tritium (Rooth and Ostlund, 1972) in the western Atlantic in 1972 (GEOSECS) and 1981 (TTO) are shown in Fig. 9-10 (Ostlund and Fine, 1979; Baes and Mulholland, 1985). In the 10 years following the atmospheric bomb tests of the early 1960s, a massive penetration of ^3H into the thermocline at all depths has occurred. Comparison of the GEOSECS and TTO data, which have a 9-year time difference, clearly shows the rapid ventilation of the North Atlantic and the value of such "transient" tracers. A similar distribution can be seen in the distribution of man-made chlorofluorocarbons, which have been released over a longer period (40 years) (Gammon *et al.*, 1982).

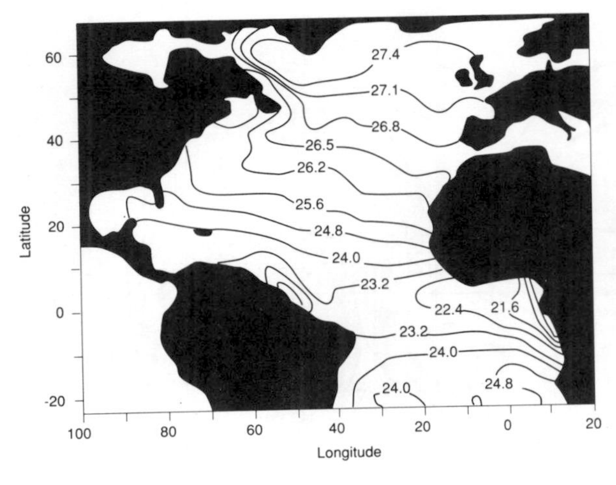

Fig. 9-9 Map of winter outcrops of isopycnal surfaces in the Atlantic Ocean. Modified from Broecker and Peng (1982) with the permission of Eldigio Press.

9.2.4 *Abyssal Circulation*

The circulation of the deep ocean below the thermocline is referred to as abyssal circulation. The currents are slow and difficult to measure but the pattern of circulation can be clearly seen in the properties of the abyssal water. For example, the water of lowest temperature in the water column is usually the densest and lies deepest. As a result, charts of the bottom water temperature have been useful in describing the pattern of the abyssal circulation (e.g. Mantyla, 1975; Mantyla and Reid, 1983). The topography of the sea floor plays an important role in constraining the circulation and much of the abyssal flow is funneled through passages such as the Denmark Straight, Gibbs Fracture Zone, Samoan Passage, Vema Channel, and the Drake Passage.

For a steady-state ocean, a requirement of the *heat balance* is that the input of new cold abyssal water (Antarctic Bottom Water and North Atlantic Deep Water) sinking in the high latitude regions must be balanced by the input of heat by geothermal heating (heat flow from the Earth), downward convection of relatively warm water (e.g. from the Mediterranean), and downward diffusion of heat across the thermocline. A general mass balance of the world's oceans requires that the water sinking in the polar regions must be exactly balanced by the upwelling of water from the abyssal ocean to the surface water. A combination of the mass and heat balances together with the forcing of the wind and the effect of a rotating Earth determine the nature of the abyssal circulation.

There are two main sources of abyssal ocean

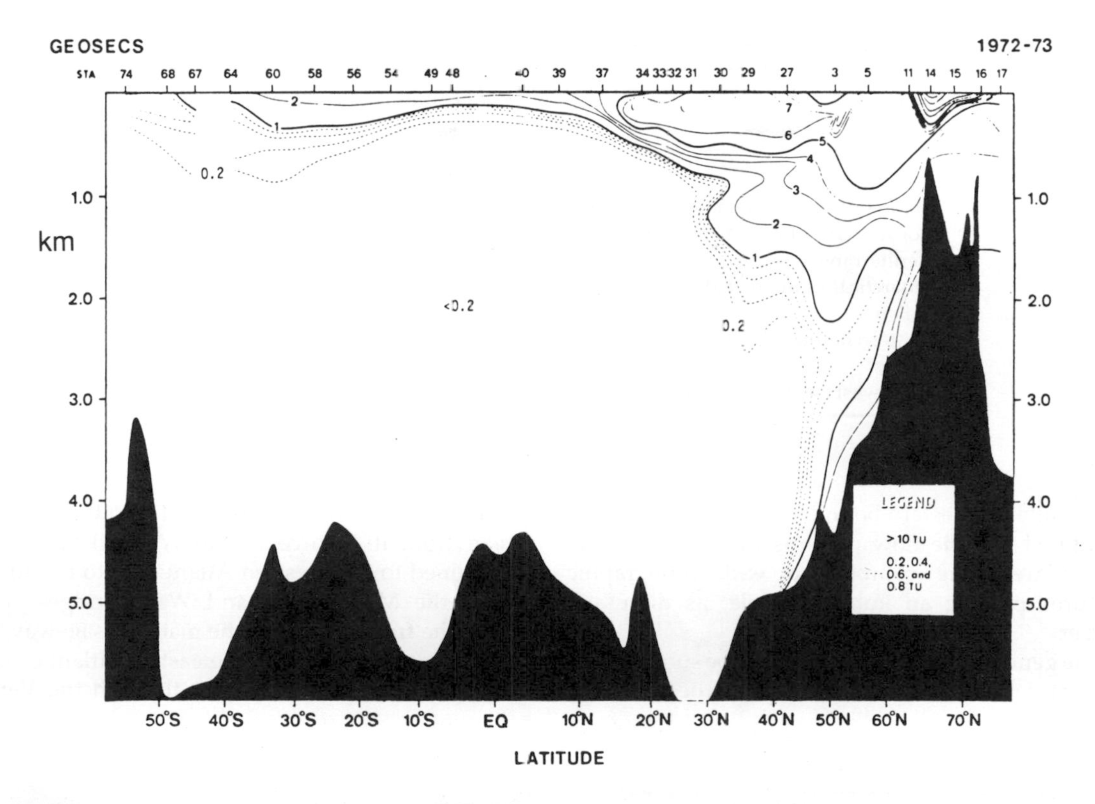

Fig. 9-10 Tritium section of the western Atlantic from 70°N to 50°S *vs* depth (km). Vertical exaggeration is 2000 : 1. Horizontal scale is proportional to cruise track. Reproduced from Ostlund and Fine (1979) with the permission of the International Atomic Energy Agency.

water. These are the Norwegian-Greenland Sea in the North Atlantic and the Weddell Sea in the Antarctic. Minor amounts of highly saline water flow from the Mediterranean and Red Seas. The low salinity water that lies at the upper boundary of the abyssal water column is formed in subpolar regions of the Antarctic and northwestern Pacific and Atlantic Oceans. The general characteristics of the major water masses are summarized in Table 9-5.

The Weddell Sea, because of its very low temperature, is the main producer of Antarctic Bottom Water. The rate of production is not well known but the best estimates are close to 38×10^6 m^3/s (Gordon, 1975). The bulk of the Antarctic Bottom Water has an initial average potential temperature of -1°C. Most of the Antarctic Bottom Water flows north into the South Atlantic and through the Vema Channel in the Rio Grande Rise into the North Atlantic. It returns southward combined with the North Atlantic Deep Water.

The formation of North Atlantic Deep Water is a complicated process that first involves the transport of relatively warm and salty North Atlantic surface water into the Norwegian-Greenland Sea by an extension of the Gulf Stream. In that region, heat is lost to the atmosphere at an extremely fast rate and the new, now cold, salty water forms the major part of the North Atlantic Deep Water. Worthington (1970) has described the Norwegian-Greenland Sea as a Mediterranean-type basin because low-density water enters the surface layers and high-density water exits at depth. The resulting North Atlantic Deep Water formation rate is about 10×10^6 m^3/s. Abyssal water does not form in a similar way in the North Pacific because the salinity is too low (Warren, 1983).

The abyssal circulation model of Stommel (1958) (Fig. 9-11) predicted that the deep waters flow most intensely along the western boundaries in all oceans and gradually mix into the interior during this flow.

Table 9-5 General characteristics of the major water masses

Water mass	Temperature (°C)	Salinity (‰)	Flow ($10^6\ m^3/s$)
1. Atlantic			
Antarctic Bottom Water (ABW)	−1.0	34.65	40
North Atlantic Deep Water (NADW)	+2.0	34.9	10
Mediterranean Water	12.0	36.6	1
Antarctic Intermediate Water (AAIW)	2.2	33.8	
2. Indian			
Western Boundary Current			5
3. Pacific			
Southwest Pacific Boundary Current	1.0	34.7	19

These intense western boundary currents have been identified but the flow patterns in the rest of the ocean are more complicated with topographic features playing an important role, as discussed earlier.

The general abyssal circulation can be summarized rather briefly. Antarctic Bottom Water, the densest of the deep waters, flows northward into the Atlantic Ocean from its source in the Weddell Sea. It is constrained to the Western Atlantic up to the equator by the Mid-Atlantic and Walvis ridges. The Romanche fracture zone is the main passageway for abyssal water to flow into the eastern Atlantic. The Antarctic Bottom Water passes through the Vema

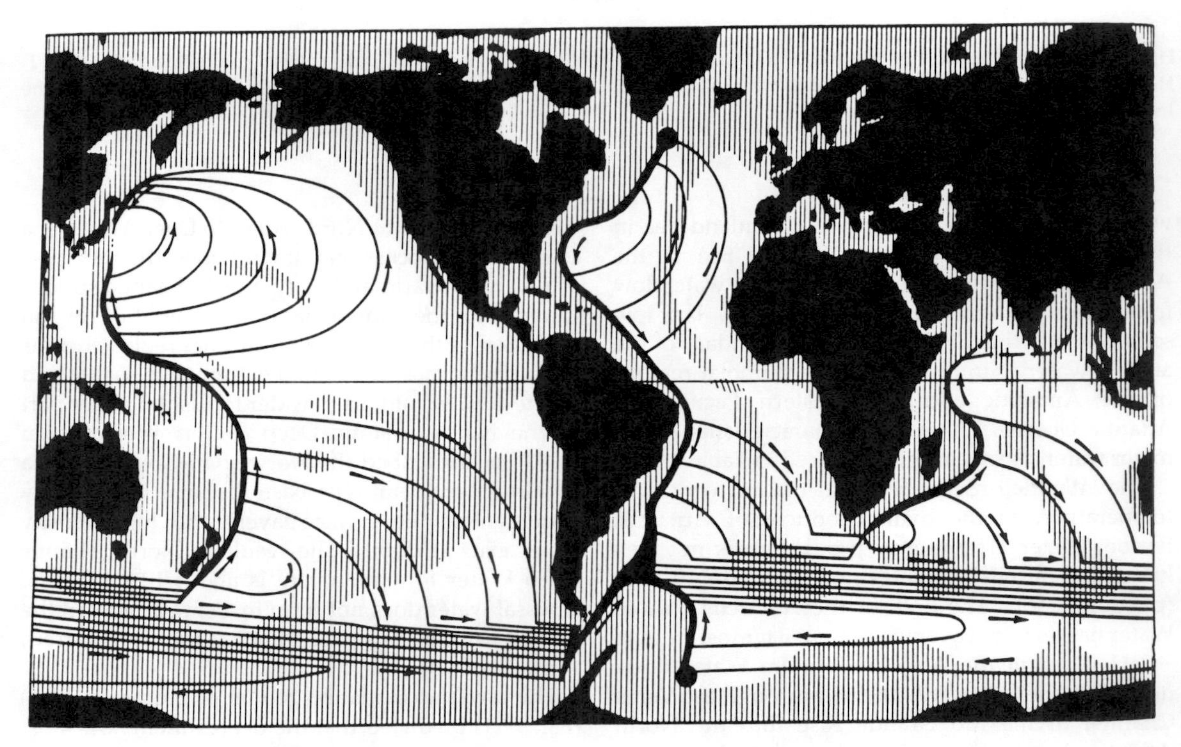

Fig. 9-11 The abyssal circulation generated by equal sources in the North Atlantic and in the Weddell Sea with uniform upwelling elswhere. Reproduced from Stommel and Arons (1958) with the permission of Pergamon Press.

Channel and remains a distinct water mass as far north as 40°N in the western North Atlantic basin.

As the Antarctic Bottom Water flows north, it gradually mixes with the southward flowing North Atlantic Deep Water, which lies immediately above. As the North Atlantic Deep Water flows to the south, it incorporates not only the Antarctic Bottom Water but also the Mediterranean Water and the Antarctic Intermediate Water which lie above. The North Atlantic Deep Water is eventually entrained into the Antarctic Circumpolar Current and flows unimpeded into the Indian and Pacific Oceans.

By the time the North Atlantic Deep Water and the Antarctic Circumpolar Water reach the Pacific Ocean, they are well mixed. The resulting water mass is referred to as the Pacific Common Water. The Pacific Common Water enters the Pacific in the southwest corner and flows north along the western boundary in the Tonga Trench. Most of the northward abyssal flow passes from the Southwest Pacific to the North Central Pacific through the Samoan Passage west of Samoa. This feature can be clearly seen in the distribution of near bottom potential temperature in the Pacific Ocean. In the North Pacific, the abyssal flow splits and goes west and east of the Hawaiian Islands. These flows meet again north of Hawaii where they mix, upwell, and flow back to the south at mid-depths. The circuit is completed by surface water flowing from the Pacific to the Indian Ocean through gaps in the Indonesian Archipelago and then into the South Atlantic through the Arghulus Reflection south of Africa.

Although the general circulation patterns are fairly well known, it is difficult to quantify the rates of the various flows. Abyssal circulation is generally quite slow and variable on short time-scales. The calculation of the rate of formation of abyssal water is also fraught with uncertainty. Probably the most promising means of assigning the time dimension to oceanic processes is through the study of the distribution of radioactive tracers. Difficulties associated with the interpretation of radioactive tracer distributions lie both in the models used, non-conservative interactions, and the difference between the time-scale of the physical transport phenomenon and the mean life of the tracer.

An example of the power of such tracers is in the "dating" of abyssal water using ^{14}C. Stuiver *et al.* (1983) have measured the ^{14}C distribution in dissolved inorganic carbon in deep samples from major ocean basins (Fig. 9-12). These data were used to calibrate a box model which indicated that the replacement times for the Atlantic, Indian, and Pacific ocean deep waters (depths > 1500 m) are 275, 250 and 510 years respectively. The calculations imply that approximately $41 \times 10^6\ m^3/s$ of bottom water must be forming in the circumpolar region (mostly in the Weddell Sea).

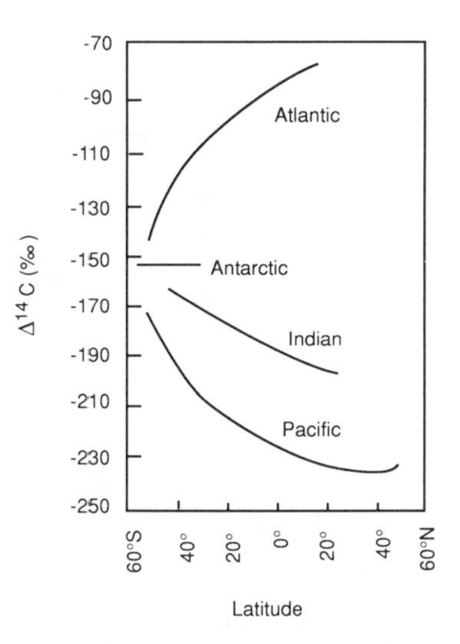

Fig. 9-12 The $\Delta^{14}C$ values of the cores of the North Atlantic, Pacific, and Indian Ocean deep waters. The oldest waters are encountered near 40 °N in the Pacific Ocean. Modified from Stuiver *et al.* (1983) with the permission of the American Association for the Advancement of Science.

9.3 Biological Processes

Almost all elements in the periodic table are involved in at least one way or another in the biological cycle of the ocean. Many elements are essential or required nutrients. Others are carried along as passive participants. In either case, the rates of biological processes need to be known.

9.3.1 The RKR Model

Essentially, all organic matter in the ocean is ultimately derived from inorganic starting materials (nutrients) converted by photosynthetic algae into biomass. A generalized model for the production of plankton biomass from nutrients in seawater was presented by Redfield *et al.* in 1963. The schematic

Table 9-6 The C, N, and P composition (average) of phytoplankton, zooplankton, and average plankton of all kinds

	C	N	P
Zooplankton	103	16.5	1
Phytoplankton	108	15.5	1
Average	106	16	1

"RKR" equation is given below:

$$106CO_2 + 16HNO_3 + H_3PO_4 + 122H_2O \text{ (light)} \rightarrow (CH_2O)_{106}(NH_3)_{16}(H_3PO_4) \text{ (plankton protoplasm)} + 138O_2$$

This equation was originally proposed for "average" plankton, a category that included both zooplankton and phytoplankton: Table 9-6. The composition of zooplankton and phytoplankton are actually slightly different. Zooplankton tend to be richer in N, perhaps because they contain more protein.

The following characteristics of the RKR reaction should be noted:

1. This is an organic redox reaction. Carbon in CO_2 and nitrogen in HNO_3 are reduced by oxygen from water as the oxygen in these compounds is oxidized to O_2. Only phosphorus does not undergo any change in oxidation state.
2. The reaction is endothermic. Energy from sunlight is stored in the form of organic biomass and O_2, the raw materials for the support of heterotrophic organisms dependent upon the food source.
3. The plankton themselves have the ability to leave the waters in which they form (by sinking or swimming).
4. This is not a reversible reaction in the strict sense and does not spontaneously seek equilibrium between products and reactants. The effective reverse reaction, respiration, occurs in a different part of phytoplankton cells or is mediated by heterotrophic organisms.
5. Inasmuch as the RKR model is a generalization, specific exceptions should be expected. The most important exceptions relate to growth conditions that can affect the stoichiometry of nutrient incorporation into plankton biomass. For example, a depletion in either N or P in seawater relative to the usual ratio results in plankton biomass poor in that element. Ryther (1969) observed that N:P ratios in plankton can vary from 3:1 to 30:1 depending on the natural availability. During respiration, the reverse reaction occurs and nutrients are regenerated. Phosphorus tends to be regenerated preferentially relative to nitrogen which is preferentially regenerated relative to carbon (see also Section 8.5). Recent interpretations of data along constant density surfaces in the Atlantic suggest that the regeneration ratios of $P:N:C:O_2$ are about 1:16:103:172, slightly different than the RKR ratios (Takahashi *et al.*, 1985).

9.3.2 *Factors Affecting the Rate of Plankton Productivity*

9.3.2.1 Nutrients

Liebig's Law of the Minimum states that under equal conditions of temperature and light, the nutrient available in the smallest quantity relative to the requirement of a plant will limit productivity. One means of showing this is to compare the nutrient composition of "average" seawater with the requirements of "average" marine plankton. As can be seen in Table 9-7, there is approximately a ten-fold excess of inorganic carbon (largely as HCO_3^-) in "average" (i.e. deep) seawater relative to the availability of phosphorus (as phosphate) and fixed nitrogen (as NO_3). These calculations suggest that nitrogen might become limiting slightly before phosphorus. They also imply that most of the carbon fixed by plankton is already present in seawater and does not come from the atmosphere.

A field test for the limiting nutrient is simply to see which nutrient first approaches zero concentrations in surface waters. In coastal waters off Long Island, nitrogen is apparently lacking (Fig. 9-13). In the central gyres of the ocean, PO_4^{3-} and $NO_3^- + NH_4^+$ concentrations are normally both extremely low (Perry, 1976). Under such conditions, growth experiments suggest that P may also be limiting. There are many areas, away from the continental margins, where N and P levels are relatively high, but chlorophyll and productivity values are not as large as would be expected. These locations include the sub-arctic North Pacific, the central Equatorial Pacific, and the circumpolar current around Antarctica. Martin and Fitzwater (1988) and Martin *et al.* (1989) have proposed that iron is really the

Table 9-7 Availability of nutrient elements in "average" seawater ($S = 34.7‰$, $T = 2°C$) and the ratios of their availability and utilization by plankton

	Availability in "average" seawater		Utilization by plankton (ratio)	Ratio of availability to utilization
	mg atoms/m^3	ratio		
Phosphorus	2.3	1	1	1.00
Nitrogen	34.5	15	16	0.94
Carbon	2340.0	1017	106	9.60
Oxygen saturation value	735.0	320	276	1.16

limiting nutrient in these cases. What little iron there is appears to be supplied by atmospheric transport.

It has been argued that phosphorus limits oceanic productivity on the million year time-scale (Broecker, 1971). The reason is that essentially all phosphorus in the ocean is introduced by rivers and thus ultimately from the weathering of continental rocks. This flux is, in effect, fixed by the rate of chemical weathering of the continents. By comparison, fixed nitrogen can be derived from atmospheric N_2 (via nitrogen fixation by marine blue green algae such as *Trichodesmium*) as well as by weathering of rocks. The reservoir of atmospheric N_2 is so large that nitrogen fixation can, over long time periods, adjust the overall supply of fixed nitrogen in seawater to the ratio needed by "average" plankton without significantly depleting the N_2 source. There is no evidence at present to support this argument for long-term phosphate limitation.

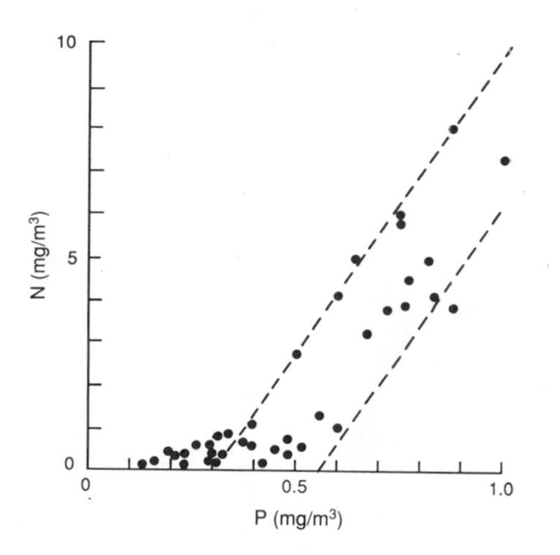

Fig. 9-13 Concentration of phosphate phosphorus and nitrate nitrogen in coastal waters south of Long Island. Slope of envelopes: $\Delta N/\Delta P = 15$. Reproduced from Redfield *et al.* (1963) with the permission of John Wiley and Sons, Inc.

Silicic acid (H_4SiO_4) is a necessary nutrient for diatoms, which build their shells from opal ($SiO_2 \cdot nH_2O$). Whether silicic acid becomes limiting for diatoms in seawater again depends on the availability of Si relative to N and P. Estimates of diatom uptake of Si relative to P range from 16:1 to 23:1. Field studies in the eastern Pacific suggest that Si can become limiting in surface waters south of the equator, with the effect becoming progressively more pronounced toward Antarctica.

If any nutrient (other than iron) is limiting growth in the ocean, it is not likely to be N. Nitrogen is primarily available as NO_3^- or NH_4^+. Nitrate is supplied to the euphotic zone from greater depth. Thus, biologists characterize the primary productivity support by nitrate as "new production" (Dugdale and Goering, 1967). The productivity supported by ammonium is called regenerated production. Under steady-state conditions, the new production, as defined above, should be equal to the particulate flux of N settling out of the euphotic zone. This "export production" can also be examined in terms of carbon (Eppley and Peterson, 1979).

9.3.2.2 *Light*

Light is always necessary for photosynthesis and becomes limiting in the winter at high latitudes. In addition, the depth profiles of productivity and light energy correlate well at locations undergoing bloom conditions. This suggests that the decline in productivity with depth reflects light penetration. The chemical constituents of seawater that can inhibit light penetration include dissolved humic

Table 9-8 Distribution of ocean productivity

Province	Percentage of ocean	Area (10^6 km^2)	Mean productivity (g C/m^2 year)	Total productivity (10^{15}g C year)
Open ocean	90.0	326	50	16.3
Coastal zone[b]	9.9	36	100	3.6
Upwelling areas	0.1	0.36	300	0.1
Total				20.0

[a] From Ryther (1969).
[b] Includes offshore areas of high productivity.

substances (Gelbstuff) and suspended particulate matter. Both factors can become important factors in estuaries and other near shore environments.

9.3.2.3 *Availability of trace metals*

Trace metals can serve as essential nutrients and as toxic substances. For example, cobalt is a component of vitamin B_{12}. This vitamin is essential for nitrogen-fixing algae. In contrast, copper is toxic to marine phytoplankton at free ion concentrations similar to those found in seawater (Sunda and Guillard, 1976). The possibility that iron availability may limit primary productivity was discussed earlier.

9.3.3 *The Geographic Distribution of Primary Productivity*

The geographic distribution of primary productivity in the ocean is shown in Fig. 9-14. High productivity is characteristic of marine zones where surface water is replenished with deeper water either by upwelling (as on western continental margins) or by deep mixing (as at high latitudes where stratification is less pronounced). Although upwelling regions are characterized by very high productivities (~300 g C/m^2 year), they together contribute less than 1% of the total ocean production (Table 9-8). Coastal regions have mean productivities of about 100 g C/m^2 year, but account for approximately 100 times the surface area of upwelling zones. These coastal regions contribute about 25% of the total primary production, with the remaining 75% coming from the wide expanses (90% of total area) of low production (50 g C/m^2 year) open ocean.

9.3.4 *Forms of Organic Matter in Seawater*

To this point, organic matter in the ocean has been treated primarily as "RKR average plankton". We

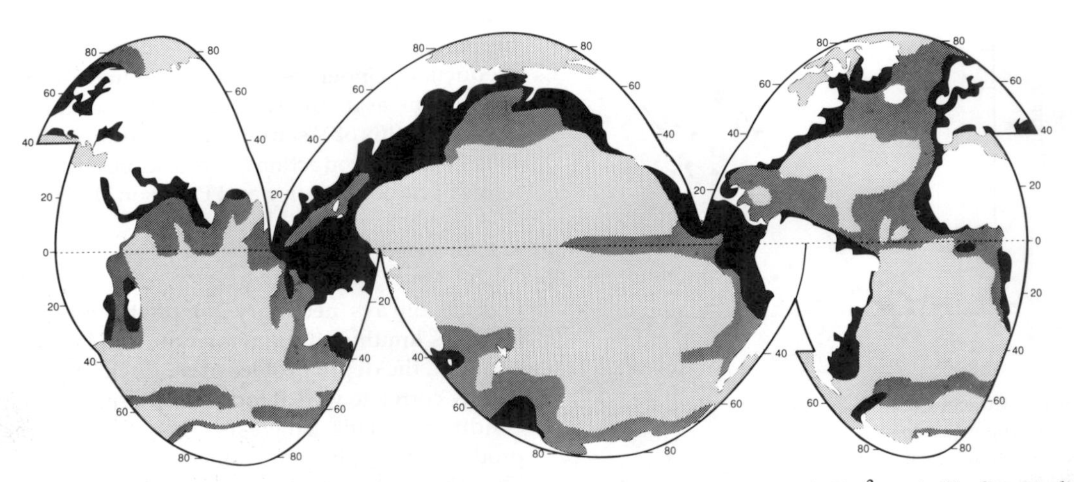

Fig. 9-14 Approximate geographical distribution of primary productivity in the oceans (gC/m^2 year). Shadings indicate productivity; light, <50 g C/m^2 year; medium, 50–100 g C/m^2 year; dark, >100 g C/m^2 year.

Table 9-9 Organic carbon reservoirs in the ocean

Depth (m)	Dissolved (10^6 tonnes C)	Approximate concentration (μg C/L)	Particulate (10^6 tonnes C)	Approximate concentration (μg C/L)	Living fraction (10^6 tonnes C)	Approximate % of particulate
0–300[a]	110 000	1000–1500	11 000	100	550[f]	5[f]
300–3800[b]	630 000	500–800	13 000	3–10	<400[g]	<3[g]
Totals	740 000[c]		24 000[c]			
	670 000[d]		14 000[d]			
	1 000 000[e]		30 000[e]			

[a] Concentrations vary widely with geographical area and with season. The depth at which concentration tends to approach a more or less constant level varies widely from 100 to 500 m.
[b] Concentrations more constant; [c] Williams (1971); [d] Menzel (1974); [e] Williams (1975); [f] phytoplankton component only (see Table 9-3 and Williams, 1975); [g] total living matter (Parsons *et al.*, 1977).

now need to focus on the fate of this biologically produced organic carbon.

There is an exceedingly broad range of size of organic material in seawater, ranging from simple organic molecules, such as dissolved glucose (scale $\sim 10^{-9}$ m), to the blue whale ($\sim 10^{2}$ m). Although the distribution curve of organic particles is smooth over the 10^{-3} to 10^{-9} m-size interval, it has become customary to divide these particles into "dissolved" and "particulate" categories on the basis of filtration through a 0.45-μm pore filter. By operational definition, "dissolved" particles pass through the filter, whereas "particulate" materials are retained.

Although somewhat arbitrary, the 0.45-μm "cut-off" between dissolved and particulate organic matter is for the most part convenient. For example, particles above about 1.0 μm are observable with a microscope and tend to settle in seawater. Particles less than 1.0 μm are submicroscopic and generally sink very slowly and disperse as a result of Brownian motion. In addition, particles less than 0.45 μm fall below the range of most living organisms (except for some viruses and small bacteria).

Essentially, all the dissolved organic matter in seawater can be assumed to be non-living. However, particulate organic matter can be either living or dead, with the latter often referenced to as "detritus".

9.3.5 Oceanic Reservoirs of Organic Carbon

The distribution of dissolved, total particulate, and living particulate organic carbon in the surface (0–300 m) and deep ocean (>300 m) is summarized in Table 9-9. The important aspects of this compilation are:

1. The organic carbon (and thus the organic matter) in seawater is predominantly in dissolved form (DOC) with an average for the whole ocean being 97%. Recent analytical advances by Sugimura and Suzuki (1988) have revealed that DOC in the upper ocean is about a factor of three larger than previously accepted. The DOC decreases with depth and correlates well with oxygen consumption.
2. Of the remaining particulate organic matter, very little is living (> 95% detritus).
3. Organic matter in general occurs at low concentrations in seawater. The average level of total organic carbon (TOC) is about 3 mg/L or 3 ppm (mass).

9.3.6 An Oceanic Budget for Organic Carbon

A simple budget for marine organic matter is given in Fig. 9-15. The two main sources of organic matter are primary productivity (200×10^{14} g C/year) and terrigenous matter introduced by rivers in both dissolved and particulate form (2×10^{14} g C/year). The main mechanism for removal of organic carbon *as such* is burial in sediments. This flux is equal to the average global sedimentation rate for marine sediments × their weight percent organic carbon. The total sink by burial (1×10^{14} g C/year) does not nearly approach the source terms. If the reservoir of total organic matter in seawater is to remain constant (i.e. steady-state), then over 99% of the input must be remineralized by respiration and decomposition either in the water column or the surface sediments. The closest estimate that can be made for the mean residence time of organic matter in the ocean is 10 000 years. This is obtained by dividing the reservoir mass by the sedimentary removal rate. This crude estimate is, however, slanted toward the refractory organic components and says little about organic carbon cycling rates in the ocean prior to sedimentation.

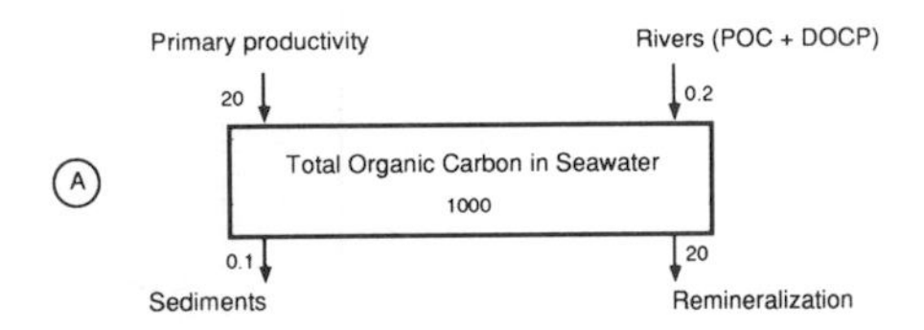

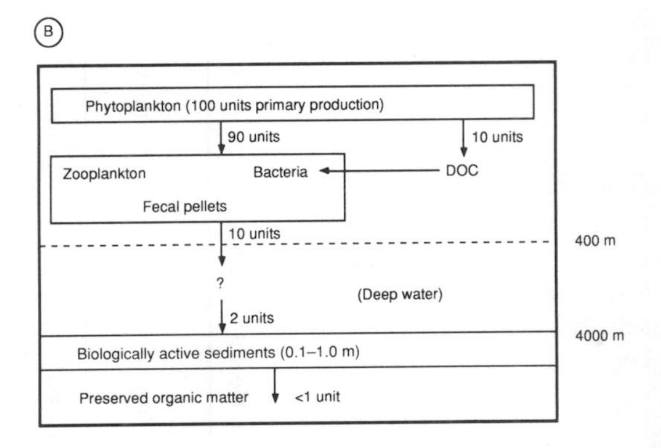

Fig. 9-15 (A) The ocean carbon budget. Units are 10^{15} g C (burdens) and 10^{15} g C/year (fluxes). (B) A rough idea where it all goes.

9.3.7 Organic Carbon Pathways in the Ocean

To understand the distribution and pathways of organic material in the ocean, the key question is: "What happens to that 99% of the phytoplankton biomass that is remineralized between photosynthesis and burial?"

A major advance in our understanding of the processes controlling the vertical transport of organic matter in the ocean has been the realization that most of the vertical flux of particulate material in the ocean water column is provided by large particles (>128 μm) which account for less than 5% of the total mass concentration. These larger particles are predominantly zooplankton fecal pellets or marine snow particles. About 90% of all phytoplankton are eaten by zooplankton and encapsulated into large (100–300 μm), fast-sinking (50–500 m/day) fecal pellets. This result is particularly significant because the large, fast-sinking particles are less likely to be collected using conventional water samplers. As a result, different types of "sediment traps" have been developed in order to collect a representative sample of the vertical particle flux.

Sediment trap studies in the open ocean show that the flux of organic carbon at any depth is directly proportional to the rate of primary productivity in the surface water and inversely proportional to the depth of the water column (Suess, 1980; see Section 8.5.1):

$$C_{flux} = \frac{C_{prod}}{0.0238Z + 0.212}$$

where C_{prod} is the primary production (g/m^2 year) and C_{flux} is the flux at depth Z (m). The original data used to calibrate this equation are shown in Fig. 9-16 as a plot of the ratio of carbon flux/primary production *vs* water depth. As can be seen, about 10% of the primary production falls to a depth of 400 m, whereas only about 1% reaches 5000 m.

This general relationship has other implications and applications. If depth in Fig. 9-16 can be transformed into time, then the slope of the plot represents a rate constant for *in situ* organic carbon loss from the sinking particles. Assuming an average settling rate of 100 m/day, the previous equation becomes:

$$2.38t = \frac{C_{prod}}{C_{flux}} - 0.212$$

where $t = Z/(\mathrm{d}Z/\mathrm{d}t)$. Thus, the estimated time for degradation of 90% of the primary production from falling particles is 4.1 days and after this time the particles would be at about 400 m.

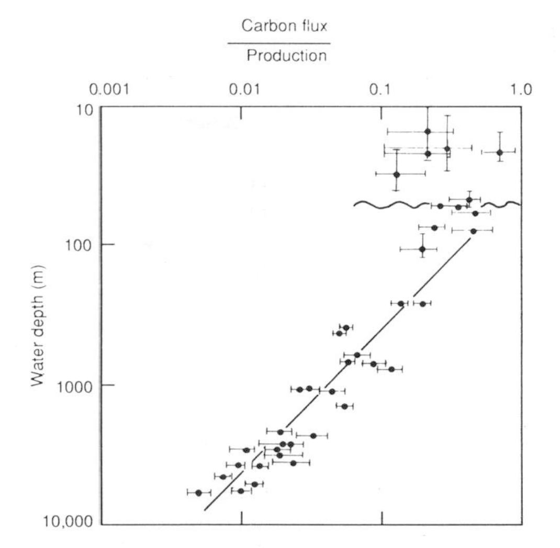

Fig. 9-16 Organic carbon fluxes with depth in the water column normalized to mean annual primary production rates at the sites of sediment trap deployment. The undulating line indicates the base of the euphotic zone; the horizontal error bars reflect variations in mean annual productivity as well as replicate flux measurements during the same season or over several seasons; vertical error bars are depth ranges of several sediment trap deployments and uncertainities in the exact depth location. Adapted from Suess (1980) with the permission of Macmillan Magazines, Ltd.

This discussion suggests a rapid and relatively direct transport of organic material vertically through the ocean water column. However, this transport is not efficient and under "average" ocean conditions (primary productivity = 100 g C/m^2 year and water depth of 4000 m) only 5% of the production can be expected to reach the ocean floor.

9.4 Chemistry of the Oceans

Chemical oceanography came to be identified as a discipline in its own right during the 1960s. The significance of chemical oceanography grew when it was realized that many of the stable and radioactive components of the ocean vary significantly in concentration and that knowledge of these variations could provide important information about natural processes. During the 1970s, the distributions of most elements and isotopes became fairly well

Table 9-10 Predicted mean oceanic concentrations

Atomic number	Element	Species	Predicted mean water concentration
1	Hydrogen	H_2O	
2	Helium	He (gas)	1.9 nmol/kg
3	Lithium	Li^+	178 μg/kg
4	Beryllium	$BeOH^+$	0.2 ng/kg
5	Boron	$B(OH)_3$	4.4 mg/kg
		$B(OH)_4^-$	
6	Carbon	ΣCO_2	2200 μmol/kg
7	Nitrogen	N_2^-, NO_2^-	590 μmol/kg
		NO_3^-, NH_4^+	30 μmol/kg
8	Oxygen	Dissolved O_2	150 μmol/kg
9	Fluorine	F^-, MgF^+	1.3 mg/kg
10	Neon	Ne (gas)	8 nmol/kg
11	Sodium	Na^+	10.781 g/kg
12	Magnesium	Mg^{2+}	1.28 g/kg
13	Aluminum	$Al(OH)_4^-$	1 μg/kg
14	Silicon	Silicate, $Si(OH)_4$	110 μmol/kg
15	Phosphorus	Reactive phosphate	2 μmol/kg
16	Sulfur	Sulfate SO_4^{2-}	2.712 g/kg
17	Chlorine	Chloride Cl^-	19.353 g/kg
18	Argon	Ar (gas)	15.6 μmol/kg
19	Potassium	K^+	399 mg/kg
20	Calcium	Ca^{2+}	415 mg/kg
21	Scandium	$Sc(OH)_3^0$	<1 ng/kg
22	Titanium	$Ti(OH)_4^0$	<1 ng/kg
28	Vanadium	$H_2VO_4^-$, HVO_4^{2-}	<1 μg/kg
24	Chromium	Cr(tot)	330 ng/kg
		$Cr(OH)_3(s)$	330 ng/kg
		CrO_4^{2-}	350 mg/kg
25	Manganese	Mn^{3+}, $MnCl^+$	10 ng/kg
26	Iron	$Fe(OH)_2^+$, $Fe(OH)_4^-$	40 ng/kg
27	Cobalt	Co^{2+}	2 ng/kg
28	Nickel	Ni^{2+}	480 ng/kg
29	Copper	$CuCO_3^0$, $CuOH^+$	120 ng/kg
30	Zinc	$ZnOH^+$, Zn^{2+}, ZnCO	390 ng/kg
31	Gallium	$Ga(OH)_4^-$	10–20 ng/kg
32	Germanium	$Ge(OH)_4^0$	5 ng/kg
33	Arsenic	$HAsO_4^{2-}$, $H_2AsO_4^-$	2 μg/kg
		Dimethylarsenate	
34	Selenium	[Se(tot)]	170 ng/kg
		SeO_3^{2-}	
		Se(IV)	—
		Se(VI)	—
35	Bromine	Br^-	67 mg/kg
37	Rubidium	Rb^+	124 μg/kg
38	Strontium	Sr^{2+}	7.8 mg/kg
			7.7 mg/kg
39	Yttrium	$Y(OH)_3^0$	13 ng/kg
40	Zirconium	$Zr(OH)_4^0$	<1 μg/kg
41	Niobium	—	1 ng/kg

Table 9-10 *Continued*

Atomic number	Element	Species	Predicted mean water concentration
42	Molybdenum	MoO_4^{2-}	11 μg/kg
44	Ruthenium	—	0.5 ng/kg
45	Rhodium	—	
46	Palladium	—	
47	Silver	$AgCl_2^-$	3 ng/kg
48	Cadmium	$CdCl_2^0$	70 ng/kg
49	Indium	$In(OH)_2^+$	0.2 ng/kg
50	Tin	$SnO(OH)_3^-$	0.5 ng/kg
51	Antimony	$Sb(OH)_6^-$	0.2 μg/kg
52	Tellurium	$HTeO_3^-$	
53	Iodine	IO_3^-, I^-	59 μg/kg (PO_4 corr.) 60 μg/kg (NO_3 corr.)
54	Xenon	Xe(gas)	0.5 nmol/kg
55	Cesium	Cs^+	0.3 ng/kg
56	Barium	Ba^{2+}	11.7 μg/kg
57	Lanthanum	$La(OH)_3^0$	4 ng/kg
58	Cerium	$Ce(OH)_3$	4 ng/kg
59	Praseodymium	$Pr(OH)_3$	0.6 ng/kg
60	Neodymium	$Nd(OH)_3$	4 ng/kg
61	Promethium	$Pm(OH)_3$	
62	Samarium	$Sm(OH)_3$	0.6 ng/kg
63	Europum	$Eu(OH)_3$	0.1 ng/kg
64	Gadolinium	$Gd(OH)_3$	0.8 ng/kg
65	Terbium	$Tb(OH)_3$	0.1 ng/kg
66	Dysprosium	$Dy(OH)_3$	1 ng/kg
67	Holmium	$Ho(OH)_3$	0.2 ng/kg
68	Erbium	$Er(OH)_3$	0.9 ng/kg
69	Thulium	$Tm(OH)_3$	0.2 ng/kg
70	Ytterbium	$Yb(OH)_3$	0.9 ng/kg
71	Lutetium	$Lu(OH)_3^0$	0.2 ng/kg
72	Hafnium	—	<8 ng/kg
73	Tantalum	—	<2.5 ng/kg
74	Tungsten	WO_4^{2-}	<1 ng/kg
75	Rhenium	ReO_4^-	4 ng/kg
76	Osmium	—	
77	Iridium	—	
78	Platinum	—	
79	Gold	$AuCl_2^-$	11 ng/kg
80	Mercury	$HgCl_4^{2-}$, $HgCl_3^-$	6 ng/kg
81	Thallium	Tl^+	12 ng/kg
82	Lead	$PbCO_3^0$	1 ng/kg
83	Bismuth	BiO^+, $Bi(OH)_2^+$	10 ng/kg
84	Polonium	PoO_3^{2-}, $PoO(OH)_2^0$	
86	Radon	Rn (gas)	
88	Radium	Ra^{2+}	
89	Actinium	—	
90	Thorium	$Th(OH)_4^0$	<0.7 ng/kg
91	Protactinium	—	
92	Uranium	$UO_2(CO_3)_2^{4-}$	3.2 μg/kg

understood. In the process, the sub-discipline of marine chemistry emerged. This field focuses on the chemical reactions and mechanisms in the ocean and at its boundaries. A summary of the observed concentrations and predicted mean oceanic concentrations are given in Table 9-10 (from Quinby-Hunt and Turekian, 1983).

9.4.1 Residence Time

As a starting point, we can view the ocean as one large reservoir to which materials are continuously added and removed (Fig. 9-17). The major sources of material include rivers and winds, which carry dissolved and particulate materials from the continents to the sea. The major removal process is the formation of marine sediments both by settling of particles through the water column as well as by precipitation of insoluble solid phases. For many elements, hydrothermal circulation through the ocean crust may be imporant.

The concept of average residence time, or turnover time, provides a simple macroscopic approach for relating the concentrations in ocean reservoirs and the fluxes between them. For the single box ocean in Fig. 9-17, the rate of change of the concentration of component n can be expressed as:

$$\left(\frac{dn}{dt}\right)_{\text{ocean}} = \sum_i \left(\frac{dn_i}{dt}\right)$$

If $(dn/dt) = 0$, we have a steady-state. This is also referred to as a state of dynamic equilibrium where the rate of input equals the rate of removal.

The concept of residence time was first introduced by Barth (1952) and given by the following expression (see also Chapter 4 and Li, 1977):

$$\tau_0 = \frac{\text{mass of element in the sea}}{\text{mass supplied (or removed) per year}}$$

$$= \frac{M}{S} = \frac{M}{Q}$$

where Q and S represent the mean total input and removal rates, respectively, and M represents the total mass of an element dissolved in the sea. For most elements, it appears that the removal rate is proportional to the total amount present or $S = kM$ where k is a first-order rate constant. At steady-state, $\tau_0 = 1/k$. This relation predicts an inverse correlation between residence time and the removal rate constant, which must be a measure of the chemical reactivity. The approach to equilibrium for this problem is discussed in Chapter 4 and by Lasaga (1980, 1981).

Chemically reactive elements should have a short residence time in seawater and a low concentration. A positive correlation exists between the mean ocean residence time and the mean oceanic concentration; however, the scatter is too great for the plot to be used for predictive purposes. Whitfield and Turner (1979) have shown that a more important correlation exists between residence time and a measure of the partitioning of the elements between the ocean and crustal rocks. The rationale behind this approach is that the oceanic concentrations have been roughly constant, while the elements in crustal rocks have cycled through the oceans. This partitioning of the elements may reflect the long-term chemical controls. The relationship can be summarized by an equation of the form:

$$\log \tau_0 = a \log K_D + b$$

where K_D is the ratio of the mean concentration in seawater/mean concentration in the crust. Appropriate values for 40 elements have been tabulated by Whitfield (1979).

Li (1981) has proposed that the distribution coefficients reflect adsorption–desorption reactions at the surface of mineral grains. To emphasize this point, Li plotted a slightly different distribution coefficient (log C_{op}/C_{sw}, where C_{op} and C_{sw} are the concentrations in oceanic pelagic clay sediments and seawater respectively *vs* the first hydrolysis constants of the metals or the dissociation constant of the oxyanion acids. The argument is that those elements that hydrolyze the strongest will adsorb the

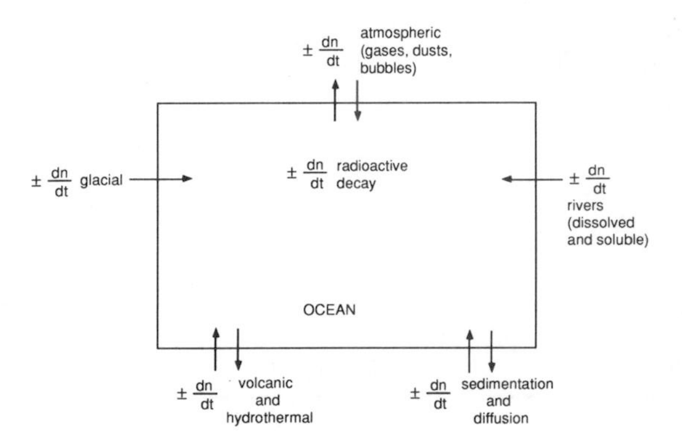

Fig. 9-17 A simplified box model of the ocean.

strongest and thus have a larger preference for the solid phase as represented by pelagic clay. For oxy-anions, the larger the acidity constant the weaker the adsorption on solid phases.

Thus, the chemical reactivity of the elements in seawater is reflected by the residence time. It is important to note, however, that while residence times tell us something about the relative reactivities, they also tell us something about the nature of the reactions. The best source of clues for understanding these reactions is to study the shape of the dissolved profiles of the different elements. When we do this we find that there are six main characteristic types of profiles as shown in Table 9-11. Notice that most of these reactions occur at the phase discontinuities between the atmosphere, biosphere, hydrosphere, and lithosphere.

9.4.2 Composition of Seawater

9.4.2.1 Major ions

The salinity of seawater is defined as grams of dissolved salt per kg of seawater. Using good technique, salinity can be reported to 0.001‰ or 1 ppm(m). By tradition, the major ions have been defined as those that make a significant contribution to the salinity. Thus, major ions are those with concentrations greater than 1 mg/kg or 1 ppm(m). By this definition, there are 11 major ions in seawater (Table 9-12).

The elements Na, K, Cl, S, Br, B, and F are the most conservative major elements. No significant variations in the ratios of these elements to chlorine have been demonstrated. Strontium has a small (<0.5%) depletion in the photic zone (Brass and Turekian, 1974), possibly due to the plankton *Acantharia* which makes its shell from $SrSO_4$ (celestite). Calcium has been known since the nineteenth century to be about 0.5% enriched in the deep sea relative to surface waters. Alkalinity (HCO_3^-) also shows a deep enrichment. These elements are controlled by the formation and dissolution of $CaCO_3$ and are linked by the following reaction:

$$CaCO_3 + CO_2 + H_2O \rightarrow Ca^{2+} + 2HCO_3^-$$

9.4.2.2 Minor elements

By definition, a minor element in seawater is one that has a concentration of less than 1 ppm(m). It is

Table 9-11 Characteristic types of profiles of elements in the ocean with example elements and probable controlling mechanisms

Type of profile	Example elements	Mechanisms
1. Similar to salinity	Na, K, Mg, SO_4, F, Br	Conservative elements of very low reactivity
2. Sea surface enrichment		Atmospheric input
	^{210}Pb, Mn	– natural
	^{90}Sr, Pb	– pollution, bomb tests
	NO	– photochemistry
	$As(CH_3)_2$, H_2, NO_2	Biological production
3. Photic zone depletion with deep ocean enrichment	Ca, Si, ΣCO_2, NO_3, PO_4, Cu, Ni	Biological uptake and regeneration
4. Mid-water maxima		
3000 m	^3He, Mn, CH_4, ^{222}Rn	Hydrothermal input
200–1000 m	^3H	Isopycnal transport
	Mn, NO_2	Redox chemistry in oxygen minimum
5. Bottom water enrichment	^{222}Rn, ^{228}Ra, Mn, Si	Flux out of the sediments
6. Deep ocean depletion	^{210}Pb, ^{230}Th, Cu	Scavenging by settling particles

Table 9-12 The major ions of seawater. The concentration at 35‰, the ratio to chlorinity, and the molar concentration[a]

Ion	g/kg at $S = 35‰$	g/kg (chlorinity ‰)	M (mol/L)
Cl^-	19.354	0.9989	5.46×10^{-1}
SO_4^{2-}	2.712	0.14	2.82×10^{-2}
Br^-	0.0673	0.00347	8.4×10^{-4}
F^-	0.0013	0.000067	6.8×10^{-5}
B	0.0045	0.000232	4.1×10^{-4}
Na^+	10.77	0.556	4.68×10^{-1}
Mg^{2+}	1.29	0.0665[b]	5.32×10^{-3}
Ca^{2+}	0.4121	0.02127	1.02×10^{-2}
K^+	0.399	0.0206	1.02×10^{-2}
Sr^{2+}	0.0079	0.00041	9.1×10^{-5}
HCO_3^-	0.142	—	2.387×10^{-3}

[a] From Wilson (1975).
[b] Recent reported values lie between 0.06612 and 0.06692.

experimentally challenging to determine the overall concentrations, much less their major chemical forms. Because early data (prior to about 1975) were so erratic, the principle of oceanographic consistency was proposed as a test for the data (Boyle and Edmond, 1975). According to this principle, the analyses of minor elements should:

1. Form smooth vertical profiles.
2. Have correlations with other elements that share the same controlling mechanism.

The concentrations of the minor elements are given in Table 9-10. There is at least one oceanographically consistent profile for most of the elements (refer to Quinby-Hunt and Turekian, 1983), for references and representative profiles for individual elements). The main mechanisms that control the distribution of minor elements are given in Table 9-11.

The chemical reactivity of minor elements in seawater is strongly influenced by their speciation (see Stumm and Brauner, 1975). For example, Cu^{2+} ion is toxic to phytoplankton (Sunda and Guillard, 1976), Uranium(VI) forms the soluble carbonate complex, $UO_2(CO_3)_3^{4-}$, and as a result uranium behaves like an unreactive conservative element in seawater (Ku *et al.*, 1977).

Although the speciation of some minor elements has been determined directly by experimental means (e.g. ion selective electrodes, polarography, electron spin resonance), most of our thinking about speciation is based on equilibrium calculations. Garrels and Thompson (1962) conducted speciation calculations for the major elements of seawater. They showed that the major cations (Na, K, Ca, Mg) and Cl are mostly (>90%) uncomplexed in seawater. The anions SO_4^{2-}, CO_3^{2-}, and HCO_3^- are tied up as complexes to a significant extent. When similar calculations are done for the minor elements in seawater, we find a different story. Most of the minor elements exist as complex ions or ion pairs. In particular, the metals form complexes with anions (ligands) such as CO_3^{2-}, Cl^-, and especially OH^-. The best estimates of the speciation of the elements in seawater are given in Table 9-10.

Stumm and Brauner (1975) conducted a Garrels and Thompson type calculation for some major and minor elements in seawater. These results are shown in Table 9-13. There are actually two calculations shown: first, there is an inorganic seawater model; then, the calculations are repeated using organic compounds with functional groups similar to those found in nature. The metals tested are listed in the left-hand column followed by a column of their concentration in seawater (as best known at that time). The inorganic and organic ligands tested are listed across the top row. Of the metals studied, only the major ions (Ca, Mg, Na, K) and Ni and Co occur mostly as the free metal ion. Complex species predominate for all other metals. When organic ligands are added to the model, only the speciation of Cu is seriously affected. Direct measurements by differential pulse anodic stripping voltammetry have shown that more than 99.7% of the total dissolved copper in surface seawater is associated with organic complexes (Coale and Bruland, 1988).

9.4.2.3 *Dissolved gases*

All deep waters of the ocean were once in contact with the atmosphere. Since over 95% of the total of all gases (except radon) reside in the atmosphere, the atmosphere dictates the ocean's gas contents. As discussed in Chapter 10, the composition of the atmosphere is nearly constant horizontally.

The solubility of many gases depends mainly on molecular weight. The heavier the molecule, the greater the solubility. Thus He is less soluble than Xe. The gases CO_2 and N_2O are exceptions to this general trend because they interact more strongly with water. Solubility also increases with decreasing temperature. Thus, high latitude surface seawater has higher gas concentrations than seawater at mid- or low latitudes.

Table 9-13 Equilibrium model: Effect of complex formation on distribution of metals [all concentrations are given as $-\log(M)$]. pH = 8.0, T = 25 °C. Ligands: pSO_4, 1.95; $pHCO_3$, 2.76; pCO_3, 4.86; pCl, 0.25. Ion changes not indicated.

A. Inorganic seawater

M	M_T[c]	Free M	Major species
Ca	1.97	2.03	$CaSO_4$, 2.94; $CaCO_3$, 3.5
Mg	1.26	1.31	$MgSO_4$, 2.25; $MgCO_3$, 3.3
Na	0.32	0.33	$NaSO_4$, 1.97; $NaHCO_3$, 3.3
K	1.97	1.98	KSO_4, 3.93
Fe(III)	8.0	18.90	$Fe(OH)_2$, 8.3; $FeSO_4$, 18.5
Mn(II)	7.5	8.1	MnCl, 7.8;[e] $MnCl_2$, 8.3[e]
Cu(II)	7.7	9.2	$CuCO_3$, 7.7; $Cu(CO_3)_2$, 9.1
Cd	8.5	10.9	$CdCl_2$, 8.7; CdCl, 9.2
Ni	7.7	7.9	NiCl, 8.3; $NiSO_4$, 8.7
Pb	8.2	9.9	$PbCO_3$, 8.6; PbOH, 8.7
Co(II)	8.3	8.5	CoCl, 9.0; $CoSO_4$, 9.1
Ag	8.7	13.1	$AgCl_2$, 8.7; AgCl, 10.0
Zn	7.2	7.8	ZnOH, 7.4; ZnCl, 8.0

B. Inorganic seawater plus soluble organic matter (2.3 mg C/L)[a]

Total concentration same as above			Organic complexes with ligands[b] and free ligand concentration					
M	Free M	Major inorganic species	Acet. (5.21)	Citr. (4.7)	Tartr. (5.41)	Glyc. (6.96)	Glut. (6.89)	Phthal. (5.20)
Ca	2.03	$CaSO_4$, 2.94; $CaCO_3$, 3.50	7.41	5.9	6.41	9.06	8.19	6.28
Mg	1.31	$MgSO_4$, 2.25; $MgCO_3$, 3.3	6.06	5.25	5.56	7.31	6.34	—[d]
Na	0.33	$NaSO_4$, 1.97; $NaHCO_3$, 3.3	—	—	—	—	—	—
K	1.98	KSO_4, 3.93	—	—	—	—	—	—
Fe(III)	18.9	$Fe(OH)_2$, 8.3; $FeSO_4$, 18.5	20.7	8.6	—	15.9	13.7	—
Mn(II)	8.1	MnCl, 7.8; $MnCl_2$, 8.3	12.8	11.4	—	13.1	12.2	—
Cu(II)	10.8	$CuCO_3$, 9.4; $Cu(CO_3)_2$, 10.5	14.3	7.7	16.7	9.6	10.6	13.0
Cd	10.9	$CdCl_2$, 8.7; CdCl, 8.7	15.1	13.1	13.5	13.5	13.4	13.6
Ni	8.0	NiCl, 8.5; $NiSO_4$, 8.8	12.5	8.4	—	9.2	9.4	11.1
Pb	9.9	$PbCO_3$, 8.6; PbOH, 8.7	13.2	11.34	11.5	11.8	—	11.7
Co(II)	8.5	CoCl, 9.0; $CoSO_4$, 9.1	12.7	26.5	11.9	10.8	10.8	14.9
Ag	13.1	$AgCl_2$, 8.7; AgCl, 10.0	17.9	26.5	—	16.7	—	—
Zn	7.8	ZnOH, 7.4; ZnCl, 8.0	11.7	11.3	10.9	8.8	9.7	10.9
		%[f]	13.0	98.6	44.9	0.7	6.6	7.5

[a] Organic matter of approximate composition $C_{13}H_{17}O_{12}N$ consists of a mixture of acetate, citrate, tartrate, glycine, glutamic acid, and phthalate, each present at 7×10^{-6} M (11 millimoles donor groups per litre).
[b] The concentrations given refer to the sum of all complexes, e.g. CuCit, CuHCit, $CuCit_2$.
[c] Total concentration of metal species: note that Fe(III) is slightly oversaturated with respect to $Fe(OH)_3(s)$; Cu(II) is over-saturated with respect to malachite but because formation is slow, precipitation of the solid has not been allowed. All other metals are thermodynamically soluble at the concentrations specified.
[d] No stability constants for such complexes are available.
[e] There is some uncertainty regarding the validity of the stability constants of chloro complexes of Mn^{2+}; according to other computations, Mn^{2+} is a major inorganic species.
[f] Percentage of total ligand bound to metal ions.

Table 9-14 Solubilities of various gases in surface ocean water

Gas	Partial pressure in dry air (atm)	Equilibrium concentration in surface seawater (cm^3/L)	
		0 °C	24 °C
H_2	5×10^{-7}		
He	5.2×10^{-6}	4.1×10^{-5}	3.4×10^{-5}
Ne	1.8×10^{-5}	1.7×10^{-4}	1.5×10^{-4}
N_2	0.781	14	9
O_2	0.209	8.8	5.5
Ar	9.3×10^{-3}	0.36	0.22
CO_2	3.2×10^{-4}	0.47	0.23
Kr	1.1×10^{-6}	8.1×10^{-5}	4.9×10^{-5}
Xe	8.6×10^{-8}	1.2×10^{-5}	0.6×10^{-5}

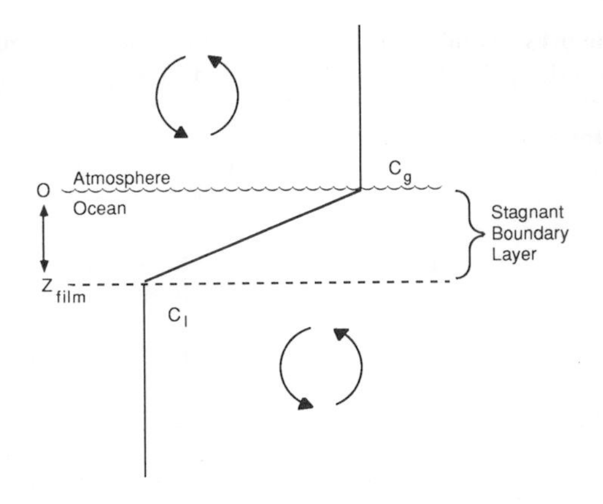

Fig. 9-18 A schematic of a stagnant boundary layer gas exchange model. C_g, gas concentration at the liquid side of the interface; C_l, gas concentration at the base of the stagnant boundary layer; Z_{film}, stagnant boundary layer thickness.

The equilibrium concentration in seawater is described by Henry's Law, which relates the partial pressure of the gas to its concentration (see Chapter 5 and Waser, 1966). Using the appropriate values of Henry's Law constant, K_H, and the partial pressures of gases in the atmosphere, the equilibrium concentrations of several gases are given in Table 9-14 for 0° C and 24 °C.

In the ocean, inert gas concentrations tend to follow the temperature solubility dependence closely. This suggests that water parcels obtain their gas signatures when they are at the sea surface close to equilibrium with the atmosphere at ambient temperature.

The process of equilibration of the atmosphere with the ocean is called gas exchange. Several models are available; however, the simplest model for most practical problems is the one-layer stagnant boundary layer model (Fig. 9-18). This model assumes that a well-mixed atmosphere and a well-mixed surface ocean are separated by a film on the liquid side of the air–water interface through which gas transport is controlled by molecular diffusion. [A similar layer exists on the air side of the interface that can be neglected for most gases. SO_2 is a notable exception (Liss and Slater, 1974).]

If transport across this film is controlled by diffusion, then from Fick's Second Law of Diffusion:

$$\frac{\partial C}{\partial t} = D\left(\frac{\partial^2 C}{\partial x^2}\right)$$

The boundary conditions are $C = C_g$ at $x = 0$ and $C = C_l$ at $x = z$. The steady-state concentration profile across the boundary layer is given by:

$$C = \frac{C_l \text{Å} C_g}{z} x + C_g$$

The steady-state flux from the atmosphere to the ocean across the layer is given by Fick's First Law:

$$F = -D\left(\frac{\partial C}{\partial x}\right) = D\,\frac{C_g - C_l}{z}$$

This treatment may be compared with that given in Chapter 4. The top of the stagnant film is assumed to have a gas concentration in equilibrium with the overlying air (i.e. $C_g = K_H P_g$). The unknown values are the flux and the thickness of the diffusive layer z.

The thickness z has been determined by analyses of isotopes (^{14}C and ^{222}Rn) that can be used to obtain the flux (Broecker and Peng, 1974; Peng *et al.*, 1979). The thickness averaged over the entire ocean has been estimated from a ^{14}C balance to be 17 μm.

Since the units of D/z are the same as velocity, we can think of this ratio as the velocity of two imaginary pistons: one moving up through the water pushing ahead of it a column of gas with the concentration of the gas in surface water (C_l), and one moving down into the sea carrying a column of gas with the concentration of the gas in the upper few molecular layers (C_g). For a hypothetical example with a film thickness of 17 μm and a diffusion coefficient of 1×10^{-5} cm^2/s, the piston velocity is 5 m/day. Thus in each day, a column of seawater 5 m thick will exchange its gas with the atmosphere.

Example: Obtain a relationship for the residence time of gases in the atmosphere with respect to gas exchange:

$$\tau_{atm} = \frac{\text{mass in atmosphere}}{\text{flux into the ocean}} = \frac{M_{atm}}{F}$$
$$= \frac{P \cdot 3 \times 10^5 \,(\text{atm})\,(\text{mol/m}^2\,\text{atm})}{PK_H(D/z)\,(\text{atm})\,(\text{mol/m}^3\,\text{atm})\,(\text{m/year})}$$
$$= \frac{3 \times 10^5}{K_H(D/z)}$$

If we assume that $Z_{film} = 17\,\mu m$ and $D = 1 \times 10^{-5}\,cm^2/s$, then the piston velocity is about 1800 m/year. The solubility of N_2, O_2, CO, H_2, NO, Ar, CH_4, etc., is about 1 mol/m^3 atm, while it is about 30 mol/m^3 atm for CO_2. Thus the residence time for most inert gases is about 160 years. For CO_2 it is about 5.4 years or a factor of 30 faster than for the other gases.

9.4.2.4 Nutrients

Oceanic surface waters are efficiently stripped of nutrients by phytoplankton. If phytoplankton biomass was not reconverted into simple dissolved nutrients, the entire marine water column would be depleted in nutrients and growth would stop. But as we saw from the carbon balance presented earlier, more than 90% of the primary productivity is released back to the water column as a reverse RKR equation. This reverse reaction is called remineralization and is due to respiration. An important point is that while production via photosynthesis can only occur in surface waters, the remineralization by heterotrophic organisms can occur over the entire water column and in the underlying sediments.

It follows that deep seawater contains nutrients from two sources. First, it may contain nutrients that were present with the water when it sank from the surface. These are called "preformed nutrients". Second, it may contain nutrients derived by the *in situ* remineralization of organic particles. These are called oxidative nutrients. The oxidative nutrients can be estimated from the RKR equation. From this model, we might expect the four dissolved chemical species (O_2, CO_2, NO_3, PO_4) to vary in seawater according to the proportions predicted. The key to understanding these remineralization reactions is the parameter Apparent Oxygen Utilization (AOU), defined as:

$$AOU = O_2' - O_2$$

where O_2' is the saturation value at the salinity and potential temperature of the sample. O_2 is the measured O_2 at the time of sampling. From the respiration form of the RKR equation, it follows that for every 138 moles of O_2 consumed *in situ*, one gets 106 moles of CO_2, 16 moles of NH_3, and 1 mole of H_3PO_4. As can be seen from Fig. 9-19, the slopes of the regression lines of AOU *vs* phosphate and nitrate closely correspond to the complete remineralization of average plankton.

The preformed nutrients are obtained by subtracting the oxidized nutrient from total nutrient (for N and P):

$$P_{total} = P_{preformed} + P_{oxidized} \qquad P_{oxidized} = {}^{1}\!/_{138} AOU$$
$$N_{total} = N_{preformed} + N_{oxidized} \qquad N_{oxidized} = {}^{16}\!/_{138} AOU$$

For silica, an additional component of silicic acid generated by inorganic dissolution also occurs. One can estimate "oxidized" Si as being 23/138 AOU, but discrimination between preformed Si and inorganic Si is not possible.

Representative profiles of PO_4, NO_3, and Si are given in Fig. 9-19. Note that after subtracting the

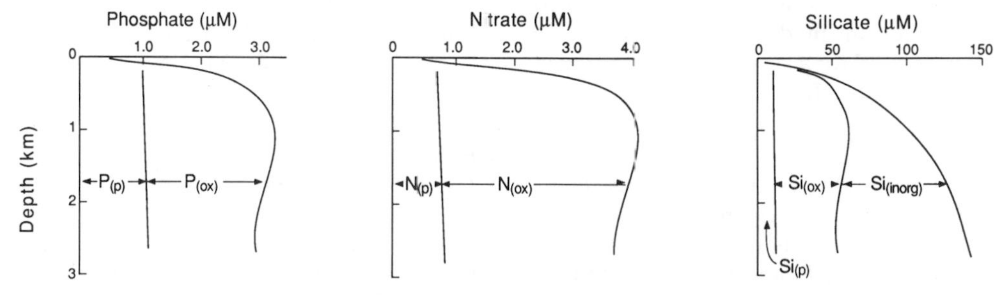

Fig. 9-19 Representative vertical distributions of various nutrients and their components. See text for explanation of designations.

oxidative nutrients the preformed values are relatively constant.

9.4.3 Equilibrium Models of Seawater

Now that we have reviewed some basic aspects of the chemical composition of the ocean, we can turn to a more fundamental question. What processes determine the composition of the ocean? Current evidence suggests that rivers are the most important contributors of dissolved substances to the ocean. Since there is geological evidence that the concentration and composition of the ocean has been relatively constant over the last ~1.5 Ga, we must conclude that river input must be balanced by removal.

9.4.3.1 Sillén's model

Sillén was a Swedish inorganic chemist who specialized in solution chemistry. In 1959, Sillén was asked as "an outsider" to give a lecture on the physical chemistry of seawater to the International Oceanographic Congress in New York (Sillén, 1961). Sillén proposed that the ionic composition of seawater might be controlled by equilibrium reactions between the dissolved ions and various minerals occurring in marine sediments. Goldschmidt (1937) had earlier proposed a schematic reaction for the geochemical balance.

$$\text{igneous rock (0.6 kg)} + \text{volatiles (1.0 kg)} \rightarrow \text{seawater (1 L)} + \text{sediments (0.6 kg)} + \text{air (3 L)}$$

This is a weathering reaction. Sillén argued that Goldschmidt's reaction could also go in the other direction. The reverse reaction would be called reverse weathering.

The framework for constructing such multicomponent equilibrium models is the Gibbs Phase Rule. This rule is valid for a system that has reached equilibrium and it states that:

$$f = c + 2 - p$$

where the number of degrees of freedom is represented by f. These are chosen from the list of all quantitatively related aspects of a system that can change. This includes T, P, and the concentrations of c components in each phase. c is the minimum number of components necessary to reproduce the system (ingredients), and p is the number of phases present at equilibrium. A phase is a domain with uniform composition and properties. Examples are a gas, a liquid solution, a solid solution, and solid phases.

In a mathematical sense, f represents the difference between the number of independent variables (including T and P) and the number of constraints (equations). If the number of equations equals the number of unknown variables, we can solve for all the concentrations using equilibrium equations. For more discussion of the phase rule, see Stumm and Morgan (1981). Sillén's approach was to mix components, pick a reasonable set of phases that might be present, and then see how many degrees of freedom there are to be fixed.

Sillén constructed his models in a stepwise fashion starting with a simplified ocean model of five components [HCl, H_2O, KOH, $Al(OH)_3$, and SiO_2] and five phases (gas, liquid, quartz, kaolinite, and potassium mica) (Sillén, 1967). His complete (almost) seawater model was composed of nine components: HCl, H_2O, and CO_2 are acids that correspond to the volatiles from the Earth; KOH, CaO, SiO_2, $NaOH$, MgO, and $Al(OH)_3$ correspond to the bases of the rocks. If there was an equilibrium assemblage of nine phases, the system would have only two independent variables. Sillén argued that a plausible set could include a gas phase and a solution phase and the following seven solid phases:

Calcite $CaCO_3$

Quartz SiO_2

Kaolinite $Al_2Si_2O_5(OH)_4$

Illite $K_{0.59}(Al_{1.38}Fe_{0.73}Mg_{0.38})(Si_{3.41}Al_{0.59})O_{10}(OH)_2$

Chlorite $Mg_3(OH)_6Mg_3Si_4O_{10}(OH)_{10}$

Montmorillonite $Na_{0.33}Al_2(Si_{3.67}Al_{0.33})O_{10}(OH)_2$

Phillipsite (zeolite) $M_3Al_3Si_4O_{16}(H_2O)_6$ (where $M = Na + K + Ca + Mg$)

If these phases all exist at equilibrium, then $f = 2$. Sillén argued that we should fix T and $[Cl^-]$ (Cl^- does not enter any of the reactions and is thus conservative.) If so, the composition of the aqueous and gas phases would be fixed. The implications are far-reaching, because these equilibria would fix P_{CO_2} of the atmosphere and thus the pH of the ocean!

9.4.3.2 *Mackenzie and Garrells' chemical mass balance between rivers and oceans*

Evaporation of river water will not make seawater. Instead, evaporation of the nearly neutral $Na^+/Ca^{2+}/HCO_3^-$ river water produces a highly alkaline $Na^+/HCO^-/CO_3^{2-}$ water such as found in the evaporite lake beds of eastern California and Nevada (Garrels and Mackenzie, 1967). In addition, on comparing the amount of material supplied to the ocean with the amount in the ocean, it may be seen that most of the elements could have been replaced many times (Table 9-15). Thus some chemical reactions must be occurring in the ocean to consume the river flux.

Mackenzie and Garrels (1966) approached this problem by constructing a model based on a river balance. They first calculated the mass of ions added to the ocean by rivers over 10^8 years. This time period was chosen because geological evidence suggests that the chemical composition of seawater has remained constant over that period. They assumed that the river input is balanced only by sediment removal. The results of this balance are shown in Table 9-16.

In this balance, SO_4 is removed by $CaSO_4$ and FeS_2 in proportion to their abundance in the sedimentary record (50/50). Calcium is removed as $CaCO_3$ with enough Mg to correspond to the natural proportions. Chloride is removed as NaCl, and enough H_4SiO_4 is removed to make opal sediments. Some Na^+ is taken up and Ca^{2+} released during ion exchange reactions in estuaries. At this point, they still had to account for 15% of the initial Na, 90% of the Mg, 100% of the K, 90% of the SiO_2, and 43% of the HCO_3^-. To remove these excess ions, Mackenzie and Garrels proposed reverse weathering type reactions of the general type:

$$\text{X-ray amorphous clays} + H_2SiO_4 + \text{cations} + HCO_3^- \rightarrow$$
$$\text{cation-rich aluminosilicates} + CO_2 + H_2O$$

In their model, they used a kaolinite-like clay for the degraded silicate and allowed Na, Mg, and K to react to form sodic montmorillonite, chlorite, and illite respectively. The balance is essentially complete with only small residuals for H_4SiO_4 and HCO_3^-. The newly formed clays would comprise about 7% of the total mass of sediments.

This last requirement has been the greatest stumbling block for accepting the Sillén and Mackenzie and Garrels equilibrium models. Most marine clays appear to be detrital and derived from the continents by river or atmospheric transport. Authigenic phases (formed in place) are found in marine sediments; however, they are nowhere near abundant enough to satisfy the requirements of the river balance. For example, Kastner (1974) calculated that less than 1% of the Na and 2% of the K transported by rivers is taken up by authigenic feldspars.

Table 9-15 Number of times river constituents have passed through the ocean in 10^8 years assuming present annual worldwide river discharge mean dissolved constituent concentration of rivers and ocean, and ocean volume of 1.37×10^{21} L (all units $= 10^{18}$ kg)

Constituent	Amount in ocean	Amount delivered by rivers to ocean in 10^8 years	No. of times constituents have been "renewed" in 10^8 years
SiO_2	0.008	42.6	5300
HCO_3^-	0.19	190.2	1000
Ca^{2+}	0.6	48.8	81
K^+	0.5	7.4	15
SO_4^{2-}	3.7	36.7	10
Mg^{2+}	1.9	13.3	7
Na^+	14.4	20.7	1.4
Cl^-	26.1	25.4	1
H_2O	1370	3 333 000	2400

Table 9-16 Mass balance calculation for removal of river-derived constituents from the ocean (all units = 10^{21} mmol)

Reaction (balanced in terms of mmol of constituent)	Constituents								Products	% of total products formed
	SO_4^{2-}	Ca^{2+}	Cl^-	Na^+	Mg^{2+}	K^+	SiO_2	HCO_3^-		
To be removed from ocean in 10^8 years	382	1220	715	900	554	189	710	3118		
Reaction 1	191	1220	715	900	554	189	710	3500	96 pyrite 287 kaolinite	3% 8%
Reaction 2	0	1029	715	900	554	189	710	3500	191 "$CaSO_4$"	5%
Reaction 3	0	1029	715	900	502	189	710	3396	52 $MgCO_3$ in magnesium calcite	2%
Reaction 4	0	0	715	900	502	189	710	1338	1029 calcite or aragonite	29%
Reaction 5	0	0	0	185	502	189	710	1338	715 "NaCl"	20%
Reaction 6	0	0	0	185	502	189	639	1338	71 silica	2%
Reaction 7	0	24	0	139	502	189	639	1338	138 sodic montmorillonite	4%
Reaction 8	0	0	0	139	502	189	639	1290	24 calcite or aragonite	1%
Reaction 9	0	0	0	0	502	189	278	1151	417 sodic montmorillonite	12%
Reaction 10	0	0	0	0	0	189	218	147	100 chlorite	3%
Reaction 11	0	0	0	0	0	0	29	−42	378 illite	11%

1: $95.5FeAl_6Si_6O_{20}(OH)_4 + 191SO_4^{2-} + 47.8CO_2 + 55.7C_6H_{12}O_6 + 238.8H_2O \rightarrow 286.5Al_2Si_2O_5(OH)_4 + 95.5FeS_2 + 382HCO_3^-$.
2: $191Ca^{2+} + 191SO_4^{2-} \rightarrow 191CaSO_4$.
3: $52Mg^{2+} + 104HCO_3^- \rightarrow 52MgCO_3 + 52CO_2 + 52H_2O$.
4: $1029Ca^{2+} + 2058HCO_3^- \rightarrow 1029CaCO_3 + 1029CO_2 + 1029H_2O$.
5: $715Na^+ + 715Cl^- \rightarrow 715NaCl$.
6: $71H_4SiO_4 \rightarrow 71SiO_{2(s)} + 142H_2O$.
7: $138Ca_{0.17}Al_{2.33}Si_{3.67}O_{10}(OH)_2 + 46Na^+ \rightarrow 138Na_{0.33}Al_{2.33}Si_{3.67}O_{10}(OH)_2 + 23.5Ca^{2+}$.
8: $24Ca^{2+} + 48HCO_3^- \rightarrow 24CaCO_3 + 24CO_2 + 24H_2O$.
9: $486.5Al_2Si_{2.4}O_{5.8}(OH)_4 + 139Na^+ + 361.4SiO_2 + 139HCO_3^- \rightarrow 417Na_{0.33}Al_{2.33}Si_{3.67}O_{10}(OH)_2 + 139CO_2 + 625.5H_2O$.
10: $100.4Al_2Si_{2.4}O_{5.8}(OH)_4 + 502Mg^{2+} + 60.2SiO_2 + 1004HCO_3^- \rightarrow 100.4Mg_5Al_2Si_3O_{10}(OH)_8 + 1004CO_2 + 301.2H_2O$.
11: $472.5Al_2Si_{2.4}O_{5.8}(OH)_4 + 189K^+ + 189SiO_2 + 189HCO_3^- \rightarrow 378K_{0.5}Al_{2.5}Si_{3.5}O_{10}(OH)_2 + 189CO_2 + 661.5H_2O$.

So while the equilibrium approach is attractive and certainly tells us the directions reactions tend to go, the experimental and empirical vertification is lacking.

9.4.4 Kinetic Models of Seawater

The failure to identify the necessary authigenic silicate phases in sufficient quantities in marine sediments has led oceanographers to consider different approaches. The current models for seawater composition emphasize the dominant role played by the balance between the various inputs and outputs from the ocean. Mass balance calculations have become more important than solubility relationships in explaining oceanic chemistry. The difference between the equilibrium and mass balance points of view is not just a matter of mathematical and chemical formalism. In the equilibrium case, one would expect a very constant composition of the ocean and its sediments over geological time. In the other case, historical variations in the rates of input and removal should be reflected by changes in ocean composition and may be preserved in the sedimentary record. Models that emphasize the role of kinetic and material balance considerations are called kinetic models of seawater. This reasoning was pulled together by Broecker (1971) in a paper entitled "A kinetic model for the chemical composition of sea water".

9.4.4.1 *Fast processes: Internal cycling*

The most obvious effects of nonequilibrium in the ocean are large variations in present day composition. These arise mainly as the result of two processes:

1. *Biological cycling*. Organisms in surface seawater take up dissolved species during their growth. The remains of these organisms sink under the influence of gravity and gradually decompose by oxidation during respiration (release of C, N, P) and corrosion of hard parts (release of Ca, C, Si, trace elements like Ba, Cd, Zn, Cr, Ni, Se). This leads to a vertical segregation in the ocean of low concentrations in the surface water and higher concentrations at depth.
2. *Oceanic circulation*. The process of ocean circulation described earlier yields an ocean circulation pattern that results in progressively older deep water as the water passes, in sequence from the Atlantic, Indian, to the Pacific Ocean. Surface water returns relatively quickly to the place of origin for the deep water.

The superposition of the biological cycle on the ocean circulation pattern leads to three general features in the distribution of the elements involved in this cycle:

1. The warm surface ocean tends to have a constant composition.
2. The surface ocean is depleted relative to the deep ocean in those elements fixed by organisms.
3. Deep ocean concentrations increase progressively as the abyssal water flows (ages) from the North Atlantic, through the Indian Ocean to the North Pacific.

The characteristics are demonstrated in Fig. 9-20. As a result, the elements influenced in this manner show various degrees of correlation. For element X, in terms of phosphorus:

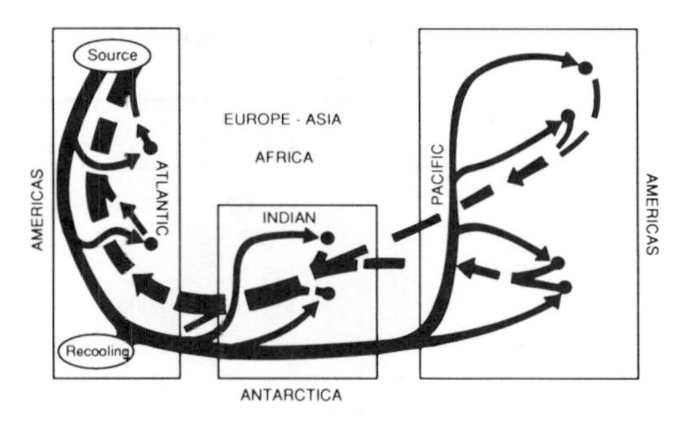

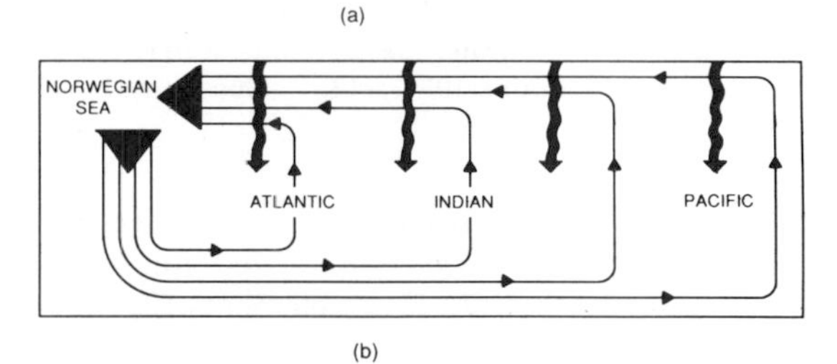

Fig. 9.20 (a) An idealized map of the patterns of deep water flow (solid lines) and surface water flow (dashed lines). The large circles designate the sinking of North Atlantic Deep Water (NADW) in the Norwegian Sea and the recooling of water along the perimeter of the Antarctic Continent; the dark circles indicate the distributed upwelling which balances this deep water generation. (b) An idealized vertical section running from the North Atlantic to the North Pacific showing the major advective flow pattern (thin lines) and the rain of particles (thick wavy lines). The combination of these two cycles leads to the observed distribution of nutrients. Adapted from Broecker and Peng (1982) with the permission of Eldigio Press.

$$[X] = a[P] + b$$

The moles X/moles P in average plankton is given by a, and b is the surface water concentration in phosphorus free water (water stripped of nutrients). In the case of P itself, the surface ocean concentration is close to zero, while the deep Pacific has a concentration of 2.5 μM. For N, the N/P ratio of plankton is 16 and the surface water concentration is 0 μM. The predicted deep sea nitrate is 40 μM. The ratio of (deep)/(surface) is greater than 10. For calcium, the Ca/P of plankton = 36 and the surface water content is 10 000 μM. The predicted deep ocean concentration

Table 9-17 Concentration distributions in the sea for elements used in significant amounts by marine organisms

Element	[S]	[DA]	[DP]	DP/S	(DP-S)/(DA-S)	Ref.
P[a]	<0.02	0.17	0.25	>10	1.7	c
N[a]	<0.2	2.1	3.3	>10	1.7	c
C[a]	205	227	248	1.25	1.9	c
Ca[a]	1000	1004	1009	1.01	~2	c
Si[b]	<100	1000	5000	>50	5	d
Ba[b]	9	12	27	3	6	d

Abbreviations: [S], [DA], and [DP] represent the concentrations of the given element in warm surface water, deep Atlantic water, and deep Pacific water respectively.
[a] Amount: 10^{-5} mol/L; [b] values in μg/L; [c] Li *et al.* (1969); [d] Wolgemuth and Broecker (1970).

is 10 090 resulting in a deep ocean enrichment ratio of 1.01. These features are summarized for several elements in Table 9-17. Three elements (P, N, Si) show nearly complete depletion in surface water, reflecting the fact that Si limits diatom growth while P and N limit the remaining organisms. Si shows the largest enrichment in the deep Pacific relative to the deep Atlantic suggesting that biological filtering is more efficient for Si than P. Hard parts of organisms undergo destruction at greater average depths than do the soft parts. Three other elements (C, Ba, Ca) show a similar deep Pacific to surface distribution but of smaller magnitude (less than 10). Note that the present cycle requires less than 100% efficient surface removal, otherwise all nutrients would be in the Pacific after a one-way trip.

There are two important consequences of this superposition of biological cycling on the ocean circulation pattern that show up in the sediments.

1. It leads to lower diatom productivity in the Atlantic relative to the Pacific.
2. It leads to a tilting of the depth of $CaCO_3$ preservation in the sediments. The deep Pacific is more corrosive to $CaCO_3$ than the deep Atlantic (more CO_2 from respiration) and thus $CaCO_3$ is found in sediments 1500 m deeper in the Atlantic than in the Pacific.

9.4.4.2 Long-term processes: Control of composition

In the previous section we considered only internal cycling. The questions we want to turn to now are:

1. What controls surface water concentrations?
2. What controls the P content of deep water and thus the deep water content of other elements?

Broecker's (1971) approach was to form groups of elements that appear to be controlled by similar processes. We will follow that approach here while examining the important factors and time-scales. The groups presented will differ from Broecker's in that we will include new information on hydrothermal processes not available at the time Broecker wrote his paper (Edmond *et al.*, 1979; McDuff and Morel, 1980). The groups used are kept as close as possible to Broecker's original list.

9.4.4.2.1 Group Ia (e.g. Cl). Elements in this group have long oceanic residence times. These elements are soluble and not reactive. The original source was degassing of the Earth's interior, which is either very slow now or complete. The main property of this group is geological removal by formation of soluble salts in evaporite deposits.

The present sources to the ocean are the weathering of old evaporites (75% of river flux) and Cl^- carried by atmospherically cycled seasalts (25% of river flux). Loss from the ocean occurs via aerosols (about 25%) and formation of new evaporites. This last process is sporadic and tectonically controlled by the closing of marginal seas where evaporation is greater than precipitation. The oceanic residence time is so long for Cl^- (~100 Ma) that an imbalance between input and removal rates will have little influence on oceanic concentrations (over periods of less than tens of millions of years).

9.4.4.2.2 Group Ib (Mg, SO_4, probably K). The key property of this group is removal during seafloor hydrothermal circulation. This fits in with Broecker's original group I, tectonically controlled elements, but enlarged by two (Mg, K).

A simple model can be used to describe this control of the concentration. In this model, the input is from rivers and the output is uptake by reactions in the ocean crust under hydrothermal systems (an application of this model is given in Section 13.5). Thus:

$$V_{riv} C_{riv} = V_{hydro}(C_{sw} - C_{exit\ fluid})$$

where V_{riv} is the river volume per time and V_{hydro} the hydrothermal circulation rate; C_{riv} is the river concentration and C_{sw} is the normal seawater concentration. Hydrothermal vents have near-zero concentrations of Mg. Therefore:

$$C_{sw} = (V_{riv}/V_{hydro})\, C_{riv}$$

The present-day best estimates are that V_{riv}/V_{hydro} is about 300. As V_{hydro} increases, e.g. faster spreading of ocean crust at ridges, C_{sw} responds. The dominant control is tectonics.

9.4.4.2.3 Group II (Ca, Na). This group includes the remaining cations with relatively long residence times. One important constraint is the charge balance of seawater:

$$2[Ca^{2+}] + [Na^+] - [HCO_3^-] = [Cl^-] + 2[SO_4^{2-}] - 2[Mg^{2+}] - [K^+]$$

Tectonic processes control the terms on the right-hand side, as already discussed. This also defines the sum of terms on the left-hand side. The control of the relative proportions of elements on the left-hand side is uncertain. The most plausible controls based on present data are:

1. Ca/Na by ion exchange in estuaries.
2. Ca/HCO_3 by calcium carbonate equilibria and control of carbon.

9.4.4.2.4 Group III (Si, P, C, N, trace elements). The common property of these elements is that they are biologically reactive and are deposited as thermodynamically unstable debris. For each of these elements, kinetic controls can be hypothesized. A first step is to describe the present-day situation.

Let us define a two-box model for a steady-state ocean as shown in Fig. 9-21. The two well-mixed reservoirs correspond to the surface ocean and deep oceans. We assume that rivers are the only source

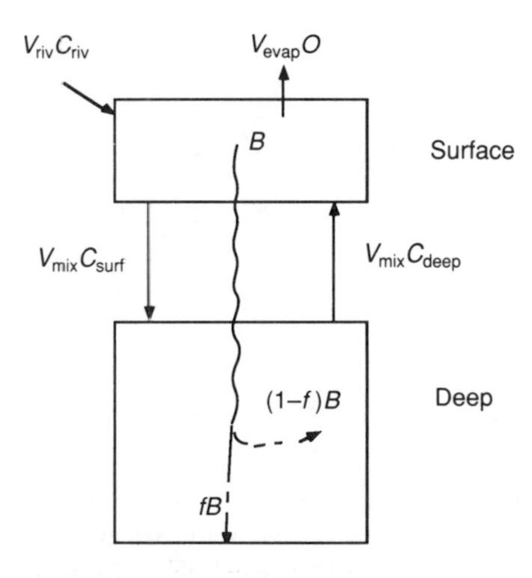

Fig. 9-21 Schematic of mass influx/outflux balance to surface/deep layers.

and sediments are the only sink. Elements are also removed from the surface box by biogenic particles (B). We also assume there is mixing between the two boxes that can be expressed as a velocity $V_{mix} = 2$ m/year and that rivers input water to the surface box at a rate of $V_{riv} = 0.1$ m/year. The resulting ratio of V_{mix}/V_{riv} is 20.

Water conservation can be expressed as:

$$V_{riv} = V_{evap}$$
$$V_{down} = V_{up} = V_{mix}$$

The mass balance for biogenic elements in the surface box is:

$$V_{riv} C_{riv} + V_{mix} C_{deep} = V_{mix} C_{surf} + B$$

For the lower box

$$V_{mix} C_{surf} + (1 - f) B = V_{mix} C_{deep}$$

where f is the fraction of the biogenic flux (B) buried and $(1 - f)$ is the fraction of the biogenic flux (B) regenerated in the deep box. For the entire ocean, the balance for biogenic elements is:

$$V_{riv} C_{riv} = fB$$

Now that we have defined the biogenic model, we can define two important properties. The first is the ratio of falling particles to the input to the surface. This property (g) is equivalent to the efficiency of bioremoval from surface waters:

$$g = \frac{B}{V_{riv}C_{riv} + V_{mix}C_{deep}}$$

$$= \frac{V_{riv}C_{riv} + V_{mix}C_{deep} - V_{mix}C_{surf}}{V_{riv}C_{riv} + V_{mix}C_{deep}}$$

$$= 1 - \frac{\left(\frac{V_{mix}}{V_{riv}}\right)\left(\frac{C_{surf}}{C_{riv}}\right)}{\left(\frac{V_{mix}}{V_{riv}}\right)\left(\frac{C_{deep}}{C_{riv}}\right) + 1}$$

As stated earlier, $V_{mix}/V_{riv} = 20$. The resulting values of g for several biogenic elements are given in Table 9-18. For Si, N, and P, at least 95% of the elements brought to the surface are removed in particulate biogenic form. Only 20% of the C is removed in this form.

The second property is the fraction of particles sinking that are preserved in the sediments:

$$f = fB/B = V_{riv}C_{riv}/B$$
$$= V_{riv}C_{riv}/(V_{riv}C_{riv} + V_{mix}C_{deep} + V_{mix}C_{surf})$$
$$= 1/(1 + (V_{mix}/V_{riv})(C_{deep}/C_{riv} - C_{surf}/C_{riv})$$

The values of f (Table 9-18) for N, P, C, and Si are 2% or less and are about 12% for Ba and Ca.

The product of $f \cdot g$ gives us the fraction of the elements that are permanently removed for each visit to the surface ocean. Conversely $1/f(f \cdot g)$ gives the number of times an element is recycled before it is permanently removed. For example, for a total ocean residence time of 1600 years, P goes through 105 cycles of 15 years each before being permanently removed.

We can end this discussion of the biogenic elements by considering the implications of this kind of cycle on the long-term stability of the concentration. Phosphorus will be used as an example. For elements like P that have a high value of g, the particulate flux responds directly to changes in the input. For P, the main input to the surface layer is upwelling and thus $B = V_{mix}P_{deep}$. The system is now operating at close to 100% efficiency. P removal to the sediments depends on the O_2 level in the deep ocean. When there is less O_2 there is less efficient regeneration of P and thus f increases. The O_2 content of the deep sea depends on the balance between the input from above ($V_{mix}O_{2surf}$) and O_2 consumption during regeneration (e.g. respiration) of the flux B. Suppose the rate of ocean circulation increased. In our simple model, we would parametrize this as saying V_{mix} increases. Because upwelling increases, productivity increases and the biogenic flux (B) increases. Now we have a situation where both the organic flux (B) and O_2 flux ($V_{mix}O_{2surf}$) to the deep ocean increase. The net effect is to stabilize the concentration of P in the deep sea. Suppose there was an increase in river input ($V_{riv}C_{riv}$) with no change in mixing. Productivity and particle flux (B) increase without a coincident change in the O_2 flux ($V_{mix}O_{2surf}$ = constant). The increased B will result in more O_2 consumption (and PO_4 release) and thus lower O_2 levels in the deep sea. As a result, there will be more efficient P removal to the sediments. Thus f and $f \cdot B$ increase, matching the increased river input. Again the net effect is to stabilize the deep P concentration. As f for P increases, the P content of the deep sea will tend to decrease, balancing the tendency to increase that which was initially caused by the increased flux of organic particles (B). The general result is that the P (or N, Si, C) content of the deep sea seeks that level where upwelling of deep water brings PO_4 to the surface at a rate such that P in organisms resistant to decomposition just balances the amount of new PO_4 entering the sea. This negative feedback model tends

Table 9-18 Parameters for element cycles within the sea

Element	[R]/[D]	[S]/[D]	g	f	$f \cdot g$
N	~0.20[a]	~0.05	0.95	0.01	0.01
P	~0.20[a]	~0.05	0.95	0.01	0.01
C	~0.10[b]	0.80	0.20	0.02	0.004
Si	~0.20	<0.05	~1.0	0.01	0.01
Ba	~2	0.30	0.75	0.12	0.09
Ca	0.04	0.99	0.01	0.12	0.001

[a] River value poorly known; [b] corrected for atmospheric recycling.

to drive the PO_4 content of the ocean toward a value where PO_4 loss equals PO_4 input.

Questions

9-1 There is some debate about what controls the magnesium concentration in seawater. The main input is rivers. The main removal is by hydrothermal processes (the concentration of Mg in hot vent solutions is essentially zero). First, calculate the residence time of water in the ocean due to (1) river input and (2) hydrothermal circulation. Second, calculate the residence time of magnesium in seawater with respect to these two processes. Third, draw a sketch to show this box model calculation schematically. You can assume that uncertainties in river input and hydrothermal circulation are 5% and 10%, respectively. What does this tell you about controls on the magnesium concentration? Do these calculations support the input/removal balance proposed above? Do any questions come to mind? Volume of ocean = 1.4×10^{21} L; river input = 3.2×10^{16} L/year; hydrothermal circulation = 1.0×10^{14} L/year; Mg concentration in river water = 1.7×10^{-4} M; Mg concentration in seawater = 0.053 M.

9-2 The flux of oxygen can be in or out of the ocean. The oxygen partial pressure in the atmosphere is 0.20 atm and Henry's Law constant is 1.26×10^{-3} M/atm. (a) As a result of photosynthesis, the nitrate concentration in seawater originally in equilibrium with the atmosphere has decreased by 20 μM. What is the new (non-equilibrium) oxygen concentration?
(b) What is the flux of oxygen due to gas exchange? Use a diffusion coefficient of 2.0×10^{-5} cm^2/s and a film thickness of 50 μm.

9-3 Hydrothermal vents have been sampled at 21° along the East Pacific Rise. The pure end member hydrothermal solutions have a temperature of 350° C and the following major ion composition (von Damm *et al.*, 1985). All concentrations are in mM and the pH is 3.4. Discuss the ways in which the vent solution differs from average seawater.

	Vent	Seawater
Na^+	432	468
K^+	23	10.2
Mg^{2+}	0	5.32
Ca^{2+}	15	10.2
Mn^{2+}	0.96	0.001
Fe^{2+}	1.7	10^{-6}
Cl^-	489	546
HCO_3^-	0	2
CO_3^{2-}	0	0.2
SO_4^{2-}	0.5	28
H_2S	7.3	

9-4 A proponent of "reverse weathering" suggested that gibbsite, kaolinite, and quartz exist in equilibrium according to the following equation:

$$Al_2Si_2O_5(OH)_4 \text{ (kaolinite)} + H_2O \rightarrow Al_2O_3(H_2O)_3 \text{ (gibbsite)} + 2SiO_2 \text{ (quartz)}$$

In equilibrium expressions for these reactions, water will appear as the *activity*, rather than concentration. The activity can be approximated by the mole fraction of water. What is the activity of water if this equilibrium is maintained? Could this equilibrium exist in seawater, where the mole fraction of water is about 0.98? ΔG^0 values (kJ/mol): gibbsite, 2320.4; kaolinite, 3700.7; quartz, 805.0; water, 228.4.

Answers can be found on p. 365.

References

Baes, C. F., A. Bjorkstrom and P. J. Mulholland (1985). Uptake of carbon dioxide by the oceans. *In* "Atmospheric Carbon Dioxide and the Global Carbon Cycle" (J. R. Trabalka, ed.) USDOE/ER-0239, pp. 81–112.

Barth, T. W. (1952). "Theoretical Petrology." John Wiley, New York.

Boyle, E. and J. M. Edmond (1975). Copper in surface waters south of New Zealand. *Nature* **253**, 107–109.

Brass, G. W. and K. K. Turekian (1974). Strontium distribution in GEOSECS oceanic profiles. *Earth Planet. Sci. Lett.* **23**, 141–148.

Brewer, P. G. (1978). Direct observation of the oceanic CO_2 increase. *Geophys. Res. Lett.* **5**, 997–1000.

Broecker, W. S. (1971). A kinetic model for the chemical composition of seawater. *Quart. Res.* **1**, 188–207.

Broecker, W. S. and T.-H. Peng (1974). Gas exchange rates between air and sea. *Tellus* **26**, 21–35.

Broecker, W. S. and T.-H. Peng (1982). "Tracers in the Sea." Eldigio Press, Palisades, N.Y.

Budyko, M. I. (1958). "The Heat Balance of the Earth's Surface" (translated by N. A. Stepanova). Office of Technical Services, Department of Commerce, Washington, D. C.

Coale, K. H. and K. W. Bruland (1988). Copper complexation in the Northeast Pacific. *Limnol. Oceanogr.* **33**, 1084–1101.

Dugdale, R. C. and J. J. Goering (1967). Uptake of new and regenerated forms of nitrogen in primary productivity. *Limnol. Oceanogr.* **12**, 196–206.

Edmond, J. M., C. Measures, R. E. McDuff, L. H. Chan, R. Collier, B. Grant, L. I. Gordon, and J. B. Corliss (1979). Ridge crest hydrothermal activity and the balance of the major and minor elements in the ocean: The Galapagos data. *Earth Planet. Sci. Lett.* **46**, 1–18.

Eppley, R. W. and B. J. Peterson (1979). Particulate organic matter flux and planktonic new production in the deep ocean. *Nature* **282**, 677–680.

Fine, R. A., H. Peterson, C. G. H. Rooth, and H. G. Ostlund (1983). Cross-equatorial tracer transport in the upper waters of the Pacific Ocean. *J. Geophys. Res.* **88**, 763–769.

Gammon, R. H., J. Cline, and D. Wisegarver (1982). Chlorofluoromethanes in the Northeast Pacific Ocean: Measured vertical distributions and application as transient tracers of upper ocean mixing. *J. Geophys. Res.* **87**, 9441–9454.

Garrels, R.M. and F. T. Mackenzie (1967). Origin of the chemical compositions of some springs and lakes. *In* "Equilibrium Concepts in Natural Water Systems" (W. Stumm, ed.), pp. 222–274. Advances in Chemistry Series No. 67. American Chemical Society, Washington, D.C.

Garrels, R.M. and M. E. Thompson (1962). A chemical model for seawater at 25° C and one atmosphere total pressure. *Amer. J. Sci.* 57–66.

Goldschmidt, V. M. (1937). The principles of distribution of chemical elements in minerals and rocks. *J. Chem. Soc.* **1937**, 655–674.

Gordon, A. L. (1975). General ocean circulation. *In* "Numerical Models of Ocean Circulation," pp. 39–53. National Academy of Science, Washington, D.C.

Jenkins, W. J. (1980). Tritium and ^{3}He in the Sargasso Sea. *J. Mar. Res.* **38**, 533–569.

Joyce, T. M., B. A. Warren, and L. D. Talley (1986). The geothermal heating of the abyssal subarctic Pacific Ocean. *Deep-Sea Res.* **33**, 1003–1015.

Kastner, M. (1974). The contribution of authigenic feldspars to the geochemical balance of alkalic metals. *Geochim. Cosmochim. Acta* **38**, 650–653.

Ku, T. L., K. G. Knauss, and G. G. Mathieu (1977). Uranium in open ocean: Concentration and isotopic concentration. *Deep-Sea Res.* **24**, 1005–1017.

Lasaga, A. C. (1980). The kinetic treatment of geochemical cycles. *Geochim. Cosmochim. Acta* **44**, 815–828.

Lasaga, A. C. (1981). Dynamic treatment of geochemical cycles: Global kinetics. *In* "Kinetics of Geochemical Processes" (A. C. Lasaga and R. J. Kirkpatrick, eds.), pp. 69–110. Mineralogical Society of America, Washington, D.C.

Leopold, L. B. (1974). "Water." W. H. Freeman, San Francisco.

Li, Y.-H. (1977). Confusion of the mathematical notation for defining the residence time. *Geochim. Cosmochim. Acta* **44**, 555–556.

Li, Y.-H. (1981). Ultimate removal mechanisms of elements from the ocean. *Geochim. Cosmochim. Acta* **45**, 1659–1664.

Li, T.-H., T. Takahashi and W. S. Broecker (1969). The degree of saturation of $CaCO_3$ in the oceans. *J. Geophys. Res.* **74**, 5507–5525.

Liss, P. S. and P. G. Slater (1974). Flux of gases across the air–sea interface. *Nature* **247**, 181–184.

McDuff, R. E. and F. M. M. Morel (1980). The geochemical control of seawater (Sillen revisited). *Environ. Sci. Technol.*, **14**, 1182–1186.

Mackenzie, F. T. and R. M. Garrels (1966). Chemical mass balance between rivers and oceans. *Amer. J. Sci.* **264**, 507–525.

Mantyla, A. W. (1975). On the potential temperature in the abyssal Pacific Ocean. *J. Mar. Res.* **33**, 341–354.

Mantyla, A. W. and J. L. Reid (1983). Abyssal characteristics of the world ocean waters. *Deep-Sea Res.* **30**, 805–833.

Martin, J. H. and S. E. Fitzwater (1988). Iron deficiency limits phytoplankton growth in the north-east Pacific subarctic. *Nature* **331**, 341–343.

Martin, J. H. and K. M. Gordon (1988). Northeast Pacific iron distributions in relation to phytoplankton productivity. *Deep-Sea Res.* **35**, 177–196.

Martin, J. H., R. M. Gordon, S. Fitzwater, and W. W. Broerkow (1989). VERTEX: Phytoplankton/iron studies in the Gulf of Alaska. *Deep-Sea Res.* **36**, 649–680.

Menzel, D. W. (1974). Primary productivity, dissolved and particulate organic matter and the sites of oxidation of organic matter. *In* "The Sea" (E. D. Goldberg, ed.), Vol. 5, pp. 659–678. John Wiley, New York.

Ostlund, H. G. and R. A. Fine (1979). Oceanic distribution and transport of tritium. *IAEA-SM-232/62*. pp. 303–314. International Atomic Energy Agency, Vienna.

Ostlund, H.G., H. G. Dorsey, and C. G. Rooth (1974). GEOSECS North Atlantic radiocarbon and tritium results. *Earth Planet. Sci. Lett.* **23**, 69–86.

Parsons, T. R., M. Takahashi, and B. Hargrave (1977). "Biological Oceanographic Processes," 2nd edn. Pergamon Press, New York.

Peng, T.-H., W. S. Broecker, G. G. Mathieu, and Y.-H. Li (1979). Radon evasion rates in the Atlantic and Pacific oceans as determined during the GEOSECS program. *J. Geophys. Res.* **84**, 2471–2486.

Perry, M. J. (1976). Phosphate utilization by an oceanic diatom in phosphorus-limited chemostat culture and in the oligotrophic waters of the central North Pacific. *Limnol. Oceanogr.* **21**, 88–107.

Pickard, G. L. (1963). "Descriptive Physical Oceanography." Pergamon Press, Oxford.

Pickard, G. L. and W. J. Emery (1982). "Descriptive Physical Oceanography," 4th edn. Pergamon Press, Oxford.

Quinby-Hunt, M. S. and K. K. Turekian (1983). Distribution of elements in sea water. *EOS* **64**, 130–131.

Redfield, A. C., B. H. Ketchum, and F. A. Richards (1963). The influence of organisms on the composition of sea-water. *In* "The Sea" (M. N. Hill, ed.), Vol. 2, pp. 26–77. Wiley-Interscience, New York.

Rooth, C. G. and H. G. Ostlund (1972). Penetration of tritium into the Atlantic thermocline. *Deep-Sea Res.* **19**, 481–492.

Ryther, J. H. (1969). Photosynthesis and fish production in the sea. *Science* **166**, 72–76.

Sarmiento, J. L., C. G. H. Rooth, and W. Roether (1982). The North Atlantic tritium distribution in 1972. *J. Geophys. Res.* **87**, 8047–8056.

Siegenthaler, U. (1983). Uptake of excess CO_2 by an outcrop-diffusion model of the ocean. *J. Geophys. Res.* **88**, 3599–3608.

Sillén, L. G. (1961). The physical chemistry of seawater. *In* "Oceanography" (M. Sears, ed.), pp. 549–581. American Association for the Advancement of Science, Washington, D.C.

Sillén, L. G. (1967). The ocean as a chemical system. *Science* **156**, 1189–1196.

Stommel, H. (1958). The abyssal circulation. *Deep-Sea Res.* **5**, 80–82.

Stommel, H. and A. Arons (1960). On the circulation of the world ocean, 1 and 2. *Deep-Sea Res.* **6**, 140–154, 217–233.

Stuiver, M., P. D. Quay, and H. G. Ostlund (1983). Abyssal water carbon-14 distribution and the age of the world oceans. *Science* **219**, 849–851.

Stumm, W. and P. A. Brauner (1975). Chemical specia-tion. *In* "Chemical Oceanography" (J. P. Riley and G. Skirrow, eds.), 2nd edn., Vol. 1, pp. 173–240. Academic Press, London.

Stumm, W. and J. J. Morgan (1981). "Aquatic Chemistry." John Wiley, New York.

Suess, E. (1980). Particulate organic carbon flux in the oceans: Surface productivity and oxygen utilization. *Nature* **288**, 260–263.

Sugimura, Y. and Y. Suzuki (1988). A high-temperature catalytic oxidation method for the determination of non-volatile dissolved organic carbon in seawater by direct injection of a liquid sample. *Mar. Chem.* **24**, 105–131.

Sunda, W. and R. R. L. Guillard (1976). The relationship between cupric ion activity and the toxicity of copper to phytoplankton. *J. Mar. Res.* **34**, 511–529.

Tai, C.-K. and C. Wunsch (1983). Absolute measurement by satellite altimetry of dynamic topography of the Pacific Ocean. *Nature* **301**, 408–410.

Takahashi, T., W. S. Broecker, and S. Langer (1985). Redfield ratio based on chemical data from osopycnal surfaces. *J. Geophys. Res.* **90**, 6907–6924.

UNESCO (1981). Background papers and supporting data on the Practical Salinity Scale 1978. UNESCO Technical Papers in Marine Science No. 37, UNESCO.

Von Damm, K. L. J. M. Edmond, B. Grant, C. I. Measures, B. Walden and R. F. Weiss (1985). Chemistry of submarine hydrothermal solutions at 21°N, East Pacific Rise. *Geochim. Cosmochim. Acta* **49**, 2197–2220.

Warren, B. A. (1983). Why is no deepwater found in the North Pacific? *J. Mar. Res.* **41**, 327–347.

Waser, J. (1966). "Basic Chemical Thermodynamics." W. A. Benjamin, New York.

Whitfield, M. (1979). The mean oceanic residence time (MORT) concept: A rationalization. *Mar. Chem.* **8**, 101–123.

Whitfield, M. and D. R. Turner (1979). Water-rock partition coefficients and the composition of seawater and rain water. *Nature* **278**, 132–137.

Williams, P. J. LeB. (1975). Biological and chemical aspects of dissolved organic matter in seawater. *In* "Chemical Oceanography" (J. P. Riley and G. Skirrow, eds.), Vol. 2, pp. 301–364, Academic Press, London.

Williams, P. M. (1971). The distribution and cycling of organic matter in the ocean. *In* "Organic Compounds in Aquatic Environments" (S. D. Faust and J. W. Hunter, eds.), pp. 143–163. Marcel-Dekker, New York.

Wilson, T. R. S. (1975). Salinity and the major elements of seawater. *In* "Chemical Oceanography" (J. P. Riley and G. Skirrow, eds.), 2nd edn., Vol. 1, pp. 365–414. Academic Press, London.

Wolgemuth, K. and W. S. Broecker (1970). Barium in seawater. *Earth Planet. Sci. Lett.* **8**, 372–378.

Worthington, L. V. (1970). The Norwegian Sea as a Mediterranean basin. *Deep-Sea Res.* **17**, 77–84.

Worthington, L. V. (1981). The water masses of the world ocean: Some results of fine-scale census. *In* "Evolution of Physical Oceanography" (B. A. Warren and C. Wunsch, eds.), pp. 42–69. MIT Press, Cambridge, Mass.

10

The Atmosphere

Robert J. Charlson

10.1 Definition

The atmosphere is a thin layer of gas uniformly covering the whole Earth. Its main constituents are nitrogen (N_2), oxygen (O_2), argon (Ar), water (H_2O gas, liquid, and solid), and carbon dioxide (CO_2). The origins of these main constituents are discussed in Chapter 2. This chapter will first concentrate on the physical properties of the atmosphere and the ways in which these influence chemical composition, particularly through diffusion and transport. Then the chemical processes of the atmosphere are discussed with emphasis on minor constituents. As will become evident, most of the chemical functioning of the atmosphere involves substances other than N_2, O_2, Ar, H_2O, and CO_2. Even though some of these are often products of, or participants in, reactions, their abundance is so great that their concentrations are not perturbed. Because of the particular importance of CO_2, it is considered in detail in Chapter 11.

Five chemical features of the atmosphere are emphasized.

1. *Altitude dependence.* The composition varies with altitude. Part of that vertical structure is due to the physical behavior of the atmosphere, whereas part is due to the influence of trace substances (notably ozone and condensed water) on thermal structure and mixing.
2. *Transport and diffusion.* With the exception of N_2, O_2, Ar, and numerous other long-lived species that are well-mixed in the bulk of the atmosphere, horizontal and vertical transport are closely coupled with chemical reactions in controlling atmospheric trace-substance concentrations.
3. *Composition.* Air is a mixture of a large number of species with concentrations varying in space and time. Of particular interest are ozone and compounds of sulfur, nitrogen, and carbon, and their chemical interactions.
4. *Role of composition in atmospheric physical processes.* The composition of the atmosphere plays a distinct set of roles in controlling and affecting certain physical processes of the atmosphere, most notably the thermal structure.
5. *Processes that occur at the upper and lower boundaries of the atmosphere.* Many atmospheric constituents are formed, and many undergo a wide range of reactions at the lower boundary. At the upper boundary, lighter elements are lost to space and some important substances are acquired.

Before setting out to discuss the vertical structure of the atmosphere, we note that it is useful to have access to conventional nomenclature. Figure 10-1, based on the thermal profile of the atmosphere, includes a number of commonly used definitions.

10.2 The Vertical Structure of the Atmosphere

10.2.1 Hydrostatic Equation

The atmosphere is very close to being in hydrostatic equilibrium in the vertical dimension. This can be described by the hydrostatic equation:

$$\frac{dP}{dz} = -\varrho g \qquad (1)$$

Global Biogeochemical Cycles
ISBN 0-12-147685-5

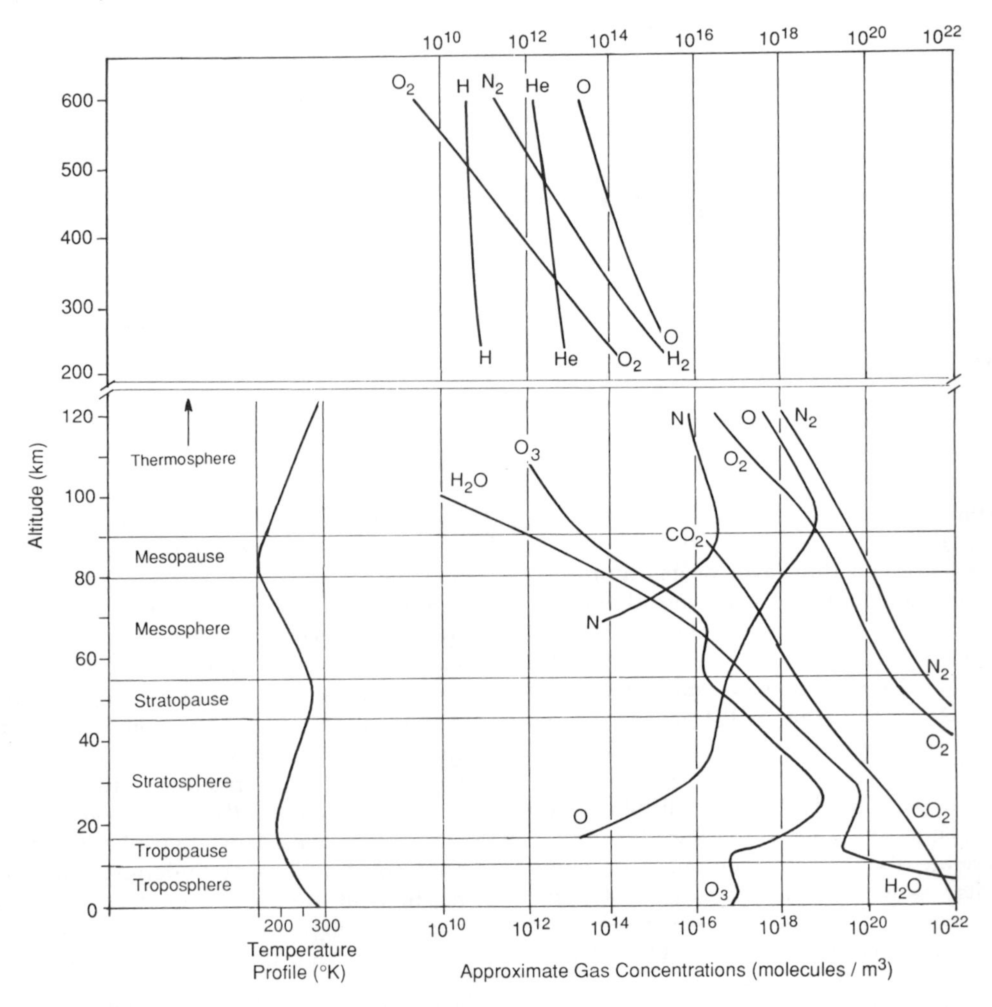

Fig. 10-1 Atmospheric vertical structure including temperature, composition, and conventional names of atmospheric layers or altitude regions.

where P is pressure, ϱ is density, g is the acceleration of gravity, and z is the vertical coordinate. We have two choices for describing dP/dz. For an ideal gas, $PV = nRT$, where V = volume, n = number of moles, R is the gas constant, and T is temperature. So,

$$P = \frac{n}{V} RT = \frac{mRT}{MV} \tag{2}$$

where m is the mass and M is the molecular mass of the gas. Since $m/V = \varrho$:

$$P = \frac{\varrho RT}{M} \tag{3}$$

The choices come in defining M for this mixture of gases. We might define M_i for each gas separately, or we might define a mean value $M = \Sigma_i X_i M_i$, where X_i is the mole fraction of component i. The use of M_i would hold in the absence of any physical mixing (e.g. by turbulence or Brownian motion), while M would be used in the case of perfect mixing.

10.2.2 *Scale Height*

If we define a molecular weight for each constituent, then we can rearrange equation (1).

Because $\varrho = PM/RT$:

$$\frac{dP}{dz} = -\frac{PMg}{RT} \quad (4)$$

so:

$$\frac{dP_i}{P_i} = -(M_i g/RT)\,dz = -dz/H_i \quad (5)$$

where $H_i \equiv RT/M_i g$ is called the scale height. In this situation, constituents with low M have large H and so pressure (and concentration) tends to fall off slowly with altitude, whereas the opposite is true of constituents with high values of M. In such cases, each gas behaves as if no other substance were present. High molecular mass gases (e.g. Xe, Kr) would be concentrated in a layer at the bottom of the atmosphere and lighter gases (H_2, He) would extend to greater altitude. This diffusive separation is not generally significant at low altitudes but occurs increasingly at altitudes above 120 km. Turbulent mixing separates the atmosphere into two layers: the mixed layer at the bottom being called the "homosphere" and the upper layer the "heterosphere". The highest reaches of the atmosphere are thus dominated by H and He, and in the heterosphere heavy unreactive gases (^{40}Ar, Xe, Kr, etc.) fall off rapidly with height. Figure 10-1 illustrates this compositional feature of the atmosphere at altitudes above *c.* 120 km.

The abundance of light elements at high altitude leads to a finite flux of these substances escaping the Earth's gravitational field. This results from a combination of a very long mean free path and a few particles having the requisite escape velocity due to the high velocity "tail" of the Boltzmann velocity distribution.

In terms of relevance to biogeochemical cycling, most of our emphasis is placed on the so-called homosphere (which really is homogeneous *only* with respect to N_2, O_2, ^{40}Ar, and other long-lived gases).

In the case of a mixed atmosphere, M cannot be defined precisely since the composition is variable (especially due to water vapor). If dry air is assumed (which is a good approximation most of the time at altitudes above about 5 km), then $M = 28.97$ g/mol. If the atmosphere is assumed to be roughly isothermal, then from equation (5) pressure falls off with altitude as:

$$P = P_0 \exp(-z/H) \quad (6)$$

Since H is constant if T and M are constant, $H \approx 8$ km if $T = 273$ K. H is the height the entire atmosphere would have if its density were constant at the sea level value throughout.

10.2.3 Lapse Rate

The atmosphere *is not* isothermal – largely due to the fact that it really is a compressible medium. In the simplest case of a dry atmosphere being mixed in the vertical direction with no addition or loss of energy, we might assume that an air parcel behaves adiabatically when it is not in contact with the ground. This implies that its enthalpy, H, is constant. If we define the geopotential at a given height, ϕ, as the work needed to move a unit mass from sea level to that height, then:

$$\phi = \int_0^z g\,dz' \quad (7)$$

For one mole of an ideal gas, $dH = C_p\,dT + M d\phi$, where C_p is the constant pressure heat capacity per mole. Now the adiabatic condition implies that:

$$d(C_p T + M\phi)$$

If we divide by dz and use equation (1), we have:

$$dT/dz = -Mg/C_p = -\Gamma_d$$

Γ_d is called the dry adiabatic lapse rate. For air, $C_p = 29.09$ J/mol K, and on Earth $g = 9.81$ m/s^2, so $\Gamma_d = 9.8$ K/km.

Another way of expressing the way temperature varies in the vertical direction involves the concept of potential temperature, θ. Potential temperature is defined as the temperature a parcel of air would reach if brought adiabatically from its existing temperature and pressure to a standard pressure, P_0. Hence:

$$\theta = T\left(\frac{P_0}{P}\right)^{R/C_p} \quad (8)$$

if T is the temperature at pressure P. This concept is widely used in meteorology, since in the absence of clouds and heat exchange it is convenient to assume that θ is constant for a parcel of air as it moves in the atmosphere.

10.2.4 Static Stability

Now, intuitively, we can consider a perfectly mixed cloudless atmosphere to have constant θ and thus it behaves in a sense like an isothermal body of water, i.e. no part of it is buoyant. If we allow for a layer of air of low θ to occur at the bottom of the atmosphere (like cold water at the bottom of a lake), it is stably stratified and an inversion is said to exist. This layer of relatively low θ acts as a barrier to vertical mixing and hence becomes a physical feature of the atmosphere that is dominant in controlling the dispersion of trace substances (see box).

Another widely used concept is that of a planetary boundary layer (PBL) in contact with the surface of the Earth above which lies the "free atmosphere". The PBL is to some degree a physically mixed layer due to the effects of shear-induced turbulence of convective overturning near the Earth's surface.

The PBL has different characteristics depending on wind speed and static stability; these can be roughly distributed between two extreme categories:

1. Cold air under warm air (inversion), such as warm air over snow or cold ocean water, coupled with low wind speed produces a thin PBL and thus a thin mixed layer. As an extreme example, in the Arctic winter, the PBL may be only 100 m deep leading to the trapping of water and pollutants near the ground and the formation of ice fog. A less extreme but well-known example is the inversion in such cities as Los Angeles, London, or Mexico City, again trapping trace substances and causing elevated concentration of pollutants.
2. The lapse rate in the PBL is unstable and vertical motion leads to the transport of significant amounts of energy upward, due to the buoyancy of air that has been in contact with the surface. A mixed layer forms up to a height where static stability of the air forms a barrier to thermally induced upward motion. This extreme occurs practically daily over the arid areas of the world and the barrier to upward mixing is often the tropopause itself. On the average in mid-latitudes, the unstable or mixed layer is typically 1–2 km deep.

Figure 10-2 shows the vertical profiles of temperature, dew point, light scattering (a measure of aerosol concentration), and the concentrations of O_3 and SO_2. Here we see that up to about 1.5 km, the temperature, dew point, light scattering, and the concentrations of O_3 and SO_2 are nearly constant. This indicates the presence of a mixed PBL. Above 1.5 km, the profiles change dramatically.

In between the two extremes of stability and instability, there are numerous near-neutral stability

Static Stability

Fluids on the Earth's surface that are in hydrostatic equilibrium may be stable or unstable depending on their thermal structure. In the case of freshwater (an incompressible fluid), density decreases with temperature above c. 4° C. Warm water lying over cold water is said to be *stable*. If warm water underlies cold, it is buoyant; it rises and is *unstable*. The buoyant force, F, on the parcel of fluid of unit volume and density ϱ' is:

$$F = (\varrho - \varrho')g$$

where ϱ is the density of the surrounding medium. Since the acceleration, a, is just F/ϱ':

$$a = \left(\frac{\varrho - \varrho'}{\varrho'}\right) g.$$

by Archimedes' principle.

Now, in the case of an ideal gas, pressure, density, and temperature are related so that:

$$a = g\left[\frac{(P/T) - (P'/T')}{(P'/T')}\right]$$

where the prime again denotes the parcel of unit volume.

But in hydrostatic equilibrium and low acceleration, $P = P'$ so:

$$a = g\left(\frac{T' - T}{T}\right)$$

If we let T' be represented by a simple Taylor's series, i.e. $T' = T_0 + (dT'/dz)\,\Delta z$, and if the rest of the ideal atmosphere has a lapse rate dT/dz, so $T = T_0 + (dT/dz)\,\Delta z$, we have for small Δz:

$$a = g\frac{\Delta z}{T_0}\left(\frac{dT'}{dz} - \frac{dT}{dz}\right)$$

If the acceleration is positive, our parcel is buoyant and spontaneous convection occurs. The atmospheric layer is said to be *unstable*. Negative acceleration implies that a small displacement, Δz, results in the parcel accelerating back toward its initial position and therefore indicates *stability*. If dT'/dz is that for an adiabatic test parcel $dT'/dz = -gM/C_p$ and dT/dz that of the existing layer, then for $dT/dz > -9.8$ K/km is *stable* and for $dT/dz < -9.8$ K/km is *unstable*. The 9.8 K/km figure then provides a simple benchmark for static stability of dry air.

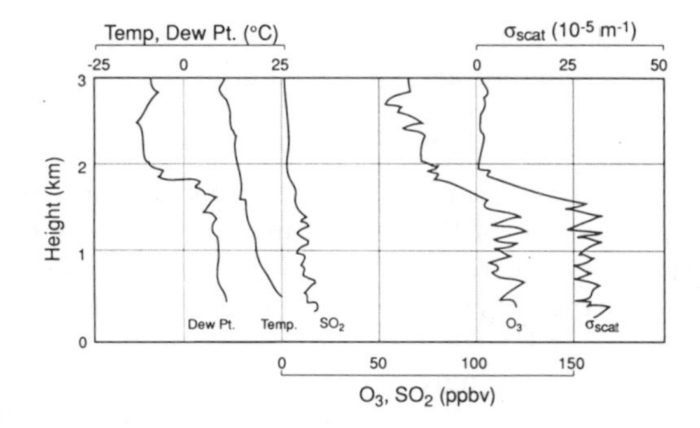

Fig. 10-2 Vertical profiles of physical (temperature, dew point, and backscatter coefficient) and chemical (ozone, sufure dioxide) variables near Scranton, Penn., during the afternoon of July 20, 1978. Modified from Mueller and Hidy (1982) with the permission of the Electric Power Research Institute.

situations, resulting in varying degrees of vertical mixing. In this regime, the mixing depends on such factors as shear-induced turbulence and vertical mixing in and by clouds.

10.3 Vertical Motions, Relative Humidity, and Clouds

When air moves vertically, its temperature changes in response to the local pressure. Indeed, the amount of temperature change is quite large for small changes in height; 1° C per 100 m, if the dry adiabatic lapse rate applies. Considering upward motion (and therefore cooling), an adiabatic decrease of pressure of only 10% due to an altitude increase from 100 to 1000 m results in a change in temperature of −9° C. This amount of cooling results in a major increase in the relative humidity (RH) due to the very strong dependence of the saturation vapor pressure of water on temperature. Details of this are discussed in Section 10.7; however, the consequence of increasing RH due to upward vertical motion is that upward motions of more than a few tens to a few hundreds of meters often cause the air to reach RH = 100% and form clouds. It is important to realize that, even though the decrease in pressure causes a decrease in the amount of water in a fixed volume of air, the temperature decrease is more significant, causing an increase in RH. Thus we see that vertical motions upward cause water clouds to develop; conversely, air that descends becomes warm, causing the RH to decrease and clouds to disappear.

Vertical motions in the atmosphere are caused by a variety of factors:

1. Convection due to the solar heating of the Earth's surface. Upward velocities of 2–20 cm/s occur.
2. Upward motion associated with convergence of horizontal motions (or vice versa, sinking due to divergence). This will be evident in the discussion of horizontal motions in Section 10.5. Again, vertical velocities of only cm/s usually are observed.
3. Horizontal motion over topographic features at the Earth's surface. A classic example of this is seen in the cap clouds associated with flow over mountains.
4. Buoyancy caused by the release of latent heat of condensation of water. As will be seen in Section 10.7, water releases a substantial amount of energy when it condenses.

Even though upward motion causes cooling of a parcel of air, the condensation of water vapor can maintain the temperature of a parcel of air above that of the surrounding air. When this happens, the parcel is buoyant and may accelerate further upward. Indeed, this is an unstable situation which can result in violent updrafts at velocities of meters per second. Cumulus clouds are produced in this fashion, with other phenomena such as lightning, heavy precipitation, and locally strong horizontal winds below the cloud (which provide the air needed to support the vertical motion).

On the average, the air over roughly half of the Earth's surface has an upward velocity and half has a downward velocity. This and the interactions of marine air with the cold ocean surface result in about half of the Earth being covered by clouds and half being clear.

10.4 The Ozone Layer and the Stratosphere

Another major feature of the vertical thermal structure of the atmosphere is due to the presence of ozone (O_3) in the stratosphere. This layer is caused by photochemical reactions involving oxygen. The absorption of solar UV radiation by O_3 causes the temperature in the stratosphere and mesosphere to be much higher than expected from an extension of the adiabatic temperature profile in the troposphere (see Fig. 10-1).

Briefly, oxygen can be photodissociated by solar UV of wavelength less than 242 nm:

$$O_2 + h\nu \rightarrow O + O \qquad (9)$$

Subsequently, the following reactions occur.

$$O + O_2 + M \rightarrow O_3 + M \tag{10}$$

$$O_3 + h\nu \rightarrow O + O_2 \tag{11}$$

The dissociation of O_3 in equation (11) occurs at longer wavelengths than for the dissociation of O_2. The progress of this reaction is halted by:

$$O + O_3 \rightarrow 2O_2 \tag{12}$$

In reality, many other chemical and photochemical processes take place leading to a sort of steady-state concentration of O_3 which is a sensitive function of height. To be accurate, it is necessary to include the reactions of nitrogen oxides, chlorine- and hydrogen-containing free radicals (molecules containing an unpaired electron). However, occurrence of a *layer* due to the altitude dependence of the photochemical processes is of fundamental geochemical importance and can be demonstrated simply by the approach of Chapman (1930).

The concentration of O_2 is approximately an exponential function of altitude:

$$\varrho_{O_2} = \varrho_{0,O_2}\, e^{-z/H} \tag{13}$$

where ϱ_{O_2} is the concentration of O_2, e.g. in molecules per unit volume, and ϱ_{0,O_2} is the concentration at $z = 0$. Now the intensity, I, of solar UV light falling on the atmosphere (in the direction of decreasing z) at an angle χ from the zenith will be attenuated as it penetrates into the atmosphere:

$$\frac{dI}{I} = A\varrho_{O_2} \sec\chi\, dz \tag{14}$$

or

$$\frac{dI}{I} = A\varrho_{0,O_2} \sec\chi\, e^{-z/H}\, dz \tag{15}$$

where A is the absorption cross-section for O_2. Integrating:

$$I = I_0 \exp\{-A\varrho_{0,O_2} H \sec\chi \exp(-z/H)\} \tag{16}$$

Now, the rate of the production of O_3 via the photolysis of O_2 is roughly given by the rate of photolysis of O_2 itself (the $O + O_2$ reaction is assumed to be fast). Thus, the rate of O_3 production as a function of altitude, $q(z)$, should be proportional to the rate of disappearance of photons as a function of altitude:

$$q(z) = \beta\left(\frac{dI}{dz}\right)\cos\chi \tag{17}$$

where β denotes proportionality. Using equation (16), we find that $q(z)$ has a maximum at a height:

$$Z_{max} = H \cdot \ln(A\varrho_{0,O_2} H \sec\chi) \tag{18}$$

The dependence of I on z results in a layer of O_3, the upper portion of the layer being controlled by the exponential decrease of ϱ_{O_2} with altitude. The lower part of the layer is controlled by the fall-off of intensity of UV light as the solar beam penetrates into the increasingly dense atmosphere. More extensive treatments of this phenomenon can be found (e.g. see Wayne, 1985, p. 117 ff.).

The resultant O_3 layer is critically important to life on Earth as a shield against UV radiation. It also is responsible for the thermal structure of the upper atmosphere and controls the lifetime of materials in the stratosphere. Many substances that are short-lived in the troposphere have lifetimes of a year or more in the stratosphere due to the limited removal by precipitation and the presence of the permanent thermal inversion and lack of vertical mixing that it causes.

Besides these features, the formation of a *layer* due to an interaction of a stratified fluid with light is itself noteworthy. Analogs to this phenomenon can be found in other media. Examples include photochemical reactions in the atmosphere near the Earth's surface, photochemical reactions in the surface water of the ocean, and biological activity near the ocean surface.

10.5 Horizontal Motions, Atmospheric Transport, and Dispersion

The horizontal motion of the atmosphere (or wind) is characterized by three spatial scales. These with their conventional names are:

- 0 to 10 km: the *micrometeorological* scale, in which turbulent dispersion of materials is dominant.
- 10 to hundreds of km: the *mesometeorological* scale, in which both advection and turbulent dispersion are effective.
- hundreds to thousands of km: the *synoptic* scale, in which motions are those of whole weather systems. Advection is the dominant transport process.
- $> 5 \times 10^3$ km: the *global* scale.

Going along with these spatial scales, we can define temporal scales as well. Micrometeorological processes tend to be important for times less than

1 h, mesoscale processes up to about 1 day, and synoptic processes a few days or more.

10.5.1 Microscale Turbulent Diffusion

An accurate description of mixing processes on each of these scales is only possible in a few selected and idealized cases. One of the best understood cases is that of a turbulent PBL over flat terrain and a point source of a trace substance. In this case, the concentration downwind of the source is often described as a plume. Figure 10-3 shows such an idealized plume. Spreading in the downward direction results from advection by the wind. Spreading at right angles to the wind results from turbulence. This description does not often hold for distances greater than a few tens of kilometers. Mixing and transport over the mesoscale is extremely hard to describe and is often dominated by local topography, presence of organized vertical motions (e.g. into clouds or "thermals" due to convection), and stable layers that are embedded in the PBL.

10.5.2 Synoptic Scale Motion: The General Circulation

The motion of substances on the synoptic scale is often assumed to be pure advection. The flux through a unit area perpendicular to the wind is simply the product of wind velocity and concentration. (If F is flux, V the velocity, and c concentration:

$$F = cV \tag{19}$$

where such transport is called *advection* (see also Chapter 4).

The motions on the largest spatial scales amount to the aggregate of the world's synoptic weather systems, often called the general circulation. Both with respect to substances that have atmospheric lifetimes of a day or more and with regard to the advection of water, it is useful to depict the nature of this general circulation. The mean circulation is described to some extent in terms of the Hadley and Ferrel cells shown in Fig. 10-4. They describe a coupled circulation driven by the large input of solar radiation near the equator. While departures from the circulation in Fig. 10-4 are substantial, this average pattern does account for major aspects of the pattern of global precipitation.

Three regions of the atmosphere are seen to have significant zonal components of flow and thus of advection. The *mid-latitude* troposphere at the surface tends to exhibit westerly flow (i.e. flow from west to east) on the average. This region contains the familiar high- and low-pressure systems that cause periodicity in mid-latitude weather. Depending on the lifetime of the substances of concern, the motion in these weather systems may be important.

The *tropical* regions of the northern hemisphere's troposphere exhibit easterly flow called the trade winds (of course, these also exist in the southern hemisphere). Finally, the *jet stream* – sometimes described as a river of air – flows with velocities of 25–50 m/s from west to east, often carrying material

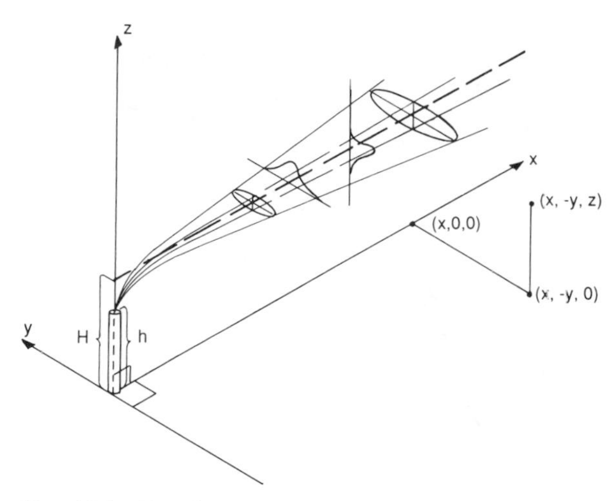

Fig. 10-3 Coordinate system showing the formation downwind from a source of Gaussian distributions of chemical concentrations in the horizontal and vertical. Ellipses denote the loci of two standard deviations.

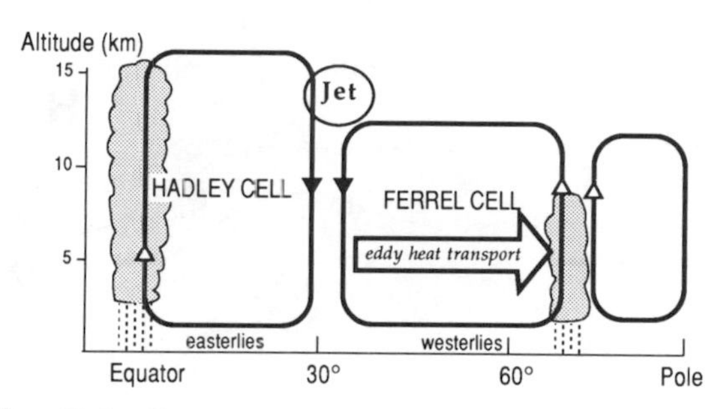

Fig. 10-4 Cross-section of the northern hemisphere atmosphere showing first-order circulation.

completely around the Earth at its altitude close to the tropopause.

10.5.3 Geostrophic Wind

Horizontal motion of the atmosphere, or wind, is a response of the air to the forces that are present. These include the force due to the pressure gradient, the Coriolis force associated with the rotation of the Earth, and frictional forces acting to retard any motion. If the acceleration of the air mass and frictional effects are small, the horizontal velocity is described by the following expression:

$$V_g = \frac{1}{\varrho f} \cdot \frac{\partial P}{\partial x} \qquad (20)$$

This describes the *geostrophic wind* ($f = 2\omega \sin \phi$, where ω is the angular velocity of the Earth and ϕ is the latitude). The air moves parallel to the isobars (lines of constant pressure). The geostrophic wind blows counterclockwise around low pressure systems in the northern hemisphere, clockwise in the southern.

At sea level, for 30°N and S, and a pressure gradient of 1 mbar per 100 km (or 1×10^{-3} Pa/m), $V_g \approx 15$ m/s. In many instances, the observed wind is indeed close to the geostrophic wind and it is often useful to have maps of isobars so that the transport trajectory can be approximated from V_g. For a complete derivation and explanation of the geostrophic wind, departures from it, and related topics, the reader is referred to textbooks on meteorology (e.g. Wallace and Hobbs, 1977).

In the mid-latitude region depicted in Fig. 10-5, the motion is characterized by "large-scale eddy transport". Here the "eddies" are recognizable as ordinary high- and low-pressure weather systems, typically about 10^3 km in horizontal dimension. These eddies actually mix air from the polar regions with air from nearer the equator. At times, air parcels with different water content, different chemical composition, and different thermodynamic characteristics are brought into contact. When cold, dry air is mixed with warm moist air, clouds and precipitation occur. A *frontal system* is said to exist. Two such frontal systems are depicted in Fig. 10-5 (heavy lines in the midwest and southeast).

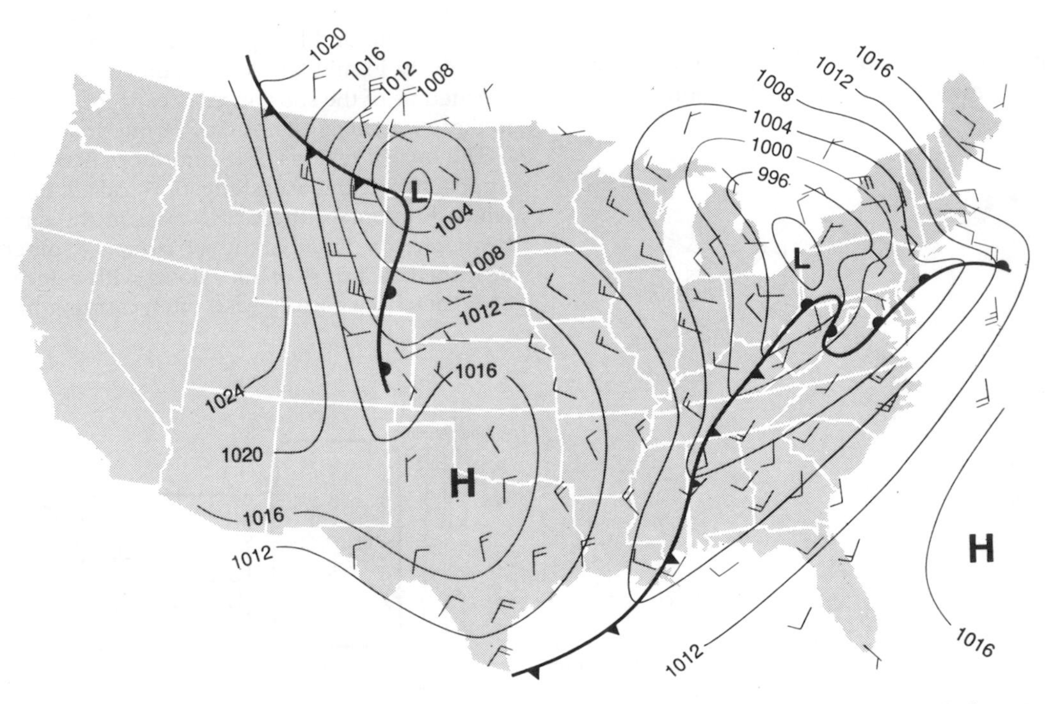

Fig. 10-5 Surface pressure map (millibars). Fronts are shown by heavy lines. H, high-pressure system; L, low-pressure system. Wind directions are shown by arrows; wind speeds correspond to the number of bars on the arrow tails.

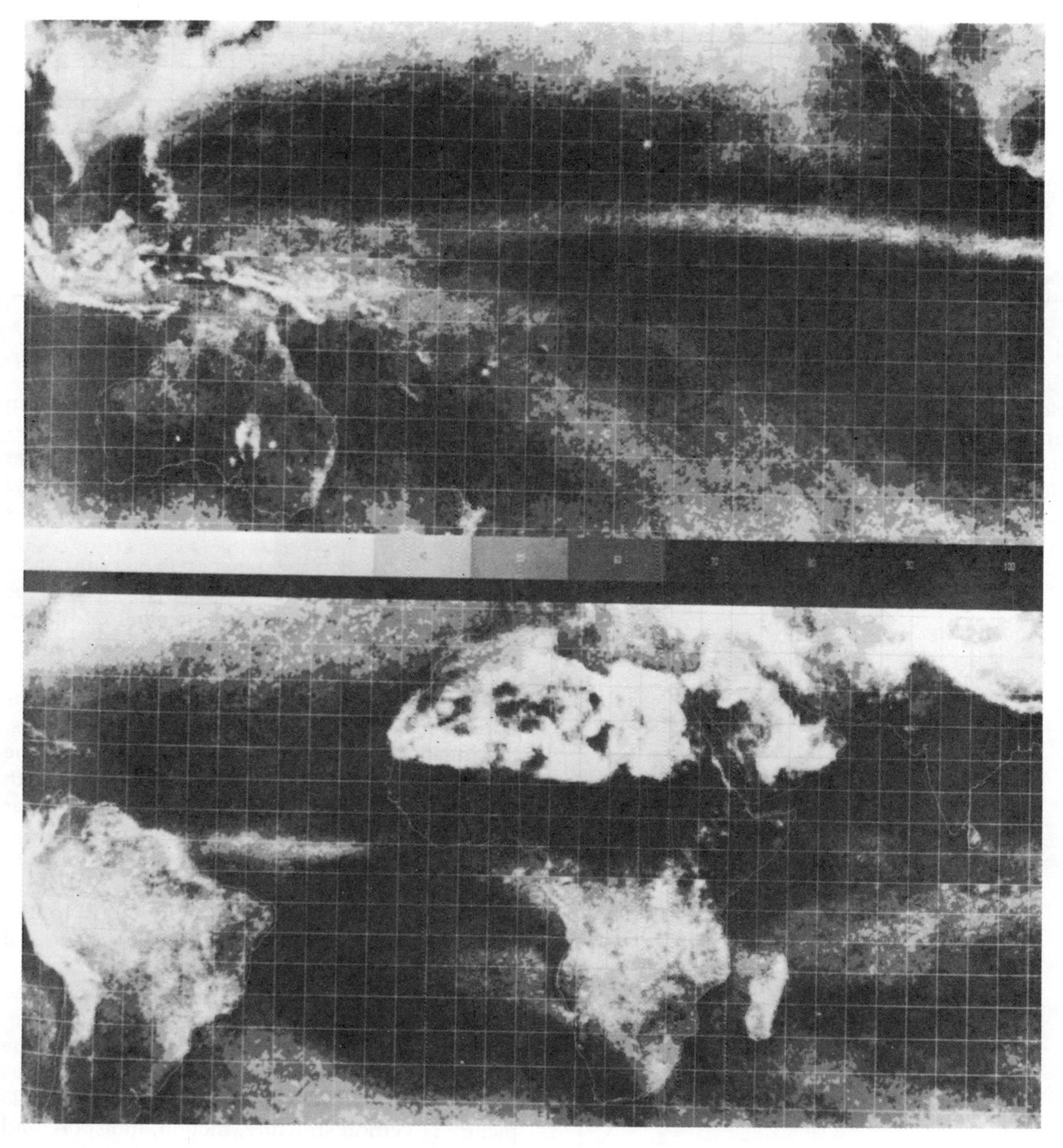

Fig. 10-6 Satellite observations of global reflectivity for January 1967–1970. White indicates areas of persistent cloudiness and relatively high precipitation, except for northern Africa where desert surface regions are highly reflective. From US Air Force and US Department of Commerce (1971).

10.5.4 Meridional Transport of Water: The ITCZ

Among the consequences of this general circulation are convergent and divergent flows in the surface wind leading to systematic vertical motions, especially those of the Hadley cell in the tropics. Upward motion (such as near the equator in the Hadley cell) often results in the formation of clouds due to adiabatic cooling, while subsidence (downward motion) results in heating and the absence of

Table 10-1 Average residence time of water vapor in the atmosphere as a function of latitude

	Latitude range (degrees)								
	0–10	10–20	20–30	30–40	40–50	50–60	60–70	70–80	80–90
Average precipitable water (g/cm^2)	4.1	3.5	2.7	2.1	1.6	1.3	1.0	0.7[a]	0.45[a]
Average precipitation (g/cm^2 year)	186	114	82	89	91	77	42	19	11
Residence time (days)	8.1	11.2	12.0	8.7	6.4	6.2	8.7	(13.4)	(15.0)

[a] Values extrapolated.

clouds. Figure 10-6 shows composite satellite photographs depicting the mean brightness of the region from 40°N to 40°S, showing clearly the bands of clouds in the intertropical convergence zone (ITCZ) and the clear areas north and south of it. The influence of the land masses on this simplified picture is also apparent, clearly underscoring the difficulties in describing air motions in the vicinity of either topographic roughness or thermal discontinuity. Just at the edge of the picture at latitude 30–40°N or S, the cloudiness of the mid-latitude weather system is apparent.

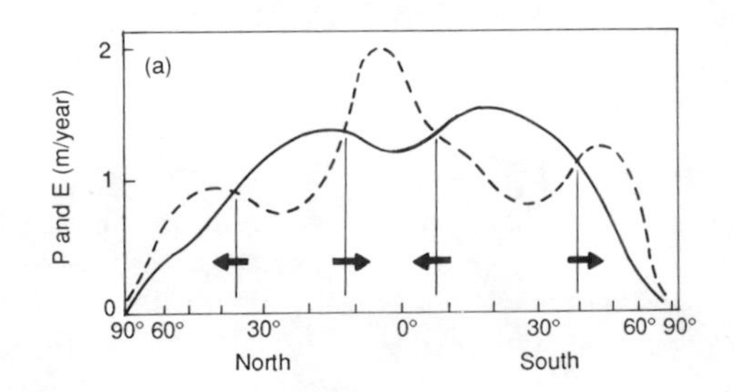

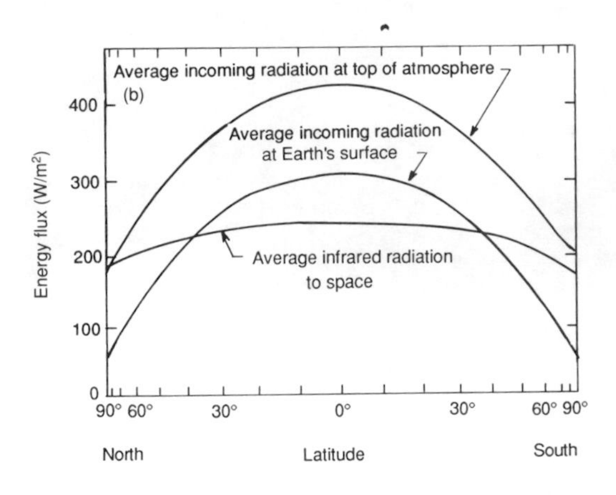

Fig. 10-7 (a) Average annual precipitation (P) and evaporation (E) per unit area latitude. Arrows represent the sense of the required water vapor flux in the atmosphere. (b) Incoming solar energy (top of atmosphere and surface) and outgoing terrestrial energy *vs* latitude.

Figure 10-7 depicts the transport of one substance, i.e. water, due to the general circulation. Here we see the overall consequence of the general circulation with its systematic pattern of vertical motions and weather systems. Water evaporates from the oceans and land surfaces at subtropical latitudes and is transported both toward the equator and the poles. Precipitation falls largely at the equator and in the mid-latitudes. Hence, the subtropics are arid, with evaporation exceeding precipitation. The polar regions likewise are arid due to water having been removed in mid-latitude weather systems prior to arrival in the Arctic and Antarctic.

The average lifetime of water vapor in the atmosphere obviously is a function of latitude and altitude. In the equatorial regions, its residence time in the atmosphere is a few days, while water in the stratosphere has a residence time of 1 year or more. Table 10-1 (Junge, 1963) provides an estimate of the average residence time for water vapor for various latitude ranges in the troposphere. Given this simple picture of vertical structure, motion, transport, and diffusion, we can proceed to examine the behavior of reactive trace substances in this dynamic milieu. However, before we can do so, it is useful to summarize briefly the overall composition of the atmosphere.

10.6 Composition

Table 10-2 includes most of the main *gaseous* constituents of the troposphere with observed concentrations. In addition to gaseous species, the *condensed* phases of the atmosphere (i.e. aerosol particles and clouds) contain numerous other species. The physical characteristics and transformations of the aerosol state will be discussed later in Section 10.10. The list of major gaseous species can be organized in several different ways. In Table 10-2, it is in order of decreasing concentration. We can see that there are five approximate categories based simply on concentration:

1. The major gases: the concentration often given as a *percent* (N_2, O_2, Ar, H_2O, CO_2).
2. Those gases having concentrations expressed as *part-per-million* (Ne, He, CH_4, CO).
3. Gases expressed as *parts-per-billion* (O_3, NO, N_2O, SO_2).
4. Gases in the *parts-per-trillion* category (CCl_2F_2, CF_4, NH_3).
5. Gases expressed in *number of atoms or molecules* per cm^3 – notably radionuclides and free radicals like OH.

Alternatively, we could organize the list by variability in which we would see that N_2, O_2, and the noble gas concentrations are extremely stable, with increasing variability for substances of low concentration and for chemically reactive substances. Both the temporal and spatial variability are influenced by the same factors: source strength and its variability, sink mechanisms and variability, and atmospheric lifetime. Close to sources (such as in a polluted urban setting), variability is likely to be dominated by proximity to and variations of the source. Urban data, for example, often show clearly the influence of temporal features of human activities like automobile traffic. However, when observations are made in more remote settings, sink mechanisms or lifetimes tend to become more evident in determining variability. Junge (1974) posed a hypothesis relating variability to residence time, suggesting that there is a geometric and inverse relationship between the relative standard deviation of concentration and residence time as indicated in Fig. 10-8.

Focusing on the chemical reactivity, we could list the noble gases, N_2, and perhaps O_2 as the least reactive. Even though the reaction of N_2 and O_2 in the presence of H_2O is favored thermodynamically, the reaction rate is very slow, so these two species do not end up as HNO_3 in the oceans. More will be said about this in Chapter 12. Reactivity being a rather unspecific term, it seems logical to organize the composition on an element-by-element basis. However, before getting to the major elements (N, S, and C), it is useful to examine H_2O as the most variable of the dominant species. In Table 10-2, we deliberately omitted water because of its variability.

Table 10-2 Major gaseous constituents of dried air

		Average concentration (volume fraction)
N_2	Nitrogen	0.78084
O_2	Oxygen	0.20946
^{40}Ar	Argon	9.34×10^{-3}
CO_2	Carbon dioxide	3.5×10^{-4}
Ne	Neon	1.8×10^{-5}
He	Helium	5.24×10^{-6}
CH_4	Methane	1.7×10^{-6}
Kr	Krypton	1.13×10^{-6}
H_2	Hydrogen	5×10^{-7}
N_2O	Nitrous oxide	3×10^{-7}
Xe	Xenon	8.7×10^{-8}
CO	Carbon monoxide	$5–20 \times 10^{-8}$
OCS	Carbonyl sulfide	5×10^{-10}
O_3	Ozone	
	Troposphere (clean)	5×10^{-8}
	Troposphere (polluted)	4×10^{-7}
	Stratosphere	1×10^{-7} to 6×10^{-6}

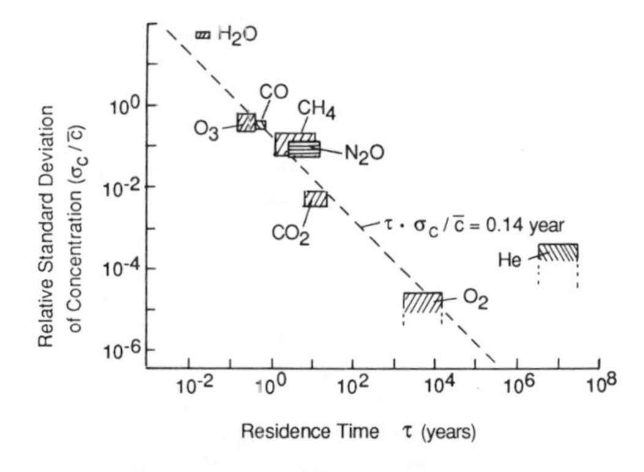

Fig. 10-8 Inverse relationship between relative standard deviation of concentration, σ_c/c, and residence time, τ, for important trace chemicals in the troposphere. Modified from Junge (1974) with the permission of the Swedish Geophysical Society.

It can range from ppmv levels in the Antarctic and the stratosphere to several percent in moist tropical air. Thus, it is necessary to reference the concentrations in the table to dry air, or to devise another measure to get around the variability of water. Another scheme would be to present all the average concentrations relative to one of the more constant constituents, e.g. to nitrogen.

10.7 Atmospheric Water and Cloud Microphysics

Water forms strong hydrogen bonds and these lead to a number of important features. First, it has a much higher boiling point than other period VI dihydrides (H_2S, H_2Se, and H_2Te). Because of the strength of these hydrogen bonds, a large amount of energy (latent heat of vaporization) is required to evaporate a unit mass of water. Similarly, the latent heat of freezing is largely due to further strong bonding in ice crystals. The surface tension (surface free energy) is also large. Table 10-3 summarizes these physical properties of water.

All three phases of water exist in the atmosphere, and the condensed phases can exist in equilibrium with the gas phase. The equilibria between these phases is summarized by the phase diagram for water (Fig. 10-9). We see from Fig. 10-9 that the partial pressures of H_2O at ordinary conditions range from very small values to perhaps 30 or 40 mbar. This corresponds to a mass concentration range up to about 25 g water/m^3. In typical clouds, relatively little of this is in the condensed phase. Liquid water contents in the wettest of cumulus clouds are around a few grams per cubic meter; ordinary mid-latitude stratus clouds have 0.3–1 g/m^3.

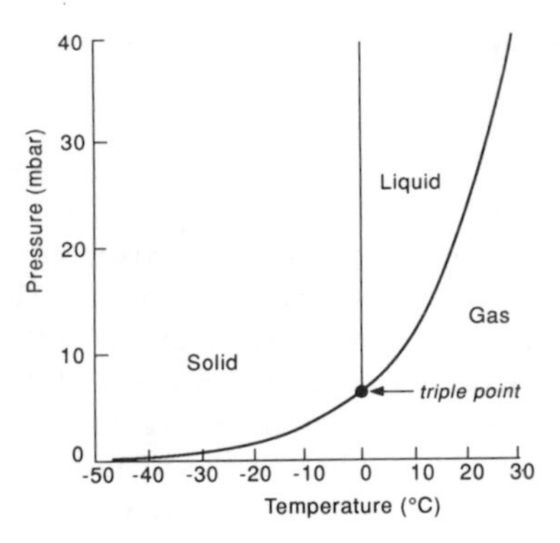

Fig. 10-9 P-T phase diagram for bulk water (based on data of the Smithsonian meteorological tables).

Water clouds play two key roles in biogeochemical cycles on the Earth:

1. They deliver water from the atmosphere to the Earth's surface as rain or snow, and are thus a key step in the hydrologic cycle.
2. They scavenge a variety of materials from the air and make them available for delivery in precipitation.

Thus we proceed to examine the physicochemical nature of the cloud nucleation process.

Only two possibilities exist for explaining the existence of cloud formation in the atmosphere. If there were no particles to act as cloud condensation nuclei (CCN), water would condense into clouds at relative humidities (RH) of around 300%. That is, air can remain supersaturated below 300% with water vapor for long periods of time. If this were to occur, condensation would occur on surface objects and the hydrologic cycle would be very different from what

Table 10-3 Properties of water

	Mass basis	Molar basis
Specific heat of water vapor at constant pressure, C_P	1952 J/kg K	35.14 J/mol K
Specific heat of water vapor at constant volume, C_V	1463 J/kg K	26.33 J/mol K
Specific heat of liquid H_2O at 0° C	4218 J/kg K	75.92 J/mol K
Latent heat of vaporization		
at 0° C	2.5×10^6 J/kg	4.5×10^4 J/mol
at 100° C	2.25×10^6 J/kg	4.05×10^4 J/mol
Latent heat of fusion, 0° C	3.3×10^5 J/kg	5.94×10^3 J/mol
Surface tension (water *vs* air)	0.073 J/m^2	

is observed. Thus, a second possibility must be the case; particles are present in the air and act as CCN at much lower RH. These particles must be small enough to have small settling velocity, stay in the air for long periods of time, and be lofted to the top of the troposphere by ordinary updrafts of cm/s velocity. Two further possibilities exist – the particles can either be water-soluble or insoluble. In order to understand why it is likely that CCN are soluble, we examine the consequences of the effect of curvature on the saturation water pressure of water.

As a result of the high surface free energy of water, the vapor pressure of a water droplet increases with decreasing radius of curvature, r, as deduced by Kelvin:

$$RT \ln \frac{P}{P_\infty} = \frac{2M\sigma}{\varrho r} \quad (21)$$

where P_∞ is the water pressure over a flat surface and σ is the surface free energy. This, combined with vapor pressure depression due to dissolved substances (Raoult's Law), results in a requirement for supersaturation (i.e. RH > 100%) as a condition for the formation of micrometer-sized droplets in clouds or fog. If the amount of soluble material is large, the supersaturation is small, while small soluble particles require higher supersaturation. Figure 10-10 shows the vapor pressure as a function of size for a variety of soluble particle masses.

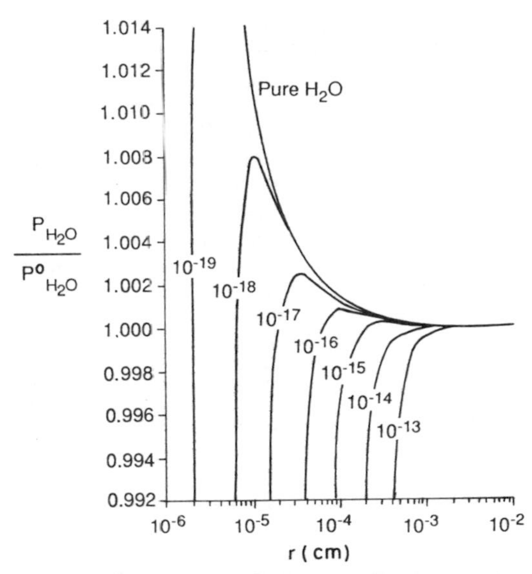

Fig. 10-10 Köhler curves calculated for the saturation ratio $P_{H_2O}/P^0_{H_2O}$ of a water droplet as a function of droplet radius r. The quantity im/M is given as a parameter for each line, where m = mass of dissolved salt, M = molecular mass of the salt, i = number of ions created by each salt molecule in the droplet.

These plots are called *Köhler curves* after their originator (Köhler, 1936). His assumptions that cloud condensation nuclei (CCN) are water-soluble materials is now widely accepted. In the past, it was often thought that NaCl particles from the ocean were the main CCN; however, more recent studies have demonstrated the frequent dominance of sulfate particles with composition between H_2SO_4 and $(NH_4)_2SO_4$.

When a droplet reaches the peak of its appropriate curve, due to being in a region of RH greater than the RH for that critical size, it will continue to grow in an uncontrolled fashion. As it gets larger, the curvature effect decreases its vapor pressure and it enters a region of increased supersaturation relative to that at the peak of the Köhler curve. A particle that turns into a droplet and passes the critical size is said to be an *activated* CCN.

Following growth by condensation, droplets grow further by collision coalescence (colliding mainly due to different fall speeds). A small amount of precipitation is produced in this fashion, recognizable as drizzle. Larger water particles and heavier precipitation occur when ice is present. Due to the ability of small droplets of liquid water to exist in a supercooled state, most cloud water is liquid. Freezing occurs due to the presence of an ice nucleating aerosol (IN), typically at temperatures of -5 to $-20°$ C. Since ice has a lower vapor pressure than supercooled water, the ice particles grow at the expense of the droplets. Ice particles that are large enough to fall can subsequently collect larger amounts via collision with droplets with resulting graupel or hail (if the particle remains frozen) or rain (if the ice melts).

The overall rainfall rate and amount depend on these microphysical processes and even more greatly on the initial amount of water vapor present, and on the vertical motions that transport water upward, cool the air, and cause supersaturation to occur in the first place. Thus the delivery of water to the Earth's surface as one step in the hydrologic cycle is controlled by both microphysical and meteorological processes. The global average precipitation amounts to about 75 cm/year or 750 L/m^2 year.

Cloud nucleation also has *chemical* consequences. The soluble material of the CCN introduces solute into cloud droplets which, in many instances, is a major and even dominant ingredient of cloud and

Table 10-4 Gaseous atmospheric sulfur compounds

Species	Concentration	Sources	Sinks
SO_2	0–0.5 ppmv (urban)	Oxidation of fossil fuel S	Direct reaction with Earth surface, oxidation to sulfate
	20–200 pptv (remote)	Oxidation of S gases	
H_2S	0–40 pptv	Biological decay of protein in anaerobic water	Oxidation to SO_2
CH_3SH	Sub-ppbv	Paper pulping	Oxidation to SO_2
CH_3CH_2SH	Sub-ppbv		Oxidation to SO_2
OCS	500 pptv		Destruction in the stratosphere
CH_3SCH_3	20–200 pptv	Oceanic phytoplankton and algae	Oxidation to SO_2
CH_3SSCH_3	Small		Oxidation to SO_2
CS_2	10–20 pptv		Destruction in the stratosphere and tropospheric OH

rainwater. A simple but useful expression for the amount of solute from CCN is:

$$[X] = \frac{\varepsilon(X)_{\text{air}}}{M_X L} \qquad (22)$$

where $[X]$ = the average molarity of the solute X in cloud water, ε = the fraction of aerosol particles of X that are activated CCN, M_X = the molecular weight of X, $(X)_{\text{air}}$ = the concentration of X in air entering the cloud (g/m^3), and L = the liquid water content of the cloud (mL/m^3). As an example, if 5 μg/m^3 of sulfate aerosol were present, and $\varepsilon \approx 1$, with $L = 1$ mL/m^3, then: $[SO_4] = 5 \times 10^{-5}$ M. This example is realistic for the industrialized areas such as eastern North America and Europe.

In addition to solute from CCN, clouds contain dissolved gases (e.g. SO_2, NH_3, HCHO, H_2O_2, HNO_3, and many more). In turn, some of these may react in the cloud droplets to form other substances which subsequently can appear in rainwater. Finally, falling raindrops can collect other materials (e.g. large dust particles) on their way to the Earth's surface. Thus, rainwater composition does not uniquely reflect the chemistry of the CCN.

10.8 Trace Atmospheric Constituents

10.8.1 Sulfur Compounds

There is a large variety of atmospheric sulfur compounds, in the gas, solid, and liquid phases. Table 10-4 lists a number of gaseous compounds, ranges of concentration, sources, and sinks (where known). As this list illustrates, a significant number of these gases contribute to the existence of oxidized sulfur in the forms of SO_2 and sulfate aerosol particles. Table 10-5 lists the oxyacids of sulfur and their ionized forms that could exist in the atmosphere. Of these, the sulfates are certainly dominant, with H_2SO_4 and its products of neutralization with NH_3 the most frequently reported forms.

Table 10-5 Oxy acids, their salts, and ionized forms which could exist in atmospheric aerosol particles

Oxy acid	Formula	Salt/ionized form
Sulfuric	H_2SO_4	HSO_4^-, SO_4^{2-}
		NH_4HSO_4
		$(NH_4)_3H(SO_4)_2$
		$(NH_4)_2SO_4$
		$MgSO_4$
		$CaSO_4$
		Na_2SO_4
		R—O—SO_3^{2-}
Sulfurous	$SO \cdot H_2O$	HSO_3^-, SO_3^{2-}
Sulfonic	R—SO_3—H	R—SO_3^-
Hydroxymethane sulfonic	$CH_2(OH)SO_3H$	$CH_2(OH)SO_3^-$
Dithionic	$H_2S_2O_6$	$S_2O_6^{2-}$
Thiosulfuric	$H_2S_2O_3$	$S_2O_3^{2-}$
Polythionic	$H_2S_nO_6$	$S_nO_6^{2-}$
Pyrosulfurous	$H_2S_2O_5$	$S_2O_5^{2-}$

As a result of the water solubility of both SO_2 and the sulfates, these compounds are frequently found in or associated with water in a condensed phase. The acidity of H_2SO_3 ($K_1 = 1.7 \times 10^{-2}$, $K_2 = 6.2 \times$

Table 10-6 Nitrogen atmospheric compounds

Species	Concentration	Source	Sink
NH_3	0–20 ppbv	Biological	Precipitation
$RNH_2 \ldots R_3N$	—	Biological	Precipitation
N_2	78.084%	Primitive volatile, denitrification	Biological nitrification
N_2O	0.1–0.4 ppmv	Biological	Photolysis in the stratosphere
(N_2O_3)	—	Reaction intermediate	—
NO	0–0.5 ppmv	Oxidation of N_2 in combustion	HNO_3
NO_2	—	NO oxidation	HNO_3
HNO_2	—	OH + NO	Precipitation
HNO_3	—	$OH + NO_2$	Precipitation

—, global values not established.

10^{-8}) can cause low pH in cloud and rainwater, although most measurements indicate that low rain pH is associated with SO_4^{2-} and NO_3^-. In any case, SO_2 and SO_4^{2-} removal from air is dominated by precipitation. These points will be amplified and quantified later in Chapter 13.

There are two dominant stable isotopes of sulfur found in atmospheric sulfur compounds, ^{32}S and ^{34}S. While it is attractive to utilize the ratio of these two for studies of atmospheric processes, source influences or sink mechanisms, no clear-cut results have yet been demonstrated. The general features of the S isotope distributions will be summarized in Chapter 13.

10.8.2 Nitrogen Compounds

Like sulfur, nitrogen has stable compounds in a wide range of oxidation states and many of them are found in the atmosphere. Again, both gaseous and particulate forms exist as do a large number of water-soluble compounds. Table 10-6 lists the gaseous forms. The nitrogen cycle is discussed in Chapter 12.

Here we see a range of oxidation states from -3 to $+5$. The reduced forms are undoubtedly the most important gaseous bases in air, while the oxides tend to produce HNO_3 as one of the two dominant strong atmospheric acids (H_2SO_4 is the other one).

In particulate form, we find the condensed phase of some of these in the form of salts, as listed in Table 10-7.

Table 10-7 Nitrogenous aerosol constituents

Species	Concentration
$(NH_4)_2SO_4$	Up to 30 $\mu g/m^3$
NH_4HSO_4	—
Amine sulfates	—
NH_4NO_3	Up to a few $\mu g/m^3$, but only in polluted air
Other nitrates	—
Organic N	—

10.8.3 Carbon Compounds

Unlike sulfur and nitrogen compounds in air where we found a wide range of oxidation states in relatively few compounds, carbon has a nearly unlimited number of compounds. These compounds fall roughly into two categories: inorganic and organic compounds. The global carbon cycle is discussed in Chapter 11.

10.8.3.1 Elemental carbon

Common soot contains significant amounts of elemental carbon in the molecular form of graphite, along with organic impurities. All forms of combustion of carbonaceous materials produce some soot, so its presence is ubiquitous in both pristine and polluted regions. The exact composition is variable and dependent on the nature of the source. All soots are, however, characterized by the presence of very small particles (sub-micrometer) and correspondingly a large ratio of surface area to mass. The graphitic component provides an exceedingly inert physical structure for such particles. Thus, soot particles can be transported large distances. They probably become coated with sulfates and other condensed material by Brownian coagulation and eventually come out of the atmosphere in rain and are eventually sequestered geologically in sediments.

The key features of soot are its chemical inertness, its physical and chemical adsorption properties, and its light absorption. The large surface area coupled with the presence of various organic functional groups allow the adsorption of many different materials onto the surfaces of the particles. This type of sorption occurs both in the aerosol phase and in the aqueous phase once particles are captured by cloud droplets. As a result, complex chemical processes occur on the surface of soot particles, and otherwise volatile species may be scavenged by the soot particles.

10.8.3.2 Carbon oxides

The oxides are gaseous and do not undergo reactions in the atmosphere that produce aerosol particles. Carbon monoxide is a relatively inert material with its main sinks in the atmosphere via reactions with free radicals, e.g.

$$OH\cdot + CO \rightarrow CO_2 + H\cdot \quad (23)$$

Other sinks – largely biological – probably exist at the Earth's surface. Seiler (1974) deduced a lifetime of *c.* 0.5 years for CO, attesting to the lack of reactivity or water solubility in comparison to sulfur and nitrogen compounds.

Carbon dioxide is likewise an inert material. As a result, its only known sinks are photosynthesis and solubility in seawater. The cycle of carbon dioxide through the atmosphere will be a major focal point in Chapter 11.

Besides its inertness, CO_2 is modestly water-soluble and in aqueous media forms carbonic acid which is a weak acid. Henry's Law for CO_2 states:

$$P(CO_2) = K_H[H_2CO_3^*] \quad (24)$$

where $K_H = 29$ atm/M [$H_2CO_3^*$ is the sum of H_2CO_3 and CO_2(aq): see Chapter 5]. Thus 350 ppmv of CO_2 results in $[H_2CO_3^*] = 10^{-4.4}$ M in otherwise pure H_2O, as is often assumed for clouds and rain. Given that $K_1 = 4.5 \times 10^{-7}$ mol/L and $K_2 = 4.7 \times 10^{-11}$ mol/L for the first and second dissociations of H_2CO_3, cloud and rainwater affected only by H_2CO_3 would have a pH of about 5.6. However, other solutes (notably H_2SO_4) usually dominate the effect of H_2CO_3, even in pristine locations.

Most CO and CO_2 in the atmosphere contain the mass 12 isotope of carbon. However, due to the reaction of cosmic ray neutrons with nitrogen in the upper atmosphere, ^{14}C is produced. Nuclear bomb explosions also produce ^{14}C. The ^{14}C is oxidized, first to ^{14}CO and then to $^{14}CO_2$ by OH· radicals. As a result, all CO_2 in the atmosphere contains some ^{14}C, currently a fraction of *c.* 10^{-12} of all CO_2. Since ^{14}C is radioactive (β^- emitter, 0.156 MeV, half-life of 5770 years), all atmospheric CO_2 is slightly radioactive. Again, since atmospheric CO_2 is the carbon source for photosynthesis, all biomass contains ^{14}C and its level of radioactivity can be used to date the age of the biological material.

10.8.3.3 Organic carbon

The remaining carbon compounds fall into the category of organic molecules. The number of identified species is large – at least several hundred – so we cannot produce an exhaustive list here. Instead, we will list molecular forms following conventional schemes for organic chemistry with a few selected samples.

Tables 10-8 through 10-12 give a sense of the range of organic molecules present in the atmosphere. Both natural sources and human activity contribute to the variety of organic molecules (Graedel, 1978). The sinks often involve *in situ* reactions.

10.8.4 Other Trace Elements

The atmosphere may be an important transport medium for many other trace elements. Lead and other metals associated with industrial activity are found in remote ice caps and sediments. The transport of iron in wind-blown soil may provide this nutrient to remote marine areas. There may be phosphorus in the form of phosphine, PH_3, although the detection of volatile phosphorus has not been convincingly or extensively reported to date.

Table 10-8 Atmospheric hydrocarbons

Class	Compound	Typical source	Probable sink	Concentration
Alkanes	Methane	Microbes, natural gas	OH	1.7 ppmv
	Ethane	Auto	OH	0–100 ppbv
	Hexane	Auto	OH	0–30 ppbv
	(Up to C_{37})	—	—	—
Alkenes	Ethene	Auto, microbes, vegetation	OH, O_3	1–1000 ppbv
	Isoprene	Trees, auto	OH	0.2–30 ppbv
Alkynes	Acetylene	Auto, microbes	OH	0.2–200 ppbv
Terpenes	α-Pinene	Trees	O_3, OH	0–1 ppbv
	Limonene	Trees	O_3, OH	0–1 ppbv
Cyclic hydrocarbons	Cyclopentane	Auto	OH	0–10 ppbv
	Cyclohexane	Auto	OH	0–10 ppbv
	Cyclopentene	Auto	O_3, OH	0–10 ppbv
Aromatic hydrocarbons	Benzene	Auto	OH, O	—
	Toluene	Auto	OH	0–100 ppbv
Polynuclear aromatic hydrocarbons	Phenanthrene	Aluminium manufacture, combustion		0–300 ng/m^3
	Benzo(a)pyrene	Auto, wood combustion		0–100 ng/m^3

Table 10-9 Oxygenated organic compounds

Class	Example	Typical source	Probable sink	Concentration
Aliphatic aldehydes	Formaldehyde		$h\nu$, OH, rain	1–100 ppbv
	Acetaldehyde	Animal waste	$h\nu$	1–10 ppbv
Olefinic aldehydes	Acrolein	Auto	O_3	0–1 ppbv
Aromatic aldehydes	4-Methyl-benzaldehyde	Auto		0–300 pptv
Aliphatic ketones	Acetone	Animal waste	$h\nu$	0–10 ppbv
Aliphatic acids	Formic acid		Rain	0–100 ppbv
	Pentanedioic acid	Cyclopentene	Rain	0–1 mg/m^3
Cyclic acids	Pinonic acid	α-Pinene + O_3	Rain	
Aromatic acids	Benzoic acid	Auto	Rain	0–400 ng/m^3
Aliphatic alcohols	Methanol CH_3OH	Animal waste	—	0–100 ppbv
Phenols	Phenol	Auto	—	0–3 ppbv

—, global values not established.

Table 10-10 Atmospheric nitrogen-containing organic compounds

Class	Example	Typical source	Probable sink	Concentration
Amines	Methylamine	Protein decay	H_2SO_4 aerosol	—
Nitrates	Peroxy acetyl nitrate	Photochemical	—	0–100 ppbv
Heterocyclic N compounds	Benzo[f]quinoline	Coal combustion	—	0–200 pg/m^3

Table 10-11 Atmospheric sulfur-containing organic compounds

Class	Example	Typical source	Probable sink	Concentration
Mercaptans	Methyl mercaptan	Microbiota	Oxidation to SO_2 by OH	0–4 ppbv
Sulfides	Dimethyl sulfide	Algae	Oxidation to SO_2 by OH	20–300 pptv

Table 10-12 Atmospheric halogen-containing organic molecules

Class	Example	Typical source	Probable sink	Concentration
Halogenated aliphatic compounds	CH_3Cl	Biological	OH	620 pptv
	CH_2Cl_2	Industrial	OH	30 pptv
	CH_3CCl_3	Solvent		140 pptv
	CCl_4	Solvent		130 pptv
	$CFCl_3$ (CFC-11)	Propellant, refrigerant	Stratosphere	220 pptv
	CF_2Cl_2 (CFC-12)	Refrigerant	Stratosphere	375 pptv
Halogenated aromatic compounds	DDT	Pesticide	—	0.009–500 ng/m^3

10.9 Chemical Interactions of Trace Atmospheric Constituents

Unlike the chemistry of simple mixtures of small numbers of reactants, the chemistry of the atmosphere involves complex interactions of large numbers of species. However, several key aspects of these interactions have been identified that account for major observable properties of the atmospheric chemical system. It is convenient to separate the description into gas phase and condensed phase interactions, not least because different chemical and physical processes are involved in these two cases.

10.9.1 Gas Phase Interactions

Figure 10-11 and its caption (Crutzen, 1983) depict the most important of the gas phase and photochemical reactions in the atmosphere. Perhaps the single most important interaction involves the hydroxyl free radical, OH·. This extremely reactive radical is produced principally from the reactions of electronically excited atomic oxygen, $O(^1D)$, with water vapor. Photodissociation of ozone produces $O(^1D)$ and also the less reactive $O(^3P)$. In the troposphere, O_3 is produced largely by photochemical reactions involving still other free radicals, including the nitrogen oxides, NO and NO_2. OH· appears to be the dominant oxidant for CO, CH_4, SO_2, and $(CH_3)_2S$, as well as the main source of HNO_3 and HNO_2.

10.9.2 Condensed Phase Interactions

Condensed phase interactions can be divided roughly into two further categories – chemical and physical. The latter involves all purely physical processes such as condensation of species of low volatility onto the surfaces of aerosol particles, adsorption, and absorption into liquid cloud and rainwater. Here, the interactions may be quite complex. For example. cloud droplets require a CCN, which in many instances is a particle of sulfate produced from SO_2 and gas-particle conversion. If this particle is strongly acidic (as is often the case), HNO_3 will not deposit on the aerosol particle; rather, it will be dissolved in liquid water in clouds and rain. Thus, even though HNO_3 is not very soluble in the concentrated H_2SO_4 of the aerosol particle, the atmospheric residence time of HNO_3 is in part determined by the physical role of H_2SO_4 particles as CCN.

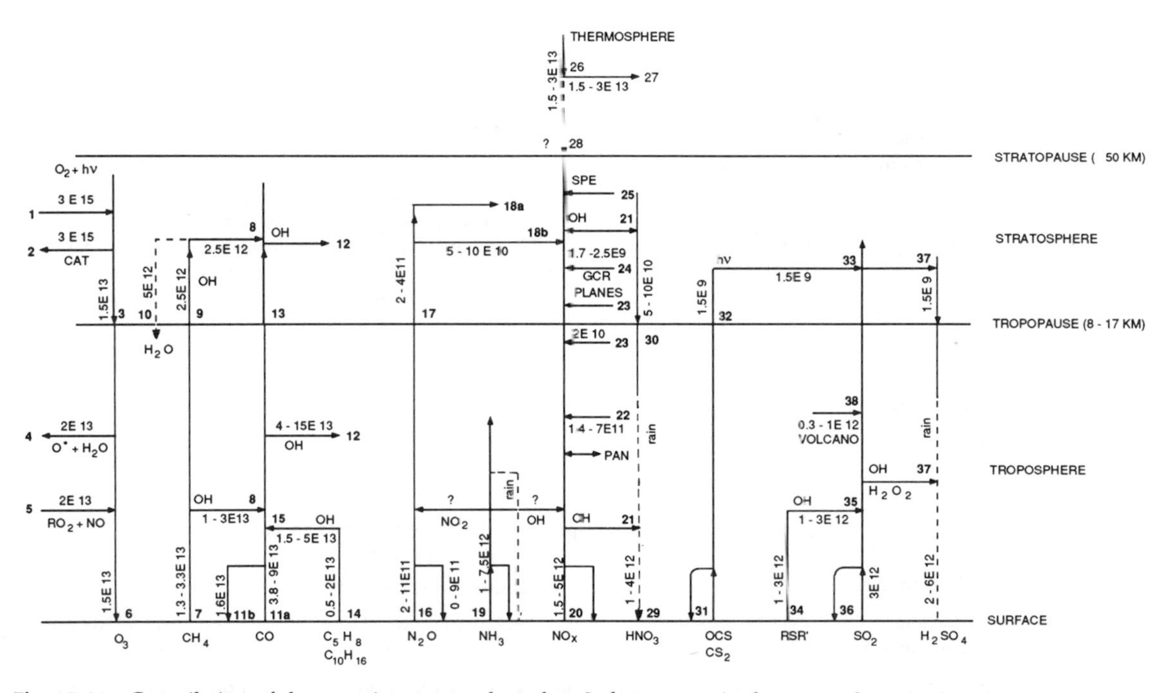

Fig. 10-11 Compilation of the most important photochemical processes in the atmosphere, including estimates of flux rates expressed in moles per year between the Earth's surface and the atmosphere and within the atmosphere.

1: $O_2 + h\nu \rightarrow 2O$

$O + O_2 + M \rightarrow O_3 + M (2\times)$

$3O_2 \rightarrow 2O_3$

2: $O_3 + h\nu \rightarrow O + O_2$

$XO + O \rightarrow X + O_2$

$X + O_3 \rightarrow XO + O_2$

$2O_3 \rightarrow 3O_2$

$X = NO, Cl, OH$

3: Downward flux to troposphere, small difference between 1 and 2

4: $O_3 + h\nu \rightarrow O(^1D) + O_2$

$O(^1D) + H_2O \rightarrow 2OH$

$HO_2 + O_3 \rightarrow OH + 2O_2$

5: $RO_2 + NO \rightarrow RO + NO_2$

$NO_2 + h\nu \rightarrow NO + O$

$O + O_2 + M \rightarrow O_3 + M$

$RO_2 + O_2 \rightarrow RO + O_3$

$R = H, CH_3$ etc from CO and hydrocarbon oxidation, e.g.

$CO + OH \rightarrow H + CO_2$

$H + O_2 + M \rightarrow HO_2 + M$

6: Ozone destruction at ground; difference between 3, 4 and 5.

7: Release of CH_4 at ground by variety of sources with range 1.3–3.3 E13 moles/year (E13 = 10^{13}). 2.5 E13 with average OH concentration of 6E 5 molecules cm^3

8: $CH_4 + OH \rightarrow CH_3 + H_2O$

$CH_3 + O_2 + H \rightarrow CH_3O_2 + M$

$CH_3O_2 + NO \rightarrow CH_3O + NO_2$

$CH_3O + O_2 \rightarrow CH_2O + HO_2$

$CH_2O + h\nu \rightarrow CO + H_2$

and other oxidation routes.

9: Flux of CH_4 to the stratosphere.

10: Flux of H_2O to the troposphere from methane oxidation.

11a: Release of CO from variety of sources, mostly man-made.

11b: Uptake of CO by microbiological processes in soils.

12: $CO + OH \rightarrow H + CO_2$.

Global loss of CO of 4–15E 13 moles/year; 7E 13 calculated with (OH) 6E 5 molecules cm^3.

15: Isoprene and terpene oxidation to CO following reaction with OH; oxidation mechanism and CO yield not well known.

16, 17: Release of N_2O to atmosphere by variety of sources; no significant sinks of N_2O in troposphere discovered; stratospheric loss estimated by model calculations.

Continued on p. 232

18a: $$N_2O + h\nu \rightarrow N_2 + O$$
$$N_2O + O(^1D) \rightarrow N_2 + O_2$$

18b: $$N_2O + O(^1D) \rightarrow 2NO$$

19: Release of NH_3 by variety of sources to atmosphere; redeposition at the ground; most ammonia removed by rain, but some NO_x loss and N_2O formation possible by reactions
$NH_2 + NO_x \rightarrow N_2O_x + H_2O$, while NO_x may be formed via
$$NH_2 + \rightarrow NH_2O + O_2$$

20: Release of NO_x at ground by variety of sources – redeposition at ground.

21: $$NO_2 + OH \rightarrow HNO_3$$
$$HNO_3 + h\nu \rightarrow OH + NO_2$$
$$HNO_3 + OH \rightarrow H_2O + NO_3$$

22: NO_x produced from lightning.

23: NO_x produced by subsonic aircraft.

24: NO_x from galactic cosmic rays.

25: NO_x from sporadic solar proton events; maximum production recorded in August 1972 event: 1 E10 moles.

26: NO production by fast photoelectrons in thermosphere and by auroral activity.

27: $$NO + N \rightarrow N_2 + O$$

28: Downward flux of NO to stratosphere; small difference between 26 and 27 may be important.

31, 32: COS destruction in stratosphere calculated with model; uptake of COS in oceans and hydrolysis may imply an atmospheric lifetime of only a few years and a source of a few times E10 moles per year. There is very little knowledge on the sources and sinks of COS and CS_2.

33: $$COS + h\nu \rightarrow S + CO$$
$$S + O_2 \rightarrow SO + O$$
$$SO + O_2 \rightarrow SO_2 + O$$

34: Release of H_2S, CH_3SCH_3 and CH_3SH by biological processes in soils and waters.

35: Oxidation of H_2S, CH_3SCH_3 and CH_3SH to SO_2 after initial attack by OH.

36: Industrial release of SO_2.

37: SO_2 oxidation to H_2SO_4 on aerosols, in cloud droplets, and by gas phase reactions following attack by OH.

38: Volcanic injections of SO_2; average over past centuries.

Modified from Crutzen (1983) with the permission of John Wiley and Sons, Inc.

Chemical interactions also occur in the condensed phases. Some of these are expected to be quite complex, e.g. the reactions of free radicals on the surfaces of or within aerosol particles. Simpler sorts of interactions also exist. Perhaps the best understood is the acid–base relationship of NH_3 with strong acids in aerosol particles and in liquid water. Often, the main strong acid in the atmosphere is H_2SO_4, and one may consider the nature of the system consisting of H_2O (liquid), NH_3, H_2SO_4, and CO_2 under realistic atmospheric conditions. Carbon dioxide is not usually important to the acidity of atmospheric liquid water (Charlson and Rodhe, 1982); the dominant effects are due to NH_3 and H_2SO_4. Figure 10-12 illustrates the sensitivity of the pH of cloud (*or* rainwater produced from it) to NH_3 and SO_4^{2-} aerosol acting as CCN in a cloud. Here, the familiar shape of a titration curve is evident, with a steep drop in pH as the anion concentration increases due to increased SO_4^{2-} aerosol input as CCN.

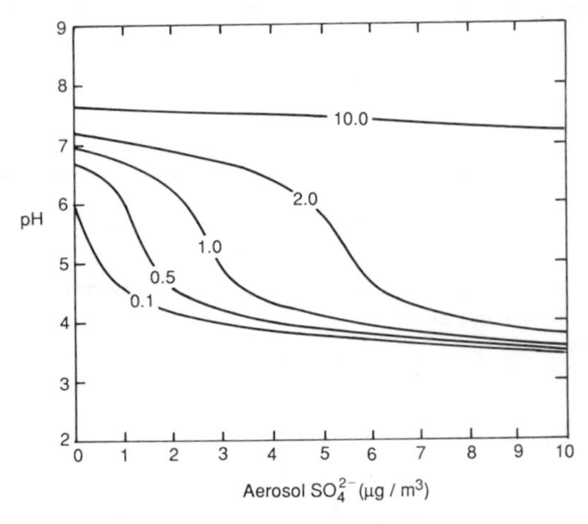

Fig. 10-12 pH sensitivity to SO_4^{2-} and NH_4^+: model calculations of expected pH of cloud water or rainwater for cloud liquid water content of 0.5 g/m³, 100 pptv SO_2, 330 ppmv CO_2, and NO_3^-. The abscissa shows the assumed input of aerosol sulfate in μg/m³ and the ordinate shows the calculated equilibrium pH. Each line corresponds to the indicated amount of total NH_3 + NH_4^+ in units of μg/m³ of cloudy air. Adapted from Vong and Charlson (1985) with the permission of the American Chemical Society.

10.10 Physical Transformations of Trace Substances in the Atmosphere

Perhaps because the unpolluted atmosphere can appear to be perfectly free of turbidity, it is not immediately obvious that it is a mixture of solid, gaseous, and liquid phases – even in the absence of clouds. Particles in the *aerosol* state (see box) comprise only a miniscule portion of the mass of the atmosphere – perhaps 10^{-9} or 10^{-10} in unpolluted cases. However, the condensed phases are important intermediates in the cycles of numerous elements, notably ammonia-N, sulfate-S, and organic C. They are also absolutely necessary participants in the hydrologic cycle (see Sections 10.7 and 10.11).

> **Explanatory Note:**
>
> *Aerosols* are solid or liquid particles, suspended in the liquid state, that have stability to gravitational separation over a period of observation. Slow coagulation is implied.

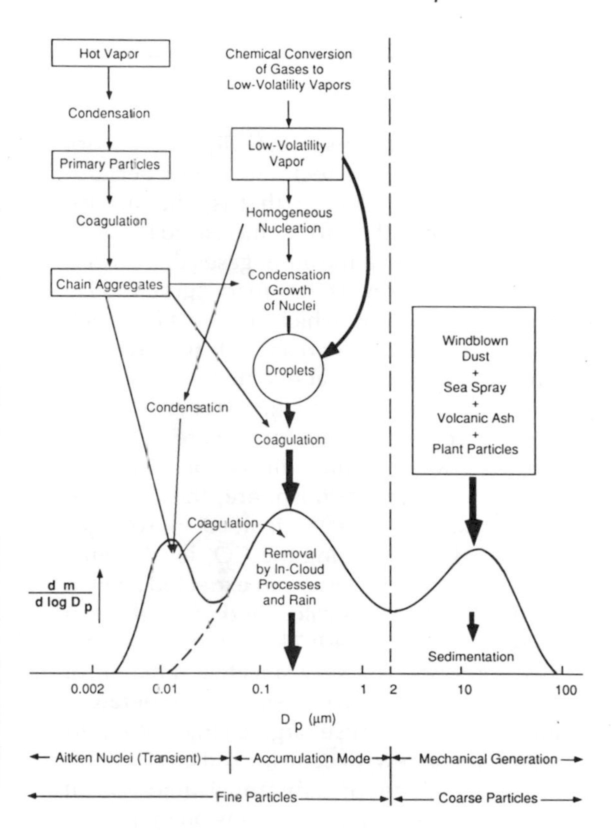

Fig. 10-13 Schematic of an atmospheric aerosol size distribution. This shows the three mass modes, the main sources of mass for each mode, and the principal processes involved in inserting mass into and removing mass from each mode (m = mass concentration, D_p = particle diameter). Adapted from Whitby and Sverdrup (1980) with the permission of John Wiley and Sons, Inc.

Figure 10-13 depicts the main physical pathways by which aerosol particles are introduced into and removed from the air. Processes that occur within the atmosphere also transform particles as they age and are transported. This form of distribution of mass with size was originally discovered in polluted air in Los Angeles, but it is now known to hold for remote unpolluted locations as well. In the latter case, the particle sizes of the accumulation mode (see Fig. 10-13) are probably somewhat smaller, perhaps by a factor of two.

Much of the fine particle aerosol is produced in the atmosphere by chemical reactions of gaseous precursors. Following the formation of very small *nuclei* (diameter less than *c.* 0.1 μm) by chemical processes (e.g. the oxidation of SO_2 to H_2SO_4), the physical process of Brownian coagulation, along with the deposition of new condensates from further reactions, causes the mass of nuclei to move up to dry diameters generally between 0.1 and 1.0 μm. The fine particle aerosol is thus composed of nuclei and larger conglomerates of material that has been accumulated. This is commonly called the *accumulation mode*. In addition, a *coarse* particle aerosol also exists, largely comprised of sea-salt and soil dust of surface derivation, with diameters usually larger than *c.* 1 μm. Except in cloud droplets, there is limited chemical contact between the coarse mode and the accumulation mode.

Because the particles in the accumulation mode are very small (most of them have diameters less than 1 μm when dry), they have very small fall speeds. Thus, they are only removed in any quantity by the formation of clouds with subsequent precipitation.

This brief description leads to Fig. 10-14, which depicts the physical transformations of trace substances that occur in the atmosphere. These physical transformations can be compared to the respective chemical transformations within the context of the

The Greenhouse Effect

Although greenhouses actually don't work the same way, this effect is so named because the result is the same; that is, the surface temperature of the Earth is increased because of the presence in the air of gases that absorb infrared radiation. The primary "greenhouse gas" is water vapor, which accounts for much of the observed increase of the average surface temperature above that expected for this planet without an atmosphere. If the albedo (reflectivity) of Earth were 0.17 (the same as Mars), and without any infrared absorption in the atmosphere, the temperature would be *c.* 260 K. Instead, it averages about 283 K. Increasing the CO_2 level from a 1980 level of 339 ppmv to an expected level in 2030 of 450 ppmv, is calculated to result in an increase of only *c.* +0.7 K. However, climate as we know it is a very sensitive function of temperature such that even a 1 K temperature increase would cause large changes if it persisted for a few decades. Predicted effects include melting of substantial amounts of continental ice, thermal expansion of the surface seawater, and increases in sea level. Thus, it is important to understand the influence of all con-tributors to the "greenhouse effect".

Figure 10-16 shows the expected temperature increase profiles for 2030 as a function of altitude. Here, another key feature of the infrared interactions is evident. In the troposphere, the temperature increase is expected to be relatively uniform with height, due largely to the vertical mixing of the troposphere by convective motions. In the stratosphere, the effect is just the opposite. The gases that absorb infrared also *emit* energy to space, such that the stratosphere becomes cooler as the concentration of greenhouse gases increases.

Finally, we might ask why it is possible for a gas like $CFCl_3$ (CFC-11) to have any effect at all when its concentration is only *c.* 0.2 ppbv? CO_2 is much more abundant (351 ppmv in 1988) and water vapor is dominant at the percent level. The reason for the sensitivity of the "greenhouse effect" to such gases lies in the details of their infrared absorption spectra. Specifically, gases that absorb strongly within that part of the infrared region where water vapor and CO_2 do not absorb strongly are the ones that can have the biggest effect. The so-called "window" region between about 7 and 12 μm wavelength is of particular importance because (a) water vapor and CO_2 do not absorb there and (b) the Earth's surface emission is a maximum at around 10 μm. Further details on the greenhouse effect may be found in Goody and Yung (1989), Goody and Walker (1972), and Ramanathan *et al.* (1985).

individual elemental cycles (e.g. sulfur). This comparison suggests that the overall lifetime of some species in the atmosphere can be governed by the chemical reaction rates, whereas others are governed by these physical processes.

10.11 Influence of Atmospheric Composition on Climate

10.11.1 Carbon Dioxide

Climate may be defined as the aggregate of all physical atmospheric properties and conditions. As such, it is absolutely clear that the chemical composition of the atmosphere as well as the physical characteristics of condensed phase trace species are of leading importance as determinants of climate. A well-known example is the increase in the temperature of the Earth's surface due to the absorption of infrared radiation from the Earth's surface by CO_2 in the air (see box). Without CO_2, the Earth's surface would be several degrees cooler than at present, depending on cloud cover, water vapor, and other controlling factors. Of course, there is substantial concern over the secular increase of CO_2, which will double from its pre-industrial level by the early to mid-twenty-first century.

10.11.2 Other "Greenhouse Gases"

Carbon dioxide is not the only gas that can influence terrestrial infrared radiation, and infrared absorption is not the only way that composition influences climate. Other gases that are important for their infrared absorption, sometimes known as "greenhouse gases", include CH_4, CCl_2F_2 (CFC-12), $CFCl_3$ (CFC-11), N_2O, and O_3. Figure 10-15 shows the relative effects of these, with the suggestion that taken together these other species are about of equal importance to CO_2. That some of these have increased dramatically due to human activities (the chlorofluorocarbons are only synthesized by man) or are increasing due to unknown causes (e.g. CH_4), suggests that the overall problem involves much more than just understanding or predicting CO_2.

10.11.3 Particles and Clouds

The condensed phases are also important to the physical processes of the atmosphere; however, their role in climate poses an entirely open set of scientific questions. The highest sensitivity of physical processes to atmospheric composition lies within the process of cloud nucleation. In turn, the albedo (or reflectivity for solar light) of clouds is sensitive to the number and properties of CCN (Twomey, 1977). At this time, it appears impossible to predict whether the temperature of the Earth might be expected to increase or decrease due to known changes in the concentrations of gases because aerosol and cloud effects cannot yet be predicted. In addition, since secular trends in the appropriate aerosol properties are not monitored very extensively, there is no way to know the degree to which changes have occurred, e.g. due to human activity.

10.12 Chemical Processes and Exchanges at the Lower and Upper Boundaries of the Atmosphere

10.12.1 CO_2, Photosynthesis, and Nutrient Exchange

The atmosphere, as a single body of gas, is in physical contact with the entire surface of the Earth. It extends upward toward space, with its density decreasing roughly a factor of ten every 16 km. Some processes of importance to geochemistry occur at the lower and upper boundaries. Evaporation of water from the Earth's surface is the first big step in the hydrologic cycle of the atmosphere. Indeed, this flux of water into and out of the atmosphere represents its largest flux and most massive cycle. Next is the exchange of CO_2 from the atmosphere to the biosphere via photosynthesis, with the return flow of CO_2 to the atmosphere via respiration, decay, and combustion. Many other trace substances also are exchanged with the biosphere in natural processes which occur at the Earth's surface, notably nitrogen and sulfur nutrient species.

10.12.2 Reactions at the Surface

Another major process at the Earth's surface not involving rapid exchange is the chemical weathering of rocks and dissolution of exposed minerals. In some instances, the key weathering reactant is H_3O^+ in rainwater (often associated with the atmospheric sulfur cycle), whereas in other cases H_3O^+ comes from high concentrations of CO_2 (e.g. in vegetated soils).

Numerous atmospheric species react with the Earth's surface, mostly in ways that are not yet chemically described. The dissolution and reaction of SO_2 with the sea surface, with the aqueous phase inside of living organisms, or with basic soils is one example. Removal of this sort from the atmosphere is usually called *dry removal* to distinguish it from removal by rain or snow. In this case, the removal flux is often empirically described by a *deposition velocity*, V_{dep}:

$$V_{dep} = \frac{F}{C} \quad (25)$$

where F = the flux per unit area and time in appropriate units and C = concentration in the atmosphere, in units to match that of the flux. At the Earth's surface, V_{dep} for many reactive species (e.g. SO_2, NO_2, O_3, HNO_3, etc.) is the order of 1 cm/s.

10.12.3 Cosmic Ray-induced Nuclear Reactions

At the top of the atmosphere, more properly at altitudes where the density is sufficiently low, high-energy cosmic ray particles cause nuclear chemical reactions with important products. The production of radioactive ^{14}C (or radiocarbon) has already been mentioned. Other radioisotopes known to be produced by cosmic rays include ^{10}Be, ^{3}H, ^{22}Na, ^{35}S, ^{7}Be, ^{33}P, and ^{32}P. Of these, ^{35}S, ^{7}Be, ^{32}P, and ^{33}P have activities that are high enough to be measured in rainwater. In several instances, notably ^{14}C and ^{7}Be, these radioactive elements are useful as tracers.

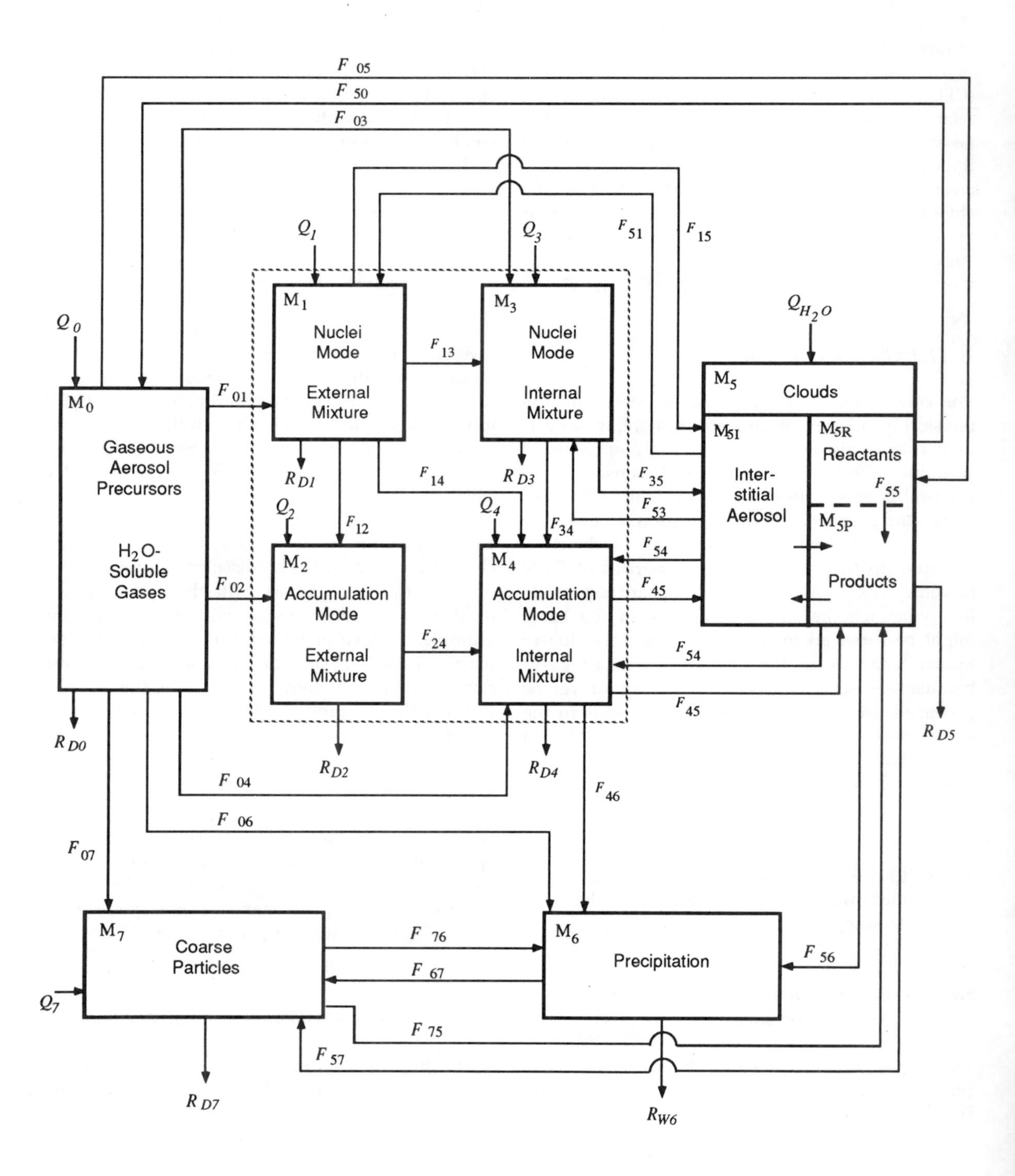

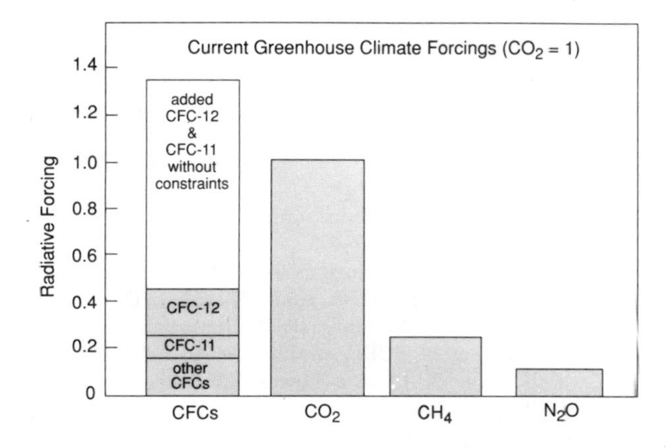

Fig. 10-15 Relative radiative forcing of the climate system (normalized to CO_2) due to increasing trace gases. Calculated radiative forcings are for smoothed 1979 annual trace gas increments. Radiative forcing for CFC-11 and CFC-12 added without constraints is based on extrapolations of the annual increase of these gases (8–11%) before 1974, when their production was severely curtailed. Adapted from Hansen *et al.* (1989) with the permission of the American Geophysical Union.

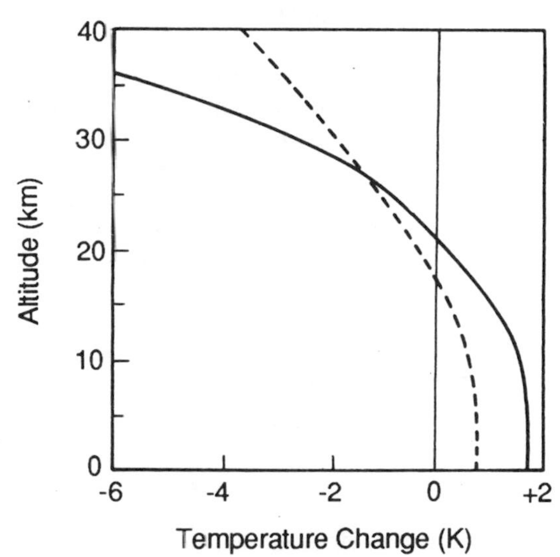

Fig. 10-16 Change in the vertical distribution of temperature expected in AD 2030 due to an increase in CO_2 alone, and CO_2 along with other radiatively important trace gases. Reproduced from Ramanathen *et al.* (1985) with the permission of the American Geophysical Union.

Fig. 10-14 Physical transformations of trace substances in the atmosphere. Each box represents a physically and chemically definable entity. The transformations are given in F_{ij} *(from the* ith to the jth box). Q_i represents sources contributing to the mass or burden, M_i, in the ith box. R_{Di} and R_{Wi} are dry and wet removals from M_i. The dashed box represents what may be called the fine particle aerosol and could be a single box instead of the set of four sub-boxes (i = 1, 2, 3, 4). The physical transformations are as follows:

F_{01}: production of new nuclei-mode particles.
F_{02}: growth of existing accumulation-mode particles by the deposition of products of chemical reactions.
F_{03}: growth of pre-existing nuclei-mode particles, as in F_{02}.
F_{04}: growth of internally mixed accumulation mode, as in F_{02}.
F_{05}: dissolution of gaseous reactants in cloud drops.
F_{50}: reverse of F_{05} evaporation or gaseous exchange or both.
F_{06}: below-cloud scavenging of gaseous reactants or reactant products.
F_{07}: interaction of gases with coarse particles, e.g. $HNO_3(g)$ + sea salt → coarse mode NO_3^-.
F_{12}: Brownian coagulation of nuclei-mode particles with themselves to produce accumulation-size (chemically) externally mixed particles.
F_{13}: adsorption, condensation.
F_{14}: coagulation.
F_{15}: cloud formation yielding nuclei-mode interstitial aerosol, coagulation with cloud droplets, or (unlikely) activation as a cloud condensation nuclei (CCN).
F_{51}: cloud evaporation releasing interstitial aerosol.
F_{57}: cloud evaporation releasing coarse particles.
F_{24}: adsorption, coagulation, condensation.
F_{34}: adsorption, coagulation, condensation.
F_{35}: cloud formation, as in F_{15}.
F_{53}: cloud evaporation, as in F_{51}.
F_{45}: cloud formation, likely to be CCN.
F_{54}: cloud evaporation, releasing CCN.
F_{55}: reactions in cloud water producing changes in solute mass.
F_{46}: below-cloud scavenging.
F_{56}: formation of precipitation.
F_{67}: evaporation of precipitation particles (raindrops) before reaching ground.
F_{75}: coarse particles acting as CCN.
F_{75}: below-cloud scavenging of coarse-mode particles.
R_{D5}: occult precipitation (deposition of cloud droplets directly to the Earth's surface, trees, etc.).

Reproduced from Charlson *et al.* (1985) with the permission of Kluwer Academic Publishers.

10.12.4 Escape of H and He

As mentioned at the beginning of this chapter, diffusive separation of low atomic or molecular weight species into space causes them to be permanently lost from the Earth. Thus, the Earth is deficient in He and H_2 relative to the best estimates of initial terrestrial composition. Some species might be accreted from space; certainly, micrometeorites represent a small but identifiable flux. Published speculations exist regarding other substances, notably water. However, these appear to be relatively unimportant at present.

References

Chapman, S. (1930). On ozone and atomic oxygen in the upper atmosphere. *Phil Mag. Ser. 7* **10** (64), 369–383.

Charlson, R. J. and H. Rodhe (1982). Factors influencing the natural acidity of rainwater. *Nature* **295**, 683.

Charlson, R. J., W. L. Chameides, and D. Kley (1985). The transformations of sulfur and nitrogen in the remote atmosphere. *In* "The Biogeochemical Cycling of Sulfur and Nitrogen in the Remote Atmosphere" (J. N. Galloway, R. J. Charlson, and M. O. Andreae, eds.), pp. 67–80. Reidel, Dordrecht, The Netherlands.

Crutzen, P. J. (1983). Atmospheric interactions: Homogeneous gas reactions of C, N and S containing compounds. *In* "The Major Biogeochemical Cycles and Their Interactions" (B. Bolin and R. Cook, eds.), pp. 67–112. John Wiley, Chichester.

Goody, R. M. and J. C. G. Walker (1972). "Atmospheres." Prentice-Hall, Englewood Cliffs, N.J.

Goody, R. M. and Y. L. Yung (1989). "Atmospheric Radiation." Oxford University Press, New York.

Graedel, T. E. (1978). "Chemical Compounds in the Atmosphere." Academic Press, New York.

Hansen, J., A. Lacis and M. Prather (1989). Greenhouse effect of chlorofluorocarbons and other trace gases. *J. Geophys. Res.* **94**, 16,417–16,421.

Junge, C. E. (1963). "Air Chemistry and Radioactivity." Academic Press, New York.

Junge, C. E. (1974). Residence variability of tropospheric trace gases. *Tellus* **26**, 477–488.

Köhler, H. (1936). The nucleus in and the growth of hygroscopic droplets. *Trans. Farday Soc.* **32**, 1152.

Mueller, P. K. and G. M. Hidy (1982). The sulfate regional documentation of SURE sampling sites. *EPRI Report EA-1901*, Vol. 3. Electric Power Research Institute, Palo Alto, Ca, USA.

Ramanathan, V., R. J. Cicerone, H. B. Singh, and J. T. Kiehl (1985). Trace gas trends and their potential role in climate change. *J. Geophys. Res.* **90**, 5547–5566.

Seiler, W. (1974). The cycle of atmospheric CO. *Tellus* **26**, 116–135.

Twomey, S. (1977). "Atmospheric Aerosols." Elsevier, Amsterdam, The Netherlands.

US Air Force and US Department of Commerce (1971). "Global Atlas of Relative Cloud Cover, 1967–1970." US Department of Commerce, Washington, D.C.

Vong, R. J. and R. J. Charlson (1985). The equilibrium pH of a cloud or rain drop: A computer-based solution for a six component system. *J. Chem. Ed.* **62**, 141–143.

Wallace, J. M. and P. V. Hobbs (1977). "Atmospheric Sciences: An Introductory Survey." Academic Press, New York.

Wayne, R. P. (1985). "Chemistry of Atmospheres." Clarendon Press, Oxford.

Whitby, K. T. and G. M. Sverdrup (1980). California aerosols: Their physical and chemical characteristics. *In* "The Character and Origins of Smog Aerosols" (P. K. Mueller, D. Grosjean, B. R. Appel, and J. J. Wesolowski, eds.), pp. 417–518. John Wiley, New York.

11

The Global Carbon Cycle

Kim Holmén

11.1 Introduction

Although many elements are essential to living matter, carbon is the key element of life on Earth. The carbon atom's ability to form long covalent chains and rings is the foundation of organic chemistry and biochemistry. The biogeochemical cycle of carbon is necessarily very complex, since it includes all life forms on Earth as well as the inorganic carbon reservoirs and the links between them. Despite being a complicated elemental cycle, it is extensively studied and, to date, probably the best understood. Lately, the possibility of global climatic changes brought about by the "greenhouse" effect of fossil fuel CO_2 in the atmosphere has also prompted much carbon-related research.

There are many review articles and books about the carbon cycle available with varying degrees of detail and points of emphasis (e.g. Bolin, 1970a,b; Keeling, 1973; Woodwell and Pecan, 1973; Woodwell, 1978; Bolin *et al.*, 1979; Revelle, 1982; Bolin and Cook, 1983; Degens *et al.*, 1984).

This chapter is an attempt to give an account of the fundamental aspects of the carbon cycle from a global perspective. An outline of the details we shall encounter is shown in Fig. 11-1. After a presentation of the main characteristics of carbon on Earth, four sections follow: a section about the carbon reservoirs within the atmosphere, the hydrosphere, the biosphere, and the lithosphere; a section covering the most important fluxes between the reservoirs; a section giving brief accounts of selected models of the carbon cycle; and a final section describing cultural influences on the carbon cycle today.

The relevant time-scales vary over many orders of magnitude, from millions of years for processes controlled by the movement of the Earth's crust, to day and even seconds for processes related to air–sea exchange and photosynthesis. Depending on the problem studied, models only include processes that work on similar time-scales. For example, most models used to study mankind's perturbation of atmospheric CO_2 exclude the geological processes working on time-scales longer than 5000 years and only include those processes that actively respond to atmospheric P_{CO_2} changes on the decadal time-scale. Although the CO_2 content in the atmosphere is modulated by changes in the exchange rate between atmosphere–ocean and atmosphere–biosphere, the level of the CO_2 concentration in the atmosphere is ultimately determined by geological processes. The P_{CO_2}-controlled erosion rate, together with volcanism, releases carbon from the lithosphere into the ocean–atmosphere–biosphere system. This system is counteracted by the sedimentation rate of carbon in the deep oceans. The balance between these two processes determines the CO_2 level in the atmosphere.

There are more than 1 million known carbon compounds, thousands of which are vital to life processes. The carbon atom's unique and characteristic ability to form long stable chains makes life itself possible. Elemental carbon is found free in nature in three allotropic forms: amorphous carbon, graphite, and diamond. Graphite is a very soft material, whereas diamond is well known for its hardness. Curiosities in nature, the amounts of elemental carbon on Earth are insignificant in a treatment of the carbon cycle. Carbon atoms have oxidation states ranging from +IV to −IV. The most common state is +IV in CO_2 and the familiar carbonate forms. Carbonate exists in two reservoirs,

Global Biogeochemical Cycles
ISBN 0-12-147685-5

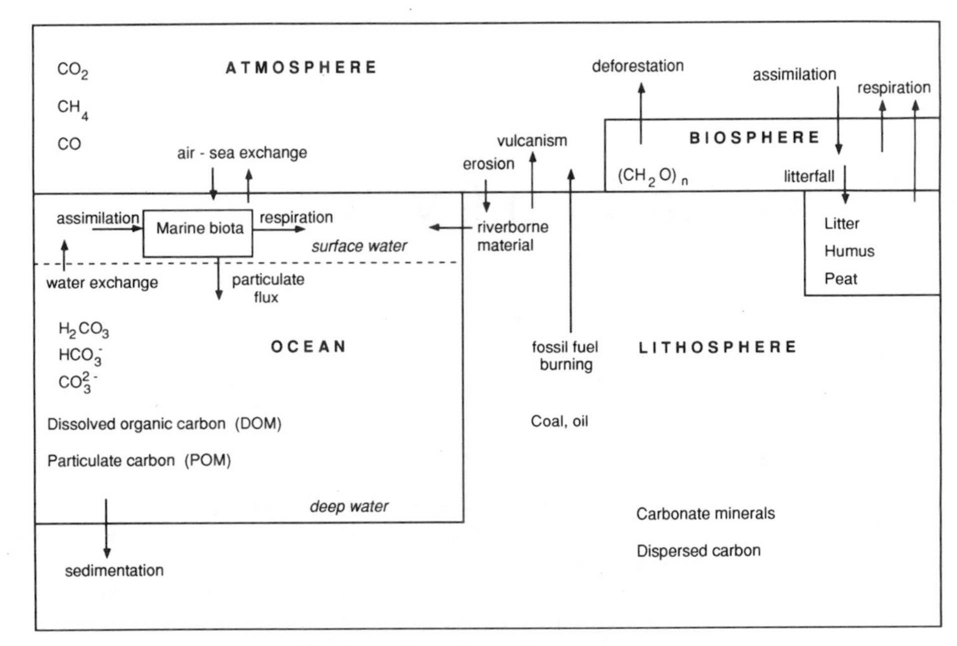

Fig. 11-1 Major reservoirs and fluxes of global carbon cycle.

in the oceans as dissolved carbon [$H_2CO_3(aq)$, $HCO_3^-(aq)$, $CO_3^{2-}(aq)$] and in the lithosphere as carbonate minerals [$CaCO_{3(s)}$, $CaMg(CO_3)_2$, $FeCO_3$]. Carbon monoxide, CO, is a trace gas present in the atmosphere with carbon in oxidation state +II. Assimilation of carbon by photosynthesis creates the reduced carbon (CH_2O) pools of the Earth. Reduced carbon is present in many forms that will be discussed separately. Methane, CH_4, another trace gas in the atmosphere, is formed by reduction of carbon by anaerobic bacteria or inorganic processes. Methane is the most reduced form of carbon with an oxidation state of −IV.

11.2 The Isotopes of Carbon

There are seven isotopes of carbon (^{10}C, ^{11}C, ^{12}C, ^{13}C, ^{14}C, ^{15}C, ^{16}C) of which two are stable (^{12}C and ^{13}C). The rest are radioactive with half-lives between 0.74 s (^{16}C) and 5726 years (^{14}C). Only the stable isotopes and ^{14}C (often referred to as "radiocarbon") are of interest in the carbon cycle.

The most abundant isotope is ^{12}C, which constitutes almost 99% of the carbon in nature. About 1% of the carbon atoms are ^{13}C. There are, however, small but significant differences in the relative abundance of the carbon isotopes in different carbon reservoirs. The differences in isotopic composition are important when estimating exchange rates between the reservoirs. Isotopic variations are caused by fractionation processes and, for ^{14}C, radioactive decay. Formation of ^{14}C takes place only in the upper atmosphere where neutrons generated by cosmic radiation react with nitrogen:

$$^{14}N + {}^{1}n \rightarrow {}^{14}C + {}^{1}p$$

The ^{14}C content of the material in a carbon reservoir is a measure of that reservoir's direct or indirect exchange rate with the atmosphere, although variations in solar activity (Stuiver and Quay, 1980, 1981) also create variations in atmospheric ^{14}C content. Geologically important reservoirs (i.e. carbonate rocks) contain no radiocarbon because the turnover times of these reservoirs are much longer than the isotope's half life. The distribution of ^{14}C is used in studies of ocean circulation and studies of the terrestrial biosphere.

Fractionation is another major process responsible for creating inhomogeneities in the isotope distribution. Physical, chemical, and biological processes may be sensitive to the molecular weights of the molecules involved. The definition of δ used to describe variations in isotope composition was introduced in Chapter 5. For ^{13}C, $\delta^{13}C$, in parts per thousand (‰) is defined by:

$$\delta^{13}C = [(^{13}R_S/^{13}R_0) - 1]\,1000. \qquad (1)$$

where $^{13}R_S$ is the $^{13}C/^{12}C$ ratio in the sample and $^{13}R_0$ is the $^{13}C/^{12}C$ ratio, the accepted standard PDB (Pee Dee belemnite, after a cretaceous belemnite from the Pee Dee formation in North Carolina). Craig (1957a) has determined R_0 to be 0.0112372.

The ^{14}C composition is described in a similar manner. The basis for $^{14}R_0$ is an oxalic acid standard of the US National Bureau of Standards normalized for ^{13}C fractionation and corrected for radioactive decay since a reference date January 1, 1950 (Stuiver and Polach, 1977). The absolute value of $^{14}R_0$ is 1.176×10^{-12} (Stuiver *et al.*, 1981).

^{14}C content is usually reported in $\Delta^{14}C$ units. The $\Delta^{14}C$ scale was originally defined by Broecker and Olson (1959). The reasoning behind the scale is that all variations in ^{14}C due to fractionation should be eliminated by correcting for the sample's observed $^{13}C/^{12}C$ ratio relative to that of postulated average terrestrial wood. The wood is assumed to have a $\delta^{13}C$ value of -25‰ and fractionation for the ^{14}C isotope is assumed to occur as the square of that for ^{13}C for all processes. The approximate expression for $\Delta^{14}C$ proposed by Broecker and Olson (1959) is:

$$\Delta^{14}C = \delta^{14}C_s - 2(\delta^{13}C_s + 0.025)(1 + \delta^{14}C_s) \qquad (2)$$

$\Delta^{14}C$ is useful in modelling, since no corrections for fractionation are necessary when modelling fluxes between reservoirs.

11.3 The Major Reservoirs of Carbon

11.3.1 The Atmosphere

Carbon is present in the atmosphere mainly as CO_2, with minor amounts present as CH_4, CO, and other gases. The CO_2 content of the atmosphere is probably the best known quantity of the global carbon cycle. Accurate measurements were begun in 1957 (Keeling *et al.*, 1976a,b; Bacastow and Keeling, 1981) with other groups following in the 1960s (Bischof, 1981) and 1970s (Pearman, 1981). The "Keeling" or "Mauna Loa" record is a classic piece of geophysical knowledge that illustrates some important aspects of carbon transfers in nature. The Mauna Loa and South Pole records are shown in Fig. 11-2. Seasonal variations are seen at both loca-

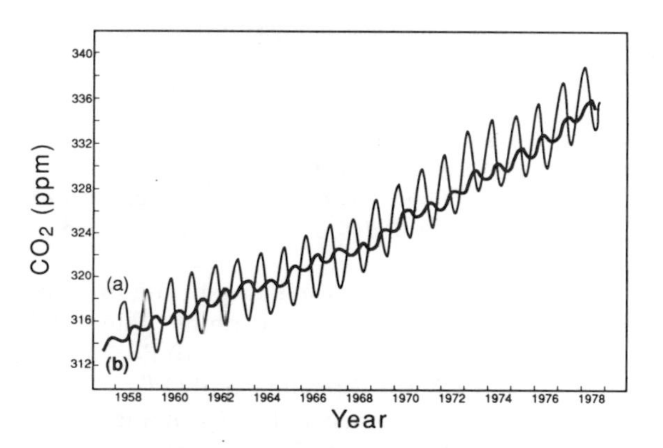

Fig. 11-2 Atmospheric CO_2 concentration (a) at Mauna Loa Observatory, Hawaii, 19.5°N, 155.6°W and (b) at the South Pole. Smoothed average of flask measurements adjusted to the 15th of each month. Modified from Bacastow and Keeling (1981) with the permission of John Wiley and Sons, Inc.

tions with a 6-month phase shift between the two hemispheres. The amplitude of the seasonal variations varies with latitude. The largest variations (10–15 ppmv) are seen at high latitudes, north of 50°N. In southerly high latitudes, the amplitude is only about 1 ppmv and along the equator there are small seasonal variations. The greater amplitude in the northern hemisphere is consistent with the extensive seasonal forests in that hemisphere. Another very obvious fact in the CO_2 record is the increasing concentration caused by mankind's perturbation of the carbon cycle. This is mainly due to the combustion of fossil fuel, but carbon mobilized from carbon pools on land, mainly oxidation of phytomass and soil organic carbon, is also significant. The atmospheric CO_2 concentration in 1988 (CDIAC, 1990) was 351.2 ppmv, which corresponds to 747 Pg (1 Pg $= 10^{15}$ g) of carbon; the 1959 concentration was 316 ppmv. Estimates of the pre-industrial CO_2 content vary considerably with values ranging between 243 and 290 ppmv. Fossil fuel emissions are well known: the trade in coal and oil has considerable economic value and is well documented. Assuming that all fossil fuel produced is oxidized within a few years, a good estimate of the total emissions can be made. For the period January 1, 1959 to January 1, 1978, the atmospheric increase was from 315.80 to 334.60 ppmv for an increased amount of 39.8 Pg C (Bacastow and Keeling, 1981), while the fossil fuel emissions were 72.573 Pg C (Rotty, 1981). This yields an observed airborne fraction of 0.55. (The observed airborne fraction is defined as the observed CO_2 increase divided by the amount produced by fossil fuel combustion. Since this quantity does not take any biospheric influences into account, it has limited value, although its use is widespread.)

There are a number of ways to estimate the pre-industrial atmospheric CO_2 content. The total emissions (from fossil fuel combustion) during the period 1850–1982 are estimated at 173 Pg C with an uncertainty of less than 10 Pg. Assuming a constant airborne fraction of 0.54, a pre-industrial atmospheric CO_2 content of 614 Pg (290 ppmv) is calculated (Bolin *et al.*, 1981). This calculation does not take into account any biospheric emissions, nor is the assumption of a constant airborne fraction perfectly sound. A recent modeling effort, where the ^{13}C variations in tree rings are used to determine magnitudes of fluxes from the different carbon reservoirs to the atmosphere, gives a historic CO_2 concentration of 243 ppmv (Peng *et al.*, 1983). Estimates of the pre-industrial CO_2 content based on the carbonate chemistry of "old" ocean (water that has yet to be contaminated by anthropogenic carbon emissions) give values ranging between 250 (Chen and Millero, 1979) and 275 $\pm$ 20 ppmv (Brewer, 1978). In a critical examination of CO_2 measurements performed in the 1880s, Wigley (1983) arrives at a CO_2 level around 260–270 ppmv. Finally, measurements of CO_2 content in air bubbles occluded in glacial ice from Antarctica and Greenland (Neftel *et al.*, 1982; Barnola *et al.*, 1983) give the least criticized data and indicate a value of 270 $\pm$ 10 ppmv.

Approximately 1% of the atmospheric carbon cycle is maintained by methane (Ehhalt, 1974). Global background levels of CH_4 are estimated at 1.7 ppmv, which corresponds to 3 Pg C (Blake and Rowland, 1988). Sources of methane are found in the sea and on land; however, the main sink for methane in the atmosphere is oxidation by the hydroxyl radical ($\cdot$OH).

Oxidation of methane is one of the sources of atmospheric CO. Another internal source of importance is the oxidation of terpenes and isoprenes emitted by forests (Crutzen, 1983). The carbon monoxide concentration in the atmosphere ranges from 0.05 to 0.20 ppmv in the remote troposphere (with considerable differences between the northern and southern hemispheres), which means that about 0.2 Pg of carbon is present as CO in the atmosphere.

Apart from CO_2, CH_4, and CO, there are many other gases in the atmosphere that contain carbon: terpenes, isoprenes, various compounds of petrochemical origin, and others. We will not discuss them further, although some, like dimethylsulfide [DMS: $(CH_3)_2S$], are of great importance in the biogeochemical cycles of other elements. The total amount of atmospheric carbon in forms other than the three discussed is estimated at 0.05 Pg C (Freyer, 1979).

11.3.2 The Hydrosphere

Oceanic carbon is mainly present in four forms: dissolved inorganic carbon (DIC), dissolved organic carbon (DOC), particulate organic carbon (POC), and the marine biota itself. The marine biota, although it is a small carbon pool with a standing crop of about 3 Pg C (De Vooys, 1979), has a profound influence on the distribution of many elements in the sea (Broecker and Peng, 1982). Primary production in the photic zone is the major input of organic carbon in the oceans (Mopper and Degens, 1979). Labile organic compounds are efficiently reoxidized in the mixed layer, whereas less than 10% of the primary production is distributed into the reservoirs of POC and DOC. The boundary separating POC and DOC is usually defined based on filtration with a 0.45-μm filter. Williams (1975) has used ^{14}C techniques to determine the average age of deep water DOC to be 3400 years. The DOC is thus clearly older than the turnover time of water in the deep oceans (100–1000 years), indicating the persistent nature of the dissolved organic compounds in the seas.

Detailed characterization of dissolved organic carbon is difficult to make; a large number of compounds have been detected, but only a small portion of the total DOC has been identified. Identified species include amino acids, fatty acids, carbohydrates, phenols, and sterols. The amount of carbon in the oceans as DOC has been estimated as 1000 Pg and the amount present as particulate organic carbon as about 30 Pg (Mopper and Degens, 1979).

DIC concentrations have been studied extensively since the appearance of a precise analytical technique (Dyrssen and Sillén, 1967; Edmond, 1970). The aquatic chemistry of CO_2 has been treated extensively; reviews can be found in Skirrow (1975), Takahashi *et al.* (1980), and Stumm and Morgan

(1981). When CO_2 dissolves in water, it may hydrate to form $H_2CO_3(aq)$, which in turn dissociates to HCO_3^- and CO_3^{2-}. The conjugate pairs responsible for most of the pH buffer capacity in seawater are HCO_3^-/CO_3^{2-} and $B(OH)_3/\ B(OH)_4^-$. Although the predominance of HCO_3^- at the oceanic pH of 8.2 actually places the carbonate system close to a pH buffer minimum, its importance is maintained by the high DIC concentration (~2 mM). Ocean water in contact with the atmosphere will, if the air–sea gas exchange rate is short compared to the mixing time with deeper waters, reach an equilibrium according to Henry's Law.

Two further reactions to be considered are the ionization of water and the borate equilibrium:

$$H_2O + B(OH)_3(aq) \leftrightarrow B(OH)_4^-(aq) + H^+(aq)$$

$$K_B = \frac{[H^+]\,[B(OH)_4^-]}{[B(OH)_3]} \tag{3}$$

In order to be able to solve for hydrogen ion concentration, we define total borate (ΣB) and total carbon ($\Sigma C \equiv$ DIC) as:

$$\Sigma B = [B(OH)_3] + [B(OH)_4^-]$$

$$\Sigma C = [H_2CO_3^*] + [HCO_3^-] + [CO_3^{2-}]$$

$[H_2CO_3^*]$ represents the sum of $CO_2(aq)$ and H_2CO_3. Alkalinity, a capacity factor, representing the acid-neutralizing capacity of the aqueous solution, is given by the following equation (see also Chapter 5):

$$\mathrm{Alk} = [OH^-] - [H^+] + [B(OH)_4^-] + [HCO_3^-] + 2[CO_3^{2-}] \tag{4}$$

Given any two of the four quantities ΣC, Alk, pH, P_{CO_2}, the other two can always be calculated provided appropriate equilibrium constants are available (the equilibrium constants depend on temperature, salinity, and pressure). The standard analytical technique for the carbonate species in seawater measures alkalinity and total carbon simultaneously in an acid titration. Hydrogen ion concentration can then be determined with the equation:

$$\mathrm{Alk} = \frac{[H^+]}{K_w} - [H^+] + \frac{\Sigma B \cdot K_B}{[H^+] + K_B} + \frac{K_1[H^+] + 2K_1K_2}{[H^+]^2 + [H^+]K_1 + K_1K_2}\Sigma C \tag{5}$$

where K_1 and K_2 are the dissociation constants for H_2CO_3. Alkalinity and ΣC are the analyzed values

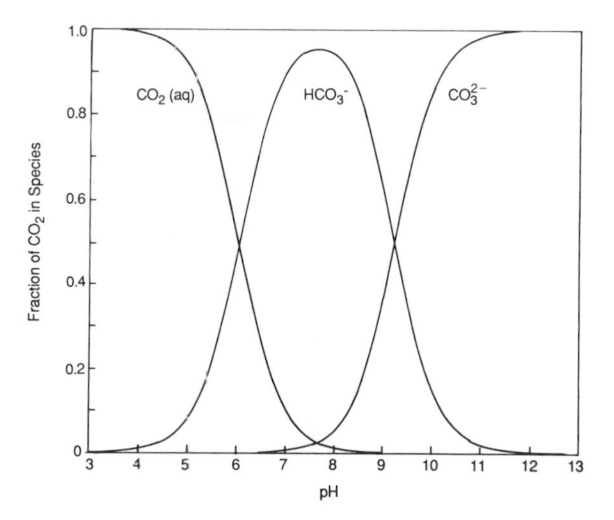

Fig. 11-3 Distribution of dissolved carbon species in seawater as a function of pH. Average oceanic pH is about 8.2. The distribution is calculated for a temperature of 15 °C and a salinity of 35‰. The equilibrium constants are from Mehrbach *et al.* (1973).

and ΣB is calculated from salinity. P_{CO_2} can then be calculated by:

$$P_{CO_2} = K_H\left(1 + \frac{K_1}{[H^+]} + \frac{K_1K_2}{[H^+]^2}\right)^{-1}\Sigma C \tag{6}$$

where K_H is the Henry's Law constant for CO_2 described in Chapter 5. At the pH of ocean water (about 8), most of the DIC is in the form of HCO_3^- and CO_3^{2-} (Fig. 11-3) with a very small proportion being $[H_2CO_3^*]$. Although $[H_2CO_3^*]$ changes in proportion to $CO_2(g)$, the ionic form changes little as a result of the various acid–base equilibria. This fact is responsible for the "buffer" factor (buffer here refers to buffering of CO_2 exchange) also known as the "Revelle" factor (Revelle and Suess, 1957; see Chapter 4 for an application of this factor). The buffer factor is defined by:

$$\beta = \frac{(\Delta P_{CO_2}/P_{CO_2})}{(\Delta\Sigma C/\Sigma C)} \tag{7}$$

Figure 11-4 shows how P_{CO_2}, H, and β are dependent on ΣC and Alk. For current atmospheric P_{CO_2}, prevailing temperature and salinity, β is about 14 in polar regions and about 10 in equatorial waters. The

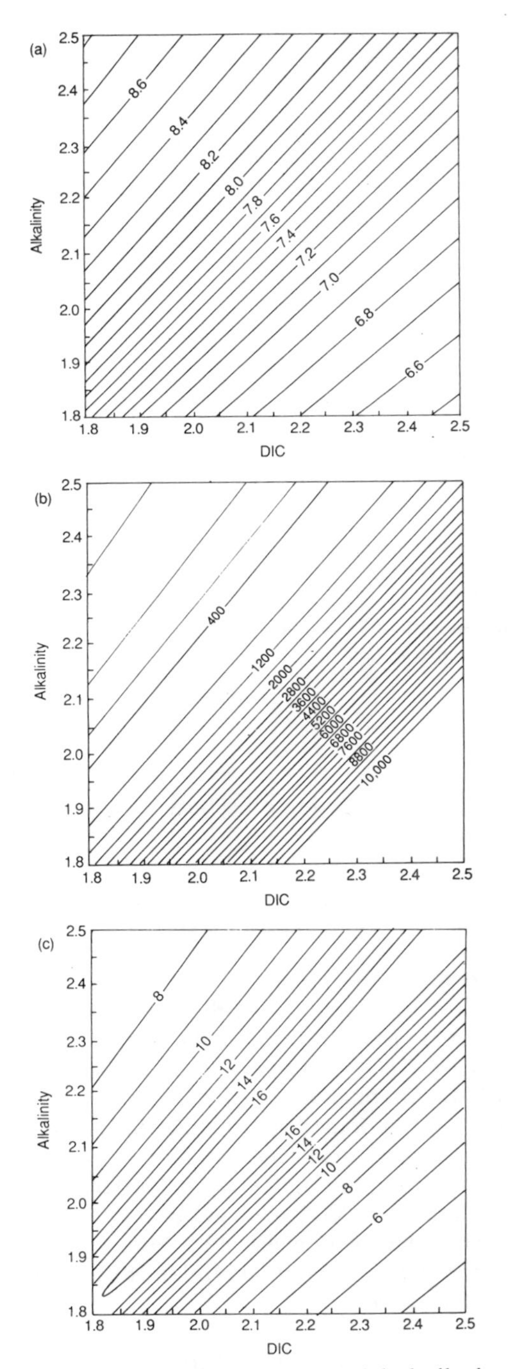

Fig. 11-4 Isolines of pH (a), P_{CO_2} (b), and the buffer factor (c) plotted as functions of DIC and alkalinity. The lines have been calculated for a temperature of 15 °C and a salinity of 35‰. The equilibrium constants for K_1 and K_2 are from Mehrbach *et al.* (1973), K_0 from Hansson (1973), and β calculated from salinity according to the formula given by Culkin (1965).

significance of $\beta \approx 10$ is that for a 10% increase in P_{CO_2}, only a 1% increase in ΣC is necessary to reach a new equilibrium. The buffer factor's large value is an important fact since it greatly constrains the ocean's ability to take up atmospheric CO_2. Average DIC and Alk concentrations for the world oceans can be seen in Fig. 11-5. With an average DIC of 2.35 mmol/kg seawater and a world oceanic volume of $1370 \times 10^6\, km^3$, the DIC carbon reservoir is estimated to be 37 900 Pg C (Takahashi *et al.*, 1981). The surface waters of the ocean contain a minor part of the dissolved inorganic carbon, 700 Pg C. Nevertheless, the surface waters play an important role as a means of communication between the atmosphere and the deep oceans. Although DIC is a large carbon reservoir with lively exchange with the atmosphere, its importance as a sink for anthropogenic CO_2 emission is restricted by several factors. The static uptake of the seawater (solubility and "buffer" factor), the degree of equilibrium between the ocean water and atmosphere, the ventilation of the deep ocean, and the oceanic sedimentation rate all impose constraints on the role of the oceans as a sink for atmospheric CO_2.

Oceanic surface water is everywhere supersaturated with respect to the two solid calcium carbonate species calcite and aragonite. Nevertheless, carbonate precipitation is exclusively controlled by biological processes, specifically the formation of hard parts (i.e. shells, skeletal parts, etc.). The very few existing accounts of spontaneous inorganic precipitation of $CaCO_{3(s)}$ (so-called "whitings") come from the Bahamas region of the Caribbean (Morse *et al.*, 1984).

The detrital rain of carbon-containing particles can be divided into two groups: the hard parts comprised of calcite and aragonite and the soft tissue containing organic carbon. The composition of the soft tissue shows surprising uniformity, the average composition being $(CH_2O)_{106}(NH_3)_{16}PO_4$ (see Chapter 9). The average composition of the particulate matter (here a composite of organic and inorganic particles) settling through the water column and subsequently being dissolved in the deep ocean, is given by $P:N:C:Ca:S = 1:15:131:26:50$ (Broecker and Peng, 1982), with a $CaCO_3$-C : Org-C ratio of 1 : 4. Calculating an average composition of the carbon that actually is deposited in sediments is more difficult since the areas of deposition are different for organic and inorganic carbon. More than 90% of the deposition of organic material takes place on the continental shelves; soft tissues falling into the deep

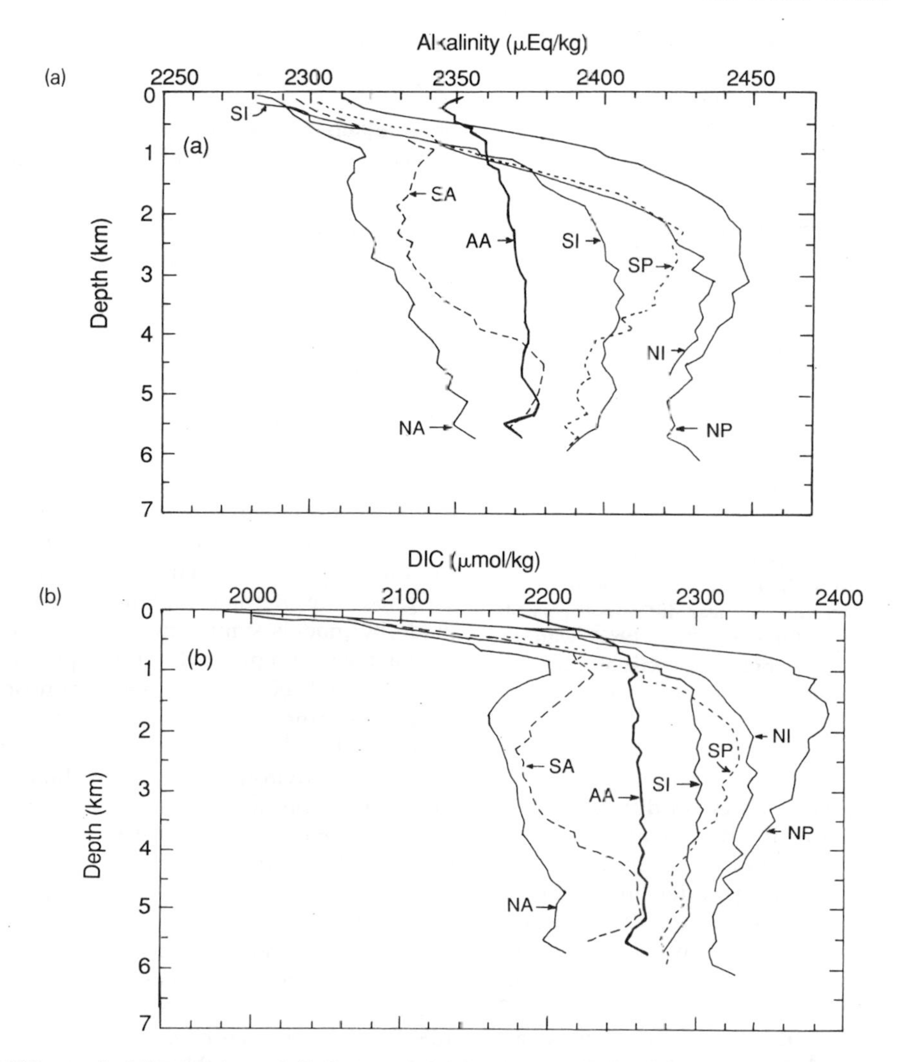

Fig. 11-5 (a) The vertical distributions of alkalinity and (b) dissolved inorganic carbon in the world oceans. Ocean regions shown are the North Atlantic (NA), South Atlantic (SA), Antarctic (AA), South Indian (SI), North Indian (NI), South Pacific (SP), and North Pacific (NP) Oceans. Modified from Takahashi *et al.* (1981) with the permission of John Wiley and Sons, Inc.

oceans are consumed by heterotrophic organisms before isolation from the water column within the sediments.

The solubility of calcite and aragonite increases with increasing pressure and decreasing temperature in such a way that deep waters are undersaturated with respect to calcium carbonate, whereas surface waters are supersaturated. The level at which the effects of dissolution are first seen on carbonate shells in the sediments is termed the lysocline and coincides fairly well with the depth of the carbonate saturation horizon. The lysocline commonly lies between 3 and 4 km depth in today's oceans. Below the lysocline is the level where no carbonate remains in the sediment – this level is termed the carbonate compensation depth.

The variations in ^{14}C seen in the deep oceans of the world (Fig. 11-6) essentially show features created by radioactive decay. The radiocarbon distribution is an important tool for determining the replacement times of the deep oceans. Great care has to be taken when interpreting the ^{14}C distribution to take into

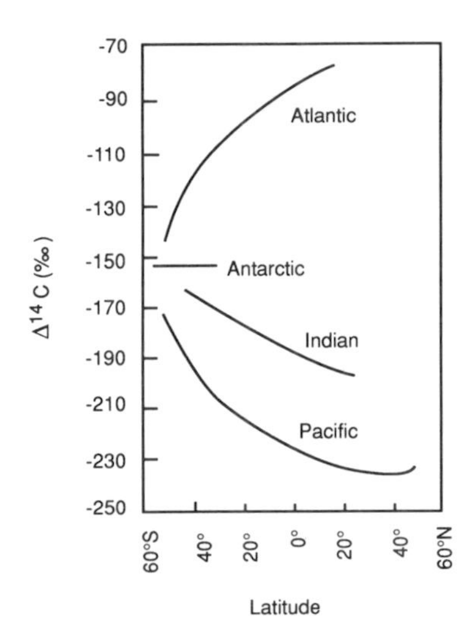

Fig. 11-6 The $\Delta^{14}C$ values of the cores of the North Atlantic, Pacific, and Indian Ocean deep waters. The oldest waters are encountered near 40°N in the Pacific Ocean. Modified from Stuiver *et al.* (1983) with the permission of the American Association for the Advancement of Science.

account mixing between waters of different origin. This is especially true in the Atlantic, since the degree of isotopic equilibrium is different in the two source areas of Atlantic deep water, the Arctic and Antarctic surface waters (Broecker, 1979). In other words, the apparent ^{14}C age in seawater is not simply a measure of the time elapsed since isolation from the atmosphere. Replacement times of 275 years for the Atlantic, 510 years for the Pacific, and 250 years for the Indian Ocean deep waters have been deduced (Stuiver *et al.*, 1983).

11.3.3 The Terrestrial Biosphere

Large amounts of carbon are found in the terrestrial ecosystems and there is a rapid exchange of carbon between the atmosphere, terrestrial biota, and soils. The complexity of the terrestrial ecosystems makes any description of their role in the carbon cycle a crude simplification, and we shall only review the most important aspects of organic carbon on land. Inventories of the total biomass of terrestrial ecosystems have been made by several researchers, and a survey of these is given by Ajtay *et al.* (1979).

Primary production maintains the main carbon flux from the atmosphere to the biota. In the process of photosynthesis, CO_2 from the atmosphere is reduced by autotrophic organisms to a wide range of organic substances. The complex biochemistry involved can be represented by the formula:

$$CO_2 + H_2O \underset{\text{respiration}}{\overset{\text{assimilation}}{\rightleftarrows}} (CH_2O)_n + O_2 \quad (8)$$

Gross primary production (GPP) is the total rate of photosynthesis including organic matter consumed by respiration during the measurement period, while net primary production (NPP) is the rate of storage of organic matter in excess of respiration. There are two main routes taken to estimate the world NPP and standing phytomass. The first method is to classify the biosphere into ecosystems in which, from measurements of estimates, values for the primary productivity and phytomass are assigned. The alternative method is to use estimates made by models simulating the effects of environmental factors on productivity and phytomass.

The possible effects of increased atmospheric CO_2 on photosynthesis are reviewed by Goudriaan and Ajtay (1979) and Rosenberg (1981). Increasing CO_2 in a controlled environment (i.e. glass house) increases the assimilation rate of some plants; however, the anthropogenic fertilization of the atmosphere with CO_2 is probably unable to induce much of this effect, since most plants in natural ecosystems are growth-limited by other environmental factors, notably light, temperature, water, and nutrients.

Estimates of terrestrial biomass vary considerably, ranging from 480 Pg C (Garrels *et al.*, 1973) to 1080 Pg C (Bazilevich *et al.*, 1970). Bazilevich *et al.* attempted to estimate the magnitude of the biomass before mankind's perturbation of the ecosystems. The major reasons for the mismatch in the estimates above is probably the difficulty in drawing sharp distinctions between biomes. For example, when does a savannah with its cover of bushes and patches of trees grade into woodlands or grasslands? The latest work that undoubtedly had the most data available estimates the total terrestrial biomass, valid as of 1970, as 560 Pg C (Olson *et al.*, 1983).

Terrestrial biomass is divided into a number of sub-reservoirs with different turnover times. Forests contain 90% of all carbon in living matter on land but their NPP is only 60% of the total. About half of the primary production in forests is in the form of twigs, leaves, shrubs, and herbs that only make up 10% of the biomass. Carbon in wood has a turnover time of the order of 50 years, whereas turnover times of carbon in leaves, flowers, fruits, and rootlets are less than a few years. When plant material becomes detached from the living plant, carbon is moved from the phytomass reservoir to litter. "Litter" can either refer to a layer of dead plant material on the soil or all plant materials not attached to a living plant. A litter layer can be a continuous zone without sharp boundaries between the obvious plant structures and a soil layer containing amorphous organic carbon. Decomposing roots are a kind of litter that seldom receives a separate treatment due to difficulties in distinguishing between living and dead roots. Total litter is estimated as 60 Pg C and total litterfall as 40 Pg C/year (Ajtay *et al.*, 1979). The average turnover time for carbon in litter is thus about 1.5 years, although caution should be observed when using this figure. For tropical ecosystems with mean temperatures above 30° C, the litter decomposition rate is greater than the supply rate and so storage is impossible. For colder climates, NPP exceeds the rate of decomposition in the soil. The average temperature at which there is balance between production and decomposition is about 25° C. The presence of peat, often treated as a separate carbon reservoir, exemplifies the difficulty in defining litter. The total amount of peat is estimated at 165 Pg C (Ajtay *et al.*, 1979).

Humus is a group of organic compounds in terrestrial ecosystems that is not readily decomposed and therefore makes up a carbon reservoir with a long turnover time. There are also significant structural differences between the marine and terrestrial substances (Stuermer and Payne, 1976). The soil organic matter of humus is often separated into three groups similar in structural characteristics but with differing solubility behavior in water solutions. Humic acids, fulvic acids, and humin are

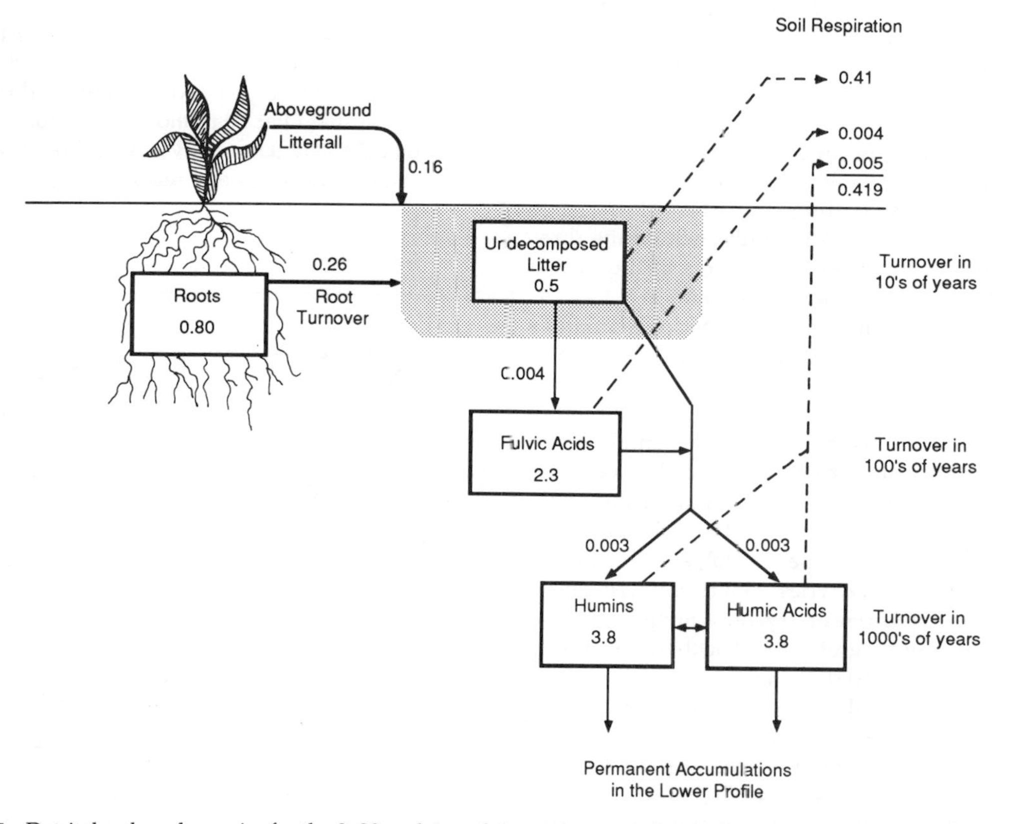

Fig. 11-7 Detrital carbon dynamics for the 0–20 cm layer of chernozem grassland soil. Carbon pools (kg C/m^2) and annual transfers (kg C/m^2 year) are indicated. Total profile content down to 20 cm is 10.4 kg C/m^2. Adapted from Schlesinger (1977) with the permission of Annual Reviews, Inc.

discussed in Chapter 7. Schlesinger (1977) presented an assessment of the various carbon pools for a temperate grassland soil (Fig. 11-7). The undecomposed litter (4% of the soil carbon) has a turnover time measured in tens of years, and the 22% of the soil carbon in the form of fulvic acids is intermediate with turnover times of hundreds of years. The largest part (74%) of the soil organic carbon (humins and humic acids) also has the longest turnover times (thousands of years).

The $\Delta^{14}C$ content of newly formed terrestrial biota is close to 0‰. Transformation of living plants into the various carbon reservoirs with longer turnover times is not associated with fractionation processes; $\delta^{13}C$ is close to -25‰ for all terrestrial organic material, whereas $\Delta^{14}C$ declines with the age of the material. Humus typically has a $\Delta^{14}C$ near -50‰.

11.3.4 The Lithosphere

Although the largest reservoirs of carbon are found in the lithosphere, the fluxes between the lithosphere and the atmosphere, hydrosphere, and biosphere are small. It follows that the turnover time of carbon in the lithosphere is many orders of magnitude longer than the turnover times in any of the other reservoirs. Many of the current modeling efforts studying the partitioning of fossil fuel carbon between different reservoirs only include the three "fast" spheres, the lithosphere's role in the carbon cycle having received much less attention.

Fossil fuel burning is an example of mankind's ability to alter fluxes between reservoirs significantly. The burning of fossil fuel transfers carbon from the vast pool of reduced carbon in the lithosphere to the atmosphere. The elemental carbon reservoir is estimated from average carbon contents in different types of rocks (ranging from 0.9% elemental carbon in shales to 0.1% in igneous and metamorphic rocks (Kempe, 1979b)) and the relative abundance of the rock types. The resulting estimate is 20 million Pg C (Hunt, 1972), a single reservoir several orders of magnitude larger than the sum of all reservoirs discussed so far. Of the 2.0×10^6 Pg of recycled elemental carbon (recycled carbon has traveled at least once through the lithospheric cycle) in the lithosphere, only 10^4 Pg make up the economically extractable reserves of oil and coal. Most of the reduced carbon species in the Earth's crust are highly dispersed and probably never will be used as fuels. The carbonate minerals distributed in sedimentary rocks represent a carbon reservoir that is even larger than the elemental carbon reservoir. About 75% of the carbon in the Earth's crust is present as carbonates. Several forms exist, the dominant biogenic forms being calcite and aragonite. Both are stoichiometrically $CaCO_3$, but calcite has six-coordinated Ca atoms and is capable of substituting several percent Mg into its lattice. Aragonite has nine-coordinated Ca atoms and several percent Sr can be incorporated into its lattice. Both forms can precipitate depending on the Ca/Mg ratio in the solution; for the present ocean Mg-calcite or aragonite are precipitated. (As mentioned earlier, this is not a widespread process today.) Dolomite [$CaMg(CO_3)_2$] is a carbonate mineral of wide importance formed by diagenetic disintegration of Mg-rich calcites. Formation of dolomite is slow today and largely confined to evaporitic settings (Holland, 1978). The invasion of the land by plants 600 million years ago at the beginning of the Phanerozoic increased the availability of CO_2 in the soil. There was a marked decrease of dolomitic sediments and an increase in limestone sediments coherent with the appearance of terrestrial vegetation (Fig. 11-8).

The amount of recycled carbonate in the lithosphere is estimated at 70 million Pg C (Hunt, 1972). Holland (1978) calculates a juvenile carbon reser-voir (carbon that never has been released from the litho-

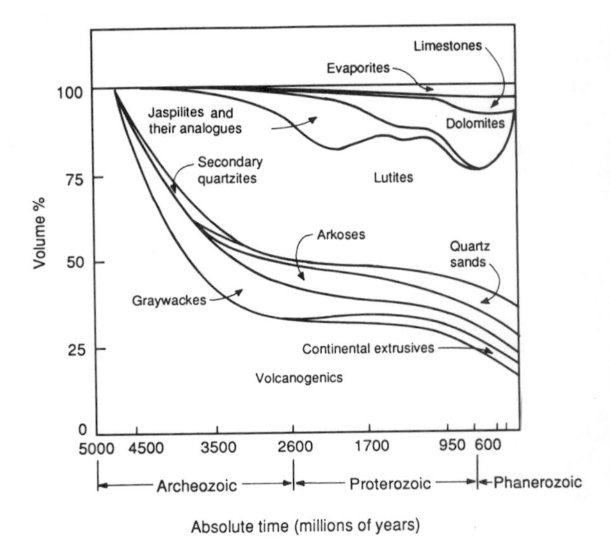

Fig. 11-8 Volume percent of sedimentary rocks as a function of age. Modified from Ronov (1964) with the permission of the American Geological Institute.

sphere) of the order of 90 million Pg C assuming that the Earth is half degassed. Although there is no ^{14}C, there are some characteristic features in the ^{13}C distribution. The formation of carbonates does not involve fractionation processes, and the carbonate minerals have $\delta^{13}C$ values close to 0‰. Non-carbonate carbon has a $\delta^{13}C$ value of -25‰. Feux and Baker (1973) have compiled ^{13}C data for various types of rocks, which they combine with the known distribution of sedimentary organic and carbonate carbon to estimate the $\delta^{13}C$ of the Earth's mantle as -7‰.

11.4 Fluxes of Carbon between Reservoirs

Carbon is released from the lithosphere by erosion and resides in the oceans 10^5 years before being deposited again in some form of oceanic sediment. It remains in the lithosphere on the average 10^8 years before again being released by erosion (Broecker, 1973). The amount of carbon in the ocean–atmosphere–biosphere system is maintained in a steady-state by geological processes; the role of biological processes is mainly to redistribute the carbon among the reservoirs.

Chemical weathering of crustal material can both add and withdraw carbon from the atmosphere. The oxidation of elemental and organic carbon releases CO_2 to the atmosphere:

$$C^0(s) + O_2 \rightarrow CO_2$$

whereas dissolution of carbonates is associated with uptake of CO_2:

$$CaCO_3(s) + CO_2(g) + H_2O \rightarrow Ca^{2+}(aq) + 2HCO_3^-(aq)$$

Silicates behave similarly to carbonates. Weathering of a non-aluminous silicate like Mg-olivine may be written:

$$Mg_2SiO_4(s) + 4CO_2 + 4H_2O \rightarrow 2Mg^{2+} + 4HCO_3^- + H_4SiO_4(aq)$$

An example of alumino silicate weathering is the reaction of the feldspar albite to a montmorillonite-type mineral:

$$2NaAlSi_3O_8(s) + 2CO_2 + 2H_2O \rightarrow Al_2Si_4O_{10}(OH)_2(s) + 2Na^+ + 2HCO_3^- + 2SiO_2(s)$$

In the weathering of carbonates, 1 mole of rock CO_2 is mobilized for each mole of atmospheric CO_2 consumed. The reverse reaction (sedimentation of carbonate in the oceans) will again release 1 mole equivalent of $CO_2(g)$. For the weathering of silicates, there is a 1 : 1 relationship between $CO_2(g)$ consumed and HCO_3^- produced, in contrast to the 1:2 relationship for carbonate weathering. Estimates of global erosion rates are based either on average river data (Livingstone, 1963; Kempe, 1979b) or on global material balance calculations performed by extrapolating the material balance from a well-documented area to the world (Kempe, 1979b).

The freshwater cycle is an important link in the carbon cycle as an agent of erosion and as a necessary condition for terrestrial life. Although the amount of carbon stored in freshwater systems is insignificant as a carbon reservoir (DeVooys, 1979; Kempe, 1979a), about 90% of the material transported from land to oceans is carried by streams and rivers.

Pure water in equilibrium with atmospheric CO_2 has a pH of about 5.6 at ambient temperatures. Although rainwater pH is affected by other airborne species with acid–base characteristics (Charlson and Rodhe, 1982), calculating the flux of carbon carried from the atmosphere to the surface by pH 5.6 rainwater is a good approximation. The flux proves to be small: 0.065 Pg C/year (Kempe, 1979a). According to the weathering reaction, half the bicarbonate in stream water originates in the atmosphere. Rainwater clearly is unimportant as a source for this carbon. Soil air is the major source of CO_2 taking part in weathering reactions with carbonates and silicates. Bacterial decomposition of organic material together with root respiration maintain high partial pressures of CO_2 in soil air (see Chapter 7; Kempe, 1979a). Approximately 0.4 Pg C/year are withdrawn from the atmosphere by weathering reactions, 0.1 Pg C/year is released by oxidation of elemental carbon, yielding a net flux of 0.3 Pg C/year from atmosphere to lithosphere (Holland, 1978).

Garrels and Mackenzie (1971) calculated global river loads based on Livingstone's (1963) data. From these figures, Kempe (1979a) deduced the following fluxes of carbon to the oceans transported by rivers:

DIC	0.45 Pg/year
PIC	0.20 Pg/year
DOC	0.12 Pg/year
POC	0.07 Pg/year
Total	0.84 Pg/year

Kempe (1979b) also estimates the flux of carbon from the lithosphere to oceans from glacial erosion

(0.033 Pg/year), global dust production (0.06 Pg/year), and marine erosion (0.0045 Pg/year) to give a combined flux of 0.1 Pg/year. Although there is a general consensus that geological processes control the amount of carbon in the ocean–atmosphere system, the mechanisms are debated. Walker (1977) assumes a steady-state model where the weathering rate is dependent on the CO_2 partial pressures in the atmosphere. An increase in P_{CO_2} will give a higher weathering rate which results in an increase of the cation content of the oceans. The increased cation concentration yields a higher precipitation rate of carbonates which removes carbon, thereby restoring the original balance. The atmospheric carbon dioxide budget is dominated by a volcanic source and consumption by the weathering of silicates. Volcanism does not depend on P_{CO_2}, but weathering does. The equilibrium level of CO_2 is therefore determined by the demand that the weathering sink must just balance the volcanic source. Holland (1978) presents somewhat different arguments but the picture is essentially the same. However, from balance calculations, he finds indications of a juvenile carbon flux from the Earth's interior of 0.08 Pg C/year.

The exchange of carbon between the terrestrial biosphere and atmosphere goes through two channels. CO_2 is the major route with CH_4 making up about 1% of the exchange. Methane is mainly produced by enteric fermentation by animals, and anaerobic production in paddy fields, freshwater lakes, swamps, and marshes. Ehhalt (1974) estimates an upper limit for the production of methane from the assumption that 10% of the dry plant matter produced on land is consumed by herbivores and the food-to-methane conversion ratio for cattle is valid for all herbivores (cattle have the highest measured ratio). The present upper limit is 0.17 Pg C/year, with a range down to 0.08 Pg C/year. Interestingly, the increase in the world cattle population since the early 1940s accounts for about 25% of this figure. The release of methane from paddy fields is estimated by Ehhalt (1974) to be 0.021 Pg C based on a series of flux measurements on paddy fields in Japan (Koyama, 1963). Methane emissions to the atmosphere from freshwater lakes, swamps, and marshes are in the range 0.15–0.22 Pg C/year. The total flux of methane to the atmosphere is thus 0.5 Pg C/year.

Carbon monoxide emissions from the terrestrial biosphere are small, but forest fires produce 0.02 Pg C/year. Degradation of chlorophyll in dying

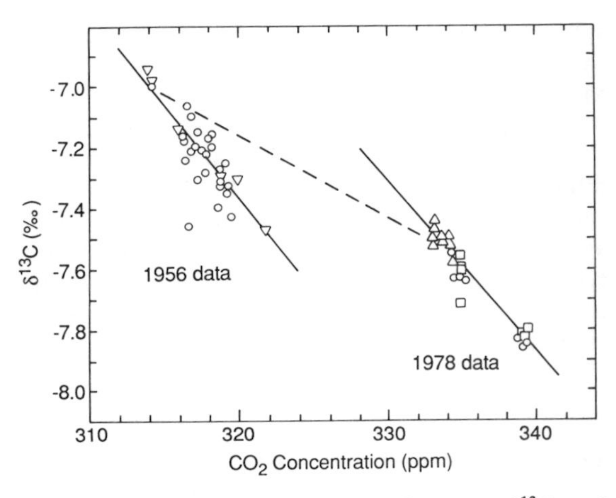

Fig. 11-9 Change in the relation between $\delta^{13}C$ and concentration of atmospheric CO_2 over 22 years. Mean change is shown as a dashed line. Solid lines show mixing relations for 1956 and 1978. Symbols refer to different data sets from western USA. Modified from Keeling *et al.* (1979) with the permission of Macmillan Magazines Ltd.

plant material seems to be the largest CO-producing mechanism at 0.04–0.2 Pg C/year (Freyer, 1979).

The exchange of CO_2 between atmosphere and terrestrial biota is one of the prime links in the global carbon cycle. This is seen by studying the variations of ^{13}C in the atmosphere. Figure 11-9 presents atmospheric $\delta^{13}C$ for the years 1956 and 1978. The lines are consistent with addition or subtraction of CO_2 with a $\delta^{13}C$ of about −27‰. This CO_2 could be derived from either fossil fuel or plants. It cannot be oceanic, since surface water DIC has a $\delta^{13}C$ of about +2‰ (Kroopnick, 1980). This confirms that the annual P_{CO_2} variations are primarily due to exchange with the terrestrial biosphere, and not caused by seasonal exchange with the oceans.

Many estimates of total terrestrial net primary production are available, with the most recent ones ranging between 45.5 Pg C/year (Lieth, 1972) and 78 Pg C/year (Bazilevich *et al.*, 1970). Ajtay *et al.* (1979) have revised the various estimates and methods involved, and they also reassess the classifications of ecosystem types and the extent of the ecosystem surface area using new data and arriving at a total NPP of 60 Pg C/year. Gross primary production is estimated to be twice net primary production, i.e. 120 Pg C/year. This implies that about 60 Pg C/year are returned to the atmosphere during the respiratory phase of photosynthesis. It is well known that carbon dioxide uptake by plants follows daily cycles; most plants take up CO_2 during the day and emit it at night. These diurnal patterns give rise to large variations in P_{CO_2} close to the vegetation sites (Fig. 11-10). The diurnal cycles that produce local changes are superimposed on the yearly cycles that give the hemispheric P_{CO_2} variations seen in Fig. 11-2.

The subsequent fate of the assimilated carbon depends on which biomass constituent the atom enters. Leaves, twigs, and the like enter litterfall, decompose, and recycle the carbon to the atmosphere within a few years, whereas carbon in stemwood has a turnover time counted in decades. In a steady-state ecosystem, the net primary production is balanced by the total heterotrophic respiration plus other outputs. The non-respiratory outputs to be considered are fires and transport of organic material to the oceans. Fires mobilize about 7 Pg C/year (Baes *et al.*, 1976), most of which is converted to CO_2. Since bacterial heterotrophes are unable to oxidize elemental carbon, the production rate of pyroligneous graphite, a product of incomplete combustion, is an interesting quantity to assess. The inability of the biota to degrade elemental carbon puts carbon into a reservoir that is effectively isolated from the atmosphere and oceans. Seiler and Crutzen (1980) estimate the production rate of graphite to be 1 Pg C/year. River transport of organic carbon, estimated earlier as 0.1 Pg C/year, brings the sum of non-respiratory outputs to 7 Pg C/year. Total respiration should therefore be around 50 Pg C/year. This figure is in agreement with estimates of soil respiration rates determined from compilations of ecosystem types and their measured soil respiration rates (Ajtay *et al.*, 1979).

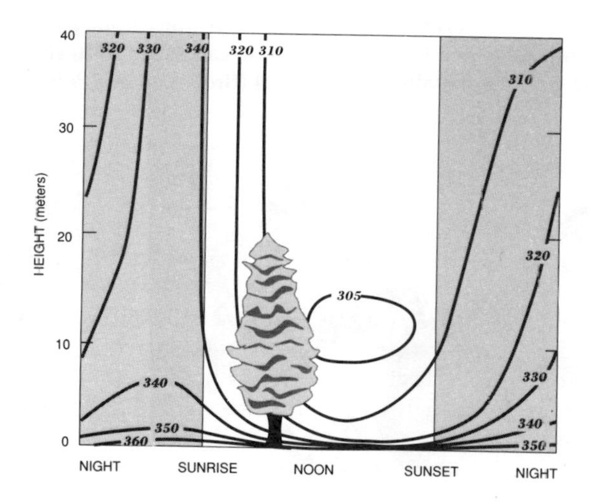

Fig. 11-10 Variation in the vertical distribution of carbon dioxide in the air around a forest with time of day. CO_2 isopleths are in ppm. Adapted from Bolin (1970c) with the permission of W. H. Freeman.

The oceans mediate some important carbon fluxes. The exchange of carbon dioxide between ocean and atmosphere has been studied extensively, since the prevailing view is that fossil fuel derived CO_2 not remaining in the atmosphere has entered the oceans. To appraise the ocean–atmosphere exchange, we make use of the radiocarbon distribution in the oceans. All ^{14}C is produced in the atmosphere; hence all radiocarbon in the oceans must have entered through the air–sea interface. Under a steady-state assumption, the net influx of ^{14}C must be balanced by the total decay within the oceans. Using our knowledge of the ^{14}C distribution in the oceans (Fig. 11-5) and the radiocarbon decay constant for ^{14}C, we can calculate the flux of carbon by the following simple relations:

$$F_{ma} = F_{am}$$

$$F_{am} R_a = F_{ma} R_m + k R_o V_o C_o$$

where F_{ma} is the flux from mixed layer to atmosphere, F_{am} is the flux from atmosphere to mixed layer, k is the ^{14}C decay rate, R_a, R_m, and R_o are the ^{14}C ratios in atmosphere, mixed layer, and average ocean respectively, V_o is the ocean volume, and C_o an average DIC concentration of the oceans. Solving for F_{am}, we obtain the following.

$$F_{am} = (k R_o V_o C_o/(R_a - R_m) \cong 6.5 \times 10^{15} \text{ mol/year} \quad (9)$$

The gross flux of carbon from atmosphere to ocean is thus 80 Pg C/year. There are several complications with the above calculation. The isotopic ratios must be steady-state values that are unavailable due to the changes resulting from atmospheric atom bomb testing. The few available pre-bomb measurements from the late 1950s (Broecker *et al.*, 1960), together with $\Delta^{14}C$ determinations in corals (Druffel and Linick, 1978), are invaluable tools for determining a steady-state R_m value. Nevertheless, we must be aware of the great sensitivity of the flux estimate to the R_m value since F_{am} is dependent on the reciprocal

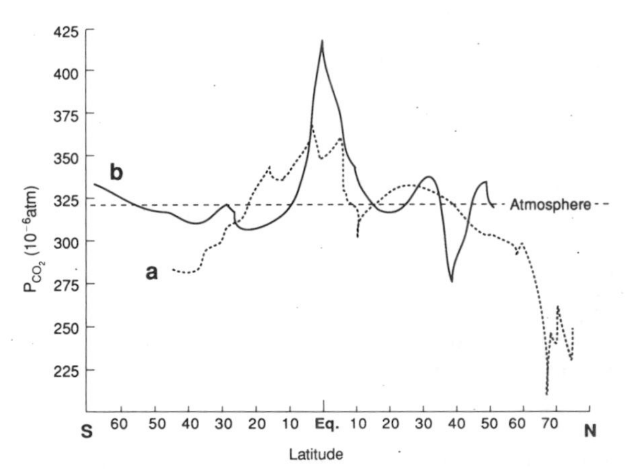

Fig. 11-11 Partial pressure of CO_2 (P_{CO_2}) in surface ocean water along the GEOSECS tracks: (a) the Atlantic western basin data obtained between August 1972 and January 1973; (b) the central Pacific data along the 180° meridian from October 1973 to February 1974. The dashed line shows atmospheric CO_2 for comparison. The equatorial areas of both oceans release CO_2 to the atmosphere, whereas the northern North Atlantic is a strong sink for CO_2. Modified from Broecker *et al.* (1979) with the permission of the American Association for the Advancement of Science.

of $R_a - R_m$, which is a small number. Equation (9) can be reformulated to include the variations of surface water P_{CO_2} and the variations of $\Delta^{14}C$ in surface waters. Figure 11-11 shows latitudinal distributions of P_{CO_2} in the Atlantic and Pacific Oceans (Broecker *et al.*, 1979). Surface water $\Delta^{14}C$ exhibits considerable variations with very low values in the Antarctic surface waters. Although large data sets of surface water P_{CO_2} measurements (Williams, 1982) and gas transfer velocities (Peng *et al.*, 1979) have been collected, estimates of CO_2 exchange based on direct flux measurements still remain to be done (see Liss, 1983, for a review on gas-exchange calculations).

The two prime mechanisms of carbon transport in the ocean are biogenic detrital rain from the photic zone to the deeper oceans and advection of dissolved carbon species. The detrital rain creates inhomogeneities of nutrients illustrated by the characteristic alkalinity profiles (Fig. 11-5). The amount of carbon leaving the photic zone as sinking particles should not be interpreted as the net primary production of the surface oceans, since most of the organic carbon is recycled within the photic zone; only about 10% settles as detritus (Bolin *et al.*, 1979). There is considerable patchiness in the rate of oceanic primary production (Fig. 11-12), with high values in areas of intense upwelling, whereas areas of slow sinking motions (the subtropical gyres) show photosynthesis rates only 10% as large. De Vooys (1979) gives a thorough account of the many methods employed and uncertainties involved in estimating the net primary production of the aquatic environments. We adopt an estimate of total primary production of 50 Pg C/year but note that the range of estimates is huge (15–126 Pg C/year).

If 10% of the total primary production settles as detrital material, about 5 Pg C leaves the photic zone annually. The $CaCO_3$ to organic carbon ratio in the detritus is usually taken as 1 : 4 (Broecker and Peng,

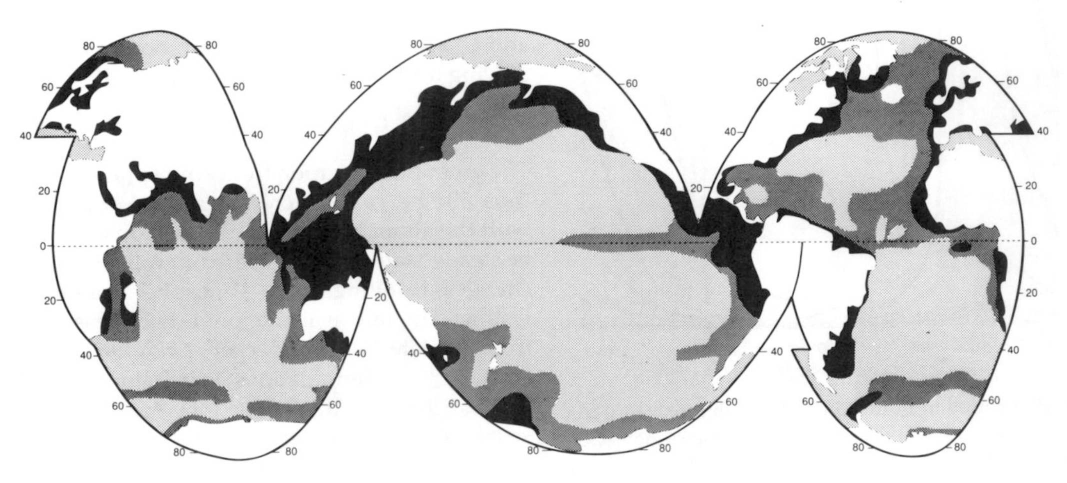

Fig. 11-12 Distribution of primary production in the World Oceans (Degens and Mopper, 1976). Shadings indicate approximate productivity. Light, <50 g C/m^2 year; medium, 50–100 g C/m^2 year; dark, >100 g C/m^2 year.

1972), which signifies a carbonate flux around 1 Pg C/year. The inorganic : organic ratio can actually vary; Chen (1978) obtained ratios between 1 : 10 and 1 : 3 in a water column study. The detrital rain is balanced by a small river input of carbon and upwelling of deep water enriched in carbon from the decomposition of the detritus at depth. To balance the carbon budget of the photic zone, oceanic circulation must provide a net transport of about 5 Pg C/year from deeper layers to the surface. Comparing Fig. 11-12 with a map of upwelling areas clearly shows that upwelling and primary production are coupled processes. We can also study Fig. 11-11 where the high P_{CO_2} values in equatorial regions are caused by the upwelling of carbon-rich and cold, deep water. Upon warming, this water is supersaturated with respect to atmospheric CO_2. The North Atlantic has low P_{CO_2} values caused by biological depletion of carbon and cooling of waters flowing northward (i.e. the Gulf Stream). The yearly circulation of carbon through the atmosphere from equatorial upwelling regions to high latitudes is around 0.01 Pg C/year (Bolin and Keeling, 1963).

In a steady-state ocean, the sedimental deposition rate of a nutrient like phosphorus ought to be balanced by river-borne influx to the oceans: 1.5–4.0 Tg P are transported to the oceans by rivers (Richey, 1983). Assuming a C : P molar ratio of 106 : 1 in the sedimental organic material, the corresponding carbon flux is in the range of 0.06–0.16 Pg C/year. The sedimentation rate of organic material is also estimated from the total annual sedimentation rate of 6.1×10^{15} g/year by applying an average organic carbon content of 0.5% (Kempe, 1979b), resulting in 0.03 Pg C/year. Kempe (1979b) estimates inorganic sedimentation at 0.09–0.22 Pg C/year based on an oceanic calcium balance calculation and the carbonate content in dated deep-sea cores.

11.5 Models of the Carbon Cycle

The descriptive account of the carbon cycle presented above is a first-order model. A variety of numerical models have been used to study the dynamics and response of the carbon cycle to different transients. This subject is an extensive field because most scientists modeling the carbon cycle develop a model tailored for their particular problem. The SCOPE 16 volume (Bolin, 1981) is useful to consult as a "state of the art" description of the field; it also includes suggested standards for notation and procedures to be used when making carbon models. We shall examine briefly some of the simplest and most frequent modeling approaches.

Box models have a long tradition (Craig, 1957b; Revelle and Suess, 1957; Bolin and Eriksson, 1959) and still receive a lot of attention. Most work is concerned with the atmospheric CO_2 increase, with the main goal of predicting global CO_2 levels during the next 100 years. This is accomplished with a model that reproduces carbon fluxes between the atmosphere and other reservoirs on time-scales of 10–100 years, but does not include deep ocean circulation or sedimentary phenomena in detail. To study changes over thousands of years, the whole oceans, terrestrial biota, and soils must be considered. Extension to even longer time-scales must deal with all geological processes. On the other hand, many processes simulated in the fossil fuel studies can then be omitted or treated as instantaneous.

Simple three-box models with the atmosphere assumed to be one well-mixed reservoir and the oceans described by a surface layer and a deep-sea reservoir have been used extensively. Keeling (1973) has discussed this type of model in detail. The two-box ocean model is refined by including a second surface box, simulating an "outcropping" (deep-water-forming) polar sea (e.g. Keeling and Bolin, 1967, 1968), and to include a better resolution of the main thermocline (e.g. Björkström, 1979). The terrestrial biota is included in a simple manner (e.g. Bolin and Eriksson, 1959) in some studies; Fig. 11-13 shows a model used by Machta (1971), where the role of biota is simulated by one reservoir connected to the atmosphere with a time lag of 20 years. Many refinements are possible. Figure 11-14 shows an example of a suggested structural framework for modeling a biome (Bolin, 1981).

The inadequacy of the two-box model of the ocean led to the box-diffusion model (Oeschger *et al.*, 1975). Instead of simulating the role of the deep sea with a well-mixed reservoir in exchange with the surface layer by first-order exchange processes, the transfer into the deep sea is maintained by vertical eddy diffusion. In its original formulation, the box diffusion model assumed that the eddy diffusivity remained constant with depth. Siegenthaler (1983) further developed the diffusion model to include polar outcropping areas. The box diffusion model is in widespread use. The most ambitious attempt (Peng *et al.*, 1983) to simulate the changes in total carbon, ^{13}C, and ^{14}C during the last 100 years uses a

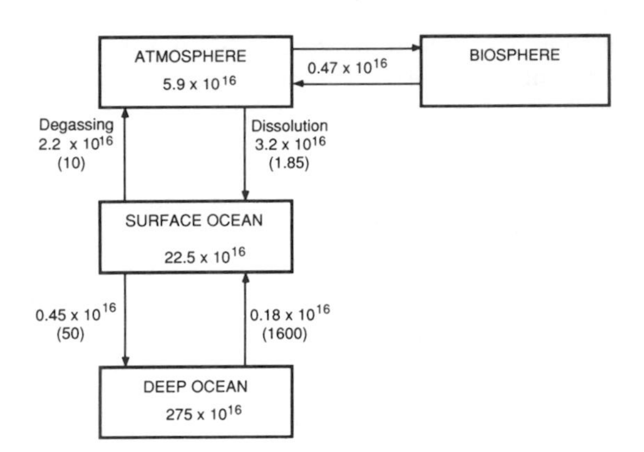

Fig. 11-13 A four-box model of the global carbon cycle. Reservoir inventories are given in moles and fluxes in mol/year. The turnover time of CO_2 in each reservoir with respect to the outgoing flux is shown in parentheses. Adapted from Machta (1971) with the permission of John Wiley and Sons, Inc.

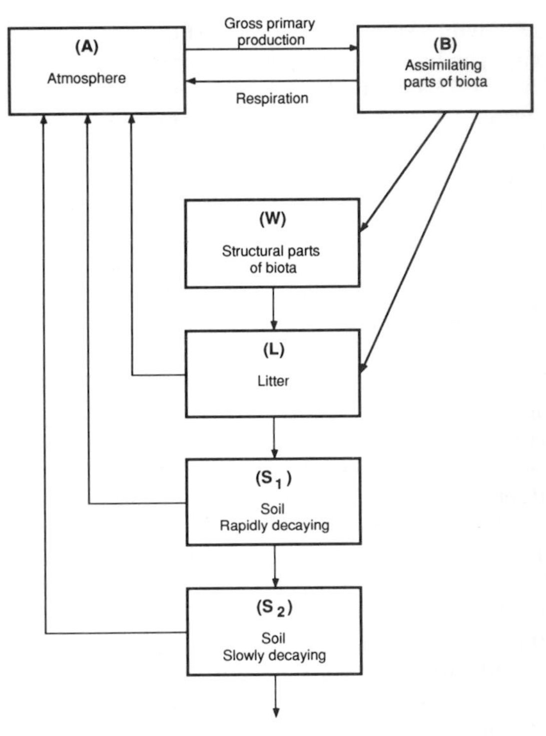

Fig. 11-14 The structural framework of a biome. Reprinted from Bolin *et al.* (1981) with the permission of John Wiley and Sons, Inc.

version of the Oeschger model combined with four boxes simulating the terrestrial system.

Box models and box diffusion models have few degrees of freedom and they must describe physical, chemical, and biological processes very crudely. They are based on empirical relations rather than on first principles. Nevertheless, the simple models have been useful for obtaining some general features of the carbon cycle and refining them to higher resolution will reveal some new aspects.

11.6 Trends in the Carbon Cycle

Throughout this chapter, many of the arguments are based on an assumption of steady-state. Before the agricultural and industrial revolutions, the carbon cycle presumably was in a quasi-balanced state. Natural variations still occur in this unperturbed environment; the Little Ice Age, 300–400 years ago, may have influenced the carbon cycle. The production rate of ^{14}C varies on time-scales of decades and centuries (Stuiver and Quay, 1980, 1981), implying that the pre-industrial radiocarbon distribution may not have been in steady-state.

Measurements of CO_2 concentrations in air bubbles trapped in glacial ice (Berner *et al.*, 1980; Delmas *et al.*, 1980) show that atmospheric P_{CO_2} was about 200 ppmv toward the end of the last glaciation 20 000 years ago. In Fig. 11-15, an estimated range of atmospheric CO_2 during the past 40 000 years is shown. The steady CO_2 level maintained during the past 10 000 years supports the steady-state assumption often invoked in modeling.

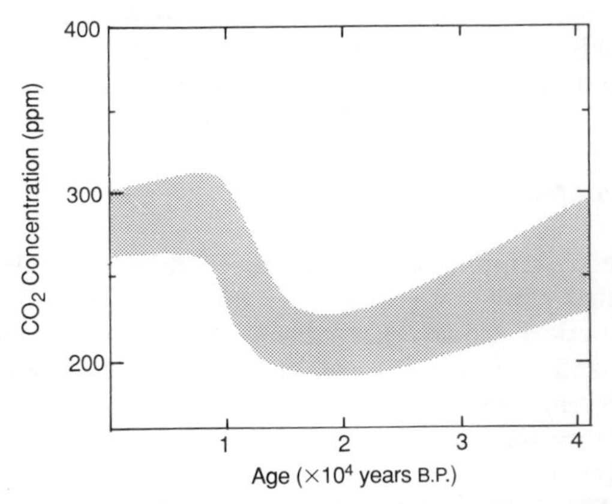

Fig. 11-15 Estimated range of atmospheric CO_2 during the past 40 000 years. Adapted from Neftel *et al.* (1982) with the permission of Macmillan Magazines Ltd.

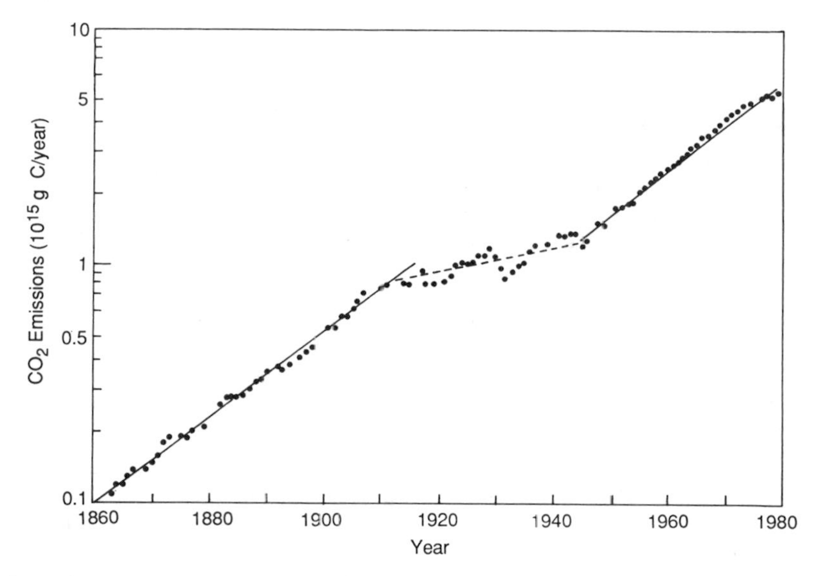

Fig. 11-16 Rate of transfer of carbon to the atmosphere due to fossil fuel combustion according to Rotty (1981).

Fossil fuel combustion undoubtedly accounts for a significant portion of the anthropogenic emissions, although it is rivaled by carbon mobilized by deforestation and land-use changes (Woodwell *et al.*, 1983). The industrial revolution marked the onset of large-scale fossil fuel combustion in the early part of the nineteenth century. Around 1860, the emissions had reached 0.1 Pg C/year (Fig. 11-16). The increase has been steady since that time, although the rate of increase has changed during the past century. In 1860, essentially only coal was used; the use of oil began at the end of the nineteenth century, followed by gas in the early decades of this century. Total fossil fuel emissions increased by 4% per year between 1860 and the beginning of the First World War, when fossil fuel consumption had reached 0.9 Pg C/year (90% of which was coal). For the next 30 years (1914–1945), the yearly increase was about 1%, but after the Second World War the growth rate returned to 4%. Figure 11-17 shows the marked changes in fossil fuel use since 1973. In this period, the annual rate of increase diminished to less than 2%, and striking changes in fuel mix occurred. The use of oil and gas increased more rapidly than coal from the turn of the century to 1973, so that oil emissions surpassed coal in the late 1960s. During the period 1973–1982, the rate of coal production increased more than that for oil. The total emissions of carbon from fossil fuels was 5.2 Pg in 1979 and had even decreased to 5.0 Pg in 1982. High oil prices, efforts to use fuels more efficiently, and an economic recession all contributed to these dramatic changes in fuel usage.

The projection of future emissions of fossil fuel

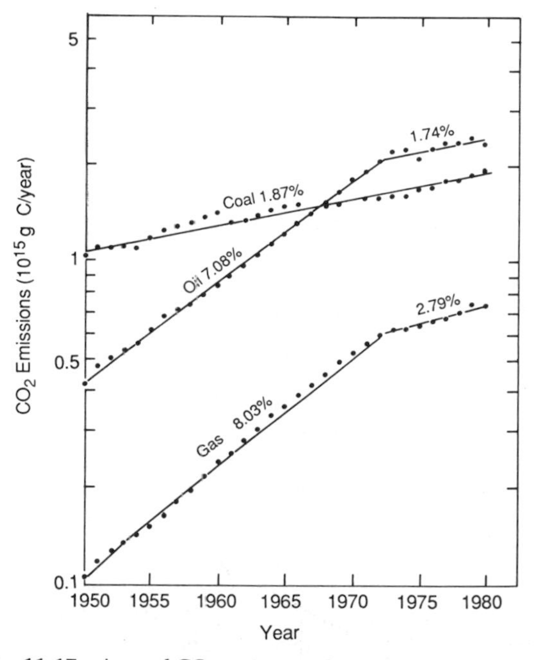

Fig. 11-17 Annual CO_2 emissions from each type of fossil fuel with growth rates (Rotty, 1981).

CO_2 is subject to a number of uncertainties. The growth rate has already shown great variations and the reserves of fossil fuels are not accurately known. Estimates of reserves range between 5000 and 10 000 Pg C, much of which would be costly to exploit with present techniques. Many models (Bolin, 1981) make use of logistic distributions of fossil fuel emissions as a function of time, as suggested by Keeling and Bacastow (1977) and Revelle and Munk (1977). Although these are crude predictors of future fossil fuel usage, they are useful for sensitivity analysis of the carbon cycle models.

A modified logistic function (Bacastow and Björkström, 1981) is:

$$\frac{dQ}{dt} = \mu(1 - (Q/Q_T)^n)\,Q \qquad (10)$$

where Q is the total amount of stable carbon added to the atmosphere prior to time t, Q_T is the total resource (5000 Pg), n is a parameter ($\sim$1), and μ the assumed growth rate (a typical value for μ is 1/22.5 per year). Equation (10) yields emissions that initially grow exponentially, but when the total consumption becomes significant in comparison to Q_T, the incremental increase decreases, becoming zero when half the resources have been used and declining thereafter. Figure 11-18 shows the emission scenarios for a number of different assumed growth rates.

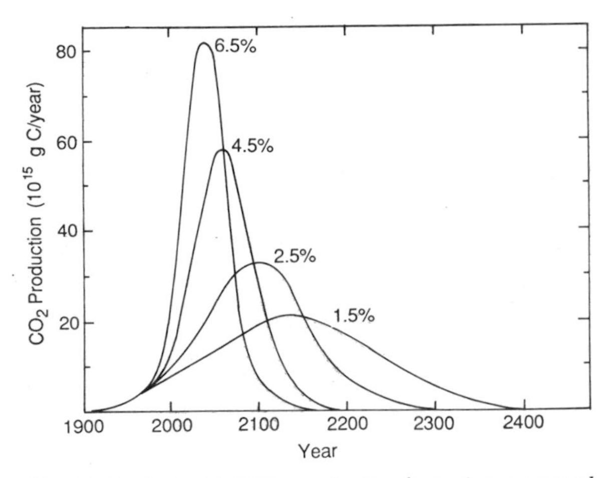

Fig. 11-18 Industrial CO_2 production for various assumed patterns of fossil fuel consumption. The figures at the different curves indicate the initial (1975) percentage increase of the annual consumption. Adapted from Keeling and Bacastow (1977) with the permission of the National Academy Press.

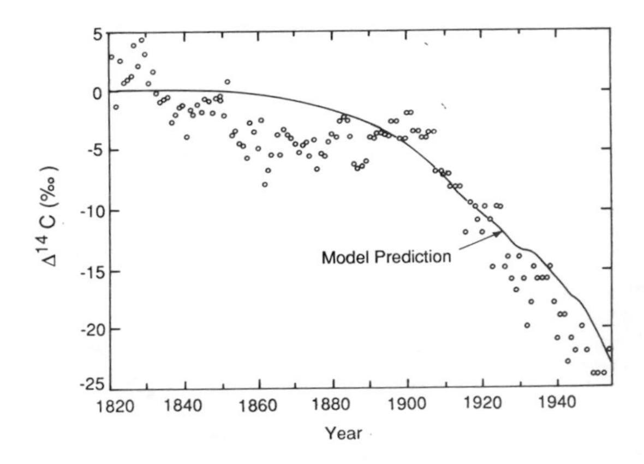

Fig. 11-19 Comparison between the Peng *et al.* (1983) model-derived Suess effect curve (solid line) and the observed $^{14}C/^{12}C$ trend (points) for atmospheric CO_2 as reconstructed by Stuiver and Quay (1981) from measurements of tree rings. Reproduced from Broecker *et al.* (1983) with the permission of the American Geophysical Union.

The atmospheric CO_2 content increased by about 1 ppmv per year during the period 1959–1978 (Bacastow and Keeling, 1981), with the South Pole P_{CO_2} increase lagging somewhat behind the Mauna Loa (19.5°N, 155.6°W) data. This difference is consistent with our knowledge of inter-hemispheric mixing times and the fact that most fossil fuel emissions occur in the northern hemisphere.

Fossil fuel emissions alter the isotopic composition of atmospheric carbon, since they contain no ^{14}C and are depleted in ^{13}C. Releasing radiocarbon-free CO_2 to the atmosphere dilutes the atmospheric ^{14}C, yielding lower $^{14}C/C$ ratios ("the Suess effect"). From 1850 to 1954, the $^{14}C/C$ ratio in the atmosphere decreased by 2.0–2.5% (see Fig. 11-19; Suess, 1965; Stuiver and Quay, 1981). The downward trend in ^{14}C was disrupted by a series of atmospheric nuclear tests. A series by the USA took place in 1958 and the USSR made extensive tests during 1960–1963. Figure 11-20 shows the atmospheric $\Delta\,^{14}C$ trend during recent years. The two superpowers have ceased performing atmospheric bomb tests, resulting in the spike-like injection of ^{14}C in the early 1960s whose further fate in the global biogeochemical cycle has proven to be an unintentional but valuable tool when deducing carbon fluxes between reservoirs.

Freyer and Balacy (1983) constructed a $^{13}C/^{12}C$

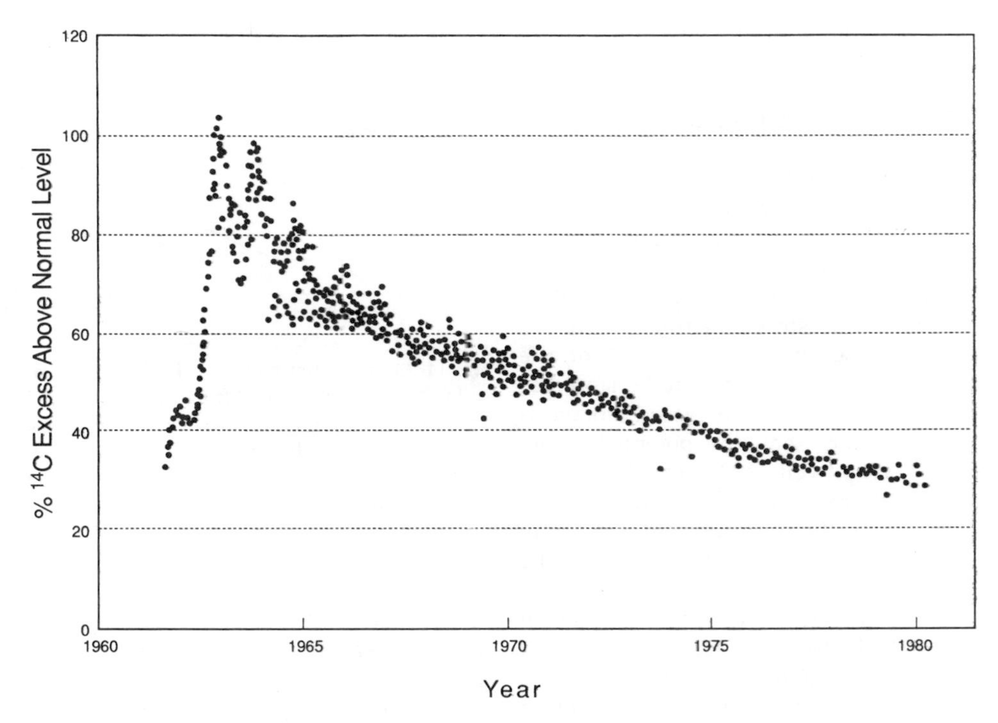

Fig. 11-20 ^{14}C in the troposphere 1962–1981. Modified from Nydal and Lovseth (1983) with the permission of the American Geophysical Union.

record from tree-ring studies (Fig. 11-21) that shows a clear decline in atmospheric ^{13}C since about 1825. An interpretation of the data in Fig. 11-21 must be based on a global carbon cycle model including ocean–atmosphere exchange. The fossil fuel CO_2 not remaining in the atmosphere is either dissolved in the oceans or incorporated into the terrestrial biosphere.

The oceanic sink for CO_2 has often been assumed (Broecker *et al.*, 1979) when analyzing the carbon cycle. Most oceanic carbon models limit the oceanic uptake of fossil fuel to about 40% of the fossil fuel emissions (Bacastow and Björkström, 1981), leading to the conclusion that there must be a "missing sink" to explain the actual observed airborne fraction of 50% (Bacastow and Keeling, 1981). Analytical limitations preclude the detection of changes in oceanic dissolved inorganic carbon directly (Broecker *et al.*, 1979). Consequently, oceanic uptake must be determined by indirect modeling approaches. A possible conclusion when tackling the "missing sink" problem is that there must be a net uptake of carbon in the terrestrial biota. Agriculture has the explicit goal of harvesting organic matter and no accumulation of carbon in agricultural ecosystems is to be expected. It is important to recognize that an increased carbon fixation rate is not equivalent to an

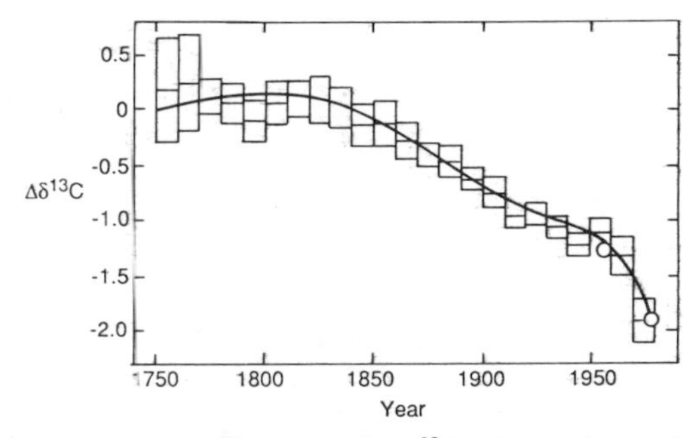

Fig. 11-21 Decade-averaged $\Delta^{13}C$ data of northern hemisphere tree ring records from 1750 to 1979 and 7th-degree polynomial fit of the data. The vertical extension of blocks represents 95% confidence limits of the mean. The open circles give the ^{13}C change of -0.65% in atmospheric CO_2 observed from 1956 to 1978 by Keeling *et al.* (1979). Adapted from Peng *et al.* (1983).

increased storage of carbon. Carbon accumulation in ecosystems is determined by the balance between net primary production and heterotrophic respiration – changes in both have to be considered in the search for the missing carbon.

Cultural eutrophication by the release of nitrogen and phosphorus in various forms could contribute significantly to the biotic storage of carbon. Anthropogenic releases of N and P correspond to the fixation of 9.6 and 17.6 Pg C/year, respectively, if all N and P released was stored as trees in forests (Houghton and Woodwell, 1983). Most of the released nutrients are not available to be stored as wood and, consequently, the increased storage is probably much smaller than the potential value. Houghton and Woodwell (1983) conclude after considering eutrophication and other alterations of the environment that there is little evidence for an increased storage of carbon in the ecosystem of the Earth.

The terrestrial biota seems unable to take up much of the excess CO_2. In fact, a careful assessment of the impact of deforestation and land-use changes indicates that the terrestrial biota has been a considerable source of CO_2 during the past century (Bolin, 1977; Woodwell *et al.*, 1983). A complex effort to deduce mankind's impact on terrestrial biota using a book-keeping model based on historical records on land use in all parts of the world (Moore *et al.*, 1981; Houghton *et al.*, 1983; Woodwell *et al.*, 1983) gives the curves in Fig. 11-22. Woodwell *et al.* (1983) arrive at a carbon release in 1980 of 1.8–4.7 Pg C/year from deforestation, which is comparable to the 5 Pg released from fossil fuels. Using a global carbon cycle model with a box-diffusion model for the oceans and a simple four-box model for soil and land biota, Peng *et al.* (1983) make use of the $\delta^{13}C$ changes in the atmosphere (Fig. 11-21) to deduce the CO_2 contribution from forest and soils. The CO_2 emissions arrived at for 1980 are about 1.5 Pg C/year, similar to the minimum values of Woodwell *et al.* (1983), but the trends during the past century differ significantly (Fig. 11-23). Rather than finding a sink for the fossil fuel CO_2 that the oceans could not accommodate, in the terrestrial biosphere we find a source! The "missing sink" problem is yet more severe. Obviously our knowledge of the global cycle of carbon is inadequate to get ends to meet.

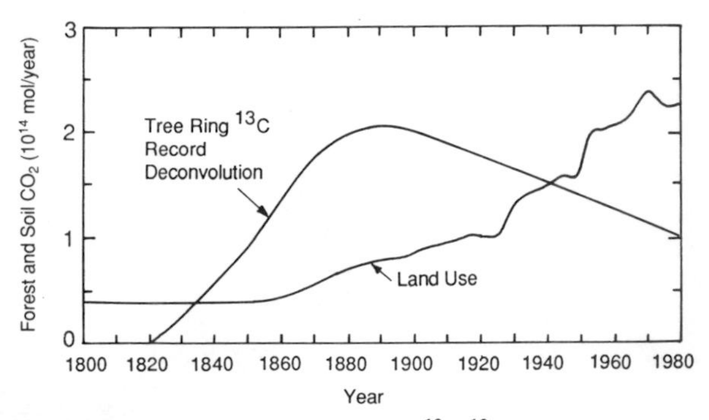

Fig. 11-23 Contrast between the $^{13}C/^{12}C$ derived forest-soil CO_2 scenario obtained from Peng *et al.* (1983) with that based on land use and stored carbon response functions obtained by Houghton *et al.* (1983). Reproduced from Broecker *et al.* (1983) with the permission of the American Geophysical Union.

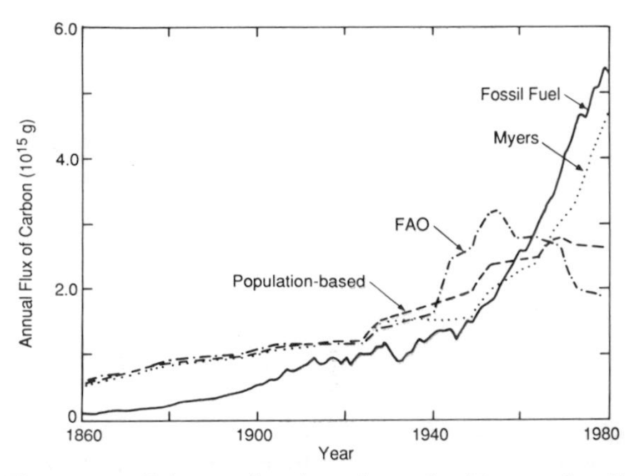

Fig. 11-22 Release of carbon from the biota and soils globally according to various estimates. The fossil fuel flux is from the data of Rotty. Modified from Woodwell *et al.* (1983) with the permission of the American Association for the Advancement of Science.

The role of carbon dioxide in the Earth's radiation budget merits this interest in atmospheric CO_2. There are, however, other changes of importance. The atmospheric methane concentration is increasing, probably as a result of increasing cattle populations, rice production, and biomass burning (Crutzen, 1983). Increasing methane concentrations are important because of the role it plays in stratospheric and tropospheric chemistry. Methane is also important to the radiation budget of the world.

As we have seen, civilization is altering the global carbon cycle in several ways. Even though our knowledge of the most relevant processes in the carbon cycle has increased considerably, it remains

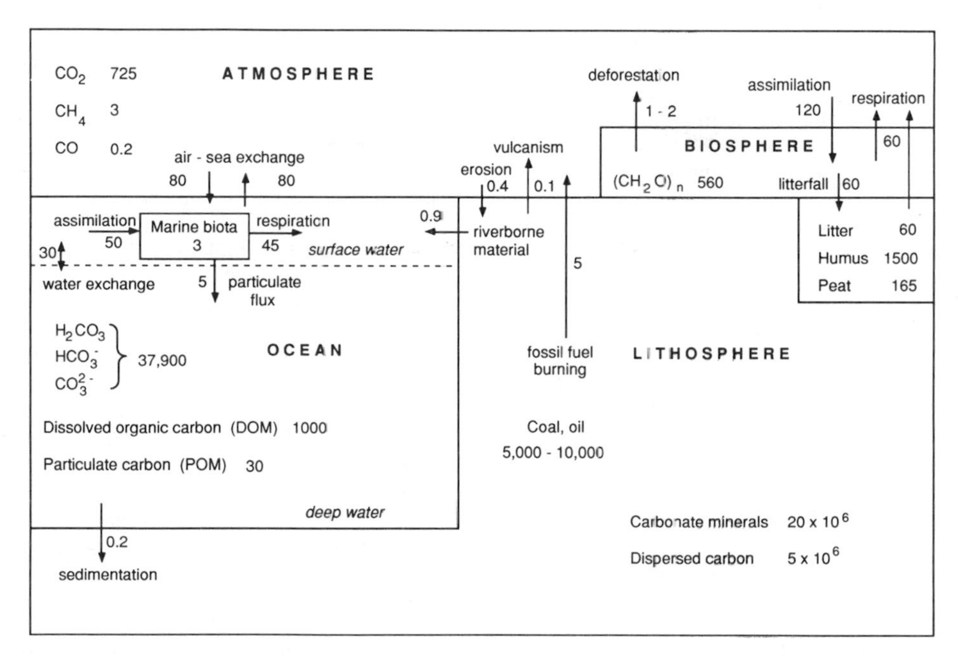

Fig. 11-24 A summary of the global carbon cycle as described in the text. Reservoir contents are given in Pg C and fluxes in Pg C/year.

limited. Figure 11-24 summarizes the carbon cycle as described in this chapter.

Studying elemental cycles one by one is a common but sometimes misleading practice since all elemental cycles on Earth are interrelated (Likens, 1981; Bolin and Cook, 1983) and since the cycles of several other elements are strongly coupled with that of carbon. Few attempts are made to arrive at internally consistent pictures for several elements simultaneously. Increased atmospheric CO_2 is associated with many processes that are poorly understood, ranging from air–sea gas exchange to changes in terrestrial biomass associated with massive land use changes. To arrive at an overall view of the carbon cycle will require much interdisciplinary research. The ability to anticipate the future for the global environment requires an understanding of the full consequences of the ongoing changes. Such wisdom can only be acquired by increasing our understanding of the most important biogeochemical cycles and their interactions.

References

Ajtay, G. L., P. Ketner, and P. Duvigneaud (1979). Terrestrial primary production and phytomass. *In* "The Global Carbon Cycle" (B. Bolin, E. T. Degens, S. Kempe, and P. Ketner, eds.), pp. 129–181. John Wiley, New York.

Bacastow, R. B. and A. Björkström (1981). Comparison of ocean models for the carbon cycle. *In* "Carbon Cycle Modelling" (B. Bolin, ed.), pp. 29–79. John Wiley, New York.

Bacastow, R. B. and C. D. Keeling (1981). Atmospheric carbon dioxide concentration and the observed airborne fraction. *In* "Carbon Cycle Modelling" (B. Bolin, ed.), pp. 103–112. John Wiley, New York.

Baes, C. F., H. E. Goeller, J. S. Olson, and R. M. Rothy (1976). "The Global Carbon Dioxide Problem" ORNL-5194, pp. 1–72. Oak Ridge National Laboratory, Oak Ridge, Tenn.

Barnola, J. M., D. Raynaud, A. Neftel, and H. Oeschger (1983). Comparison of CO_2 measurements by two laboratories on air from bubbles in polar ice. *Nature* **303**, 410–413.

Bazilevich, N. I., L. Rodin, L. Ye and N. N. Roznov (1970). Geographical Aspects of Biological Productivity. *Pap. V Congr. USSR Geogr. Soc.*, USSR, Leningrad.

Berner, W., H. Oeschger, and B. Stauffer (1980). Information on the CO_2 cycle from ice core studies. *Radiocarbon* **22**, 227–235.

Bischof, W. (1981). The CO_2 content of the upper polar troposphere between 1963–1979. *In* "Carbon Cycle Modelling" (B. Bolin, ed.), pp. 113–116. John Wiley, New York.

Björkström, A. (1979). A model of CO_2 interaction between atmosphere, oceans, and land biota. *In* "The Global Carbon Cycle" (B. Bolin, E. T. Degens, S. Kempe, and P. Ketner, eds.), pp. 403–457. John Wiley, New York.

Blake, D. R. and F. S. Rowland (1988). Continuing worldwide increase in tropospheric methane 1978 to 1987. *Science* **239**, 1129–1131.

Bolin, B. (1970a). Changes of land biota and their importance for the carbon cycle. *Science* **196**, 613–615.

Bolin, B. (1970b). The carbon cycle. *Sci. Amer.* **223** (3), 124–132.

Bolin, B. (1970c). The carbon cycle. *In* "The Biosphere", pp. 47–56. W. H. Freeman, New York.

Bolin, B. (1977). Changes in land biota and their importance for the carbon cycle. *Science* **196**, 613–615.

Bolin, B. (ed.) (1981). "Carbon Cycle Modelling." John Wiley, New York.

Bolin, B. and R. Cook (eds.) (1983). "The Major Biogeochemical Cycles and Their Interactions." John Wiley, New York.

Bolin, B. and E. Eriksson (1959). Changes of the carbon dioxide content of the atmosphere and sea due to fossil fuel combustion. *In* "Atmosphere and Sea in Motion" (B. Bolin, ed.), pp. 130–142. The Rockefeller Institute Press, New York.

Bolin, B. and C. D. Keeling (1963). Large-scale atmospheric mixing as deduced from the seasonal and meridional variations of carbon dioxide. *J. Geophys. Res.* **68**, 3899–3920.

Bolin, B., E. T. Degens, S. Kempe, and P. Ketner (eds.) (1979). "The Global Carbon Cycle." John Wiley, New York.

Bolin, B., A. Björkström, C. D. Keeling, R. Bacastow, and U. Siegenthaler (1981). Carbon cycle modelling. *In* "Carbon Cycle Modelling" (B. Bolin, ed.), pp. 1–28. John Wiley, New York.

Brewer, P. G. (1978). Direct observation of the oceanic CO_2 increase. *Geophys. Res. Lett.* **5** (12), 997–1000.

Broecker, W. S. (1973). Factors controlling CO_2 content in the oceans and atmosphere. *In* "Carbon and the Biosphere" (G. M. Woodwell and E. V. Pecan, eds.), pp. 32–50. United States Atomic Energy Commission, Springfield, Virginia.

Broecker, W. S. (1979). A revised estimate for the radiocarbon age of North Atlantic Deep Water. *J. Geophys. Res.* **84** (C6), 3218–3226.

Broecker, W. S. and E. A. Olson (1959). Lamont Radiocarbon Measurements VI. *Am. J. Sci. (Radiocarbon Suppl.)* **1**, 111–132.

Broecker, W. S. and T.-H. Peng (1982). "Tracers in the Sea." Eldigio Press, New York.

Broecker, W. S., R. Gerard, M. Ewing, and B. C. Heezen (1960). Natural radiocarbon in the Atlantic Ocean. *J. Geophys. Res.* **65** (9), 2903–2931.

Broecker, W. S., Takahashi, T., Simpson, H. J. and T.-H. Peng (1979). Fate of fossil fuel carbon dioxide and the global carbon budget. *Science* **206**, 409–418.

CDIAC (1990). *Carbon Dioxide Information Analysis Center Communications,* Spring. Oak Ridge National Laboratory, Oak Ridge, Tenn.

Charlson, R. J. and H. Rodhe (1982). Factors controlling the acidity of natural rainwater. *Nature* **295**, 683–685.

Chen, C.-T. (1978). Decomposition of calcium carbonate and organic carbon in the deep oceans. *Science* **201**, 735–736.

Chen, C.-T. and F. J. Millero (1979). Gradual increase of oceanic CO_2. *Nature* **277**, 205–206.

Craig, H. (1957a). Isotopic standards for carbon and correction factors for mass-spectrometric analysis of carbon dioxide. *Geochim. Cosmochim. Acta* **12**, 133–149.

Craig, H. (1957b). The natural distribution of radiocarbon and the exchange time of carbon dioxide between atmosphere and sea. *Tellus* **9**, 1–17.

Crutzen, P. J. (1983). Atmospheric interactions: Homogeneous gas reactions of C, N, and S containing compounds. *In* "The Major Biogeochemical Cycles and Their Interactions" (B. Bolin, ed.), pp. 67–112. John Wiley, New York.

Culkin, F. (1965). The major constituents of sea water. *In* "Chemical Oceanography" (J. P. Riley and G. Skirrow, eds.), 1st edn., pp. 121–161. Academic Press, London.

Degens, E. T. and K. Mopper (1976). Factors controlling the distribution and early diagenesis of organic material in marine sediments. *In* "Chemical Oceanography," Vol. 6 (J. P. Riely, Ed.), pp. 59–113. Academic Press, New York.

Degens, E. T., S. Kempe, and A. Spitzy (1984). Carbon dioxide: A biogeochemical portrait. *In* "The Handbook of Environmental Chemistry" (O. Hutzinger, ed.), Vol. 1, Part C, pp. 127–215. Springer-Verlag, New York.

Dlemas, R. J., J.-M. Ascencio, and M. Legrand (1980). Polar ice evidence that atmospheric CO_2 20 000 yr BP was 50% of present. *Nature* **284**, 155–157.

De Vooys, C. G. N. (1979). Primary production in aquatic environments. *In* "The Global Carbon Cycle" (B. Bolin, E. T. Degens, S. Kempe, and P. Ketner, eds.), pp. 259–292. John Wiley, New York.

Druffel, E. M. and T. W. Linick (1978). Radiocarbon in annual coral rings of Florida. *Geophys. Res. Lett.* **5**, 913–916.

Dryssen, D. and L. G. Sillén (1967). Alkalinity and total carbonate in sea water: A plea for P-T-independent data. *Tellus* **19** (1), 113–120.

Edmond, J. M. (1970). High precision determination of titration alkalinity and total carbon dioxide concentration of sea water by potentiometric titration. *Deep-Sea Res.* **17**, 737–750.

Ehhalt, D. H. (1974). The atmospheric cycle of methane. *Tellus* **26**, 58–70.

Feux, A. N. and D. R. Baker (1973). Stable carbon isotopes in selected granitic, mafic, and ultramafic igneous rocks. *Geochim. Cosmochim. Acta* **37**, 2509–2521.

Freyer, H.-D. (1979). Atmospheric cycles of trace gases containing carbon. *In* "The Global Carbon Cycle" (B. Bolin, E. T. Degens, S. Kempe, and P. Ketner, eds.), pp. 101–128. John Wiley, New York.

Freyer, H. D. and N. Belacy (1983). $^{13}C/^{12}C$ records in Northern Hemispheric trees during the past 500 years: Anthropogenic impact and climatic superpositions. *J. Geophys. Res.* **88**, 6844–6852.

Garrels, R. M. and F. T. Mackenzie (1971). "Evolution of Sedimentary Rocks." W. W. Norton, New York.

Garrels, R. M., F. T. Mackenzie, and C. Hunt (1973). "Chemical Cycles and the Global Environment." W. Kaufmann, Los Altos, Calif.

Goudriaan, J. and G. L. Atjay (1979). The possible effects of increased CO_2 on photosynthesis. *In* "The Global Carbon Cycle" (B. Bolin, E. T. Degens, S. Kempe, and P. Ketner, eds.), pp. 237–249. John Wiley, New York.

Hansson, I. (1973). A new set of acidity constants for carbonic acid and boric acid in sea water. *Deep-Sea Res.* **20**, 461–478.

Holland, H. D. (1978). "The Chemistry of the Atmosphere and Oceans." Wiley-Interscience, New York.

Houghton, R. A. and G. M. Woodwell (1983). Effect of increased C, N, P, and S on the global storage of C. *In* "The Major Biogeochemical Cycles and Their Interactions" (B. Bolin and R. B. Cook, eds.), pp. 327–343. John Wiley, New York.

Houghton, R. A., J. E. Hobbie, J. M. Melillo, B. Moore, B. J Peterson, G. R. Shaver, and G. M. Woodwell (1983). Changes in the carbon content of terrestrial biota and soils between 1860 and 1980: A net release of CO_2 to the atmosphere. *Ecol. Monogr.* **53**, 235–262.

Hunt, J. M. (1972). Distribution of carbon in crust of Earth. *Bull. Amer. Assoc. Pet. Geol.* **56**, 2273–2277.

Keeling, C. D. (1973). The carbon dioxide cycle: Reservoir models to depict the exchange of atmospheric carbon dioxide with the oceans and land plants. *In* "Chemistry of the Lower Atmosphere" (S. Rasool, ed.), pp. 251–329. Plenum Press, New York.

Keeling, C. D. and R. B. Bacastow (1977). Impact of industrial gases on climate. *In* "Energy and Climate," pp. 72–95. National Academy of Sciences, Washington, D.C.

Keeling, C. D. and B. Bolin (1967). The simultaneous use of chemical tracers in oceanic studies. I. General theory of reservoir models. *Tellus* **19**, 566–581.

Keeling, C. D. and B. Bolin (1968). The simultaneous use of chemical tracers in oceanic studies II. A three-reservoir model of the North and South Pacific Oceans. *Tellus* **20**, 17–54.

Keeling, C. D., R. B. Bacastow, A. E. Bainbridge, C. A. Ekdahl, P. R. Guenther, L. S. Waterman, and J. F. S. Chin (1976a). Atmospheric carbon dioxide variations at Mauna Loa Observatory, Hawaii. *Tellus* **28** (6), 538–551.

Keeling, C. D., J. A. Adams, C. A. Ekdahl, and P. R. Guenther (1976b). Atmospheric carbon dioxide variations at the South Pole. *Tellus* **28** (6), 552–564.

Keeling, C. D., W. G. Mook, and P. P. Tans (1979). Recent trends in the $^{13}C/^{12}C$ ratio of atmospheric carbon dioxide. *Nature* **277**, 121–123.

Kempe, S. (1979a). Carbon in the freshwater cycle. *In* "The Global Carbon Cycle" (B. Bolin, E. T. Degens, S. Kempe, and P. Ketner, eds.), pp. 317–342. John Wiley, New York.

Kempe, S. (1979b). Carbon in the rock cycle. *In* "The Global Carbon Cycle" (B. Bolin, E. T. Degens, S. Kempe, and P. Ketner, eds.), pp. 343–377. John Wiley, New York.

Koyama, T. (1963). Gaseous metabolism in lake sediments and paddy soils and the production of atmospheric methane and hydrogen. *J. Geophys. Res.* **68**, 3971–3973.

Kroopnick, P. (1980). The distribution of ^{13}C in the Atlantic Ocean. *Earth Planet. Sci. Lett.* **49**, 469–484.

Lieth, H. (1972). Über die primarproduktion der erde. *Z. Angewandte Bot.* **46**, 1–37.

Likens, G. E. (ed.) (1981). "Some Perspectives of the Major Biogeochemical Cycles." John Wiley, New York.

Liss, P. S. (1983). Gas transfer: Experiments and geochemical implications. *In* "Air–Sea Exchange of Gases and Particles" (P. S. Liss and W. G. N. Slinn, eds.), pp. 241–298. NATO ASI Series. Reidel, Dordrecht, The Netherlands.

Livingstone, D. A. (1963). Chemical composition of rivers and lakes. *In* "Data of Geochemistry" (M. Fleischer, ed.), 6th edn., pp. 1–61. U.S. Geol. Surv. Prof. Pap. 440G.

Machta, L. (1971). The role of the oceans and biosphere in the carbon dioxide cycle. *In* "The Changing Chemistry of the Oceans" (D. Dryssen and D. Jagner, eds.), pp. 121–145. John Wiley, London.

Mehrbach, C., C. H. Culberson, S. E. Hawley, and R. M. Pytkowicz (1973). Measurement of the apparent dissociation constants of carbonic acid in seawater at atmospheric pressure. *Limnol. Oceanogr.* **18** (6), 897–907.

Moore, B., R. D. Boone, J. E. Hobbie, R. A. Houghton, J. M. Melillo, B. J. Peterson, G. R. Shaver, C. J. Vorosmarty, and G. M. Woodwell (1981). A simple model for analysis of the role of terrestrial ecosystems in the global carbon budget. *In* "Carbon Cycle Modelling" (B. Bolin, ed.), pp. 365–385. John Wiley, New York.

Mopper, K. and E. T. Degens (1979). Organic carbon in the ocean: nature and cycling. *In* "The Global Carbon Cycle" (B. Bolin, E. T. Degens, S. Kempe, and P. Ketner, eds.), pp. 293–316. John Wiley, New York.

Morse, J. W., F. J. Millero, V. Thurmond, E. Brown, and H. G. Ostlund (1984). The carbonate chemistry of Grand Bahama Bank Waters: After 18 years another look. *J. Geophys. Res.* **89** (C3), 3604–3614.

Neftel, A., H. Oeschger, J. Schwander, B. Stauffer, and R. Zumbrunn (1982). Ice core sample measurements give atmospheric CO_2 content during the past 40 000 yr. *Nature* **295**, 220–223.

Nydal, R. and K. Lovseth (1983). Tracing bomb ^{14}C in the atmosphere 1962–1980. *J. Geophys. Res.* **88** (C6), 3621–3642.

Oeschger, H., U. Siegenthaler, U. Schotterer, and A. Gugelmann (1975). A box diffusion model to study the carbon dioxide exchange in nature. *Tellus* **27**, 168–192.

Olson, J. S., J. A. Watts, and L. J. Allison (1983). "Carbon in

Live Vegetation of Major World Ecosystems." United States Department of Energy, TR004.
Pearman, G. (1981). The CSIRO (Australia) atmospheric CO_2 monitoring program. *In* "Carbon Cycle Modelling" (B. Bolin, ed.), pp. 117–120. John Wiley, New York.
Peng, T.-H., W. S. Broecker, G. G. Mathieu, and Y.-H. Li (1979). Radon evasion rates in the Atlantic and Pacific Oceans as determined during the GEOSECS program. *J. Geophys. Res.* **84**, 2471–2486.
Peng, T.-H., W. S. Broecker, H. D. Freyer, and S. Trumore (1983). A deconvolution of the tree ring based $\delta^{13}C$ record. *J. Geophys. Res.* **88** (C6), 3609–3620.
Revelle, R. (1982). Carbon dioxide and world climate. *Sci. Amer.* **247** (2), 33–41.
Revelle, R. and W. Munk (1977). The carbon dioxide cycle and the biosphere. *In* "Energy and Climate," pp. 140–158. National Academy of Sciences, Washington, D.C.
Revelle, R. and H. E. Suess (1957). Carbon dioxide exchange between atmosphere and ocean, and the question of an increase of atmospheric CO_2 during the past decades. *Tellus* **9**, 18–27.
Richey, J. E. (1983). C, N, P, and S cycles: Major reservoirs and fluxes; the phosphorus cycle. *In* "The Major Biogeochemical Cycles and Their Interactions" (B. Bolin and R. B. Cook, eds.), pp. 51–56. John Wiley, New York.
Ronov, A. B. (1964). Common tendencies in the chemical evolution of the earth's crust, ocean, and atmosphere. *Geochem.* **8**, 715–743.
Rosenberg, N. J. (1981). The increasing CO_2 concentration in the atmosphere and its implications on agricultural productivity. I. Effects on photosynthesis, transpiration and water use efficiency. *Climatic Change* **3**, 265–279.
Rotty, R. M. (1981). Data for global CO_2 production from fossil fuels and cement. *In* "Carbon Cycle Modelling" (B. Bolin, ed.), pp. 121–125. John Wiley, New York.
Schlesinger, W. H. (1977). Carbon balance in terrestrial detritus. *Ann. Ecol. Syst.* **8**, 51–81.
Seiler, W. and P. J. Crutzen (1980). Estimates of gross and net fluxes of carbon between the biosphere and the atmosphere from biomass burning. *Climatic Change* **2**, 226–247.
Siegenthaler, U. (1983). Uptake of excess CO_2 by an outcrop-diffusion model of the ocean. *J. Geophys. Res.* **88** (C6), 3599–3608.
Skirrow, J. (1975). The dissolved gases: Carbon dioxide. *In* "Chemical Oceanography" (J. P. Riley and G. Skirrow, eds.), 2nd edn., Vol. 2, pp. 1–192. Academic Press, London.
Stuermer, D. H. and J. R. Payne (1976). Investigation of seawater and terrestrial humic substances with carbon-13 and proton nuclear magnetic resonance. *Geochim. Cosmochim. Acta* **40**, 1109–1114.
Stuiver, M. (1980). ^{14}C distribution in the Atlantic Ocean. *J. Geophys. Res.* **85**, 2711–2718.
Stuiver, M. and H. A. Polach (1977). Discussion reporting of ^{14}C data. *Radiocarbon* **19**, 355–363.
Stuiver, M. and P. D. Quay (1981). Atmospheric ^{14}C changes resulting from fossil fuel CO_2 release and cosmic ray flux variability. *Earth Planet. Sci. Lett.* **53**, 349–362.
Stuiver, M., H. G. Ostlund, and T. A. McConnaughey (1981). GEOSECS Atlantic and Pacific ^{14}C distribution. *In* "Carbon Cycle Modelling" (B. Bolin, ed.), pp. 201–221. John Wiley, New York.
Stuiver, M., P. D. Quay, and H. G. Ostlund (1983). Abyssal water carbon-14 distribution and the age of the world oceans. *Science* **219**, 849–851.
Stumm, W. and J. J. Morgan (1981). "Aquatic Chemistry," 2nd edn. John Wiley, New York.
Suess, H. E. (1965). Secular variations of the cosmic-ray-produced carbon-14 in the atmosphere and their interpretation. *J. Geophys. Res.* **70**, 5937–5952.
Takahashi, T., W. S. Broecker, S. R. Werner, and A. E. Bainbridge (1980). Carbonate chemistry of the surface waters of the world oceans. *In* "Isotope Marine Chemistry" (E. P. Goldberg, Y. Horibe, and K. Sarubashi, eds.), pp. 291–326. Uchida Rokakuho, Tokyo.
Takahashi, T., W. S. Broecker, and A. E. Bainbridge (1981). The alkalinity and total carbon dioxide concentration in the world oceans. *In* "Carbon Dioxide Modelling" (B. Bolin, ed.), pp. 271–286. John Wiley, New York.
Walker, J. C. G. (1977). "Evolution of the Atmosphere." Macmillan, New York.
Wigley, T. M. L. (1983). The pre-industrial carbon dioxide level. *Climatic Change* **5**, 315–320.
Williams, P. J. le B. (1975). Biological and chemical aspects of dissolved organic material in sea water. *In* "Chemical Oceanography (J. P. Riley and G. Skirrow, eds.), 2nd edn., Vol. 2, pp. 301–363. Academic Press, London.
Williams, R. T. (1982). "Transient Tracers in the Ocean." Preliminary Hydrographic Report, Vol. 1–4. Scripps Institution of Oceanography. La Jolla. Calif.
Woodwell, G. M. (1978). The carbon dioxide question. *Sci. Amer.* **238**, 38–43.
Woodwell, G. M. and E. V. Pecan (eds.) (1973). "Carbon and the Biosphere." United States Atomic Energy Commission, Washington, D.C.
Woodwell, G. M., J. E. Hobbie, R. A. Houghton, J. M. Melillo, B. Moore, B. J. Peterson, and G. R. Shaver (1983). Global deforestation: Contribution to atmospheric carbon dioxide. *Science* **222**, 1081–1086.

12

The Nitrogen Cycle

Daniel A. Jaffe

12.1 Introduction

The nitrogen cycle offers a rich variety of important biological and abiotic processes that involve many important compounds in the gas, liquid, and solid phases. Compounds of nitrogen also play important roles in a wide range of contemporary environmental issues, from the perturbation of stratospheric ozone to the contamination of groundwater. Global nitrogen cycles also offer some very interesting puzzles Why does the Earth have a predominantly nitrogen atmosphere? What factors are responsible for the striking increase in atmospheric nitrous oxide?

There are several reasons for trying to come to grips with the nitrogen cycle. First, much nitrogen is coupled with other elements of living matter – C, S, and P. Understanding nitrogen will help us understand the role of living matter in biogeochemical cycles. Second, nitrogen is implicated in several of the ways humankind impacts the natural environment. Photochemical smog, acid precipitation, and nitrate pollution of groundwater are all related to compounds of nitrogen. Ammonia plays an important role in atmospheric aerosols and may have been a significant greenhouse gas in the early atmosphere.

In this chapter, we will look at the parts of the nitrogen cycle, and then integrate these parts into a single comprehensive view of the nitrogen cycle. The first step in this process is to take a look at the many varied nitrogen compounds that exist in natural systems. We then consider the biological processes that transform nitrogen. Next, we will consider abiotic transformations (mainly inorganic reactions occurring in the atmosphere). We can then consider the physical processes that serve to transport nitrogen from one reservoir to another; for example the process by which N_2O reaches the stratosphere, or NH_3 is volatilized from decomposing organic matter.

12.2 Chemistry

Nitrogen has five valence electrons and can take on oxidation states between +5 and −3. Most of the nitrogen compounds we will discuss either have nitrogen bonded to carbon and hydrogen, in which case the oxidation state of the nitrogen is negative (N is more electronegative than either C or H), or have nitrogen bonded to O, in which case the nitrogen has a positive oxidation state.

Table 12-1 lists the most common nitrogen compounds that exist in the natural world, by oxidation state. In addition, it also lists the boiling point (b.p.) for each compound as well as its heat of formation [$\Delta H^0(f)$] and free energy of formation [$\Delta G^0(f)$]. For comparison, the data on H_2O are also included.

To fully understand some of the major players in the nitrogen cycle, we should also consider some of the industrial and social implications of these compounds:

1. *HNO_3*. Nitric acid is a very strong acid; about 6.8 million metric tons per year are manufactured for industrial purposes in the USA. Most of it is produced from ammonia by the catalytic oxidation to NO, which is then further oxidized to NO_2. Addition of water and oxidation forms HNO_3. Most of the

Global Biogeochemical Cycles
ISBN 0-12-147685-5

Table 12-1 Chemical data on important nitrogen compounds

Oxidation state	Compound	b.p. (°C)	$\Delta H^0(f)$ (kJ/mol, 298 K)	$\Delta G^0(f)$
+5	N_2O_5 (g)	11	115	
	HNO_3 (g)	83	−135	−75
	$Ca(NO_3)_2$ (s)		−900	−720
	HNO_3 (aq)		−200	−108
+4	NO_2 (g)	21	33	51
	N_2O_4		9	98
+3	HNO_2 (g)		−80	−46
	HNO_2 (aq)		−120	−55
+2	NO (g)	−152	90	87
+1	N_2O (g)	−89	82	104
0	N_2 (g)	−196	0	0
−3	NH_3 (g)	−33	−46	−16.5
	NH_4^+ (aq)		−72	−79
	NH_4Cl (s)		−201	−203
	CH_3NH_2 (g)		−28	28
	H_2O (g)	100	−242	−229

nitric acid produced is used in the manufacture of fertilizers and explosives.

In the troposphere, nitrogen oxides also produce HNO_3, but in general the oxidants are free radicals produced photochemically, such as HO_2, RO_2, and OH. The HNO_3 produced in this manner is an important contributor to "acid rain".

In its pure form, nitric acid is a liquid with a high vapor pressure (47.6 torr at 20° C), so that in the lower atmosphere, HNO_3 exists as a gas, in an aerosol or in a cloud droplet. When nitric acid reacts with a base a nitrate salt is produced. If the atmospheric base is ammonia, NH_4NO_3 is the result:

$$NH_3(g) + HNO_3(g \text{ or } aq) \rightarrow NH_4NO_3(s \text{ or } aq)$$

If the reaction is between two gas phase species, this reaction could be a source of cloud condensation nuclei, or simply a means to neutralize an acidic aerosol. Although there are some questions concerning the measurement of atmospheric HNO_3 (Lawson, 1988), most measurements indicate that gaseous HNO_3 concentrations predominate over particle NO_3^-.

In addition, there are numerous other nitrate salts. These are all high melting, colorless solids, and are very soluble in water. Many of these, such as KNO_3 and $NaNO_3$, are mined in large quantities and used in fertilizers and explosives. Prior to the industrial production of HNO_3 from ammonia, the mining of these salts was the major means for producing explosives. During the First World War, Germany's supply of $NaNO_3$ from Chile was cut off. The Haber process ($N_2 + 3H_2 = 2NH_3$), developed by Fritz Haber only a few years earlier, allowed Germany to produce ammonia, and therefore nitric acid to make nitrate salts for explosives.

2. *NO_2*. Very little NO_2 is produced industrially as an end-product, at least not intentionally. NO_2 is a brown/yellow gas at room temperature due to its light absorption at wavelengths shorter than 680 nm. NO_2 dimerizes into the colorless N_2O_4 (and indeed there is a very small amount of N_2O_4 found in urban atmospheres). Nitrogen dioxide has a very irritating odor, and is quite toxic. It is produced by the oxidation of NO, so that the concentrations of these two gases are coupled in the atmosphere. Since NO is a byproduct of virtually all combustion processes, NO and NO_2 are generally found in much higher

concentrations in urban areas than in the natural background, and this is a significant source of photochemical smog.

3. *NO*. Nitric oxide, or nitrogen monoxide, is a colorless gas at room temperature. As we have already seen, it is industrially produced by the oxidation of ammonia. However, in the urban nitrogen cycle, a more significant process is the equilibrium process discussed in Chapter 5.

4. *N_2O*. Nitrous oxide, or dinitrogen oxide, is also a colorless gas at room temperature. Its principal uses are as an anaesthetic and as an aerosol propellant, but it is not a major industrial chemical. It is generally produced (in a lab or industrial setting) by the carefully controlled thermal decomposition of ammonium nitrate (NH_4NO_3):

$$NH_4NO_3(s) \rightarrow N_2O(g) + 2H_2O(g)$$

Great care is necessary, as ammonium nitrate is also explosive when heated.

In many respects, N_2O is analogous to CO_2. It has the same linear structure, the same number of electrons (isoelectronic), and a similar (low) reactivity; however, CO_2 is more soluble in water as a result of the acid–base reaction of CO_2 and water. It is the low reactivity of N_2O that results in a long tropospheric lifetime, and therefore its eventual transport to the stratosphere, where it is believed to be a primary control on the concentration of ozone in the stratosphere. Concentrations of N_2O in the troposphere are increasing, and this has raised concerns that anthropogenically produced N_2O could decrease stratospheric ozone concentrations (McElroy *et al.*, 1976; Söderlund and Svensson, 1976; Weiss, 1981).

5. *N_2*. Dinitrogen, or simply nitrogen, is a colorless gas at room temperature. It is generally considered to be a very stable molecule; however, it is not its thermodynamic stability, but rather its chemical kinetic inertness that accounts for its low reactivity. This is shown by the values of the thermodynamic parameters given in Table 12.1. In the presence of oxygen, N_2 is thermodynamically unstable with respect to aqueous NO_3^-, but the high activation energy necessary to break the N_2 triple bond results in its chemical inertness. If we lived in a world dominated by equilibrium chemistry, most atmospheric N_2, and all of the O_2, would be consumed, yielding an ocean containing approximately 0.1 M HNO_3 (Lovelock, 1979). As a result of its nonpolar nature, N_2 has a low solubility in water, but with its high partial pressure in the atmosphere it is the most prevalent nitrogen species in the ocean. Nitrogen constitutes some 78% by volume of the atmosphere, and is industrially separated by the liquefication and distillation of air.

6. *NH_3*. Ammonia is a colorless gas. It is a strong base, forms hydrogen bonds, is soluble in water, and is a fairly reactive molecule. Each year, 12.4 million metric tons are manufactured by the Haber process ($N_2 + 3H_2 \rightarrow 2NH_3$ at 400° C and 250 atmospheres), principally for nitric acid production, use in fertilizers and explosives. As a fertilizer, ammonia can be utilized in three ways: first, by direct injection of the boiling (−33° C) liquid. This method works only because most soils are moist and are acidic, so that the NH_3 dissolves in the wet soil before it evaporates. The second method is to use ammonia in ammonium salts, such as NH_4Cl (s) or NH_4NO_3 (s). The third method is to oxidize the ammonia to HNO_3, and use it in a nitrate salt. In the atmosphere, ammonia is the primary gaseous base, so that it will react with acids either in the gas or aqueous phase to produce an ammonium salt, as in these reactions:

$$NH_3(g) + HCl(g) \rightarrow NH_4Cl(s)$$

$$2NH_3(g) + H_2SO_4(l) \rightarrow (NH_4)_2SO_4(s)$$

A significant proportion of the total reduced nitrogen in the atmosphere exists as aqueous or aerosol ammonium ion, with lesser amounts of gaseous ammonia (Quinn *et al.*, 1988). Ammonia has a very irritating odor, and is toxic at low concentrations. Ammonium salts contain the tetrahedral NH_4^+ ion. Some ammonium salts, like NH_4NO_3 and $(NH_4)_2SO_4$, are manufactured on a large scale, 6.0 and 1.8 million metric tons per year respectively. Ammonium sulfate is used principally as a fertilizer, whereas ammonium nitrate is also used in explosives. Since plants can utilize nitrogen in both the −3 and +5 oxidation states, ammonium nitrate is particularly well suited for use as a fertilizer; however, it must be handled with caution. In April 1947, a ship loaded with ammonium nitrate exploded and killed 576 people in Texas City, Texas.

There are also many significant nitrogen compounds that are a necessary part of all organisms. Most of these have nitrogen in a formal −3 oxidation state.

7. *Amines*. Amines (R—NH_2) are an organic derivative of ammonia, where an alkyl group (abbreviated by an —R) replaces one or more of the

hydrogens. The simplest amine is methyl amine, $CH_3—NH_2$. An amine can have one, two, or three alkyl groups, as in trimethyl amine, $(CH_3)_3—N$. Like ammonia, amines are fairly basic and are good hydrogen bonders; low molecular weight amines are quite water soluble. As the alkyl groups get larger, the amine begins to show more properties of an organic molecule than of ammonia, and water solubility decreases. The amines are generally odorous compounds with relatively high boiling points, due to hydrogen bonding. Methyl amine, and other amines, are often found in the flesh of rotting fish, and this could represent a pathway into the atmosphere for these compounds. Amines are bases, and their reaction with acids in the atmosphere is probably their principal removal mechanism (as for ammonia):

$$CH_3—NH_2(g) + H_3O^+(aq) \rightarrow CH_3—NH_3^+(aq) + H_2O$$

The alkyl ammonium ion is quite soluble in water. Like ammonia, amines may also be oxidized. As would be expected from their structure, amines are fairly polar. Amines (including aromatic amines) are quite common in the biological world. For example, many vitamins (such as thiamine and niacinamide) and all alkaloids (e.g. caffeine, cocaine, nicotine, and lysergic acid) contain an amine functional group. Many of these groups are found in ring systems, with pyridine rings (C_5H_5N) being quite common.

8. *Amides*. Amides occur quite frequently in nature, most significantly in urea and proteins. Urea, $NH_2—CO—NH_2$, is an important carrier of nitrogen between animals and plants. Animals metabolize proteins and amino acids, and excrete large amounts of urea. Plants break urea down into ammonia, which they can utilize. For this reason, urea is also an excellent fertilizer, and 5 million metric tons are manufactured each year through the high temperature reaction of CO_2 and NH_3. Animal excrement and urea fertilizers are thought to be significant sources of atmospheric ammonia (Freney *et al.*, 1983).

Proteins are also important nitrogen compounds. They constitute much of the cell materials, and are present in every type of organism known. In humans, muscle tissue, skin, and hair is mostly protein, about half of the dry weight of our bodies. From a chemical point of view, proteins are polymers of amino acids, alpha amine derivatives of carboxylic acids. Only about 20 different amino acids are actually found in proteins. It is the large number of variations in the protein chain, using only these 20 amino acids, that gives rise to the great diversity of proteins. Numerous other organic nitrogen compounds are found in natural systems in small amounts, some of which are very toxic or carcinogenic. These include certain types of nitro compounds, cyano compounds, or nitrosamines, for example.

Having considered many of the different type of compounds we will be discussing, we should briefly consider some of the utility of the data presented in Table 12-1. The thermodynamic relationships discussed in Chapter 5 may be used to calculate equilibrium concentrations for such gases as NO and NO_2 in air. This only requires the relationship between ΔG^0 and the equilibrium constant. A more complex analysis would include the contribution of photolytically driven reactions, to derive steady-state concentrations for many trace species. Holland (1978) has presented such an analysis for an abiotic world, including the input energy from sunlight. His results show that virtually all nitrogen compounds are currently found at much higher levels than the calculated steady-state value, indicating that other important processes have been ignored. This can only happen if the chemical reactions are slow compared to the rates of other processes. Lovelock (1979) has stated that on Earth, the disagreement between abiotic steady-state calculated concentrations and measured atmospheric concentrations indicates the presence of life, and suggests that atmospheric chemical probes can be used as a simple means to detect life on any planetary system. If chemical measurements of a planet's atmosphere show that it is close to thermodynamic equilibrium, substantial numbers of living organisms are probably not present. In natural systems, some thermodynamic equilibria do exist, but in many cases, environmental systems are in some type of dynamic non-equilibrium steady-state. On Earth, the presence of life significantly alters the atmospheric concentrations of many trace species.

12.3 Biological Transformation of Nitrogen Compounds

Our next task is to consider the various ways that nitrogen is processed by the biosphere. These mechanisms are the primary mover in the terrestrial and oceanic nitrogen cycles. It is important to remember that even though much of the discussion of the nitrogen cycle revolves around transfer of nitrogen between the major global reservoirs (atmosphere, aquatic, biosphere, lithosphere), these fluxes represent only a small portion of the nitrogen transferred within the biosphere–soil and biosphere–aquatic systems. Rosswall (1976) estimates that on a global basis, the "internal" biological nitrogen cycle accounts for 95% of all nitrogen fluxes. The important processes are indicated schematically in Fig. 12-1. The many new terms are easier to identify if one remembers that they are defined from the perspective of the organism:

1. *Nitrogen fixation* is any process in which N_2 in the atmosphere reacts to form *any* nitrogen compound. Biological nitrogen fixation is the enzyme-catalyzed reduction of N_2 to NH_3, NH_4^+, or any organic nitrogen compound.
2. *Ammonia assimilation* is the process by which NH_3 or NH_4^+ is taken up by an organism to become part of its biomass in the form of organic nitrogen compounds.
3. *Nitrification* is the oxidation of NH_3 or NH_4^+ to NO_2^- or NO_3^- by an organism, as a means of producing energy.
4. *Assimilatory nitrate reduction* is the reduction of NO_3^-, followed by uptake of the nitrogen by the organism as biomass.
5. *Ammoniafication* is the breaking down of organic nitrogen compounds into NH_3 or NH_4^+.
6. *Denitrification* is the reduction of NO_3^- to any gaseous nitrogen species, generally N_2 or N_2O.

All of these processes are mediated by various types of microorganisms. Some of these processes are energy-producing, and some of these occur in symbiotic relationships with other organisms. It is

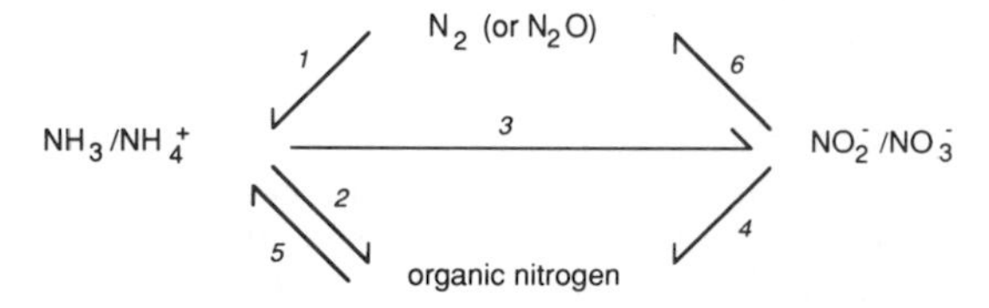

Fig. 12-1 Biological transformations of nitrogen compounds. The numbers refer to processes described in the text.

appropriate to begin our discussion of these processes with nitrogen fixation, since this is the only means by which nitrogen can be brought into natural biosystems (in the absence of artifical fertilization). Similarly, it is principally denitrification that removes nitrogen from the biosphere.

Biological nitrogen fixation is the ultimate source of nitrogen in all living organisms, in the absence of industrial fertilizers. It can be done by a variety of bacteria and algae, both symbiotic and free-living, although, in general, the symbiotic organisms are thought to be quantitatively more significant. There are two major limitations to biological nitrogen fixation. The first is that the process takes a large amount of input energy to overcome the high activation energy of the nitrogen triple bond. Despite the fact that ΔG^0 for the production of NH_3 from N_2 and H_2 is negative at 25° C, only those organisms with highly developed catalytic systems are able to fix nitrogen. The second limitation is that nitrogen fixation is a reductive process, highly sensitive to the presence of O_2, and so only those organisms that live in anaerobic environments or can provide an anaerobic environment will fix nitrogen. In terrestrial ecosystems, plants are often nitrogen-limited, so that from an agricultural point of view, understanding nitrogen fixation is important for optimizing food yields. Burns and Hardy (1975) present a good review of biological nitrogen fixation. Anthropogenic contributions to soil and aquatic nitrogen occur in several ways and, as we shall see, are of similar magnitude as biological fixation.

In terrestrial systems, the symbiotic bacteria, particularly strains of the genus *Rhizobium*, are a significant source of nitrogen fixation. These bacteria are found on the roots of many leguminous plants (clover, soybeans, chickpeas, etc.), and have been used agriculturally as a means of replenishing soil nitrogen. Many of these organisms are anaerobes, or have developed mechanisms to maintain an anaerobic environment (Granhall, 1981). Many other symbiotic diazotrophs (nitrogen-fixing organisms) exist, but the *Rhizobium* have been the most extensively studied (see, e.g. Postgate, 1982).

In freshwater aquatic systems, cyanobacteria are believed to account for up to 78% of the total fixed nitrogen (Mague, 1977). These symbiotic organisms have been found to vary with season and often to precede, and be part of, algal blooms. This can lead to eutrophication of an aquatic system, by supplying NH_3, one of the nutrients essential to all plant life. In marine systems, little is known about the nitrogen-fixing capacity. Cyanobacteria are widespread so that significant nitrogen fixation is possible, and this could be a major, global, source of fixed nitrogen. Since phytoplankton are often nitrogen-limited and the ocean represents 71% of the Earth's area, this question is of great importance.

Most diazotrophic (nitrogen-fixing) organisms utilize the nitrogenase enzyme. This enzyme has been the focus of intensive research in recent years due to the possibilities of utilizing it to improve industrial nitrogen fixation (Postgate, 1982). It has been isolated from 20 to 30 different prokaryotic organisms and appears to have very similar properties regardless of the source. Nitrogenase consists of two metalloproteins, one a Mo-Fe protein which serves to bind the N_2 to the enzyme, probably at the metal site, the other an Fe-protein, which is the source of electrons for the reduction. Both of these metalloproteins are very sensitive to O_2. It is interesting to note that, whereas microorganisms can fix N_2 at a partial pressure of 0.8 atm and 20° C, the industrial fixation requires 250 atm and 400° C!

In the broadest possible definition, nitrogen fixation includes any activites that produce nitrogen compounds from gaseous N_2. Anthropogenic nitrogen fixation can be classified into three types: direct intentional industrial production of NH_3 and HNO_3, unintentional production of NO_x (NO and NO_2) by combustion (industrial and biomass), and biological nitrogen fixation as a result of agricultural management practices (e.g. planting clover so as to replenish nitrogen on farmlands). The values for the fluxes of direct industrial fixation and combustion are probably the most reliably known of any in the nitrogen cycle. Referring to the table of fluxes (Table 12.3), the sum of the first two items listed above, direct fixation and combustion (60 Tg/year), is within an order of magnitude of the biological fixation (140 Tg/year), and the last item, biological fixation due to agriculture, is rarely estimated.

It has been suggested (Delwiche, 1981; Söderlund and Svensson, 1976) that in order to account for the estimated burden and residence time for atmospheric N_2O, the global rate of nitrogen fixation, and consequently nitrification and denitrification, would have to be about an order of magnitude greater than current estimates. Numerous measurements have found that the ocean, at certain times and places, is supersaturated with respect to N_2O (Hahn, 1981), suggesting a flux of nitrogen out of the ocean to balance the overall budget. However, other data suggest that the ocean is not a significant source of N_2O (McElroy *et al.*, 1977; Cohen, 1978). Nitrogen fixation in the oceans is not well understood, and this represents a significant gap in the understanding of the nitrogen cycle.

Once nitrogen has been fixed in the soil or aquatic system as NH_3 or NH_4^+, there are two major pathways it can follow. It can be oxidized to NO_2^-/NO_3^- or assimilated by an organism to become part of its biomass. The latter process is termed ammonia assimilation, and for those organisms that can directly utilize ammonia, this represents a significant nitrogen source. Since many plants obtain most of their nitrogen from nitrate, via reductive assimilation, direct ammonia assimilation yields significant energy savings, and therefore gives those organisms a competitive advantage. Free ammonium ion does not exist for long in aerobic soils before nitrification occurs, and so NO_3^- is the prevalent form of nitrogen in aerobic soil and aquatic environments (Delwiche, 1981). Since ammonia assimilation does not involve nitrogen transfer to other reservoirs, and quantitatively is less significant than nitrification, we will not consider it further here.

Nitrification consists of two energy-yielding steps: the oxidation of ammonium to nitrite, and the oxidation of nitrite to nitrate. These equations are generally represented as follows.

$$NH_4^+ + 3/2O_2 \rightarrow NO_2^- + H_2O + 2H^+ \qquad \Delta G^0 = -290 \text{ kJ/mol}$$

$$NO_2^- + 1/2O_2 \rightarrow NO_3^- \qquad \Delta G^0 = -82 \text{ kJ/mol}$$

There are several organisms that utilize nitrification as an energy source (Delwiche, 1981). The first step in the process is principally done by bacteria of the genus *Nitrosamonas*, and the second by *Nitrobacter*, both autotrophic organisms. These organisms utilize CO_2 as a carbon source, and obtain their energy from the oxidation of NH_4^+. Heterotrophic bacteria, which utilize organic compounds rather than CO_2, can also perform nitrification; however, these are thought to be quantitatively much less significant than the autotrophs (Bremner and Blackmer, 1981). In the oxidation of NH_4^+ to NO_3^-, hydroxylamine (NH_2OH) and other less stable compounds are likely intermediates, and NO and N_2O are almost certainly intermediates, probably as enzyme-bound complexes. These intermediates have brought attention to nitrification as a possible source of atmospheric constituents, particularly N_2O.

Bremner and Blackmer (1981) presented a review of recent experimental results on soil nitrification. Their data show that in soils under nitrifying conditions (soil well aerated), a significant amount of N_2O production can occur, and that under these conditions the rate of N_2O production is independent of the concentration of NO_3^- (and therefore not due to denitrification) and increases with application of NH_4^+. The conclusion from these experiments is that nitrification may be responsible for the production and release to the atmosphere of significant amounts of N_2O, though quantitative fluxes are not yet possible.

As in terrestrial systems, nitrous oxide in the ocean has been assumed to result primarily from the denitrification process (Söderlund and Svensson, 1976). However, recent authors have suggested that the oxidative production of NO_3^- (nitrification) is likely to be a significant source as well (Cohen and Gordon, 1979; Hahn, 1981). The data of Cohen and Gordon show that in many measurements, N_2O is correlated with NO_3^-, and negatively correlated with O_2. From these data, they draw the conclusion that nitrous oxide is produced by the oxidation of reduced nitrogen species to nitrate, consuming O_2. They estimate N_2O production in the oceans due to nitrification at 4–10 Tg/year. Since these authors feel marine nitrification is at least as significant as marine denitrification, they claim that most estimates of oceanic N_2O fluxes are too high. For example, Söderlund and Svensson (1976) give a range of 20–80 Tg/year for oceanic N_2O production.

Once in a soil or aquatic system, nitrate has two major pathways: it can serve as a terminal electron acceptor under anaerobic conditions (denitrification), or it can be simultaneously reduced and assimilated into an organism's biomass. The latter process is termed assimilatory nitrate reduction, and is likely to be dominant when reduced nitrogen is in low supply, as during aerobic conditions. This represents a primary input of nitrogen for most plants and many microorganisms. Most plants can assimilate both NH_4^+ and NO_3^-, even though there is an energy cost in first reducing the NO_3^-. After the NO_3^- has been taken up by the root system of a plant and reduced, it then follows the same pathway by which NH_4^+ is incorporated into its biomass (Kikby, 1981).

Besides nitrogen fixation, the only other major source of reduced nitrogen is the decomposition of soil or aquatic organic matter. This process is termed ammoniafication. Heterotrophic bacteria are principally responsible for this. These organisms utilize organic compounds from dead plant or animal matter as a carbon source, and leave behind NH_3 and NH_4^+. In some instances, they may incorporate a complete organic molecule into their own biomass. The majority of the reduced nitrogen produced in this way stays within the biosphere; however, a small portion of it will be volatilized. The volatilization of ammonia can occur either as a result of the breakdown of animal excreta, or by the microbial decomposition of organic matter. It is generally assumed that volatilization from animal excreta is the more significant process; however, few data are available to support this conclusion. Söderlund and Svensson (1976) estimate that as much as 77% of the global ammonia flux to the atmosphere may be due to ammoniafication. Freney *et al.* (1983) present a good review on the overall ammonia volatilization process.

Denitrification is the only process in which the major end-product is removed from the internal biological nitrogen cycle. It is the principal means of balancing the input flux from biological nitrogen fixation. Generally, N_2 is the end-product of denitrification; however, NO, and particularly N_2O, are also common. Microorganisms use NO_3^- as a terminal electron sink (oxidant) in the absence of O_2, as in waterlogged anaerobic soils. The overall process of oxidizing an organic compound, and reducing the NO_3^-, is an energy-producing process for the microorganisms. There are approximately 17 genera of facultative anaerobic bacteria that can utilize NO_3^- as an oxidizing agent, and these are thought to be widespread. Denitrification occurs via a well-known series of intermediates, including NO_2^-, NO, and N_2O. The ratio of $N_2:N_2O$ production during denitrification is a topic of current concern. Generally, the major product is N_2, accounting for 80–100% of the nitrogen released (Delwiche, 1981). Söderlund and Svensson (1976) used a global terrestrial average of 16 : 1 for $N_2:N_2O$ in calculating the N_2O flux to the atmosphere. Under certain environmental conditions, nitrous oxide can become a major product. Generally, conditions that increase the amount of N_2O production also decrease the overall rate of denitrification. For example, at lower pH values and higher O_2 concentrations, the proportion of N_2O increases, but the overall rate of denitrification decreases.

In terrestrial systems, it is generally agreed that

denitrification is a significant natural source of both N_2 and N_2O. There is speculation, however, that increasing use of nitrogen fertilizers may be contributing to an increasing flux of N_2O from terrestrial systems to the atmosphere, either due to nitrification, denitrification, or both, and that this is the cause of the increasing tropospheric concentration of N_2O (Weiss, 1981; Prinn *et al.*, 1983; Robinson *et al.*, 1984). The increase in tropospheric N_2O is predicted to contribute to ozone depletion in the stratosphere (Söderlund and Svensson, 1976; McElroy *et al.*, 1977). With respect to agricultural utilization of nitrogen fertilizers, the subject of gaseous losses due to denitrification has been extensively studied. Rolston (1981) presents a good review of this topic, and gives data suggesting that anything from 10–75% of fertilizer nitrogen may be lost by this process (typically in the range of 20%). Various crop management practices have been developed to counter this problem. For example, having large amounts of NO_3^-, organic carbon, and water will increase denitrification, whereas limiting one of these factors will decrease it.

By comparison to terrestrial systems, denitrification in the oceans is very poorly understood. There are different views on whether the oceans are a source or sink of nitrous oxide. Most, but not all, data indicate that the ocean is, on average, supersaturated with respect to N_2O, and that N_2O supersaturations are positively correlated with NO_3^- and negatively correlated with O_2 (Hahn, 1981; Cohen and Gordon, 1979). However, the interpretations of these results vary. As previously mentioned, Cohen and Gordon (1979) claim that these data support the notion that N_2O is produced during the oxidative production of NO_3^- from reduced species, where the O_2 is depleted because it is the oxidant in this process. Other authors interpret the same data as being a result of denitrification (Söderlund and Svensson, 1976), assuming that it is the low concentrations of O_2 that encourages the microbial utilization of NO_3^- as the terminal electron acceptor. In addition, Cohen (1978) has suggested that nitrous oxide can also be consumed by denitrification in anaerobic environments, and this could represent a portion of the "missing sink" for N_2O (as will be discussed in a later section). Overall, it is clear that there are large gaps in our understanding of denitrification in the oceans.

From a biogeochemical cycle perspective, the internal biological nitrogen cycle is not a simple system. It still appears to be true that the largest nitrogen input to the biosphere is biological fixation, and the primary loss from the biosphere is denitrification. From a global perspective, it is not the exchange of N_2 that plays a major role in the biogeochemical cycle, but rather the exchange of various trace gases, such as NO, NO_2, N_2O, and NH_3. It now appears that these trace gases may be released into the atmosphere at various stages in the "internal" biological nitrogen cycle. To understand the behavior of these trace gases, it is necessary to consider a wide range of biological processes. Some of the topics that are in need of further understanding include: the nature and quantification of nitrogen fixation in the ocean and nitrification in soils; agricultural practices that can inhibit denitrification in soils; and the nature and quantification of all processes relating to nitrous oxide in the oceans.

12.4 Abiotic Processes

12.4.1 Homogeneous Gas Phase Reactions

Although abiotic nitrogen fluxes are quantitatively smaller than the biological fluxes, they are quite important with respect to tropospheric aerosols, tropospheric and stratospheric ozone, removal of numerous trace species from the atmospheric reservoirs, and therefore the global climate. These abiotic processes can be categorized into gas phase reactions or heterogeneous reactions. In the atmosphere, photochemistry plays a significant role in producing reactive species (such as OH radicals), which are available to perform transformations that otherwise would not occur. The photochemistry of the atmosphere is quite complex and will not be dealt with in great detail here. For an in-depth review on tropospheric photochemistry, the reader is referred to Logan *et al.* (1981), Finlayson-Pitts and Pitts (1986), Crutzen and Gidel (1983), or Crutzen (1988). The fundamentals of chemical kinetics are discussed in Chapter 5.

In most cases, the direct reaction of N_2 with O_2 is slow under ambient conditions. It is the presence of numerous odd electron species (for example, OH, HO_2, and RO_2 radicals) that are photochemically produced and responsible for most of the oxidizing reactions of nitrogen species in the atmosphere.

Some of the important reactions are shown below:

$$O_3 + h\nu \rightarrow O_2 + O(^1D)$$

where $O(^1D)$ is an electronically excited oxygen atom. It can decay back to a ground state oxygen atom (^3P) (which will regenerate an ozone molecule), or else it can react with water to produce two OH radicals:

$$O(^1D) + H_2O \rightarrow 2OH$$

The OH radical is a primary oxidizer in the atmosphere, oxidizing CO to CO_2 and CH_4 and higher hydrocarbons to CH_2O, CO, and eventually CO_2. OH and other radical intermediates can oxidize CH_4 and NO in the following sequence of reactions:

$$OH + CH_4 \rightarrow H_2O + CH_3$$
$$M + CH_3 + O_2 \rightarrow CH_3O_2 + M$$
$$CH_3O_2 + NO \rightarrow NO_2 + CH_3O$$
$$CH_3O + O_2 \rightarrow HO_2 + CH_2O$$
$$HO_2 + NO \rightarrow NO_2 + OH$$

net reaction:

$$CH_4 + 2O_2 + 2NO \rightarrow CH_2O + 2NO_2 + H_2O$$

Hydroxyl, OH, acts as a catalyst for the oxidation of NO to NO_2. NO_2 molecules can react with the OH radical to produce HNO_3, which may be removed in precipitation. This is how most tropospheric NO_x eventually gets removed, either in wet or dry deposition:

$$NO_2 + OH + M \rightarrow HNO_3 + M$$

NO_2 may also photolyze and produce a ground state O atom, which will go on to produce ozone:

$$NO_2 + h\nu \rightarrow NO + O \qquad (\lambda < 410 \text{ nm})$$
$$O_2 + O + M \rightarrow O_3 + M$$

These reactions are important in a cycle that oxidizes CO and hydrocarbons and produces ozone, in the presence of sufficient NO_x. In photochemical smog, ozone can build up to unhealthy levels of several hundred parts per billion (ppb) as a result of these reactions. There are many other reactions that occur, some of which may be significant at various times, including the destruction of O_3 by NO, production and loss of HONO (nitrous oxide) and peroxyacetyl nitrate (PAN), and further oxidation of CH_2O. These reactions, and many more, represent a complex set of chemical interactions. For our purposes here, it is only necessary to note the major features:

1. The oxidation of CO and all hydrocarbons to CO_2 is indirectly driven by ozone and sunlight via the OH radical.
2. In the presence of sufficient NO_x (roughly 30 parts per trillion), this oxidation produces ozone.
3. Most NO and NO_2 eventually gets removed as HNO_3.
4. The lifetime of gaseous NO_x in the troposphere is in the order of 1–30 days (Söderlund and Svensson, 1976; Garrels, 1982; Crutzen, 1988).

Ammonia is also oxidized by OH radicals, and this has been proposed as a source and a sink for NO_x (Logan, 1983), although it is a minor sink for NH_3. The reaction sequence for this oxidation begins as follows:

$$NH_3 + OH \rightarrow NH_2 + H_2O$$

The next step is not clear, but it appears that the NH_2 can react with either an O_3, NO, or NO_2:

$$NH_2 + NO \text{ or } NO_2 \rightarrow N_2 \text{ or } N_2O \qquad \text{(several steps)}$$

$$NH_2 + O_3 \rightarrow NO \text{ and other products} \qquad \text{(several steps)}$$

Since the NO_x reaction above is 1000 times faster than the O_3 reaction, at an O_3 concentration of 60 ppbv the rates of these two reactions will be comparable when the NO_x concentration is 60 pptv. At NO_x concentrations lower than this, there will be NO_x production, and at levels higher than 60 pptv there will be NO_x destruction. Stedman and Shetter (1983) suggest that this reaction sequence could account for a small production of NO_x in regions of the globe with low NO_x concentrations, and a major sink for it in areas heavily impacted by anthropogenic NO_x emissions. However, this mechanism remains highly speculative.

Nitrogen oxides also play a significant role in regulating the chemistry of the stratosphere. Aspects of oxygen atom and ozone chemistry in the stratosphere are discussed in Chapters 5 and 10. In the stratosphere, ozone is formed by the same reaction as in the troposphere, the reaction of O_2 with an oxygen atom. However, since the concentration of O atoms in the stratosphere is much higher (O is produced from photolysis of O_2 at wavelengths less than 242 nm), the concentration of

O_3 in the stratosphere is much higher:

$$O_2 + h\nu \rightarrow 2O$$
$$O_2 + O + M \rightarrow O_3 + M$$

This ozone production is balanced by various ozone destruction reactions. Principal among these is the catalytic reaction of O_3 with NO:

$$O_3 + NO \rightarrow NO_2 + O_2$$
$$NO_2 + O \rightarrow NO + O_2$$
$$\text{net reaction:} \quad O_3 + O \rightarrow 2O_2$$

The principal source of NO_x in the stratosphere is the slow upward diffusion of N_2O, and its subsequent reaction with O atoms, or photolysis (McElroy *et al.*, 1976):

$$N_2O + h\nu \rightarrow N_2 + O$$
$$N_2O + O(^1D) \rightarrow 2NO$$

The first of these reactions will result in the generation of a single ozone molecule. The second reaction produces the NO_x that leads to catalytic ozone destruction. The relative rates of these two reactions is in an approximate ratio of 9:1, favoring the first. Since NO is a catalyst for O_2 destruction (a single NO will destroy many ozone molecules before being removed), N_2O is believed to exert a significant control on stratospheric O_3 concentrations.

A ground level source, and stratospheric sink for N_2O, is consistent with the observed vertical concentration gradient. McElroy *et al.* (1976) estimate that an ozone reduction of up to 20% is possible, based upon the above catalytic cycle and increasing anthropogenic emissions of N_2O, assuming no other sinks for atmospheric nitrous oxide are found. More recent model predictions suggest a total column ozone reduction of 4% based on the current increase in tropospheric N_2O concentrations (National Research Council, 1984).

12.4.2 Heterogeneous Reactions

Heterogeneous processes play a role in the following ways: gas–particle conversions and cloud chemistry in the atmosphere; exchange of gases into or from the oceans; exchange of gases into or from soil; weathering, degassing, mineralization and burial of nitrogen compounds into or from rocks. In the atmosphere, precipitation results in the removal of a large fraction of the nitrogen compounds as NH_4^+ or NO_3^-. These ions are produced by reactions such as the following:

$$NH_3(g) + HNO_3(g) \rightarrow NH_4NO_3(s)$$
$$NH_3(g) + HNO_3(aq) \rightarrow NH_4^+(aq) + NO_3^-(aq)$$
$$NH_3(g) + H_2O(l) \rightarrow NH_3(aq)$$
$$NH_3(g) + H_2SO_4(l) \rightarrow NH_4(HSO_4)(s)$$
$$NH_3(g) + NH_4(HSO_4)(s) \rightarrow (NH_4)_2SO_4(s)$$

All of these species are very soluble in a raindrop or cloud drop. These reactions represent an important source of atmospheric aerosols. For ammonia and ammonium, the condensed phase represents approximately $^2/_3$ of the total atmospheric burden, whereas for nitric acid and nitrates, about $^2/_3$ is in the gas phase (Söderlund and Svensson, 1976).

The oceans represent a large potential reservoir for any gaseous compound. The solubility of a particular compound is governed by its chemical structure, temperature, pH and other chemical properties of the solution, as well as its atmospheric concentration. For some atmospheric gases, such as NH_3 and NO_x, the oceans probably are a net sink (through the mechanism of precipitation), since virtually all measurements of oceanic air show significantly lower concentrations than is found over continental regions. This seems reasonable in light of the fact that most sources for NH_3 and NO_x are in continental areas. However, recent measurements of gaseous and aerosol ammonia in seawater and air indicate that the oceans are supersaturated with respect to ammonia, and thus are a likely source of gaseous ammonia, while at the same time being a *net* sink for it (Quinn *et al.*, 1988).

For atmospheric N_2O, it is not clear whether the net flow is to or from the oceans. As mentioned previously, some authors (Cohen and Gordon, 1979; Hahn, 1981) believe that the numerous measurements of oceanic supersaturation are good evidence for marine export of N_2O. On the other hand, McElroy *et al.* (1976) claim that the oceans are a net sink for atmospheric N_2O, though they base this result on scanty data. Also for atmospheric N_2, it is not clear whether the net flow is to or from the oceans, although it is generally assumed that denitrification will balance the inputs from all other sources (Söderlund and Svensson, 1976). As will be seen in Section 12.5.3, this may not be the case.

Volatilization of gaseous nitrogen compounds from soils can represent a major pathway into the atmosphere. As in the aquatic systems, parameters that play an important role in this process include: the nature of the compound; soil temperature, water content, pH, aeration of the soil; and a concentration gradient of the gas in question. Ammonia has as its major sources ammoniafication of organic matter in soils, and volatilization of urea in animal feces. These two processes are difficult to distinguish experimentally, and are often grouped together. In measurements to determine ammonia fluxes to the atmosphere, an additional complication arises due to the rapid recycling of ammonia gas in the plant cover at ground level (Denmead *et al.*, 1976). The majority of the ammonia released from soil or animal wastes is quickly recycled and does not mix in the atmosphere. Since gaseous ammonia concentrations drop off rapidly with altitude (Hoell *et al.*, 1980), it seems clear that its principal source in the atmosphere is at the ground. While little is known about amines, these may also prove to have a significant terrestrial or oceanic source due to volatilization.

Since none of the other gaseous nitrogen compounds have as high a solubility in water as ammonia, it seems appropriate to consider these fluxes as direct biota-atmosphere processes (for instance NO or N_2O losses during nitrification).

The quantitative understanding of fluxes to and from rocks and sediments is poorly known, but in some instances these may represent a significant component of the nitrogen cycle. For example, in the cycling of organic nitrogen compounds, it is estimated (Söderlund and Svensson, 1976) that 38 Tg/year are permanently buried principally by sedimentation of organic particulate matter. There are numerous other instances where weathering or mineralization occur; however, little is known about the global fluxes for these processes.

12.5 The Global Nitrogen Cycle

The global nitrogen cycle is often referred to as the nitrogen cycles, since we can view the overall process as the result of the interactions of various biological and abiotic processes. Each of these processes, to a first approximation, can be considered as a self-contained cycle. We have already considered the biological cycle from this perspective (Fig. 12-1), and now we will look at the other processes, the ammonia cycle, the NO_x cycle, and the fixation/denitrification cycle.

12.5.1 Nitrogen Inventories

We will consider the inventories of nitrogen in the following compartments: crustal, terrestrial, oceanic, and atmospheric. In general, there is more agreement among researchers over the values for the nitrogen burdens than for the fluxes. In considering these inventories, it is significant to recall that 99.96% of the non-crustal nitrogen exists as uncombined atmospheric N_2 and it is this fact that causes nitrogen to often be a limiting nutrient in the condensed phase.

The principal form of nitrogen in terrestrial systems is as dead soil organic matter, with biomass accounting for only about 4%, and inorganic nitrogen about 6.5%, on a global average (Söderlund and Svensson, 1976). There is, however, a large difference in the distribution of nitrogen in the tropics and the polar regions, with tropical regions having a larger proportion of nitrogen contained as biomass. Most of the reservoirs for organic nitrogen and biomass have been estimated by knowledge of the carbon content of soils and biomass and an appropriate ratio of carbon to nitrogen. The inventories of inorganic forms of nitrogen have a higher relative uncertainty.

Dissolved N_2 is the principal form of nitrogen in the oceans, accounting for 95% of the total oceanic nitrogen. The remainder of the oceanic nitrogen is principally NO_3^- and dead organic matter. The oceans hold about 0.5% of the total non-crustal nitrogen (as N_2). In contrast to CO_2, where the oceans are a significant reservoir, the oceans contain only about 15% of the total N_2O, due to its lower solubility. As in the terrestrial compartment, organic nitrogen compounds are estimated from knowledge of the carbon content and an appropriate C : N ratio, and the inorganic nitrogen inventories have a higher relative uncertainty.

In the atmosphere, N_2 is the principal nitrogen component and over 99% of the remaining nitrogen is found as N_2O. The other trace nitrogen species all have reactivities and removal mechanisms that result in residence times of less than 1 year and low atmospheric concentrations. Gaseous ammonia, for example, has been shown (Hoell *et al.*, 1980) to decrease significantly with altitude due to its removal by acidic gases, acidic aerosols, and liquid water. The accuracies of the inventories of many of the trace species, particularly NH_3 and NO_x, are limited by scanty data, especially in the southern

Table 12-2 Nitrogen inventories in Tg N

Reservoir	Söderlund and Svensson (1976)	McElroy *et al.* (1976)	Stedman and Shetter (1983)
Crustal			
Rocks	—	—	—
Sediments	4.0 E8	6.0 E8	—
Coal dep.	1.2 E5	—	—
Terrestrial			
Soil:			
organic	3.0 E5	6.0 E4	—
inorganic	—	1.0 E4	—
Biomass:			
plants	1.1–1.4 E4	1.0 E4	—
animals	2.0 E2	—	—
Oceanic			
Dissolved N_2	2.2 E7	—	—
Dissolved N_2O	2.0 E4	—	—
NO_3^-	5.7 E5	—	—
NH_4^+	7.0 E3	—	—
NO_2^-	5.0 E2	—	—
Total inorganic	—	6.0 E5	—
Organic matter:			
dissolved	5.3 E5	—	—
particulate	0.3–2.4 E4	—	—
total organic	—	2.0 E5	—
Biomass:			
plants	3.0 E2	—	—
animals	1.7 E2	—	—
total biomass	—	8.0 E2	—
Atmospheric			
Gaseous:			
N_2	3.9 E9	4.0 E9	3.9 E9
N_2O	1.3 E3	1.1 E3	1.4 E3
NH_3	0.9	—	0.3
NO_x	1.0–4.0	—	0.2
Aerosol:			
NH_4^+	1.8	—	0.6
NO_3^-	0.5	—	0.2
Total N(-III)	—	—	—
Total $NO_x + NO_3^-$	—	—	—
Total reactive (excludes $N_2 + N_2O$)	—	3.0	—
Total organics	1.0	—	—

Note: The E notation is used to represent powers of 10.

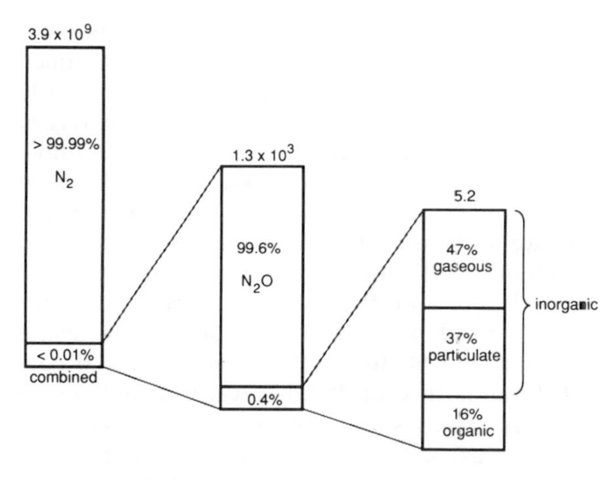

Fig. 12-2 Partitioning of the various forms of nitrogen in the atmosphere. Units are Tg N. Reprinted from Söderlund and Rosswall (1982) with the permission of Springer-Verlag.

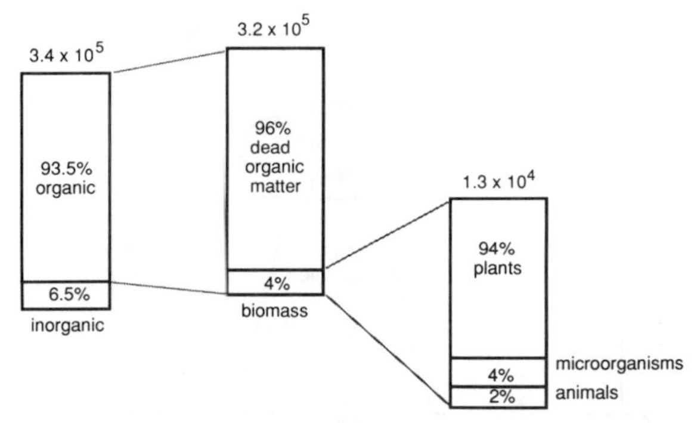

Fig. 12-4 Partitioning of the global inventories of nitrogen in the terrestrial system. Units are Tg N. Reprinted from Söderlund and Rosswall (1982) with the permission of Springer-Verlag.

hemisphere and remote regions of the globe. Table 12-2 presents inventories of nitrogen. Schematics of the distributions of nitrogen in the atmosphere, the ocean, and terrestrial biota are shown in Figs. 12-2, 12-3, and 12-4.

12.5.2 Fluxes of Nitrogen

The largest source of atmospheric ammonia is ammoniafication and volatilization from animal excreta (Freney *et al.*, 1983). Direct anthropogenic emissions, including combustion and fertilizers, are probably much smaller; however, an indirect anthropogenic contribution from domestic animals may be a significant fraction of the total. A small part of this ammonia is oxidized to nitrogen oxides, but the majority of it is returned as NH_4^+ in precipitation or as NH_3(g) via dry deposition. The ammonia is then available again to the biosphere, and the cycle is repeated (Fig. 12-5). Some of the better studied phenomena in the ammonia cycle are the wet deposition processes in populated regions; however, there are few quantitative data available for many of the other processes.

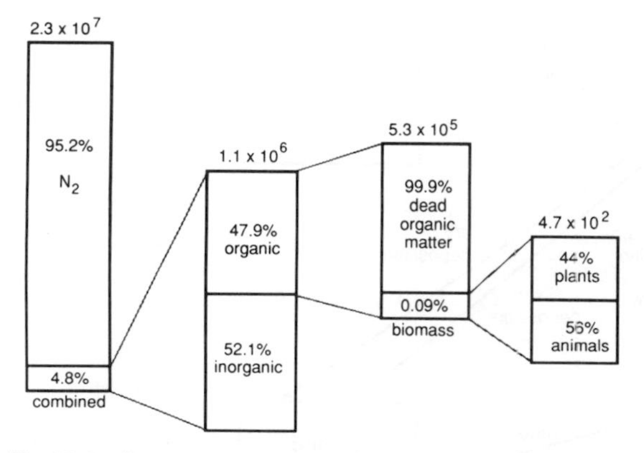

Fig. 12-3 Partitioning of the global inventories of nitrogen in the aquatic system. Units are Tg N. Reprinted from Söderlund and Rosswall (1982) with the permission of Springer-Verlag.

In the NO_x cycle (Fig. 12-6), gaseous emissions of NO and NO_2 are balanced by wet and dry deposition of NO_3^-, NO, and NO_2. The principal sources of NO_x species are thought to be anthropogenic combustion (both fossil fuels and biomass). Microbial processes in soils, lightning, and natural forest fires, probably, are minor NO_x sources. However, the fluxes of the natural NO_x production processes are poorly known, whereas the anthropogenic contributions are among the best known values in the nitrogen cycle. In the atmosphere, NO_x is converted to HNO_3 via photochemical oxidation, and therefore has a short residence time, on the order of a few days. Wet deposition occurs mainly as NO_3^- in precipitation, and since the anthropogenic emissions of NO_x occur mainly in urban areas, HNO_3 is a significant contributor to acid precipitation in adjacent regions (within a few thousand km). Anthropogenic acid deposition results mainly from HNO_3 and H_2SO_4. Anthropogenic emissions of NO_x are increasing and

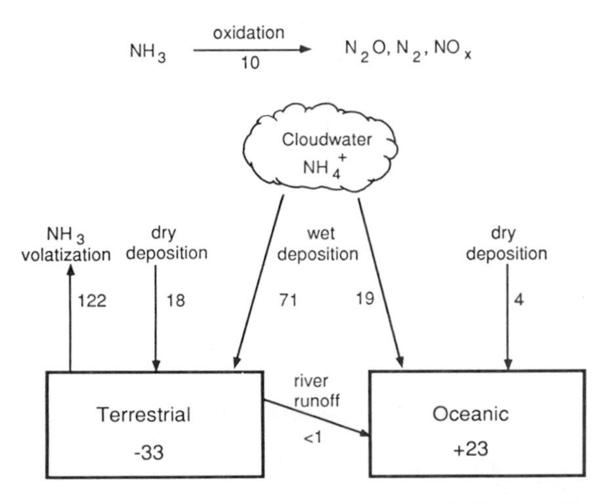

Fig. 12-5 The ammonia/ammonium cycle (Tg N/year). The numbers are fluxes and changes in burdens.

contributing an increasing amount of HNO_3 to acidic deposition (Mayewski *et al.*, 1986), whereas anthropogenic sulfur emissions appear to be leveling off. Once nitrate is deposited back to the terrestrial or oceanic systems, it can be incorporated into biomass, enter the fixation/denitrification cycle, or accumulate in the ocean.

The fixation/denitrification cycle (Fig. 12-7) is perhaps the most heavily perturbed by humans, but the effects of these perturbations are poorly understood. In the atmosphere, the principal components are N_2 and N_2O, both produced by denitrification. There is also evidence that N_2O can be consumed during denitrification. In addition, as previously mentioned, there is great uncertainty with regard to the oceanic component of the denitrification/fixation cycle. In general, it is assumed that both the oceans and the terrestrial systems are in approximate balance with respect to total nitrogen fluxes, and denitrification, being the least understood process, is used to balance the cycle. The value for denitrification used here has been obtained by using Stedman and Shetter's (1983) "pre-industrial" denitrification flux of 124.5 Tg/year, based on an assumed steady-state prior to anthropogenic influences, plus an anthropogenic contribution. The anthropogenic contribution to denitrification has been estimated to be equal to 35% of the nitrogen in fertilizer lost due to denitrification (Delwiche, 1981; Rolston, 1981), plus an additional amount due to agricultural management practices (such as planting legumes) which increase nitrogen fixation and denitrification. The denitrification due to agricultural management practices has been estimated as 25% of Burns and Hardy's (1975) estimated anthropogenic biological fixation, to give a total anthropogenic increase in denitrification of 23 Tg/year.

The data of Weiss (1981) suggest that the fixation–denitrification cycle is not balanced, which he attributes to anthropogenic influences. In his extensive data set of gaseous N_2O concentrations, several facts are apparent: First, mean N_2O con-

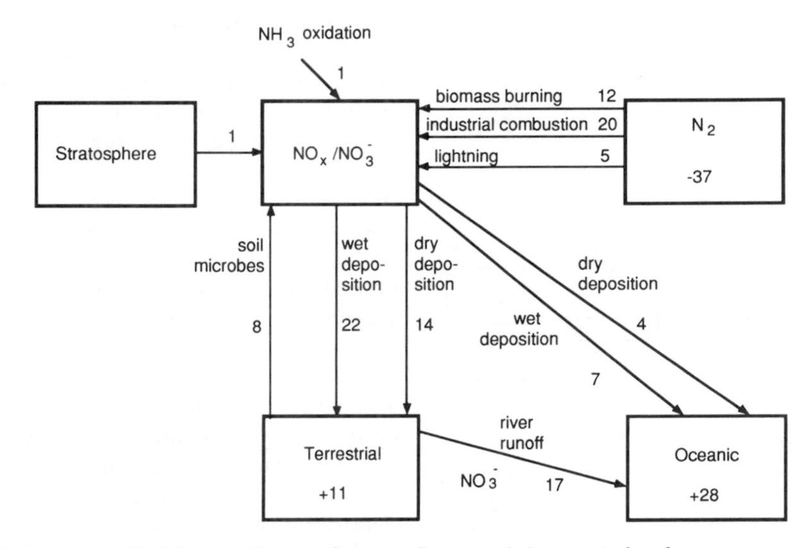

Fig. 12-6 The NO_3/NO_x cycle (Tg N/year). The numbers are fluxes and changes in burdens.

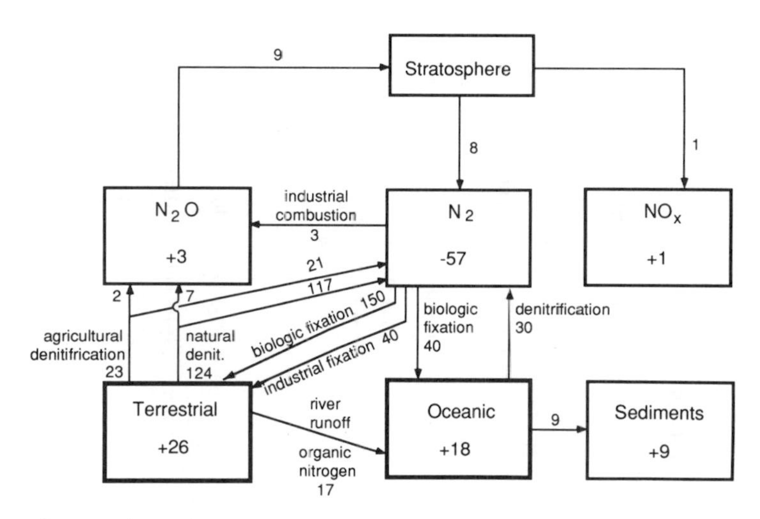

Fig. 12-7 The nitrogen fixation–denitrification cycle (Tg N/year). The numbers are fluxes and changes in burdens.

centrations in the northern hemisphere are about 300.2 ppbv, whereas southern hemispheric concentrations are lower by 0.8 ppbv. Second, global N_2O concentrations are increasing at a rate of about 0.2% per year. By fitting his data to a global box model for N_2O with a residence time of 150 years, and an exponentially increasing source strength, he calculates that the increase can be accounted for by a net unbalanced flux of N_2O into the atmosphere of 3 Tg/year. Thus flux could be accounted for simply by anthropogenic combustion, or by a combination of combustion and agricultural practices. The global increase in N_2O concentrations has been confirmed by Robinson *et al.* (1984), Prinn *et al.* (1983), and NOAA (1987).

Stedman and Shetter (1983) claim that the magnitude of the northern–southern hemispheric difference in N_2O concentrations is consistent with a major terrestrial source and a large oceanic sink, as suggested by McElroy *et al.* (1976), since the largest land masses are in the northern hemisphere. The N_2O model presented by Stedman and Shetter (1983) includes large soil source and sink terms, and gives an N_2O residence time of about 20 years. Their model is consistent with much of the soil data (presented in the earlier sections of this paper) which suggest that soils are both a major source and sink for N_2O (see also Keller *et al.*, 1983). However, Stedman and Shetter's model is inconsistent with the data and calculations of Weiss (1981), the data of Hahn and Crutzen (1982), the residence time for atmospheric N_2O as calculated by Pierotti *et al.* (1978), and soil studies by Seiler and Conrad (1981), which shows that N_2O can be both produced and consumed simultaneously in a soil profile (and, consequently, that most estimates for soil N_2O production are too high). A model for fixation–denitrification is presented that includes a total source strength for N_2O of 12 Tg/year, and a sink to the stratosphere of 9 Tg/year, giving a net unbalanced source of 3 Tg/year (Fig. 12-7). This model is consistent with the small source–sink terms suggested by Seiler and Conrad (1981), and gives an atmospheric lifetime for N_2O of about 100 years. It is interesting to note that in the flux estimates of various authors (Table 12-3), the range of values for the microbial production of N_2O falls into two non-overlapping categories. With respect to understanding the anthropogenic perturbations to the nitrogen cycle, it is very important to know which of these models is the more accurate. If the sources and sinks for N_2O are large (*c.* 100 Tg/year), an unidentified surface sink of N_2O must be in operation so that a smaller fraction of anthropogenic N_2O will reach the stratosphere and, consequently, the changes in the ozone layer will be smaller in magnitude. However, if the sources and sinks for N_2O are on the order of 10 Tg/year, then an anthropogenic contribution of only 3 Tg/year (and increasing by about 3% per year) (Weiss, 1981) becomes very significant, since the only important sink is to the stratosphere.

The fluxes in Table 12-3 have been divided into the

Table 12-3 N fluxes in Tg N/year

	Burns and Hardy (1975)	Söderlund and Svensson (1976)	McElroy *et al.* (1976)	Delwiche (1977)	Sweeney *et al.* (1978)	Stedman and Shetter (1983)	Other	Jaffe, this work	Range
Crustal-oceanic									
Sedimentation	15	38	0 (net)	0.2	10	14	10–40[h]	14	0–40
Weathering	5					14		5	5–14
Crustal-atmospheric									
Outgassing of N_2	5					0.1–2.0		1	0.1–5
Terrestrial-oceanic									
River run-off	15	14–24	20	35	30		15–40[h]	34	14–40
Terrestrial-atmospheric									
Biological fixation	139	139	170	99	100	110	140[e]	150	99–170
Microbial NO_x prod.		21–89				10	8[f]	8	8–89
Microbial N_2O prod. (nat.)		16–69[b]	40[b] net			38[b]	0.7–7[e]	7 net	0.7–69
Microbial N_2O sink						50			
Denitrification	140	107–161	243	120	90	124.5		147[d]	90–243
Ammonia volatilization	165	113–244	150	75		82[a]	16–131[e]	122	75–244
Industrial fixation	30	36	40	40	35		58[e]	40	30–58
Anthropogenic N_2O prod.						11	3–9,[e] 3[g]	5[d]	3–11
Total NH_3 deposition		138.5		65				89	65–138
Total NO_x deposition		57.5		25				36	25–57
Terrestrial + Oceanic-atmospheric[c]									
Total N deposition	200		240		220	109[a]			
Oceanic-atmospheric									
Biological fixation	36	30–130	10	30	15–90	40	60–100[h]	40	10–130
Denitrification	70	25–179	106	40	50–125	30.5		30	25–179
Microbial N_2O prod.		20–80[b]	10[b]			0 (net)	1–10[e] net	0 (net)	40–80
Microbial N_2O sink		0	50						
Total NH_3 deposition		34.5		14				23	14–34
Total NO_x deposition		22		9				11	9–22
Atmospheric-atmospheric									
N_2O loss to the strat.		10	10		20	9		9	9–20
Lightning	10		10	7	0.5–3	3	8[f]	5	0.5–10
Industrial combustion	20	19	40	18	15	20	21[f]	20	15–40
Biomass burning						5	12[f]	12	15–12

[a] Only northern hemispheric values for NH_3 are included.
[b] These values are also included as denitrification.
[c] Values given only where separate terrestrial and oceanic fluxes were not estimated.
[d] These values include 2 Tg/year from agricultural soils.
[e] Hahn and Crutzen (1982); [f] Logan (1983); [g] Weiss (1981); [h] Hahn (1981).

four principal systems: crustal, terrestrial, oceanic, and atmospheric. For those atmospheric gases that have residence times of less than 1 year (NH_3, NO_x, and any aerosol species), it is assumed that their atmospheric cycles are in steady-state; however, for those gases that have residence times much greater than 1 year (N_2 and N_2O), no assumptions about steady-state have been made. In most cases, recent estimates relate back to the review by Söderlund and Svensson (1976), and this in turn references the data of much earlier studies.

12.5.3 Anthropogenic Perturbations

Having looked at the major nitrogen cycles, we have already discussed some of the significant anthropogenic perturbations. It is useful to include a summary of some of the likely or possible consequences of these. We consider three perturbations: tropospheric NO_x/O_3 smog chemistry, stratospheric N_2O/O_3 chemistry, and possible consequences of increases in nitrogen fixation. In considering these, we will proceed from the relatively well-understood tropospheric "smog" chemistry, to the very speculative, but potentially very significant, question of increased nitrogen fixation.

12.5.3.1 Tropospheric NO_x and O_3 chemistry

The chemistry of tropospheric ozone production is one of the most extensively studied phenomena of environmental chemistry. However, even this subject has its controversial aspects, particularly with regard to how best to meet US Environmental Protection Agency mandated air quality standards for O_3 and NO_2. The essential reactions have already been presented in Section 12.4 and will only be discussed briefly here.

In all high-temperature combustion processes, particularly power plants and automobiles, NO_x is produced by the direct reaction of $N_2 + O_2$, and from nitrogen-containing fuels. It can then be oxidized by either HO_2, RO_2, or O_3 to NO_2. In the presence of NO_x and sunlight, the oxidation of CO, CH_4, and other hydrocarbons results in substantial ozone production. In an urban environment, the diurnal cycle of these trace species will generally exhibit a characteristic pattern of concentration maxima first in NO, then NO_2, followed by O_3 around midday (National Academy of Sciences, 1977). This pattern has been successfully modeled (Niki *et al.*, 1972). Whether it is necessary to control both hydrocarbons and NO_x, or just hydrocarbons, is a matter of some concern (Pitts *et al.*, 1983). Recent evidence indicates that natural hydrocarbons and anthropogenically produced NO_x are important precursors to urban and rural ozone (Liu, S.C. *et al.*, 1987; Liu, X. *et al.*, 1988). This suggests that control of anthropogenic NO_x will be required in order to reduce ozone concentrations.

As a result of anthropogenic NO_x emissions, tropospheric ozone in remote northern hemispheric locations has shown a large increase since pre-industrial times (Crutzen and Gidel, 1983; Crutzen, 1988; Oltmans and Komhyr, 1986; Logan, 1985; Volz and Kley, 1988). Chemical models suggest that ozone production per unit of NO_x is most efficient at low NO_x levels (Liu *et al.*, 1987, 1988). Since tropospheric ozone plays an important role in the oxidizing nature of the troposphere, is a greenhouse gas, and is toxic to humans and vegetation at levels not far above ambient, understanding the role of anthropogenic NO_x emissions in determining global ozone concentrations is of some significance (Dickinson and Cicerone, 1986; Logan, 1986).

As seen in Table 12-4, global NO_x production is probably dominated by anthropogenic sources; in an urban environment, virtually all NO_x is of anthropogenic origin.

12.5.3.2 Stratospheric N_2O/O_3 chemistry

As discussed previously, N_2O serves to destroy stratospheric ozone catalytically. The rate of diffusion of tropospheric N_2O to the stratosphere is a function of its concentration gradient and the eddy diffusivity. In considering anthropogenic influences on the stratosphere, one must first understand N_2O in the troposphere. This relates back to our models for the denitrification–fixation cycle. If the large source and sink term model is correct, then the anthropogenic emissions represent only a small fraction of the total flux, and loss mechanisms other than to the stratosphere must remove a substantial fraction of the total tropospheric N_2O. If, on the other hand, the model presented here is the more correct, then anthropogenic emissions are a major fraction of the total (41%), the only significant sink is to the stratosphere, and we would predict that tropospheric N_2O concentrations would rise in almost direct proportion to the total anthropogenic source strength. This suggests one method for

Table 12-4 Summary of anthropogenic perturbation to the nitrogen cycle

Natural	Anthropogenic	A/N[a]	Reference
Tropospheric flux of NO_x (Tg/year)			
4–22	19	0.9–4.8	Söderlund and Svensson (1976)
3–19	20–60	1–20	Hahn and Crutzen (1982)
15	25	1.7	Stedman and Shetter (1983)
17	33	1.9	Logan (1983)
Atmospheric flux of N_2O (Tg/year)			
36–149	?	—	Söderlund and Svensson (1976)
?	3	—	Weiss (1981)
2–17	5–11	0.3–5.5	Hahn and Crutzen (1982)
48–58	11	0.19–0.23	Stedman and Shetter (1983)
7	5	0.71	Jaffe (1985)
Global N fixation (Tg/year)			
131	74 (44)	0.56	Burns and Hardy (1975)
169–269	55	0.20–0.33	Söderlund and Svensson (1976)
190	80	0.42	McElroy *et al.* (1976)
136	58	0.43	Delwiche (1977)
109–209	180 (89)	0.9–1.6	NAS (1978)
154	132 (80)	0.86	Stedman and Shetter (1983)

[a] A/N is the ratio of the anthropogenic flux to the natural flux.

resolving this controversy: by making an extensive historical catalog of anthropogenic N_2O emissions from all possible sources (direct fertilizer use, other agricultural practices, fossil fuel combustion, sewage, and other possible sources) and comparing this to the increasing N_2O concentration, one could determine if there is a direct relationship. Of course, the question of increased emissions due to anthropogenic influences is quite open, but as more agricultural and atmospheric data become available, it may be possible to make this comparison. In dealing with the question of increased emissions due to anthropogenic influences, one must be careful to consider data that leave open the possibility of an anthropogenic sink for N_2O. In other words, much of the soil data currently available considers only the soil as either a source or a sink, not both simultaneously.

Estimates for the ratio of anthropogenic/natural N_2O production range from 0.19–5.50. As mentioned previously, total ozone reductions in the stratosphere have been calculated to be 4% at the present rate of increase of tropospheric N_2O (National Research Council, 1984), and could be much larger if N_2O concentrations grow at a faster rate. This loss is not balanced by ozone production in the troposphere. Since a decrease in the ozone column would result in a significant increase in the number of cases of skin cancer per year, might alter the global climate, and could have a major effect on the biosphere, this question is very important.

The chemistry of stratospheric ozone is complex and closely tied to nitrogen species. Data from Antarctica suggest that anthropogenically produced chlorofluorocarbons fragment and result in substantial ozone depletion in the stratosphere. This process is accelerated in the extremely cold vortex that forms over Antarctica each winter. Measurements suggest that the nitrogen oxides, which would generally remove Cl fragments, are tied up in ice particles in polar stratospheric clouds, which only form at very low temperatures. Once the nitrogen oxides are "frozen out", the Cl fragments can go on to destroy ozone catalytically. Although the complex chemistry is not fully understood at this time, it is clear that N_2O and NO_x in the stratosphere play a critical role in the chemistry of ozone (Toon *et al.*, 1986; Solomon and Schoeberl, 1988).

12.5.3.3 Nitrogen fixation

The question of impacts that might arise from

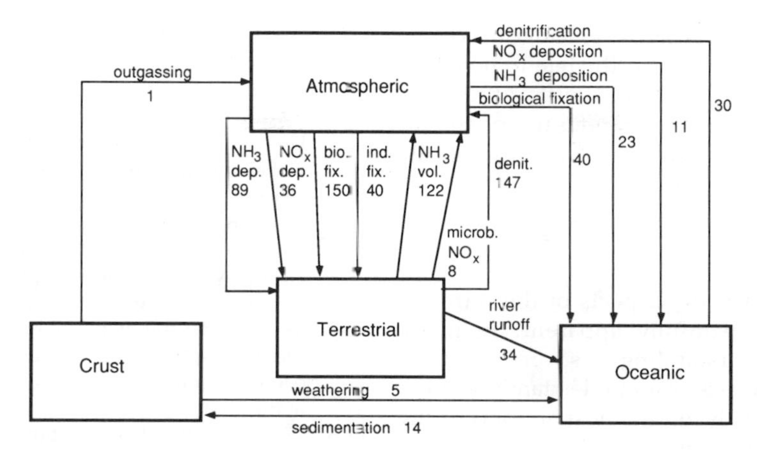

Fig. 12-8 The nitrogen cycle (Tg N/year).

anthropogenic fixation of nitrogen, apart from any of the already mentioned concerns, has been rarely considered. Of course, it is impossible to separate these effects without looking at the whole picture. For example, one could look at the N_2O question as one of the many inevitable changes in the global nitrogen cycle that result from anthropogenic nitrogen fixation, which is of the same magnitude as natural fixation. A schematic for the nitrogen cycle as a whole is presented in Fig. 12-8, and Fig. 12-9 shows the net fluxes from the various reservoirs. This model indicates that there is a net transfer of nitrogen from the atmosphere to the terrestrial, oceanic, and crustal reservoirs of 81 Tg/year. This transfer is of the same order of magnitude as is the difference in anthropogenic fixation (100 Tg/year) and increased denitrification (23 Tg/year). The increased nitrogen burden is distributed between the terrestrial and oceanic systems; however, as Stedman and Shetter (1983) point out, there is little evidence for an increase in the total nitrogen content of present-day biomass, and it may have decreased by as much as 25%. This leads to the conclusion that the net flow of nitrogen is probably to the oceans, although there is no clear evidence for this. It seems unlikely that a change in the tropospheric N_2 burden could be significant on any time-scale, so that it is the potential increase in oceanic NO_3^- concentration that could conceivably alter global patterns of nitrogen utilization. This change, approximately 0.01% of the nitrate content of the oceans per year, should be observable on the time-scale of 10–20 years. One possible effect of this perturbation is an increased output of N_2O by the microbes that undergo denitrification in the oceans in regions where nitrate is a limiting nutrient. A second possible consequence

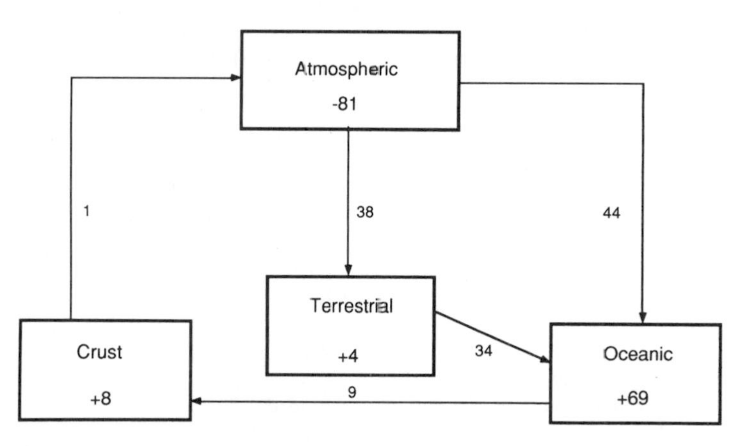

Fig. 12.9 Net flows in the nitrogen cycle (Tg N/year).

of increased NO_3^- in the oceans is a change in oceanic species. In any event, it is worthwhile to note that cultural effects are probably a significant force in shifting nitrogen between the various reservoirs and forms, and the consequences of this can only be speculated upon.

12.5.3.4 Conclusions

Having considered many aspects of the nitrogen cycle, one fact is painfully apparent. In many instances, our understanding is severely limited by a lack of data, or lack of an understanding of the basic processes involved. In most instances, the magnitudes of the anthropogenic fluxes are approximately known, but it is difficult to recognize the scale of the problem given the large error bars of the natural fluxes. Some of the most significant gaps in our knowledge include:

1. The cycle of nitrogen in the marine system, including fixation, denitrification, sources and sinks for N_2O, and changes in the concentration of NO_3^-.
2. The natural cycle for terrestrial N_2O.
3. Anthropogenic perturbations to tropospheric N_2O concentrations, including both sources and sinks.
4. The significance of natural ammonia volatilization, including both ammoniafication and volatilization from animal excrement.
5. The concentrations of most of the trace species in the remote troposphere.
6. The role of anthropogenic NO_x emissions in increasing tropospheric ozone in the northern hemisphere.
7. The concentrations of trace species and the associated chemistry in the stratosphere.
8. Deposition patterns for nitrogen species in remote regions.

It is hoped that in dealing with many of these uncertainties, and continued lively debate, we will improve our understanding of the nitrogen cycle and the anthropogenic influences on it.

Questions

12-1 How would the nitrogen cycle change if life on Earth were suddenly absent? What would be the time-scale for these changes?

12-2 If, as a result of anthropogenic activities, nitrogen is being removed from the atmospheric reservoir (as N_2) to the oceanic reservoir (as NO_3^-), how long would it take to detect this change? Is this a thermodynamically favorable process?

12-3 How have agriculture and deforestation changed the global rates of nitrogen fixation and denitrification?

12-4 Discuss the importance of atmospheric N_2O. Why is it important to know something about its natural and anthropogenic sinks? What role might atmospheric N_2O play in the control of planetary climate? (See, for instance, Lovelock, 1979.)

12-5 Describe the trends in the ozone concentrations in the troposphere and stratosphere, and the total ozone column. What roles do nitrogen oxides play in these changes?

12-6 What are the key reactions that result in the formation of photochemical smog? How do *increases* in NO_x emissions lead to *lower* peak ozone concentrations in some areas? Would you advocate the lowering of NO_x emission standards in some areas?

References

Bremner, J. M. and A. M. Blackmer (1981). Terrestrial nitrification as a source of atmospheric nitrous oxide. *In* "Denitrification, Nitrification, and Atmospheric Nitrous Oxide" (C. C. Delwiche, ed.), pp. 151–170. John Wiley, New York.

Burns, R. C. and R. W. F. Hardy (1975). "Nitrogen Fixation in Bacteria and Higher Plants." Springer-Verlag, New York.

Cohen, Y. (1978). Consumption of dissolved nitrous oxide in an anoxic basin, Saanich Inlet, B.C. *Nature* **272**, 235–237.

Cohen, Y. and L. Gordon (1979). Nitrous oxide production in the ocean. *J. Geophys. Res.* **84** (C1), 347–353.

Crutzen, P. J. (1988). Tropospheric ozone: An overview. *In* "Tropospheric Ozone-Regional and Global Scale Interactions" (I. S. A. Isaksen, ed.), pp. 3–32. NATO ASI Series C, Vol. 227. Reidel, Boston, Mass.

Crutzen, P. J. and L. T. Gidel (1983). A 2-dimensional photochemical model of the atmosphere. 2: The tropospheric budgets of the anthropogenic chlorocarbons CO, CH_4, CH_3Cl and the effects of various NO_x sources on tropospheric O_3. *J. Geophys. Res.* **88**, 6641–6661.

Delwiche, C. C. (1977). Energy relations in the global nitrogen cycle. *Ambio* **6**, 106–111.

Delwiche, C. C. (1981). The nitrogen cycle and nitrous oxide. *In* "Denitrification, Nitrification, and Atmospheric Nitrous Oxide" (C. C. Delwiche, ed.), pp. 1–15. John Wiley, New York.

Denmead, O. T., J. R. Freney, and J. R. Simpson (1976). A closed ammonia cycle within a plant canopy. *Soil Biol. Biochem.* **8**, 161–164.

Dickinson, R. E. and R. J. Cicerone (1986). Future global warming from atmospheric trace gases. *Nature* **319**, 109–115.

Finlayson-Pitts, B. and J. Pitts (1986). "Atmospheric Chemistry: Fundamentals and Experimental Techniques." John Wiley, New York.

Freney, J. R., J. R. Simpson, and O. T. Denmead (1983). Volatilization of ammonia. *In* "Gaseous Loss of Nitrogen from Plant–Soil Systems" (J. R. Freney and J. R. Simpson, eds.), pp. 1–32. Martinus Nijhoff/W. Junk, Boston, Mass.

Garrels, R. M. (1982). Introduction: Chemistry of the troposphere – some problems and their temporal frameworks. *In* "Atmospheric Chemistry" (E. D. Goldberg, ed.), pp. 3–16. Springer-Verlag, New York.

Granhall, U. (1981). Biological nitrogen fixation in relation to environmental factors and functioning of natural ecosystems. *In* "Terrestrial Nitrogen Cycles" (F. E. Clark and T. Rosswall, eds.). *Ecol. Bull.* **33**, 131–145.

Hahn, J. (1981). Nitrous oxide in the oceans. *In* "Denitrification, Nitrification, and Atmospheric Nitrous Oxide" (C. C. Delwiche, ed.), pp. 191–240. John Wiley, New York.

Hahn, J. and P. J. Crutzen (1982). The role of fixed nitrogen in atmospheric photochemistry. *In* "The Nitrogen Cycle" (W. D. P. Stewart and T. Rosswall, eds.), pp. 219–239. The Royal Society, London.

Hoell, J. M., C. N. Harward, and B. S. Williams (1980). Remote infrared heterodyne radiometer measurements of atmospheric ammonia profiles. *Geophys. Res. Lett* **7**, 325–328.

Holland, H. D. (1978). "The Chemistry of the Atmosphere and Oceans." John Wiley, New York.

Keller, M., T. J. Goreau, S. C. Wofsy, W. A. Kaplan, and M. B. McElroy (1983). Production of nitrous oxide and consumption of methane by forest soils. *Geophys. Res. Lett.* **10**, 1156–1159.

Kikby, E. A. (1981). Plant growth in relation to nitrogen supply. *In* "Terrestrial Nitrogen Cycles" (F. E. Clark and T. Rosswall, eds.). *Ecol. Bull.* **33**, 249–271.

Lawson, D. R. (1988). The nitrogen species methods comparison study: An overview. *Atmos. Environ.* **22**, 1517.

Liu, S. C., M. Trainer, F. C. Fehsenfeld, D. D. Parrish, E. J. Williams, D. W. Fahey, G. Hubler, and P. C. Murphy (1987). Ozone production in the rural troposphere and the implications for regional and global ozone distributions. *J. Geophys. Res.* **92**, 4191–4207.

Liu, X., M. Trainer, and S. C. Liu (1988). On the nonlinearity of the tropospheric ozone production. *J. Geophys. Res.* **93**, 15,879–15,888.

Logan, J. (1983). Nitrous oxides in the troposphere: Global and regional budgets. *J. Geophys. Res.* **88**, 10,785–10,807.

Logan, J. A. (1985). Tropospheric ozone: Seasonal behavior, trends, and anthropogenic influences. *J. Geophys. Res.* **90**, 10,463–10,482.

Logan, J., M. J. Prather, S. C. Wofsy, and M. B. McElroy (1981). Tropospheric chemistry: A global perspective. *J. Geophys. Res.* **86**, 7210–7254.

Lovelock, J. (1979). "Gaia: A New Look at Life on Earth." Oxford University Press, New York.

McElroy, M. B., J. W. Elkins, S. C. Wofsy, and Y. L. Yung (1976). Sources and sinks for atmospheric N_2O. *Rev. Geo. Space Phys.* **14** (2), 143–150.

McElroy, M. B., S. C. Wofsy, and Y. L. Tung (1977). The nitrogen cycle: Perturbations due to man and their impact on atmospheric N_2O and O_3. *Phil. Trans. Roy. Soc. Lond.* **B277**, 159–181.

Mague, T. H. (1977). Ecological aspects of dinitrogen fixation by blue-green algae. *In* "A Treatise on Dinitrogen Fixation" (R. W. F. Hardy and A. H. Gibson, eds.), pp. 85–140. John Wiley, New York.

Mayewski, P. A., W. B. Lyons, M. J. Spencer, M. Twickler, W. Dansgaard, B. Koci, C. I. Davidson, and R. E. Honrath (1986). Sulfate and nitrate concentrations from a South Greenland ice core. *Science* **232**, 975–977.

National Academy of Sciences (1977). "Nitrogen Oxides." National Academy of Sciences, Washington, D.C.

National Research Council (1984). "Causes and Effects of Changes in Stratospheric Ozone: Update 1983." National Academy Press, Washington, D.C.

Niki, H., E. E. Daby, and B. Weinstock (1972). Mechanisms of smog reactions. *In* "Photochemical Smog and Ozone Reactions" (R. F. Gould, ed.), pp. 16–57. American Chemical Society, Washington, D.C.

NOAA, US Department of Commerce (1987). "Geophysical Monitoring for Climatic Change," No. 15, Summary Report 1986, pp. 85–90, Boulder, Colorado.

Oltmans, S. and W. P. Komhyr (1986). Surface ozone distributions and variations from 1973–1984 measurements at the NOAA geophysical monitoring for climatic change baseline observations. *J. Geophys. Res.* **91**, 5229–5236.

Pierotti, D., R. A. Rasmussen, and R. Chatfield (1978). Continuous measurements of nitrous oxide in the troposphere. *Nature* **274**, 574–576.

Pitts, J. N., A. M. Winer, R. Atkinson, and W. P. L. Carter (1983). Comment on "Effect of Nitrogen Oxide Emissions on Ozone Levels in Metropolitan Regions," "Effect of NO_x Emission Rates on Smog Formation in the California South Coast Air Basin," and "Effect of Hydrocarbon and NO_x on Photochemical Smog Formation under Simulated Transport Conditions." *Environ. Sci. Technol.* **17**, 54–57.

Postgate, J. R. (1982). Biological nitrogen fixation: Fundamentals. *In* "The Nitrogen Cycle" (W. D. P. Stewart and T. Rosswall, eds.), pp. 73–83. The Royal Society, London.

Prinn, R. G., P. G. Simmonds, R. A. Rasmussen, R. A. Rosen, F. N. Alyea, C. A. Cardilino, A. J. Crawford, D. M. Cunnold, P. J. Frasser, and J. E. Lovelock (1983). The atmospheric lifetime experiment 1: Introduction, instrumentation, and overview. *J. Geophys. Res.* **88**, 8353–8367.

Quinn, P. K., R. J. Charlson, and T. S. Bates (1988). Simultaneous observations of ammonia in the atmosphere and ocean. *Nature* **335**, 336–338.

Robinson, E., W. L. Bamesberger, F. A. Menzia, A. S. Waylett, and S. F. Waylett (1984). Atmospheric trace gas measurements at Palmer Station, Antarctica: 1982–83. *J. Atmos. Chem.* **2**, 65–81.

Rolston, D. E. (1981). Nitrous oxide and nitrogen gas production in fertilizer loss. *In* "Denitrification, Nitrification, and Atmospheric Nitrous Oxide" (C. C. Delwiche, ed.), pp. 127–149. John Wiley, New York.

Rosswall, T. (1976). The internal nitrogen cycle between microorganisms, vegetation, and soil. *In* "Nitrogen, Phosphorus and Sulfur – Global Cycles" (B. H. Svensson and R. Soderlund, eds.). *Ecol. Bull.* **22**, 157–167.

Seiler, W. and R. Conrad (1981). Field measurements of natural and fertilizer induced N_2O release rates from soils. *J. Air Poll. Control Assoc.* **31**, 767–772.

Söderlund, R. and T. Rosswall (1982). The nitrogen cycles. *In* "The Natural Environment and the Biogeochemical Cycles" Vol. 1, Part B (O. Hutzinger, ed.), pp. 61–81. Springer-Verlag, New York.

Söderlund, R. and B. H. Svensson (1976). The global nitrogen cycle. *In* "Nitrogen, Phosphorus and Sulfur – Global Cycles" (B. H. Svensson and R. Söderlund, eds.). *Ecol. Bull.* **22**, 23–73.

Solomon, S. and M. R. Schoeberl (1988). Overview of the polar ozone issue. *Geophys. Res. Lett.* **15**, 845–846.

Stedman, D. H. and R. E. Shetter (1983). The global budget of atmospheric nitrogen species. *In* "Trace Atmospheric Constituents" (S. E. Schwartz, ed.), pp. 411–454. John Wiley, New York.

Sweeney, R. E., K. K. Liu and I. R. Kaplan (1978). Oceanic nitrogen isotopes and their uses in determining the sources of sedimentary nitrogen. *In* Stable Isotopes in the Earth Sciences (B. W. Robinson, ed.), pp. 9–26. DSIR Bulletin #220. New Zealand Department of Science and Industrial Research.

Toon, O. B., P. Hamill, R. P. Turco, and J. Pinto (1986). Condensation of HNO_3 and HCl in the winter polar stratosphere. *Geophys. Res. Lett.* **13**, 1284–1287.

Volz, A. and D. Kley (1988). Evaluation of the Montsouris series of ozone measurements made in the nineteenth century. *Nature* **332**, 240–242.

Weiss, R. F. (1981). The temporal and spatial distribution of tropospheric nitrous oxide. *J. Geophys. Res.* **86**, 7185–7195.

13

The Sulfur Cycle

R. J. Charlson, T. L. Anderson, and R. E. McDuff

13.1 Introduction

Sulfur, the 14th most abundant element in the Earth's crust, plays a variety of important roles in the chemical functioning of the Earth. In its reduced oxidation state, sulfur is a key nutrient to life providing, for example, structural integrity to protein-containing tissues. In its fully oxidized state, sulfur exists as sulfate, SO_4^{2-}, the second most abundant anion in rivers (after bicarbonate, HCO_3^-) and in seawater (after chloride, Cl^-) and is the major cause of acidity in both natural and polluted rainwater. This link to acidity makes sulfur a key player in the natural weathering of rocks and such environmental problems as "acid rain". Sulfate in the atmosphere has been identified as the dominant component of cloud condensation nuclei in both remote and polluted settings (Bigg *et al.*, 1984). Thus, it has important interactions with clouds (and perhaps the global radiative energy balance which is sensitive to clouds) and with the hydrological cycle. Finally, of the major elemental cycles (i.e. C, N, O, P, S), the sulfur cycle is one of the most heavily perturbed by human activity. It is estimated that anthropogenic emissions of sulfur into the atmosphere (largely from coal combustion) are currently of about equal magnitude to natural sulfur emissions (Andreae, 1985).

With the exception of ionic sulfides formed from highly electropositive elements (i.e. Na, K, Ca, Mg), sulfur bonding in natural environments is covalent. When fully oxidized, however, the covalently bonded sulfur atom exists within the sulfate ion which forms either sulfuric acid (a gas or liquid) or ionic compounds such as $CaSO_4 \cdot 2H_2O$ (gypsum) and $(NH_4)_2SO_4$ (ammonium sulfate). Covalent bonding occurs in organosulfur compounds. Hence, the chemistry of sulfur involves chemical complexity not found in elements at the edges of the periodic table.

Sulfur exists naturally in several oxidation states, and its participation in oxidation–reduction reactions has important geochemical consequences. For example, when an extremely insoluble material, FeS_2, is precipitated from seawater under conditions of bacterial reduction, Fe and S may be sequestered in sediments for periods of hundreds of millions of years. Sulfur can be liberated biologically or volcanically with the release of H_2S or SO_2 as gases.

There are nine known isotopes of sulfur of which four are stable:

Isotope	Average crustal abundance (%)
^{32}S	95.0
^{33}S	0.76
^{34}S	4.22
^{36}S	0.014

The prevalence of sulfur's second most abundant isotope, ^{34}S, along with the fractionation known to occur in many biogoechemical processes, make isotopic studies of sulfur a potentially fruitful method of unraveling its sources and sinks within a given reservoir.

13.2 Oxidation States of Sulfur

Table 13-1 includes many of the key naturally occurring molecular species of sulfur, subdivided by

Global Biogeochemical Cycles
ISBN 0-12-147685-5

Table 13-1 Some naturally occurring sulfur compounds

Oxidation state	Gas	Aerosol	Aqueous	Soil	Mineral	Biological
−II	H_2S, RSH	—	H_2S, HS^-, S^{2-}	S^{2-}, HS^-	S^{2-}	Methionine
	RSR		RS^-	MS	HgS	$CH_3S(CH_2)_2CHNH_2$
	OCS					Cysteine
	CS_2				CuS_2, etc.	$HSCH_2CHNH_2COOH$
						Dicysteine
−I	RSSR	—	RSSR	SS^{2-}	FeS_2	—
0	$CH_3SOCH_3^+$	—	—	S_8	—	—
II			$S_2O_3^{2-}$	—		
IV	SO_2	$SO_2 \cdot H_2O$	$SO_2 \cdot H_2O$			
		HSO_3^-	HSO_3^-		—	—
			SO_3^{2-}	SO_3^{2-}		
			$HCHO \cdot SO_2$			
VI	SO_3	H_2SO_4, HSO_4^-	SO_4^{2-}	$CaSO_4$	$CaSO_4 \cdot H_2O$	
		SO_4^{2-}	HSO_4^-, SO_4^{2-}		$MgSO_4$	
			$CH_3SO_3^-$			
		$(NH_4)_2SO_4$, etc.				
		Na_2SO_4				
		CH_3SO_3H				

oxidation state and reservoir. The most reduced forms, S(−II), are seen to exist in all except the aerosol form, despite the presence of free O_2 in the atmosphere, ocean, and surface waters. With the exception of H_2S in oxygenated water, these species are oxidized very slowly by O_2. The exception is due to the dissociation in water of H_2S into $H^+ + HS^-$. Since HS^- reacts quickly with O_2, aerobic waters may contain, and be a source to the atmosphere of RSH, RSR, etc., but not of H_2S itself. Anaerobic waters, as in swamps or intertidal mudflats, can contain H_2S and can, therefore, be sources of H_2S to the air.

S(−II and −I) also are found in minerals, notably in metal and metalloid sulfides. As many as 95 sulfide minerals appear in standard lists, where sulfur is bound to a wide variety of other elements: Ag, Fe, Cd, Hg, Mn, Ca, Te, Se, As, Sn, Cu, Pb, Pt, Sb, Co, Ni, Mo, Rn, W. Of these, FeS_2 (pyrite) is the most abundant.

Highly oxidized forms of organic sulfur exist in folic acid and sulfolipids, but the major form of sulfur in organisms is in amino acids. There are two sulfur-containing amino acids: cysteine and methionine. Although sulfur bonds assume a variety of oxidation states in living organisms, we list the amino acid sulfur as oxidation state −II because the sulfur is generally bonded to carbon or hydrogen. Thus, methionine has a —SCH_3 bond like dimethyl sulfide, and cysteine is a mercaptan which can convert to smaller gaseous molecules like CH_3SH. A key biological function of sulfur is to provide disulfide (—SS—) linkages between amino acids within protein molecules, thus giving three-dimensional structure to proteins and strength or mechanical structure to tissues composed of proteins. Sulfur in methionine is crucial to the methylation reaction, a biosynthetic process. The amounts of sulfur in organisms varies but is typically of the same order of magnitude as phosphorus, e.g. about 0.25% (by dry wt.). Even though sulfur is an essential element for biota, its abundance, 29 mmol/kg in seawater, is often so large that other elements (e.g. N and P) provide limits to growth while S does not.

The most oxidized form of sulfur, S(+VI), is predominantly sulfate, SO_4^{2-}. Sulfate particles ranging in composition from pure sulfuric acid (H_2SO_4) to fully neutralized ammonium sulfate [$(NH_4)_2SO_4$] are ubiquitous constituents of the atmosphere (see Chapter 10).

In the ocean, sulfate exists as a free ion, SO_4^{2-}. In sedimentary rock, sulfate is found in evaporite minerals (i.e. minerals produced by the evaporation of seawater), with gypsum ($CaSO_4 \cdot 2H_2O$) being the most common. By contrast, the intermediate oxidation state, S(+IV), has only a transitory existence in the atmosphere and in some volcanic and industrial emissions to the atmosphere. Gaseous SO_2 is soluble in water, leading to the presence of HSO_3^- and SO_3^{2-} ions, which are unstable under aerobic conditions, producing SO_4^{2-} as the stable end-product (Chapter 5). Oxidation of SO_2 to sulfate occurs in the gas phase as well due to the presence of the strong oxidizing agent, OH·.

Elemental sulfur also occurs naturally, with production either by biological or inorganic processes. In either case, it appears that a higher oxidation state of sulfur may react with the −II (sulfide) state to yield a zero oxidation state product, or that S(O) is an intermediate in the oxidation of S(−II). There are several genera of sulfate-reducing bacteria, two of which are widely recognized: *Desulfovibrio* and *Desulfotomaculum*. These organisms utilize sulfur in the sulfate ion as an electron acceptor and produce H_2S. This H_2S is then available to react with iron and form the insoluble precipitate, FeS_2 (pyrite). Other organisms can utilize H_2S and produce elemental sulfur, S(0). Elemental sulfur also is produced inorganically, for example by the following reaction in volcanoes:

$$2H_2S + SO_2 \rightarrow 2H_2O + 3S$$

For completeness, we mention the existence of compounds of mixed oxidation states. Here, two or more atoms of sulfur exist in the molecule or ion, each having a different oxidation state. Numerous examples of these species are known but details of their natural existence are obscure. It is suggested (Grinenko and Ivanov, 1983) that thiosulfate, $S_2O_3^{2-}$, ion can be produced by bacteria in waterlogged soils, paddy fields, and the like. The formal oxidation state for the two sulfur atoms is II, but the two sulfur atoms are chemically different. Such species may play important roles as intermediates between the major species such as S^{2-} and SO_4^{2-} (Jorgensen, 1990).

Returning to Table 13-1, we see that sulfur is found in gaseous, aerosol, aqueous, soil, mineral, and biological forms. The gaseous forms, which are found in the atmosphere, are generally of the lower oxidation states −II or +IV, while most of the aqueous form is S(+VI) as SO_4^{2-}. The vapor pressures of non-ionic S(−II) compounds are large

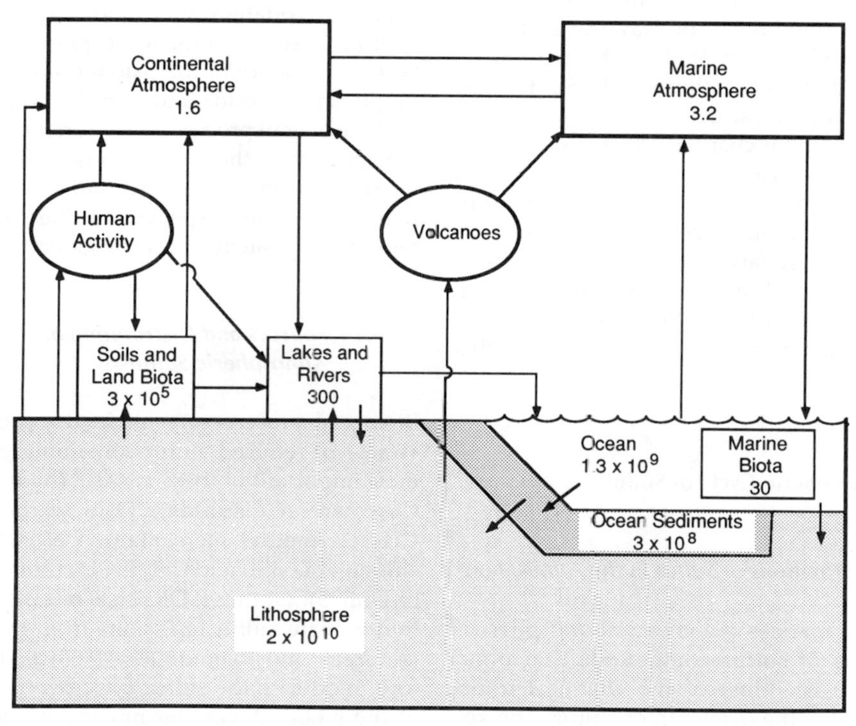

Fig. 13-1 Major reservoirs and burdens of sulfur. Units are Tg (10^{12} g) S.

enough that transfer of these species occurs from bodies of water to the atmosphere. Sulfate ion in solution is relatively inert, and much of its processing is due to transport by water or evaporation of the water to form solid evaporites. The sulfate minerals (evaporites like $CaSO_4$) are water-soluble, whereas the sulfide minerals (such as FeS_2) tend to be highly insoluble.

13.3 Sulfur Reservoirs

The preceding discussion of oxidation states demonstrates the existence of sulfur in solid, liquid, and gaseous forms and in living organisms. It is thus convenient to organize the remainder of this discussion along these same lines. Figure 13-1 is a box diagram showing the key reservoirs and the approximate burdens of sulfur in each (Freney *et al.*, 1983).

The vast majority of sulfur at any given time is in the lithosphere. The atmosphere, hydrosphere, and biosphere, on the other hand, are where most transfer of sulfur takes place. The role of the biosphere often involves reactions that result in the movement of sulfur from one reservoir to another. The burning of coal by humans (which oxidizes fossilized sulfur to SO_2 gas) and the reduction of seawater sulfate by phytoplankton which can lead to the creation of another gas, dimethyl sulfide (CH_3SCH_3), are examples of such processes.

The remainder of this chapter, which discusses the cycling of sulfur, is divided into an atmospheric part and an oceanic/solid earth part. The amount of sulfur in the atmosphere at any given time is small, even though the fluxes are large, because the lifetime of most sulfur compounds in air is relatively short (e.g. days). Sulfur in the ocean as SO_4^{2-} is cycled much more slowly, and the primary interactions in that cycle are with the solid earth.

13.4 The Atmospheric Cycle of Sulfur

13.4.1 Transformations of Sulfur in the Atmosphere

Figure 13-2 summarizes the chemical and physical transformations of sulfur compounds that occur in the atmosphere. Most of the chemical transformations involve the oxidation of sulfur. The key oxidizing agents are thought to be the OH· radical, for the gas phase, and H_2O_2, O_3, and OH· in the aqueous phase. Many of these transformations, however, can only be identified as "multistep" processes – that is, the detailed chemistry is not currently understood. The rates at which most of the transformations occur are also poorly understood and have been estimated only semi-quantitatively. The amounts of sulfur within the various reservoirs (Fig. 13-1) are better known because they can be measured directly, although these data are greatly complicated by the patchy and episodic nature of the distribution of atmospheric sulfur species. Among the fluxes (see Fig. 13-6 and Table 13-2), the best data are available for anthropogenic SO_2 emissions and SO_4^{2-} deposition to the surface in rainwater. Recent improvements have also been made in quantifying the natural emissions of reduced sulfur gases, especially from the world oceans. In sum, the qualitative picture of the atmospheric sulfur cycle now appears to be in good focus, although many quantitative details remain to be filled in.

The most important pathway of sulfur through the atmosphere involves injection as a low oxidation state gas and the removal as oxidation state VI sulfate in rainwater (see Fig. 13-2, paths 1, 4, 5, 6, 7, 8, 9, 10, 12, and 13). Since this pathway involves a change in chemical oxidation state and physical phase, the lifetime of sulfur in the atmosphere is governed by both the kinetics of the oxidation reactions and the frequency of clouds and rain. We will argue below that the overall process is fast – on the order of days – meaning that the atmospheric sulfur cycle is a regional phenomenon and that the distribution of nearly all sulfur species in the atmosphere is necessarily "patchy" over the globe.

13.4.2 Sources and Distribution of Atmospheric Sulfur

Biological processes result in the production of a variety of reduced sulfur-containing gases. The six most important of these are H_2S (hydrogen sulfide), CS_2 (carbon disulfide), OCS (carbonyl sulfide), CH_3SH (methyl mercaptan), CH_3SCH_3 (dimethyl sulfide or DMS), and CH_3SSCH_3 (dimethyl disulfide or DMDS). Of these, DMS has recently been shown to dominate sulfur emissions from the open ocean (Andreae and Raemdouck, 1983) and may, therefore, modulate the sulfur cycle over a large portion of the globe. A varying mixture of all six of these gases is found over terrestrial ecosystems (Adams

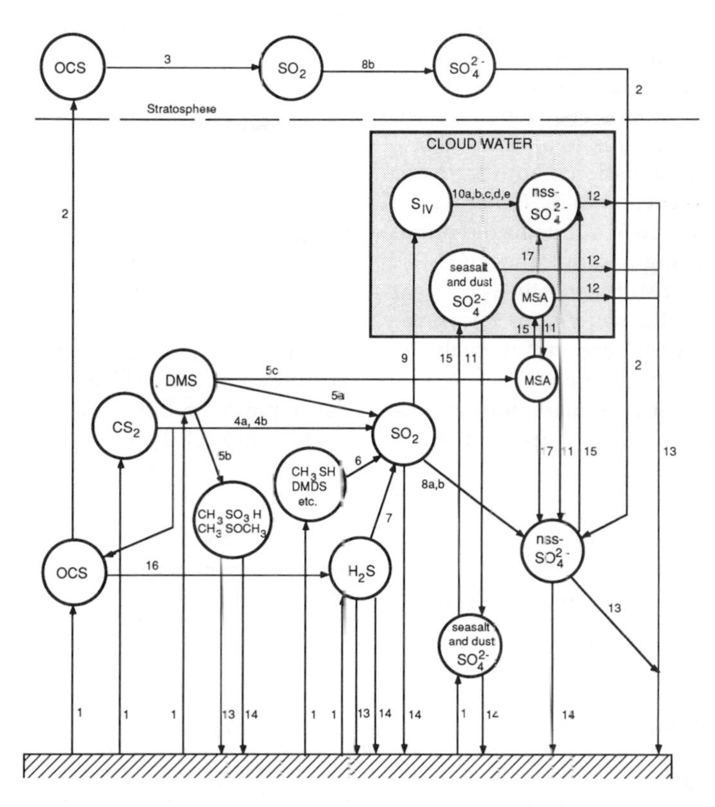

Fig. 13-2 The chemical and physical transformations of sulfur in the atmospheric cycle. Circles are chemical species, the box represents cloud–liquid phase. DMS = CH_3SCH_3; DMDS = CH_3SSCH_3; $S_{IV} = (SO_2)_{aq} + HSO_3^- + SO_3^{2-} + CH_2OHSO_3^-$; and MSA (methane sulfonic acid) = CH_3SO_3H. The chemical transformations are as follows:

1. Surface emissions
2. Tropospheric/stratospheric exchange
3. $OCS + h\nu \rightarrow S + CO$
   $S + O_2 \rightarrow SO + O$
   $SO + O_2 \rightarrow SO_2 + O$

4a. $CS_2 + OH \rightarrow CS_2OH$
   $CS_2OH \rightarrow$ multistep $\rightarrow OCS + SO_2$

4b. $CS_2 + h\nu \rightarrow CS_2^*$
   $CS_2^* + O_2 \rightarrow CS + SO_2$
   $CS + O_2 \rightarrow OCS + O$
   $CS + O_3 \rightarrow OCS + O_2$

5a. $CH_3SCH_3 + OH \rightarrow$ multistep $\rightarrow SO_2$

5b. $CH_3SCH_3 + OH \rightarrow$ multistep $\rightarrow CH_3SOCH_3$

5c. $CH_3SCH_3 + OH \rightarrow$ multistep $\rightarrow CH_3SO_3H$

6. $CH_3SH + OH \rightarrow$ multistep $\rightarrow SO_2$
7. $H_2S + OH \rightarrow$ multistep $\rightarrow SO_2$

8a. $SO_2 + OH \rightarrow HSO_3$
   $HSO_3 + O_3 \rightarrow \rightarrow HO_2 + SO_3$
   $SO_3 + H_2O \rightarrow H_2SO_4$

8b. $SO_2 \rightarrow SO_4^{2-}$ (heterogeneous reaction)

9. $(SO_2)_g \leftrightarrow (SO_2)_{aq}$
   $(SO_2)_{aq} + H_2O \rightarrow HSO_3^- + H^+$
   $HSO_3^- \leftrightarrow H^+ + SO_3^{2-}$

10a. $CH_2(OH)_2 + HSO_3^- \leftrightarrow H_2O + CH_2OHSO_3^-$
   $(H_2O_2)_g \leftrightarrow (H_2O_2)_{aq}$

10b. $HSO_3^- + (H_2O_2)_{aq} \rightarrow$ multistep $\rightarrow H^+ + SO_4^{2-}$
   $(O_3)_g \leftrightarrow (O_3)_{aq}$

10c. $HSO_3^- + (O_3)_{aq} \rightarrow$ multistep $\rightarrow H^+ + SO_4^{2-}$
   $(HO_2)_g \leftrightarrow (HO_2)_{aq}$
   $(HO_2)_{aq} \rightarrow H^+ + O_2^-$
   $(HO_2)_{aq} + O_2 \xrightarrow{H_2O} (H_2O_2)_{aq} + OH^-$

10d. $HSO_3^- + (H_2O_2)_{aq} \rightarrow$ multistep $\rightarrow 2H^+ + SO_4^{2-}$
   $(OH)_g \rightarrow (OH)_{aq}$
   $HSO_3^- + (OH)_{aq} \rightarrow$ multistep $\rightarrow H^+ + SO_4^{2-}$

10e. $HSO_3^- + O_2 \rightarrow$ multistep $\rightarrow H^+ + SO_4^{2-}$

11. Evaporation
12. Cloud water $\rightarrow$ rainwater
13. Washout, rainout
14. Dry deposition
15. Cloud nucleation
16. $OCS + OH \rightarrow$ multistep $\rightarrow H_2S$
17. $MSA \rightarrow SO_4^{2-}$ by some mechanism

Modified from Charlson *et al.* (1985) with the permission of Kluwer Academic Publishers.

et al., 1981). Referring to Table 13-2, we see that the global sum of biogenic sulfur emissions is currently estimated at 57 Tg S/year (add fluxes F4 and F14), with an uncertainty of −50% to +100%. Marine DMS accounts for 39 Tg of this total, and most of the uncertainty involves emissions (both terrestrial and marine) of H_2S (Andreae, 1985). Other natural sources of low oxidation state sulfur to the atmosphere are biomass burning and volcanoes (28 Tg S/year), each of which emit sulfur mostly as SO_2 gas. Volcanoes are a sporadic source, of secondary importance to the troposphere sulfur cycle but capable of causing huge fluctuations locally and in the stratospheric reservoir. Biomass burning has both anthropogenic and natural components (often hard to separate). The sulfur output from this source is highly uncertain; it is not included in the fluxes in Table 13-2.

Emissions of sulfur to the atmosphere by humans are almost entirely in the form of SO_2. The main sources are coal-burning and sulfide ore smelting. The total anthropogenic flux is estimated to be about 80 Tg S/year (Ivanov, 1983) and is thus essentially equal in magnitude to the natural flux of low oxidation state sulfur to the atmosphere. Clearly, the atmospheric sulfur cycle is intensely perturbed by human activity. To estimate the spatial extent of this perturbation, we will need some idea of the residence time of sulfur in the atmosphere.

The simplest definition of residence time is total

Table 13-2 Major sulfur fluxes (Tg S/year)

Flux	Description	Kellogg *et al.* (1972)	Friend (1973)	Granat *et al.* (1976)	Moller (1984a, b)	Ivanov (1983)		This work (natural only)
						Natural	Anthro.	
1. Continental part of the cycle								
F1a	Emission to atmosphere from fossil fuel burning and metal smelting	50	65	65	75	—	113	—
F1b	Effluents from chemical industry and mining	—	—	—	—	—	29	—
F1c	Soils to rivers (anthro. portion is pollution of rivers from fertilizers)	—	26	—	—	—	28	8[a]
F2	Aeolian emission (dust)	—	—	0.2	—	20	—	20
F3a	Volcanic emissions to continental atmosphere	1	1	1.5	1	14	—	9[b]
F4	Biogenic gases (land)	—	58	5	35	18	—	18
F5	Gravitational settling of large (aeolian) particles to land	—	—	—	—	12	—	12
F6	Washout and dry deposition of gases and fine particles to land	111	121	71	—	25	47	46
F7	Transport to oceanic atmosphere	5	8	18	—	35	66	13
F8	Transport from oceanic atmosphere	4	4	17	—	20	—	24
F9	Weathering to soil	—	42[c]	66[c]	—	114[c]	—	26
F10	Weathering to rivers	—	—	—	—	—	—	93
F11	Burial of sulfur in sediments from continental water bodies	—	—	—	—	—	—	35
F12	River run-off to oceans	—	136	122	—	104	104	104

Table 13-2 *Continued*

Flux	Description	Kellogg *et al.* (1972)	Friend (1973)	Granat *et al.* (1976)	Moller (1984a, b)	Ivanov (1983)		This work (natural only)
						Natural	Anthro.	
2. Oceanic part of the cycle (see also F7, F8, and F12 above)								
F3b	Volcanic emissions to marine atmosphere	1	1	1.5	1	14	—	19[b]
F13	Aeolian emissions (seasalt)	47	44	44	175	140	—	140
F14	Biogenic gases (marine)	—	48	27	35	23	—	39[d]
F15	Washout and dry deposition of gases and particles to oceans	72	96	73	—	258	—	187
F16	Burial of sulfur in oceanic sediments	—	—	—	—	139	—	69
F17	Deposition of marine sulfate via thermal vent reactions at mid-ocean ridges	—	—	—	—	—	—	43[e]
F18	Lithification of marine sediments	—	—	—	—	—	—	69[f]

[a] Deduced to balance budget for "soils and land biota" reservoir.
[b] Ivanov's (1983) global total for volcanic emissions reapportioned as ⅓ continental and ⅔ marine.
[c] Number shown represents the sum of fluxes F9 and F10.
[d] Andreae and Raemdonck (1983).
[e] Deduced to balance budget for "ocean" reservoir.
[f] Deduced to balance budget for "ocean sediments" reservoir.

burden within a reservoir divided by the flux out of that reservoir: $\tau = M/S$ (see Chapter 4). A typical value for the flux of non-seasalt sulfate (nss-SO_4^{2-}) to the ocean surface via rain is 0.11 g S/m^2 year (Galloway, 1985). Using this value, we may consider the residence time of nss-SO_4^{2-} itself and of total non-seasalt sulfur over the world oceans. Appropriate vertical column burdens (derived from the data review of Toon *et al.*, 1987) are 460 μg S/m^2 for nss-SO_4^{2-} and 1700 μg S/m^2 for the sum of DMS, SO_2, and nss-SO_4^{2-}. These numbers yield residence times of about 1.5 days for nss-SO_4^{2-} and 5.6 days for total non-seasalt sulfur. We might infer that the oxidation process is frequently slower than the rain removal process of particulate sulfate, although within the accuracy of these estimates it is safer simply to note that the two processes seem to be of similar duration. Direct estimates of the residence time of DMS with respect to oxidation by the OH· radical are in the range of 1.5–2 days (Andreae, 1985 and references therein).

As shown in Fig. 13-4, the flux of sulfate in rainwater over polluted industrial regions is of order 1 g S/m^2 year or about ten times the remote marine flux. Since natural sources of sulfur are probably much weaker over continents, this indicates a massive human perturbation of the sulfur cycle within industrialized areas and a potentially strong perturbation over much larger areas depending on how far atmospheric sulfur is transported on average before being deposited. Figures 13-3 and 13-4 give the impression of a regional phenomenon with a horizontal scale of about 1000 km. This scale is confirmed by sulfur budget studies in polluted regions. For example, Ottar (1978) estimated that 80% of European sulfur emissions are deposited over Europe. For mid-latitude weather, horizontal transport over a spatial scale of the order 1000 km corresponds to a time-scale of a few days. Studies of the rate of oxidation of SO_2 to sulfate in polluted regions indicate that a conversion rate of about 1% per hour is a good average. This corresponds to a residence time for SO_2 gas of about 2 days. Once converted to sulfate, washout by rain should occur on a time-scale of a few days.

For both polluted and remote conditions, there-

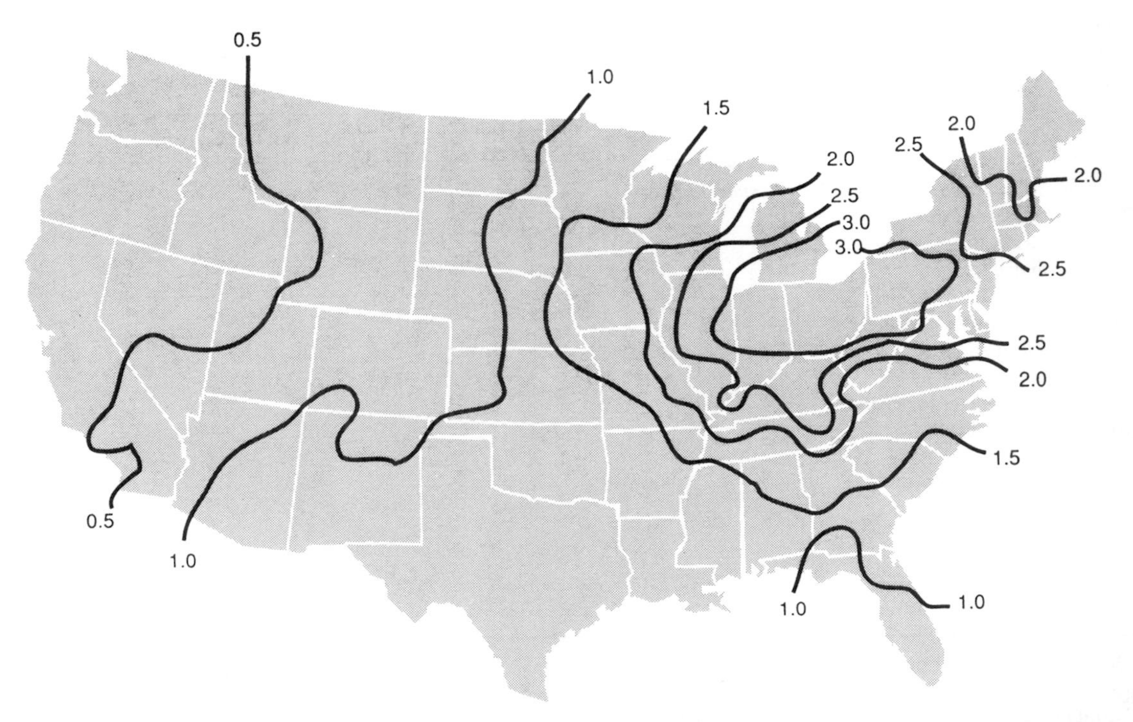

Fig. 13-3 Precipitation-weighted concentration of SO_4^{2-} (mg/L) over the United States, 1987. Modified with permission from the National Atmospheric Deposition Program, 1988.

fore, the cycling of sulfur from low oxidation state gas to sulfate particles and then back to the surface in rain takes place on a time-scale of a few days.

Some sulfur gases are directly absorbed at the surface, and particulate sulfur (aerosol) has some flux to the surface due to Brownian motion and gravitational settling. Together, these processes are referred to as "dry deposition" (to distinguish them from the rain removal mechanism). Dry deposition may be an important removal process for SO_2 in polluted areas and is probably fairly minor for SO_2 and the reduced sulfur gases over the oceans (Galloway, 1985). The dry deposition of sulfate particles (other than seasalt and dust sulfate) is also minor. In any case, these additional sinks for atmospheric sulfur can only have the effect of shortening its residence time.

An interesting exception to the patchiness of atmospheric sulfur compounds is carbonyl sulfide (OCS). This compound, which may be emitted directly or produced by the oxidation of CS_2, is highly stable against further oxidation (until it reaches the stratosphere) and so is unavailable for rapid wet removal. As a result, OCS has a long residence time (~1 year) as well as a large and fairly uniform concentration (~500 pptv) in the troposphere. Being inert in the troposphere, this compound, like the chlorofluorocarbons, is available for gradual mixing into the stratosphere. Here it can be broken apart by ultraviolet radiation and oxidized to SO_2 and sulfate (see top of Fig. 13-2). Through this mechanism, OCS is thought to be the major source of sulfate particles in the stratosphere during volcanically quiescent periods (Crtuzen, 1976). In the troposphere, while chemically unimportant, it is actually the largest reservoir of sulfur. This situation is analogous to that of nitrogen, where the dominant species is N_2, which being so inert, is usually ignored in discussions of atmospheric nitrogen chemistry.

Finally, there is a major flux of sulfur through the atmosphere in both seasalt particles (~140 Tg S/year) and terrestrial dust (~20 Tg S/year). In each case, the form of sulfur is sulfate, originating mostly as the mineral gypsum in the case of dust and as sulfate ion in seawater in the case of seasalt. Already in its fully oxidized, stable state, this component of

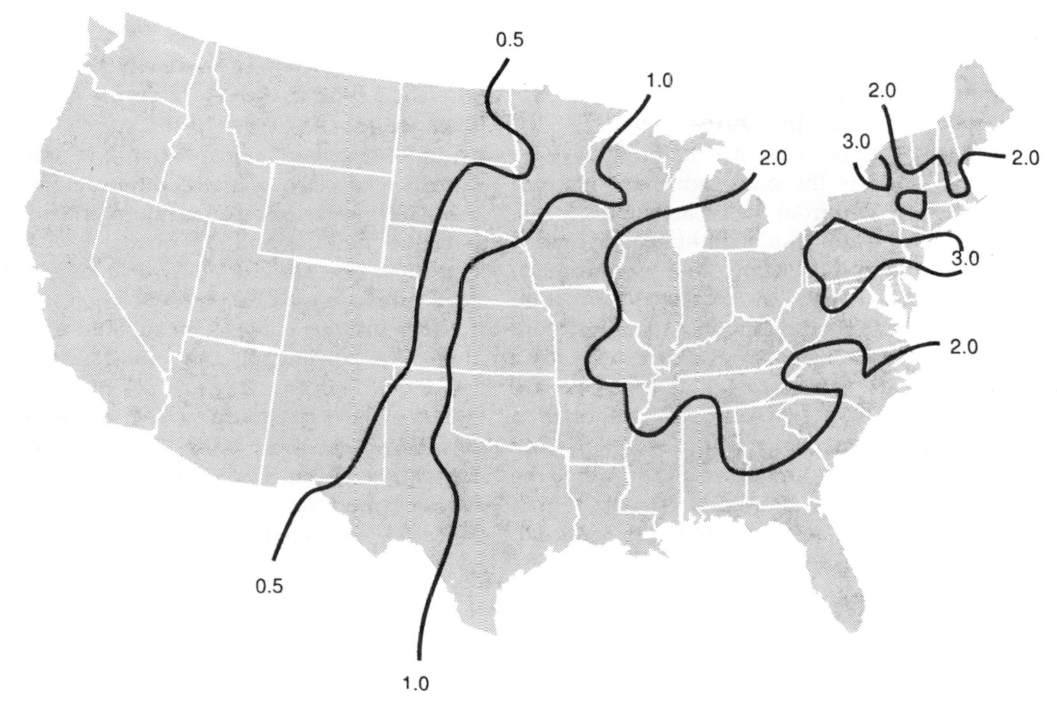

Fig. 13-4 Wet sulfate ion deposition (g/m^2 year), 1987. Modified with permission from the National Atmospheric Deposition Program, 1988.

atmospheric sulfur does not participate in atmospheric redox reactions, nor does it contribute to the acidity of cloud or rainwater. Instead, it is simply returned to the surface, via dry or wet removal, in the same chemical form in which it was emitted. Sulfate in both dust and seasalt particles is best viewed as an inert, secondary constituent, merely "along for the ride", with little geochemical consequence. (It is very important to measure this component, however, so that it may be subtracted from total sulfate measurements in order to derive the non-seasalt sulfate quantity of chemical interest.) Since it does not need to be oxidized prior to wet removal and has, in addition, a large sink via dry deposition, the residence time of seasalt and dust sulfate is significantly shorter than that of other sulfur species. Because both seasalt and dust particles are relatively large, they are subject to significant removal by gravitational settling.

We have seen that, except for OCS, sulfur species in the atmosphere have residence times that are short (days) such that their geographical distribution is patchy. This perspective on the atmospheric sulfur cycle has important implications. While human emissions certainly constitute an overwhelming perturbation within heavily industrialized regions, there may be even larger areas of the globe in which the human influence on the sulfur cycle is relatively unimportant, e.g. much of the southern hemisphere (see Section 13.4.3). In addition, we see that the sulfur cycle can only sensibly be studied on a regional basis and that the calculation of global budgets will necessarily be a painstaking process of making myriad measurements over a wide range of regions and seasons to allow accurate averaging.

13.4.3 The Remote Marine Atmosphere

Let us turn now to a detailed, box model investigation of a regional sulfur cycle. The discussion so far suggests that the sulfur cycle over much of the ocean should be largely uninfluenced by human or other continental input. The absence of complex, polluted-air chemistry, along with a high degree of horizontal spatial homogeneity, should provide the simplest possible system for studying the transformations introduced in Fig. 13-2. We begin by assuming that the atmosphere is in steady-state (fluxes into and out of each box are equal) and that the remote marine environment is a closed system with respect to sulfur (no fluxes from or to the continents). We ignore seasalt sulfate. Finally, we assume that DMS is the only significant reduced sulfur species emitted from the ocean surface.

Figure 13-5 is the box model of the remote marine sulfur cycle that results from these assumptions. Many different data sets are displayed (and compared) as follows. Each box shows a measured concentration and an estimated residence time for a particular species. Fluxes adjoining a box are calculated from these two pieces of information using the simple formula, $S = M/\tau$. The flux of DMS out of the ocean surface and of nss-SO_4^{2-} back to the ocean surface are also quantities estimated from measurements. These are converted from surface to volume fluxes (i.e. from μg S/m^2 h to ng S/m^2 h) by assuming the effective scale height of the atmosphere is 2.5 km (which corresponds to a reasonable thickness of the marine planetary boundary layer, within which most precipitation and sulfur cycling should take place). Finally, other data are used to estimate the factors for partitioning oxidized DMS between the MSA and SO_2 boxes, for SO_2 between dry deposition and oxidation to sulfate, and for nss-SO_4^{2-} between wet and dry deposition.

We begin our analysis by comparing the surface fluxes. According to the indicated partitioning factors, 67% of the 11μg DMS-S/m^2 h emitted from the ocean surface should be returned as nss-SO_4^{2-} in rain. This leads to a predicted wet deposition flux of nss-SO_4^{2-} of 7.4μg S/m^2 h, which is 43% lower than the measured flux of 13μg S/m^2 h (Galloway, 1985). Since the estimated accuracy of the DMS emission

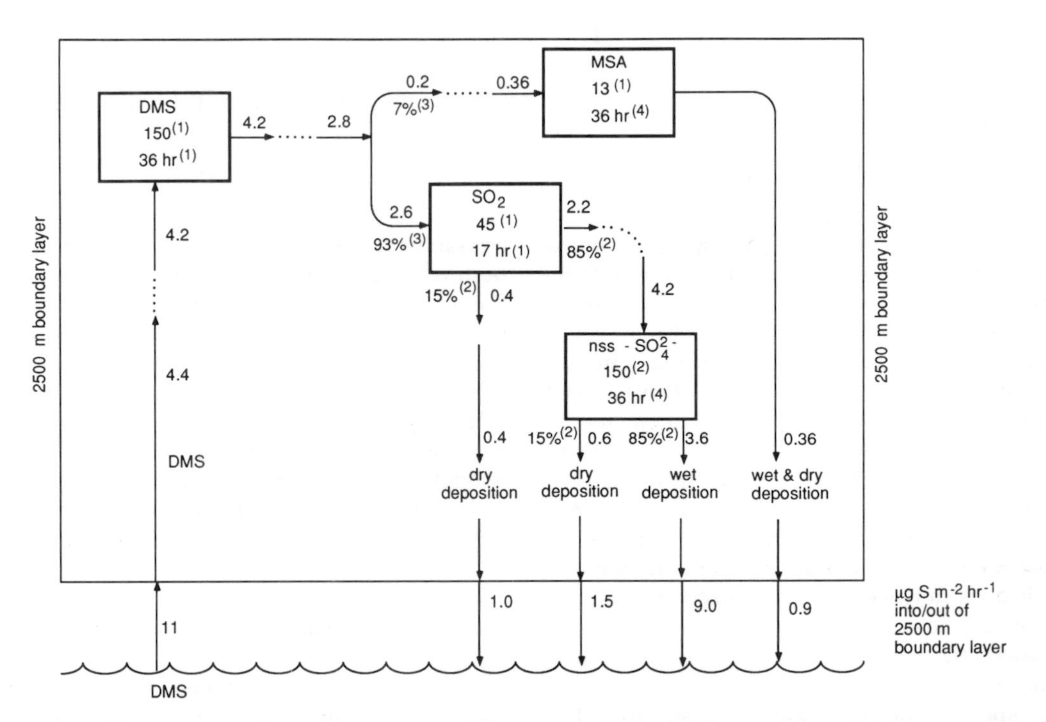

Fig. 13-5 The sulfur cycle in the remote marine boundary layer. Within the 2500 m boundary layer, burden units are ng S/m^3 and flux units are ng S/m^3 h. Fluxes within the atmospheric layer are calculated from the burden and the residence time. Dots indicate that calculations based on independent measurements are being compared. The measured wet deposition of nss-SO_4^{2-} (not shown) is 13 ± 7 μg S/m^2/h[(2)]. Inputs and outputs roughly balance, suggesting that a consistent model of the remote marine sulfur cycle within the planetary boundary layer can be constructed based on biogenic DMS inputs alone. Data: (1) Andreae (1986); (2) Galloway (1985); (3) Saltzman *et al.* (1983); (4) sulfate aerosol lifetime calculated earlier in this chapter based on marine rainwater pH; the same lifetime is applied to MSA aerosol. Modified from Crutzen *et al.* (1983) with the permission of Kluwer Academic Publishers.

flux is ±50% (Andreae, 1986), this is about as good agreement as can be expected. It indicates that our "closed system" assumption is at least a reasonable first approximation.

The two estimates of the flux into the DMS box are in excellent agreement, tending to support our 2.5-km assumed boundary layer height. However, the flux out of the DMS box is about 50% larger than necessary to support the fluxes through the MSA and SO_2 boxes. This might suggest a missing sink for DMS. Could DMSO, another known oxidation product of DMS whose concentrations and lifetimes have not been carefully studied, fill this gap? An even larger discrepancy exists between the two estimates of the flux from SO_2 to nss-SO_4^{2-}, which differ by almost a factor of 2. The fact that the flux out of the DMS box (multiplied by the appropriate partitioning factors) provides better agreement with the flux into the nss-SO_4^{2-} box, suggests that the error may lie in the measured SO_2 concentration (too low) or the estimated SO_2 lifetime (too high). In any case, the various data sets are in reasonable agreement and all of the above comparisons suggest further measurements which would help to refine our understanding of these atmospheric processes.

13.4.4 The Global Atmospheric Sulfur Budget

Figure 13-5 is an example of the direct application of modeling concepts from Chapter 4 to the tropospheric portion of the sulfur cycle. As such, it illustrates two important aspects of the cycle approach:

1. This approach allows independent data sets to be compared.
2. It suggests which measurements would be most helpful to more precisely determine the main features of the system.

Another way in which the cycle approach is often used is in the development of global atmospheric budgets. Here, the only assumption made is that the atmosphere, taken as a whole, is in steady-state. Thus, if the total burden of S compounds in the air is not increasing or decreasing, the global fluxes into the atmosphere must be equal to the total of fluxes out of the atmosphere. A comparison of the flux out (from rainwater analyses) with known fluxes from industrial production led Eriksson (1959, 1960) and many others to suggest that a "missing source" was needed to balance the budget. Indeed, this notion has provided substantial impetus over the past 20 years to study natural sources, including emission of reduced sulfur from swamps (Adams *et al.*, 1981), emission from volcanoes (Warneck, 1988), and biological production in and emission from the oceans (Andreae and Raemdonck, 1983).

Figure 13-6a (Ivanov, 1983) is a depiction of the natural global sulfur budget. Figure 13-6b depicts the budget with natural and anthropogenic sources. Table 13-2 serves to explain Fig. 13-6 and includes the wide range of estimates of various fluxes, and demonstrates the degree of uncertainty inherent in such approaches.

Comparison of Figs. 13-6a and 13-6b clearly demonstrates the degree to which human activity has modified the cycle of sulfur, largely via an atmospheric pathway. The influence of this perturbation can be inferred, and in some cases measured, in reservoirs that are very distant from industrial activity. Ivanov (1983) estimates that the flux of sulfur down the Earth's rivers to the ocean has roughly doubled due to human activity. Included in Table 13-2 and Fig. 13-6 are fluxes to the hydrosphere and lithosphere, which leads us to these other important parts of the sulfur cycle.

13.5 The Hydrospheric Cycle of Sulfur

The ocean plays a central role in the hydrospheric cycling of sulfur, since the major reservoirs of sulfur on the Earth's surface are related to various oceanic depositional processes. In this section, we consider the reservoirs and the fluxes, focusing on the cycling of sulfur through this oceanic node.

There are three major sulfur reservoirs at the Earth's surface (Table 13-3): as S(−II) in sedimentary shales, as S(+VI) in evaporite deposits, and as S(+VI) in seawater. The rate of cycling through these reservoirs is closely related to the fluxes of sulfur to and from the ocean. However, sulfur is unusual among the *major* constituents of natural waters in that its fluxes have been significantly modified by human activities. Thus, a useful starting point is the pre-industrial cycle of Fig. 13-6a. The main features are (1) a tightly closed loop through which sulfate is carried by sea spray into the atmosphere, but quickly returned to the ocean, and (2) a larger, slower loop in which sulfate is derived by weathering, carried by rivers to the ocean, and returned to the continents through the cycling of rocks (Chapter 6). Over geologic periods of time, the ocean composition is

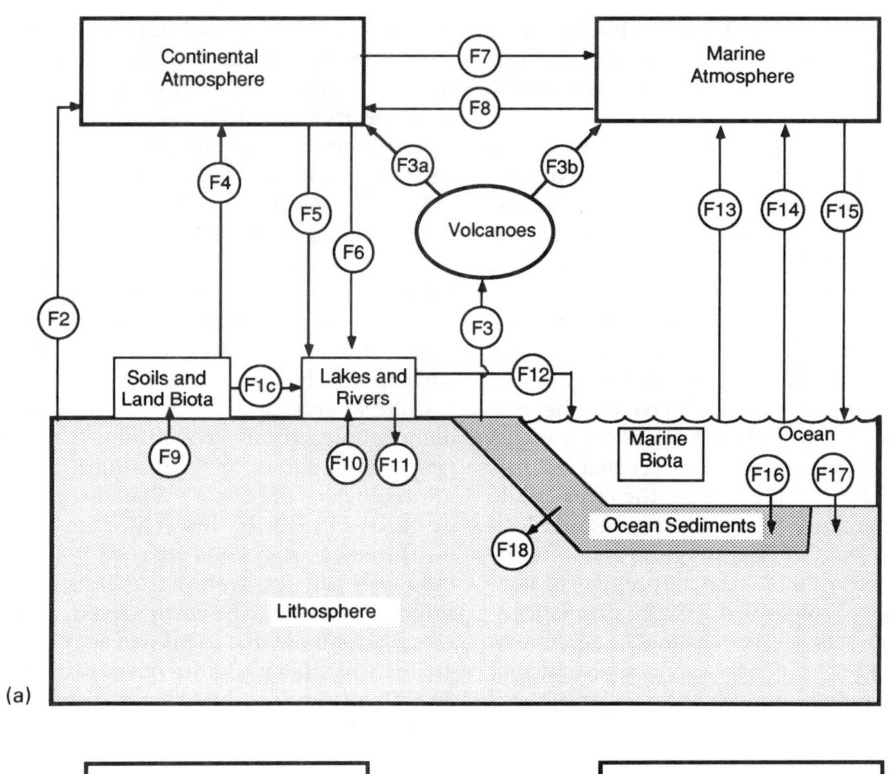

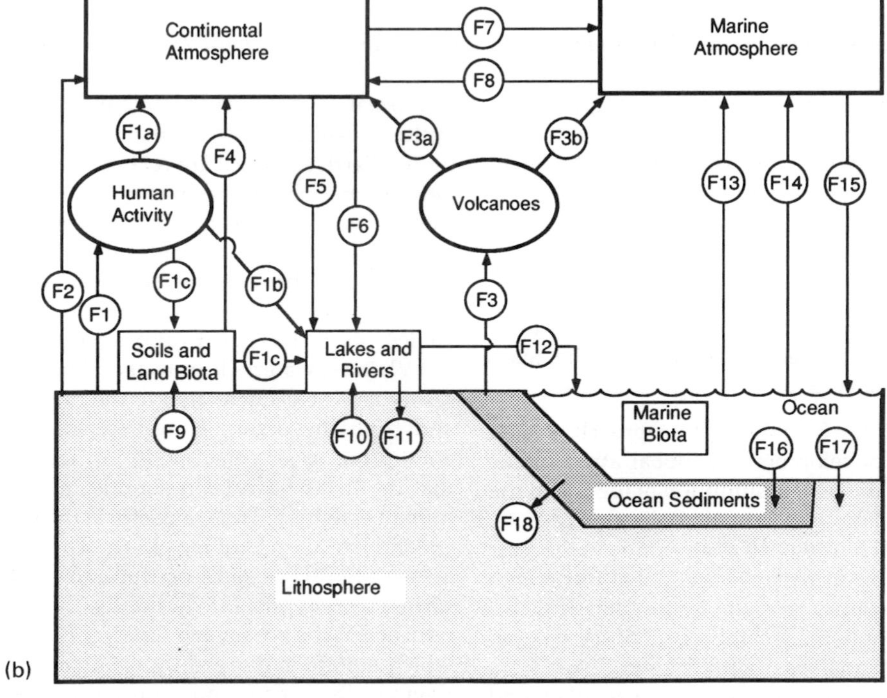

Fig. 13-6 Major fluxes of the global biogeochemical sulfur cycle excluding (a) and including (b) human activity. Numbers in circles designate fluxes described in Table 13.2. (Modified with permission from John Wiley and Sons, Inc.)

Table 13-3 Major sulfur reservoirs (Tg S)

Reservoir	Major form(s)	Burden
Continental atmosphere[a]	OCS, SO_4^{2-}, SO_2, DMS, H_2S	1.6
Marine atmosphere[a]	OCS, SO_4^{2-}, SO_2, H_2S, DMS	3.2
Soils and land biota[a]	Reduced	3×10^5
Lakes and rivers[d]	SO_4^{2-}	300
Marine biota[e]	Reduced	30
Seawater[a,b]	SO_4^{2-}	1.3×10^9
Ocean sediments[a,c]	Gypsum ($CaSO_4$), pyrite (FeS_2)	3×10^8
Rest of lithosphere[a,c]	Pyrite (FeS_2), gypsum ($CaSO_4$)	2.4×10^{10}

[a] Freney *et al.* (1983); [b] Volkov and Rozanov (1983); [c] Migdisov *et al.* (1983).
[d] Calculated from mass of surface freshwater (1.3×10^{20} g H_2O) and average sulfate concentration (2.5×10^{-6} g S/g H_2O).
[e] Calculated from carbon in ocean biota (3000 Tg) and approximate mass ratio for sulfur to carbon of 1 : 100.

thought to be relatively constant, with river inputs to the ocean balanced by depositional processes. What processes maintain this balance?

13.5.1 Oceanic Outputs

There are both uncertainties and controversies concerning the removal of sulfur from the ocean. FeS_2 (pyrite) in sedimentary shales is most often formed in reducing marine sediments, particularly when these sediments underlie waters of high biological productivity. In these areas, the flux of (dead) organic carbon to the sediments greatly exceeds the rates at which oxygen can diffuse into the sediments (Chapter 8) and so iron (III) hydroxides and then sulfate become the terminal electron acceptors for the oxidation of organic matter, thereby providing a source of Fe^{2+} and HS^- for pyrite formation.

Various workers have estimated the rate of pyrite formation. Berner (1972) summed the sulfur accumulation rates of various sediment types in proportion to their areal coverage and found a flux of about 10% of the river flux. Li (1981) carried out a similar calculation and finds 30% of the river flux, probably indicative of the uncertainty of the approach. Toth and Lerman (1977) established that the decrease of sulfate with depth in sediment pore waters is a function of sedimentation rate. This information was used to estimate the diffusive flux of sulfur into sediments driven by pyrite formation, again a value about 10% of the river flux. Apparently, pyrite formation, while measureable, is not the dominant removal process.

Evaporite deposition is a much more episodic process and thus difficult to quantify. Because seawater is significantly undersaturated with respect to common evaporitic minerals, like gypsum and halite, evaporites are only formed when restricted circulation develops in an ocean basin in which evaporation exceeds precipitation. A geologically recent example is the Mediterranean Sea of 5–6 Ma ago. At this time, excess evaporation exceeded the supply of ocean water through shallow inlet(s) from the Atlantic Ocean. As salinity increased, first $CaSO_4$, then NaCl, precipitated. Over time, salt deposits 2–3 km thick formed. This thickness represents about 40 desiccations of the entire volume of the Mediterranean. How rapidly would the Mediterranean have to evaporate to remove all the sulfate introduced by rivers to the ocean? The timescale τ is given by $\tau = V_{Med} C_{oc}/F_r$, where F_r is the river flux of sulfate, V_{Med} is the volume of the Mediterranean, and C_{oc} is the concentration of sulfate in the ocean. This time is about 25 000 years, which could be readily achieved (2000 m of water depth in 25 000 years is only 8 cm/year of excess evaporation). A second point to consider is that the basin will eventually fill. One meter of seawater evaporates to a layer of solid just over 1 cm thick. Evaporite formation could only be sustained in the Mediterranean for 2–3 Ma, removing only about ⅓ of the ocean's sulfate in this time. Evaporite formation can match the river flux, but it must be taking place on a large scale.

A solution, still controversial, has recently been proposed. This is the loss of sulfate from seawater during hydrothermal circulation through mid-ocean ridges (Edmond *et al.*, 1979). The flow of water through these systems is estimated to be about 1.4×10^{14} L/year, about 0.4% of the flow of rivers. However, sulfate is quantitatively removed, yielding a flux of 125 Tg S/year capable of balancing the river flux. The controversy is whether the chemistry involved in removing sulfate is the formation of anhydrite ($CaSO_4$) or reduction and subsequent precipitation as metal sulfides. Anhydrite forms on simple heating of seawater. As the ocean crust ages and cools it would redissolve, creating an equal and opposite flux. In contrast, metal sulfides are much more resistant to subsequent interaction with seawater.

13.5.2 Oceanic Inputs

Referring again to Fig. 13-6a, the materials that constitute the oceanic sinks for sulfur are recycled over time by exposure to weathering on the continents. The rates at which these processes occur help to regulate the flux of sulfur into rivers.

The evaporite source is characterized by covariation of sulfate (from gypsum) and chloride (from halite). That elements can be recycled from the ocean to land by movement of salt-bearing aerosols (so-called "cyclic salts") has confused the interpretation of river flux data somewhat. While this cycling generally follows the ratio of salts in the sea, the S : Cl ratio is an exception. Taking the S : Cl ratio of the cyclic component to be 2 (based on compositional data for marine rains) and assuming that all chloride in rivers is cyclic, an upper limit for the cyclic influence can be calculated:

$$F_{r,Cl}(SO_4^{2-}/Cl)_{seawater} = (310\ \text{Tg Cl/year})(0.046\ \text{g S/g Cl}) = 29\ \text{Tg S/year}$$

However, not all the chloride is cyclic, a fact first appreciated in recent years. An example comes from a detailed study of river geochemistry conducted in the Amazon Basin. In the inland regions, rains typically have a chloride content of 10 μM, whereas major inland tributaries have chloride contents of 20–100 μM. These data suggest that only 25% of the Cl is cyclic, whereas 75% is derived by weathering of evaporites. Indeed, 90% of this 75% can be shown to have its origin in the Andean headwaters, derived from evaporites that make up only 2% of the area of the Amazon Basin (Stallard and Edmond, 1981). As the ratio of sulfate to chloride in evaporite deposits is generally much higher than seawater [gypsum ($CaSO_4 \cdot 2H_2O$) precipitates first as seawater is evaporated], dissolution of evaporites represents one of the principal sources of riverine sulfate.

The other principal source is weathering of sedimentary or igneous sulfides, mainly pyrite, by the oxidation:

$$FeS_2 + {}^{15}\!/_4 O_2 + {}^{7}\!/_2 H_2O \rightarrow 2SO_4^{2-} + Fe(OH)_3(s) + 4H^+$$

What are the relative contributions of these two sources? Two approaches have been taken. One is to establish the geology and hydrology of a basin in great detail. This has been carried out for the Amazon (Stallard and Edmond, 1981) with the result that evaporites contribute about twice as much sulfate as sulfide oxidation. The other approach is to apply sulfur isotope geochemistry. As mentioned earlier, there are two relatively abundant stable isotopes of S, ^{32}S and ^{34}S. The mean 34/32 ratio is 0.0442. However, different source rocks have different ratios, which arise from slight differences in the reactivities of the isotopes. These deviations are expressed as a difference from a standard; in the case of sulfur, the standard being a meteorite found at Canyon Diablo, Arizona.

Evaporitic sulfur has a range of sulfur isotopic composition from +10‰ to +30‰ whereas sedimentary sulfur is depleted in the heavy isotope and has a range of isotopic composition of about −40‰ to +10‰. Most of this variation reflects systematic changes with geologic age. The source fractions of a river water can be estimated from an isotopic mass balance:

$$\delta^{34}S_{evap}X_{evap} + \delta^{34}S_{pyrite}X_{pyrite} = \delta^{34}S_{mean\ river}$$

where X_i is the fraction from that source. This approach has been applied to the Volga River, which has a $\delta^{34}S$ of +6‰, suggesting again about $^2\!/_3$ from evaporites and $^1\!/_3$ from sulfides (calculated taking $\delta^{34}S_{evap}$ as +15‰ and $\delta^{34}S_{pyrite}$ as −15‰).

Accepting these relative proportions from evaporites ($^2\!/_3$) and sulfides ($^1\!/_3$), the characteristic times (τ_i) of cycling of the evaporite sulfur and sulfide sulfur reservoirs can be estimated from the reservoir sizes (R_i) in Table 13-3 and the river flux of sulfur. For evaporites:

$$\tau_{evaporite} = (93 \times 10^6\ \text{Tg})/(2 \cdot 104/3\ \text{Tg/year}) = 140\ \text{Ma}$$

and for sulfides:

$$\tau_{pyrite} = (48 \times 10^6 \text{ Tg})/(104/3 \text{ Tg/year}) = 140 \text{ Ma}$$

The characteristic times of sulfur cycling through the two reservoirs are nearly identical. Sulfur cycling through evaporites is on a time-scale similar to the cycling of the evaporites themselves (about 200 Ma; Garrels and Mackenzie, 1971), suggesting that physical factors limit the rate of weathering of this very soluble component. Sulfur cycling through shales is considerably faster than the cycling of the overall shale reservoir (about 600 Ma).

Questions

13-1 What would be the approximate sulfate concentration of rainwater globally for the following cases (assume that rainfall is uniformly 75 cm/year):

(a) The only source of sulfur were DMS and it was uniform over the globe.
(b) In addition to (a) consider the sulfur from industry uniformly distributed over the globe.
(c) Same as (b), but assume that all industry is in the northern hemisphere and consider the hemispheres separately.
(d) Compare these concentrations to the data for eastern North America (Fig. 13-4).

13-2 Sulfur and oxygen are in the same column of the periodic table. List their chemical similarities and differences and consider the biogeochemical consequences of each.

13-3 Estimate the total amount of oxidation (moles/year) caused by the reduction of SO_4^{2-} to S(−II). Compare this to the total amount of oxidation by O_2.

13-4 After considering the Redfield, Ketchum and Richards ratio (see Chapter 9) for C : N : P, consider the analog for sulfur in land and marine biota. What key biochemical species are the major determinants of the C : S relationship in biota?

13-5 *Hypothetical problem for chemists*: Consider the global cycle of selenium which has many chemical similarities to sulfur. Construct a box diagram for the global selenium cycle based on known similarities and differences of Se and S.

13-6 This problem is a first-order attempt to quantify the possible anthropogenic perturbation of the northern hemisphere (NH) marine sulfur cycle. First, assume that present-day anthropogenic sulfur emissions result in 20 Tg S/year being transported from North America to the atmosphere over the NH Atlantic and 10 Tg S/year being transported from Asia to the atmosphere over the NH Pacific. Assume a uniform concentration in the N–S direction, average westerly wind speeds of 10 m/s, that both ocean regions are 4000 km in N–S extent, that the NH Atlantic is 5500 km from east to west, and that the NH Pacific is 8500 km from east to west. Next, assume that biogenic emissions of DMS from the ocean to the atmosphere are uniformly 0.1 g S/m^2 year (the average value from Andreae and Raemdonck, 1983). Now, if we use a 2-day residence time (or e-folding time – see Chapter 4) for sulfur in the atmosphere, what percentage of total atmospheric sulfur would be anthropogenic sulfur at the middle and western edge of each ocean?

References

Adams, D. F., S. O. Farwell, E. Robinson, M. R. Park, and W. L. Bamsberger (1981). Biogenic sulfur source strengths. *Environ. Sci. Technol.* **15**, 1493–1498.

Andreae, M. O. (1985). The emission of sulfur to the remote atmosphere. *In* "The Biogeochemical Cycling of Sulfur and Nitrogen in the Remote Atmosphere" (J. N. Galloway, R. J. Charlson, M. O. Andreae, and H. Rodhe, eds.), pp. 5–25. Reidel, Dordrecht, The Netherlands.

Andreae, M. O. (1986). The ocean as a source of atmospheric sulfur compounds. *In* "The Role of Air–Sea Exchange in Geochemical Cycling" (P. Buat-Menard, ed.), pp. 331–362. Reidel, Dordrecht, The Netherlands.

Andreae, M. O. and H. Raemdonck (1983). Dimethyl sulfide in the surface ocean and the marine atmosphere: A global view. *Science* **221**, 744–747.

Berner, R. A. (1972). Sulfate reduction, pyrite formation and the oceanic sulfur budget. *In* "The Changing Chemistry of the Oceans" (D. Dyrssen and D. Jagner, eds.), pp. 347–361. Wiley-Interscience, New York.

Bigg, E. K., J. L. Gras, and C. Evans (1984). Origin of Aitken particles in remote regions of the Southern Hemisphere. *J. Atmos. Chem.* **1**, 203–214.

Charlson, R. J., W. L. Chameides, and D. Kley (1985). The transformations of sulfur and nitrogen in the remote atmosphere. *In* "The Biogeochemical Cycling of Sulfur and Nitrogen in the Remote Atmosphere" (J. N. Galloway, R. J. Charlson, and M. O. Andreae, eds.), pp. 67–80. Reidel, Dordrecht, The Netherlands.

Crutzen, P. J. (1976). The possible importance of CSO for the sulphate layer of the stratosphere. *Geophys. Res. Lett.* **3**, 73–76.

Crutzen, P. J., D. M. Whelpdale, D. Kley, and L. A. Barrie (1985). The cycling of sulfur and nitrogen in the remote atmosphere. *In* "The Biogeochemical Cycling of Sulfur and Nitrogen in the Remote Atmosphere" (J. N. Galloway, R. J. Charlson, and M. O. Andreae, eds.), pp. 203–212. Reidel, Dordrecht, The Netherlands.

Edmond, J. M., C. Measures, R. E. McDuff, L. H. Chan, R. W. Collier, B. Grant, L. I. Gordon, and J. B. Corliss

(1979). Ridge crest hydrothermal activity and the balance of the major and minor elements in the ocean: The Galapagos data. *Earth Planet. Sci. Lett.* **46**, 1–18.

Eriksson, E. (1959). The yearly circulation of chloride and sulfur in nature: Meteorological, geochemical and pedological implications, Part I. *Tellus* **11**, 375–603.

Eriksson, E. (1960). The yearly circulation of chloride and sulfur in nature: Meteorological, geochemical and pedological implications, Part II. *Tellus* **12**, 63–109.

Freney, J. R., M. V. Ivanov, and H. Rodhe (1983). The sulphur cycle. *In* "The Major Biogeochemical Cycles and Their Interactions" (B. Bolin and R. B. Cook, eds.), pp. 56–61. John Wiley, Chichester.

Friend, J. P. (1973). The global sulfur cycle. *In* "Chemistry of the Lower Atmosphere" (S. I. Rasool, ed.), pp. 177–201. Plenum Press, New York.

Galloway, J. N. (1985). The deposition of sulfur and nitrogen from the remote atmosphere. *In* "The Biogeochemical Cycling of Sulfur and Nitrogen in the Remote Atmosphere" (J. N. Galloway, R. J. Charlson, M. O. Andreae, and H. Rodhe, eds.), pp. 143–175. Reidel, Dordrecht, The Netherlands.

Garrels, R. M. and F. T. Mackenzie (1971). "Evolution of Sedimentary Rocks." W. W. Norton, New York.

Granat, L., H. Rodhe and R. O. Hallberg (1976). The global sulphur cycle. *In* "Nitrogen, Phosphorus and Sulphur – Global Cycles" (B. H. Svensson and R. Soderlund, eds.) pp. 89–134. SCOPE Report 7, Ecol. Bull., Stockholm.

Grinenko, V. A. and M. V. Ivanov (1983). Principal reactions of the global biogeochemical cycle of sulphur. *In* "The Global Biogeochemical Sulfur Cycle" (M. V. Ivanov and J. R. Freney, eds.), pp. 1–23. John Wiley, Chichester.

Ivanov, M. V. (1983). Major fluxes of the global biogeochemical cycle of sulphur. *In* "The Global Biogeochemical Sulphur Cycle" (M. V. Ivanov and J. R. Freney, eds.), pp. 449–463. John Wiley, Chichester.

Jorgensen, B. B. (1990). A thiosulfate shunt in the sulfur cycle of marine sediments. *Science* **249**, 152–154.

Kellogg, W. W., R. D. Cadle, E. R. Allen, A. L. Lazrus, and E. A. Martell (1972). The sulfur cycle. *Science* **175**, 587–596.

Li, Y. H. (1981). Geochemical cycle of elements and human perturbation. *Geochim. Cosmochim. Acta* **45**, 2037–2084.

Migdisov. A. A., A. B. Ronov, and V. A. Grinenko (1983). The sulphur cycle in the lithosphere, Part I: Reservoirs. *In* "The Global Biogeochemical Sulphur Cycle" (M. V. Ivanov and J. R. Freney, eds.), pp. 25–95. John Wiley, Chichester.

Moller, D. (1984a). Estimation of the global man-made sulphur emission. *Atmos. Environ.* **18**, 19–27.

Moller, D. (1984b). On the global natural sulphur emission. *Atmos. Environ.* **18**, 29–39.

National Atmospheric Deposition Program (1988). NADP/NTN Annual Data Summary. Precipitation Chemistry in the United States, 1987. National Resources Ecology Laboratory, Colorado State University, Fort Collins, Colorado, 353 pp.

Ottar, B. (1978). As assessment of the OECD study on long range transport of air pollutants (LRTAP). *Atmos. Environ.* **12**, 445–454.

Saltzman, E. S., D. L. Savoie, R. G. Zika and J. M. Prospero (1983). Methane sulfonic acid in the marine atmosphere. *J. Geophys. Res.* **88**, 10 897–10 902.

Stallard, R. F. and J. M. Edmond (1981). Geochemistry of the Amazon 1. Precipitation chemistry and the marine contribution to the dissolved load at the time of peak discharge. *J. Geophys. Res.* **86**, 9844–9858.

Toon, O. B., J. F. Kasting, R. P. Turco, and M. S. Liu (1987). The sulfur cycle in the marine atmosphere. *J. Geophys. Res.* **92**, 943–963.

Toth, D. J. and A. Lerman (1977). Organic matter reactivity and sedimentation rates in the ocean. *Amer. J. Sci.* **277**, 265–285.

Volkov, I. I. and A. G. Rozanov (1983). The sulphur supply in oceans, Part I: Reservoirs and fluxes. *In* "The Global Biogeochemical Sulphur Cycle" (M. V. Ivanov and J. R. Freney, eds.), pp. 357–423. John Wiley, Chichester.

Warneck, P. (1988). "Chemistry of the Natural Atmosphere." Academic Press, New York.

14

The Phosphorus Cycle

Richard A. Jahnke

Phosphorus is one of the most important elements on Earth. It participates in or controls many of the biogeochemical processes occurring in the biosphere. To understand the interaction between P and other biogeochemical processes and elemental distributions, it is necessary to understand the distribution of P on the Earth's surface and the processes that control its distribution. The strategy of this chapter, therefore, is to (1) discuss the chemical forms in which P is present in the environment; (2) describe the processes that control its distribution in terrestrial, aquatic, and oceanic systems; and (3) define the major P reservoirs on the Earth's surface and the rate at which P is exchanged between these reservoirs. Because all of these subjects must be addressed within this single chapter, the discussion is somewhat superficial and intended to expose the reader to the individual topics rather than to provide a thorough discussion of each. References in each section provide more detailed presentations of the individual topics.

14.1 Occurrence of Phosphorus

The global occurrence of P differs from that of the other major biogeochemical elements, C, N, S, O, and H, in several very important aspects. First, while gaseous forms of P can be produced in the laboratory, none have ever been found in significant quantities in the natural environment. Thus, although some P is transported within the atmosphere on dust particles and dissolved in rain and cloud droplets, the atmosphere generally plays a minor role in the global P cycle. It should be noted, however, that at certain locations, this small atmospheric source of P can be important. An example is the surface waters in the central gyres of the oceans where extremely low standing stocks of P are observed and the transport of P from other potential sources is very slow.

The second significant difference between P and the other major biogeochemical elements is that oxidation–reduction reactions play a very minor role in controlling the reactivity and distribution of P in the natural environment. While several oxidation states for P are chemically possible, these forms are generally restricted to controlled laboratory settings. In natural systems, therefore, P is almost exclusively present in the V oxidation state where it is found as phosphate (PO_4^{3-}), a tetrahedral oxy-anion. Nearly all dissolved and particulate forms of P are combined, complexed, or slightly modified forms of this ion. In general, the biogeochemical cycle of P is synonymous with that of phosphate.

Finally, P also differs from other elements in that it is overwhelmingly dominated by a single isotopic form containing 15 protons and 16 neutrons. There are only two naturally occurring radioactive forms of P, ^{32}P and ^{33}P. Cosmic rays produce these radionuclides in the atmosphere by nuclear reactions with argon. A small amount of ^{32}P is also contributed by ^{32}Si decay. Because these isotopes have extremely short half-lives (^{32}P half-life, 14.3 days; ^{33}P half-life, 25.3 days), their activities in the environment are always very low and account for a minute portion of the total P in the Earth. While these isotopes may be useful tracers for processes occurring near the atmospheric source such as the uptake of P in the photic zone of lakes and oceans (Lal and Lee, 1988), they are of little use in deciphering the longer-term aspects of the P cycle.

Global Biogeochemical Cycles
ISBN 0-12-147685-5

14.1.1 Dissolved Inorganic Forms of Phosphorus

Phosphate, PO_4^{3-}, is the fully dissociated anion of triprotic phosphoric acid, H_3PO_4:

$$H_3PO_4 \leftrightarrow H^+ + H_2PO_4^- \leftrightarrow 2H^+ + HPO_4^{2-} \leftrightarrow 3H^+ + PO_4^{3-}$$

The dissociation constants for these equilibria for freshwater and seawater are listed in Table 14-1. The proportion of the individual protonated species in distilled water and seawater over the pH range of 2–10 is shown in Fig. 14-1. At the pH of freshwater systems (very roughly 6–7), $H_2PO_4^-$ is the dominant phosphate species. The high ionic strength of seawater and the presence of cations such as Ca^{2+}, Mg^{2+}, and Na^+, which form ion pairs with the PO_4^{3-} species, significantly alter the dissociation of phosphoric acid. In seawater at a pH of 8, HPO_4^{2-} dominates. The importance of ion pairs on the PO_4^{3-} activity in seawater is further demonstrated in Fig. 14-2. It is clear that ion pairs with dissolved cations play a central role in controlling the aqueous PO_4^{3-} speciation and that free PO_4^{3-} ions constitute a very small proportion of the total present (0.01% at standard seawater conditions). The chemical reactivity of PO_4^{3-} in aqueous systems is therefore highly dependent on the composition and pH of the solution. Acid–base and complexation reactions are not only important in seawater systems, but also influence the reactivity of PO_4^{3-} in groundwater and freshwater systems.

Another important class of inorganic PO_4^{3-} compounds are the condensed or polyphosphates. In these compounds, two or more phosphate groups bond together via P—O—P bonds to form chains or in some cases cyclic compounds. Although polyphosphates generally account for only a small portion of the total P present in natural waters, they are extremely reactive compounds and are routinely used in industrial and commercial applications. Condensed phosphates form soluble complexes with many metal cations and are used, therefore, as water softeners.

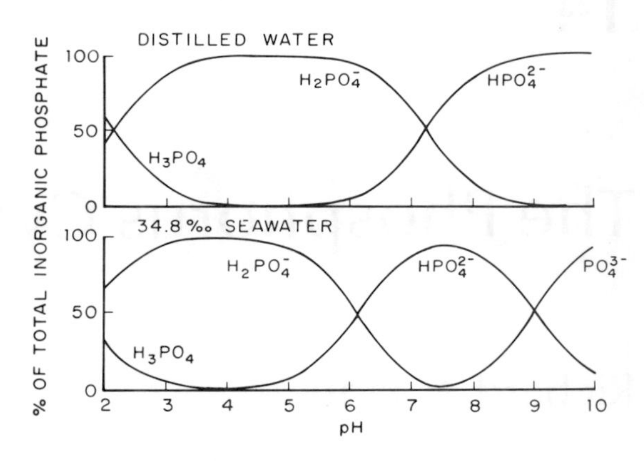

Fig. 14-1 Distribution of phosphoric acid species as a function of pH in distilled water and seawater (Atlas, 1975).

Table 14-1 Dissociation constants of phosphoric acid at 25° C

	Distilled[a] (pK)	Sea water[b] (pK)
$H_3PO_4 \leftrightarrow H^+ + H_2PO_4^-$	2.2	1.6
$H_2PO_4^- \leftrightarrow H^+ + HPO_4^{2-}$	7.2	6.1
$HPO_4^{2-} \leftrightarrow H^+ + PO_4^{3-}$	12.3	8.6

[a] Stumm and Morgan (1981); [b] Atlas (1975).

14.1.2 Particulate Forms of Phosphorus

Phosphorus is the tenth most abundant element on Earth with an average crustal abundance of 0.1% and may be found in a wide variety of mineral phases. There are approximately 300 naturally occurring minerals in which PO_4^{3-} is a required structural component. Phosphate may also be present as a trace component in many minerals (Nriagu and Moore, 1984; Slansky, 1986), either by the substitution of

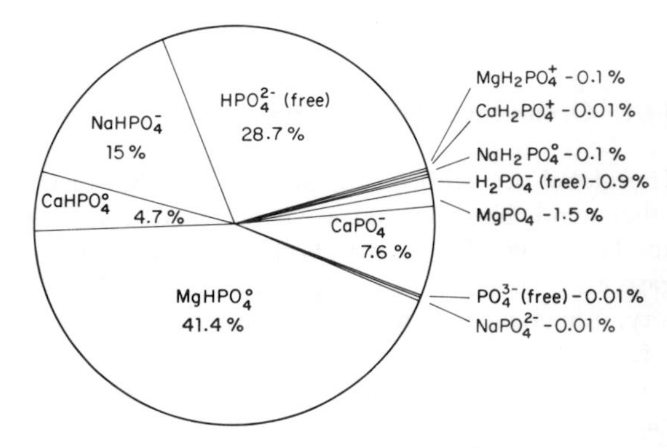

Fig. 14-2 Calculated speciation of PO_4^{3-} in seawater of 34.8‰ at 20 °C and a pH of 8.0 (Atlas, 1975).

small quantities of PO_4^{3-} into the crystal structure or by the adsorption of PO_4^{3-} onto the mineral surface.

By far the most abundant phosphate mineral is apatite, which accounts for more than 95% of all P in the Earth's crust. The basic composition of apatite is listed in Table 14-2. Apatite exhibits a hexagonal crystal structure with long open channels parallel to the "c" axis. In its pure form, F^-, OH^-, or Cl^- occupy sites along this axis to form fluorapatite, hydroxyapatite, or chlorapatite, respectively. However, because of the "open" nature of the apatite crystal lattice, many substitutions are possible and "pure" forms of apatite as depicted by the general formula in Table 14-2 are essentially never found. Of the possible substituting ions, carbonate ion is by far the most important followed by Na^+, SO_4^{2-}, and Mg^{2+}. The most common form of natural apatite is francolite, a highly substituted form of carbonate fluorapatite deposited in marine systems. The substitution of CO_3^{2-} ions into the mineral lattice has a substantial effect on apatite solubility (Jahnke, 1984). More studies are required, however, before the effects of all substituting ions are understood and an accurate assessment of the solubility of complex, natural apatites can be made.

The importance of this mineral is perhaps best demonstrated by the diversity of its sources. Assuming that nearly all of the P present is in the form of apatite, igneous rocks contain between 0.02 and 1.2% apatite. Apatite is also produced by organisms (including man) as structural body parts such as teeth, bones, and scales. After an organism dies, these components tend to accumulate in sediments and soils. In some locations, these constituents are reworked and concentrated by physical processes to form economically important deposits. By far the largest accumulations of P on the Earth's surface are massive sedimentary apatite deposits (phosphorites). The mining of these deposits provides 82% of the total world PO_4^{3-} production and 95% of the total remaining reserves (Howard, 1979). Phosphate rock may also form by the accumulation of bird or bat droppings and subsequent diagenetic alteration and crystallization. The mining of guano can be important locally, although this comprises a negligible fraction of the world's phosphate rock production.

Table 14-2 The chemical formula of apatites

General formula $Ca_{10}(PO_4)_6X_2$

X =	
F^-	Fluorapatite
OH^-	Hydroxyapatite
Cl^-	Chlorapatite

Possible substitutes for Ca^{2+}:
Na^+, K^+, Ag^+, Sr^{2+}, Mn^{2+}, Mg^{2+}, Zn^{2+}, Cd^{2+}, Ba^{2+}, Sc^{3+}, Y^{3+}, rare earth elements, Bi^{3+}, U^{4+}

Possible substitutes for PO_4^{3-}:
CO_3^{2-}, SO_4^{2-}, CrO_4^{2-}, AsO_4^{3-}, VO_4^{3-}, $F \cdot CO_3^{3-}$, $OH \cdot CO_3^{3-}$, SiO_4^{4-}

In general, the major phosphorite deposits are of marine origin and occur as sedimentary beds ranging from a few centimeters to tens of meters in thickness. The biogenic matter produced in the water column settles to the sediment surface and decomposes, releasing PO_4^{3-} to the seawater and pore waters. Thus, the P required to form large deposits is thought to be supplied through this decomposition. Once released to the pore waters, the PO_4^{3-} is available to be incorporated into apatite. In coastal Peru waters, the precipitation of modern apatites is associated with high dissolved PO_4^{3-} in the near-surface pore waters. This PO_4^{3-}, although ultimately transported to the sediments as organic matter, is apparently concentrated by chemical processes occurring at redox boundaries or by biological processes associated with bacterial mats that form at the sediment surface in this area (Froelich *et al.*, 1988). The mechanisms by which precipitation occurs are not yet understood, although mathematical models suggest that apatite precipitation is an early diagenetic event, occurring near the sediment–water interface (Van Cappellen and Berner, 1988). The major sedimentary phosphorite deposits appear to have formed either by the direct precipitation of a PO_4^{3-} phase from seawater or pore waters or by the replacement of CO_3^{2-} by PO_4^{3-} in biogenic calcium carbonate (Sheldon, 1981). These deposits are found on all continents at a wide range of latitudes. Deposits that appear to be forming today, however, tend to be located on continental margins underlying regions of major or local upwelling and high biological primary production.

The occurrence of other phosphate minerals is also important in certain locations. In sediments, the formation of vivianite, $Fe_3(PO_4)_2 \cdot 8H_2O$, and struvite, $NH_4MgPO_4 \cdot 6H_2O$, has been reported (Emerson and Widmer, 1978; Murray *et al.*, 1978). Minerals that may form as secondary weathering products in PO_4^{3-} deposits include crandallite, $CaAl_3(PO_4)_2(OH)_5 \cdot H_2O$, brushite, $CaHPO_4 \cdot 2H_2O$, whitlockite, $Ca_3(PO_4)_2$, vivianite, and others. Nriagu

(1976) has described the weathering of PO_4^{3-} minerals with thermodynamic arguments. His calculations suggest that in neutral to acidic soil systems, calcareous PO_4^{3-} minerals may be weathered to forms richer in aluminum such as crandallite, wavellite, $Al_3(PO_4)_2(OH)_3 \cdot 5H_2O$, or montgomeryite, $Ca_2Al_2(PO_4)_3OH \cdot 7H_2O$, along with others.

Phosphate is ubiquitous as a minor component within the crystal lattices of other minerals or adsorbed onto the surface of particles such as clays, calcium carbonate, or ferric oxyhydroxides (Berner, 1973). Therefore, in general, transport of these other particulate phases represents an important transport pathway of P as well.

14.1.3 Organic Forms of Phosphorus

Many of the most fundamental biochemicals required for life contain P generally linked to long, complicated organic molecules by phosphate ester bonds. Among the impressive list of compounds in which phosphate is a necessary constituent are the nucleic acids, DNA, and RNA (see Chapter 3). In these compounds (Fig. 14-3a), phosphates covalently link the mononucleotide units forming long polymers which, depending on the composition of the attached base, encode all necessary genetic information.

Phosphate also plays a central role in the transmission and control of chemical energy within the cells primarily via the hydrolysis of the terminal phosphate ester bond of the adenosine triphosphate (ATP) molecule (Fig. 3b). In addition, phosphate is a necessary constituent of phospholipids, which are important components in cell membranes and as mentioned before, of apatite, which forms structural body parts such as teeth and bones. It is not surprising, therefore, that the cycling of P is closely linked with biological processes. This connection is, in fact, inseparable as organisms cannot exist without P, and their existence controls, to a large extent, the natural distribution of P.

Because these P-containing compounds are abundant in all organisms, P is one of the major components of all organisms. In general, marine microorganisms contain 105–125 carbon atoms for every P atom (Redfield *et al.*, 1963; Peng and Broecker, 1987). Because of the increased abundance of structural parts not involving phosphorus, the average C:P ratio of terrestrial plants is much higher, approximately 800:1 (Deevey, 1970).

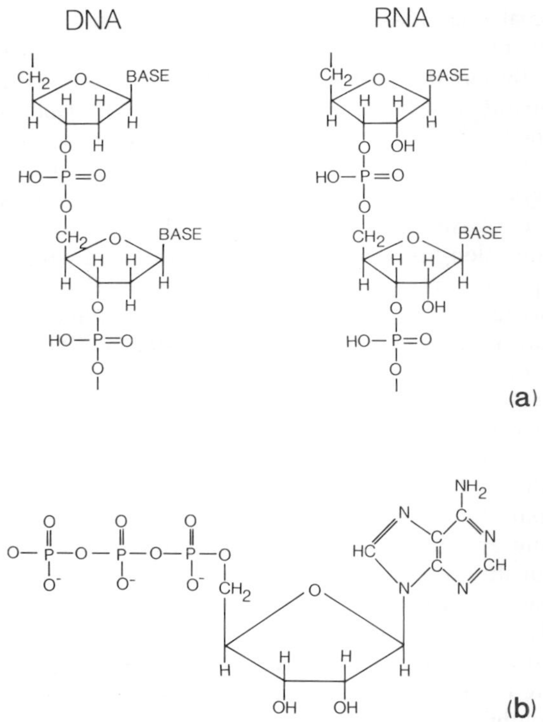

DNA RNA

(a)

(b)

Fig. 14-3 Structure of RNA and DNA (a) and ATP (b).

Organic P also occurs in dissolved forms, particularly in euphotic surface waters, where it may exceed inorganic phosphate concentrations (Smith *et al.*, 1986). The labile fraction of dissolved organic phosphorus may be the most rapidly cycled marine phosphate pool (Orrett and Karl, 1987; Jackson and Williams, 1985). This pool may also provide a "buffer", providing utilizable phosphate to phytoplankton between episodic inputs from other sources (Jackson and Williams, 1985).

14.2 Sub-global Phosphorus Cycles

A global representation of the P cycle will, by necessity, be general and will parameterize intricate processes and feedback mechanisms into simple first-order transfers and will combine a wide variety of P-containing components into relatively few reservoirs. To appreciate the rationale behind the construction of such a model and to understand its limitations, the transfers of P within a hypothetical terrestrial ecosystem and in a generalized oceanic system will be discussed first.

14.2.1 Freshwater Terrestrial Ecosystems

The dominant processes controlling the movements of P through terrestrial ecosystems are schematically presented in Fig. 14-4. In a general way, the overall movement of P on the continents may be envisioned as the constant erosion of P from continental rocks and transport in both dissolved and particulate form by rivers to the ocean, stopping occasionally along this pathway to interact with biological and mineralogical systems.

Physical and chemical erosion of continental rocks, represented by the arrow labeled "1" on Fig. 14-4, introduces particulate and dissolved P to the soil system. The majority (90%) of the P eroded from rocks remains trapped in the mineral lattices of the particulate matter. This P will be transported with the suspended material or bedload downstream until it eventually reaches the estuaries and the oceans, never having entered the biological cycles. The small proportion of the P that is leached from the minerals into solution, however, is readily available to enter biological cycles (2) and to react with inorganic soil particles (3).

Dissolved P in groundwaters is, of course, taken up by plants. Although many elements are required for plant life, in many ecosystems P is the least available and, therefore, limits overall primary production (Schindler, 1977; Smith *et al.*, 1986). Thus, in many instances, the availability of P influences or controls the cycling of other bioactive elements. When organisms die, the organic P compounds decompose and the P is released back into the soil–groundwater system.

Inorganic reactions in the groundwaters also influence dissolved P concentrations. These reactions include the dissolution or precipitation of P-containing minerals or the adsorption and desorption of P onto and from mineral surfaces. As discussed above, the inorganic reactivity of phosphate is strongly dependent on pH. In alkaline systems, apatite solubility should limit groundwater phosphate, whereas in acidic soils, aluminum phosphates should dominate. Adsorption of phosphate onto mineral surfaces, such as iron or aluminum oxyhydroxides and clays, is favored by low solution pH and may influence groundwater concentrations. Phosphorus will be exchanged between organic materials, groundwaters, and mineral phases many times on its way toward the ocean.

Lakes (4) also constitute an important component of the terrestrial P system. Because much of mankind's activities occur on or adjacent to lakes and because P availability has such a major influence on the biological community, P cycling in lakes has been studied extensively. The importance of understanding the transport of P through the lake system may be demonstrated by considering the hypo-

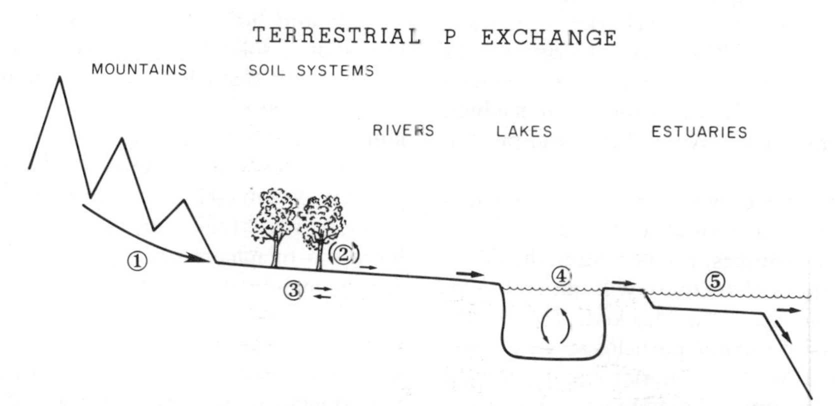

Fig. 14-4 Schematic representation of the transport of P through the terrestrial system. The dominant processes considered in this description are: (1) mechanical and chemical weathering of rocks, (2) incorporation of P into terrestrial biomass and its return to the soil system through decomposition, (3) exchange reactions between groundwaters and soil particles, (4) cycling in freshwater lakes, and (5) transport through the estuaries to the oceans of both particulate and dissolved P.

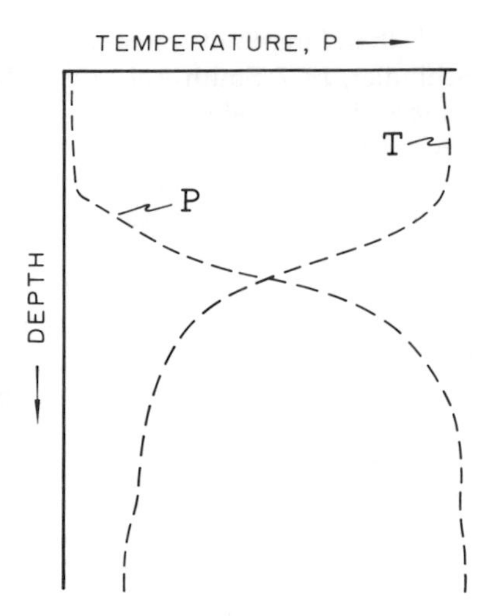

Fig. 14-5 Hypothetical distribution of P and temperature in a temperate lake in summer. Thermal stratification restricts exchange between surface and deep waters. Phosphorus is depleted in the surface waters by the sinking of biologically produced particles.

thetical P and temperature profiles for a temperate lake in summer displayed in Fig. 14-5. Briefly, in summer, warming of the surface layers produces strong stratification which restricts exchange between the lighter, warm surface water and the colder, denser deep water. During photosynthesis, the dissolved P in the photic zone is incorporated into plants and is eventually transported below the thermocline on sinking particles. Because of the stratification, this P is transported back to the surface photic zone very slowly. The constant stripping of P from the surface layers by sinking particles results in extremely low levels of P in the surface waters which, therefore, generally limit overall biological productivity.

In this context, it is easy to envision the potential influence of increased P input to the surface layers via anthropogenic sources. If P is continually added to the photic zone, productivity will not be limited but will simply continue unchecked. This results in large amounts of organic particles sinking below the thermocline. As this material decomposes, it consumes oxygen. Since exchange with the surface layer is restricted, the deep waters will become depleted in oxygen (cf. Lehman, 1988). If enough organic materials sink into the deep waters, the oxygen will be thoroughly consumed via decomposition. In extreme conditions, this can result in the formation of anoxic deep waters and fish kills.

As cooling occurs in the late fall and early winter, the thermal stratification breaks down, permitting mixing of the deep and surface layers. This allows the surface layers to be replenished with P. During the winter months, biological productivity in a temperate lake is limited by the availability of light rather than nutrients.

In the hypothetical terrestrial system depicted in Fig. 14-4, the P eroded from the land is eventually transported to the estuaries. As in lakes, soils, and rivers, many chemical and biological processes act to control the transport of P within and from the estuary (Lucotte and d'Anglejan, 1988; Jonge and Villerius, 1989). Dissolved P may be removed from solution onto the particulate phase and deposited in the sediments. On the other hand, the change in the solution composition may cause P to be released from the particulate load. The P that is transported from the estuaries to the ocean in particulate form will rapidly settle to the sea floor and be incorporated into the sediments. The dissolved P will enter the surface ocean and participate in the biological cycles. Determining what proportion of P that is transported out of the estuary is reactive is a critical step in the elucidation of the marine P budget (Froelich *et al.*, 1982).

14.2.2 The Oceanic System

Over much of the ocean (exclusive of upwelling regions and high latitude areas) the vertical distribution of dissolved PO_4^{3-} is represented by the shape of the profile displayed in Fig. 14-6, which is similar to the shape observed for the temperate lake in summer. Also included on the figure are the major processes responsible for controlling this shape. In general, dissolved P is near undetectable levels in the euphotic zone (generally the upper 20–100 m) and increases to maximum concentrations of 1–3 μM at approximately 1000 m. The distribution can best be envisioned as the balance between the incorporation of P into organisms with the eventual sinking of some fraction of this P from the surface waters and the constant slow rate of return of P by physical processes to the surface layer. The majority of ocean deep water is formed in the North Atlantic and slowly spreads sequentially to the South Atlantic, Indian, and finally to the Pacific Oceans. Because of

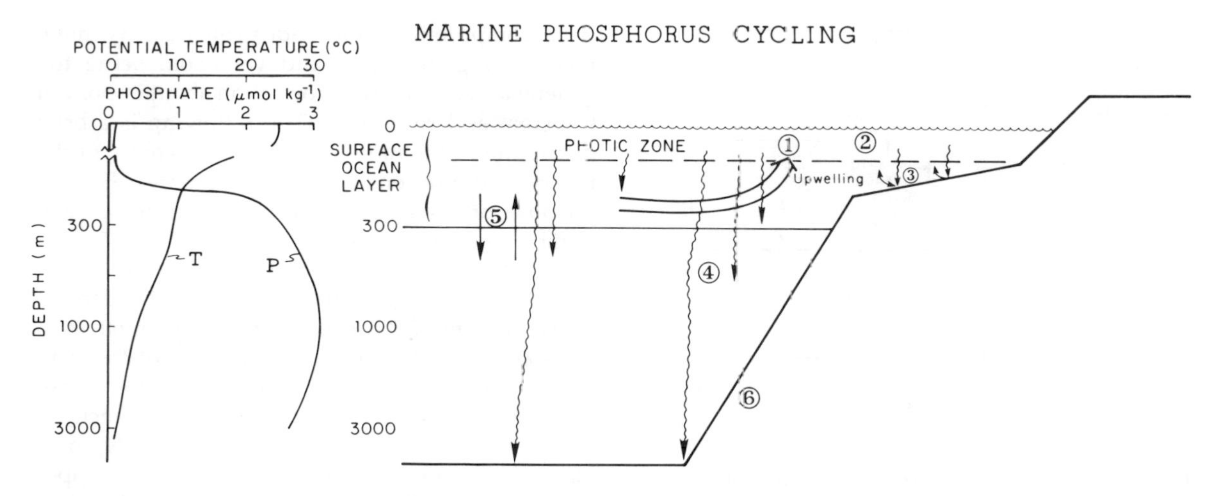

Fig. 14-6 Profiles of potential temperature and phosphate at 21°21′N, 122°15′W in the Pacific Ocean (J. W. Murray, unpublished results) and a schematic representation of the oceanic processes controlling the P distribution. The dominant processes considered are: (1) upwelling of nutrient-rich waters, (2) biological productivity and the sinking of the biogenic particles produced, (3) regeneration of nutrients by the decomposition of organic matter in surface waters and shallow sediments, (4) decomposition of particles below the main thermocline, (5) slow exchange between surface and deep waters, (6) incorporation of P into the bottom sediments.

the continuous rain of particulate P into the deep waters from the surface layers, the deep water PO_4^{3-} concentration increases progressively from the North Atlantic to the Pacific.

Unlike the temperate lake, stratification in the ocean does not completely break down in the winter. Except for the polar regions of the ocean, processes other than deep convective overturn are responsible for returning the P stored in the deep waters to the surface. A relatively slow exchange occurs between water layers everywhere in the ocean and this supplies some P to the surface ocean (process 5 in Fig. 14-6) from the intermediate layers below. More important sources of P to the photic zone are the major upwelling regions generally located adjacent to the western, sub-tropical continental margins (1) and in equatorial divergence zones. In the western margins, the prevailing winds tend to transport surface water offshore. This water is replaced by nutrient-rich water from below. At these locations, P input (along with other required nutrients) is not limited by slow diffusive transport processes but is enhanced many-fold by the upward advection of water. For this reason, upwelling areas are capable of supporting extremely high rates of biological primary production and abundant populations of higher organisms. Thus, the major fisheries of the world are concentrated in upwelling regions such as off Peru. A significant amount of P is also returned to the surface ocean in cold, high-latitude regions where less stratification results in greater vertical mixing than in the temperate and equatorial regions.

Once in the photic zone, P is readily incorporated into biogenic particles (2) via the photosynthetic activities of plants and begins to sink. The majority of these particles decompose in the surface layer or in shallow sediments and the P is recycled directly back into the photic zone (3) to be reincorporated into biological particles. A small portion of the particles produced in the surface layers, however, does escape the surface layers and sinks into the deep ocean. Most of these particles eventually decompose (4), and the cycle is repeated. A very small fraction of these particles, however, escapes decomposition and is incorporated into the sediments (6).

The dominant forms in which P is removed from seawater appear to be: (1) the burial of organic P compounds that escape decomposition; (2) the burial of P adsorbed onto $CaCO_3$ surfaces; and (3) the formation of apatite in continental margin sediments. To a lesser degree, the decomposition of fish debris and the adsorption of P onto iron oxyhydroxide phases also contribute to the removal of P from seawater (Froelich *et al.*, 1977, 1982, 1983).

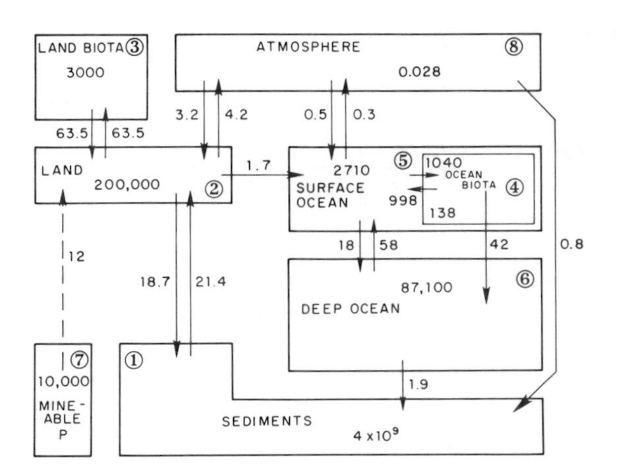

Fig. 14-7 The global phosphorus cycle. The values shown are in millions of metric tons. Reprinted from Lerman *et al.* (1975) and Graham (1977) with the permission of the Geological Society of America. See also Fig. 4-7.

14.3 The Global Phosphorus Cycle

The global P cycle proposed by Lerman *et al.* (1975) is presented in Fig. 14-7. This representation has been modified slightly to include the atmosphere (Graham, 1977). For clarity, the mass of P estimated to be contained in each of these reservoirs is listed in Table 14-3 along with the method by which it was calculated.

In choosing these reservoirs to describe the P cycle, compromises were made to maintain a general focus and global scale and yet avoid being too general and hence lose information about important transfers and reservoirs. The following is a brief discussion of the rationale behind the choice of the reservoir definitions and their estimates. For the purpose of discussion, the reservoirs have been numbered as presented in Lerman *et al.* (1975) with the addition of the atmosphere (reservoir 8).

The reservoir representing the land (2) is defined as the amount of P contained in the upper 60 cm of the soil. This rather narrow definition of the land reservoir is made because it is through the upper portions of the soil system that the major interactions with the other P reservoirs occur. Specifically, most plants receive their nutritive P needs from the upper soil horizons and the return of P to the soil system by the decomposition of plant matter is also concentrated in this upper soil zone. Similarly, the major interactions with the atmosphere, groundwaters, and rivers occur near the soil surface. And, finally, phosphate in the form of fertilizer is applied directly to the soil surface. Thus, in attempting to represent the land and its interaction with other reservoirs, the surface soil horizon most directly interacts with all components and best represents the dynamical nature of this reservoir. Phosphorus in soils deeper than 60 cm and in crustal rocks is included in the sediment reservoir (1). This reservoir accounts for all of the particulate P that exchanges with the other reservoirs only on very long timescales.

Table 14-3 The mass of P contained in each geochemical reservoir (values are given in metric tons of phosphorus)[a]

Reservoir	P content	References and comments
1. Sediments	4.00×10^{15}	Van Wazer (1961)
2. Land	2.00×10^{11}	Computed from land area of 133×10^6 km^2, assumed soil thickness of 60 cm, density of 2.5 g/cm^3 and a mean P content of 0.1% (Taylor, 1964)
3. Land biota	3.00×10^9	Computed from an estimate of the N in land biota (12×10^4 tons N; Delwiche, 1970); and a mean P : N atomic ratio in land plants (1.8 : 16; Deevey, 1970)
4. Oceanic biota	1.38×10^8	Computed from N in ocean biota (1×10^9 tons N; Vaccaro, 1965) and a mean P : N atomic ratio of 1 : 16 (Redfield *et al.*, 1963)
5. Surface ocean	2.71×10^9	Computed from assumed mean concentration of dissolved P of 25 mg/m^3, 300 m thick surface layer, and an area of 3.61×10^8 km^2
6. Deep ocean	8.71×10^{10}	Computed from assumed mean deep P concentration of 80 mg/m^3, a water depth of 3000 m, and the same surface area as above
7. Mineable P	1.00×10^{10}	Mean of values reported in Stumm (1973), Ronov and Korzina (1960), and Van Wazer (1961)
8. Atmospheric P	28 000	Graham (1977)

[a] From Lerman *et al.* (1975).

The land biota reservoir (3) represents the phosphorus contained within all living terrestrial organisms. The dominant contributors are forest ecosystems with aquatic systems contributing only a minor amount. Phosphorus contained in dead and decaying organic materials is not included in this reservoir. It is important to note that although society most directly influences and interacts with the P in lakes and rivers, these reservoirs contain little P relative to soil and land biota and are not included in this representation of the global cycle.

The ocean system is separated into three major reservoirs which best represent the dominant pools and pathways of P transport within the ocean. The surface ocean reservoir (5) is defined as the upper 300 m of the oceanic water column. As discussed in Section 14.2.2 and displayed in Fig. 14.6, the surface layer roughly corresponds to the surface mixed layer where all the photosynthetic uptake of P and the majority of the decomposition and release of P from sinking organic matter occur. Therefore, the most active exchange of P between solution and ocean biota occurs in this zone. Also, the 300-m depth roughly corresponds to the top of the main thermocline which restricts exchange between surface and deep waters, thus representing a natural boundary. Additionally, dissolved PO_4^{3-} brought to the ocean via rivers is introduced directly to the surface layer.

The oceanic biota reservoir (4) is also within the surface layers. Although organisms reside at all depths within the ocean, the overwhelming majority reside within the photic zone where phytoplankton dominate. The oceanic biota reservoir only contains 1/30 as much P as the land biota reservoir. This is primarily because oceanic biomass is composed of relatively short-lived organisms, while land biomass is dominated by massive long-lived forests.

The deep ocean (6) is the portion of the water column from 300 to 3300 m and is the largest ocean reservoir of dissolved P. However, since the deep ocean is devoid of light, this P is not significantly incorporated into ocean biota. It is simply stored in the deep waters until it is eventually transported back into the photic zone via upwelling or eddy diffusive mixing.

The sediment reservoir (1) represents all phosphorus in particulate form on the Earth's crust that is (a) not in the upper 60 cm of the soil and (b) not mineable. This includes unconsolidated marine and freshwater sediments and all sedimentary, metamorphic, and volcanic rocks. The reason for this choice of compartmentalization has already been discussed. In particulate form, P is not readily available for utilization by plants. The upper 60 cm of the soil system represents the portion of the particulate P that can be relatively quickly transported to other reservoirs or solubilized by biological uptake. The sediment reservoir, on the other hand, represents the particulate P that is transported primarily on geologic time-scales.

The atmospheric reservoir (8) represents P contained on dust particles. Because the mean residence time of dust in the air is very short, the standing stock of P in the atmosphere is relatively small. Mineable P (7) is simply an estimate of the total amount of P stored in economic deposits.

14.3.1 Fluxes between Reservoirs

A summary of the estimated fluxes between reservoirs and the methods of calculation are presented in Table 14-4. The fluxes between reservoirs are chosen to represent the principal pathways by which P is transported between reservoirs. The notation used here is the same as that presented by Lerman *et al.* (1975), with the first number representing the reservoir from which the P originates and the second number representing the receiving reservoir. It is important for the reader to understand that the evaluation of the fluxes is extremely difficult. This is, in part, caused by the introduction of P from anthropogenic sources obscuring natural levels. Also, as stated earlier, most P is present in particulate form and is not biologically active. Thus, the evaluation of the P that is actively transferred between the inorganic reservoirs and the biota or as the dissolved component in rivers must always be made in the presence of a large inert background. Small exchanges between the relatively non-reactive particulate P and reactive (primarily dissolved) P could significantly alter the flux estimate.

The transfer of P from land to terrestrial biota (F_{23}) represents the sum of terrestrial biological productivity. There is no significant gaseous form of P, nor is there a major transfer of living organisms between the freshwater–terrestrial system and the oceans. The terrestrial biota system is, therefore, essentially a closed system where the flux of P to the biota (F_{23}) is balanced by the return of P to the land from the biota (F_{32}) due to the decay of dead organic materials.

Table 14-4 Summary of the flux of P between reservoirs in metric tons/year[a]

Flux		References and comments
F_{12}	2.00×10^7	Computed from combined rates of mechanical and chemical denudation of continents (2×10^9 tons/year; Garrels and MacKenzie, 1971) and a mean P content of crustal material of 0.1% (Taylor, 1964)
F_{23}, F_{32}	6.30×10^7	Computed from Bolin's (1970) total C fixed annually ($20–30 \times 10^9$ tons) and a mean P : C atomic ratio in land biota of 1.8 : 1480 (Deevey, 1970)
F_{25}	1.70×10^6	Garrels *et al.* (1973), excluding agricultural and other anthroprogenic contributions
F_{54}	1.04×10^9	Computed from rate of N fixation of oceanic biota of 7.5×10^9 tons N/year (Vaccaro, 1965) and a mean P : N atomic ratio in ocean biota of 1 : 16 (Redfield *et al.*, 1963)
F_{45}	9.98×10^8	Computed assuming that 96% of oceanic biota recycled within upper 300 m thick layer
F_{46}	4.20×10^7	Difference between fluxes F_{54} and F_{45}
F_{56}	1.80×10^7	Computed from P content of surface layer given in Table 14-3 and a water exchange rate between surface and deep ocean of 2 m/year (Broecker, 1971)
F_{65}	5.80×10^7	Computed as for F_{56} using the P content of the deep ocean given in Table 14-3
F_{61}	1.90×10^6	Calculated assuming the ocean is in steady-state
F_{72}	1.20×10^7	Stumm (1973)
F_{21}	1.87×10^7	Computed from the difference between F_{12} and $F_{25} + F_{28}$
F_{28}	4.20×10^6	Graham (1977)
F_{82}	3.20×10^6	Graham (1977)
F_{58}	3.00×10^5	Graham (1977)
F_{85}	5.00×10^5	Graham (1977)
F_{81}	8.00×10^5	Graham (1977)

[a] From Lerman *et al.* (1975).

The transfer of P from the continents to the ocean is separated into two distinct pathways. The flux of reactive P (F_{25}) is estimated via measurements of dissolved organic and inorganic P in rivers. This P is transported directly to the surface ocean and is available for biological uptake. The other pathway by which P is transported to the oceans is that associated with particulate materials, either suspended in river waters or simply transported to the ocean as bedload. The P in these materials is considered locked in the solid structure and not available for biological uptake. In general, these particles rapidly settle to the ocean bottom and are incorporated into the sediments. This removal is rapid enough that this flux is represented as the direct transport of P from the land reservoirs to the sediments (F_{21}).

Separating "reactive" and "unreactive" P in the estuarine system is extremely difficult. Since the majority of P is associated with the particles, a small exchange between the particulate and the dissolved fraction would alter the "reactive" flux (F_{25}) significantly. Kaul and Froelich (1984) have investigated the fate of dissolved and particulate P in a small estuary. They determined that although P is readily incorporated into plants in the estuary, all of the dissolved P brought to the estuary by the river is eventually transported to the ocean. In addition, they found that ⅓ of the "reactive" P that leaves the estuary is derived from fluvial particulate matter. Thus, estimates of the flux of "reactive" P from the land to the surface ocean based on dissolved P concentration data may be 50% too low. Contradicting this conclusion are reports that estuaries are a sink for oceanic PO_4^{3-} (Jonge and Villerius, 1989). Thus, there is significant uncertainty in estimates of the exchange of PO_4^{3-} between terrestrial and oceanic systems.

A small flux is shown between the land and atmosphere. This represents the transport of dust particles to the atmosphere (F_{28}) and the deposition of these particles back on land either as dry deposition or associated with atmospheric precipitation (F_{82}). Similarly, fluxes that represent the transport of sea salt from the surface ocean to the atmosphere (F_{58}) and the deposition of soluble (F_{85}) and insoluble (F_{81}) atmospheric forms are also shown. As already discussed for the river fluxes, the insoluble particulate flux is represented as a direct transport of P to the sediment reservoir.

The natural circulation of the oceans also exchanges waters between the deep and surface ocean reservoirs. Because biological uptake constantly strips P from the surface layers, the P concentration is much less in the surface reservoir than the deep reservoir. Thus, although the continuity of water demands equal volumes of water to be exchanged between the reservoirs, far more P is carried by the upwelling waters (F_{65}) than by the downwelling waters (F_{56}).

By far the largest exchange of P between reservoirs occurs between the surface ocean and ocean biota. These large numbers attest to the tremendous primary production that occurs in the ocean. However, the relatively small standing stock also attests to the short life-span and rapid turnover of these organisms. Most of the recycling occurs in the surface waters and, hence, the flux of P into the biota (F_{54}) is nearly equaled by the release of P from the organisms (F_{45}). However, a small fraction (4%) of the P incorporated into the biota sinks out of the surface layers before being solubilized by decomposition. Although this particulate flux to the deep ocean reservoir (F_{46}) is a small fraction of the total amount incorporated into oceanic biota, ocean productivity is so great that this flux is quantitatively quite large.

Nearly all of the detrital particles sinking into the deep ocean decompose and release the associated P. A small percentage (approximately 5%), however, does survive and accumulate on the sea floor. This P is then buried in the sediments (F_{61}) and represents the ultimate removal of P from the ocean.

The basic cycle of P transport from continents to oceans to sediments is then completed by the slow geologic processes that eventually return marine sediments to the continents (F_{12}). One additional flux is considered which represents the P mined and placed on agricultural fields in the form of fertilizers (F_{72}). Notice that there are no fluxes into the reservoir marked mineable P. This is undoubtedly incorrect. However, at time-scales important to human beings, the formation of mineable P deposits is assumed to be too slow to be included. Notice also that because of this, the mineable P reservoir cannot achieve steady-state.

14.3.2 The Steady-state Cycle

The response time and average residence time for the reservoirs are listed in Table 14-5. As discussed in Chapter 4, the residence time of an element within a reservoir reflects the reactivity of that element. A short residence time suggests that removal processes or reactions are rapid and significant over short time-scales compared to the amount in the reservoir.

From Table 14-5, it is obvious that the residence time of P in the atmosphere is extremely short. This does not represent chemical reaction and removal of P from the atmosphere but rather the rapid removal of most phosphorus-containing particulate matter that enters the atmosphere.

More informative is the comparison between the residence times of P in the land and ocean biota. Although there is 16 times more biological incorporation of P in the oceans, the standing stock is only 5% of that on land. The residence time of a P atom incorporated into oceanic biota is relatively short (48 days) compared to the 50-year residence time of P in land biota. This difference represents

Table 14-5 Summary of reservoir amounts, fluxes, and residence times

Reservoir	A (Tg)	Σ flux (Tg/year)	Residence time, τ (years)
Atmosphere	0.028	4.5	0.006 (53 h)
Land biota	3000	63.5	47.2
Land	200 000	88.1–100.1	2270–1998
Surface ocean	2710	1058	2.56
Ocean biota	138	1040	0.1327 (48 days)
Deep ocean	87 100	60	1452
Sediments	4×10^9	21.4	1.87×10^8
Total ocean system	89 810	1.9	47 270

a fundamental difference in the types of organisms in the two reservoirs. Whereas oceanic biota are dominated by single-cell, short-lived planktonic plants, terrestrial biomass is dominated by forests.

The residence time of P in the deep ocean is 1400 years and is dependent primarily on the rate at which P is transported to the surface waters via upwelling. The ocean system also demonstrates an important characteristic of global reservoir models. The residence times of P in the oceanic biota, surface ocean, and deep ocean are 48 days, 2.5 years, and 1400 years, respectively. These relatively short times suggest that P is recycled rapidly throughout the ocean. However, since this cycling is almost exclusively between oceanic reservoirs and not to outside reservoirs, the residence time of P in the entire ocean system is quite long, i.e. 50 000–100 000 years. Thus, the average P atom is cycled 50 times between the deep water and surface waters before being removed to the sediments. Also, each time a P atom reaches the surface waters, it is cycled between the oceanic biota and the dissolved organic pool 25 times before being transported to the deep water. Thus, the average P atom is incorporated into the ocean biota a total of 1250 times during its stay in the ocean.

14.3.3 Perturbations

One of the main goals in establishing a global understanding of P cycling is to evaluate the influence of changing conditions on the P distribution. Perhaps the most obvious and visible mechanism by which the P cycle may be altered is through anthropogenic input. This includes P applied to the soil in the form of fertilizers as well as P used in detergents and various industrial applications. The amount of P added to the environment in this manner may be estimated from the amount of P mined each year. We have all witnessed or have read in newspapers the detrimental effects of large P inputs on some freshwater systems. The influence of this added P on the global cycle is less obvious.

Lerman *et al.* (1975) considered several cases in which mankind's activities perturbed the natural cycle. If we assume that all mined P is supplied to the land as fertilizer and that all of this P is incorporated into land biota, the mass of the land biota will increase by 20%. This amount is small relative to the P stored in the land reservoir. Since P incorporated into land biota must first decompose and be returned to the land reservoir before being transported further, there is essentially no change in the other reservoirs. Thus, although such inputs would significantly alter the freshwater–terrestrial ecosystem locally where the P release is concentrated, the global cycle would be essentially unaffected.

A greater perturbation results if one assumes that the rate at which P is mined doubles every 10 years and that the dissolved river-borne flux to the ocean increases in proportion to the mining rate. In this case, mineable P will be exhausted in 60 years. By that time, the P contained in the surface ocean will have increased by 38% and the P present in ocean biota will have increased by 30% (assuming no other limiting factors for biological production). The other reservoirs will not change significantly. Since, in this case, the mineable P is completely consumed in 60 years, the increased P input will then cease. The system will then return to the present-day levels after 150 years.

There may also be natural fluctuations within this cycle that occur over time-scales ranging from thousands of years (glacial–interglacial) to millions of years as demonstrated in Fig. 14-8. In this figure, the abundance of P in known phosphorite deposits is plotted as a function of geologic time. Notice that this is a semi-log plot, so that the amount of P stored in phosphorites in different time intervals varies by

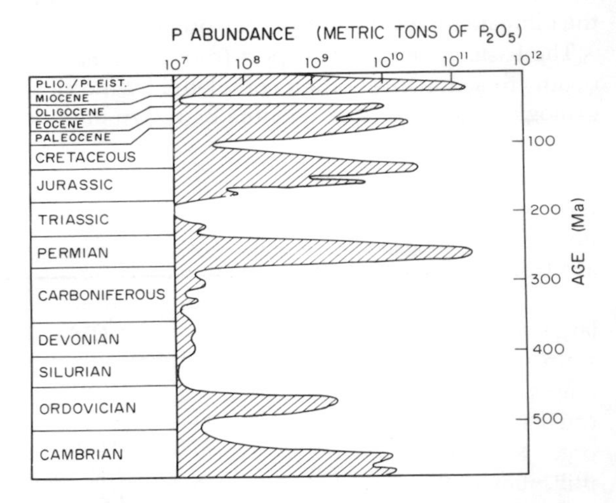

Fig. 14-8 The phosphate abundance in sedimentary deposits as a function of geologic age. Reprinted from Cook and McElhinney (1979) with the permission of The Economic Geology Publishing Co.

more than three orders of magnitude. Very dynamic processes must be involved to alter the rate at which phosphorite-P is incorporated into the geologic record to this extent. Clearly, to evaluate the influence of anthropogenic P inputs on the global cycle, the natural fluctuations must be assessed.

Because P is such an important element in biological systems, the P cycle may influence other biogeochemical cycles and processes. Perhaps one of the most important examples of such a relationship is the potential link between the P content of deep ocean water and atmospheric CO_2. Atmospheric CO_2 levels are increasing rapidly today due to the burning of fossil fuels and deforestation. Because CO_2 influences the efficiency with which solar radiation is absorbed in the atmosphere and, hence, the temperature at the Earth's surface, there is great concern as to how this increase will influence the global climate. However, recent analyses of gas pockets trapped in ice cores taken from Greenland and Antarctica suggest that there have been natural fluctuations in the concentration of atmospheric CO_2, and that during the last glacial period, atmospheric CO_2 was 80 ppmv less than the modern pre-industrial value (Neftel *et al.*, 1982). Thus, to predict the consequences of the build-up of CO_2 in the atmosphere, one must understand the natural variations. Broecker (1982) has pointed out that one mechanism by which atmospheric CO_2 may be altered is that of changing the PO_4^{3-} concentration in the deep waters of the ocean. His argument is as follows.

The present average PO_4^{3-} concentration of deep ocean water is 2.2 μmol/kg. When a parcel of deep water is transported to the photic zone, this PO_4^{3-} is completely incorporated into plants. Note that this assumes that net primary productivity is not limited by the availability of other micronutrients. In short-term laboratory studies, this assumption is clearly not true in that it has been demonstrated that the availability of fixed N limits production. Because certain organisms are capable of fixing N from the large N_2 gas pool, whereas there is no alternative source or substitute for PO_4^{3-}, it is likely that on longer time-scales, PO_4^{3-} limits productivity. Because the overall chemical composition of marine organisms is relatively constant, the complete utilization of the upwelled PO_4^{3-} also determines the amount of dissolved inorganic carbon and alkalinity that is removed from the surface waters and transported downward on sinking particles. This, in turn, alters the solution chemistry and partial pressure of CO_2 in the surface waters and, in the long term, the P_{CO_2} of the atmosphere as well. If the PO_4^{3-} concentration of the water entering the photic zone were to change, the resulting P_{CO_2} would also be affected. In fact, Broecker (1982) speculates that if the PO_4^{3-} concentration in the deep water during the last glacial period averaged 3.2 μmol/kg, the resulting atmospheric CO_2 would be 80 ppmv less than the modern pre-industrial value. Thus, natural variations in the transfer rates of P between reservoirs which would alter the abundance of P in the deep ocean reservoir may have profound effects on other geochemical cycles and climate. The elucidation of such interactions between cycles is the focus of many exciting research programs currently underway.

Questions

14-1 A farmer in the Imperial Valley has been irrigating and fertilizing her fields for many years. Because of the generally hot conditions, much of this irrigation water is lost to evaporation, increasing the salt content of the soil and groundwater. Assuming that the composition of the salt is roughly similar to that of seawater, how might this salt build-up influence the availability of P to the plants? Will the farmer need to increase the amount of fertilizer she uses?

14-2 Because of a primitive sewage system and the use of fertilizers, the citizens of a small community have added a significant amount of P to a nearby lake. Recognizing that this will stimulate biological production and may cause anoxia in the deep waters when the lake is stratified, the community decides to install a pumping system to continually exchange the deep water with the surface water. What will this do to the productivity of the lake? Will this prevent the deep water from becoming anoxic?

14-3 In recent years, many of the world's forests have been cut down and replaced with short-lived crops. What effect, if any, might this have on: (1) the P stored in the land biota reservoir, (2) the exchange rate of P between the land biota and the land reservoirs, and (3) the exchange rate between the land reservoir and the surface ocean?

References

Atlas, E. L. (1975). Phosphate equilibria in seawater and interstitial waters. Ph.D. Thesis, Oregon State University, Oregon.

Berner, R. A. (1973). Phosphate removal from sea water by adsorption on volcanic ferric oxides. *Earth Planet. Sci. Lett.* **18**, 77–86.

Bolin, B. (1970). The carbon cycle. *Sci. Amer.* **223**, 125–132.

Broecker, W. S. (1971). A kinetic model for the chemical composition of sea water. *Quat. Res.* **1**, 188–207.

Broecker, W. S. (1982). Ocean chemistry during glacial time. *Geochim. Cosmochim. Acta* **46**, 1689–1706.

Cook, P. J. and M. W. McElhinney (1979). A reevaluation of the spatial and temporal distribution of sedimentary phosphate deposits in the light of plate tectonics. *Econ. Geol.* **74**, 315–330.

Deevey, E. S., Jr. (1970). Mineral cycles. *Sci. Amer.* **223**, 149–158.

Delwiche, C. C. (1970). The nitrogen cycle. *Sci. Amer.* **223**, 137–146.

Emerson, S. and G. Widmer (1978). Early diagenesis in anaerobic lake sediments II. Equilibrium and kinetic factors controlling the formation of iron phosphate. *Geochim. Cosmochim. Acta* **42**, 1307–1316.

Froelich, P. N., M. L. Bender, and G. R. Heath (1977). Phosphorus accumulation rates in metalliferous sediments on the East Pacific Rise. *Earth Planet. Sci. Lett.* **34**, 351–359.

Froelich, P. N., M. L. Bender, N. A. Luedtke, G. R. Heath, and T. DeVries (1982). The marine phosphorus cycle. *Amer. J. Sci.* **282**, 474–511.

Froelich, P. N., K. H. Kim, R. A. Jahnke, W. C. Burnett, A. Soutar, and M. Deakin (1983). Pore water fluoride in Peru continental margin sediments: Uptake from seawater. *Geochim. Cosmochim. Acta* **47**, 1605–1612.

Froelich, P. N., M. A. Arthur, W. C. Burnett, M. Deakin, V. Hensley, R. Jahnke, L. Kaul, K.-H. Kim, K. Roe, A. Soutar, and C. Vathakanon (1988). Early diagenesis of organic matter in Peru continental margin sediments: Phosphate precipitation. *Mar. Geol.* **80**, 309–343.

Garrels, R. M. and F. T. MacKenzie (1971). "Evolution of Sedimentary Rocks." W. W. Norton, New York.

Garrels, R. M., F. T. MacKenzie, and C. A. Hunt (1973). "Chemical Cycles and the Global Environment." W. Kaufmann, Los Altos, Calif.

Graham, W. F. (1977). Atmospheric pathways of the phosphorus cycle. Ph.D. Thesis, University of Rhode Island, R.I.

Howard, P. F. (1979). Phosphate. *Econ. Geol.* **74**, 192–194.

Jackson, G. A. and P. M. Williams (1985). Importance of dissolved organic nitrogen and phosphorus in biological nutrient cycling. *Deep-Sea Res.* **32**, 223–235.

Jahnke, R. A. (1984). The synthesis and solubility of carbonate fluorapatite. *Am. J. Sci.* **284**, 58–78.

Jonge, V. N. de and L. A. Villerius (1989). Possible role of carbonate dissolution in estuarine phosphate dynamics. *Limnol. Oceanogr.* **34**, 332–340.

Kaul, L. W. and P. N. Froelich, Jr. (1984). Modeling estuarine nutrient geochemistry in a simple system. *Geochim. Cosmochim. Acta* **48**, 1417–1434.

Lal, D. and T. Lee (1988). Cosmogenic ^{32}P and ^{33}P used as tracers to study phosphorus recycling in the upper ocean. *Nature* **333**, 752–754.

Lehman, J. T. (1988). Hypolimnetic metabolism in Lake Washington: Relative effects of nutrient load and food web structure on lake productivity. *Limnol. Oceanogr.* **33**, 1334–1347.

Lerman, A., F. T. MacKenzie, and R. M. Garrels (1975). Modeling of geochemical cycles: Phosphorus as an example. *Geol. Soc. Amer. Mem.* **142**, 205–218.

Lucotte, M. and B. d'Anglejan (1988). Seasonal changes in the phosphorus–iron geochemistry of the St. Lawrence estuary. *J. Coast. Res.* **4**, 339–349.

Murray, J. W., V. Grundmanis, and W. M. Smethie, Jr. (1978). Interstitial water in the sediments of Saanich Inlet. *Geochim. Cosmochim. Acta* **42**, 1011–1026.

Neftel, A., H. Oeschger, J. Swander, B. Stauffer, and R. Zumbrunn (1982). New measurements on ice core samples to determine the CO_2 content of the atmosphere during the last 40 000 years. *Nature* **295**, 220–223.

Nriagu, J. O. (1976). Phosphate–clay mineral relations in soils and sediments. *Can. J. Earth Sci.* **13**, 717–736.

Nriagu, J. O. and P. B. Moore (1984). "Phosphate Minerals." Springer-Verlag, New York.

Orrett, K. and D. M. Karl (1987). Dissolved organic phosphorus production in surface seawater. *Limnol. Oceanogr.* **32**, 383–395.

Peng, T. H. and W. S. Broecker (1987). C : P ratios in marine detritus. *Global Biogeochem. Cycles* **1**, 155–162.

Redfield, A. C., B. H. Ketchum, and F. A. Richards (1963). The influence of organisms on the composition of seawater. *In* "The Sea" (M. N. Hill, ed.), Vol. 2, pp. 26–77. Wiley Interscience, New York.

Ronov, A. B. and G. A. Korzina (1960). Phosphorus in sedimentary rocks. *Geochemistry* **8**, 805–829.

Schindler, D. W. (1977). Evolution of phosphorus limitation in lakes. *Science* **195**, 260–262.

Sheldon, R. P. (1981). Ancient marine phosphorites. *Ann. Rev. Earth Planet. Sci.* **9**, 251–284.

Slansky, M. (1986). "Geology of Sedimentary Phosphates." Elsevier, New York.

Smith, S. V., W. J. Kimmerer, and T. W. Walsh (1986). Vertical flux and biogeochemical turnover regulate nutrient limitation of net organic production in the North Pacific Gyre. *Limnol. Oceanogr.* **31**, 161–167.

Stumm, W. (1973). The acceleration of the hydrogeochemical cycling of phosphorus. *Water Res.* **7**, 131–144.

Stumm, W. and J. J. Morgan (1981). "Aquatic Chemistry." Wiley-Interscience, New York.

Taylor, S. R. (1964). Abundance of chemical elements in the continental crust: A new table. *Geochim. Cosmochim. Acta* **28**, 1273–1285.

Vaccaro, R. F. (1965). Inorganic nitrogen in sea water. *In* "Chemical Oceanography" (J. P. Riley and G. Skirrow, eds.), Vol. 1, pp. 365–408. Academic Press, New York.

Van Cappellen, P. and R. A. Berner (1988). A mathematical

model for the early diagenesis of phosphorus and fluorine in marine sediments. *Amer. J. Sci.* **288**, 289–333.

Van Wazer, F. (ed.) (1961). "Phosphorus and Its Compounds," Vol. 2. Wiley Interscience, New York.

15

Trace Metals

Mark M. Benjamin and Bruce D. Honeyman

15.1 Introduction

Industrialized society is built upon the use of metals. Unlike many of the synthetic organic compounds used in industry, medicine, and agriculture, metals are part of natural biogeochemical cycles. Human activity influences their cycling in two interrelated ways: by altering the rate at which metals are transported among different reservoirs and by altering the form of the metals from that in which they were originally deposited.

Metals and other elements of economic interest are deposited when geochemical conditions reduce their mobility. Deposits range in quality from the nearly pure element, such as native copper, to highly disseminated deposits of marginal economic value. In addition, there are natural background levels of nearly all the elements in what constitutes the average crustal rocks: the shales, sandstones, igneous and metamorphic rocks that make up the continents and ocean floors. In the absence of human activities, elements are released to terrestrial and aquatic environments at rates corresponding to natural chemical and mechanical erosion times. Mining, construction, and large-scale changes to the natural environment alter the rate of release of elements to that part of the biogeochemical environment we call the ecosphere. These alterations, in turn, have a cascading effect on the rate at which metals are exchanged among various reservoirs in the ecosphere. In this way, the release of metals from, for example, the crustal reservoir, can affect biota in aquatic systems far from the original deposit site.

One of the characteristics of the cycle of metal mobilization and deposition is that the form of the metal is changed. This change in speciation of a metal has a profound effect on its fate. The link between metal speciation and fate is the central theme of this chapter.

This chapter differs from previous ones in that it describes the cycling of several chemical elements. These elements have many similar properties and can be considered as a group, but each also has properties that make it unique. One of the most important properties that distinguishes metals from other elements is their tendency to bond reversibly with a very large number of compounds. The availability and nature of such compounds in a system can control the transport and fate of metals. In view of the importance of these reactions, this chapter starts with a broad overview of global metal cycling, which is then interpreted in the context of the nature of metals and their chemical reactions. Finally, these concepts are applied in some detail to the natural and perturbed biogeochemical cycling of two specific metals.

15.2 Metals and Geochemistry

15.2.1 *Metal Abundance and Availability*

The average composition of the Earth's crust is essentially the composition of igneous rocks, since metamorphic and sedimentary rocks constitute a relatively insignificant portion of the total crustal mass. Eight elements – O, Si, Al, Fe, Ca, Na, K, and Mg – make up nearly 99% of the total elemental mass; the remaining elements are differentiated throughout the crust according to their particular

Global Biogeochemical Cycles
ISBN 0-12-147685-5

chemical properties. Geochemical differentiation based upon chemical affinities was first described by Goldschmidt (1954) in his proposal for a general geochemical classification scheme. In his framework, elements are considered to be siderophiles, chalcophiles, lithophiles, or atmophiles depending on their relative affinities for minerals containing iron, sulfide, or silicate, or for the atmosphere, respectively. Such a classification scheme was a significant advance in our understanding of the distribution of elements, making it possible to relate the general geochemical character of elements to their position in the Periodic Table and to fundamental chemical properties such as electronegativity and ion size.

With regard to geochemical cycling (as well as for economic considerations), it is important to distinguish between the abundance of an element and its availability. The availability of an element is related not only to its relative abundance on Earth but also the stability of minerals in which it is a major constituent. Thus, a number of elements (e.g. copper, mercury, tin, and arsenic) that are scarce in terms of their average crustal abundance are easily isolated due to their ability to form mineral deposits. The most unavailable elements are those that form no major minerals of their own. Many of the rarer elements are available for economic use only to the extent that they are obtained as byproducts of the extraction of more abundant elements. Tellurium, for example, is produced during the electrolytic refining of copper.

15.2.2 Metal Mobilization

The availability of a metal describes one aspect of its potential to cycle among biogeochemical reservoirs. The initiation of the cycling process is called mobilization. Metals may be mobilized, i.e. made available for transport away from their region of deposition, when the geochemical character of the depositional environment changes. These changes may be due to either natural or anthropogenic causes.

Natural mobilization includes chemical, mechanical, and biological weathering and volcanic activity. In chemical weathering, the elements are altered to forms that are more easily transported. For example, when basic rocks are neutralized by acidic fluids (such as rainwater acidified by absorption of CO_2), the minerals contained in them dissolve, releasing metals to aqueous solution. Several examples of the involvement of the atmosphere in the mobilization of metals follow:

$$Al(OH)_3 + 3H_2CO_3 \rightarrow Al^{3+} + 3HCO_3^- + 3H_2O$$
$$Fe(OH)_3 + 3H_2CO_3 \rightarrow Fe^{3+} + 3HCO_3^- + 3H_2O$$
$$ZnS + 2O_2 \rightarrow Zn^{2+} + SO_4^{2-}$$
$$PbCO_3 + H_2CO_3 \rightarrow Pb^{2+} + 2HCO_3^{2-}$$

Biological and volcanic activity also have roles in the natural mobilization of elements. Plants can play multiple roles in this process. Root growth breaks down rocks mechanically to expose new surfaces to chemical weathering, while chemical interactions between plants and the soil solution affect solution pH and the concentration of salts, in turn affecting the solution–mineral interactions. Plants also aid in decreasing the rate of mechanical erosion by increasing the stability of the land to erosion. These factors are discussed more fully in Chapters 6 and 7.

Volcanic activity has a significant effect on the mobilization of metals, particularly the more volatile ones, e.g. Pb, Cd, As, and Hg. The effects of volcanism are qualitatively different from those of the weathering and other near-surface mobilization processes mentioned above. Volcanism transports materials from much deeper in the crust and may inject elements into the atmospheric reservoir.

15.2.3 Human Activities as Geochemical Processes

That humans have significantly altered the biogeochemical cycles of many metals is no longer an arguable point. What is uncertain is the magnitude of the effects from these alterations, particularly in the long term. Given that the flux of energy and materials through the biosphere is self-regulating, at least within certain limits, the issue becomes one of evaluating the ability of the biosphere to assimilate anthropogenic metal inputs and predicting the rate and types of changes that will occur as a new steady-state condition is approached. These sorts of predictions are hampered not only by the state of our understanding of the complex interactions that occur between metals and the environment, but also because the determination of background metal concentrations in uncontaminated environments is so difficult. In view of the worldwide dissemination of anthropogenic materials that has already occurred, locating an uncontaminated site is extremely difficult. For example, concentrations of mercury, selenium, and sulfur above background levels have been found in arctic ice sheets. These elemental "signals" correlate with the beginning of worldwide industrialization (Weiss *et al.*, 1971a, b). In addition, there are problems associated with preparing and analyzing samples containing extremely low concentrations of metals without contaminating them.

Despite the difficulties, there have been many efforts in recent years to evaluate trace metal concentrations in natural systems and to compare trace metal release and transport rates from natural and anthropogenic sources. There is no single parameter that can summarize such comparisons. Frequently, a comparison is made between the composition of atmospheric particles and that of average crustal material to indicate whether certain elements are enriched in the atmospheric particulates. If so, some explanation is sought for the enrichment. Usually, the contribution of sea spray to the enrichment is estimated, and any enrichment unaccounted for is attributed to other natural inputs (volcanoes, low-temperature volatilization processes, etc.) or anthropogenic sources.

A second approach is to compare total mining production of a metal to an estimate of its total natural flux, making the implicit assumption that all mined materials will be released to the environment in the near future (a reasonable assumption when comparing with geologic processes).

Finally, some authors have computed metal loading to the environment from specific human activities, such as discharges of wastewater, and compared these with natural release rates. While the details of the computations and conclusions vary, the general observation for many metals is that anthropogenic contributions to metal ion transport rates and environmental burdens are approaching and in many cases have already exceeded natural contributions. A few such comparisons are provided in Tables 15-1 to 15-4.

The amount of metals released as the byproduct of a single activity, the burning of coal, illustrates the potential importance of anthropogenic sources. Inorganic, non-combustible materials are present in coal, and these materials constitute the ash that remains after combustion. Fly ash (the ash that leaves the furnace and is collected by flue-scrubbing

Table 15-1 Natural and anthropogenic sources of atmospheric emissions[a]

Element	Natural rate (Gg/year)	Anthropogenic rate (Gg/year)	Anthropogenic/ natural ratio
Al	48 900	7 200	0.15
Ti	3 500	520	0.15
Sm	4.1	1.2	0.29
Fe	27 800	10 700	0.39
Mn	605	316	0.52
Co	7	4.4	0.63
Cr	58	94	1.6
V	65	210	3.2
Ni	28	98	3.5
Sn	5.2	43	8.2
Cu	19	263	13.6
Cd	0.3	5.5	19.0
Zn	36	840	23.5
As	2.8	78	27.9
Se	0.4	14	33.9
Sb	1	38	38.0
Mo	1.1	51	44.7
Ag	0.06	5	83.3
Hg	0.04	11	27.5
Pb	5.9	2 030	34.6

[a] From Lantzy and Mackenzie (1979).

Table 15-2 Comparison between artificial and natural rates of global metal injection into the oceans and atmosphere[a]

	Input from industrial world's municipal waste water (Gg/year)	Input from combustion[b] (Gg/year)	Natural weathering[c] (Gg/year)
Cd	3	—	36
Cr	55	1.5	50
Cu	42	2.1	250
Fe	440	1400	24 000
Pb	15	3.6	110
Mn	7.4	7.0	250
Ni	17	3.7	11
Ag	2.3	0.07	11
Zn	100	7	720

[a] From Galloway (1979); [b] Bertine and Goldberg (1971); [c] Turekian (1971).

devices) is particularly enriched in metals, and in 1975 approximately 33 million tonnes of fly ash were produced in the USA (Theis and Wirth, 1977). This represents an average of 270 000 tonnes of fly ash per year for a typical 1000-MW power plant. A trace element with a concentration of 1 mg/kg (ppm) in the fly ash would be produced as waste at an average rate of 270 kg per 1000-MW plant per year. The composition of a few fly ash samples is shown in Table 15-5. As, Cd, Co, Cr, Cu, Hg, Pb, Se, V, and Zn are present in concentrations ranging from 1 to 1000 ppm(mass). Furthermore, several trace metals including As, Pb, Cd, Se, Cr, and Zn are concentrated on the surfaces of the smallest particles, which have the greatest likelihood to escape the plant and be transported significant distances in the atmosphere. When these ashes are ponded or exposed to rain, the pH of the water can decrease to less than 4 or increase to greater than 13, depending on the chemistry of the fly-ash matrix. The fraction of the total metal solubilized under these conditions is a sensitive function of pH and, although this fraction is usually 10% or less, it can exceed 50%.

Table 15-3 Calculated present-day fluxes of heavy metals into the sediments of Lake Erie[a]

Element	Anthropogenic flux ($\mu g/cm^2$ year)	Natural flux ($\mu g/cm^2$ year)
Cd		
Stn 1	0.36	0.16
Stn 2	0.02	0.02
Stn 7	0.54	0.09
Cu		
Stn 1	12.0	7.8
Stn 3	0.15	0.47
Stn 7	8.8	4.8
Pb		
Stn 1	11.8	4.3
Stn 3	0.33	0.44
Stn 7	10.9	2.4
Zn		
Stn 1	36.2	14.8
Stn 3	1.0	0.68
Stn 7	30.6	5.9

[a] From Nriagu (1979).

Once the anthropogenic release rates of metals are established, the next critical step is to evaluate their fate upon discharge to receiving waters. Sediment analyses are often useful in this regard because changes in metal concentration as a function of depth in the sediment can indicate historical trends. Also, the concentrations of metals are typically much greater in sediments than in the water column and are therefore easier to analyze and evaluate. Bruland *et al.* (1974) used metal : aluminum ratios in sediments to determine the magnitude of the anthropogenic component of the heavy metal transport rate to a Southern California basin. Assuming that aluminum has had a uniform rate of transport to

Table 15-4 Inventory of sources and sinks of heavy metals in Lake Erie[a]

Source	Flux (tonnes/year)			
	Cadmium	Copper	Lead	Zinc
Detroit River (import from Upper Lakes)	—	1640	630	5220
Tributaries, USA	—	100	52	271
Tributaries, Ontario	—	31	19	140
Sewage discharges	5.5	448	283	759
Dredged spoils	4.2	42	56	175
Atmospheric inputs	39	206	645	903
Shoreline erosion	7.9	190	221	308
Total, all sources	—	2477	1906	7776
Export, Niagara River and Welland Canal	—	1320	660	4440
Retained in sediments	—	1157	1246	3376

[a] From Nriagu (1979).

the sediments over the past century from crustal rock sources, they concluded that Pb, Cr, Cd, Zn, Cu, Ag, V, and Mo are now accumulating at higher rates than a century or more ago. For all of these metals, the anthropogenic component represented at least ⅓ of the natural emission rate, and for Pb, Ag, and Mo, the anthropogenic rate exceeded the natural rate. A qualitatively similar conclusion applies to trace metal transport into the northern portion of Chesapeake Bay (Helz, 1976).

In a subsequent section, trace metal movement through some other aquatic systems will be reviewed and analyzed in the context of specific chemical reactions. First, though, it is appropriate to summarize and review the most important chemical reactions of metal ions.

Table 15-5 Comparison of elemental concentrations in size-classified fly-ash fraction[a]

Element	Concentration (μg/g)			
	Fraction 1 (18.5 μm)[b]	Fraction 2 (6.0 μm)	Fraction 3 (3.7 μm)	Fraction 4 (2.4 μm)
Cr	28	53	64	68
Ni	25	37	43	40
Zn	68	189	301	590
Cu	56	89	107	137
Cd	0.4	1.6	2.8	4.6
Pb	73	169	226	278
As	13.7	56	87	132
Se	19	59	78	198
V	86	178	244	327

[a] From Coles *et al.* (1979); [b] mass median diameters determined by centrifugal sedimentation.

15.3 An Overview of Metal Ion Chemistry

15.3.1 Introduction

Metals, like many of the elements discussed in previous chapters, can exist in nature in several different oxidation states. When bonded to other elements, metal ions are almost always assigned a positive oxidation number and are somewhat electrophilic. Because of this, they are stabilized by association with electron-rich atoms. In particular, atoms that have a free electron pair can "donate" some of their electron density to the metal to form a bond. The most common and environmentally important donor atoms are oxygen, nitrogen, and sulfur. The bonds they form with metal ions range in strength from relatively weak associations such as those between a dissolved metal ion and water to very strong covalent bonds. These types of bonds are significant in both aqueous phase reactions and in the formation of insoluble compounds.

The characteristic affinity of metals for electron-rich donor atoms leads to an important distinction between the geochemical behavior of metal ions and that of the elements discussed earlier in this text. Specifically, metal ions in a single oxidation state can bind to donor atoms from a variety of cationic, anionic, or neutral molecules. Thus, the metals can be found in and can move through the environment as parts of molecules spanning the complete range of charge, molecular weight, bioavailability, and other chemical characteristics. In addition, since metals comprise an extremely small fraction of the composition of biological organisms, the connection between metal transport through the environment and biological activity is generally less direct and of less quantitative importance than for, say, oxygen or phosphorus. (However, recent evidence is pointing to a more significant role for biota in controlling metal concentrations and transport in the ocean.) In a relative sense, then, purely chemical reactions are of more importance for metals than for the elements discussed earlier. The strength of the chemical bonding between the donor atoms O, N and S and metals is the overriding factor controlling the geochemical cycling of metals. Once these reactions are understood, the behavior of metals, both with respect to transport and biological impacts, can be discussed in a more logical framework.

15.3.2 Oxidation–Reduction Reactions

Metals exist in nature primarily in positive oxidation states, and many form stable compounds in more than one oxidation state. The formal oxidation number of the most common form can range from +1 to +6. The stable form in a given environment depends on the oxidation potential and chemical composition of that environment. Often the stable form at the Earth's surface in the presence of molecular oxygen is different from that which is stable in anoxic sediments or waters.

Thermodynamically, virtually all metals in the elemental form are unstable with respect to redox reactions in environments where they are exposed to air and water, i.e. virtually all environments where they are used. Those metals least likely to oxidize (corrode) were long ago given the distinguished title of "noble metals". Efforts to prevent metals from corroding, and the cost of repairing and replacing metal structures that have done so runs into billions of dollars annually. Thus, one characteristic feature of society's use of metals is that the metals are continuously, albeit slowly, "degrading" to a less useful form from the moment they are put into use.

As noted previously, many metals have more than one potentially stable positive oxidation state. These different oxidation states can have dramatically different chemical properties, which in turn affec their biogeochemical forms and significance. For example, almost 4 g/L ferrous iron, Fe(II), can dissolve in distilled water maintained at pH7.0. However, if the water is exposed to air and the iron is oxidized to Fe(III), essentially all the iron will precipitate, reducing the soluble Fe concentration by over eight orders of magnitude. Oxidation state can also affect a metal ion's toxicity. For instance, the toxicity of As(III) results from its ability to inactivate enzymes, while As(V) interferes with ATP synthesis. The former is considerably more toxic to both aquatic organisms and humans.

The thermodynamically stable oxidation state of a metal in a given environment is a function of the prevailing oxidation potential. The key relationship, which is described in Chapter 5 and in most water chemistry textbooks, is known as the Nernst equation:

$$\mathrm{Eh} = \mathrm{Eh}^0 - (2.3RT/nzF) \log [\mathrm{product}]/[\mathrm{reactant}]$$

Using this relationship, one can compute the potential for each redox half-reaction that can occur in the system. If the potentials for two half-reactions are different, electrons will transfer from the reduced species at the lower potential to the oxidized species at the higher potential. The process continues until all half-reactions have the same potential. Water chemistry texts describe rapid graphical or computerized approaches to solve for the concentrations of all species once this equilibrium condition has been attained.

While these calculations provide information about the ultimate equilibrium conditions, redox reactions are often slow on human time-scales, and sometimes even on geologic time-scales. Furthermore, the reactions in natural systems are complex and may be catalyzed or inhibited by the solids or trace constituents present. There is a dearth of information on the kinetics of redox reactions in such systems, but it is clear that many chemical species commonly found in environmental samples would not be present if equilibrium were attained. Furthermore, the conditions at equilibrium depend on the concentration of other species in the system, many of which are difficult or impossible to determine analytically. Morgan and Stone (1985) have reviewed the present state of knowledge regarding the kinetics of many environmentally important reactions. They point out that a determination of whether or not an equilibrium model is appropriate in a given situation depends on the relative time constants of the chemical reactions of interest and the physical processes governing the movement of material through the system. This point is discussed in some detail in Section 15.3.8. In the absence of detailed information with which to evaluate these time constants, chemical analyses for metals in each of their oxidation states, rather than equilibrium calculations, must be conducted to evaluate the current state of a system and the biological or geochemical importance of the metals it contains.

To summarize, an evaluation of the oxidation state of metals in an environment is central to determining their probable fate and biological significance. Redox reactions can lead to orders of magnitude changes in the concentration of metals in various phases, and hence in their mode and rate of transport. While equilibrium calculations are a valuable tool for understanding the direction in which changes are likely to occur, field measurements of the concentrations of metals in their various oxidation states are always needed to evaluate metal speciation, since chemical equilibrium is often not attained in natural systems.

15.3.3 Volatilization

15.3.3.1 Volatilization from the solid state

The extent to which any chemical species is volatilized is governed by its vapor pressure, which is sensitive to temperature. Most metals and their compounds have very low vapor pressures at normal temperatures, low enough that their tendency to vaporize can be ignored. The major exceptions are metallic Hg and organometallic compounds. Nevertheless, in some environments, significant quantities of metals can be volatilized either as elements or inorganic compounds such as oxides or carbonates. The most obvious such environments are high-temperature furnaces such as in smelters or fossil fuel-burning power plants and in regions of geothermal activity or vulcanism. While the oxides, sulfates, carbonates, and sulfides of a metal all have somewhat different volatilities, the most volatile metals, regardless of the anion with which they are associated, are Hg, As, Cd, Pb, and Zn. Metallic Hg and organometallic compounds may be transported significant distances as gases and eventually be removed from the atmosphere by dissolution in rain droplets. However, most volatilized metals condense rapidly as they cool and fall to the surface associated with particulate matter, either as dry deposition or scavenged by precipitation. Some of the condensed particles are light enough to be carried long distances, and those that fall to Earth may be resuspended by wind action or washed into a water body by surface run-off. Particles produced by high-temperature combustion processes are mostly in the $<2\ \mu m$ size range and typically have atmospheric residence times of 7–14 days, whereas those generated by soil erosion are larger ($>5\ \mu m$) (Hardy *et al.*, 1985). Anthropogenic particles are typically enriched in trace metals (normalized to the concentration of Al) by a factor of 100–10 000 compared with atmospheric particulates generated by natural erosion and wind action. As an indication of the importance of volatilization of metals to their overall biogeochemical budgets, Galloway (1979) estimated that volatization of As, Hg, and Se overwhelms total dust and volcanic emanation rates by factors of 7.5, 625, and 7.3, respectively.

15.3.3.2 Volatilization from solution

The equilibrium volatility of a species dissolved in water is characterized by its equilibrium constant for the reaction $X(g) \rightarrow X(aq)$, i.e. its Henry's Law constant. These equilibrium constants are related, but are not directly comparable to vapor pressures of the corresponding dry solids because of the effect of the solvent, water. [The most appropriate direct comparison involves the calculation of fugacities. The fugacity, or escaping tendency of a chemical species from the environment in which it exists, depends upon both the concentration of the chemical species of interest and the strength of its interactions with the surrounding (solvent) molecules. Details of the calculation are provided in chemical thermodynamics textbooks and in a number of articles by Mackay and co-workers (e.g. Mackay, 1979).] Suffice it to say that when volatile metal species such as methylmercury are present in water, there virtually always exists a driving force to strip them out of the aqueous phase, since their partial pressures in air are essentially zero. Thus, equilibrium between gaseous and aqueous phases is rarely a limiting factor. For volatilization of these species, the limiting factor is usually the rate at which they can move through the water column to and across the water–air interface.

15.3.3.3 *Volatilization of mercury and lead*

Mercury and lead are two metals of environmental significance for which volatilization is of dominant importance. Lead is used as a gasoline additive. This situation is unusual in that lead is added to gasoline as an organo-metal compound, tetraethyl lead. In internal combustion engines, most of the lead is converted to inorganic forms and released in the exhaust. The widespread use of this compound as an anti-knock agent in gasoline engines has led to its dispersion everywhere automobiles travel, and from there to very remote locations. As with metals near smelters, lead concentrations in the soil near major roadways decrease rapidly with distance from the source, but significant amounts of the metal are transported by wind, either directly after being emitted or by resuspension after a period of deposition.

The ratio of anthropogenic emissions to total natural emissions is highest for the atmophilic elements Sn, Cu, Cd, Zn, As, Se, Mo, Hg, and Pb (Lantzy and Mackenzie, 1979). In the case of lead, atmospheric concentrations are primarily the consequence of leaded gasoline combustion. Atmospheric fluxes of lead in the USA rose steadily from the first decades of this century, reaching a maximum in the early 1970s (see Eiseneich *et al.*, 1986 and references therein). Passage of the Clean Air Act of 1972 and its subsequent amendments resulted in dramatic reductions in atmospheric lead concentrations, although lead fluxes worldwide still remain 10–1000 times above background levels (Settle *et al.*, 1982; Settle and Patterson, 1982).

The second metal for which volatilization is a dominant transport mode is mercury, which is the most volatile metal in its elemental state. As with lead, a key reaction that can increase the volatility of mercury is formation of an organometallic compound. In this case, the reactions take place in water and are primarily biological, being mediated by bacteria commonly found in the upper levels of sediments. These reactions and their importance in the global mercury cycle are discussed in some detail later in the chapter.

To summarize, metals can be transferred into the gas phase in high-temperature processes either in their elemental form or as inorganic compounds, and these compounds can then be transported long distances as gases or in other physical/chemical forms. Many organometallic compounds are also volatile. In a few cases, natural organometallic compounds may be formed that are volatile. These compounds are formed by microorganisms in mildly reducing aquatic environments and are then transported to the surface and across the air–water interface to enter the gas phase. Methylmercury compounds are probably the most important and certainly have been the most widely studied of these because of their central role in the bioaccumulation of mercury; others, such as methylarsine and organotin compounds, are environmentally important as well. Anthropogenic release to the atmosphere overwhelms natural sources for Hg, Cd, Cu, Ag, Zn, Pb, As, and Se, and possibly other metals.

15.3.4 Complexation Reactions

The stability of liquid water is due in large part to the ability of water molecules to form hydrogen bonds with one another. Such bonds tend to stabilize the molecules in a pattern where the hydrogens of one water molecule are adjacent to oxygens of other water molecules. When chemical species dissolve, they must insert themselves into this matrix, and in the process break some of the bonds that exist between the water molecules. If a substance can form strong bonds with water, its dissolution will be thermodynamically favored, i.e. it will be highly soluble. Similarly, dissolution of a molecule that breaks water-to-water bonds and replaces these with weaker water-to-solute bonds will be energetically unfavorable, i.e. it will be relatively insoluble. These principles are presented schematically in Fig. 15-1.

Metal ions form strong bonds with water molecules. By orienting themselves in such a way that the metal "faces" an oxygen atom, the negative charge on the oxygen is partially distributed onto the metal, forming the analog of a strong hydrogen bond. Similarly, some of the charge on the metal ion is neutralized (recall the electrophilic nature of metal ions). These bonds are strong enough that, in most aqueous environments, most metal ions are surrounded by an "inner hydration sphere" of 4–8 strongly bound water molecules, as well as a loosely attached "outer hydration sphere" of variable size. Although chemical convention represents dissolved metal ions as Me^{n+}, a more accurate designation is $Me(H_2O)_x^{n+}$. The strength of the metal-to-water bond increases with decreasing size and increasing charge of the metal ion, and also depends on the distribution of electrons around it.

The water molecules in the inner hydration sphere can undergo dissociation reactions just as water molecules far from a dissolved metal ion do, but the presence of the metal ion changes the equilibrium constant for this reaction. The magnitude of this alteration is directly related to the strength of the metal-to-water bond. If the metal attracts a large portion of the electron density from the oxygen, it weakens the oxygen-to-hydrogen bonds and enhances the tendency for the water molecule to dissociate. This leads to dissolved species that are typically designated hydrolyzed metal ions, $Me(OH)_y^{n-y}$, but which are more accurately portrayed as $Me(H_2O)_x(OH)_y^{n-y}$. As an extreme example, we can think of a metal like molybdenum bonding so strongly to four oxygen atoms that they lose their ability to bond to any hydrogen at all, forming MoO_4^{3-}. In such a situation, the oxygens are not considered part of the hydration sphere at all, but as covalently bonded to the molybdenum. Nevertheless, in a qualitative sense, this reaction is no different from the hydrolysis reactions described above. Comparison of a metal hydrolysis reaction with hydrolysis of pure water indicates that if the equilibrium constant for metal hydrolysis is larger than 10^{-14} (the dissociation constant of pure water), the metal acts as an acid, i.e. it increases the tendency for hydrogen ions to be released to solution (Fig. 15-2b). Furthermore, since metal ions are surrounded by several waters of hydration, they can act as multi-protic acids, releasing four, five, or even six hydrogen ions to solution and acquiring a net negative charge in the process.

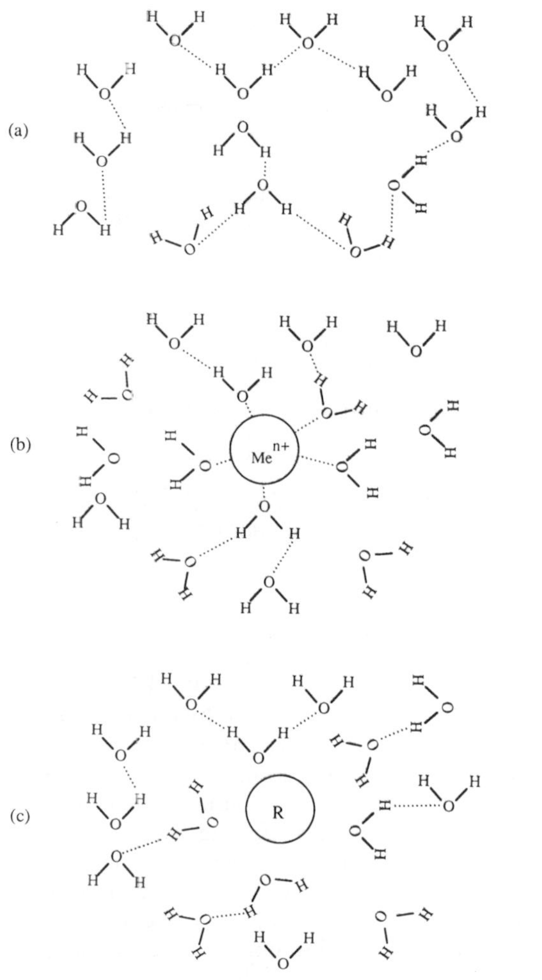

Fig. 15-1 Schematic representation of the change in water structure (water molecule orientation) due to the presence of a charged (hydrophilic) solute. (a) Pure water; (b) a solute forming strong bonds with water (dissolution favorable); (c) a solute forming weak bonds with water (dissolution unfavorable).

The comparisons between pure water and a solution containing dissolved metal ions can be extended by considering the behavior of other

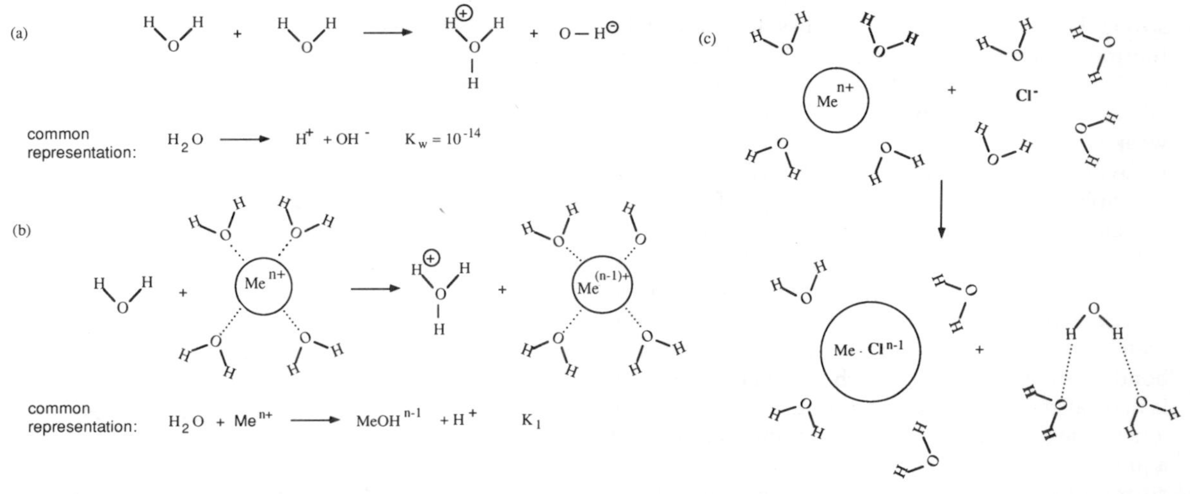

(a)

common representation: $H_2O \longrightarrow H^+ + OH^-$ $K_w = 10^{-14}$

(b)

common representation: $H_2O + Me^{n+} \longrightarrow MeOH^{n-1} + H^+$ K_1

(c)

Fig. 15-2 Comparison of water dissociation in bulk solution (a) and in the hydration sphere of a metal ion (b). Exchange of water of hydration for a chloride ion (c) forms the Me–Cl complex (from Manahan, 1979).

dissolved ions such as Cl^-, SO_4^{2-}, HCO_3^-, S^{2-}, and dissolved organic molecules. Just as metal ions do, these ions tend to orient themselves in a way that maximizes the bond strength between them and other constituents of the solution. For instance, the water molecules surrounding a chloride ion are somewhat structured by the negative charge of the chloride, so that the positive (hydrogen) ends of the water molecule are closer to the ion than is the oxygen. Typically, this tendency to attract and orient a hydration sphere is much weaker for anions than for metals. If a metal ion is dissolved in the same solution as the chloride, then an additional possibility presents itself. The chloride ion could replace one of the waters of hydration surrounding the metal, thereby exchanging a water–metal bond and a water–chloride bond for a metal–chloride bond and a water–water bond (Fig. 15-2c). Equilibrium constants for such reactions are called stability constants, and are tabulated in a number of sources. Values for these constants can range considerably. Very large values imply that metals and ligands are virtually certain to form complexes, even if both ions are present in very low concentrations.

The exchange of water molecules in the hydration sphere for other dissolved species can be extended to include "mixed ligand complexes", i.e. those in which water molecules have been replaced by two or more different types of ligands, and "multi-dentate" complexes, those in which a single molecule binds to the metal through more than one atom, and therefore causes more than one water of hydration to be released (Fig. 15-3). As might be expected, the metal-ligand binding strength typically decreases as ligands are added, and the bond strength of a multi-dentate ligand is typically greater than that of mono-dentate ligands, as a result of the multiple bonds formed. When a ligand forms a strong multi-dentate complex, it is referred to as a chelating agent, and the complex is called a chelate.

It is important to recognize that although many ligands are anions and hence have an electrostatic attraction for the metal ion, this is not a requirement. For instance, the neutral ammonia molecule (NH_3) is a strong complexing agent for many metals. If electrostatic attraction were critical to formation of these types of bonds, then complexation would cease once the positive charge on the metal was neutralized. However, complexation can continue until several of the molecules in the hydration sphere are replaced, yielding a chemical species consisting of a positively charged central metal ion surrounded by as many as six negatively charged ligands, and carrying a net negative charge of -3 or -4.

Complexation reactions are of crucial importance in the biogeochemical cycling of metals because a large fraction of the total dissolved metal may be complexed. In addition, biological effects can be

(a)

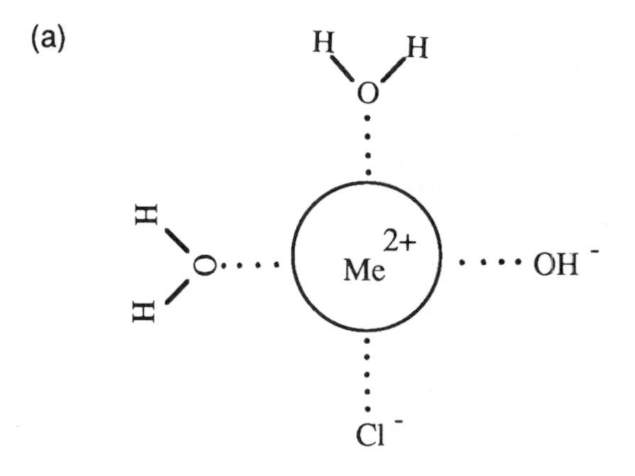

(b)

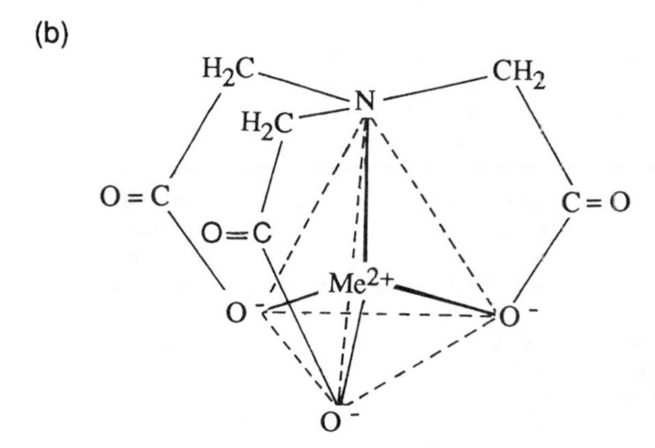

Fig. 15-3 Mixed ligand and multi-dentate complexes: (a) A hypothetical $Me(OH)Cl^0$ complex; (b) nitrilotriacetate chelate of a divalent metal ion in a tetrahedral configuration (from Manahan, 1979).

sensitive to the types of complexes present. In many cases, free metal ions (fully hydrated metal ions) are more toxic than complexed ones, especially those complexed by multi-dentate organic ligands. There is also evidence that some organisms secrete chelating agents in response to high metal concentrations in their environment, presumably as a defense against metal ion toxicity. On the other hand, organisms can also produce chelating agents to acquire metals that are necessary for certain metabolic functions. These chelating agents are often extremely specific for a given metal and are used to "collect" metals from solution or maintain a desired concentration of metals inside the cell.

In addition to those complexing agents produced by organisms to regulate their environment, numerous complexing and chelating agents are present in aquatic systems as a result of the normal exchange of metabolites between organisms and their environment, cell death and decay, or as a result of anthropogenic activities. The high molecular weight compounds known as fulvic and humic acids are particularly important. These acidic polymers are primarily the residue from organic decay processes and do not have a well-defined chemical structure. However, like synthetic chelating agents, they contain high concentrations of electron-rich donor atoms, especially oxygen in carboxylic and phenolic groups, which can complex or chelate metal ions. Sulfur- and nitrogen-containing portions of the molecules may also be important in complexing such metals. These molecules are especially effective chelators since they are large and flexible, and have many complexing sites on each molecule. At least in theory, they may be able to arrange themselves in such a way that several donor groups are at the optimal positions to form very strong bonds.

The speciation (ignoring organic complexes) of selected metals in various aquatic environments is described in Table 15-6. The relative strengths of the complexes of a given metal, and differences among metals in the strength of binding to a given ligand, are apparent. Also apparent is the fact that the dominant form of a given metal often changes when the local environment changes pH or ionic strength (as in an estuary). Important changes also occur when the environment changes from aerobic to anaerobic (for instance, in the bottom of a seasonally anoxic lake), or when the concentration of dissolved organic matter changes, such as near a sewage outfall.

Summarizing this section, metal ions may be present in aqueous solution surrounded by and bonded to water molecules or a wide variety of inorganic and organic complexing agents. The bonds formed by complexation are often reversible, and their strength varies widely depending on the chemical and physical properties of the metal and ligand. As a result of complexation reactions, concentrations of total dissolved metal in an aquatic system can be orders of magnitude higher than those of free metal ions. Thus, if the concentration of free metal ion in a system is limited by virtue of low solubility, complexes can still cause large amounts of that metal to dissolve and be transported among the various geochemical reservoirs.

Table 15-6 Model results for metal speciation in natural waters[a,b]

	Freshwater			Seawater	
	Inorganic (pH 6)	Inorganic (pH 9)	Inorganic and organic (pH 7)	Inorganic (pH 8.2)	Inorganic and organic (pH 8.2)
Ag^{+}	72, Cl	65, Cl, CO_3	65, Cl	<1, Cl	<1, Cl
Al^{3+}	<1, OH, F	<1, OH		<1, OH	
Cd^{2+}	96, Cl, SO_4	47, CO_3, OH	87, org, SO_4	3, Cl	1, Cl
Co^{2+}	98, SO_4	20, CO_3, OH		58, Cl, CO_3, SO_4	63, Cl, SO_4
Cr^{3+}	<1, OH	<1, OH		<1, OH	
Cu^{2+}	93, CO_3, SO_4	<1, CO_3, OH	<1, org	9, CO_3, OH, Cl	<1, org, CO_3
Fe^{2+}	99	27, CO_3, OH		69, Cl, CO_3, SO_4	
Fe^{3+}	<1, OH	<1, OH	<1, org, OH	<1, OH	<1, OH, org
Hg^{2+}	<1, Cl, OH	<1, OH^{-}		<1, Cl	
Mn^{2+}	98, SO_4	62, CO_3	91, SO_4	58, Cl, SO_4	25, Cl, SO_4
Ni^{2+}	98, SO_4	9, CO_3		47, Cl, CO_3, SO_4	50, org, Cl, SO_4
Pb^{2+}	86, CO_3, SO_4	<1, CO_3, OH	9, CO_3, org	3, Cl, CO_3, OH	2, CO_3, OH
Zn^{2+}	98, SO_4	6, OH, CO_3	95, SO_4, org	46, Cl, OH, SO_4	25, OH, Cl, org

[a] Data for inorganic freshwater and inorganic seawater from Turner *et al.* (1981). Data for systems with inorganics and organics from Stumm and Morgan (1981). Six organic ligands are included corresponding to 2.3 mg/L total soluble organic carbon. Stability constants and inorganic composition of model water were not identical in the two studies, and so comparisons are qualitatively valid but may have minor quantitative inconsistencies.

[b] Each entry has the % of total metal present as the free hydrated ion, then the ligands forming complexes, in decreasing order of expected concentration. For instance, in inorganic freshwater at pH 9, Ag is present as the free aquo ion (65%), chloro-complexes (25%), and carbonato-complexes (9%).

15.3.5 Precipitation Reactions

If metal concentrations in solution become large, soluble complexes that contain more than one metal ion can form. Eventually, these can grow so large that a three-dimensional network is created in which most of the ions in the interior are not in contact with the bulk solution, and a separate phase is formed. In systems where large soluble polymeric metal complexes are stable, the exact point at which the new phase forms is open to question and is somewhat a matter of definition. In most practical situations, however, these large polymers have very limited stability and the system undergoes a dramatic change from a completely soluble state containing monomeric and relatively small polymeric species to one containing an identifiable second phase. The activation energy required to form this new phase is large, and solutions must be highly supersaturated for precipitation to be initiated in particle-free systems. On the other hand, the activation energy for growth of solids once a solid–liquid boundary has been established is much smaller, so once precipitation has started it can often proceed rapidly. In natural aquatic systems, the problem of high activation energies for formation of new phases is usually circumvented by new solids forming on the surfaces of existing solids with a compatible crystal structure.

Solubility equilibria are described quantitatively by the equilibrium constant for solid dissolution, K_{SP} (the solubility product). Formally, this equilibrium constant should be written as the activity of the products divided by that of the reactants, including the solid. However, since the activity of any pure solid is defined as 1.0, the solid is commonly left out of the equilibrium constant expression. The activity of the solid is important in natural systems where the solids are frequently not pure, but are mixtures. In such a case, the activity of a solid component that forms part of an "ideal" solid solution is defined as its mole fraction in the solid phase. Empirically, it appears that most solid solutions are far from ideal, with the dilute component having an activity considerably greater than its mole fraction. Nevertheless, the point remains that not all solid components found in an aquatic system have unit activity, and thus their solubility will be less than that defined by the solubility constant in its conventional form.

Metal precipitates of biogeochemical significance are primarily oxides, hydroxides, carbonates, and sulfides. Not surprisingly, these are also the ligands with which many metals form strong complexes. However, the correspondence between the strength of bonding in a solid phase and that in a soluble complex is far from exact, because of the effects of the bonds with water in the latter cases. All of these anions have concentrations that are strongly dependent on pH. Because of this, a change in pH is one of the most important driving forces for precipitation or dissolution reactions. The other dominant factor is the oxidation–reduction potential of the solution. Metal sulfides tend to be extremely insoluble, but they can only exist in environments where the redox potential is sufficiently reducing for sulfur to exist in the −II oxidation state (H_2S, HS^-, or S^{2-}). Thus, some metals are soluble in the oxidized layers of sediments but are precipitated in lower anoxic layers, where sulfate (SO_4^{2-}) is reduced to sulfide.

Model calculations that illustrate these points have been performed by Morel *et al.* (1975) for the speciation of a sewage–seawater mixture. They concluded that sulfides of Cu, Cd, Pb, Zn, Ag, Hg, and Co would be the predominant form of those metals in the sewage, along with oxidized precipitates $Cr(OH)_3$ and Fe_2O_3, cyano-complexes of Ni^{2+}, and free Mn^{2+}. However, as dilution and oxidation of the sewage by seawater proceed and the S(−II) concentration decreases, the sulfides dissolve and all the metals except Fe eventually enter solution. Their calculations show that $CoCO_3(s)$, $ZnCO_3(s)$, and CuO(s) may form as intermediates between the initial (sulfidic solid) and final (soluble) end-states.

In addition to effects on the concentration of anions, the redox potential can affect the oxidation state and solubility of the metal ion directly. The most important examples of this are the dissolution of iron and manganese under reducing conditions. The oxidized forms of these elements [Fe(III) and Mn(IV)] form very insoluble oxides and hydroxides, while the reduced forms [Fe(II) and Mn(II)] are orders of magnitude more soluble [in the absence of S(−II)]. The oxidation or reduction of the metals, which can occur fairly rapidly at oxic–anoxic interfaces, has an important "domino" effect on the distribution of many other metals in the system due to the importance of iron and manganese oxides in adsorption reactions. In an interesting example of this, it has been suggested that arsenate accumulates in the upper, oxidized layers of some sediments by diffusion of As(III), Fe(II), and Mn(II) from the deeper, reduced zones. In the aerobic zone, Fe and Mn are oxidized by oxygen, and precipitate. The solids can then oxidize As(III) to As(V), which is subsequently immobilized by sorption onto other Fe or Mn oxyhydroxide particles (Takamutsu *et al.*, 1985).

While the solubility constants for various potential solids can indicate which solid is thermodynamically stable under a given set of conditions, reactions involving precipitation or dissolution of a solid are typically more subject to kinetic limitations than are reactions that take place strictly in solution. As a result, a thermodynamically unstable solid (often referred to as "metastable") may form and remain in place for geologic time periods. Such is frequently the case when an insoluble but relatively less stable solid phase precipitates, and the thermodynamically stable phase has a different crystal structure. Often the only way for the more stable solid to form is by dissolution of ions from the first solid and reprecipitation of the second. Since the first solid is quite insoluble itself, dissolution occurs very slowly and since this causes the solution to be only very slightly supersaturated with respect to the second solid, the driving force for its formation is also small. Thus, for example, goethite (α-FeOOH) is more stable than lepidocrocite (γ-FeOOH), but lepidocrocite forms more rapidly than goethite when iron sulfide is oxidized by oxygen, and lepidocrocite can be found in geological formations long after the oxidation has taken place.

An important result of the concepts discussed in this section and the preceding one is that precipitation and complexation reactions exert joint control over metal ion solubility and transport. Whereas precipitation can limit the dissolved concentration of a specific species (Me^{n+}), complexation reactions can allow the total dissolved concentration of that metal to be much higher. The balance between these two competing processes, taking into account kinetic and equilibrium effects, often determines how much metal is transported in solution between two sites.

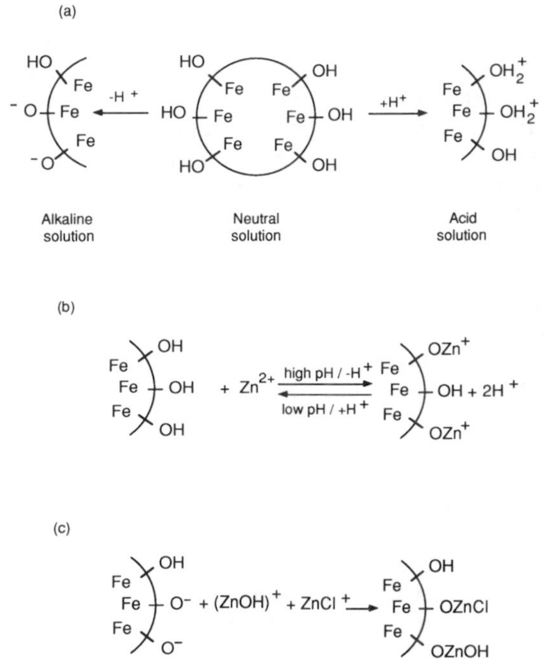

Fig. 15-4 Analogy between dissolved ligands and adsorbents (surface-bound ligands): (a) Surface acid–base reactions; (b) surface complexation of free metals; (c) formation of "mixed-ligand" surface complexes.

15.3.6 Adsorption

The importance of one other type of reaction that metal ions undergo has been recognized and studied extensively in the past 30 years. This reaction is adsorption, in which metal ions bind to the surface of particulate matter and are thereby transported as part of a solid phase even though they do not form an identifiable precipitate. Conceptually, these reactions can be thought of as hybrids between complexation and precipitation reactions. Most studies of these reactions have used metal oxides or hydroxides as the solid (adsorbent) phase, and they will be discussed in that context here, although sorption onto other solids may be important as well.

At a solid–solution interface, atoms are fixed in place by their attachment to the solid, but are freer to move and react with solution components than are other atoms in the solid. Put another way, these atoms are half in solution and half out. To the extent that their bonding requirements are not completely met by the bond to the bulk solid, they are energetically driven to react with solution components. In the case of a ferric oxide, for instance, an oxygen atom may be attached to an iron atom on the solid side, and may bind to a hydrogen ion on the solution side, to form the analog of an FeOH complex. If we treat the FeO^- surface group as a reactant, we can write an equilibrium reaction describing this interaction. Furthermore, as shown in Fig. 15-4a, we can postulate that the surface group might react with a second hydrogen ion to form the analog of a hydrated Fe ion. The result is that each surface Fe atom can behave as a diprotic acid, binding or releasing hydrogen ions in response to changes in solution pH. A second result is that the surface acquires a net charge, which may be positive or negative depending on the pH and the affinity of the surface for hydrogen ions. This charge is neutralized by ions of opposite charge accumulating in solution near the surface. This structure of the oxide–solution interface – a net surface charge balanced by a swarm of ions in the solution near the surface – is called the electrical double layer. This effect is also discussed in Chapter 8.

Consider the situation of a different metal ion, say Zn^{2+}, dissolved in a solution containing FeOOH. Just as in the cases described earlier, the zinc ion may appear as a free aquo ion, as a hydroxo- or other complex, or in the present case as a complex with the surface oxide site. The empirical evidence is that in many cases the most stable arrangement is for the Zn or other dissolved metal ion to replace hydrogen ions at the surface and hence be preferentially adsorbed (Fig. 15-4b). As would be expected, since the reaction is effectively a competition between the metal and hydrogen ions to bind to the oxide, metal ion adsorption is enhanced at high pH and diminished at low pH (Fig. 15-5a). This conceptual model for adsorption is also consistent with the observation that the adsorptive bond strength of

Fig. 15-5 Comparative adsorption of several metals onto amorphous iron oxyhydroxide systems containing 10^{-3} M Fe_T and 0.1 M $NaNO_3$. (a) Effect of solution pH on sorption of uncomplexed metals. (b) Comparison of binding constants for formation of soluble Me–OH complexes and formation of surface Me–O–Si complexes (i.e. sorption onto SiO_2 particles). (c) Effect of solution pH on sorption of oxyanionic metals. (a) and (c) adapted from Manzione and Merrill (1989) with the permission of the Electric Power Research Institute; (b) adapted from Balistrieri *et al.* (1981) with the permission of Pergamon Press.

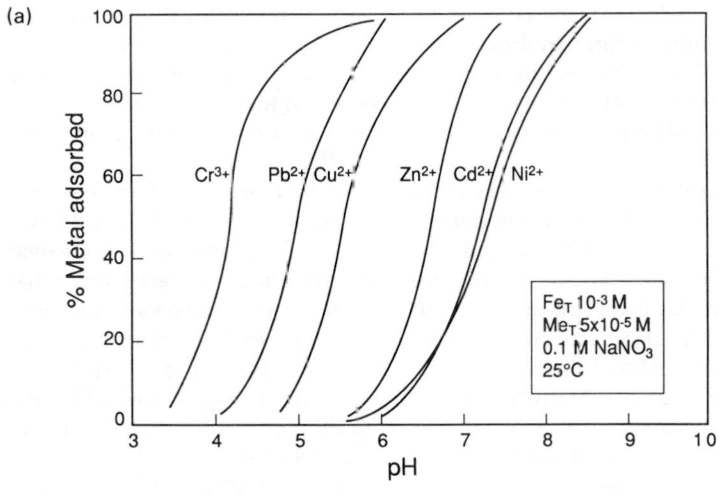
(a)
% Metal adsorbed
pH
Cr3+
Pb2+
Cu2+
Zn2+
Cd2+
Ni2+
FeT 10-3 M
MeT 5x10-5 M
0.1 M NaNO3
25°C

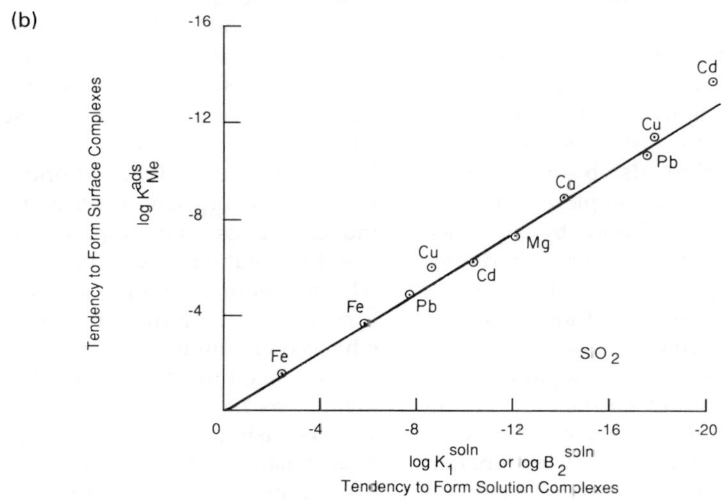
(b)
Tendency to Form Surface Complexes
log K ads Me
log K1 soln or log B2 soln
Tendency to Form Solution Complexes
Cd
Cu
Pb
Ca
Mg
Cu
Cd
Fe
Pb
Fe
SiO2

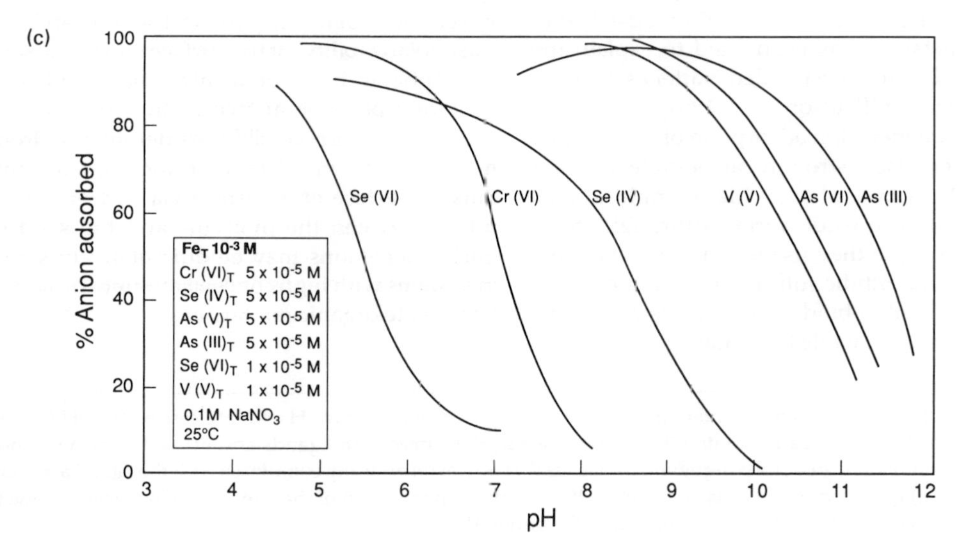
(c)
% Anion adsorbed
pH
Se (VI)
Cr (VI)
Se (IV)
V (V)
As (VI)
As (III)
FeT 10-3 M
Cr (VI)T 5 x 10-5 M
Se (IV)T 5 x 10-5 M
As (V)T 5 x 10-5 M
As (III)T 5 x 10-5 M
Se (VI)T 1 x 10-5 M
V (V)T 1 x 10-5 M
0.1M NaNO3
25°C

most metal ions onto oxide surfaces is strongly correlated with their ability to form hydroxo complexes in solution (Fig. 15-5b). That is, those metals that bind strongly to the oxygen atom of water molecules in solution also bond strongly to surface-bound oxygen atoms (e.g. Fe(III)).

As with the complexation reactions described earlier involving only dissolved species, the electrostatic attraction between a negatively charged surface and a positively charged metal ion can enhance the attraction of these reactants for one another, but much of the driving force for the reaction is provided by the formation of a chemical bond between the surface and the metal. Because of this, many metals can bind even to a positively charged surface, and in other cases can cause the surface charge to change sign from negative to positive as a result of their adsorption.

Dissolved metal complexes such as $MeCl_x^{n-x}$ or $Me(HCO_3)_y^{n-y}$ are also potential adsorbing species and can form the analog of mixed ligand complexes at the surface (Fig. 15-4c). In most cases, these simple inorganic complexes tend to adsorb somewhat less strongly than aquo or hydroxo complexes. This may be partially accounted for simply by statisitical factors: when one of the metal's waters of hydration is replaced by another ligand, there are fewer waters of hydration that can be exchanged for the surface oxide ligand. The bond strength of the complex-to-surface bond may be less than the aquo metal-to-surface bond, since the ligand satisfies some of the bonding requirements of the metal ion. In any case, complexation by simple inorganic ligands generally acts to retain metal ions in solution at the expense of metal ion adsorption. Formation of Cd–Cl complexes, for instance, has been cited to explain the desorption of Cd from suspended matter as it passes through an estuary (Paulson *et al.*, 1984).

The effects on metal ion adsorption of ligands that can themselves adsorb strongly can be quite different from that described above. Many multi-atomic ligands can bond to oxide surfaces through atoms different from those they use to bind to metal ions. For example, the sulfidic sulfur of thiosulfate ($—S—S—O_3^{2-}$) is thought to bind to polarizable metal ions such as Ag^+ or Cd^{2+}, while the sulfate group is more likely to bond to an oxide surface site. In such a case, the ligand may simultaneously bind to both the surface and the metal, thereby acting as a bridge between them. To explain the interactions in such a system, it is necessary to discuss ligand adsorption briefly.

The adsorption of anions, including metalloids such as SeO_4^{2-}, MoO_4^{2-}, and CrO_4^{2-}, anionic ligands such as PO_4^{3-} and SO_4^{2-}, and fulvic acids, is similar to that of cationic metals, except that it usually has the inverse dependence on solution pH. That is, the sorption reaction involves competition with OH^- ions for the surface, and ligand adsorption is therefore strong at low pH, where competition is weak (Fig. 15.5c). The complete set of equilibria describing the interactions in a system containing dissolved metal and ligand molecules and a solid that can absorb either of these is complicated, but conceptually involves a balance among all the relatively simple individual reactions discussed thus far. These interactions are summarized schematically in Fig. 15-4. Low pH favors adsorption of the free ligand and ligand-bridged metals. High pH favors adsorption of free metal ions and metal-bridged ligands. The complexation of the metal and ligand in solution may or may not be a function of pH. The net interaction of all these factors can lead to solutions where the adsorption of metal ions is enhanced or diminished compared to that in ligand-free systems. Interesting examples of some of these interactions have been provided by Davis and coworkers, using both synthetic ligands and natural organic matter as the complexing agents (Fig. 15-6).

In addition to the interactions discussed above, which all depend in part on the ionizability, or at least polarizability, of the surface and the adsorbates, hydrophobic parts of ligands may bind to corresponding parts of surfaces. Thus, if a metal ion is complexed or irreversibly bonded to a hydrophobic molecule, the metal may be incorporated into the bulk or surface of a particle via hydrophobic interaction between the molecule and the solid phase. Such interactions may be quantitatively significant in systems with high concentrations of dissolved and particulate organic matter.

Fig. 15-6 Adsorbing, complexing ligands can enhance sorption of metals at low pH and interfere with it at high pH. This is shown (a) in a strictly inorganic solution (line shows adsorption curve with ligands absent) and (b) in a solution with natural organic matter as the complexing agent. (c) Metals that do not form strong complexes with the ligand are unaffected by its presence. (a) modified from Davis and Leckie (1978) with the permission of the American Chemical Society; (b) and (c) adapted from Davis (1984) with the permission of Pergamon Press.

(a)

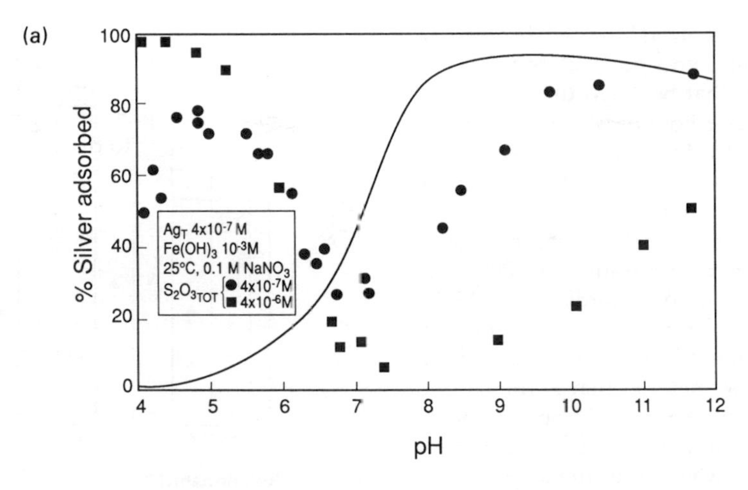
% Silver adsorbed
pH
Ag_T 4x10^{-7} M
$Fe(OH)_3$ 10^{-3}M
25°C, 0.1 M $NaNO_3$
S_2O_{3TOT} 4x10^{-7}M 4x10^{-6}M

(b)

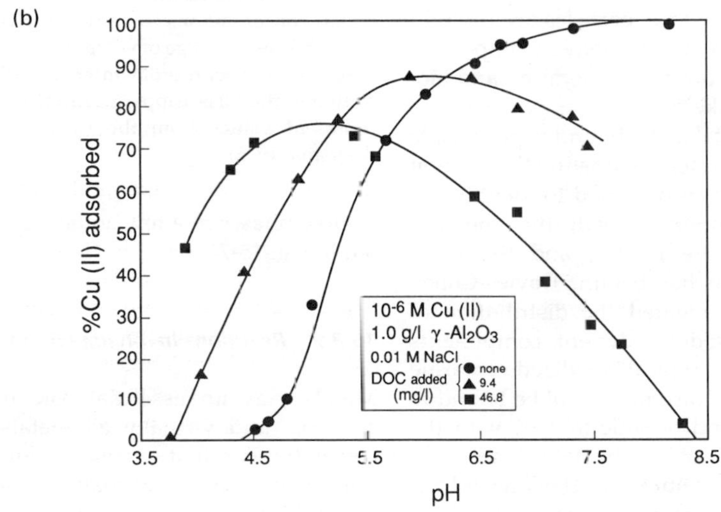
%Cu (II) adsorbed
pH
10^{-6} M Cu (II)
1.0 g/l γ-Al_2O_3
0.01 M NaCl
DOC added (mg/l) none 9.4 46.8

(c)

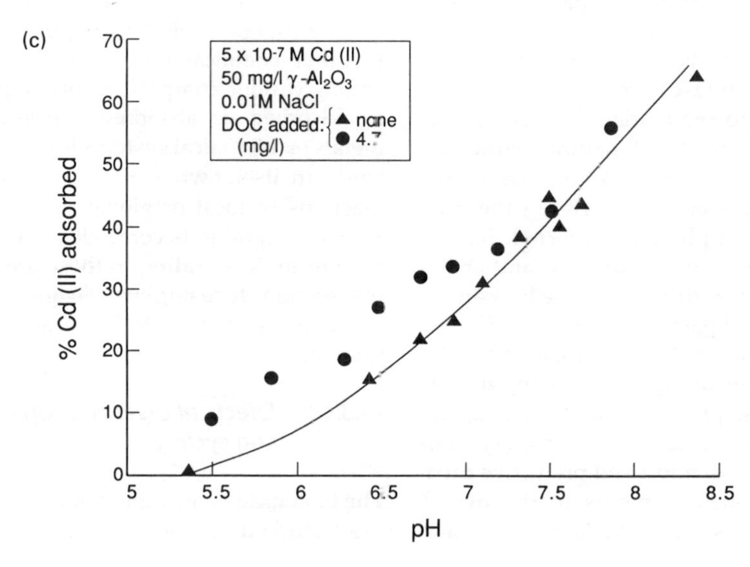
% Cd (II) adsorbed
pH
5 x 10^{-7} M Cd (II)
50 mg/l γ-Al_2O_3
0.01M NaCl
DOC added: (mg/l) none 4.7

A related phenomenon is the adsorption of organic matter, metals, and particulate matter at another interface, i.e. that between the bulk water and air. At the air–water boundary, a microlayer of approximately 50 μm thickness is established with physical/chemical properties different from those of the bulk water. These properties cause metals, particles, microorganisms, and dissolved organic matter to accumulate in concentrations 5–1000 times greater than in bulk solution (Hardy *et al.*, 1985). While this layer does not contain a significant fraction of the total metal in the aquatic system, it comprises much of the material that is transferred to the atmosphere as spray and represents an important region through which substances must pass to enter the bulk water from the atmosphere. During the several hours that particles are thought to spend in this enriched environment, important dissolution, complexation, photochemical, and biological reactions may take place.

The fraction of the total metal bound to particulate matter can be large, at times dominating the fraction in bulk solution. Research designed to identify the form of the particle-bound metal, the chemical nature of the particulate matter, and the bond strength between them has begun. Davies-Colley *et al.* (1984) have investigated the distribution of Cd among various model sediment components and concluded that for "typical" oxidized estuarine sediments, most of the trace metal will be bound to hydrous iron oxides and organic matter, with the organic-bound fraction of Cu being greater than that of Cd. Manganese oxides and clays contribute negligibly to the total binding capacity for both metals, according to their work.

Another approach to assess the partitioning of metals among the phases comprising natural particulate matter is to sequentially and selectively extract or dissolve portions of natural particulate matter. Based on the release of trace metals accompanying each step, associations between the trace metal and the extracted phase are inferred. Both of the above approaches have drawbacks, and at this time it is impossible to predict in advance how and to what extent metals and particulate matter will bond to one another in a natural system. Despite the uncertainties, empirical results can often be interpreted using the framework provided here, offering insights into the behavior of metals in complex systems and ultimately leading to improved predictive capability. Experimental measurements of the overall fractionation of metals between particulate and soluble phases in a few aquatic systems are presented in Fig. 15-7.

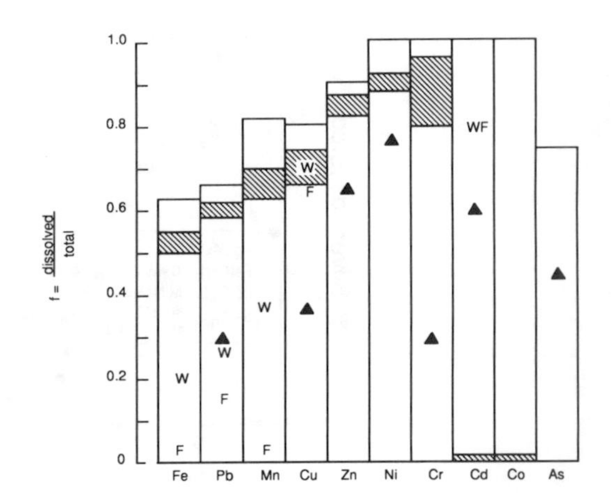

Fig. 15-7 Relationship between dissolved and total heavy metal concentrations in several rivers. Cross-hatched bands represent range of values from the Ruhr (Imhoff *et al.* 1980); W and F represent winter and fall values at a selected station in the Mississippi (Eisenreich *et al.*, 1980); triangles represent values from the Rhine river (de Groot and Allersma, 1975).

15.3.7 Reactions Involving Organisms

Metals play an essential role in many enzyme systems, and virtually all metals can be toxic at concentrations that exceed the levels at which they are required or are normally found in the environment. Microorganisms play a central role in converting inorganically and organically bound metals to other chemical forms and transporting metals among various compartments of aquatic ecosystems as adsorbed or absorbed species. The effects of metals on biological systems is an established field of study in itself, while the importance of biological reactions in local or global metal cycling has only recently begun to become clear. An overview of the current understanding in these areas is provided in this section, focusing once again on the importance of metal speciation to the understanding of the system.

15.3.7.1 Effects of organisms on metal speciation and cycling

The biological contribution to metal cycling has been most studied for those metals for which biological reactions contribute significantly to the total global flux, or for which biological transformations have particularly significant implications for humans. In this category fall mercury and arsenic, whose biological cycles were first described clearly by Wood (1974). He emphasized that these metals, and perhaps many others, have natural biological cycles even though the metals themselves are not biologically essential. Many biological conversions of metals may be evolutionary responses that microorganisms have developed as detoxification mechanisms. However, at times, the "detoxification" reactions of one group of organisms actually intensify the toxicity of the metal to higher organisms. Such is the case with methylation of mercury and arsenic.

In the case of mercury (Fig. 15-8), Wood suggests that reduction of Hg^{2+} to Hg^0 and alkylation to form methyl- or dimethylmercury can both be viewed as detoxification reactions, because all of the products are volatile and can be lost from the aqueous phase. Organisms can also convert the methylated forms to Hg^0, which is more volatile and less toxic. However, both the methylated and the reduced forms are more toxic to humans and other mammals than is Hg^{2+}.

The arsenic cycle in ocean waters and sediments also has important biological steps (Andreae, 1979). Arsenate, As(V), can be biologically converted into arsenite, As(III), and at least eight different organoarsenic compounds, all presumably representing detoxification processes mediated by bacteria in

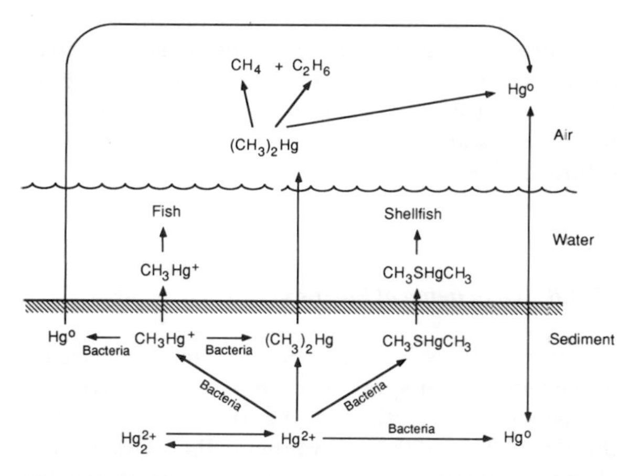

Fig. 15-8 The mercury cycle, demonstrating the bioaccumulation of mercury in fish and shellfish. Adapted from the National Academy of Sciences (1978) with permission.

reduced sediments or by algae in the water column. On the other hand, biologically catalyzed demethylation and perhaps oxidation of arsenite serve to return arsenate to the water. In the bulk water below 400 m, one study found the ratio of arsenite to arsenate to be at least 12 orders of magnitude greater than that expected from equilibrium calculations, indicating the importance of biological reactions.

Even if microorganisms do not mediate a reaction with the metal directly, e.g. by changing its oxidation state or forming organometallic compounds, they still may play an important role in the cycling of many metals. For instance, Wood pointed out that bacteria can facilitate the mobilization of mercury from mineral deposits by oxidizing sulfide and thereby allowing mercury which had been sequestered in the extremely insoluble solid cinnabar (HgS) to dissolve. The same mechanism can be important in the release of many metal sulfides, and the use of bacteria to release metals from ores in mining operations is under investigation. Additionally, recent evidence has shown that the cycling of several metals in the open ocean is closely tied to the cycling of particulate organic matter, i.e. microorganisms (Jones and Murray, 1984). Figure 15-9 shows correlations of cadmium and phosphate in the ocean off the coast of California. Like phosphate, Cd is depleted in the surface waters due to biological uptake and released to the aqueous phase in deeper water as the sinking organic solids decay. It is not yet known whether the metal uptake is active and carried out only by living organisms, or is passive and involves adsorption on dead cells as well. Similar correlations have been reported for Ni, Zn, and Cu.

The range of processes that must be considered in the cycle of metals is described in Fig. 15-10 (Nelson *et al.*, 1977). Both the complexity of metal cycle analysis in a real system and the importance of speciation are well-stated by Andreae (1979) in his overview of the arsenic cycle in seawater:

> The biological cycle of arsenic in the surface ocean involves the uptake of arsenate by plankton, the conversion of arsenate to a number of as yet unidentified organic compounds, and the release of arsenite and methylated species into the seawater. Biological demethylation of the methylarsenicals and the oxidation of arsenite by as yet unknown mechanisms serve to regenerate arsenate. The concentrations of the arsenic species are then controlled primarily by the relative rates of biologically mediated reactions, superimposed on processes of physical transport and mixing. The

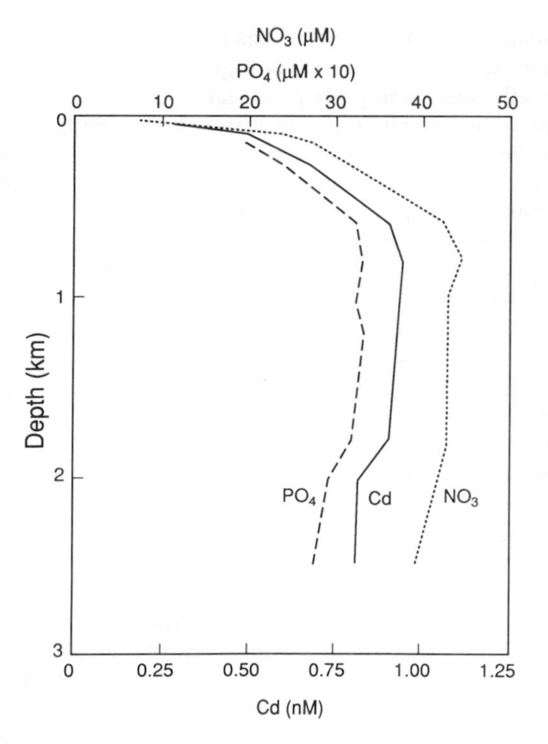

Fig. 15-9 Depth profiles for NO_3^-, PO_4^{3-}, and Cd observed at station 64 off the coast of California in April 1977. Adapted from Bruland *et al.* (1978) with the permission of the Society for Limnology and Oceanography.

presence of arsenite in the deep ocean at concentrations far from thermodynamic equilibrium further emphasizes the importance of kinetic restraints on redox equilibration in the ocean.

15.3.7.2 Effects of metals on organisms: Toxicity and bioavailability

In the last decade or so, recognition of the importance of metal speciation has had an enormous impact in the area of toxicity studies. Previously, data were often difficult to interpret even from a single study, let alone from a range of investigations each with slightly different conditions. Recent analysis of speciation has led to a remarkably consistent conclusion: for all metals that have been studied (Cu, Cd, and Zn have been investigated the most), the toxicity of metals to algae, various aquatic invertebrates, and fish is strongly related to the activity of the free metal ion, regardless of total metal in solution. This means, for example, that a solution containing 10 mg/L total Cu and enough organic matter so that 99% of the copper is strongly complexed, will be less toxic than an otherwise comparable solution with only 0.5 mg/L total dissolved Cu and little or no complexing organic matter. One recent investigation reached the same conclusion with respect to metal ion availability as an essential nutrient (in this case Cu).

Strongly complexed metals, especially those complexed by chelating agents such as EDTA or by natural humic material, appear to be completely unavailable and non-toxic. A typical experiment showing such a result is summarized in Fig. 15-11. Weaker complexes formed with monodentate ligands such as chloride or carbonate also provide some protection against toxicity, although it is not yet clear whether these complexes are as innocuous as the chelates. Of course, the overall effect of potentially toxic metals depends not only on metal speciation but on all aspects of solution chemistry and on the identity and previous exposure history of the test organism. For example, one of the unresolved issues in this field has to do with the effects of water hardness (conferred primarily by dissolved Ca and Mg ions) and alkalinity on metal toxicity. In some studies, these variables are negatively correlated with toxicity, and in others they seem to have no effect. Undoubtedly, the results depend partially on the ability of the hardness ions to compete with the toxic ones for specific binding sites and on other responses of the organism to changes in the concentrations of these ions, quite apart from the response to the toxin. Regardless of the ultimate resolution of the remaining issues, it is clear that consideration of speciation has been the key step in rationalizing studies of toxicity and bioavailability in recent years and will be a crucial part of any such future studies.

15.3.8 Geochemical Kinetics

To this point, we have emphasized that the cycle of mobilization, transport, and redeposition involves changes in the physical state and chemical form of the elements, and that the ultimate distribution of an element among different chemical species can be described by thermochemical equilibrium data. Equilibrium calculations only describe the potential for change between two end states; only in certain

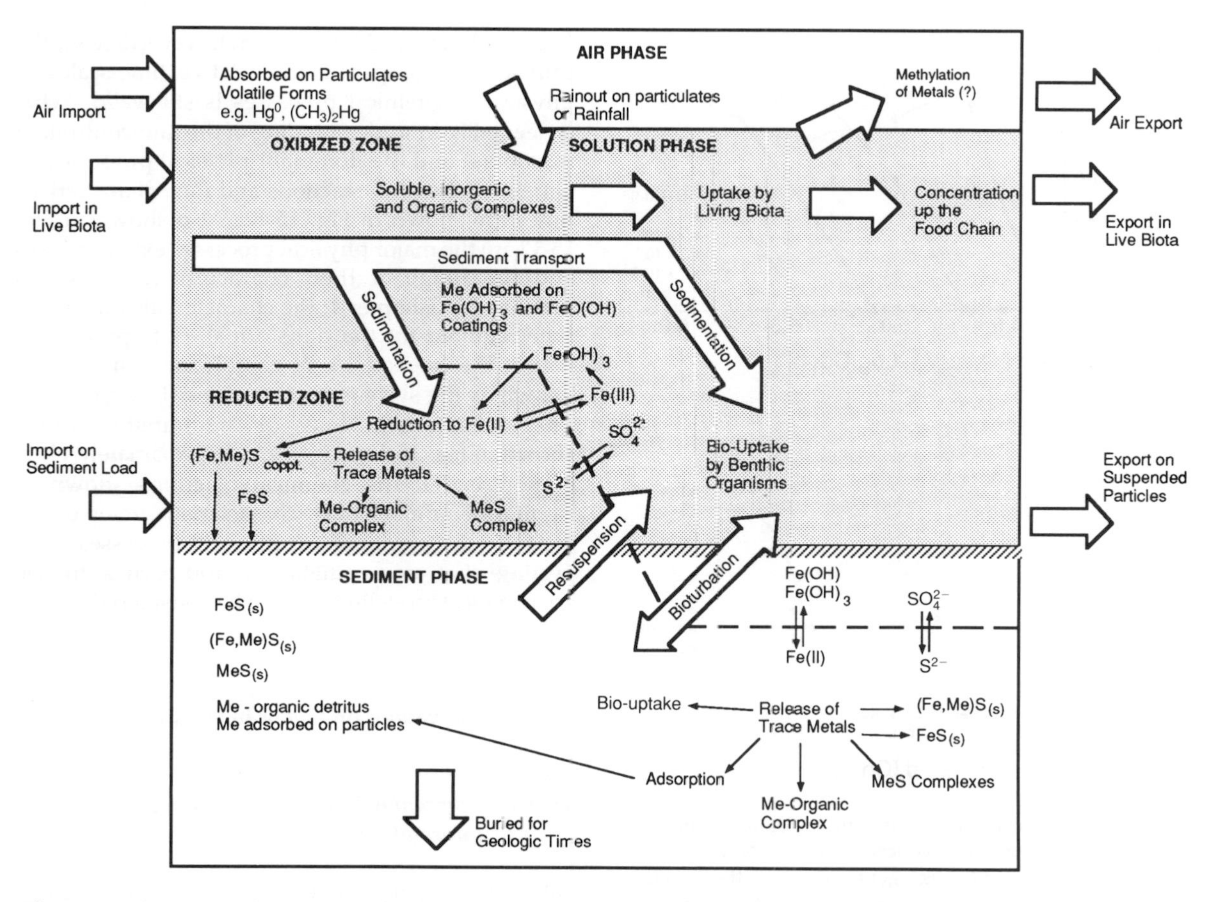

Fig. 15-10 Summary of reactions and processes important in metal biogeochemical cycling (after Nelson *et al.*, 1977).

cases can they provide information about rates (Hoffman, 1981). In analyzing and modeling a geochemical system, a decision must be made as to whether an equilibrium or non-equilibrium model is appropriate. The choice depends on the time-scales involved, and specifically on the ratio of the rate of the relevant chemical transition to the rate of the dominant physical process within the physico-chemical system.

Comparisons of time-scales for various physical processes have been discussed by Lerman (1979) and Schwartzenbach and Imboden (1984). In general. physical processes in lakes have characteristic times in the range of 10^{-4} to 10^{2} years. Oceanic processes span times from days to thousands of years (Broecker, 1974) and geological events may occur catastrophically (e.g. earthquakes or landslides exposing new weathering surfaces) or very slowly, as in the case of continental subduction. If the characteristic chemical reaction times, τ_{chem}, are short compared to the characteristic times of the dominant physical processes, τ_{phys}, then it is probably appropriate to consider the chemical transition in terms of equilibrium concepts. If τ_{chem} is large compared to τ_{phys}, then reactions will proceed only slightly, if at all, from their initial conditions. When τ_{chem} is of the same order as τ_{phys}, then quasi-equilibrium or kinetic descriptions may be employed (Morgan and Stone, 1985; Morel, 1983; Keck, 1978). Thus, a description of the chemical transitions occurring in each reservoir of a biogeochemical system must consider the relative physical and chemical time-scales characteristic of that reservoir. For example, the hydrolysis of Fe(III) has a charac-teristic time (τ_{chem}) of 3.2×10^{-7} s (Hemmes *et al.*, 1971). Since most natural physical processes are much slower than this rate, they can be

(a)

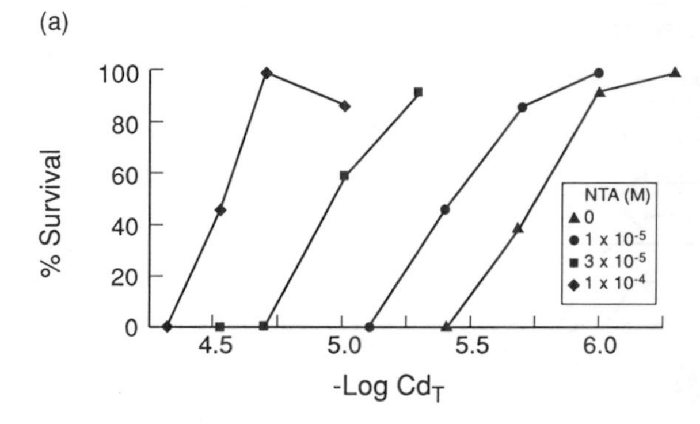

(b)

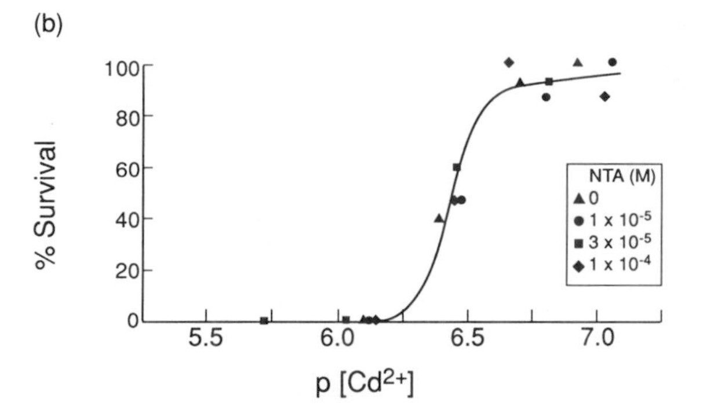

Fig. 15-11 Effects of strong complexation on metal ion toxicity. (a) Increasing concentration of NTA, a strong multi-dentate complexing agent, decreases the toxicity of Cd to grass shrimp. All systems have equal concentrations of total Cd. (b) When the results are replotted showing survival as a function of Cd^{2+} concentration, the data for all concentrations of NTA collapse to a single curve. Adapted from Sunda *et al.* (1978) with the permission of the American Chemical Society.

considered to be at equilibrium with respect to Fe(III) hydrolysis. In contrast, the dissolution of silica proceeds at a much slower rate, with a characteristic time of approximately 8 years. The oxygenation of Mn(II) is more ambiguous. It has characteristic times ranging from weeks to tens of years depending on whether the reaction is homogeneous, mediated by bacteria or catalyzed by the presence of metal-oxide surfaces (Morgan and Stone, 1985).

The suite of chemical reactions taking place in a geochemical compartment, such as the atmosphere or a lake, are often quite complex, involving higher-order reaction kinetics and multiple, linked reactions (e.g. Pankow and Morgan, 1981). Nevertheless, the principle of comparing the relative time-scales of physical *vs* chemical processes is still valid. What is needed is an understanding of the rate-controlling reaction(s) and the dominant physical processes. A variety of chemical reactions and their characteristic times are shown in Fig. 15-12. Also shown in Fig. 15-12 are the major physical processes extant in lakes and the range of their characteristic times. As described in Chapter 4, the characteristic time is the time required for the transformation to proceed to within the fraction 1/e of completion. This is also known as the system response time (Lasaga, 1980). The characteristic times of the transformations shown in Fig. 15-12 range from approximately 1 h to 6 months. Of the chemical reactions shown, a significant number have characteristic times of the order of the predominant physical processes, suggesting that non-equilibrium approaches to the geochemical modeling need to be considered.

15.4 Observations on Metals in Natural Systems

15.4.1 Combining Physical and Chemical Information

All the factors mentioned in the previous sections play a role in the movement of metals through their overall biogeochemical cycle: injection into the atmosphere, deposition onto land or water surfaces, transport via rivers and estuaries to the oceans, and sedimentation and ultimate burial in the sediments. The physical/chemical form in which each metal is transported in the aquatic phases of the cycle will depend on the specific metal and its interactions with other dissolved and suspended constituents in accord with the principles discussed above. However, it is important to keep in mind that the physical processes of fluid flow and sediment transport must be combined with the chemical reactions to gain insight into the functioning of the complete system. This point has been well-stated by Turekian (1977) in a discussion of metal concentrations in the ocean:

> Why are the oceans so depleted in these trace metals? Certainly it is not for the lack of availability from rock weathering or because of constraints imposed by the solubility of any unique compound of these elements. The reason must lie in the dynamics of the system of delivery of the metals to the oceans and their sub-

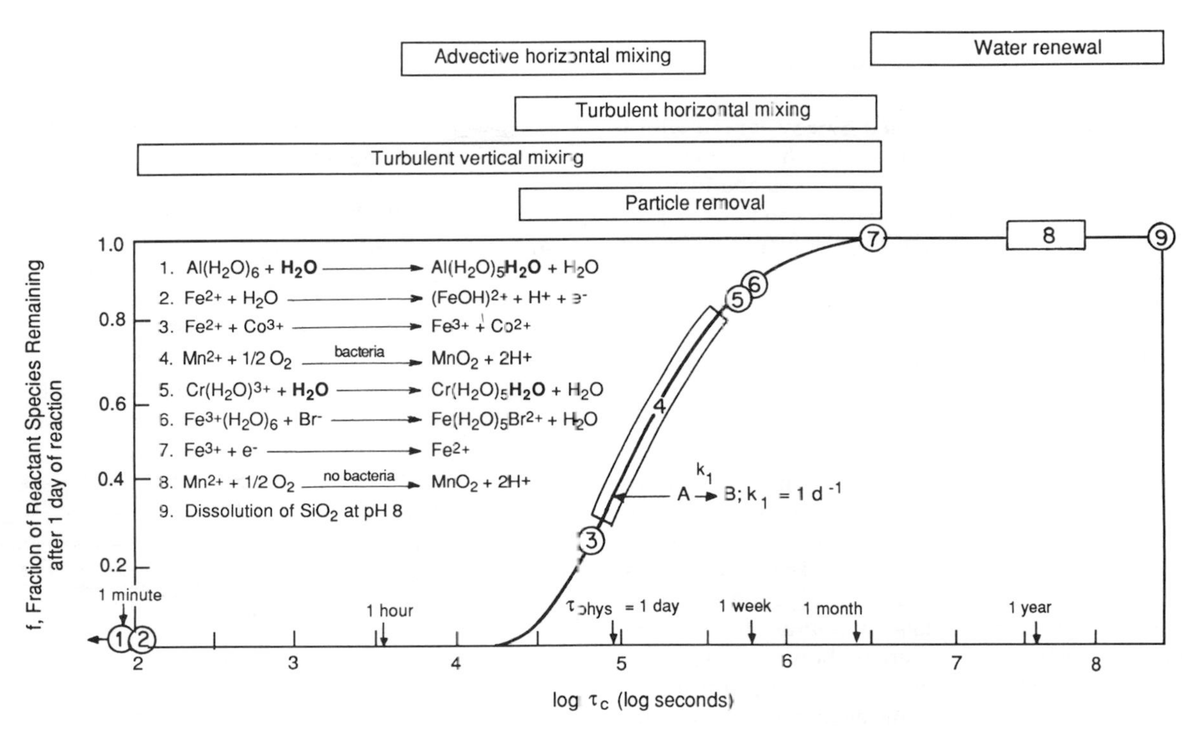

Fig. 15-12 Examples of characteristic times for lake transport processes and chemical reactions, and extent of reaction of several environmentally significant reactions in a system with $\tau_{phys} = 1$ day. Numbers in circles and boxes refer to example reactions.

sequent behavior in an ocean that cannot be simulated by simple *in vitro* experiments involving homogeneous reaction kinetics.

In this section, an overview of the net transport of metals through rivers, estuaries, and the oceans provides an example of how these chemical and physical forces interact.

Recent studies have reported that from 10% to more than 90% of the trace metal load of streams is carried in the particulate fraction, with the fraction of particle-bound Fe, Pb, Cu, As, and Zn typically being greater than that of Cd, Co, and Ni. As noted earlier, the exact balance will depend on the concentration of inorganic and organic complexing ligands, the type and quantity of particulate matter available, and the concentrations of other ions competing for the binding sites. In interpreting these data, one must bear in mind that the fraction defined as being "particulate" does not represent an absolute measure; rather, it is defined by the separation technique employed. Often any material that will not pass a 0.45-μm filter is defined as particulate. However a significant fraction of the "soluble" trace metals Fe and Mn, and organic material that does pass the filter, can subsequently be removed using filters with smaller pores. Furthermore, much of the 0.45-μm filterable material is converted to non-filterable matter (still using 0.45-μm filters) when the concentrations of Ca^{2+}, Mg^{2+}, and Na^{+} increase to levels they attain in estuaries. This process and its implications merit some special attention.

15.4.2 Particle and Metal Interactions in Estuaries

Estuaries exhibit physical and chemical charactristics that are distinct from oceans or lakes. In estuaries, water renewal times are rapid (10^{-3} to 10^{-4} years compared to 1–10 years for lakes and 10^4 years for oceans), redox and salinity gradients are often transient, and diurnal variations in nutrient concentrations can be significant. The biological productivity of estuaries is high and this, coupled with accumulation of organic debris within estuary boundaries, often produces anoxic conditions at the sediment–water interface. Thus, in contrast to the relatively constant chemical composition of the oceans, the chemical environment in an estuary varies in time and space.

An estuary can be defined as a system in which ocean waters have been diluted by freshwater from land drainage. Many ions are more abundant in coastal or oceanic waters than river water, e.g. Na, K, Mg, Ca, Cl, (SO_4), and (HCO_3). However, nutrients (N, P, and Si) and many transition series metals, including Fe, are more concentrated in river water. Dissolved and particulate organic matter (DOM and POM) are 1–2 orders of magnitude more concentrated in rivers than in ocean water. As end-member waters mix, dissolved constituents originally in each of the end-member waters may remain in solution during mixing (i.e. behave conservatively), interact with sinking particulate matter, or precipitate out of solution. The wide variety of mixing regimes and end-member waters makes it difficult to generalize about the estuarine behavior of many metals of interest to environmental geochemists. Generally, however, the major components of seawater – Na, Ca, K, and SO_4 – behave conservatively. Some elements, such as zinc, may behave conservatively in some estuaries but be rapidly removed in others. Iron, by contrast is removed rapidly and efficiently in nearly all estuarine environments, i.e. its behavior is highly non-conservative.

As river water (Fig. 15-13a) high in dissolved and particulate iron and other inorganics (PIM) and organic matter enters an estuarine mixing zone (Fig. 15-13b), rapid flocculation of the DOM and coagulation of particulate matter occurs (Fig. 15-13c). The extent to which this occurs is a function of the salinity of the water mixture: flocculation and coagulation generally increase with increasing salinity up to a salinity of about 15‰, with greater salinities having little added effect. While systems containing only colloidal iron oxides have been shown to coagulate as salinity increases, the presence of organic matter increases the coagulation rate and leads to the formation of Fe oxide–organic aggregates. The efficiency of metal removal by the flocculation process generally follows the relative order of trace metal–DOM complex stabilities: $Fe^{3+} > Al^{3+} > Cu^{2+} \sim Ni^{2+}$, etc. Reactions that may initiate the iron–organic removal process include coagulation of pre-existing colloidal particles or precipitation of previously soluble humic acids, and adsorption of trace metals onto particles before or after either of the above processes.

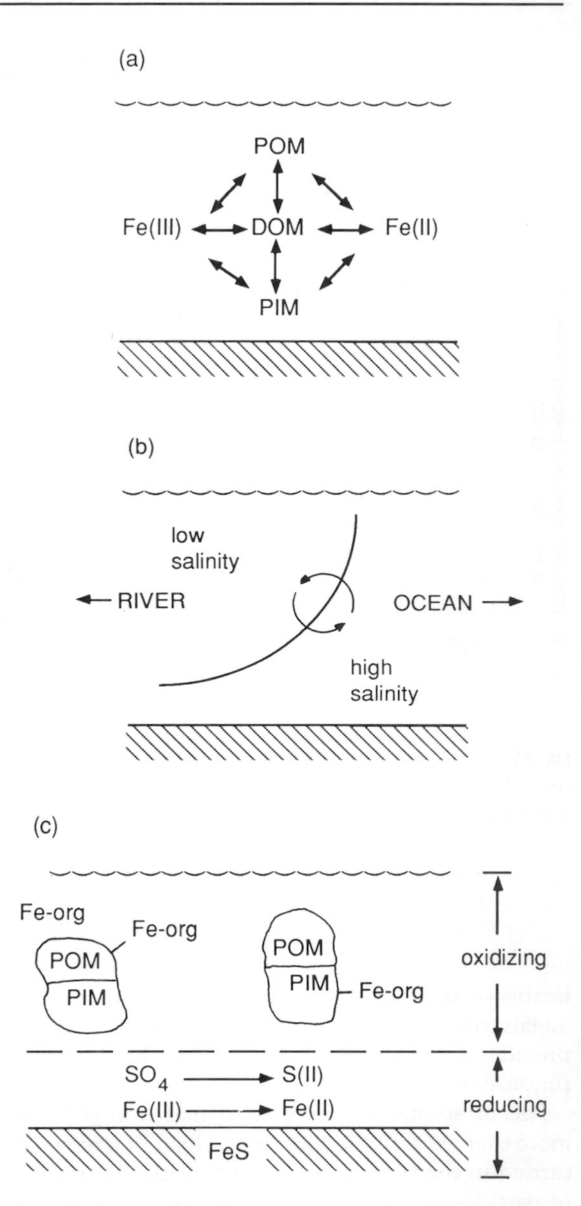

Fig. 15-13 Schematic representation of Fe and organic matter interactions in an estuary. POM, particulate organic matter; DOM, dissolved organic matter; PIM, particulate inorganic matter.

The efficient removal of iron in estuaries allows very little of the initial river-borne iron to escape the estuary to the coastal waters. Nearly all the iron entering the estuary is transported in particulate form to sediments on the estuary floor where, due to the high accumulation of organic matter, reducing conditions are often met. Fe(III) is reduced in anaerobic sediments and, since S is often also accumulating in the sediments, Fe(II) may be precipitated as iron sulfides. Resuspension of sediments may reintroduce iron to the oxygenated layer of the estuary where it can be rapidly oxidized, forming new colloidal particles, which can sorb other metals, associate with suspended organic matter, and return to the sediments in repeated episodes of internal cycling.

The ultimate result is that not only Fe, but most metals that interact strongly with organic matter or oxide adsorbents, are likely to settle out of the water column in estuaries. Not only does this process reduce the metal flux reaching the ocean, but 50% of the time the current at the bottom of an estuary actually facilitates the movement of metals upstream.

While Fe- and Mn-oxides certainly are important in the binding and transport of the trace metals in the estuarine system, organic matter appears to play a dominant role in the coagulation step. For instance, adjustment of the solution pH to values where humics normally precipitate (pH = 1–3) causes all the components (Fe and Mn, organics, and trace metals) to coagulate into filterable particles. By contrast, adjustment of the pH upward, which would normally favor precipitation of Fe- and Mn-oxides, has no effect. In view of this, it is not surprising that the tendency of trace metals to be removed from estuarine waters corresponds roughly with their tendency to bind to organics, with Cu being the most strongly affected. In one study, for instance, approximately 40% of the dissolved Cu, 15% of the dissolved Cd and Ni, and less than 5% of the dissolved Co and Mn were filterable after coagulation was induced by addition of Ca and Mg to river water (Sholkovitz and Copland, 1981). As noted earlier, desorption of metals owing to the formation of soluble Me–Cl complexes may partially or completely counteract this process for some metals.

To summarize, even though some Fe- and Mn-oxides may dissolve in the anoxic sediments of estuaries and thereby release trace metals, re-oxidation of the Fe and Mn in the upper layers of the sediments, coagulation of Fe–Mn–humic–trace metal mixtures in the water column, and circulation patterns within the estuary all combine to return trace metals to the upstream particulate fraction. Turekian has suggested that while these processes allow significant internal cycling of trace metals in the estuary, they prevent any more than a small fraction of the total trace metal load from reaching the ocean. When he considered the role that soils and rocks play in attenuating the movement of metals from the land to water bodies and the scavenging of metals from the water column by particles in the ocean in combination with the estuarine processes, Turekian (1977) claimed to have identified "the great particle conspiracy", which maintains soluble trace metal concentrations at extremely low and remarkably consistent levels throughout the world's open oceans.

15.4.3 Application of Chemical Principles to Evaluate Field Data

Examples were given earlier of a few systems for which the theoretical speciation of metals was calculated, and a few others in which it was analyzed experimentally, at least into some sub-groups if not specific species. These approaches have rarely been combined to evaluate data or hypotheses critically regarding metal transport through a real system. One study, already noted earlier in this chapter, can serve as a model for combining theoretical considerations with field data. In that study, Morel *et al.* (1975) used data for the partitioning of metals between the soluble and particulate fractions to estimate the redox potential of primary-treated sewage. Knowing the composition of the Southern California coastal waters into which the sewage was discharged, they were then able to interpret the relative changes in the concentrations of several metals as a function of dilution and oxidation. Combining chemical analyses with data for sedimentation rates of metals and organic carbon, they suggested that most metals are not mobilized in the near-field deposition zone, and they offered plausible explanations for the deviation of Pb and Cd from this pattern. As the authors noted, extension of their results to other systems is not warranted, but extension of the methodology, i.e. the combination of experimental data with chemical speciation models to test hypotheses and suggest profitable research directions, certainly is.

15.5 Examples of Global Metal Cycling

In this final section, the global cycles of two metals, mercury and copper, are reviewed. These metals

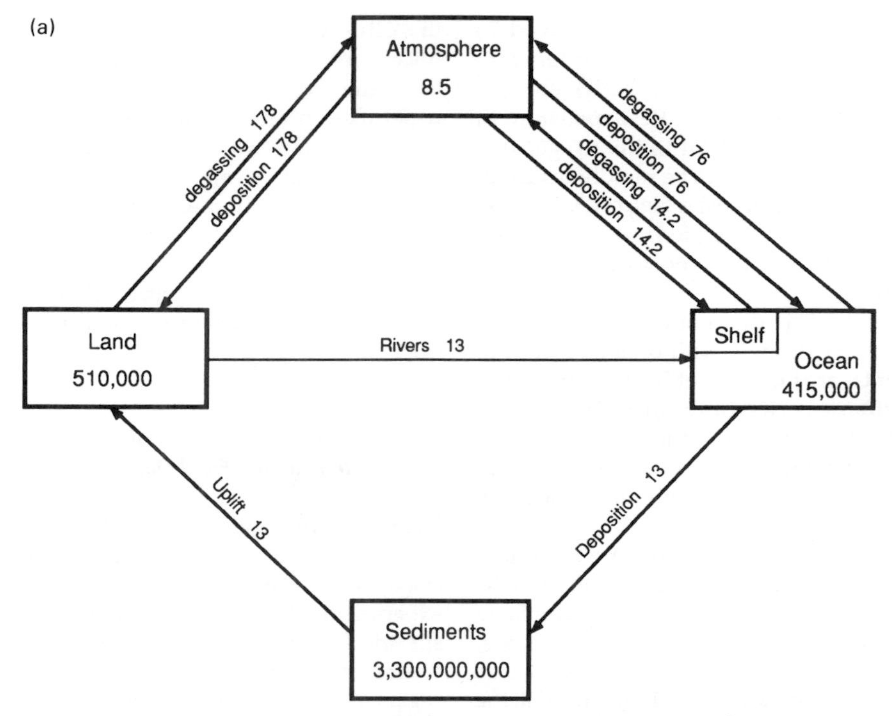

Fig. 15-14 The (a) "pre-man" and (b) present-day global cycles for mercury. Units are 10^8 g (burdens) and 10^8 g Hg/year (fluxes). Adapted from the National Academy of Sciences (1978) with permission.

were chosen because their geochemical cycles have been studied extensively, and their chemical reactions exemplify the full gamut of reactions described earlier. In addition, the chemical forms of the two metals are sufficiently different from one another that they behave differently with respect to dominant transport modes and pose different risks to organisms.

15.5.1 Mercury

There have been several reviews of various aspects of the environmental behavior of mercury. Much of the material in this section was derived from a review published by the National Academy of Sciences (1978). The reader is referred to that publication and the original references cited therein for further information.

Mercury occupies a unique (and infamous) place in environmental history because it was the first chemical for which a direct connection was proven between relatively low concentrations in a natural water system, bioaccumulation up the food chain, and a serious health impact on a human population at the top of the food chain. Within a relatively few years, such epidemics were documented at two sites in Japan, and somewhat less definitive evidence suggesting mercury poisoning was reported in Canada. In addition, consumption of animals that had been fed grain seed contaminated by methylmercury fungicide led to acute poisoning of a family in the USA, and, in one of the worst cases of acute human exposure to an environmental hazard, over 6000 Iraquis were hospitalized with symptoms of mercury poisoning after consuming homemade bread made from seed wheat treated with this same type of fungicide. In the latter case, over 500 deaths were reported by hospitals, and it is likely that many other affected individuals did not report to the hospital.

15.5.1.1 Global mercury cycling

Unlike most heavy metals, the natural and anthropogenic cycles of mercury are dominated by atmospheric transport. Metallic mercury has the highest vapor pressure of any heavy metal, and it is re-

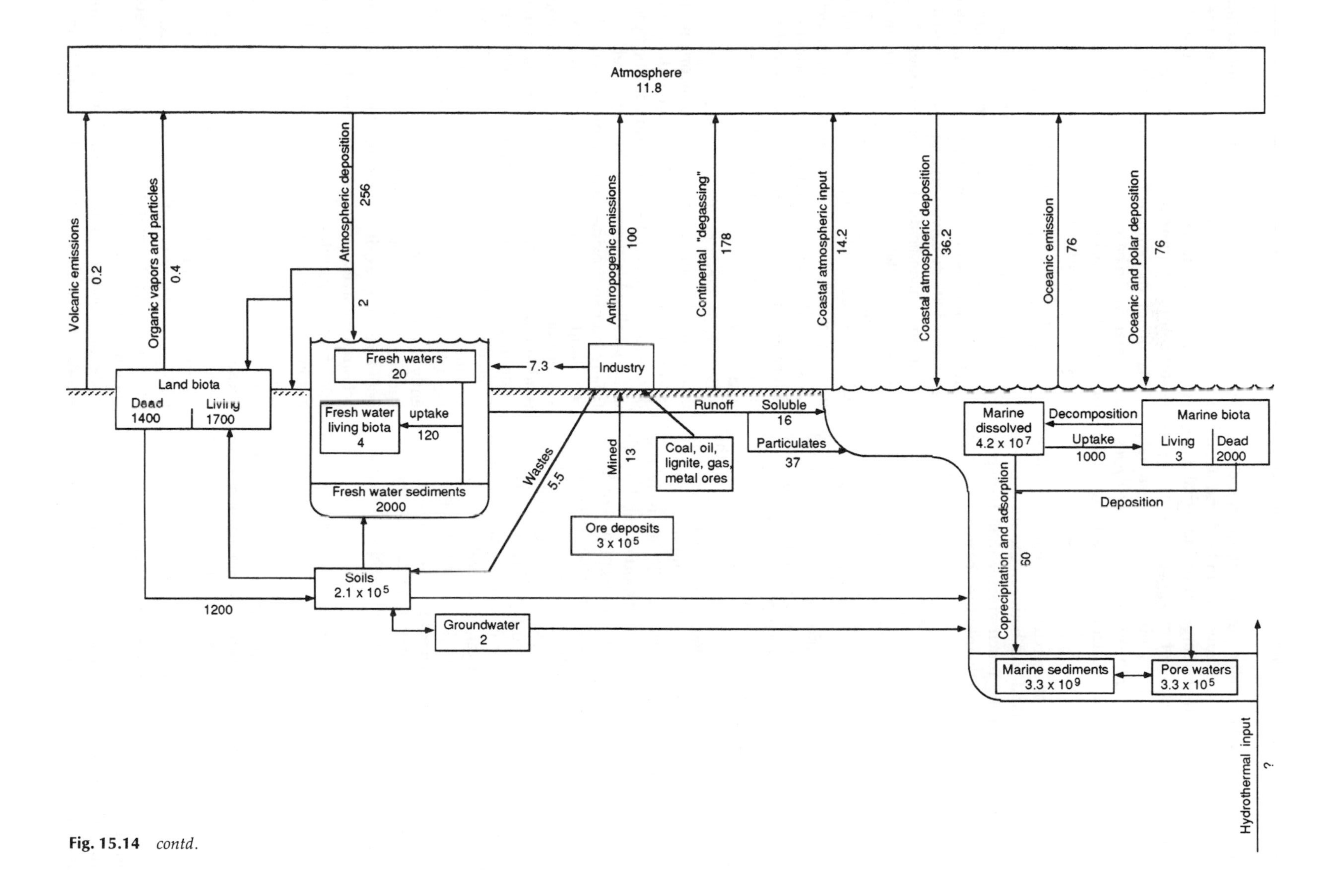

Fig. 15.14 *contd.*

leased in geochemically significant quantities by volcanic activity, volatilization from land and ocean surfaces, and in high-temperature industrial processes such as smelting of minerals and burning of fossil fuels. Figure 15-14 provides a 1978 estimate of the natural and anthropogenic mercury cycles. It demonstrates many salient points, among which are the following:

1. Exchange rates of mercury between the atmosphere and the land and between the atmosphere and the ocean are much greater than transport from the land directly into the ocean via riverine discharge.
2. While the natural exchange of mercury between the land and atmosphere and the atmosphere and oceans is balanced, human activity has tipped this balance so that there is now net transport to the atmosphere. It has also led to an increase of about 40% (3.3/8.5) in the amount of mercury stored in the atmospheric reservoir and a roughly equivalent increase (100/268) in the annual mercury flux through the atmosphere.
3. The transport rate of mercury flowing from the land to the oceans in rivers has been increased by a factor of about 4 (53/13) by human activity. While the increased rate is still relatively less important than the total transport of Hg through the atmosphere, it can represent a significant stress on the exposed organisms, particularly since the increased flux is unevenly distributed. That is, human activity has created local environments where the transport of mercury or its concentration in a river or estuary is many tens of times higher than background levels.
4. The average residence times for mercury in the atmosphere, terrestrial soils, oceans, and oceanic sediments are approximately 11 days, 1000 years, 3200 years, and 2.5×10^8 years, respectively.

15.5.1.2 *Environmentally important forms of mercury*

There are several environmentally significant mercury species. In the lithosphere, mercury is present primarily in the +II oxidation state as the very insoluble mineral cinnabar (HgS), as a minor constituent in other sulfide ores, bound to the surfaces of other minerals such as oxides, or bound to organic matter. In soil, biological reduction apparently is primarily responsible for the formation of mercury metal, which can then be volatilized. Metallic mercury is also thought to be the primary form emitted in high-temperature industrial processes. The insolubility of cinnabar probably limits the direct mobilization of mercury from this mineral, but oxidation of the sulfide in oxygenated water can allow mercury to become available and participate in other reactions, including bacterial transformations.

In aqueous solution, mercury forms strong complexes with both organic and inorganic ligands. In particular, mercury is strongly complexed by compounds containing reduced sulfur atoms or carboxyl groups, by hydroxide, and by chloride. As a result, much of the dissolved mercury in freshwater is associated with dissolved organic matter, most likely humic and fulvic acids. Even in extremely "pure" river water, containing few dissolved minerals or organic compounds, mercury exists not as the divalent free aquo metal ion, but as a neutral or anionic hydroxo complex, depending on the solution pH. In seawater, dissolved mercury exists primarily as chloro and possibly organic complexes and mixed ligand complexes of chloride, bromide, and hydroxide.

Dissolved mercury also reacts to form strong complexes with inorganic and probably organic particulate matter. Figure 15-15 shows the distribution of mercury between dissolved and adsorbed phases in an idealized system consisting of water, an artificially prepared oxide, and various concentrations of chloride. The mercury is strongly sorbed in the absence of the chloride, but as more of this ligand is added to the system, the mercury remains in solution at pH values where it had previously sorbed. Chemically, the effect of the chloride is to convert the mercury from Hg–OH complexes to Hg–Cl or Hg–Cl–OH or complexes. These complexes apparently do not sorb as strongly as the Cl-free mercury species. While the conditions in any natural water are more complex, it is clear that this type of interaction may have a significant effect on the form and fate of mercury under changing environmental conditions, for instance in water moving through an estuary.

Finally, in addition to being an unusual metal because of its high volatility, mercury is unusual because of the importance of biological reactions in altering its form. It has been pointed out that biological activity is critical in converting oxidized (+2) mercury to metallic (0) mercury in soils, allowing subsequent volatilization. Mercury also undergoes biomethylation, which is of overriding significance with respect to its toxicity. In this

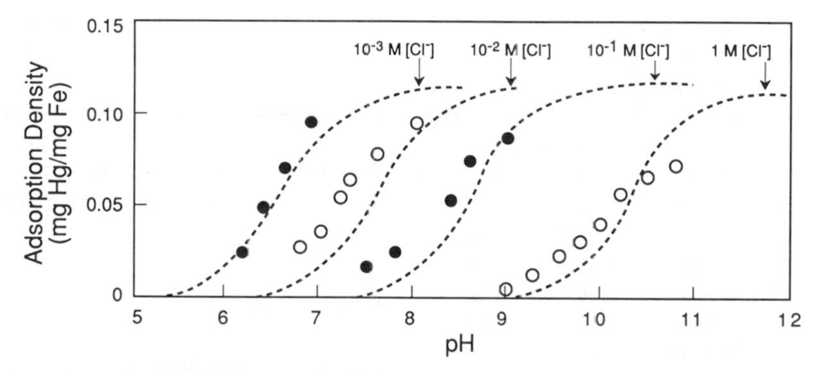

Fig. 15-15 The effect of chloride on adsorption of mercury by hydrous iron oxide at constant total mercury concentration of 3.4×10^{-5} M. The lines represent the predicted adsorption assuming that Hg–Cl complexes do not sorb at all. Adapted from Avotins (1975) with permission.

reaction, which takes place in the upper layers of fresh- and saltwater sediments, divalent mercury forms covalent bonds with organic methyl groups. The organo-mercury compounds are more volatile and more easily bioaccumulated than their inorganic precursors. They may sorb to organic matter in the sediment, but once released they are rapidly taken up by organisms and are concentrated up the food chain, eventually becoming available for human consumption. This is the path that led to the first recognized epidemic of mercury poisoning affecting humans, at Minamata Bay, Japan. The rate of biomethylation of mercury is proportional to the concentration of inorganic mercury in the system and the concentration of organisms capable of carrying out the reaction. Thus, it has been suggested that limiting the nutrient supply discharged to receiving waters may be an effective method of limiting biomethylation in contaminated areas. Biodemethylation has also been observed and may be a microorganismal response to mercury intoxication.

15.5.1.3 Mercury movement in aquatic systems: Two case studies

Although there have been numerous efforts to describe the global mercury flux, studies in which fluxes of the individual species in a well-defined natural system were evaluated are rare. Two studies that have been reported – one for a highly contaminated saltwater system, the other for a relatively less contaminated stretch of a river – provide an interesting synthesis of the above information. In the former case, the movement of mercury out of Minamata Bay over a period of about 25 years was estimated (Kudo and Miyahara, 1983). The results support a remarkably strong affinity between the discharged mercury and bottom sediments, so strong in fact that the majority of the mercury discharged to the Bay remained there, attached to bottom sediments, 25 years after all Hg discharges had ceased. This result is all the more surprising considering the rapid flushing rate of the Bay: the residence time of water in the Bay is only 2.5 days due to vigorous tidal action. Even so, it is estimated that between 1960 and 1975 the rate of mercury elimination from the Bay to the sea was only 0.4% per year, which could easily be accounted for by sediment transport. The rate of movement of mercury from the Bay to the sea increased by approximately a factor of 10 in the period 1975–78, which the authors of the report hypothesize was caused by increased sediment movement generated by an increase in the traffic of large ships in the Bay during this period. As supporting evidence for the speed and tenacity with which mercury binds to sediments in the Bay, the investigators point out that the sea sediments just outside of Minamata Bay were not contaminated by the initial mercury discharges, indicating that sorption or other processes immobilizing mercury in the Bay must have occurred very quickly, on a time-scale of hours. Thus, whatever the immobilization process is, it appears to be rapid, efficient, and long-lasting.

A detailed study of mercury in a 3-mile (4.9-km) section of the Ottawa River (Kudo, 1983), represents one of the few comprehensive studies in which the transport of mercury in several physical and chemical forms was evaluated simultaneously. This stretch of river had received mercury as an industrial discharge for many years, a practice that was stopped 3 years prior to the start of the study. In addition, some mercury is derived from natural weathering reactions in the river's headwaters region. The vast majority of the total mercury in the test section (>96%) was associated with the bed sediments, and in conformity with the study of Minamata Bay described above, this mercury was not easily released to the solution phase. Thus, despite the relatively large reservoir of mercury in the sediments, slow bed sediment velocities limited the significance of sediment-mediated mercury transport. Although the water contained only 13 ng/L total dissolved Hg, it accounted for 58% of the total mercury flux through the system, while 41% could be attributed to suspended soil transport. Of the total mercury in the system, less than 6% was methylmercury. This mercury species was apparently less strongly sorbed to suspended sediments than was inorganic mercury, and as a result the aqueous phase accounted for 80% of its transport, with the remaining 20% moving with the suspended sediments. While the exact percentages would undoubtedly vary with such parameters as suspended sediment load, average water velocity, and sediment composition, these results are useful as a reference point with which other systems can be compared. This study is also representative of the type of analysis necessary for an evaluation of heavy metal movements in water systems and of the risks they pose to aquatic and human populations.

Mercury provides an excellent example of the importance of metal speciation in understanding biogeochemical cycling and the impact of human activities on these cycles. Mercury exists in solid, aqueous, and gaseous phases, and is transported among reservoirs in all these forms. It undergoes precipitation/dissolution, volatilization, complexation, sorption, and biological reactions, all of which alter its mobility and its effect on exposed populations. The effect of all these reactions on the environmental behavior of mercury has been indicated in this section. The importance of analyzing speciation is perhaps most evident from the fact that the most toxic form of mercury and the only form that has been identified as having affected large populations of humans, methylmercury, is relatively insignificant in the global mercury balance. In the next sub-section, a similar summary will be presented for copper. Despite significant differences from mercury with respect to certain chemical properties and environmental transport modes, it will once again be shown that an understanding of chemical speciation is critical to an appreciation of the metal's environmental behavior.

15.5.2 Copper

The biogeochemical cycling of copper has been extensively studied. While some questions remain regarding the gross inventories of copper in various environmental reservoirs, studies dating from as early as the turn of the century provide considerable insight into the flux rates among the reservoirs. The speciation of copper in natural waters has also been well-studied, and in recent years a rough consensus seems to have been reached regarding the dominant copper species in "typical" environments, although the exact values of the stability constants of some important copper complexes are still somewhat uncertain.

Unlike mercury, copper is known to be a metabolically essential element for virtually all organisms. It displays a well-established property of many of the heavy metals, being essential to growth at low concentrations and toxic at high concentrations. The requirement for Cu stems from its inclusion in several proteins, in which it is always coordinated to N, S, or O ligands (Moore and Ramamoorthy, 1984). At the other extreme, copper's toxic properties have been exploited to reduce growth of unwanted organisms such as algae and fungus in water bodies and in soils. Copper biogeochemistry has been discussed and reviewed frequently in recent years. In this section, only an outline of the literature will be provided, with emphasis on how the speciation of copper controls its behavior.

15.5.2.1 Global cycling

The global cycling of copper has been reviewed by Nriagu (1979) and is described schematically in Fig. 15-16. Like most heavy metals and in contrast to mercury, the flux of copper from terrestrial to oceanic reservoirs is dominated by transport in rivers. Copper reaching the oceans by atmospheric transport is of the same order of magnitude as that by three strictly anthropogenic sources: direct discharge

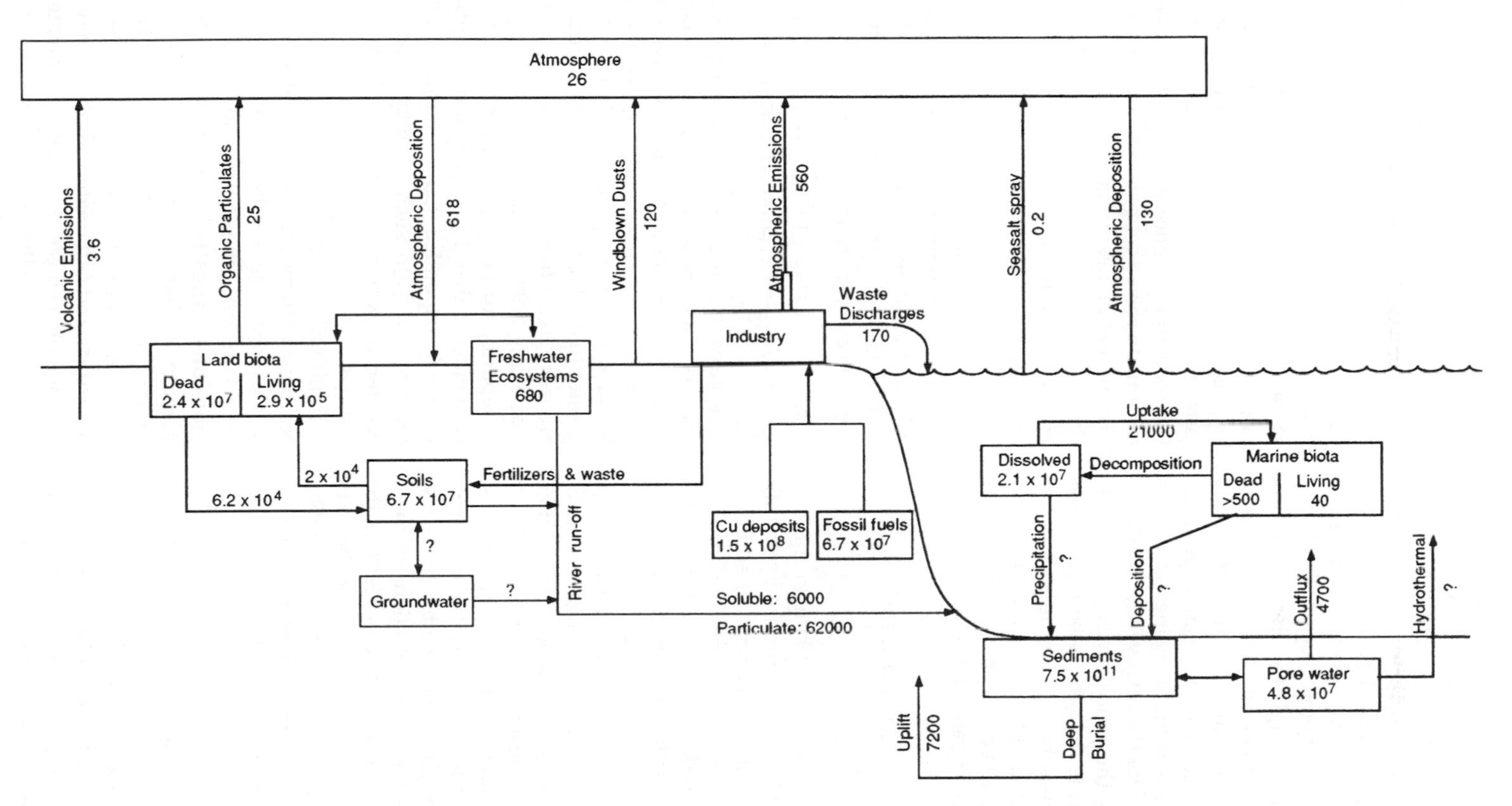

Fig. 15-16 The global copper cycle. Units are 10^8 g Cu (burdens) and 10^8 g Cu/year (fluxes). Adapted from Nriagu (1979) with the permission of Wiley-Interscience.

of wastes and sludge into the oceans, discharge of domestic and industrial wastes into rivers which eventually discharge into the ocean, and the use of copper-containing anti-fouling paints. Each of these sources is several hundred times less than the copper flux carried to the oceans by riverine run-off as part of the natural biogeochemical cycle. The transport that does occur through the atmosphere is essentially all via particulate matter, not transport of gaseous copper species, and roughly 90% of this particulate matter is injected into the atmosphere as a result of smelting, fossil fuel burning, and other human activities. Despite the relative insignificance of human inputs to the global cycle of copper, atmospheric transport is thought to be responsible for most of the transport to some inland water systems, e.g. the Great Lakes, and there is no question that in localized ecosystems anthropogenic inputs can dominate over natural ones.

15.5.2.2 *Copper in rocks, minerals, and atmospheric dust*

Copper exists in crustal rocks at concentrations ranging from about 10 to a few hundred ppm(m), with 70 ppm(m) being about average. In addition, at least 20 copper minerals have been identified, containing copper in the 0, +I, or +II oxidation state. These are primarily sulfides, hydroxides, and carbonates, of which chalcopyrite ($CuFeS_2$) is most common. Copper is also found in relatively high concentrations in deep-sea ferromanganese nodules, in many cases at concentrations greater than 0.5% and occasionally greater than 1.0%.

Solids containing oxidized anions (carbonates, sulfates, hydroxides, and oxides) are the dominant forms of Cu in airborne particulate matter. In the few studies that have addressed the reactions of these particles in atmospheric washout, about 50% of the copper has been found to be soluble. Since the solubility is strongly dependent on pH, acid precipitation and acidification of receiving waters may have a significant effect on the form and fate of airborne copper.

15.5.2.3 *Reactions of copper in aquatic systems*

15.5.2.3.1 Redox reactions and complexation. Copper in solution, as in solids, can carry either a +1 or +2 charge. The divalent form is the stable one in oxygenated water and is the prevalent form in natural waters. The distinction between the two forms is particularly significant to copper transport because they tend to form complexes with different ligands. Cuprous (Cu^+) ions form strong bonds with Cl^-, which probably controls its speciation in marine systems, but these have been little studied since Cu(I) is rarely detected in such systems other than in anoxic sediments. By contrast, cupric ions (Cu^{2+}) form strong complexes with hydroxide, carbonate, phosphate, and ammonia, and its complexes with organic molecules are typically the second strongest (after mercury) of any of the heavy metals. Even in waters with low concentrations of dissolved organic carbon, dissolved copper is associated primarily with organic matter, i.e. humic or fulvic acids. Turner *et al.* (1981) computed that in seawater with a total humic acid concentration of 10^{-6} M as carbon, organic complexes could account for 47% of the total dissolved copper, 2% of the lead, and less than 0.1% of the Mn, Co, Ni, Zn, Cd, or Hg. Metals other than Cu are prevented from complexing with the organics by a combination of competition for the ligand by Ca^{2+} and Mg^{2+} and complexation of the metals themselves by inorganic ligands:

$$Me^{n+} + (Cl^- \text{ or } CO_3^{2-}) \longleftrightarrow \text{Me/anion complex}$$

$$(Ca^{2+} \text{ or } Mg^{2+}) + \text{Hum} \longleftrightarrow (\text{CaHum or MgHum})$$

In seawater, most of the non-organically bound Cu^{2+} is complexed with inorganic carbon. In freshwater, because both types of competition shown above are less significant than in seawater, and because humic and fulvic acid concentrations are greater than in seawater, all metals, including Cu^{2+}, are more likely to exist as organic complexes. Thus, frequently >90% of the dissolved copper in rivers is reported to be organically bound. It should be remembered that this may include significant quantities of suspended Cu–Fe–humate colloids. The tendency to form such complexes would be even greater in waters contaminated by organic discharges.

Equilibrium complexation constants for Cu reactions with natural organic matter and the details of Cu speciation are bound to remain somewhat uncertain, since the composition of the complexing molecules varies from site to site. What is not in dispute is that the fraction of dissolved copper present as free aquo Cu^{2+} is probably very small in any natural water. In extremely pristine waters, hydroxide and carbonate complexes may dominate, but organic complexes usually dominate in waters containing more than a few tenths of a mg/L organic carbon.

15.5.2.3.2 Adsorption. Considering the similarities between adsorption and complexation reactions, it is not surprising that Cu^{2+} is among the most strongly sorbing of the heavy metals, and that it sorbs onto both inorganic and organic solids. Sorption of copper onto oxides and clays has been investigated extensively, both because of its importance to biogeochemical cycling and as part of studies of the availability of soil-bound copper to plants. Typical results showing the relative adsorptive strength of copper and several other metals in organic systems were presented in Fig. 15-5.

In a completely inorganic system, binding of Cu^{2+} to suspended and bottom sediments would remove much of the Cu from the dissolved phase. The formation of Cu-organic complexes and the presence of organic solids complicates the analysis, however. In essence, free Cu^{2+} has a strong tendency to react with many components of the system, and when these components are tested one at a time, each is able to sequester most of the Cu in the system. In a real system, these reactants compete and interact with one another, leading to complex speciation patterns that depend on the relative concentrations of each component and solution pH. The general picture that emerges from studies of the speciation of Cu in real and simulated natural waters is that natural organic complexing agents can bind Cu^{2+} so strongly that direct copper-to-surface bonding is effectively inhibited. Rather, the copper remains bonded to the organic matter under almost all conditions realistically expected in natural waters, and is adsorbed only under conditions where the organics sorb. Copper bound to organic particulates, or to organic matter adsorbed onto inorganic particulates, is thus the most likely form of Cu to reach bottom sediments, and speciation studies have consistently found Cu to be primarily associated with organics in aerobic sediments. The role of coagulation of colloidal matter in facilitating transport of Cu to estuarine sediments has already been discussed. In anaerobic sediments, reduction of SO_4^{2-} to S^{2-} leads to precipitation of extremely insoluble CuS or mixed metal sulfides containing Cu.

15.5.2.4 Biological effects

The strength of the copper–organic bond and the well-established dual potential of copper as a biological stimulant or inhibitor has led to many studies of the effect of copper speciation on growth. A few of these studies, showing that strong organic and inorganic complexes of most metals, including Cu, are non-toxic, were referred to earlier. The general conclusion that hydrated Cu^{2+} appears to be more biologically active than chelated or certain organically or inorganically complexed forms, bears repeating. This fact becomes more significant when put in the context of the very small ratio of Cu^{2+} to total Cu in most natural waters. More than for any other metal, it has been demonstrated for copper that toxicity does not correlate well with total copper or even total soluble copper in a system. Some uncertainty remains regarding the toxicity of certain inorganic complexes (particularly the hydroxo- and carbonato-complexes) and the roles of alkalinity, hardness, and other heavy metals in modifying copper toxicity. Nevertheless, studies of Cu toxicity interpreted in terms of speciation have been extremely successful and have helped establish the value of such an approach. In the past decade, analysis of speciation has shifted from the state-of-the-art to the mainstream in toxicity studies and will undoubtedly continue to shed light on the biological effects of all heavy metals.

15.6 Summary

We have tried to give an overview of the important processes and reactions controlling biogeochemical cycling of metals. While each metal has some unique properties, several generalizations can be made. The most important of these have to do with metal speciation, i.e. the physical/chemical form in which the metal exists in a given environment. Because metals tend to undergo a greater number and variety of relatively rapid, reversible reactions than the elements disussed in previous chapters, they are found in the environment in solid, aqueous, or gaseous phases, associated with literally thousands of different compounds. These reactions often reflect the affinity of metal ions for other atoms with free electron pairs, in particular O, N, or S.

The critical processes controlling global metal cycling are volatilization (exchange between aqueous and gaseous phases), adsorption/precipitation/dissolution (exchange between aqueous and solid phases), and complexation (conversion among various dissolved forms of the metal). For most metals, transport as a gaseous species is of little quantitative importance except in very high-temperature environments. A few metals, most notably mercury, can exist as gases at ambient temperatures, and several metals form volatile organometallic compounds that may dominate transport of the metal in local environments. Transport of particles suspended in the air is an important process for distributing many metals to regions far from their sources.

Transport in solution or aqueous suspension is the major mechanism for metal movement from the land to the oceans and ultimately to burial in ocean sediments. In solution, the hydrated metal ion and inorganic and organic complexes can all account for major portions of the total metal load. Relatively pure metal ores exist in many places, and metals from these ores may enter an aquatic system as a result of weathering. For most metals, a more common sequence is for a small amount of the ore to dissolve, for the metal ions to absorb onto other particulate matter suspended in flowing water, and for the metal to be carried as part of the particulate load of a stream in this fashion. The very insoluble oxides of Fe, Si, and Al (including clays), and particulate organic matter, are the most important solid adsorbents on which metals are "carried".

The distribution of metals between dissolved and particulate phases in aquatic systems is governed by a competition between precipitation and adsorption (and transport as particles) *vs* dissolution and formation of soluble complexes (and transport in the solution phase). A great deal is known about the thermodynamics of these reactions, and in many cases it is possible to explain or predict semi-quantitatively the equilibrium speciation of a metal in an environmental system. Predictions of complete speciation of the metal are often limited by inadequate information on chemical composition, equilibrium constants, and reaction rates.

Metals act upon and are acted upon by biota in important ways. Through their effect on the chemical environment in soils, sediments, or open bodies of water, organisms can help dissolve, complex, or precipitate metals. Additionally, as noted above, suspended organic particulates can be important adsorbents for metals, carrying them through an aquatic system to its outlet or through a water column to the bottom sediments. Organisms can also directly mediate reactions involving metals, such as by formation of organometal compounds.

Metals, in turn, can either stimulate or inhibit biological activity. In this regard, the different effects of dissolved metal species are especially important. For most metals, the hydrated metal ion and simple inorganic complexes have a much greater effect on organisms (in both the stimulatory and inhibitory ranges) than large organic complexes or adsorbed metal ions.

The full appreciation of the overriding importance of metal speciation in evaluating the transport and effects of metals in an environment is a relatively recent event. As more information is gathered on the forms in which metals exist and are transported through various environmental compartments, it will become possible to predict more accurately the response of the biological communities exposed to the metals and hopefully avert or mitigate the adverse effects.

Questions

15-1 Imagine that an industrial spill allows an acidic solution containing chromate ion (CrO_4^{2-}) to enter a turbid, slow-moving stream containing organic-rich run-off. The streamwater pH is lowered to about 5 by the spill. As the water moves downstream, some of the particles settle, eventually being buried in the sediment. Other particles remain suspended, and as additional tributaries mix in, the pH of the water returns to a normal level of around 7.5. Describe the various fates that chromate ions might experience in this system.

15-2 Critique the statement, "All metals oxidize (corrode) when exposed to air and water. Therefore over a period of years they oxidize completely." Discuss whether the statement is formally valid, whether the second statement follows from the first, and how the statements apply to such mundane settings as steel or copper water pipes or aluminium lawn furniture.

15-3 In the treatment of water for public consumption, iron(III) chloride is frequently added to the water. After pH adjustment, the iron forms a precipitate of $Fe(OH)_3$, which is thought to enhance coagulation and subsequent settling of suspended colloids. Discuss what you think would happen to dissolved copper in the water when this iron solid dissolves. Consider two "types" of water supply: a relatively pristine supply generated by snowmelt from a remote area, and a supply high in natural organic matter (humic matter) generated by percolation of rainwater through organic-rich vegetated zones.

15-4 Using the equilibrium constants below, calculate the concentrations of free (uncomplexed) cadmium ion in a freshwater with a chloride concentration of 15 mg/L, and in seawater containing 17 000 mg/L chloride. Ignore complexation with other ions.

$$Cd^{2+} + Cl^- \leftrightarrow CdCl^+ \quad \log K_1 = 2.0$$

$$CdCl^- + Cl^- \leftrightarrow CdCl_2 \quad \log K_2 = 0.7$$

$$CdCl_2 + Cl^- \leftrightarrow CdCl_3^- \quad \log K_3 = 0.0$$

References

Andreae, M. O. (1979). Arsenic speciation in seawater and interstitial waters: The role of biological–chemical interactions on the chemistry of a trace element. *Limnol. Oceanogr.* **24**, 440–452.

Avotins, P. V. (1975). Adsorption and coprecipitation studies of mercury on hydrous iron oxides. Ph.D. Dissertation, Stanford University, Stanford, Calif.

Balistrieri, L., P. G. Brewer, and J. W. Murray (1981). Scavenging residence times of trace metals and surface chemistry of sinking particles in the deep ocean. *Deep-Sea Res.* **28A,** 101–121.

Bertine, K. K. and E. D. Goldberg (1971). Fossil fuel combustion and the major sedimentary cycle. *Science* **173**, 233–235.

Broecker, W. S. (1974). "Chemical Oceanography." Harcourt Brace Jovanovich, New York.

Bruland, K. W., K. Bertine, M. Koide, and E. D. Goldberg (1974). History of metal pollution in Southern California coastal zone. *Environ. Sci. Technol.* **8**, 425–432.

Bruland, K. W., G. A. Knauer, and J. H. Martin (1978). Cadmium in Northeast Pacific waters. *Limnol. Oceanogr.*, **23**, 618–625.

Coles, D. G., R. C. Ragaini, J. M. Ondor, G. L. Fisher, D. Silberman, and B. A. Prentice (1979). Chemical studies of stack fly ash from a coal-fired power plant. *Environ. Sci. Technol.* **13**, 455–459.

Davies-Colley, R. J., P. O. Nelson, and K. J. Williamson (1984). Copper and cadmium uptake by estuarine sedimentary phases. *Environ. Sci. Technol.* **18**, 491–499.

Davis, J. S. (1984). Complexation of trace metals by adsorbed natural organic matter. *Geochim. Cosmochim. Acta* **48**, 679–691.

Davis, J. S. and J. O. Leckie (1978). Effect of adsorbed complexing ligands on trace metal uptake by hydrous oxides. *Environ. Sci. Technol.* **12**, 1309–1315.

de Groot, A. J. and E. Allersma (1975). Field observations on the transport of heavy metals in sediments. *In* "Heavy Metals in the Aquatic Environment" (P. A. Krenkel, ed.) pp. 85–95. Pergamon Press, Oxford.

Eisenreich, S. J., M. R. Hoffman, D. Rastetter, E. Yost, and W. J. Maier (1980). Metal transport phases in the upper Mississippi River. *In* "Particulates in Water" (M. Kavanaugh and J. O. Leckie, eds.), Advances in Chemistry, Vol. 189, pp. 135–176. American Chemistry Society, Washington, D.C.

Eisenreich, S. J., N. A. Metzer, and N. R. Urban (1986). Response of atmospheric lead to decreased use of lead in gasoline. *Environ. Sci. Technol.* **20**, 171–174.

Galloway, J. N. (1979). Alteration of trace metal geochemical cycles due to the marine discharge of wastewater. *Geochim. Cosmochim. Acta* **43**, 207–218.

Goldschmidt, V. M. (1954). "Geochemistry." Oxford University Press, Fairlawn, N.J.

Hardy, J. T., C. W. Apts, E. A. Crecelius, and G. W. Fellingham (1985). The sea–surface microlayer. Fate and residence times of atmospheric metals. *Limnol. Oceanogr.* **30**, 93–101.

Helz, G. R. (1976). Trace element inventory for the Northern Chesapeake Bay with emphasis on the influence of man. *Geochim. Cosmochim. Acta* **40**, 573–580.

Hemmes, P., L. D. Rich, D. L. Cole, and E. M. Eyring (1971). Kinetics of hydrolysis of ferric ion in dilute aqueous solution. *J. Phys. Chem.* **75**, 929–932.

Hoffman, M. R. (1981). Thermodynamic, kinetic and extra-thermodynamic considerations in the development of equilibrium models for aquatic systems. *Environ. Sci. Technol.* **15**, 345–353.

Imhoff, K. R., P. Koppe, and F. Dietz (1980). Heavy metals in the Ruhr River and their budget in the catchment area. *Prog. Water Technol.* **12**, 735–749.

Jones, C. J. and J. W. Murray (1984). Nickel, cadmium and copper in the northeast Pacific off the coast of Washington. *Limnol. Oceanogr.* **29**, 711–720.

Keck, J. C. (1978). Rate-controlled constrained equilibrium method for treating reactions in complex systems. *In* "Maximum Entropy Formalism" (R. D. Levine and M. Tribus, eds.), pp. 219–245. MIT Press, Cambridge, Mass.

Kudo, A. (1983). Physical/chemical/biological removal mechanisms of mercury in a receiving stream. *In* "Toxic Materials – Methods for Control" (N. E. Armstrong and A. Kudo, eds.), pp. 325–285. The Center for Research in Water Resources, University of Texas at Austin, Austin, Texas.

Kudo, A. and S. Miyahara (1983). Migration of mercury from Minamata Bay. *In* "Toxic Materials – Methods for Control" (N. E. Armstrong and A. Kudo, eds.), pp. 265–285. The Center for Research in Water Resources, University of Texas at Austin, Texas.

Lantzy, R. J. and F. T. Mackenzie (1979). Atmospheric trace metals: Global cycles and assessment of man's impact. *Geochim. Cosmochim. Acta* **43**, 511–525.

Lasaga, A. (1980). The kinetic treatment of geochemical cycles. *Geochim. Cosmochim. Acta* **44**, 815–828.

Leckie, J. O., A. R. Appleton, N. B. Ball, K. F. Hayes, and B. D. Honeyman (1989). Adsorptive removal of trace elements from fly-ash pond effluents onto iron oxyhydroxide. *In* Field Evaluation. EPRI Report GS-6438. Electric Power Research Institute, Palo Alto, Calif.

Lerman, A. (1979). "Geochemical Processes: Water and Sediment Environments." Wiley-Interscience, New York.

Mackay, D. (1979). Finding fugacity feasible. *Environ. Sci. Technol.* **13**, 1218–1223.

Manahan, S. E. (1979). "Environmental Chemistry," 3rd edn. Willard Grant Press, Boston, Mass.

Manzione, M. A. and D. T. Merrill (1989). Trace Metal Removal by Iron Coprecipitation: Field Evaluation. Electric Power Research Institute Report EPRI GS-6438, Palo Alto, CA, USA.

Moore, J. W. and W. Ramamoorthy (1984). "Heavy Metals in Natural Waters: Applied Monitoring and Impact Assessment." Springer-Verlag, New York.

Morel, F. M. M. (1983). "Principles of Aquatic Chemistry." Wiley-Interscience, New York.

Morel, F. M. M., J. C. Westall, C. R. O'Melia, and J. J. Morgan (1975). Fate of trace metals in Los Angeles County wastewater discharge. *Environ. Sci. Technol.* **9**, 756–761.

Morgan, J. J. and A. T. Stone (1985). Kinetics of chemical processes of importance in lacustrine environments. *In* "Chemical Processes in Lakes" (W. Stumm, ed.), pp. 389–426. John Wiley, New York.

National Academy of Sciences (1978). "An Assessment of Mercury in the Environment." National Academy of Sciences, Washington, D. C.

Nelson, M. B., J. A. Davis, M. M. Benjamin, and J. O. Leckie (1977). The role of iron sulfides in controlling trace heavy metals in anaerobic sediments: oxidative dissolution of ferrous monosulfides and the behavior of associated trace metals." Report #CEEDO-TR-77-13 Civil and Environmental Engineering Development Office, Air Force Systems Command, Tyndall AFB, Florida, USA.

Nriagu, J. O. (1979). "Copper in the Environment, Part I; Ecological Cycling" Wiley-Interscience, New York.

Pankow, J. F. and J. J. Morgan (1981). Kinetics for the aquatic environment. *Environ. Sci. Technol.* **15**, 1155–1164.

Paulson, A. J., R. A. Feely, H. C. Curl, and J. F. Gendron (1984). Behavior of Fe, Mn, Cu, and Cd in the Duwamish River estuary downstream of a sewage treatment plant. *Water Res.* **18**, 633–641.

Schwartzenbach, R. P. and P. M. Imboden (1984). Modelling concepts for hydrophobic pollutants in lakes. *Ecol. Model.* **22**, 171.

Settle, D. M. and C. C. Patterson (1982). Magnetudes and sources of precipitation and dry deposition fluxes of industrial and natural leads to the North Pacific at Enewetak. *J. Geophys. Res.* **87**, 8857–8869.

Settle, D. M., C. C. Patterson, K. K. Turekian, and J. K. Cochran (1982). Lead precipitation fluxes at tropical ocean sites determined from ^{210}Pb measurements. *J. Geophys. Res.* **87**, 1239–1245.

Sholkovitz, E. R. and D. Copland (1981). The coagulation, solubility, and adsorption properties of Fe, Mn, Cu, Ni, Cd, Co, and humic acids in a river water. *Geochim. Cosmochim. Acta* **45**, 181–189.

Stumm, W. and J. J. Morgan (1981). "Aquatic Chemistry." 2nd edn. Wiley-Interscience, New York.

Sunda, W. G., D. W. Engel, and R. M. Thuotte (1978). Effect of chemical speciation on toxicity of cadmium to grass shrimp. *Palaemonetes pugio:* Importance of free cadmium ion. *Environ. Sci. Technol.* **12**, 409–413.

Takamatsu, T., M. Kawashima, and M. Koyama (1985). The role of Mn-rich hydrous manganese oxide in the accumulation of arsenic in lake sediments. *Water Res.* **19**, 1029–1032.

Theis, T. L. and J. L. Wirth (1977). Sorptive behavior of trace metals on fly ash in aqueous systems. *Environ. Sci. Technol.* **11**, 1096–1100.

Turekian, K. K. (1971). Rivers, tributaries and estuaries. *In* "Impingement of Man on the Ocean" (D. W. Wood, ed.), pp. 9–73. John Wiley, New York.

Turekian, K. K. (1977). The fate of metals in the oceans. *Geochim. Cosmochim. Acta* **41**, 1139–1144.

Turner, D. R., M. Whitfield, and A. G. Dickson (1981). The equilibrium speciation of dissolved components in freshwater and seawater at 25° C and 1 atm pressure. *Geochim. Cosmochim. Acta* **45**, 855–881.

Weiss, H. V., M. Koide, and E. D. Goldberg (1971a). Selenium and sulfur in a Greenland ice sheet: Relation to fossil fuel combustion. *Science* **172**, 261–263.

Weiss, H. V., M. Koide, and E. D. Goldberg (1971b). Mercury in a Greenland ice sheet: Evidence of recent input by man. *Science* **174**, 692–694.

Wood, J. (1974). Biological cycles for toxic metals in the environment. *Science* **183**, 1049–1052.

SCOTT W. STARRATT

16

Human Modification of Global Biogeochemical Cycles

Robert J. Charlson, Gordon H. Orians, Gordon V. Wolfe, and Samuel S. Butcher

Chapter 1 began by posing several examples of global, societally important environmental issues, the solution or management of which requires an understanding and integrated treatment of biogeochemical cycles. These issues appear frequently throughout the book, particularly in the references to anthropogenic fluxes of substances that significantly perturb individual elemental cycles. In this brief concluding chapter, we revisit these issues to call attention to their biogeochemical nature.

16.1 Global Climate Change

Many biogeochemical and physical processes are involved in determining the climate of the Earth, and some of these are being significantly perturbed by human activity. Some of these perturbations can be assessed quantitatively, some have been discovered recently and can be described only qualitatively, and still others are yet to be discovered. Of particular importance is the atmosphere, through which all energy enters and leaves the Earth. The physical and chemical composition of the atmosphere determines the transmission, absorption, and reflection of incoming solar radiation and outgoing terrestrial radiation, and the resulting energy balance determines surface temperature. The biogeochemical cycles of sulfur – a crucial component of clouds and most aerosols – and of carbon and nitrogen, which form radiatively important trace gases, are central to the radiative properties of the atmosphere. The cycles of these three elements are also severely perturbed by human activity.

Figure 16-1 sketches the interplay of these cycles and climate. Here we denote observable quantities by boxes and the processes that affect them by ovals. This figure illustrates the two major processes by which chemical cycles affect climate: the greenhouse effect and aerosol/cloud formation. We see, for example, the radiatively important natural atmospheric trace gases water vapor, carbon dioxide, methane, and nitrous oxide, as well as the radiatively important anthropogenic chlorofluorocarbons. This group of gases, produced from a variety of natural and human processes affecting the cycles of water, carbon, nitrogen, and halocarbons, absorb infrared radiation in the atmosphere, changing the global heat balance, i.e. the greenhouse effect (see Section 10.11). The other important climate-affecting process, aerosol and cloud formation, appears to be dominated by the sulfur cycle: the production of sulfur gases that are oxidized to sulfuric acid in the atmosphere, forming new aerosol particles. Some of these have direct radiative effects and some may act as cloud condensation nuclei (CCN) to produce clouds and affect cloud albedo. Therefore, the sulfur cycle influences the short-wave radiation properties of the atmosphere, and the cycles of water, carbon, nitrogen, and trace halocarbons help determine the long-wave properties.

Of course, such a flow diagram cannot accurately portray the complete climate–biogeochemical cycle system. Rather, diagrams such as this are intended

Global Biogeochemical Cycles
ISBN 0-12-147685-5

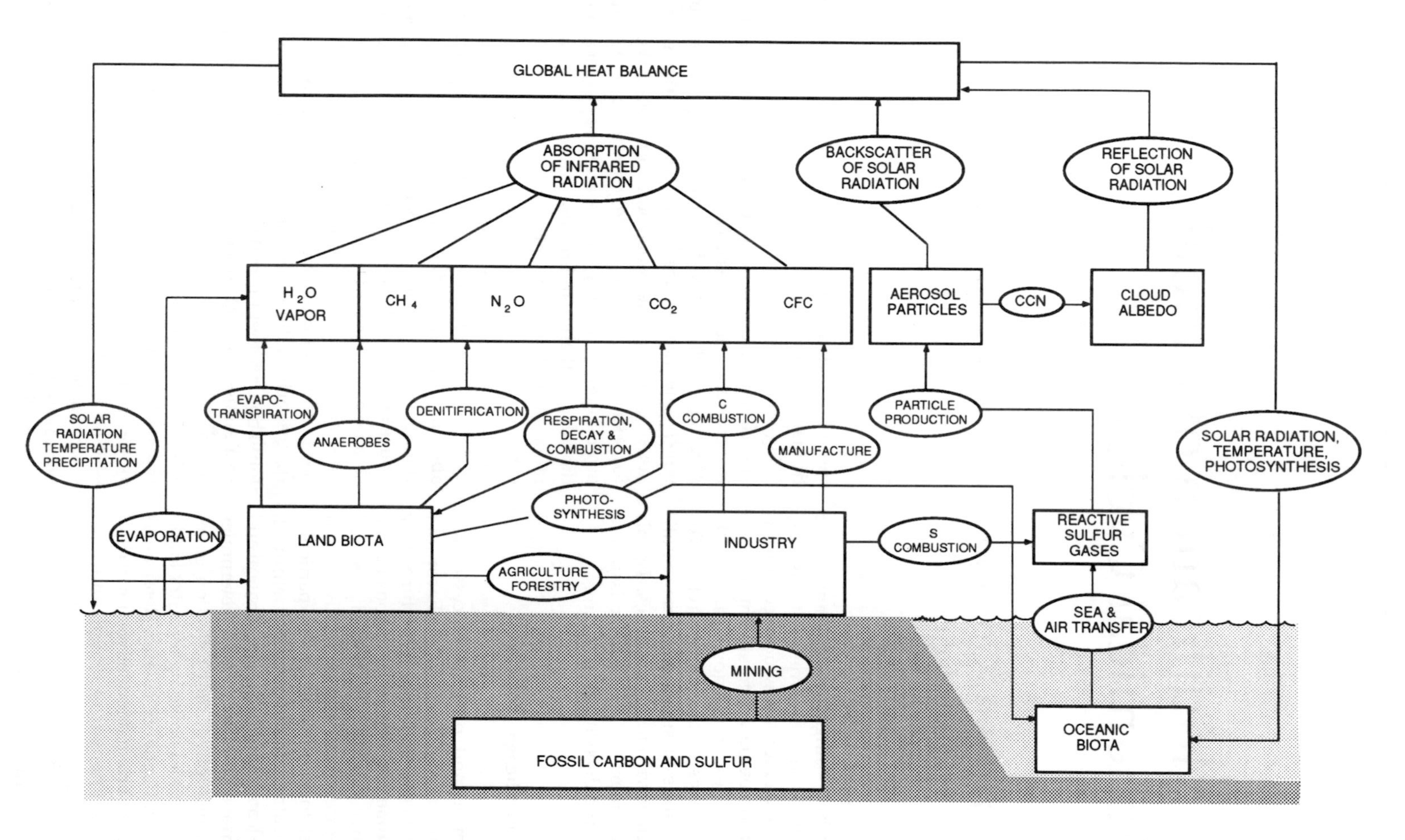

Fig. 16-1 Schematic of the processes that connect global biogeochemical cycles and climate. Boxes denote observables and ovals indicate processes that affect them.

Table 16-1 Radiatively important trace species in the atmosphere: Percent change in flux measured relative to the pre-industrial age

Cycle change	Species	% change
Long-wave absorbers		
Water	H_2O (vapor)	Not known
Carbon	CO_2	+50%
	CH_4	>+65%
Nitrogen	N_2O	+25%
Halogens	Chlorofluorocarbons	+∞%
Short-wave reflectors		
Sulfur	SO_4^{2-}	+230%

to provide integration and, most especially, the definition of key quantities and processes to be observed and measured. Ultimately, our goal is to include a full quantitative model of the biogeochemical fluxes, their geographical variations and changes, along with a complete physical model of the climate.

Table 16-1 demonstrates that, with the exception of water vapor, all of these cycles have been severely perturbed by human activity. Of course, all of these cycles are also linked in many ways. For example, the combustion of fossil fuel has increased the fluxes of carbon and sulfur to the atmosphere. Denitrification, the production of N_2O, is linked with the production of CO_2 during respiration and decay. And, in addition, other important cycles are involved which are not depicted here.

16.2 Acid Precipitation

The combustion of fossil fuels, and the consequent oxidation of nitrogen in combustion air (see Chapters 5 and 12) has greatly modified the natural atmospheric cycles of C, N, and S. Although the change of atmospheric CO_2 has little effect on the chemical composition of precipitation, sulfur- and nitrogen-containing acids have a major impact on the chemical composition of rain and snow. These acids in turn perturb the cycles of important minor elements, such as aluminum, through the weathering process in rocks and soils (Chapters 6 and 7). In developing a simple model of this phenomenon, both as a way to define key processes and to understand and quantify them, it is important to recognize the fundamental chemical nature of the acid–base balance of aqueous solutions – in this case, rainwater and melted snow water. As shown in Fig. 10-12, the pH of rainwater is especially sensitive to the addition of small amounts of strong acids because the natural system is like a titration near its endpoint.

We have only recently understood the phenomena that control rainwater pH in the natural, unpolluted environment. As pointed out in Section 10.9, these appear to be mainly the cycles of sulfur and nitrogen compounds. A model of the unperturbed system is necessary in order to understand and predict the changes that occur when strong sulfur- and nitrogen-acids are added, as well as to foresee the complex effects that perturbed rainwater pH has on other biogeochemical cycles and ecosystem processes.

16.3 Food Production

Few limitations have more profound implication for human welfare than the availability of food and water. Food production, whether by agriculture or hunting and gathering, is strongly influenced by climate, by the availability of key nutrients in adequate amounts during the growing season, by the presence of toxic materials, and by the physical, chemical, and microbiological properties of the soils. The biogeochemical cycles of C, N, P, and S are central to food production. A supply of minor elements is also important under some soil conditions.

Food production is influenced directly by biogeochemistry via precipitation chemistry and changes in soil properties induced by it. The potential benefit of increased NO_x in precipitation is an increased availability of nitrogen, a limiting nutrient in most agricultural regions. The potential negative consequence is increased soil acidity. The outcomes are certain to vary with soil types and climates. Better information on the interrelationships between precipitation chemistry and soil chemistry will help agriculture adapt to changes induced by human perturbations of biogeochemical cycles.

Agriculture, the major human activity directly exposed to climate, has always been sensitive to climate change. Droughts and floods have depressed food production. Desperate farmers have fled from formerly productive areas to seek a better life. Despite these difficulties, agricultural scientists have developed many different strains of crops adapted to different climates that have helped to buffer agricultural productivity and keep pace with increasing demands for food. However, in poorer countries, agricultural infrastructure is poorly developed and technical services are poor or lacking. Where climate change may occur, these countries are likely to experience severe difficulties in adapting to the change.

Predictions of changes in productivity are difficult in part because a major cause of probable climate change is increasing atmospheric carbon dioxide, which is also the raw material for photosynthesis. Enriching the atmosphere with CO_2 could potentially speed rates of photosynthesis and, thereby, increase agricultural production. Indeed, laboratory experiments in which CO_2 concentrations were increased from 300 to 600 ppm, increased photosynthetic rates by 20% in maize and 60% in wheat (Akita and Moss, 1973). Laboratory experiments also show that enrichment with CO_2 increases the growth of root shoots and increases the efficiency of water use.

What will happen under field conditions is, however, highly uncertain. Laboratory experiments have been conducted under conditions of abundant nutrients and water, ideal temperatures, and no competition among experimental plants. Such conditions are rare in the field. The best prediction is that, whereas increases similar to those found in the laboratory are unlikely under field conditions, increased concentrations of CO_2 are likely to ameliorate detrimental effects of climate change. However, field scale experiments under a variety of soils and climates and with several crop plants are needed to provide information on the effects of climate change, accompanied by higher concentrations of CO_2, on agricultural productivity.

16.4 Stratospheric Ozone Depletion

The Antarctic "ozone hole" is one of the most dramatic indications of anthropogenic environmental change. Depletion of stratospheric ozone via the catalytic mechanisms described in Chapters 5, 10, and 12 was first detected in a surprising fashion by British observers in Antarctica (Farman *et al.*, 1985). They measured increases of springtime solar UV radiation penetrating the atmosphere at wavelengths that normally are absorbed by O_3. The results showed almost a factor of 2 depletion of ozone between 1956 and 1985, with most of the change occurring after 1976, as seen in Fig. 16-2. Subsequent re-analysis of satellite data revealed that most of this depletion was limited to a geographical area perhaps 1000 km across, fortunately situated over an unpopulated place with no agriculture and very few people (Stolarski, 1988).

The chemical reactions in the oxygen-only mechanism (see Sections 5.3 and 10.4) substantially underestimate the ozone destruction rate:

$$O_2 + h\nu \rightarrow O + O$$
$$O + O_2 + M \rightarrow O + M$$
$$O_3 + h\nu \rightarrow O + O_2$$
$$O + O_3 \rightarrow 2O_2$$

Crutzen (1971) and Molina and Rowland (1974) showed that a second class of catalytic processes exist that result in the destruction of ozone:

$$X + O_3 \rightarrow XO + O_2$$
$$XO + O \rightarrow X + O_2$$
$$\text{net: } O_3 + O \rightarrow 2O_2$$

Here X may be H, OH, NO, Cl, or Br. Cl from the photodissociation of chlorofluorocarbons such as

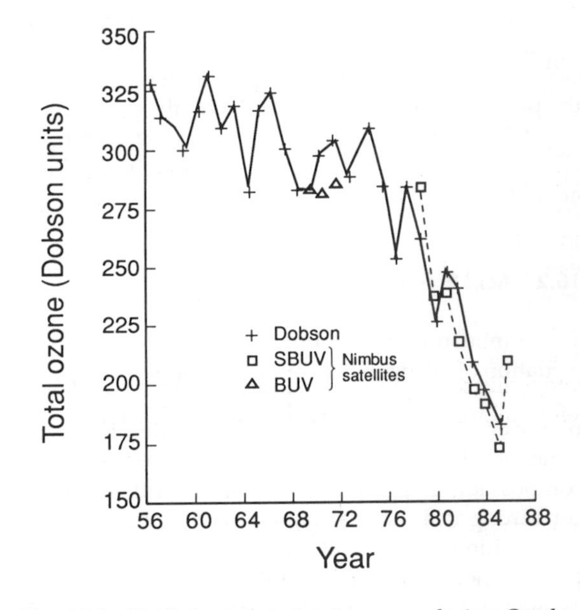

Fig. 16-2 Depletion of Antarctic ozone during October between 1956 and 1985. Adapted from Stolarski (1988). Dobson data are from ground-based observations with the permission of John Wiley and Sons, Inc.

CFC-12 (CCl_2F_2) is the main known catalyst currently acting in the so-called "ozone hole" in the Antarctic spring.

It now appears that both the extreme magnitude and geographic limitations of the Antarctic ozone depletion are due to meteorological patterns peculiar to the South Pole. The decrease, which exceeds the small reduction in the rest of the stratosphere apparently involves the circulation of the polar vortex, a complex interaction of Cl with oxides of nitrogen, their physical trapping in extremely cold ($T < -80°C$) clouds, and preferential removal of some species by precipitation.

Although the unique circumstance of the Antarctic ozone hole does not extend to heavily populated parts of the world, there is disturbing evidence that stratospheric ozone levels worldwide have also begun to drop. Certainly, the distribution of CFCs, which are extremely long-lived compounds, is global, and their atmospheric burdens are ever increasing.

The depletion of stratospheric ozone is an example of the interaction of chemical cycles that was not predicted to occur when the supposedly inert CFCs were invented. It is, in fact, the extremely low reactivity of CFCs which made their use so popular and which is responsible for their very long lifetimes in the atmosphere. Therefore, one lesson of the ozone hole has been that we must have a more complete understanding of the natural chemical cycles in order to better predict the effects and environmental implications of our novel compounds. Examples abound of our failure to predict the interactions of products of industrialization with natural chemical cycles: leaded gasolines, halogenated pesticides such as DDT, the biovolatilization of waste selenium and mercury, and so on. Currently, we develop, license, and use hundreds of new compounds each year. The greater our understanding of biogeochemical cycles, the greater the chances that potentially damaging compounds will be recognized before they cause new environmental disasters.

16.5 Oxidative Capacity of the Global Troposphere

Because of the abundance of molecular oxygen in the atmosphere, surface waters, and soils, oxidation processes are common. Often, the reduced materials that enter into oxidation reactions are produced by biota, such as in photosynthesis. Because the atmospheric reservoir of O_2 is a major feature of this global redox system, its concentration and involvement in chemical reactions are important. Although no major changes in atmospheric O_2 as a result of human activity are forecast, changes in the oxidative capacity of the atmosphere are possible. Ordinarily, the atmosphere is a self-cleansing system due to the abundance of O_3, OH, NO_2, and other reactive species. For example, hydrocarbon emissions from biota (such as terpenes) are oxidized in a matter of hours or days to CO and then on to CO_2. Alternatively, carboxylic acids may be formed and then transferred to the hydrosphere or pedosphere by rain. The atmosphere acts much like a low-temperature flame, converting numerous reduced compounds to oxidized ones that are more readily removed from the air. The limit to the rate of oxidation can be defined by the concentration of OH radicals and other trace species, such as hydrogen peroxide.

The increase of several reduced species in air (e.g. CH_4, CO, N_2O, SO_2, non-methane hydrocarbons) suggests that the oxidative capacity of the atmosphere may be decreasing. The increase in these reduced gases may even be the cause of a decrease in oxidative capacity, by lowering the steady-state concentrations of OH and other oxidizing agents. Rasmussen and Khalil (1984) describe the mysterious increase of CH_4 starting around A.D. 1800 and accelerating in recent years to almost 1% per year (Fig. 16-3e). This increase is still a mystery because there is no clear indication of its main cause. Studies of the presence of the reduced species in air strongly suggest that several of the major biogeochemical cycles are involved in controlling the oxidative capacity of the atmosphere. The nitrogen cycle (which is heavily perturbed by the human activities of combustion and fertilizer production) is a key factor through its role in atmospheric photochemistry. The sulfur cycle is involved through the presence of SO_2 from combustion of fossil sulfur in coal and oil. Carbon monoxide as a part of the carbon cycle is also implicated. It is clearly impossible to separate out any one chemical species as the single controlling factor; rather, it is necessary to consider the entire system as a unit.

16.6 Life and Biogeochemical Cycles

The evolution of life on Earth has depended on a sustained supply of nutrients provided by the

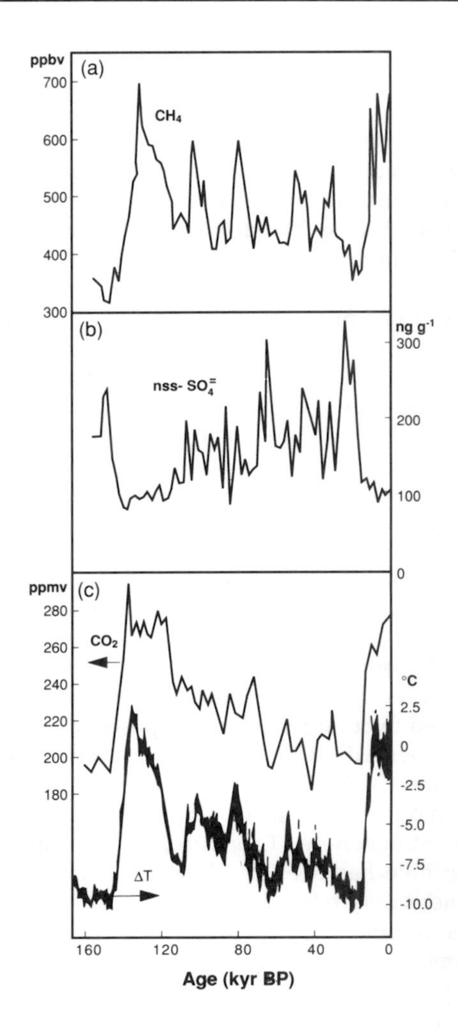

Fig. 16-3 Illustration of the linked behavior of radiatively important trace species concentrations over different time-scales. (a, b, c) Concentrations of methane, nss-SO_4^{2-}, and CO_2 over the past 160 000 years found in ice cores from Vostok, Antarctica (temperature deduced from $^{18}O_2$ also shown). (d, e, f) Secular trends in nitrous oxide, methane, and CO_2 over the past 300 years. (g, h, i, j, k) Changes in October stratospheric ozone column burden over Antarctica, and chloroflourocarbon, methane, sulfate, and nitrate from south Greenland ice, and carbon dioxide concentrations over the past 30 years. Figures adapted from: (a) Chappellaz *et al.* (1990) with the permission of Macmillan Magazines Ltd.;(b) Legrand *et al.* (1988) with the permission of Macmillan Magazines Ltd.; (c) Barnola *et al.* (1987) with the permission of Macmillan Magazines Ltd.; (d) Khalil and Rasmussen (1988) with the permission of the International Glaciological Society; (e) Stauffer *et al.* (1985) with the permission of the American Society for the Advancement of Science; (f) Siegenthaler and Oeschger (1987) with the permission of the Swedish Geophysical Society; (g) Stolarski (1988) with the permission of John Wiley and Sons, Inc. (h) Cunold *et al.* (1986) with the permission of the American Geophysical Union; (i) Blake and Rowland (1988) with the permission of the American Association for the Advancement of Science; (j) Mayewski *et al.* (1990) with the permission of Macmillan Magazines, Ltd.; (k) Keeling (1989) with the permission of the American Geophysical Union.

physical environment. Life, in turn, has profoundly influenced the availability and cycling of these nutrients; hence the inclusion of *bio* in biogeochemical cycles. The involvement of the biosphere with biogeochemical cycles has been determined by the evolution of life's biochemical properties in the context of the physical and chemical properties of planet Earth.

Not surprisingly, only about 20 of the chemical elements found on Earth are used by living organisms (Chapters 3 and 7). Most of them are common elements. Rare elements are used, if at all, only at extremely low concentrations for specialized functions. An example of the latter is the use of molybdenum as an essential component of nitrogenase, the enzyme that catalyzes the fixation of elemental dinitrogen. Because they are composed of common elements, living organisms exert their most profound effects on the cycles of those elements.

Finding and extracting elements from the environment is a potentially costly process. Organisms have evolved powerful mechanisms for scavenging nutrients from rock, soil, water, and air. In addition, they re-use most materials extensively in more or less cyclic patterns. The more abundant elements leak from these cycles because organisms are unable to use everything present. Also, some elements are released rather than being recycled when the cost of recycling exceeds the cost of scavenging new molecules. Thus, amino acids and carbohydrates are discarded when deciduous leaves are dropped from plants because the costs of breakdown and reabsorption of those molecules are too high. None the less, the amount of leakage from ecosystems may be low because microorganisms and the roots of plants quickly capture and re-incorporate the elements.

Organisms also evolved powerful detoxifying mechanisms that remove toxic materials or convert them to non-toxic forms or nutrients. Examples of alterations to non-toxic forms are the conversions of hydrogen sulfide to sulfate and nitrite to nitrate. The prime example of development of the ability to use a toxic substance is the evolution of aerobic metabolism, which converted a serious and wide-

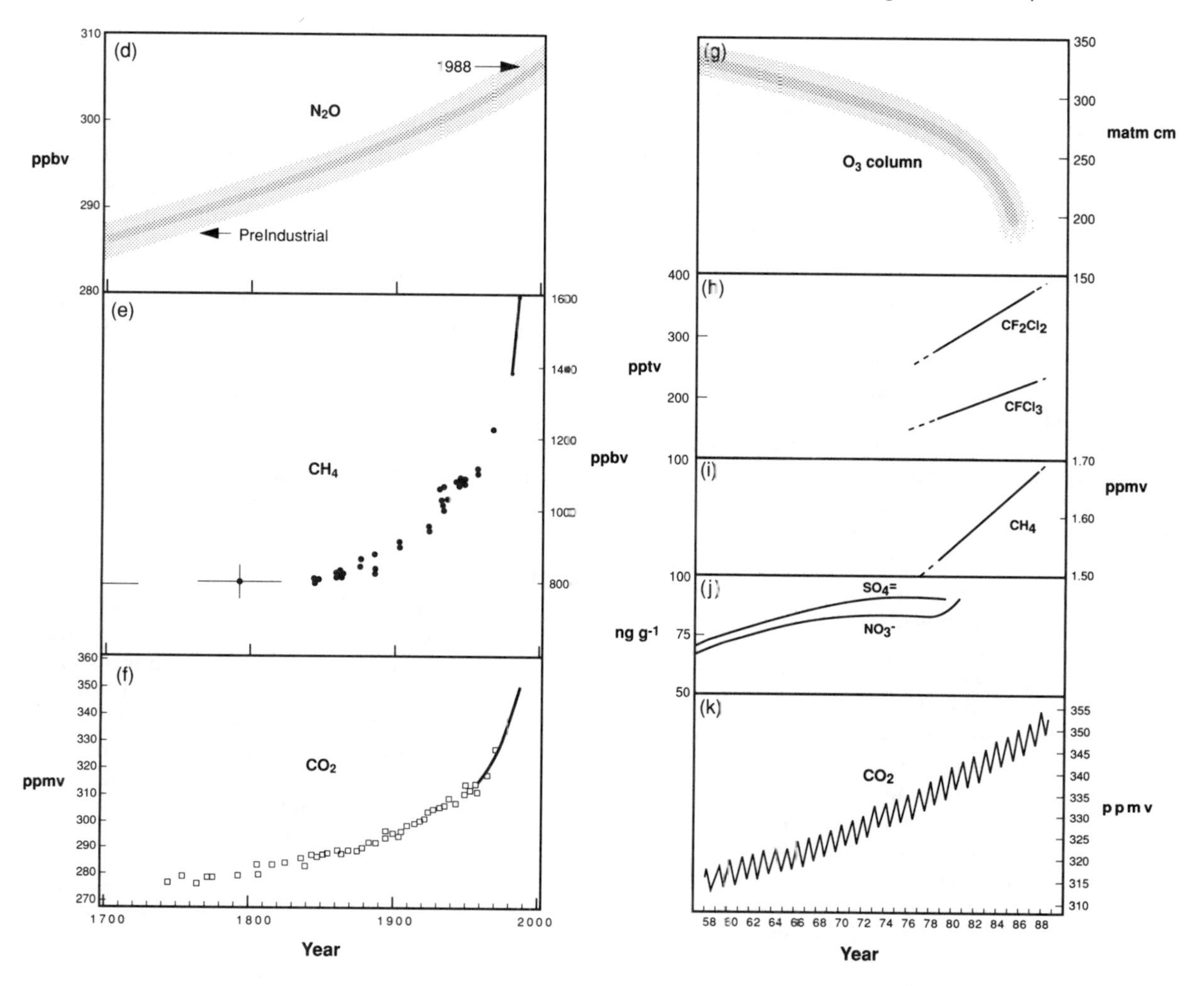

spread toxin, oxygen, into a major resource. This development, as we have seen, greatly increased the productivity of the biosphere and generated the oxygen-rich atmosphere of today.

How the biosphere will respond to human-induced climate changes is uncertain. Atmospheric concentrations of CO_2, which have been increasing for many decades, have apparently been accompanied by increased rates of photosynthesis at mid-latitudes and high latitudes during the summer. Evidence for this is provided by the increased amplitude of the high-latitude seasonal fluctuations in CO_2 concentrations. Whether or not the biosphere is, or will continue to be, a sink for CO_2 has profound implications for future climate change, productivity of agriculture and forestry, and the functioning of natural ecosystems. If increases in CO_2 concentrations are accompanied by modest and slow changes in climate, the biosphere may be able to respond by increasing productivity and sequestering more carbon. However, if climate changes are rapid, the adaptive flexibility of long-lived plants may be exceeded, lowering productivity and, hence, storage of carbon. Rapid, or even modest, climate changes are likely to shift climates latitudinally and longitudinally faster than slow-growing trees with large seeds have migrated in the past. Implications for the biological productivity and survival of species in the face of such changes are profound, lending importance to the need to better understand and predict likely climate changes. Climate change also has ramifications for the animals that feed on plants. Although animals are more mobile than plants, it is also important to take action now to establish corridors that will facilitate future migrations of species.

16.7 Conclusion

We have learned much about the individual parts and processes of the Earth's atmospheric, oceanic, continental, physical, chemical, and biological systems. However, we have just started to understand the linkages and feedbacks that make these systems function as a single entity. James Lovelock is well known as the inventor of the electron capture detector for gas chromatography, which has permitted the measurement of many environmental trace substances of anthropogenic origin, particularly those containing halogens. He is also the author of the Gaia hypothesis (Lovelock, 1979). Lovelock suggests that a new scientific discipline is emerging which he calls *geophysiology*. Other scientists have used the terms *earth system science* or *global geoscience* to describe this field of study, noting that it in no way replaces the established reductionist sciences. Rather it complements them and uses their knowledge and methods.

The degree to which this new discipline develops will depend upon the ways that world society chooses to respond to the many global processes that presently are changing. It would be speculative even to suggest that this new scientific discipline will emerge with a particular lexicon, focus, and practice. However, we do know that the global system *is* changing as a result of known processes, that it *has* changed continuously throughout the geological past and in recent times, and that current changes are large compared to natural ones in the past.

The characteristics of the changes in several species discussed in this chapter are shown in Fig. 16-3. This figure depicts the changing composition of the atmosphere on three time-scales. Figures 16-3a–c show the simultaneous variation of CO_2, CH_4, temperature, and SO_4^{2-} from the Vostok ice core. These records also clearly demonstrate that the Earth functions as a coupled system. Temperature, CO_2, and CH_4 are positively correlated with one another, but each is negatively correlated with SO_4^{2-} (for reasons that are not yet known). This time period covers 160 000 years including the present interglacial climate, the last glacial period (known in the USA as the Wisconsin ice age), the previous interglacial time, and the last bit of the penultimate ice age.

Figures 16-3d–f depict both the recent data from ice cores and the contemporary records of N_2O, CH_4, and CO_2 during the most recent 300 years. These illustrate the profound changes that have occurred since the industrial revolution. Although the exact causes of the increases of N_2O and CH_4 are not yet fully agreed upon, there is no debate regarding the relationship of the increase of CO_2 to the burning of fossil carbon and deforestation. In the case of CH_4 and CO_2, there is also excellent agreement between the ice-core records and the records from direct sampling of the atmosphere, which began in 1957 for CO_2 and about 1973 for CH_4.

Finally, Figs. 16-3g–k illustrate the accelerated rate of change that has occurred for some of these atmospheric chemical variables for the three decades from 1957 to the present. The O_3 column data (g) are for the month of October at Halley Bay, Antarctica; SO_4^{2-} and NO_3^- are for recent snow in Greenland; while the remainder are global mean values.

We want to draw two final conclusions from these figures:

1. The oldest records (a–c) clearly show a strong degree of temporal correlation between three atmospheric components and climate (as indicated by temperature). Because there is a sound physical basis for the involvement of all three in climatic processes, it is necessary to study, view, and understand these variables and climate as linked components of a system. They are all *dependent* variables and cannot be viewed as independent with climate being imposed as an exogenous factor.
2. Humans have so modified the main biogeochemical cycles of Earth that the chemical composition of the atmosphere exhibits differences that are approaching the magnitude of the changes that occurred between ice ages and interglacial periods. For example, the change in atmospheric CO_2 from the Wisconsin ice age to the preindustrial value was from 190 to 280 ppmv, an increase of about 50%. The increase from A.D 1800 until now is from 280 to 350 ppmv or an additional 25%. It appears to be inevitable that CO_2 will continue to increase, doubling from the preindustrial value by *c* A.D. 2025–2050. That the Earth's heat balance will change is also inevitable.

As much as we know about the increase in CO_2, the forecast of climatic response is unclear, largely because the current changes are unprecedented. It is likely that the near-term climatic future will be affected by human activity. Understanding the response of climate to our actions will depend upon an understanding of the biogeochemical functioning of the Earth as a linked system.

References

Akita, S. and D. N. Moss (1973). Photosynthetic responses to carbon dioxide and light by maize and wheat leaves adjusted for constant stomatal apertures. *Crop Sci.* **13**, 234–237.

Barnola, J. M., D. Raynaud, Y. S. Korotkevich, and C. Lorius (1987). Vostok ice core provides 160,000 year record of atmospheric CO_2. *Nature* **329**, 408–414.

Blake, D. R. and F. S. Rowland (1987). Continuing worldwide increase in tropospheric methane 1978 to 1987. *Science* **239**, 1129–1131.

Chappellaz, J., J. M. Barnola, D. Raynaud, Y. S. Korotkevich, and C. Lorius (1990). Ice-core record of atmospheric methane over the past 160,000 years. *Nature* **345**, 127–131.

Crutzen, P. J. (1971). Ozone production rates in an oxygen, hydrogen, nitrogen oxide atmosphere. *J. Geophys. Res.* **76**, 7311–7327.

Cunold, D. M., R. G. Prinn, R. A. Rasmussen, *et al*. (1986). Atmospheric lifetime and annual release estimates for $CHCl_3$ and CF_2Cl_2 from 5 years of ALE data. *J. Geophys. Res.* **91**, 10,797–10,817.

Farman, J. C., B. G. Gardiner, and J. D. Shouklin (1985). Large losses of total ozone in Antarctica reveal seasonal ClO_x/NO_x interaction. *Nature* **315**, 207–210.

Keeling, C. D. (1989). A three dimensional model of atmospheric CO_2 transport based on observed winds: 1. Analysis of observational data. *Geophys. Monogr.* **55**, 165–236.

Khalil, M. A. K. and R. A. Rasmussen (1988). Nitrous oxide: Trends and global mass balance. *Ann. Glaciol.* **10**, 73–79.

Legrand, M. R., R. J. Delmas, and R. J. Charlson (1988). Climate forcing implications from Vostok ice core sulphate data. *Nature* **334**, 418–420.

Lovelock, J. E. (1979). "Gaia: A New Look at Life on Earth." Oxford University Press, Oxford.

Mayewski, P. A., W. B. Lyons, M. J. Spencer, M. S. Twickler, C. F. Buck, and S. Whitlow (1990). An ice-core record of atmospheric response to anthropogenic sulfate and nitrate. *Nature* **346**, 554–556.

Molina, M. J. and F. S. Rowland (1974). Stratospheric sink for chlorofluoromethanes: Chlorine-catalyzed destruction of ozone. *Nature* **249**, 810–812.

Rasmussen, R. A. and M. A. K. Khalil (1984). Atmospheric methane in recent and ancient atmospheres: Concentrations, trends, and interhemispheric gradient. *J. Geophys. Res.* **89**, 11,599–11,604.

Siegenthaler, U. and H. Oeschger (1987). Biospheric CO_2 emissions during the past 200 years reconstructed by deconvolution of ice core data. *Tellus* **39B**, 140–154.

Stauffer, B., G. Fischer, A. Neftel, and H. Oeschger (1985). Increase of atmospheric methane record in Antarctic ice core. *Science* **229**, 1386–1388.

Stolarski, R. S. (1988). Changes in ozone over the Antarctic. *In* "The Changing Atmosphere" (F. S. Rowland and I. S. A. Isaksen, eds.), pp. 105–120. John Wiley, New York.

Glossary

This glossary includes some of the terms not defined in the text. For definitions of other terms, refer to the index for the location of discussion in the text

Adiabatic A process occurring without exchange of heat with the surroundings

Advection Flux of a quantity (e.g. heat, chemical substance) by organized fluid motion (*see also* Diffusion)

Aerobic Conditions involving the presence of oxygen as an oxidizing agent

Aerosol particles Solid or liquid particles in the atmosphere, other than cloud droplets, of diameter smaller than 10 μm

Albedo The fraction of solar energy reflected from the Earth

Alluvial Pertaining to materials moved by and deposited from flowing water

Anaerobic Conditions involving the presence of an oxidant other than O_2

Anion Ion with a negative charge

Anthropogenic Caused by human activities

Authigenic Pertaining to new minerals of clastic sedimentary rocks formed at their place of deposition, in contrast to minerals formed at other places prior to erosion, transportation, and deposition

Autotrophic Refers to organisms that can use carbon dioxide as a sole source of carbon for growth

Beta decay The nuclear decay process in which an electron is emitted and a neutron is converted to a proton in the nucleus

Biogeochemical cycle The description of the processes and quantities in the global circulation of elements

Bioturbation Mixing of sediment material by worms and other organisms

Brownian motion The random movement of a particle as a result of collisions with fluid molecules or other particles

Catalyst A substance that increases the rate of a chemical reaction without being consumed. Catalysts usually work by providing a lower activation energy pathway

Cation An ion with a positive charge

Cation exchange capacity In the past, this property has been expressed in milliequivalents per 100 g of material (meq/100 g). In SI units, it is expressed in moles of positive charge (p^+) per unit mass. Often in centimole per kilogram, cmol (p^+)/kg. The old and new units are numerically equivalent

CFC Chlorofluorocarbon: the class of industrially produced compounds containing carbon, chlorine, and fluorine

Chelate A dissolved species, typically organic, with two or more functional groups that are capable of sharing pairs of electrons with a metal cation

Clastic Material composed of pieces of pre-existing rocks

Clay Rock or mineral with a grain size less than 4 μm (for geologists: 2 μm for soil scientists), often consisting of hydrous aluminum silicates

Cloud-condensation nucleus (CCN) An atmospheric aerosol particle that serves as a nucleation site for a cloud droplet

Congruent dissolution Reaction in which a mineral forms dissolved substances

Conservative Refers to a quantity that is not changed in the course of a reaction

Craton A continental area not subject to tectonic processes

Cycle Processes and quantities in the circulation of materials through a system of reservoirs

Deposition velocity The ratio between the flux of a substance to a surface and its concentration in the medium above the surface

Dew point The temperature of a gaseous mixture at which water vapor begins to condense

Diagenesis Physical and chemical change associated with, or taking place shortly after, solidification of a sediment in the near-surface part of the Earth's crust

Diffusion Flux of a substance as a result of irregular motions of molecules (molecular diffusion) or irregular motions of fluid elements (turbulent diffusion)

DMS Dimethyl sulfide, $(CH_3)_2S$: a reduced sulfur gas produced by marine microbiota

Dry deposition The flux of material from the atmosphere to the surface by processes other than precipitation. Examples are the gravitational settling of large particles and absorption of gases such as CO_2 and SO_2 at the ocean surface

Electrophilic The tendency to attract electrons and form bonds with atoms having non-bonded electrons

Eolian Usually referring to surface material that has been transported and deposited by wind

Euphotic zone Region with enough light for photosynthesis

Eustatic Pertaining to sea-level change associated with a change in the amount of water in the oceans

Evaporite Secondary deposits, composed principally of gypsum ($CaSO_4 \cdot 2H_2O$) and halite NaCl), that are formed in ocean basins and some lakes with restricted circulation that experience climate conditions where evaporation exceeds precipitation

Fluvial Pertaining to processes and features resulting from flowing surface water

Free radical A molecule with one or more unpaired electrons

Hadley circulation The organized circulation of air rising near the equator and sinking in the sub-tropics

Heat capacity The amount of heat required to bring about a unit temperature change in a given amount of substance. Often expressed as joule/mol K or joule/kg K. Older sources may use calorie instead of joule

Heterotrophic Refers to organisms that require organic compounds for growth

Hydrophilic The tendency of a substance or a portion of a molecule to associate with water

Hydrophobic Tendency of a substance or a portion of a molecule to not associate with water. Common in hydrocarbon chains

Hypsographic A graph or map of heights or depths

Incongruent dissolution Reaction in which a mineral reacts with dissolved substances to form another mineral

Intertidal The region between the extremes of high and low tide

Isomorphous substitution Replacement of one atom with another of similar size in the structure of a crystal

Isostatic When the surface of the crust or its movement is defined by the densities of different elements of the crust

Karst Referring to topography due to the action of water on soluble rock

Lacustrine Pertaining to development in lakes

Mesic This term has a very broad meaning. In ecology, it refers to moderate conditions with respect to both temperature and moisture. In soil, specifically in soil taxonomy, mesic is used to represent a soil temperature (mean annual) that falls in the range 8–15° C

Mole fraction The ratio of the moles of a substance to the total number of moles in the sample. In the atmosphere this is the same as the volume fraction

Non-seasalt (nss) The amount of an element or compound in the bulk aerosol mass that is in excess of its seawater ratio with sodium or chloride. Often applied to sulfate

Obligatory anaerobes Organisms restricted to life in anaerobic environments

Oxidation The reaction in which a substance gives up electrons (*see also* Reduction)

Phosphorite Sedimentary deposits of calcium phosphate minerals

Phytoplankton Single-celled photosynthetic algae

Podzolization A process commonly occurring in humid-temperate environments and tropical regions. The process involves the participation of soluble organic compounds that complex Fe, Al, and other metals and transport them to depth, where they are arrested. The result is a soil that displays a dramatic segregation of horizons

Podzols Soils affected by podzolization that show a thick organic layer followed by a gray leached mineral horizon (E) and by horizons enriched in humus (Bh) and iron (Bs)

Polygenetic soil A soil that has formed as a result of more than one genetic process. For example, a soil that had originally formed under a given plant assemblage that was later replaced by another, e.g. the replacement of a prairie by a forest

Primary minerals Minerals present as they were formed in the original rock

Pyroclastic Pertaining to surface deposits from explosive volcanic eruptions

Reduction The reaction in which a substance gains electrons (see Oxidation)

Refractoriness The resistance of a substance to degradation. Often applied to organic matter

Relative humidity The ratio of the partial pressure of water to the saturation vapor pressure at that temperature

Reservoir An amount of material defined by certain chemical, physical, or biological characteristics that, under particular conditions, can be considered as reasonably homogeneous

Residence time The time that an individual atom, particle, or organism spends in a reservoir

Response time The time-scale characterizing the adjustment of a reservoir to equilibrium following a sudden change in boundary conditions, inputs, or outputs

Saprolite Finely divided weathered rock that has not been transported away from the site of weathering

Secondary minerals Minerals that are formed from primary minerals by weathering, diagenesis, and other post-consolidation reactions

Sedimentation The downward flux, due to gravity, of material suspended in a fluid

Stereoisomers Compounds that have the same atoms connected in the same sequence, but in which the spatial arrangements of the atoms differ. This term includes optical isomers and *cis–trans* isomers

Stratosphere The portion of the atmosphere extending from the top of the troposphere (8–15 km) up to about 50 km. The stratosphere is characterized by poor mixing and active photochemistry

Troposphere The portion of the atmosphere extending from the surface to a height that ranges from 8 to 15 km. The troposphere is characterized by the presence of much of the Earth's weather and by rapid mixing

Turbulent flux Flux of a quantity due to irregular motion of fluid elements (*see also* Diffusion and Advection)

Turnover time The ratio of the content of a reservoir to the sum of its sinks (or sources)

Ultraxeric Refers to extremely dry conditions, as experienced by the inland edge of the Transantarctic Mountains (Antarctica), where precipitation (solid) is less than 50 mm/year

Wet deposition Flux of a substance to the surface as a result of precipitation

Xeric Referring to dry conditions (*see also* Ultraxeric)

Zenith The imaginary line extending vertically from the Earth's surface

Answers

Answers are given for some of the numerical questions in the text.

Chapter 4

4-1 If steady state is assumed ($Q_1 + Q_2 = S$), Q_2 can be estimated as $S - Q_1 = 25$. If the uncertainties in S and Q_1 are independent, the uncertainty of the Q_2 estimate is from -25 to $+75$.

4-2 Atmosphere: 3.5 year; Surface water: 4.3 year; Shortlived biota (land): 1.2 year, etc.

4-3 $\tau_0 = \tau_r = 2\tau_a$

4-4 The response time of the system is 0.83 year.

4-5 The response time is $1/k$. However, the turn-over time τ_0 is $M_0/k(M_0 - M_1)$ and thus different from the response time. The turnover time depends on the steady state content M_0. If M_0 is just a little bit larger than the threshold value, τ_0 will be very large.

Chapter 5

5-1 $\Delta G^0 = +157.6$ kJ for the reaction $3/2\ O_2(g) \rightarrow O_3(g)$. K_{eq} is thus 1.1×10^{-38} and equilibrium O_3 pressure is 4×10^{-42} bar.

5-2 The Nernst equation for the reduction of SO_4^{2-} is $Eh = +0.2486 - (0.059/8) \log ([HS^-][SO_4^{2-}]) - (0.059/8)$ 9 pH, and Eh for this couple is -0.316. Eh must be determined by redox couples that react much more rapidly.

5-3 $[Fe^{2+}]/[Fe^{3+}] = 1.5 \times 10^{14}$.

5-4 By equating the rates of reactions 1 and 2, Cl/ClO $= 1.25 \times 10^{-4}$, or Cl and ClO are 6×10^4 and 5×10^8, respectively. The number of cycles is approximately the ratio of the rates of catalytic and removal processes. This is $1.12 \times 10^6/50 = 2 \times 10^4$.

5-5 The average residence time for CH_3CCl_3 is given by $\tau = 1/k[OH]$. Taking $T = 288$ K for the troposphere, $k = 9.8 \times 10^{-15}$ cm^3/molec s and $[OH] = 3 \times 10^5$ molec/cm^3.

5-6 7.5×10^{17} of O_2 will be consumed, or the pressure of O_2 will decrease by 0.06%. The decrease in O_2 will reduce the rate of decomposition of fixed carbon, increasing the rate of carbon burial and increasing atmospheric oxygen.

Chapter 9

9-1 τ_{water} (river input) = 43 750 year; τ_{water} (hydrothermal output) = 1.4×10^7 year; τ_{Mg} (river input) = 1.36×10^7 year; τ_{Mg} (hydrothermal) = 1.4×10^7 year.

9-2 The flux is 6.88×10^{-6} mol cm^{-2} s^{-1} out of the ocean.

9-4 The activity of water would be 0.60. This is substantially different from the activity of water in seawater and thus could not be in equilibrium with water in seawater. The spontaneous reaction in seawater will be to the right.

Index

Abbreviations, Covers
Abyssal circulation, 184–187
Abyssal water age, 187
Accumulation mode, 233
Acetaldehyde reactions atmosphere, 229
Acetylene reactions atmosphere, 229
Acid deposition, 275
Acid mine water, 51
Acid precipitation, 355
  neutralization, 124
  plants, 124
  sulfur, 285
Acid rain, *see* Acid precipitation
Acids soils, 135
Activation energy, 82
Acute toxins, 45
Adenosine triphosphate (ATP), 27, 42
  function, 27
  structure, 27, 304
Adiabatic, 363
Adiabatic lapse rate, 215
Advection, 66, 219, 363
Aerobic, 363
Aerosol, 233
  climate, 235
  composition, 226, 227
  particles, 363
  reactions, 233
  size distribution, 232
Age, 59
Age frequency function, 59
Air-sea exchange, 68
Air-water boundary metals, 334
Aitken nuclei, 233
Alanine structure, 31
Albedo, 52, 234, 363
  clouds, 235
Alfisols, 144
Aliphatic acids soils, 135
Alkalinity, 76, 243, 244
  distribution in ocean, 245
  surface water, 110
Allophane, 130
  formation, 147
  properties, 129
Alluvial, 363
Alpha reactions, 13
Aluminosilicates, 130
  weathering, 140
Aluminum
  acid precipitation, 355
  acidity soils, 139
  aqueous, 76
  complexation, 126, 133, 340
  concentration ocean, 194
  crustal abundance, 127
  mobilization, 318
  reaction rates, 339
  solar abundance, 11
  sources, 319, 328
  speciation ocean, 194
Aluminum-26 heat source, 14, 19
Aluminum cycles soils, 145
Aluminum hydroxides, 131
Aluminum oxides, 130, 131
Amazon River
  denudation rate, 109
  dissolved solids, 109
  water chemistry, 103, 110
Amides, 266
Amines, 265
  reactions atmosphere, 227, 229
Amino acids, 29–31
  chondrites, 29
  hydrophilic, 31
  hydrophobic, 31
  sediments, 168
  soils, 135
  structure, 31
  synthesis, 26
Ammonia, 265
  assimilation, 267, 268
  atmosphere, 273
  clouds, 272
  condensation sequence, 15
  fluxes, 275
  formation sediments, 169
  metals, 326
  oceans, 272
  oxidation, 271
  reactions atmosphere, 227
  sources, 273
  thermodynamic data, 264
Ammonia cycle, 276
Ammoniafication, 49, 267, 269
  sediments, 168
Ammonium
  diffusion constant sediments, 163
  thermodynamic data, 264
Ammonium chloride thermodynamic data, 264
Amphibole
  properties, 127
  weathering, 95
  weathering soils, 138
Amphoteric character, 76
Anaerobic, 363
Anatexis, 57
Animal defence, 45
Animal kingdom, 40
Anion, 363
Antarctic Bottom Water (ABW), 181, 184
  properties, 186
Antarctic Intermediate Water (AIW), 181
  properties, 186
Antarctica weathering, 142
Anthropogenic, 363
Antimony
  concentration ocean, 195
  solar abundance, 11
  sources, 319
  speciation ocean, 195
Apatite, 169
  formation, 303
  properties, 127
  structure, 303
Apparent Oxygen Utilization, 201
Archaebacteria, 27
Archean Eon, 25
Arginine structure, 31
Argon
  concentration ocean, 194
  solar abundance, 11
  solubility seawater, 200
Argon-40 concentration atmosphere, 223
Aridisols, 145
Arrhenius equation, 82
Arsenic
  adsorption effects, 331
  concentration ocean, 194
  organisms, 335
  solar abundance, 11
  solubility, 329
  sources, 319
  speciation, 334
  speciation ocean, 194
Asparagine structure, 31
Aspartic acid structure, 31
Assimilation
  animal types, 48
  mean values, 48
Assimilatory nitrate reduction, 267
Asteroid impacts, 19
Atlantic Ocean
  area, 177
  mean depth, 177
  water balance, 178
Atmophiles, 318

Atmosphere, 213–238
carbon compounds, 227, 241
carbon dioxide, 228
carbon monoxide, 228
climate, 234
composition, Covers, 223, 226–230
dispersion in, 218
elemental carbon, 227
evolution, 19
general circulation, 219
horizontal motion, 218
nitrogen compounds, 227
organic compounds, 229
oxy acids, 226
phosphorus, 309
stability, 215
sulfur compounds, 226
surface reactions, 235
synoptic scale motion, 219
trace elements, 228
trace substances figure, 237
transformations figure, 235
turbulent diffusion, 219
ultraviolet radiation, 218
vertical motions, 217
vertical structure, 213
water, 224
water transport, 221
window region, 233
Atmospheric gases biological sources, 52
Atmospheric transport, 218
Attapulgite exchange capacity, 161
Authigenic, 363
Autotrophic, 363
Autotrophic bacteria, 165
Average age, 59

Baltic Sea
area, 177
mean depth, 177
volume, 177
Banded iron formations, 28
Barium
concentration ocean, 195
crustal abundance, 127
solar abundance, 11
use by organisms, 206
Beachrock, 162
Beryllium
concentration ocean, 194
element formation, 12
solar abundance, 11
speciation ocean, 194
Beta decay, 13, 363
Bicarbonate
amount from rivers, 203
amount in ocean, 203
concentration ocean, 198
river water, 148
Big Bang, 10
Bimolecular process, 83
Biogeochemical cycle, 363
defined, 56
Biogeochemical cycles
anthropogenic effects, 353–360
effect on life, 357–360
soils, 145
Biogeochemistry description, 6
Biological processes ocean, 187
Biome structure, 254
Biota
carbon flux, 258
chemical data, Covers
Biotite
properties, 127
weathering soils, 138
Bioturbation, 167, 363
Bismuth
concentration ocean, 195
solar abundance, 11
speciation ocean, 195
Borate ocean, 243
Boron
concentration ocean, 194, 198
solar abundance, 11
speciation ocean, 194
Boundary layer, 200, 216
Bowen's reaction series, 94, 95
Box model ocean, 196
Box models, 55–66
Box-diffusion model, 253
Bromide
concentration ocean, 194, 198
solar abundance, 11
Brownian motion, 363
Budget defined, 56
Buffer capacity seawater, 160
Buffer factor, 65, 243, 244

C3 photosynthetic mechanism, 86
C4 photosynthetic mechanism, 86
Cadmium
adsorption effects, 331, 333
complexation, 331, 332, 338
concentration ocean, 195
organisms, 335
solar abundance, 11
sources, 319, 321
speciation, 328, 334
speciation ocean, 195
surface water, 335
Calcite
properties, 127
weathering, 95
Calcium
amount from rivers, 203
amount in ocean, 203
complexation, 331
concentration ocean, 194, 198
crustal abundance, 127
diffusion constant sediments, 163
plants, 125
river water, 148
solar abundance, 11
sulfate formation, 298
surface water, 103
use by organisms, 206
Calcium carbonate
seawater, 244
shells, 44
Calcium cycles soils, 145
Calcium nitrate thermodynamic data, 264
Carbohydrates, 30, 32, 33
oxidation, 42
structure, 32, 33
Carbon, 239–259
atmospheric, 241
burial, 88
carbonate rocks, 248
concentration ocean, 194
crustal abundance, 127
detritus, 244, 247
element formation, 12
flux, 165, 193, 249–253
biota, 258
detritus, 252
oceanic, 251
rainwater, 249
hydrosphere, 242
isotopes, 240
lithosphere, 248
sedimentation rate, 253
sediments, 248
shales, 248
soils, 247
solar abundance, 11
speciation ocean, 194
terrestrial biota, 246, 258
time scale of change, 239
use by organisms, 206
Carbon-13
atmosphere, 250
standard, 77
tree rings, 257
Carbon-14
carbon flux measurement, 251
deep water, 246
formation, 240
ocean, 252, 257
scale, 241
standard, 241
Carbon compounds
atmosphere, 227
oxidative capacity of atmosphere, 357
Carbon cycle, 58, 65, 66
fires, 251
models, 253
organisms, 49, 50
reservoirs, 240
summary figure, 259
tree rings, 257
trends, 254
Carbon dioxide
acid-base chemistry, 243
air borne fraction, 242
aqueous equilibrium, 76
atmosphere, 228
change in atmospheric flux, 355
concentration in atmosphere, 223, 241
concentration profile atmosphere, 214
control, 239
effect on photosynthesis, 246
equilibrium seawater, 243
food production, 356
greenhouse effect, 234, 235
non-steady-state, 61
oceanic, 252
oceanic sink, 257
past 40,000 years, 254

past concentrations, 358, 359
preindustrial, 241, 242
sediments, 159
soil effect of plants, 248
soils, 134, 135
solubility
in water, 228
in seawater, 200
variation forest, 251
weathering, 95, 156, 249
Carbon disulfide, 288
reactions atmosphere, 226, 289
Carbon monoxide
atmosphere, 228, 242
concentration atmosphere, 223
equilibrium in nebula, 17
formation, 250
oxidation, 271
oxidative capacity of atmosphere, 357
Carbonate weathering, 140
Carbonates
distribution in seawater, 243
soils, 144
weathering, 249
Carbonic acid
equilibrium, 76
weathering, 126
Carbonyl sulfide, 288
concentration atmosphere, 223
reactions atmosphere, 226, 289
residence time, 292
stratosphere, 292
Carnivores, 46
Catalyst, 83, 363
Cation, 363
Cation exchange capacity (CEC), 128, 129, 161, 363
Cell walls, 29
Cellobiose structure, 32
Cells, 29
Cellulose structure, 33
Cementation, 162
Cerium
concentration ocean, 195
solar abundance, 11
speciation ocean, 195
Chalcophiles, 318
Challenger Deep, 176
Chapman mechanism, 83, 218
Charge balance soils, 146
Chelate, 363
Chelation, 327, 336
Chemical kinetics, 82–86
Chemical signals, 43
Chemical weathering, 94–98, 135
atmosphere, 236
carbon, 249
Chitin, 44, 46
production, 46
Chloride
amount from rivers, 203
amount in ocean, 203
concentration ocean, 198
conservative element, 206
diffusion constant sediments, 163
surface water, 110
Chlorinated compounds reactions atmosphere, 230
Chlorine
concentration ocean, 194
solar abundance, 11
Chlorite, 130
exchange capacity, 161
properties, 127, 129
weathering, 95
Chlorofluorocarbon, 84, 363
change in atmospheric flux, 355
greenhouse effect, 234, 235, 236
non-steady-state, 61
ozone, 357
stratosphere, 280
Chondrite, 15
carbonaceous, 17
Chromium
adsorption effects, 331
concentration ocean, 194
reaction rates, 339
solar abundance, 11
sources, 319
speciation, 328, 334
speciation ocean, 194
Chromosome, 37
CISES, *see* Closed *in situ* experimental systems
Clastic, 363
Clay, 363
Clay aggregate, 158
Clay minerals
formation, 131
soil, 128
Climate, 234
change, 353, 354
change biosphere response, 359, 360
cloud condensation nuclei, 353
clouds, 235
greenhouse gases, 353
particles, 235
sulfur, 353
trace species, 355
water chemistry, 147
Closed *in situ* experimental systems (CISES), 169
Cloud condensation nuclei (CCN), 224, 363
climate, 353
Cloud droplets solutes, 225, 226
Clouds, 217
albedo, 235
ammonia, 272
climate, 235
formation, 224, 231
global view, 221
key roles, 224
nitrates, 272
Coarse particles atmosphere, 233
Cobalt
concentration ocean, 194
nutrient, 190
solar abundance, 11
sources, 319
speciation, 328, 334
speciation ocean, 194
Comets
impacts, 19
source of volatiles, 18
Compaction, 162
Complex formation oceans, 199
Composition Earth's crust, 127
Composition profile atmosphere, 214
Condensation sequence, 15
Congruent dissolution, 140, 363
Congruent weathering, 94
Conservation of charge, 76
Conservative, 363
Conservative elements seawater, 206
Conservative ions, 76
Constants, Covers
Continental margin, 175
Continental rise, 176
Continental shelf, 176
Copper
adsorption, 349
adsorption effects, 331
adsorption pH effect, 333
complexation, 331, 340, 348
concentration ocean, 194
crustal rocks, 348
global cycle, 346–349
organisms, 349
reactions, 348
redox reaction, 348
solar abundance, 11
sources, 221, 319
speciation, 328, 334
speciation ocean, 194
Coriolis force
atmosphere, 220
ocean, 182
Cosmic abundance, 9, 10
Cosmic rays, 235
Coulomb behavior, 109
Coupled behavior in biosphere, 358–360
Coupled cycles, 66, 89
atmosphere, 230
Coupled reservoirs, 61
Craton, 105, 363
erosion, 111
Creosote bush defensive mechanism, 45
Cretaceous-tertiary extinction, 19
Crust composition, Covers, 127
Crustal deformation, 108
Crystallization, 57
Cyanobacteria, 28
Cycle, 363
approach general, 55
defined, 56
Cycle time, 62
Cycles coupled, 89
Cysteine structure, 31

Dating geological samples, 14
Davis cycle, 111
Deamination sediments, 168
Decomposition, 50
Deep ocean water phosphorus, 307
Deep water age, 246
Defensive molecules, 44
Denitrification, 267–270

Denudation, *see also* Erosion, Weathering
Denudation rate, 107, 109
Deoxyribonucleic acid (DNA), 29, 30, 35
  structure, 304
Deoxyribose structure, 32
Deposition velocity, 69, 235, 363
Desert soils, 141–144
Desulfovibrio, 51
Detrital rain, 244, 252
Detritivores, 46
Deuterium standard, 77
Dew point, 363
  vertical profile atmosphere, 217
Diagenesis, 57, 161, 363
Diffusion, 363
  ocean surface, 200
  sediments, 162
Diffusion constants sediments, 163
Diffusivity, 67
Digestibility-reducing substances, 45
Dimethyl sulfide (DMS), 51, 288, 363
  flux, 294
  production, 49
  reactions atmosphere, 226, 289
Dinitrogen, *see* Nitrogen gas
Dinitrogen pentoxide thermodynamic data, 264
Dinitrogen tetroxide thermodynamic data, 264
Dissolved inorganic carbon (DIC), 242, 244
  distribution in ocean, 245
Dissolved organic carbon (DOC), 242
Dolomite
  properties, 127
  weathering, 95
Donnan equilibrium, 161
Dry deposition, 69, 292, 363
  marine boundary layer, 294
Dry removal, 235
Dysprosium
  concentration ocean, 195
  solar abundance, 11
  speciation ocean, 195

e-folding time, 60
Earth
  atmosphere, 20
  data, Covers
  evolution, 19
  geological history, 24
  unique factors, 9
Earth system science, 360
Ecological organization, 46
Ecological zones productivity, Covers
Ecosystem, 46
Eddy diffusivity ocean, 183
Eddy transport, 220
Eh-pH diagrams, 79–81
Eigenvalue, 62
Eigenvector, 62
Ekman spiral, 182
Electron activity, 78
Electron transport schematic, 37
Electrophilic, 363
Element
  formation, 10
  partitioning weathering, 102
Elemental carbon atmosphere, 227
Elemental cycles organisms, 49
Elements
  concentration in ocean table, 194, 195
  origin, 9
  speciation ocean table, 194, 195
Elevation of terraces, 108
Endosymbiotic theory, 40
Energy storage molecules, 42
Enthalpy of formation, 74
Entisols, 142
Entropy, 73
Enzymes, 36
Eolian, 363
Equilibrium, 73–77
  condensed phase, 75
Equilibrium constant, 74
Erbium
  concentration ocean, 195
  solar abundance, 11
  speciation ocean, 195
Erosion
  and orogeny, 108
  bedrock, 93
  craton, 111
  ice sheets, 115
  regimes, 99
  sea level change, 105
  summary, 94
  tectonically active areas, 108
  transport-limited, 114
  uplift rates, 107
  weathering-limited, 114
Escape from atmosphere, 215, 237
Essential elements, 127
Estuaries
  iron, 340
  metals, 339–341
  phosphorus, 310
Ethane reactions atmosphere, 229
Ethene reactions atmosphere, 229
Eukaryote, 37, 38
Eukaryotic cell properties, 39
Eukaryotic organisms, 40
Euphotic zone, 363
Europium
  concentration ocean, 195
  solar abundance, 11
  speciation ocean, 195
Eustatic, 363
Eustatic sea level change, 105
Evaporite, 364
  formation, 297
  minerals weathering, 96
  sulfur cycle, 298
Exchange factors, Covers
Exchange time atmosphere, 71
  ocean, 71
Expandability of minerals, 129
Extinction cretaceous-tertiary, 19

Fatty acid, 30
  structure, 34
Fecal pellets sediments, 159
Feldspar
  properties, 127
  weathering, 95, 140
  weathering soils, 138
Fermentation 26
Fertilizer nitrogen, 265
Fick's laws, 200
First Law of Thermodynamics, 73
Fischer-Tropsch reaction, 17
Five-kingdom system, 37
Floodplain development, 104
Fluorinated compounds reactions atmosphere, 230
Fluorine
  concentration ocean, 194, 198
  crustal abundance, 127
  solar abundance, 11
  speciation ocean, 194
Fluvial, 364
Flux defined, 56
Flyash metals, 320
Food production, 355, 356
Forest soils, 143
Formaldehyde reactions atmosphere, 229
Formation factor sediments, 162
Formic acid reactions atmosphere, 229
Fossil carbon addition, 66
Fossil fuel
  emissions, 242
  resources, 248
  use record, 255, 256
Fossils origin of life, 22
Francolite, 303
Free energy, 73, 74
Free radical, 230, 364
Frontal system, 220
Frost wedging, 134, 141, 142
Fructose structure, 32
Fulvic acid, 133
  complexation, 133
Fungi, 39
Fusion sequence solar, 12

Gadolinium
  concentration ocean, 195
  solar abundance, 11
  speciation ocean, 195
Gaia hypothesis, 12, 360
Galactose structure, 32
Gallium
  concentration ocean, 194
  solar abundance, 11
  speciation ocean, 194
Gas exchange boundary layer, 200
Gas solubility, 200
Gas transfer, 68
Geochemical cycle
  defined, 56
  figure, 57
Geologic history of Earth, 24
Geological ages, 24
Geological samples dating, 14
Geophysiology, 360
Geostrophic currents, 182
Geostrophic wind, 220

Germanium
  concentration ocean, 194
  solar abundance, 11
  speciation ocean, 194
Gibbs free energy, 73
Gibbs Phase Rule, 202
Glacial erosion model, 116
Glacial record trace substances, 358, 359
Glaciers, 115
  erosion, 115
  volume, 178
Gley horizon, 142
Glucose structure, 32
Glutamic acid structure, 31
Glutamine structure, 31
Glyceraldehyde structure, 32
Glycerol, 30
Glycine structure, 31
Glycolysis, 27
Gold
  concentration ocean, 195
  solar abundance, 11
  speciation ocean, 195
Grassland soils, 144
Gravitational collapse, 14
Green sulfur bacteria, 28, 51
Greenhouse effect, 234
  temperature profile, 237
  trace gases, 237
Greenhouse gases climate, 353
Gross flux, 60
Gross primary production, 47, 246
Ground water volume, 178
Guano, 303
Gypsum weathering, 95
Gyre, 182

Hadean Eon, 25
Hadley circulation, 219, 364
Hafnium
  concentration ocean, 195
  solar abundance, 11
Half cell conventions, 78
Half reaction, 78
Halite weathering, 95
Halloysite exchange capacity, 161
Halmyrolysis, 161
Halogenated organic compounds
    atmosphere, 230
Heat capacity, 364
Helium
  Big Bang, 10
  concentration
    atmosphere, 223
    ocean, 194
  concentration profile atmosphere, 214
  solar abundance, 11
  solubility seawater, 200
Henry's Law, 75, 200, 324
Herbivores, 46
Heterosphere, 215
Heterotrophic, 364
Heterotrophic bacteria, 165
Hexane reactions atmosphere, 229
High pressure system, 220
Histidine structure, 31
History of Earth, 24

Holmium
  concentration ocean, 195
  solar abundance, 11
  speciation ocean, 195
Homosphere, 215
Hudson Bay
  area, 177
  mean depth, 177
  volume, 177
Humic acid, 133
Humic substance, 126, 133, 247
  properties, 133
Humin, 133
Hydration sphere, 326
Hydrogen
  Big Bang, 10
  crustal abundance, 127
  solar abundance, 11
Hydrogen-2 standard, 77
Hydrogen-3, *see* Tritium
Hydrogen atoms concentration profile
    atmosphere, 214
Hydrogen bonds, 30
Hydrogen gas
  concentration atmosphere, 223
  solubility seawater, 200
Hydrogen sulfide, 288
  reactions atmosphere, 226, 289
Hydrologic cycle, 1
Hydrophilic, 364
  character, 30
  clay minerals, 161
Hydrophobic, 364
Hydrostatic equation, 213
Hydrothermal circulation
  removal mechanism, 207
  sulfur, 298
Hydroxides soils, 130
Hydroxyl radical (OH), 230, 242, 271
  concentration change, 357
Hypsographic, 364

Ice caps volume, 178
Ice nucleating aerosol (IN), 225
Ice sheets erosion, 115
Illite exchange capacity, 161
Imogolite, 130
  formation, 147
  properties, 129
Inceptisols, 142, 144
Incongruent dissolution, 140, 364
Incongruent weathering, 95
Indian Ocean
  area, 177
  mean depth, 177
  volume, 177
  water balance, 178
Indium
  concentration ocean, 195
  solar abundance, 11
  speciation ocean, 195
Information storage molecules, 42
Intertidal, 364
Intertropical convergence zone (ITCZ),
    221, 222
Inversion, 216

Iodine
  concentration ocean, 195
  solar abundance, 11
  speciation ocean, 195
Ion exchange sediments, 165
Ionic strength, 77
Iridium solar abundance, 11
Iron
  acidity soils, 139
  complexation, 126, 133, 331, 340
  concentration ocean, 194
  condensation sequence, 15
  crustal abundance, 127
  diffusion constant sediments, 163
  estuaries, 340
  mobilization, 318
  oxidation, 141
  pH effects, 330
  reaction rates, 339
  redox
    organisms, 49
    reactions, 322
    soils, 139, 142
  solar abundance, 11
  solubility, 329
  sources, 319
  speciation, 328, 334
    ocean, 194
Iron cycle soils, 145
Iron hydroxides adsorption, 330
Iron oxides, 130–132
  formation, 145
Iron oxidizing bacteria, 51
Iron sulfide sediments, 167
Isoleucine structure, 31
Isomorphous substitution, 128, 364
Isoprene reactions atmosphere, 229
Isopycnal surfaces, 184
Isostatic, 364
Isotope effect
  kinetic, 85
  photosynthesis, 86
  thermodynamic, 77
Isotope fractionation, 241
Isotopic composition, 77

Japanese pampas grass soils, 124–126
Jet stream, 219

Kaolinite
  exchange capacity, 161
  properties, 129
  structure, 158
  weathering, 140
Kaolinite minerals, 128
Karst, 364
Kelvin effect, 76, 225
Kinetic isotope effect, 85
King model, 111
Kjeldahl method, 168
Kohler curves, 225
Krebs cycle, 36
Krypton
  concentration atmosphere, 223
  solar abundance, 11
  solubility seawater, 200

Lactic acid bacteria, 26
Lactose structure, 32
Lacustrine, 364
Lake Erie metals, 320, 321
Lakes
  deposition of organic matter, 306
  phosphorus, 305
  stratification, 306
Lanthanum
  concentration ocean, 195
  solar abundance, 11
  speciation ocean, 195
Lapse rate, 215
Layer silicates, 128
Lead
  adsorption effects, 331
  complexation, 331
  concentration ocean, 195
  gasoline additive, 324
  mobilization, 318
  solar abundance, 11
  sources, 221, 319
  speciation, 328, 334
    ocean, 195
  tetraethyl, 324
  volatilization, 324
Lecithin structure, 34
Leibig's Law of the Minimum, 188
Leucine structure, 31
Life
  biogeochemical cycles, 357–360
  major divisions, 37
  requirements, 26
Light ocean productivity, 189
Lignins, 44
Linear systems models, 61
Linoleic acid structure, 34
Lipids, 30, 34
  structure, 34
Lithium
  concentration ocean, 194
  solar abundance, 11
Lithological factor sediments, 162
Lithophiles, 318
Litter, 132
  amount, 247
Llanos, 104
Logistic function, 256
Logistical growth, 65
Low pressure system, 220
Lutetium
  concentration ocean, 195
  solar abundance, 11
  speciation ocean, 195
Lysine structure, 31

Magnesium
  amount from rivers, 203
  amount in ocean, 203
  complexation, 133, 331
  concentration ocean, 194, 198
  condensation sequence, 15
  crustal abundance, 127
  diffusion constant in sediments, 163
  plants, 125
  solar abundance, 11
  surface water, 103
Magnesium cycles soils, 145
Magnetite properties, 127
Major ions ocean, 197
Maltose structure, 32
Manganese
  concentration ocean, 194
  crustal abundance, 127
  diffusion constant in sediments, 163
  estuaries, 341
  reaction rates, 339
  redox organisms, 49
  reduction soils, 139
  solar abundance, 11
  solubility, 329
  sources, 319
  speciation, 194, 328, 334
Mannose structure, 32
Mantle plume, 114
Maries' fir soils, 124–126
Marine boundary layer sulfur cycle, 294
Mars, 19
  atmosphere, 20
Mass balance ocean composition, 203, 204
Matrix method, 62
Mauna Loa record, 241
Mechanism of reaction, 83
Mediterranean Sea
  area, 177
  evaporites, 297
  mean depth, 177
  volume, 177
Mediterranean Water (MW), 181
  properties, 186
Melanization, 144
Membrane, 29
Mercaptan reactions in atmosphere, 230
Mercury (element)
  anthropogenic, 342, 344
  case studies, 345
  concentration ocean, 195
  contamination, 342
  fish, 335
  global cycle, 342–346
  Minimata Bay, 345, 346
  organisms, 335, 344, 345
  Ottawa River, 345, 346
  residence time, 344
  solar abundance, 11
  sources, 319
  speciation, 195, 328, 344
  volatilization, 324
Mercury (planet) atmosphere, 20
Mesic, 364
Metal cycles organisms, 49, 51
Metal oxides soils, 130
Metal sulfides, 329
Metals, 317–350
  abundance, 317
  adsorption, 330–334
    pH effect, 332, 333
  aerosols, 323
  air-water boundary, 334
  anthropogenic effects, 318–321
  bioavailability, 336
  complexation, 324–328, 332, 338
  coordination with water, 326
  estuaries, 339–341
  flyash, 320
  global cycling, 341–350
  importance of speciation, 317
  ion chemistry, 322
  Lake Erie, 320, 321
  mobilization, 318
  natural systems, 338
  ocean productivity, 190
  organic matter, 341
  organisms, 334
  partitioning, 334
  precipitation, 328–330
  rates of processes, 337
  redox reactions, 322, 323
  sediments, 320
  speciation organisms, 334
    table, 328
  soils, 139
  solubility, 328
  toxicity, 327, 336–338
  volatilization, 323
  water, 325
  weathering, 318
Metamorphism, 57
Metastable solids, 329
Methane
  atmosphere, 242
  change in atmospheric flux, 355
  concentration atmosphere, 223
  condensation sequence, 15
  equilibrium in nebula, 17
  formation, 250
  greenhouse effect, 235, 236
  non-steady-state, 61
  oxidative capacity of atmosphere, 357
  past concentrations, 358–359
  production by organisms, 4, 49, 50
  reactions, 4
  reactions atmosphere, 229
  sediments, 159
Methane sulfonic acid (MSA), 294
  reactions atmosphere, 289
Methanogenic bacteria, 27
Methanol reactions atmosphere, 229
Methionine structure, 31
Methyl amine thermodynamic data, 264
Methyl mercaptan, 288
  reactions atmosphere, 226
Methylmercury, 335
Mica, 128
  properties, 127, 129
  structure, 158
  weathering, 95, 140
Michaelis-Menten model, 82, 84
Miller experiment, 25
Mineral compartment, 136, 146, 147
Mineralization in soils, 139
Minerals
  properties, 129
  weathering reactions, 138
Minimata Bay mercury, 355, 346
Minor elements ocean, 197
Mitochondrion, 37, 40
Mixing times, 70
Modeling cycles, 55–72
Models matrix methods, 62

Mole fraction, 364
Molecular diffusion, 66
Molecular processes, 82, 83
Molecules functional types, 42
Mollisols, 144
Molybdenum
bonding, 325
concentration ocean, 195
solar abundance, 11
sources, 319
speciation ocean, 195
Monera, 37
Monosaccharide, 30
Montmorillonite, 130
exchange capacity, 161
formation, 145
properties, 129
structure, 158
Moon formation, 18
Mosses, 39
Mountainous regions water chemistry, 110
Multiplying prefixes, Covers
Muscovite, 128
weathering soils, 138

Natural selection, 42
Natural systems redox, 81
Neodymium
concentration ocean, 195
solar abundance, 11
speciation ocean, 195
Neon
concentration atmosphere, 223
concentration ocean, 194
solar abundance, 11
solubility seawater, 200
Nernst equation, 78, 163, 322
Net flux, 60
Net primary production, Covers, 47, 246
global, 250
Neutron capture, 13
Niche formation sediments, 159
Nickel
adsorption effects, 331
complexation, 340
concentration ocean, 194
solar abundance, 11
sources, 319
speciation, 328, 334
speciation ocean, 194
Niobium
concentration ocean, 194
solar abundance, 11
Nitrate
clouds, 272
diffusion constant sediments, 163
nutrient ocean, 201
oceans, 189
reactions atmosphere, 230
reduction, 267, 269
reduction soils, 139, 142
Nitric acid, 263
reactions atmosphere, 227
thermodynamic data, 264
Nitric oxide, 265, *see also* Nitrogen oxides
formation, 82
reactions atmosphere, 227
thermodynamic data, 264
Nitrification, 49, 267, 268
soils, 139
Nitrobacter, 268
Nitrogen, 263–282
abiotic processes, 270
anthropogenic effects, 279
concentration ocean, 194
fertilizer, 265
flux sediments, 166
fluxes, 275, 273
inventories, 273, 274
partitioning in reservoirs, 275
plants, 125
sediments, 167, 273
soils, 273
solar abundance, 11
speciation ocean, 194
use by organisms, 206
Nitrogen atoms concentration profile atmosphere, 214
Nitrogen compounds
acid precipitation, 355
aquatic systems, 275
atmosphere, 227, 275
biological transformations, 266–270
description, 263
gas phase reactions, 270
heterogeneous reactions, 272
terrestrial systems, 275
Nitrogen cycle
net flows, 281
organisms, 49, 50
overall description, 281
Nitrogen dioxide, 264, *see also* Nitrogen oxides
reactions atmosphere, 227
thermodynamic data, 264
Nitrogen fixation, 49, 267
anthropogenic effects, 268, 280
bacteria, 267
cycle, 277
Nitrogen gas, 263
atmosphere, 273
concentration atmosphere, 223
concentration profile atmosphere, 214
reactions atmosphere, 227
solubility seawater, 200
thermodynamic data, 264
Nitrogen monoxide, *see* Nitric oxide and Nitrogen oxides
Nitrogen oxide cycle, 276
Nitrogen oxides
catalytic effect, 272
fluxes, 275
food production, 355
formation, 86
lightning, 87
reactions, 271
removal atmosphere, 271
shock waves, 87
stratosphere, 271
tropospheric ozone, 271
tropospheric chemistry, 279
Nitrogenase, 268
Nitrosamonas, 268
Nitrous acid
reactions atmosphere, 227
thermodynamic data, 264
Nitrous oxide, 265
atmosphere, 273
change in atmospheric flux, 355
concentration, 277
concentration atmosphere, 223
global increase, 277
greenhouse effect, 237
oceans, 272
oxidative capacity of atmosphere, 357
past concentrations, 359
production, 269
reactions atmosphere, 227
source, 277
stratosphere, 272, 279
thermodynamic data, 264
Non-conservative ions, 76
Non-equilibrium systems, 86
Non-ideal behavior, 76
Non-linear models, 64
Non-seasalt (nss), 364
Non-seasalt (nss) sulfate, 291
Non-steady state, 60
North Atlantic Deep Water (NADW), 181, 184, 15
properties, 186
Nuclear bomb testing, 178, 184
carbon-14, 256
Nutrients changes in, 358
Nutrients ocean productivity, 188

Obligatory anaerobes, 364
Obliquity Earth orbit, 18
Ocean basins data, 177
Ocean margin profile, 176
Oceans, 175–209
acid-base chemistry, 243
area, 177
biological processes, 187
borate equilibrium, 243
box model, 196
chemistry, 193
circulation, 178, 205
composition, Covers, 197
characteristic profiles, 197
classes of elements, 206, 207
complex formation, 199
control, 205
dissolved gases, 198
kinetic models, 204
major ions, 198
mass balance, 203, 204
models, 202
nutrients, 201
density, 179
depth, 177
elements in table, 194, 195
evolution, 19
organic acid, 191–193
organic matter, 190
particulate carbon, 191–193
phosphorus, 309
productivity, 188–190
salinity, 179

Oceans (*cont.*)
surface currents, 182
temperature, 179
water masses characteristics, 186
volume, 177
Oleic acid structure, 34
Oligosaccharide, 30
Onlap sequences, 105
Order of reaction, 82
Organic acids soils, 135
Organic carbon oceans, 191–193
Organic compounds atmosphere, 229
Organic matter
metals, 341
oceans, 190
refractoriness, 165
sediments, 165
soils, 132
Organic molecules, 30
Organic-mineral compartment, 136, 146
Organic synthesis primeval Earth, 25
Organisms
classification, 37–40
copper, 349
first, 26
metals, 335
Origin of life, 26
historical, 21
Orinoco River
denudation rate, 109
dissolved solids, 109
erosion, 113
water chemistry, 103, 110
Orthophosphate, *see* Phosphate
Osmium solar abundance, 11
Ottawa River mercury, 345, 346
Oxidation, 77–81, *see also* Redox
soils, 139
Oxidative capacity troposphere, 357
Oxides soils, 130
Oxisols, 145
Oxy acids atmosphere, 226
Oxygen, *see also* Oxygen gas
crustal abundance, 127
solar abundance, 11
Oxygen-18 standard, 77
Oxygen atoms concentration profile
atmosphere, 214
Oxygen cycle, 88
Oxygen gas
concentration atmosphere, 214, 223
concentration ocean, 194
constancy, 88
Earth atmosphere, 20
evolution, 28
formation, 87, 88
impact on biota, 29
removal, 88
solubility seawater, 200
utilization, 201
weathering, 96, 156
Oxygen-only mechanism, 218
Ozone
catalytic destruction, 85
concentration atmosphere, 214, 223
depletion, 356, 357
layer, 217, 218
remote troposphere, 279
stratosphere, 83, 279
trend in, 359
troposphere, 279
vertical profile atmosphere, 217

Pacific Ocean
area, 177
depth, 177
volume, 177
water balance, 178
Palladium solar abundance, 11
Palmitic acid structure, 34
Panama water chemistry, 103
Particle concentration profile
atmosphere, 217
Particle transfer, 69
Particles
atmosphere, *see also* Aerosols
defined, 192, 339
Particulate carbon oceans, 191–193
Particulate organic carbon (POC), 242
pE, 78
Peat amount, 247
Pedosphere definition, 123
Permafrost, 142
Persian Gulf
area, 177
mean depth, 177
volume, 177
Phenols soils, 135
Phenylalanine structure, 31
Phosphate Baltic sediment, 171
Phosphate
diffusion constant in sediments, 163
formation sediments, 170
minerals, 303
nutrient ocean, 201
oceans, 189
Pacific Ocean, 307
removal sediments, 170
seawater, 302
water, 302
Phosphatidate structure, 34
Phosphoric acid, 302
dissociation constants, 302
Phosphorite, 364
Phosphorus, 301–313
adsorption, 305, 307
anthropogenic, 308, 312
atmosphere, 309
burdens table, 311
concentration ocean, 194
coupling with other cycles, 313
crustal abundance, 127
dissolved, 302
distribution in oceans, 307
estuaries, 310
flux sediments, 166
fluxes, 309–311
freshwater, 305
geological time, 312
global cycle, 308–312
figure, 308
table, 308
isotopes, 301
lakes, 305
limiting nutrient, 313
mining, 64
oceans, 306, 307
organic forms, 304
particulate forms, 302
redox reactions, 301
removal from oceans, 307
residence times, 311
river water, 310
sediments, 169–171, 303
time history, 312
solar abundance, 11
speciation, 302
surface water, 309
terrestrial, 305
unreactive, 310
use by organisms, 206
weathering, 305
Phosphorus compounds, 301
Phosphorus cycle, 63, 308–312
organisms, 51
sub-global, 304
soils, 145
Photochemical processes atmosphere,
231
Photolysis, 82
Photosynthesis, 28
early, 28
isotope effect, 86
total amount, 46
Physical constants, Covers
Physical weathering, 94–98, 134
Phytoalexins, 45
Phytoplankton, 364
composition, 188
Piston velocity, 200
Plagioclase
properties, 127
weathering soils, 138
Planet
accretion, 18
atmospheres, 20
bulk composition, 17
condensation, 15
formation, 15–19
Planetary boundary layer, 216
Plankton
composition, 188
productivity, 188–190
Plant kingdom, 39
Plants
elements in, Covers
mass, Covers
phosphorus, 308
Platinum solar abundance, 11
Plume spreading, 219
Podzol, 364
Podzolization, 364
Polonium speciation ocean, 195
Polygenetic soils, 364
Polysaccharide, 30
Polysaccharides soils, 144
Polywater, 159
Potassium
amount from rivers, 203
amount in ocean, 203

concentration ocean, 194, 198
crustal abundance, 127
diffusion constant sediments, 163
plants, 125
solar abundance, 11
surface water, 103
Potassium-40 heat source, 14, 19
Potassium cycles soils, 145
Potential temperature, 215
Praeseodymium
concentration ocean, 195
solar abundance, 11
speciation ocean, 195
Precipitation average, 222
Prefixes metric, Covers
Primary minerals, 364
soils, 127
Primary production distribution, 252
Primary productivity oceans, 190
Production
animal types, 48
mean values, 48
Productivity ecological zones, Covers
Prokaryote, 37, 38
Prokaryotic cell, 24
properties, 39
Proline structure, 31
Promethium speciation ocean, 195
Protein, 29, 30
Proterozoic Eon, 25
Protista, 39
Proton donors soils, 135
Proton-proton chain, 12
Purple sulfur bacteria, 28, 51
Pyrite formation, 89, 297
Pyroclastic, 364
Pyroxene properties, 127
Pyruvate oxidation, 36

Quartz
properties, 127
weathering, 95, 96
weathering soils, 138

R process, 13
Radiation effects trace species, 355
Radioactive heat source, 14, 19
Radium speciation ocean, 195
Rain acid-base chemistry, 231, 232
Rainforest soils, 145
Rainwater carbon flux, 249
Rate
molecular processes, 82
temperature dependence, 82
Rate constant, 82
Rate law, 82
Rate limiting step, 84
Reaction mechanism, 83
Reaction order, 82
Reaction rates, 82
Red beds, 28
Red giant, 12
Red Sea
area, 177
mean depth, 177
volume, 177
Redfield model, 187
Redfield ratio, 47, 165, 188, 244
Redox potential sediments, 159
Redox reactions
coupling, 166
natural systems, 81
sediments, 163
soils, 139–143
Redoxcline, 164
Reduced sulfur energy source, 28
Reduction, 77–81, 364, *see also* Redox
Reduction sequence
sediments, 164
soils, 143
Refractoriness, 364
Refractory elements, 15
Relative humidity, 217, 364
Reproduction, 35
Reservoir, 364
defined, 56
Residence time, 58, 196, 364, *see also* Transit time
and concentration, 196
and variability, 223
mercury, 344
Respiration
animal types, 48
mean values, 48
root, 98
Response time, 60, 338, 364
nitrogen oxide formation, 87
Rhenium
concentration ocean, 195
solar abundance, 11
speciation ocean, 195
Rhizobium, 50
Rhodium solar abundance, 11
Ribonucleic acid (RNA), 29, 30, 35
structure, 304
Ribose structure, 32
Ribosome, 41
River water
carbon flux, 249
phosphorus, 310
RKR model, 187
Root respiration, 98
Rubidium
concentration ocean, 194
solar abundance, 11
Rubidium-87 dating, 14
Run-off weathering, 98
Runaway greenhouse, 19
Ruthenium
concentration ocean, 195
solar abundance, 11

S process, 13
Salinity
average ocean, 181
defined, 179
Salt weathering, 134, 141
Samarium
concentration ocean, 195
solar abundance, 11
sources, 319
speciation ocean, 195
Samarium-147 dating, 14
Saprolite, 364
Savannah soils, 145
Scale height, 214
Scale of motion atmosphere, 218
Scale spatial, 57
Scandium
concentration ocean, 194
solar abundance, 11
speciation ocean, 194
Scattering coefficient profile atmosphere, 217
Sea level change, 105
Seasalt particles, 292
Second Law of Thermodynamics, 73
Secondary minerals, 364
soils, 128
Sediment layers, 155
Sedimentary cycle, 157
Sedimentation, 364
transport, 68
Sedimentation rate carbon, 253
Sediments, 155–171
cementation, 162
classification, 157
compaction, 162
diagenesis, 161
diffusion, 162
elemental composition, 111
equilibrium model, 160
formation, 155
gas phase, 159
liquid phase, 159
liquid-solid interactions, 160
metals, 320
nitrogen, 167
organic matter, 165
phosphorus, 169–171, 309
processes in, 160–165
reactions in, 165
redox reactions, 163
storage, 104
structure, 157
sulfur, 167
tectonic recycling, 108
transport, 345, 346
weathering, 155
Selection natural, 42
Selenium
adsorption effects, 331
concentration ocean, 194
solar abundance, 11
sources, 319
speciation ocean, 194
Serine structure, 31
Sexual reproduction, 35
Siderophiles, 18, 318
Silica
amount from rivers, 203
amount in ocean, 203
diffusion constant sediments, 163
dissolution reaction rates, 339
Silicate layer structure, 158
Silicates weathering, 249
Silicic acid
concentration, 147
nutrient, 189

Silicic acid (*cont.*)
river water, 148
soil reactions, 126
surface water, 110
Silicon
concentration ocean, 194
condensation sequence, 15
crustal abundance, 127
nutrient ocean, 201
solar abundance, 11
speciation ocean, 194
use by organisms, 206
Silicon cycles soils, 145
Silicon oxides, 130, 132
Sillen model, 160, 202
Silver
adsorption pH effect, 333
concentration ocean, 195
solar abundance, 11
sources, 319
speciation, 328
speciation ocean, 195
Sink defined, 56
Smectite, 130
formation, 147
Sodium
amount from rivers, 203
amount in ocean, 203
concentration ocean, 194, 198
crustal abundance, 127
diffusion constant sediments, 163
plants, 125
solar abundance, 11
surface water, 103
Soil
charge balance, 146
compartments, 146, 147
composition, 126–135
formation, 134, 141
factors, 124
global view, 136, 137
minerals, 127
water flow, 98
Soil depth weathering, 101
Soil horizon, 125, 136, 137, 146, 147
Soil profile, 124–126
Soils, 123–148
biogeochemical cycles, 145
effects of plants, 124–126
essential elements, 127
nitrogen, 273
phosphorus, 308
water chemistry, 146
weathering, 96, 100
Solar abundance, 9, 11
Solar luminosity change, 12
Solar system origin, 14, 15
Solubility equilibrium, 75
Source defined, 56
Southwest Pacific Boundary Current properties, 186
Specialization organisms, 41
Spodosols, 143
Spontaneity, 74
Spontaneous generation, 21
Stability
atmospheric, 215
minerals, 95
Stability constants, 326
Stability diagrams, 79–81
Standard hydrogen electrode, 78
Standard state, 74
Starch structure, 33
Static stability, 215, 216
Stearic acid structure, 34
Steppe soils, 144
Stereoisomers, 364
Stratosphere, 214, 217, 364
exchange, 71
ozone, 356, 357
Stromatolites, 22
Strontium
concentration ocean, 194, 198
crustal abundance, 127
solar abundance, 11
Structural molecules, 44
Subsurface water volume, 178
Sucrose structure, 32
Suess effect, 256
Sulfate
amount from rivers, 203
amount in ocean, 203
concentration ocean, 198
reduction, 167
surface water, 110
Sulfate compounds, 286
reactions atmosphere, 289
Sulfate particles
change in atmospheric flux, 355
past concentrations, 358, 359
Sulfate reducing bacteria, 84, 170, 287
Sulfate reduction
kinetics, 84
organisms, 49
soils, 142
Sulfides, 286
dissolved in sediments, 167
reactions atmosphere, 230
Sulfur, 285–299
amino acids, 286
anthropogenic, 290, 291
atmospheric reactions, 288
atmospheric sources, 288
bacteria, 28, 51
elemental, 287
evaporites, 297
flux sediments, 166
fluxes, 290
global atmospheric budget, 295
hydrospheric budget, 295–299
hydrothermal circulation, 298
isotope effects, 86
isotopes, 285, 298
major reservoirs, 287, 288
metals, 329
oceanic sinks, 297
oceanic sources, 298
oxidation organisms, 49
oxidation states, 285
precipitation, 292
rainwater, 291
redox, 81
remote marine atmosphere, 293–295
reservoirs table, 297
residence time, 290, 299
sediments, 167
solar abundance, 11
speciation ocean, 194
structure of molecules, 30
volcanoes, 287
Sulfur compounds, 286
acid precipitation, 355
atmosphere, 226, 288
concentration ocean, 194
crustal abundance, 127
reactions in atmosphere, 289
Sulfur cycle
deposition, 292
figure, 296
organisms, 49, 51
Sulfur dioxide
oxidation rate, 291
reactions atmosphere, 226, 289
residence time, 291
sources, 290
vertical profile atmosphere, 217
Sulfur species free energy, 80
Sun fusion reaction, 12
Supernovae, 13
Superwater, 159
Surface area
clays, 128
minerals, 129
Surface water volume, 178

Taiwan
uplift rate, 107
water chemistry, 103
Tantalum
concentration ocean, 195
solar abundance, 11
Tectonic processes, 104
map, 106
Tellurium
solar abundance, 11
speciation ocean, 195
Temperature
average ocean, 181
past values, 358
profile atmosphere, 214, 217
Temperature-salinity (T-S) diagram, 179
Tepuis, 112
Terbium
concentration ocean, 195
solar abundance, 11
speciation ocean, 195
Termolecular process, 83
Terrace elevation, 108
Terrestrial biomass, 246
Tetraethyl lead, 324
Thallium
concentration ocean, 195
solar abundance, 11
speciation ocean, 195
Thermal expansion weathering, 134, 141
Thermocline, 179, 306
circulation, 183
Thermodynamics, 73–77
Thiobacillus, 51
Thiosulfate, 287

Thorium
concentration ocean, 195
heat source, 14, 19
solar abundance, 11
speciation ocean, 195
Threonine structure, 31
Thulium
concentration ocean, 195
solar abundance, 11
speciation ocean, 195
Tides Moon orbit, 18
Tilt Earth axis, 18
Time scale, 57
processes in water, 337–339
Tin
concentration ocean, 195
solar abundance, 11
sources, 319
speciation ocean, 195
Titanium
concentration ocean, 194
crustal abundance, 127
solar abundance, 11
sources, 319
speciation ocean, 194
Titanium oxides, 130–132
Toxicity metals, 336
Toxins, 45
Trace elements atmosphere, 228
Trace substances atmosphere figure, 237
Transit time frequency function, 59
Transport processes, 66–70
Transport time, 70
Transport weathering interaction, 101
Tricarboxylic acid cycle (TCA), 36
Triglyceride structure, 34
Tritium ocean, 185
Trophic level, 47
Tropics weathering, 93
Troposphere, 214, 364
exchange, 71
Tryptophan structure, 31
Tundra soils, 142
Tungsten
concentration ocean, 195
solar abundance, 11
speciation ocean, 195
Turbulent diffusivity, 71
Turbulent flux, 66, 364
Turgor pressure, 44
Turnover time, 57, 364
defined, 56
Tyrosine structure, 31

Ultraviolet radiation (UV), 218
Ultraxeric, 364
Unimolecular process, 82
Upwelling regions phosphorus, 307
Uranium
concentration ocean, 195
heat source, 14, 19
solar abundance, 11
speciation ocean, 195

Valine structure, 31
van't Hoff equation, 75
Vanadium
adsorption effects, 331
concentration ocean, 194
solar abundance, 11
sources, 319
speciation ocean, 194
Vegetation weathering, 100
Venus, 19
atmosphere, 20
Vermiculite, 130
exchange capacity, 161
properties, 129
Vertical motion atmosphere, 217
Vertisols, 145
Vivianite, 169
Volatiles accretion, 17
Volcanic glass properties, 127
Volcanoes
early Earth, 26
metals, 318
Vostok record, 358

Warm Surface Water (WSW), 181
Water
amount from rivers, 203
amount in ocean, 203
change in atmospheric flux, 355
chemistry
climate, 147
mountainous regions, 110
weathering, 103
concentration profile atmosphere, 214
condensation sequence, 15
evaporation energy flux, 235
isotope effect, 77
Mars, 19
phase diagram, 224
properties, 224
redox, 78, 79
source abyssal, 184
thermodynamic data, 264
transport atmosphere, 221, 222
various reservoirs, 178
Venus, 19
Water cycle, 1
Water vapor residence time, 222
Watershed dynamics, 148
Weathering, 94–98
atmospheric gases, 95
bedrock, 93
biota, 135
carbon dioxide, 249
element partitioning, 102
grain surfaces, 96
horizon, 156
landforms, 104
metals, 318
phosphorus, 305
reaction control, 96
runoff, 98
sediments, 155
slope, 98, 100
soil depth, 101
soil fluids, 97
soils, 96, 100, 134
sulfur, 298
temperature, 98
transport control, 96
transport interaction, 101
tropics, 93–115
vegetation, 100
water chemistry, 103
Weathering rate, 96
Weathering reactions soils, 138, 140
Wet deposition, 69, 364
marine boundary layer, 294
Whitings, 244
Wind-driven circulation ocean, 182

Xenon
concentration atmosphere, 223
concentration ocean, 195
solar abundance, 11
solubility seawater, 200
Xeric, 364

Ytterbium
concentration ocean, 195
solar abundance, 11
speciation ocean, 195
Yttrium
concentration ocean, 194
solar abundance, 11
speciation ocean, 194

Zeldovich mechanism, 82
Zenith, 364
Zinc
adsorption effects, 331
concentration ocean, 194
mobilization, 318
solar abundance, 11
sources, 319, 221
speciation, 328, 334
speciation ocean, 194
Zirconium
concentration ocean, 194
solar abundance, 11
speciation ocean, 194
Zooplankton composition, 188

International Geophysics Series

Edited by

RENATA DMOWSKA

Division of Applied Science
Harvard University

JAMES R. HOLTON

Department of Atmospheric Sciences
University of Washington
Seattle, Washington

Volume 1 Beno Gutenberg. Physics of the Earth's Interior. 1959*

Volume 2 Joseph W. Chamberlain. Physics of the Aurora and Airglow. 1961*

Volume 3 S. K. Runcorn (ed.). Continental Drift. 1962*

Volume 4 C. E. Junge. Air Chemistry and Radioactivity. 1963*

Volume 5 Robert G. Fleagle and Joost A. Businger. An Introduction to Atmospheric Physics. 1963*

Volume 6 L. Dufour and R. Defay. Thermodynamics of Clouds. 1963*

Volume 7 H. U. Roll. Physics of the Marine Atmosphere. 1965*

Volume 8 Richard A. Craig. The Upper Atmosphere: Meteorology and Physics. 1965*

Volume 9 Willis L. Webb. Structure of the Stratosphere and Mesosphere. 1966*

Volume 10 Michele Caputo. The Gravity Field of the Earth from Classical and Modern Methods. 1967*

Volume 11 S. Matsushita and Wallace H. Campbell (eds.). Physics of Geomagnetic Phenomena. (In two volumes.) 1967*

Volume 12 K. Ya. Kondratyev. Radiation in the Atmosphere. 1969

Volume 13 E. Palmén and C. W. Newton. Atmospheric Circulation Systems: Their Structure and Physical Interpretation. 1969

Volume 14 Henry Rishbeth and Owen K. Garriott. Introduction to Ionospheric Physics. 1969*

Volume 15 C. S. Ramage. Monsoon Meterology. 1971*

Volume 16 James R. Holton. An Introduction to Dynamic Meterology. 1972*

Volume 17 K. C. Yeh and C. H. Liu. Theory of Ionospheric Waves. 1972

Volume 18 M. I. Budyko. Climate and Life. 1974

Volume 19 Melvin E. Stern. Ocean Circulation Physics. 1975

Volume 20 J. A. Jacobs. The Earth's Core. 1975*

Volume 21 David H. Miller. Water at the Surface of the Earth: An Introduction to Ecosystem Hydrodynamics. 1977

*These titles out of print

Volume 22 Joseph W. Chamberlain. Theory of Planetary Atmospheres: An Introduction to Their Physics and Chemistry. 1978*

Volume 23 James R. Holton. Introduction to Dynamic Meterology, Second Edition. 1979

Volume 24 Arnett S. Dennis. Weather Modification by Cloud Seeding. 1980

Volume 25 Robert G. Fleagle and Joost A. Businger. An Introduction to Atmospheric Physics, Second Edition. 1980

Volume 26 Kuo-Nan Liou. An Introduction to Atmospheric Radiation. 1980

Volume 27 David H. Miller. Energy at the Surface of the Earth: An Introduction to the Energetics of Ecosystems. 1981

Volume 28 Helmut E. Landsberg. The Urban Climate. 1981

Volume 29 M. I. Budyko. The Earth's Climate: Past and Future. 1982

Volume 30 Adrian E. Gill. Atmosphere–Ocean Dynamics. 1982

Volume 31 Paolo Lanzano. Deformations of an Elastic Earth. 1982

Volume 32 Ronald T. Merrill and Michael W. McElhinny. The Earth's Magnetic Field: Its History, Origin and Planetary Perspective. 1983

Volume 33 John S. Lewis and Ronald G. Prinn. Planets and Their Atmospheres: Origin and Evolution. 1983

Volume 34 Rolf Meissner. The Continental Crust: A Geophysical Approach. 1986

Volume 35 M. U. Sagitov, B. Bodri, V. S. Nazarenko, and Kh. G. Tadzhidinov. Lunar Gravimetry. 1986

Volume 36 Joseph W. Chamberlain and Donald M. Hunten. Theory of Planetary Atmospheres: An Introduction to Their Physics and Chemistry, Second Edition. 1987

Volume 37 J. A. Jacobs. The Earth's Core, Second Edition. 1987

Volume 38 J. R. Apel. Principles of Ocean Physics. 1969

Volume 39 Martin A. Uman. The Lighting Discharge. 1987

Volume 40 Davod G. Andrews, James R. Holton, and Conway B. Leovy. Middle Atmosphere Dynamics. 1987

Volume 41 Peter Warneck. Chemistry of the Natural Atmosphere. 1988

Volume 42 S. Pal Arya. Introduction to Micro-meteorology. 1988

Volume 43 Michael C. Kelley. The Earth's Ionosphere: Plasma Physics and Electrodynamics. 1989

Volume 44 William R. Cotton and Richard A. Anthes. Storm and Cloud Dynamics. 1989

Volume 45 William Menke. Geophysical Data Analysis: Discrete Inverse Theory. 1989

Volume 46 S. George Philander. El Niño, La niña, and the Southern Oscillation. 1990

Volume 47 Robert A. Brown. Fluid Mechanics of the Atmosphere. 1991

Volume 48 James R. Holton. An Introduction to Dynamic Meterology, Second Edition. 1992

Volume 49 A. Kaufman. Geophysical Field Theory and Method, Volume 1. 1992

Volume 50 Samuel Butcher, Robert J. Charlson, Gordon H. Orians and Gordon V. Wolfe. Global Biogeochemical Cycles. 1992

Volume 51 B. Evans and T. Wong. Fault Mechanics and Transport Properties of Rock. 1992